Multiscale Entropy Approaches and Their Applications

Multiscale Entropy Approaches and Their Applications

Editor

Anne Humeau-Heurtier

MDPI • Basel • Beijing • Wuhan • Barcelona • Belgrade • Manchester • Tokyo • Cluj • Tianjin

Editor
Anne Humeau-Heurtier
University of Angers
France

Editorial Office
MDPI
St. Alban-Anlage 66
4052 Basel, Switzerland

This is a reprint of articles from the Special Issue published online in the open access journal *Entropy* (ISSN 1099-4300) (available at: https://www.mdpi.com/journal/entropy/special_issues/multiscale_entropy_ii).

For citation purposes, cite each article independently as indicated on the article page online and as indicated below:

LastName, A.A.; LastName, B.B.; LastName, C.C. Article Title. *Journal Name* **Year**, *Article Number*, Page Range.

ISBN 978-3-03943-340-7 (Hbk)
ISBN 978-3-03943-341-4 (PDF)

Cover image courtesy of J.-L. Heurtier.

Contents

About the Editor . . . . **ix**

Anne Humeau-Heurtier
Multiscale Entropy Approaches and Their Applications
Reprinted from: *Entropy* **2020**, *22*, 644, doi:10.3390/e22060644 . . . . **1**

Antoine Jamin and Anne Humeau-Heurtier
(Multiscale) Cross-Entropy Methods: A Review
Reprinted from: *Entropy* **2020**, *22*, 45, doi:10.3390/e22010045 . . . . **7**

Dae-Young Lee and Young-Seok Choi
Multiscale Distribution Entropy Analysis of Short-Term Heart Rate Variability
Reprinted from: *Entropy* **2018**, *20*, 952, doi:10.3390/e20120952 . . . . **23**

Xiaojun Zhao, Chenxu Liang, Na Zhang and Pengjian Shang
Quantifying the Multiscale Predictability of Financial Time Series by an Information-Theoretic Approach
Reprinted from: *Entropy* **2019**, *21*, 684, doi:10.3390/e21070684 . . . . **39**

Xiefeng Cheng, Pengfei Wang and Chenjun She
Biometric Identification Method for Heart Sound Based on Multimodal Multiscale Dispersion Entropy
Reprinted from: *Entropy* **2020**, *22*, 238, doi:10.3390/e22020238 . . . . **53**

Xinzheng Dong, Chang Chen, Qingshan Geng, Zhixin Cao, Xiaoyan Chen, Jinxiang Lin, Yu Jin, Zhaozhi Zhang, Yan Shi and Xiaohua Douglas Zhang
An Improved Method of Handling Missing Values in the Analysis of Sample Entropy for Continuous Monitoring of Physiological Signals
Reprinted from: *Entropy* **2019**, *21*, 274, doi:10.3390/e21030274 . . . . **75**

Abhishek Tiwari, Isabela Albuquerque, Mark Parent, Jean-François Gagnon, Daniel Lafond, Sébastien Tremblay and Tiago H. Falk
Multi-Scale Heart Beat Entropy Measures for Mental Workload Assessment of Ambulant Users
Reprinted from: *Entropy* **2019**, *21*, 783, doi:10.3390/e21080783 . . . . **89**

Antonio Davalos, Meryem Jabloun, Philippe Ravier and Olivier Buttelli
On the Statistical Properties of Multiscale Permutation Entropy: Characterization of the Estimator's Variance
Reprinted from: *Entropy* **2019**, *21*, 450, doi:10.3390/e21050450 . . . . **109**

Dragana Bajic, Tamara Skoric, Sanja Milutinovic-Smiljanic and Nina Japundzic-Zigon
Voronoi Decomposition of Cardiovascular Dependency Structures in Different Ambient Conditions: An Entropy Study
Reprinted from: *Entropy* **2019**, *21*, 1103, doi:10.3390/e21111103 . . . . **125**

Hamed Azami, Alberto Fernández and Javier Escudero
Multivariate Multiscale Dispersion Entropy of Biomedical Times Series
Reprinted from: *Entropy* **2019**, *21*, 913, doi:10.3390/e21090913 . . . . **147**

Aurora Martins, Riccardo Pernice, Celestino Amado, Ana Paula Rocha, Maria Eduarda Silva, Michal Javorka and Luca Faes
Multivariate and Multiscale Complexity of Long-Range Correlated Cardiovascular and Respiratory Variability Series
Reprinted from: *Entropy* **2020**, *22*, 315, doi:10.3390/e22030315 . . . . **169**

Katarzyna Harezlak and Pawel Kasprowski
Application of Time-Scale Decomposition of Entropy for Eye Movement Analysis
Reprinted from: *Entropy* **2020**, *22*, 168, doi:10.3390/e22020168 . . . . **189**

Ben-Yi Liau, Fu-Lien Wu, Chi-Wen Lung, Xueyan Zhang, Xiaoling Wang and Yih-Kuen Jan
Complexity-Based Measures of Postural Sway during Walking at Different Speeds and Durations Using Multiscale Entropy
Reprinted from: *Entropy* **2019**, *21*, 1128, doi:10.3390/e21111128 . . . . **207**

Nurul Retno Nurwulan, Bernard C. Jiang and Vera Novak
Development of Postural Stability Index to Distinguish Different Stability States
Reprinted from: *Entropy* **2019**, *21*, 314, doi:10.3390/e21030314 . . . . **219**

Ian M. McDonough, Sarah K. Letang, Hillary B. Erwin and Rajesh K. Kana
Evidence for Maintained Post-Encoding Memory Consolidation Across the Adult Lifespan Revealed by Network Complexity
Reprinted from: *Entropy* **2019**, *21*, 1072, doi:10.3390/e21111072 . . . . **235**

Sreevalsan S. Menon and K. Krishnamurthy
A Study of Brain Neuronal and Functional Complexities Estimated Using Multiscale Entropy in Healthy Young Adults
Reprinted from: *Entropy* **2019**, *21*, 995, doi:10.3390/e21100995 . . . . **251**

Ofelie De Wel, Mario Lavanga, Alexander Caicedo, Katrien Jansen, Gunnar Naulaers and Sabine Van Huffel
Decomposition of a Multiscale Entropy Tensor for Sleep Stage Identification in Preterm Infants
Reprinted from: *Entropy* **2019**, *21*, 936, doi:10.3390/e21100936 . . . . **271**

Herbert F. Jelinek, David J. Cornforth, Mika P. Tarvainen and Kinda Khalaf
Investigation of Linear and Nonlinear Properties of a Heartbeat Time Series Using Multiscale Rényi Entropy
Reprinted from: *Entropy* **2019**, *21*, 727, doi:10.3390/e21080727 . . . . **287**

Mohammed El-Yaagoubi, Rebeca Goya-Esteban, Younes Jabrane, Sergio Muñoz-Romero, Arcadi García-Alberola and José Luis Rojo-Álvarez
On the Robustness of Multiscale Indices for Long-Term Monitoring in Cardiac Signals
Reprinted from: *Entropy* **2019**, *21*, 594, doi:10.3390/e21060594 . . . . **301**

David Perpetuini, Antonio M. Chiarelli, Daniela Cardone, Chiara Filippini, Roberta Bucco, Michele Zito and Arcangelo Merla
Complexity of Frontal Cortex fNIRS Can Support Alzheimer Disease Diagnosis in Memory and Visuo-Spatial Tests
Reprinted from: *Entropy* **2019**, *21*, 26, doi:10.3390/e21010026 . . . . **323**

Soheil Keshmiri, Hidenobu Sumioka, Ryuji Yamazaki and Hiroshi Ishiguro
Multiscale Entropy Quantifies the Differential Effect of the Medium Embodiment on Older Adults Prefrontal Cortex during the Story Comprehension: A Comparative Analysis
Reprinted from: *Entropy* **2019**, *21*, 199, doi:10.3390/e21020199 . . . . **337**

Chao Xu, Chen Xu, Wenjing Tian, Anqing Hu and Rui Jiang
Multiscale Entropy Analysis of Page Views: A Case Study of Wikipedia
Reprinted from: *Entropy* **2019**, *21*, 229, doi:10.3390/e21030229 . . . . **353**

Tzu-Kang Lin and Yi-Hsiu Chien
Performance Evaluation of an Entropy-Based Structural Health Monitoring System Utilizing Composite Multiscale Cross-Sample Entropy
Reprinted from: *Entropy* **2019**, *21*, 41, doi:10.3390/e21010041 . . . . **367**

Mao Ge, Yong Lv, Yi Zhang, Cancan Yi and Yubo Ma
An Effective Bearing Fault Diagnosis Technique via Local Robust Principal Component Analysis and Multi-Scale Permutation Entropy
Reprinted from: *Entropy* **2019**, *21*, 959, doi:10.3390/e21100959 . . . . **389**

Haikun Shang, Feng Li and Yingjie Wu
Partial Discharge Fault Diagnosis Based on Multi-Scale Dispersion Entropy and a Hypersphere Multiclass Support Vector Machine
Reprinted from: *Entropy* **2019**, *21*, 81, doi:10.3390/e21010081 . . . . **415**

About the Editor

Anne Humeau-Heurtier received her Ph.D. degree in Biomedical Engineering in France. She is currently a full professor in Engineering at the University of Angers, France. Her research interests include signal and image processing, mainly multiscale and entropy-based analyses and data-driven methods.

Editorial

Multiscale Entropy Approaches and Their Applications

Anne Humeau-Heurtier

LARIS—Laboratoire Angevin de Recherche en Ingénierie des Systèmes, University of Angers, 49035 Angers, France; anne.humeau@univ-angers.fr

Received: 28 May 2020; Accepted: 2 June 2020; Published: 10 June 2020

Keywords: multiscale entropy; multivariate data; entropy

1. Introduction

Multiscale entropy (MSE) measures have been proposed from the beginning of the 2000s to evaluate the complexity of time series, by taking into account the multiple time scales in physical systems. Since then, these approaches have received a great deal of attention and have been used in a large range of applications. Multivariate approaches have also been developed.

The algorithms for a MSE approach are composed of two main steps: (i) a coarse-graining procedure to represent the system's dynamics on different scales; and (ii) the entropy computation for the original signal and for the coarse-grained time series to evaluate the irregularity for each scale. Moreover, different entropy measures have been associated with the coarse-graining approach, each one having its advantages and drawbacks: approximate entropy, sample entropy, permutation entropy, fuzzy entropy, distribution entropy, dispersion entropy, etc.

In this Special Issue, we gathered 24 papers focusing on either the theory or applications of MSE approaches. These papers can be divided into two groups: papers that propose either new developments on entropy-based measures or improve the understanding of existing ones (nine papers); and papers that propose new applications of existing entropy-based measures (14 papers), as described below. Moreover, one paper proposes a review on cross-entropy methods and their multiscale approaches [1].

2. New Developments in Entropy-Based Measures

Lee et al. proposed a multiscale distribution entropy based on a moving averaging multiscale process and distribution entropy to study short-term heart rate variability (HRV) [2]. The authors show that the new entropy-based measure outperforms MSE and multiscale permutation entropy as it is insensitive to the length of signals. The new measure shows a decrease in the complexity of HRV with aging and for congestive heart failure patients.

Zhao et al. proposed the multiscale entropy difference (MED) to assess the predictability of nonlinear financial time series on several time scales [3]. MED quantifies the contributions of the past values by reducing the uncertainty of the forthcoming values in signals on several time scales. The algorithm has been validated on simulated data and then applied to the analysis of Chinese stock markets.

Cheng et al. proposed a method based on multimodal multiscale dispersion entropy for the biometric characterization of heart sounds [4]. The work relies on the use of the improved complete ensemble empirical mode decomposition with adaptive noise (ICEEMDAN) and refined composite multiscale dispersion entropy. The authors show that the proposed method is effective for heart sound biometric recognition.

Dong et al. proposed a method, KeepSampEn, to minimize the error due to missing values in sample entropy calculation [5]. For this purpose, they modified the computation process but not the data. The results reveal that KeepSampEn shows a consistent lower average percentage error than other methods as skipping the missing values, linear interpolation and bootstrapping.

Tiwari et al. investigated the multiscale features of the mental workload for ambulant users [6]. Features that outperform benchmark ones are proposed and they exhibit complementarity when used in combination. Thus, the authors reported that composite coarse-graining via a new second moment moving average scaling method, combined with the modified permutation entropy method, outperforms other combinations.

From a Taylor series expansion, Dávalos et al. developed an explicit expression for the multiscale permutation entropy (MPE) estimator's variance as a function of the time scale and ordinal pattern distribution [7]. They also determined the Cramér–Rao lower bound of the MPE. The results show that MPE variance is related to the MPE measurement and increases linearly with time scale, but not when the MPE measure reaches its maximum value. Moreover, for short time scales compared to the signal length, the MPE variance resembles the MPE Cramér–Rao lower bound.

Bajic et al. proposed a method that enables an application of MSE to an arbitrary number of signals [8]. The authors also wanted to test whether their method recognizes the changes of the dependency level (coupling strength, level of interaction) of joint multivariate signals in different biomedical experiments. For this purpose, they use the copula density to determine the coupling strength. Moreover, the authors apply the composite MSE to the systolic blood pressure, the pulse interval, and the body temperature of rats exposed to different ambient temperatures.

Azami et al. introduced the multivariate multiscale dispersion entropy (mvMDE) to quantify the complexity of multivariate time series [9]. When applied to different kinds of signals, the results show that mvMDE has some advantages over multivariate multiscale entropy (mvMSE) and multivariate multiscale fuzzy entropy (mvMFE).

Martins et al. introduced a new method to assess the complexity of multivariate time series [10]. This new method takes into account the presence of short-term dynamics and long-range correlations and uses vector autoregressive fractionally integrated (VARFI) models. This leads to a linear parametric representation of the vector's stochastic processes. Then, an analytical formulation is obtained to derive the MSE measures. The authors tested this new approach on cardiovascular and respiratory signals to assess the complexity of the heart period, the systolic arterial pressure, and the respiration variability in different physiological conditions. The results show that, by taking into account long-range correlations, the method proposed by the authors overcomes the existing ones as it captures significant variations in the complexity that are not observed with standard existing methods.

3. Applications of Existing Entropy-Based Measures

In this Special Issue, 14 papers propose to use existing entropy-based measures for different kinds of applications, as mentioned below.

Harezlak et al. studied eye movement signal characteristics [11]. For this purpose, the authors used several methods: approximate entropy, fuzzy entropy, and the largest Lyapunov exponent. For these three methods, multilevel maps are defined. The results show better accuracy for saccadic latency and saccade, than previous studies using eye movement dynamics.

Liau et al. evaluated the changes in the complexity of the center of pressure (COP) during walking at different speeds and for different durations [12]. For this purpose, the MSE was used. The authors show that both the walking speed and walking duration factors significantly affect the complexity of COP.

Based on ensemble empirical mode decomposition (EEMD) and MSE and using an accelerometer, Nurwulan et al. proposed a measure, the postural stability index (PSI), to distinguish different stability states in healthy subjects [13]. PSI is able to discriminate between normal walking and walking with obstacles in healthy subjects.

McDonough et al. were interested by post-encoding memory consolidation mechanisms in a sample of young, middle-aged and older adults [14]. For this purpose, they tested a novel measure of information processing, network complexity and studied if it was sensitive to these post-encoding mechanisms. Network complexity was determined by assessing the irregularity of brain signals within a network over time. This was performed through MSE. The results show that network complexity is sensitive to post-encoding consolidation mechanisms that enhance memory performance.

Menon and Krishnamurthy mapped neuronal and functional complexities from the MSE of resting-state functional magnetic resonance imaging (rfMRI) blood oxygen-level dependent (BOLD) signals and BOLD phase coherence connectivity [15].

De Wel et al. proposed a novel unsupervised method to discriminate quiet sleep from non-quiet sleep in preterm infants, from the decomposition of a multiscale entropy tensor [16]. This was performed according to the difference in the electroencephalography (EEG) complexity between the neonatal sleep stages.

Jelinek et al. investigated the efficacy of applying multiscale Renyi entropy on heart rate variability (HRV) to obtain information on the sign, magnitude, and acceleration of the signals with time [17]. The results show that their quantification using multiscale Renyi entropy leads to statistically significant differences between the disease classes of normal, early cardiac autonomic neuropathy (CAN), and definite CAN.

El-Yaagoubi et al. studied the dynamics, the consistency and the robustness of MSE, multiscale time irreversibility (MTI), and multifractal spectrum in HRV characterization in long-term scenarios (7 days) [18]. The results show that congestive heart failure (CHF) and atrial fibrillation (AF) populations show significant differences at long-term and very long-term scales (thus, MSE is higher for AF while MTI is lower for AF).

For an early Alzheimer's disease (AD) diagnosis, Perpetuini et al. used sample entropy and the MSE of functional near infrared spectroscopy (fNIRS) in the frontal cortex of early AD and healthy controls during three tests that were used to assess visuo-spatial and short-term-memory abilities [19]. A multivariate analysis revealed promising results (good specificity and sensitivity) in the capabilities of fNIRS and complexity for an early diagnosis.

Keshmiri et al. studied the effect of the physical embodiment on older people's prefrontal cortex (PFC) activity when they are listening to stories [20]. For this purpose, they used MSE. Their results show that, in older people, physical embodiment leads to a significant increase of MSE for PFC activity. Moreover, this increase reflects the perceived feeling of fatigue.

Xu et al. used the short-time series MSE (sMSE) to study the complexities and temporal correlations of Wikipedia page views of four selected topics [21]. The goal was to understand the complexity of human website searching activities. The results show that sMSE is useful to analyze the temporal variations of the complexity of page view data for some topics. Nevertheless, the regular variations of sample entropy cannot be accepted as is when different topics are compared.

Lin et al. developed an entropy-based structural health monitoring system to solve the problem of unstable entropy values observed when multiscale cross-sample entropy was used to assess damage in laboratory-scale structure [22]. The results could be interesting for long-term monitoring.

Ge et al. proposed a bearing fault diagnosis technique using the local robust principal component analysis (to remove background noise: it decomposed the signal trajectory matrix into multiple low-rank matrices) and multiscale permutation entropy that identified the low-rank matrices corresponding to the bearing's fault feature [23]. The latter matrices are then combined into a one-dimensional signal and represents the extracted fault feature component.

Shang et al. used variational mode decomposition and multiscale dispersion entropy to propose a novel feature extraction method for partial discharge fault analysis [24]. Moreover, a hypersphere multiclass support vector machine was used for partial discharge pattern recognition.

Let us now hope that these papers will bring other interesting applications and lead to new ideas to further improve the study of the irregularity and complexity of data (1D, 2D, *n*-D).

Funding: This research received no external funding.

Acknowledgments: I express my thanks to the authors of the above contributions and to the *Entropy* Editorial Office and MDPI for their support during this work.

Conflicts of Interest: The author declares no conflict of interest.

References

1. Jamin, A.; Humeau-Heurtier, A. (Multiscale) Cross-entropy methods: A Review. *Entropy* **2020**, *22*, 45. [CrossRef]
2. Lee, D.Y.; Choi, Y.S. Multiscale distribution entropy analysis of short-term heart rate variability. *Entropy* **2018**, *20*, 952. [CrossRef]
3. Zhao, X.; Liang, C.; Zhang, N.; Shang, P. Quantifying the multiscale predictability of financial time series by an information-theoretic approach. *Entropy* **2019**, *21*, 684. [CrossRef]
4. Cheng, X.; Wang, P.; She, C. Biometric identification method for heart sound based on multimodal multiscale dispersion entropy. *Entropy* **2020**, *22*, 238. [CrossRef]
5. Dong, X.; Chen, C.; Geng, Q.; Cao, Z.; Chen, X.; Lin, J.; Jin, Y.; Zhang, Z.; Shi, Y.; Zhang, X.D. An improved method of handling missing values in the analysis of sample entropy for continuous monitoring of physiological signals. *Entropy* **2019**, *21*, 274. [CrossRef]
6. Tiwari, A.; Albuquerque, I.; Parent, M.; Gagnon, J.F.; Lafond, D.; Tremblay, S.; Falk, T.H. Multi-Scale Heart Beat Entropy Measures for Mental Workload Assessment of Ambulant Users. *Entropy* **2019**, *21*, 783. [CrossRef]
7. Dávalos, A.; Jabloun, M.; Ravier, P.; Buttelli, O. On the statistical properties of multiscale permutation entropy: Characterization of the estimator's variance. *Entropy* **2019**, *21*, 450. [CrossRef]
8. Bajic, D.; Skoric, T.; Milutinovic-Smiljanic, S.; Japundzic-Zigon, N. Voronoi Decomposition of Cardiovascular Dependency Structures in Different Ambient Conditions: An Entropy Study. *Entropy* **2019**, *21*, 1103. [CrossRef]
9. Azami, H.; Fernández, A.; Escudero, J. Multivariate multiscale dispersion entropy of biomedical times series. *Entropy* **2019**, *21*, 913. [CrossRef]
10. Martins, A.; Pernice, R.; Amado, C.; Rocha, A.P.; Silva, M.E.; Javorka, M.; Faes, L. Multivariate and multiscale complexity of long-range correlated cardiovascular and respiratory variability series. *Entropy* **2020**, *22*, 315. [CrossRef]
11. Harezlak, K.; Kasprowski, P. Application of time-scale decomposition of entropy for eye movement analysis. *Entropy* **2020**, *22*, 168. [CrossRef]
12. Liau, B.Y.; Wu, F.L.; Lung, C.W.; Zhang, X.; Wang, X.; Jan, Y.K. Complexity-based measures of postural sway during walking at different speeds and durations using multiscale entropy. *Entropy* **2019**, *21*, 1128. [CrossRef]
13. Nurwulan, N.R.; Jiang, B.C.; Novak, V. Development of postural stability index to distinguish different stability states. *Entropy* **2019**, *21*, 314. [CrossRef]
14. McDonough, I.M.; Letang, S.K.; Erwin, H.B.; Kana, R.K. Evidence for maintained post-encoding memory consolidation across the adult lifespan revealed by network complexity. *Entropy* **2019**, *21*, 1072. [CrossRef]
15. Menon, S.S.; Krishnamurthy, K. A Study of brain neuronal and functional complexities estimated using multiscale entropy in healthy young adults. *Entropy* **2019**, *21*, 995. [CrossRef]
16. De Wel, O.; Lavanga, M.; Caicedo, A.; Jansen, K.; Naulaers, G.; Van Huffel, S. Decomposition of a multiscale entropy tensor for sleep stage identification in preterm infants. *Entropy* **2019**, *21*, 936. [CrossRef]
17. Jelinek, H.F.; Cornforth, D.J.; Tarvainen, M.P.; Khalaf, K. Investigation of linear and nonlinear properties of a heartbeat time series using multiscale Rényi entropy. *Entropy* **2019**, *21*, 727. [CrossRef]
18. El-Yaagoubi, M.; Goya-Esteban, R.; Jabrane, Y.; Muñoz-Romero, S.; García-Alberola, A.; Rojo-Álvarez, J.L. On the robustness of multiscale indices for long-term monitoring in cardiac signals. *Entropy* **2019**, *21*, 594. [CrossRef]
19. Perpetuini, D.; Chiarelli, A.M.; Cardone, D.; Filippini, C.; Bucco, R.; Zito, M.; Merla, A. Complexity of frontal cortex fNIRS can support Alzheimer disease diagnosis in memory and visuo-spatial tests. *Entropy* **2019**, *21*, 26. [CrossRef]

20. Keshmiri, S.; Sumioka, H.; Yamazaki, R.; Ishiguro, H. Multiscale entropy quantifies the differential effect of the medium embodiment on older adults prefrontal cortex during the story comprehension: A Comparative Analysis. *Entropy* **2019**, *21*, 199. [CrossRef]
21. Xu, C.; Xu, C.; Tian, W.; Hu, A.; Jiang, R. Multiscale entropy analysis of page views: A case study of Wikipedia. *Entropy* **2019**, *21*, 229. [CrossRef]
22. Lin, T.K.; Chien, Y.H. Performance evaluation of an entropy-based structural health monitoring system utilizing composite multiscale cross-sample entropy. *Entropy* **2019**, *21*, 41. [CrossRef]
23. Ge, M.; Lv, Y.; Zhang, Y.; Yi, C.; Ma, Y. An effective bearing fault diagnosis technique via local robust principal component analysis and multi-scale permutation entropy. *Entropy* **2019**, *21*, 959. [CrossRef]
24. Shang, H.; Li, F.; Wu, Y. Partial discharge fault diagnosis based on multi-scale dispersion entropy and a hypersphere multiclass support vector machine. *Entropy* **2019**, *21*, 81. [CrossRef]

Review

(Multiscale) Cross-Entropy Methods: A Review

Antoine Jamin [1,2,*] and Anne Humeau-Heurtier [2]

1 COTTOS Médical, Allée du 9 novembre 1989, 49240 Avrillé, France
2 LARIS – Laboratoire Angevin de Recherche en Ingénierie des Systèmes, Univesity Angers, 62 avenue Notre-Dame du Lac, 49000 Angers, France; anne.humeau@univ-angers.fr
* Correspondence: antoine.jamin@cottos.fr; Tel.: +33-252605954

Received: 9 December 2019; Accepted: 26 December 2019; Published: 29 December 2019

Abstract: Cross-entropy was introduced in 1996 to quantify the degree of asynchronism between two time series. In 2009, a multiscale cross-entropy measure was proposed to analyze the dynamical characteristics of the coupling behavior between two sequences on multiple scales. Since their introductions, many improvements and other methods have been developed. In this review we offer a state-of-the-art on cross-entropy measures and their multiscale approaches.

Keywords: cross-entropy; multiscale cross-entropy; asynchrony; complexity; coupling; cross-sample entropy; cross-approximate entropy; cross-distribution entropy; cross-fuzzy entropy; cross-conditional entropy

---

1. Introduction

To quantify the asynchronism between two time series, Pincus and Singer have adapted the approximate entropy algorithm to a cross-approximate entropy (cross-ApEn) method [1]. Then, other cross-entropy methods—that improve the cross-ApEn—have been developed [2–7]. Furthermore, additional cross-entropy methods have been introduced to quantify the degree of coupling between two signals, or the complexity between two cross-sequences [8–10]. Cross-entropy methods have recently been used in different research fields, including medicine [5,11,12], mechanics [13], and finance [7,10].

The multiscale approach of entropy measures was proposed by Costa et al. in 2002 to analyze the complexity of a time series [14]. In 2009, Yan et al. proposed a multiscale approach for cross-entropy methods to quantify the dynamical characteristics of coupling behavior between two sequences on multiple scale factors [15]. Then, other multiscale procedures have been published with different cross-entropy methods [16,17]. Multiscale cross-entropy methods have recently been used in different research fields, including medicine [18–21], finance [6,9], civil engineering [22], and the environment [23].

Cross-entropy methods and their multiscale approaches are used to obtain information on the possible relationship between two time series. For example, Wei et al. applied percussion entropy to the amplitude of digital volume pulse signals and changes in R-R intervals of successive cardiac cycles for assessing baroreflex sensitivity [18]. Results showed that the method is able to identify the markers of diabetes by the nonlinear coupling behavior of the two cardiovascular time series. Moreover, Zhu and Song computed cross-fuzzy entropy on a vibration time series to assess the bearing performance degradation process of motor [13]. Results showed that the method detects trend for bearing degradation process over the whole lifetime. In addition, Wang et al. applied multiscale cross-trend sample entropy to analyze the asynchrony between air quality impact factors (fine particulate matters, nitrogen dioxide, …), and air quality index (AQI) in different regions of China [23]. Results showed that the degree of synchrony between fine particulate matter and AQI is higher than the other air quality impact factor which reveals that fine particulate matter has become the main source of air pollution in China.

Our paper presents a state-of-the-art in three sections: First, the cross-entropy methods are introduced. We detail, in the second section, different multiscale procedures. A multiscale cross-entropy generalization is presented and other specific multiscale cross-entropy algorithms are proposed in the third section.

2. Cross-Entropy Methods

In this section, we classify cross-entropy methods according to their entropy measures: Cross-approximate entropy, cross-sample entropy, and cross-distribution entropy. Other methods that use different cross-entropy-based measures are also detailed. Table 1 shows the twelve measures that are detailed in this section.

Table 1. Cross-entropy measures, in chronological order, that are presented in this review. Authors, year, reference, and section location are indicated for each item.

Method	Authors	Year	Ref.	Section
Cross-approximate entropy	Pincus and Singer	1996	[1]	Section 2.1.1
Cross-conditional entropy	Porta et al.	1999	[8]	Section 2.4.1
Cross-sample entropy	Richman and Moorman	2000	[2]	Section 2.2.1
Cross-fuzzy entropy	Xie et al.	2010	[3]	Section 2.4.2
Modified cross-sample entropy	Yin and Shang	2015	[4]	Section 2.2.2
Binarized cross-approximate entropy	Škorić et al.	2017	[5]	Section 2.1.2
Modified cross-sample entropy based on symbolic representation and similarity	Wu et al.	2018	[6]	Section 2.2.3
Kronecker-delta based cross-sample entropy	He et al.	2018	[7]	Section 2.2.4
Permutation based cross-sample entropy	He et al.	2018	[7]	Section 2.2.5
Cross-distribution entropy	Wang and Shang	2018	[9]	Section 2.3.1
Permutation cross-distribution entropy	He et al.	2019	[10]	Section 2.3.2
Cross-trend sample entropy	Wang et al.	2019	[23]	Section 2.2.6
Joint permutation entropy	Yin et al.	2019	[24]	Section 2.4.3

2.1. Cross-Approximate Entropy-Based Measures

2.1.1. Cross-Approximate Entropy

Cross-approximate entropy (cross-ApEn), introduced by Pincus and Singer [1], allows to quantify asynchrony between two time series. For two vectors $\mathbf{u}$ and $\mathbf{v}$ of length N, cross-ApEn is computed as:

$$\text{cross-ApEn(m,r,N)}(\mathbf{v}||\mathbf{u}) = \Phi^m(r)(\mathbf{v}||\mathbf{u}) - \Phi^{m+1}(r)(\mathbf{v}||\mathbf{u}), \tag{1}$$

where $\Phi^m(r)(\mathbf{v}||\mathbf{u}) = \frac{1}{N-m+1}\sum_{i=1}^{N-m+1} \log C_i^m(r)(\mathbf{v}||\mathbf{u})$ and $C_i^m(r)(\mathbf{v}||\mathbf{u})$ is the number of sequences, of m consecutive points, of $\mathbf{u}$ that are approximately (within a resolution r) the same as sequences, of the same length, of $\mathbf{v}$. One major dawback of this approach is that $C_i^m(r)(\mathbf{v}||\mathbf{u})$ should not be equal to zero. This is why cross-ApEn is not really adapted for a short time series. Furthermore, it is direction-dependent because often $\Phi^m(r)(\mathbf{v}||\mathbf{u})$ is generally not equal to its direction conjugate $\Phi^m(r)(\mathbf{u}||\mathbf{v})$ [2]. The value of cross-ApEn computed from two signals can be interpreted as a degree of synchrony or mutual relationship.

2.1.2. Binarized Cross-Approximate Entropy

Binarized cross-approximate entropy (XBinEn), introduced by Škorić et al. [5] in 2017, is an evolution of cross-ApEn to quantify the similarity between two time series. It has the advantage of being faster than cross-ApEn. XBinEn encodes a time series divided into vectors of length m. For two vectors $\mathbf{u}$ and $\mathbf{v}$ of length N, the XBinEn algorithm follows these six steps:

1. Binary encoding series are obtained as:

$$x_i = \begin{cases} 0 & \text{if } u_{i+1} - u_i \leqslant 0 \\ 1 & \text{if } u_{i+1} - u_i > 0 \end{cases}, \quad y_i = \begin{cases} 0 & \text{if } v_{i+1} - v_i \leqslant 0 \\ 1 & \text{if } v_{i+1} - v_i > 0 \end{cases}, \tag{2}$$

where $i = 1, 2, ..., N-1$, $x_i \in \mathbf{X}_m^{(i)} = [x_i, x_{i+t}, ..., x_{i+(m-1)t}]$, and $y_i \in \mathbf{Y}_m^{(i)} = [y_i, y_{i+t}, ..., y_{i+(m-1)t}]$. The time lag t allows a vector decorrelation to be performed;

2. Vector histograms $N_X^{(m)}(k)$ and $N_Y^{(m)}(n)$ are computed as:

$$N_X^{(m)}(k) = \sum_{i=1}^{N-(m-1)t} I\{\sum_{l=0}^{m-1} x_{i+l\cdot t} \times 2^l = k\}, \quad N_Y^{(m)}(n) = \sum_{j=1}^{N-(m-1)t} I\{\sum_{l=0}^{m-1} y_{j+l\cdot t} \times 2^l = n\}, \tag{3}$$

where $k, n = 0, 1, ..., 2^m - 1$, and $I\{\cdot\}$ is a function that is equal to 1 if the indicated condition is fulfilled;

3. The probability mass functions are obtained as:

$$P_X^{(m)}(k) = \frac{N_X^{(m)}(k)}{N-(m-1)t}, \quad P_Y^{(m)}(n) = \frac{N_Y^{(m)}(n)}{N-(m-1)t}, \tag{4}$$

where $k, n = 0, 1, ..., 2^m - 1$;

4. A distance measure is applied:

$$d(\mathbf{X}_m^{(i)}, \mathbf{Y}_m^{(j)}) = \sum_{k=0}^{m-1} I\{x_{i+k\cdot t} \neq y_{j+k\cdot t}\}, \tag{5}$$

where $i, j = 1, ..., N-(m-1)t$;

5. The probability $p_k^m(r)$ that a vector is within the distance r from a particular vector is estimated:

$$p_k^m(r) = \Pr\{d(\mathbf{X}_m^{(k)}, \mathbf{Y}_m) \leqslant r\}; \tag{6}$$

6. XBinEn is finally obtained as:

$$\text{XBinEn}(m, r, N, t) = \Phi^{(m)}(r, N, t) - \Phi^{(m+1)}(r, N, t), \tag{7}$$

where $\Phi^{(m)}(r, N, t) = \sum_{k=0}^{2^m-1} P_X^{(m)}(k) \cdot \ln(p_k^m(r))$.

This method gives almost the same results as cross-ApEn for a non-short time series. However, it is computationally more efficient than cross-ApEn. Its main disadvantage is that it cannot identify small signal changes. XBinEn is adapted to environments where processor resources and energy are limited but it is not a substitute to cross-ApEn [5]. It is proposed when the cross-ApEn procedure cannot be applied. The value of XBinEn computed from two signals can be interpreted as a degree of relationship between a related pair of time series.

2.2. Cross-Sample Entropy-Based Measures

2.2.1. Cross-Sample Entropy

Cross-sample entropy (cross-SampEn) quantifies the degree of asynchronism of two time series. This method was introduced by Richman and Moorman in 2000 to improve the cross-ApEn limitations (see Section 2.1.1) [2]. Cross-SampEn is a conditional probability measure that quantifies the probability that a sequence of m consecutive points (called sample) of a time series $\mathbf{u}$—that matches another sequence of the same length of another time series **v**—will still match the other sequence when their length is increased by one sample ($m+1$). For two vectors **u** and **v**, cross-SampEn is computed as:

$$\text{cross-SampEn}(m,r,N)(\mathbf{v}||\mathbf{u}) = -\ln \frac{A^m(r)(\mathbf{v}||\mathbf{u})}{B^m(r)(\mathbf{v}||\mathbf{u})}, \tag{8}$$

where m is the sample length, N is the vectors (**u** and **v**) length, $A^m(r)(\mathbf{v}||\mathbf{u})$ and $B^m(r)(\mathbf{v}||\mathbf{u})$ are, respectively, the probability that a sequence of **u** and a sequence of **v** will match for $m+1$ and m points (within a tolerance r).

For two time series **u** and **v** of length N, cross-SampEn can also be described as:

$$\text{cross-SampEn}(\mathbf{u},\mathbf{v},m,r,N) = -\ln \frac{n^{(m+1)}}{n^{(m)}}, \tag{9}$$

where $n^{(m)}$ represents the total number of sequences of m consecutive points of **u** that match with other sequences of m consecutive points of **v**.

The main difference between cross-ApEn and cross-SampEn is that cross-SampEn shows relative consistency whereas cross-ApEn does not. Unlike cross-ApEn, cross-SampEn is not direction-dependent. However, cross-SampEn generates, sometimes, undefined values for short time series. The value of cross-SampEn computed from two time series can be interpreted as a measure of similarity of the two time series.

2.2.2. Modified Cross-Sample Entropy

Modified cross-sample entropy (mCSE), introduced by Yin and Shang in 2015, has been developed to detect the asynchrony of a financial time series [4]. Inspired by the generalized sample entropy, proposed by Silva and Murta, Jr. [25], the authors proposed to adapt this method to cross-SampEn. The method combines cross-SampEn and nonadditive statistics. For two vectors **u** and **v** of length N, mCSE is computed as:

$$\text{mCSE}(m,r,N) = -\log_q \frac{\sum_{i=1}^{N-m} n_i^{(m+1)}}{\sum_{i=1}^{N-m} n_i^{(m)}}, \tag{10}$$

where m is the sample length, q is the entropic index, and $n_i^{(m)}$ is the number of times that the distance between vectors $\mathbf{y}_m = \{v(i), v(i+1), ..., v(i+m-1) : 1 \leqslant i \leqslant N-m+1\}$ and $\mathbf{x}_m = \{u(i), u(i+1), ..., u(i+m-1) : 1 \leqslant i \leqslant N-m+1\}$ is less than or equal to the tolerance r. The distance is calculated with $d(x_m(i), y_m(i)) = \max\{|u(i+k) - v(j+k)| : 0 \leqslant k \leqslant m-1\}$.

The value of mCSE computed from two time series can be interpreted as a degree of synchrony between the two time series and it can illustrate some intrinsic relations between the two time series.

2.2.3. Modified Cross-Sample Entropy Based on Symbolic Representation and Similarity

Modified cross-sample entropy based on symbolic representation and similarity (MCSEBSS), introduced by Wu et al. in 2018, has been developed to quantify the degree of asynchrony of two financial time series with various trends (stock markets from different areas [6]). In comparison with cross-SampEn, this method reduces the probability of including undefined entropies and it is more robust to noise. For two vectors **u** and **v** of length N, MCSEBSS is computed as:

$$\text{MCSEBSS}(\mathbf{u},\mathbf{v},m,r,N) = -\ln \frac{n^{(m+1)}}{n^{(m)}}, \tag{11}$$

where m is the sample length and $n^{(m)}$ is the number of template matches by comparing $s(\mathbf{u}_m(i), \mathbf{v}_m(j))$ and r. For $\mathbf{u}_m = \{u(i+k)\}$ and $\mathbf{v}_m = \{v(i+k)\}$ $(0 \leqslant k \leqslant m-1$ and $1 \leqslant i \leqslant N-m)$, the similarity function $s(u_m(i), v_m(j))$ is calculated as:

$$s(u_m(i), v_m(j)) = \frac{\#\text{ of 1 in count}(i,j)}{m}, \qquad 1 \leqslant i,j \leqslant N-m, \tag{12}$$

where $count(i,j)$ is obtained by the function f defined as:

$$f = \begin{cases} 1 & \text{if } u_m(i+k) = v_m(j+k) \\ 0 & \text{if } u_m(i+k) \neq v_m(j+k) \end{cases}, \quad 0 \leqslant k \leqslant m-1. \tag{13}$$

The parameter r must be fixed between $\frac{m-n}{m+1}$ and $\frac{m-n}{m}$, where n is the maximum number of zeros obtained with $count(i,j)$ to consider $\mathbf{u}$ and $\mathbf{v}$ similar.

The value of MCSEBSS computed from two time series can be interpreted as a degree of asynchrony of the two time series. A low cross-entropy value indicates a strong synchrony between two signals.

2.2.4. Kronecker-Delta-Based Cross-Sample Entropy

The Kronecker-delta-based cross-sample entropy (KCSE), introduced by He et al. in 2018, has been developed to define the dissimilarity between two time series [7]. KCSE is based on the Kronecker-delta function $\delta_{x,y}$ that returns 1 if two variables are equal and 0 otherwise. For two vectors $\mathbf{u}$ and $\mathbf{v}$ of length N, KCSE is calculated as:

$$\text{KCSE}(m) = -\ln \frac{B^{m+1}}{B^m}, \tag{14}$$

where $B^m = \frac{\sum_{i=1}^{N-m+1} \text{KrD}_{u_m(i),v_m(i)}}{N-m+1}$ and $B^{m+1} = \frac{\sum_{i=1}^{N-m} \text{KrD}_{u_{m+1}(i),v_{m+1}(i)}}{N-m}$. The dissimilarity, between $\mathbf{u}_m(i) = [u(i), u(i+1), ..., u(i+m-1)]$ and $\mathbf{v}_m(i) = [v(i), v(i+1), ..., v(i+m-1)]$, is calculated as:

$$\text{KrD}_{u_m(i),v_m(i)} = \frac{\delta_{u_m(i),v_m(i)} + \delta_{u_m(i+1),v_m(i+1)} + \cdots + \delta_{u_m(i+m-1),v_m(i+m-1)}}{n}. \tag{15}$$

Authors show that KCSE is better to classify financial data than multidimensional scaling based on the Chebyshev distance method [7]. The value of KSCE computed from two time series can be interpreted as a degree of irregularity between the two time series.

2.2.5. Permutation-Based Cross-Sample Entropy

The permutation-based cross-sample entropy (PCSE), introduced by He et al. in 2018, is quite similar to KCSE (see Section 2.2.4) [7]. A permutation step has only been added. For two vectors $\mathbf{u}$ and $\mathbf{v}$ of length N, PCSE is calculated as:

$$\text{PCSE}(m) = -\ln \frac{B^{m+1}}{B^m}, \tag{16}$$

where $B^m = \frac{\sum_{i=1}^{N-m+1} \text{KrD}_{\text{permuX}_m(i),\text{permuY}_m(i)}}{N-m+1}$ and $B^{m+1} = \frac{\sum_{i=1}^{N-m} \text{KrD}_{\text{permuX}_{m+1}(i),\text{permuY}_{m+1}(i)}}{N-m}$. The KrD function is defined in Section 2.2.4. The two vectors $\mathbf{permuX}_m(i)$ and $\mathbf{permuY}_m(i)$ are obtained by a permutation algorithm defined with the permutation entropy [26]. The Video S1 shows an example of a permutation algorithm.

PCSE shows better results than KCSE for synthetic data (ARFIMA model) [7]. However, the two approaches give the same results for financial data [7]. Authors show that KCSE is better to classify financial data than multidimensional scaling based on the Chebyshev distance method [7]. The value of PCSE computed from two time series can be interpreted as degree of irregularity between the two time series.

2.2.6. Cross-Trend Sample Entropy

Inspired by MCSEBSS (see Section 2.2.3), Wng et al. developed the cross-trend sample entropy (CTSE) to quantify the synchronism between two time series with strong trends [23]. For two time series **u** and **v** of length N, CTSE is calculated with the following four steps algorithm:

1. The two time series are symbolized as:

$$U(j) = \begin{cases} 1 & \text{if } \tilde{u}(j) > u(j) \\ 0 & \text{otherwise} \end{cases}, \qquad V(j) = \begin{cases} 1 & \text{if } \tilde{v}(j) > v(j) \\ 0 & \text{otherwise} \end{cases}, \qquad 1 \leqslant j \leqslant N, \tag{17}$$

   where $\tilde{\mathbf{u}}$ and $\tilde{\mathbf{v}}$ are, respectively, the trend of **u** and **v** obtained by polynomial fitting (linear, quadratic or higher order).

2. The template vectors $\mathbf{u}_m$ and $\mathbf{v}_m$ are constructed as:

$$\mathbf{u}_m(i) = \{U(i+k)\}, \qquad \mathbf{v}_m(i) = \{V(i+k)\}, \tag{18}$$

   where $0 \leqslant k \leqslant m-1$ and $1 \leqslant i \leqslant N-m$.

3. The similarity between $x_m(i)$ and $y_m(i)$ is calculated as:

$$d(x_m(i), y_m(i)) = \frac{\#\text{ of 1 in } C_m(i)}{m}, \qquad 1 \leqslant i \leqslant N-m, \tag{19}$$

   where the i-th symbol vector C_m is determined with f, a symbolic function between two template vectors $\mathbf{u}_m$ and $\mathbf{v}_m$, as:

$$f = \begin{cases} 1 & \text{if } u_m(i+k) = v_m(i+k) \\ 0 & \text{otherwise} \end{cases}, \qquad 0 \leqslant k \leqslant m-1. \tag{20}$$

4. CTSE is finally computed as:

$$\text{CTSE}(u, v, r, N) = -\ln \frac{n^{(m+1)}}{n^{(m)}}, \tag{21}$$

   where $n^{(m)}$ is obtained by comparing $d(x_m(i), y_m(j))$ within a tolerance r for $1 \leqslant i \leqslant N-m$.

CTSE has two advantages over MCSEBSS: It is more sensitive to the difference of dynamical characteristic between two signals, and it works well with signals with trends (linear, quadratic, cubic, and sinusoidal) [23]. The value of CTSE computed from two time series can be interpreted as an indicator of dynamical structure regarding the two time series with potential trends.

2.3. Cross-Distribution Entropy-Based Measures

2.3.1. Cross-Distribution Entropy

In 2018, Wang and Shang introduced the cross-distribution entropy (cross-DistEn) to quantify the complexity between two cross-sequences [9]. To generalize the standard statistical mechanics, the authors replaced the standard distribution entropy (DistEn) based on Shannon entropy by DistEn based on Tsallis entropy [9]. The authors showed that cross-DistEn better illustrates the relationships between two vectors than cross-SampEn does [9]. For two times series **u** and **v** of length N cross-DistEn follow these four steps:

1. The state-space is reconstructed by building $(N-m+1)$ vectors $\mathbf{X}(i)$ and $\mathbf{Y}(i)$ with $\mathbf{X}(i) = \{u(i), u(i+1), ..., u(i+m-1)\}$, $1 \leq i \leq N-m$, and $\mathbf{Y}(j) = \{v(j), v(j+1), ..., v(j+m-1)\}$, $1 \leq j \leq N-m$. m is the intended size of the vectors $\mathbf{X}(i)$ and $\mathbf{Y}(i)$;
2. The distance matrix is built by defining the distance matrix $\mathbf{D} = \{d_{i,j}\}$ with $d_{i,j}$ being the Chebyshev distance between the vectors $\mathbf{X}(i)$ and $\mathbf{Y}(j)$ defined as:

$$d_{ij} = \max\{|u_{i+k}^{(\tau)} - v_{j+k}^{\tau}|, 0 \leqslant k \leqslant m-1\}; \tag{22}$$

3. The probability density is estimated by computing the empirical probability density function of the matrix $\mathbf{D}$ by applying the histogram approach. If the histogram has M bins, the probability of each bin will be P_t with $1 \leq t \leq M$;
4. The cross-distribution entropy based on the Tsallis entropy is computed as:

$$crossDistEn(\mathbf{u}, \mathbf{v}) = \frac{1}{\ln(a)} \frac{1}{q-1}(1 - \sum_{t=1}^{M} P_t^q), \tag{23}$$

where q is the order of the Tsallis entropy and a the logarithm base of the entropy computation.

The main advantage of cross-DistEn is that it is adapted for short time series. With financial data, cross-DistEn illustrates better the relationship between signals than cross-SampEn [9]. The value of cross-DistEn computed from two time series can be interpreted as a degree of linkage of the two time series.

2.3.2. Permutation Cross-Distribution Entropy

The permutation cross-distribution entropy (PCDE), introduced by He et al. in 2019, is a variant of cross-DistEn (see Section 2.3.1) [10]. The permutation allows to characterize fluctuations and prevents the impact of spatial distances on results. The PCDE algorithm is the same as the one of cross-DistEn, detailed in Section 2.3.1. However, an additional step is added before step 2 to permute $\mathbf{X}(i)$ and $\mathbf{Y}(j)$ with the permutation algorithm mentioned in Section 2.2.5. The distance matrix is therefore constructed with the permuted vectors. The value of PCDE computed from two time series can be interpreted as a degree of dissimilarity between the two time series.

2.4. Other Cross-Entropy-Based Measures

2.4.1. Cross-Conditional Entropy

Cross-conditional entropy (CCE), introduced by Porta et al. in 1919, quantifies the degree of coupling between two signals [8]. A corrected conditional entropy has been introduced to improve the approximate entropy that suffers from limitations when a finite number of sample is considered [27]. CCE is an adaptation of the corrected conditional entropy. For two signals $\mathbf{u} = \{u(i), \quad i = 1, ..., N\}$ and $\mathbf{v} = \{v(i), \quad i = 1, ..., N\}$, CCE is computed as:

$$\text{CCE}_{\mathbf{v}/\mathbf{u}}(L) = -\sum_{L-1} p(\mathbf{u}_{L-1}) \sum_{i/L-1} p(v(i)/\mathbf{u}_{L-1}) \times \log p(v(i)/\mathbf{u}_{L-1}), \tag{24}$$

where L is the length of the pattern extracted to be compared, $p(\mathbf{u}_{L-1})$ is the joint probability of the pattern $\mathbf{u}_{L-1}(i) = (u(i), u_{L-1}(i-1))$, and $p(v(i)/\mathbf{u}_{L-1})$ is the probability of the sample $v(i)$ given the pattern $\mathbf{u}_{L-1}(i)$. If a mixed pattern, composed by $L-1$ samples, of $\mathbf{u}$ and $\mathbf{v}$: $(v(i), u(i), ..., u(i-L+2)) = (v(i), \mathbf{u}_{L-1})$, is defined and with the Shannon entropy $E(\mathbf{u}_L) = -\sum_L p(\mathbf{u}_L) \log p(\mathbf{u}_L)$, CCE can also be described as:

$$\text{CCE}_{\mathbf{v}/\mathbf{u}}(L) = E(v(i), \mathbf{u}_{L-1}) - E(\mathbf{u}_{L-1}). \tag{25}$$

For a limited amount of samples, the approximation of CCE always decreases to zero while increasing L. To solve this problem, a modification has been introduced as:

$$\mathrm{CCE}_{\mathbf{v}/\mathbf{u}}(L) = \widehat{\mathrm{CCE}_{\mathbf{v}/\mathbf{u}}}(L) + \mathrm{perc}_{\mathbf{v}/\mathbf{u}}(L) \times \widehat{E}(\mathbf{v}), \tag{26}$$

where $\mathrm{perc}_{\mathbf{v}/\mathbf{u}}$ is the ratio of mixed patterns found only once over the total number of mixed patterns, $\widehat{\mathrm{CCE}_{\mathbf{v}/\mathbf{u}}}(L)$ and $\widehat{E}(\mathbf{v})$ are, respectively, the estimates of the $\mathrm{CCE}_{\mathbf{v}/\mathbf{u}}(L)$ and $E(\mathbf{v})$ based on the considered limited dataset.

CCE can be defined as a measure of unpredictability of one signal when the second is observed because it quantifies the amount of information carried by one signal which cannot be derived from the other. It is not fully a measure of synchronization. The main disadvantage of CCE is that it is not totally adapted for short time series.

2.4.2. Cross-Fuzzy Entropy

Cross-fuzzy entropy (C-FuzzyEn), introduced by by Xie et al. in 2010 [3], is an adaptation of fuzzy entropy, introduced by Chen et al. [28], that quantifies the synchrony or similarity of patterns between two signals [3]. C-FuzzyEn is an improvement of cross-SampEn that is more adapted for short time series and more robust to noise. For two times series **u** and **v** of length N, C-FuzzyEn is obtained with the following three steps algorithm:

1. The distance d_{ij}^m between $\mathbf{X}_i^m$ and $\mathbf{Y}_j^m$ is computed as:

$$d_{ij}^m = d[\mathbf{X}_i^m, \mathbf{Y}_j^m] = \max_{k \in (0, m-1)} |u(i+k) - \overline{u}(i) - v(j+k) - \overline{v}(i)|, \tag{27}$$

   where m is the number of consecutive data to compare, $\mathbf{X}_i^m = \{u(i), u(i+1), ..., u(i+m-1)\} - \overline{u}(i)$, and $\mathbf{Y}_j^m = \{v(i), v(i+1), ..., u(v+m-1)\} - \overline{v}(i)$. $\overline{u}(i)$ and $\overline{v}(i)$ are calculated as: $\overline{u}(i) = \frac{1}{m}\sum_{l=0}^{m-1} u(i+l)$, and $\overline{v}(i) = \frac{1}{m}\sum_{l=0}^{m-1} v(i+l)$;
2. The synchrony or similarity degree D_{ij}^m is computed as: $D_{ij}^m = \mu(d_{ij}^m, n, r)$, where $\mu(d_{ij}^m, n, r)$ is the fuzzy function obtained as:

$$\mu(d_{ij}^m, n, r) = \exp -\frac{(d_{ij}^m)^n}{r}, \tag{28}$$

   where r and n determine the width and the gradient of the boundary of the exponential function, respectively;
3. Finally, C-FuzzyEn is computed as:

$$\text{C-FuzzyEn}(m, n, r) = \ln \Phi^m - \ln \Phi^{m+1}, \tag{29}$$

   where $\Phi^m = \frac{1}{N-m}\sum_{i=1}^{N-m}(\frac{1}{N-m}\sum_{j=1}^{N-m} D_{ij}^m)$, and $\Phi^{m+1} = \frac{1}{N-m}\sum_{i=1}^{N-m}(\frac{1}{N-m}\sum_{j=1}^{N-m} D_{ij}^{m+1})$.

The value of C-FuzzyEn computed from two time series can be interpreted as the synchronicity of patterns.

2.4.3. Joint-Permutation Entropy

Joint permutation entropy (JPE), introduced by Yin et al. in 2019, quantifies the synchronism between two time series. It is based on permutation entropy that consists of comparing neighboring values of each point and mapping them to ordinal patterns to quantify the complexity of a signal [26]. For two signals **u** and **v**, JPE is computed as the Shannon entropy of the $d! \times d!$ distinct motif combinations $\{(\pi_i^{d,t}, \pi_j^{d,t})\}$:

$$\mathrm{JPE}(d, t) = -\sum_{i,j:(\pi_i^{d,t}, \pi_j^{d,t})} p(\pi_i^{d,t}, \pi_j^{d,t}) \cdot \ln p(\pi_i^{d,t}, \pi_j^{d,t}), \tag{30}$$

where d is the embedded dimension and $p(\pi_i^d, \pi_j^d)$ is the joint probability of $\{(\pi_i^{d,t}, \pi_j^{d,t})\}$ appearing in the $\mathbf{X}_l^{d,t} = \{u_l, u_{l+t}, ..., u_{l+(d-1)t}\}$ and $\mathbf{Y}_l^{d,t} = \{v_l, v_{l+t}, ..., v_{l+(d-1)t}\}$ and it is defined as:

$$p(\pi_i^{d,t}, \pi_j^{d,t}) = \frac{||l : l \leqslant T, \text{type}(X_l^{d,t}, Y_l^{d,t}) = (\pi_i^{d,t}, \pi_j^{d,t})||}{T}, \tag{31}$$

where $T = N - (d-1)t$, type$(\cdot)$ corresponds to the map from pattern space to symbol space, and $||\cdot||$ corresponds to the cardinality of a set.

The main advantages of JPE are the simplicity, the robustness, and the low computational cost. The value of JPE computed from two time series can be interpreted as a degree of correlation between the two time series [29].

3. Multiscale Procedures

To study entropy or cross-entropy measures of time series across scales, a multiscale procedure can be used. In this part we detail, in chronological order, three multiscale methods: The coarse-grained, the time-shift, and the composite coarse-grained approaches.

3.1. Coarse-Graining Procedure

In 2002 Costa et al. introduced the coarse-graining procedure to analyze the complexity, defined by the analysis of the irregularity through scale factors [14]. This method is an improvement, more adapted for a biological time series, of the coarse-graining procedure introduced by Zhang [30]. This procedure has been used in multiscale entropy and cross-entropy methods [6,9,15,20,31–33]. For each scale factor, this procedure derives a set of vectors illustrating the system dynamics. For a monovariate discrete signal $\mathbf{x}$ of length N, the coarse-grained time series $\mathbf{y}^{(\tau)}$ is calculated as:

$$y_j^{(\tau)} = \frac{1}{\tau} \sum_{i=(j-1)\tau+1}^{j\tau} x_i, \tag{32}$$

where τ is the scale factor and $1 \leqslant j \leqslant \frac{N}{\tau}$. The length of the coarse-grained vector is $\frac{N}{\tau}$. An example of coarse-graining procedure is presented in Figure 1A.

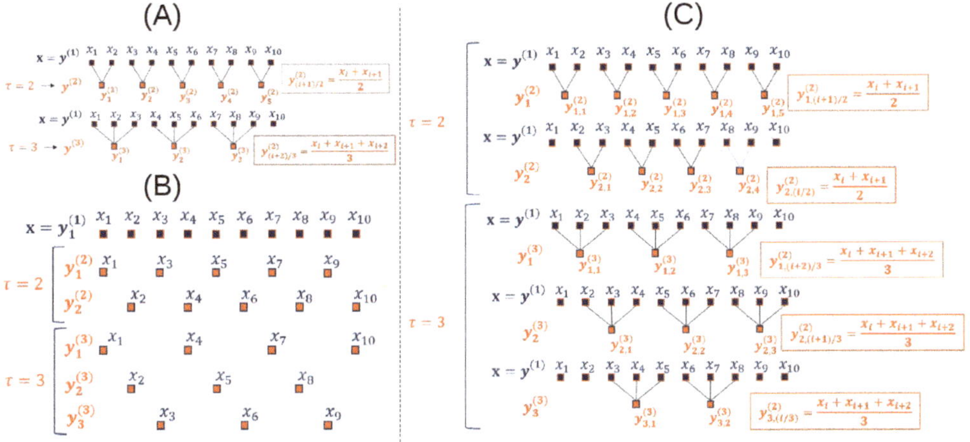

Figure 1. Examples of multiscale procedures for the ten first points of a time series $\mathbf{x}$. (**A**) represents the coarse-graining procedure (modified from [34]), (**B**) shows the time-shift procedure, and (**C**) illustrates the composite coarse-graining procedure (modified from [35]).

3.2. Time-Shift Procedure

As for the coarse-grained procedure, the time-shift procedure is used to decompose a signal through different scale factors and to perform a multiscale analysis. While coarse-graining procedure uses the averaging of time series on several interval scales, the time-shift procedure applies time shifting in time series. The main disadvantage of a coarse-graining procedure is the loss of pattern information hidden in the time series. To overcome this limitation, Pham used the Higuchi's fractal dimension (HFD) [36] and proposed a new multiscale analysis [37]. The time-shift procedure illustrates the fractal dimension of a signal. This method has been recently used with entropy and cross-entropy measures [17,37–39]. HFD shows stable numerical results for stationary, non-stationary, deterministic, and stochastic time series [40]. For a monovariate discrete signal $\mathbf{x}$ of length N, the β time-shift signal $\mathbf{y}_\beta^{(\tau)}$ is calculated as:

$$\mathbf{y}_\beta^{(\tau)} = (x_\beta, x_{\beta+\tau}, ..., x_{\beta+\lfloor\frac{N-\beta}{\tau}\rfloor\tau}). \tag{33}$$

For each time scale τ, β time-shift time series are computed ($\beta = 1, 2, ..., \tau$). An illustration of the time-shift procedure is presented in Figure 1B.

3.3. Composite Coarse-Graining Procedure

The coarse-graining procedure, introduced by Costa et al. [14], increases the variance of estimated entropy values at large scale. To overcome this limitation, by Wu et al. introduced in 2013 a composite coarse-graining procedure [35]. This method has been used with entropy and cross-entropy measures [16,32]. For a monovariate discrete signal $\mathbf{x}$ of length N, the k-th composite coarse-grained time series $\mathbf{y}_k^{(\tau)}$ is computed as:

$$y_{k,j} = \frac{1}{\tau} \sum_{i=(j-1)\tau+k}^{j\tau+k-1} x_i, \tag{34}$$

where $1 \leqslant j \leqslant \frac{N}{\tau}$. For each time scale τ, k composite coarse-grained time series are computed ($1 \leqslant k \leqslant \tau$). An illustration of the composite coarse-graining procedure is presented in Figure 1C.

4. Multiscale Cross-Entropy Methods

4.1. Generalization

Multiscale cross-entropy (MCE) methods consist of applying a cross-entropy measure for each scale factor obtained by a specific procedure. For each scale factor τ, MCE is computed as:

$$MCE(\mathbf{X}^{(\tau)}, \mathbf{Y}^{(\tau)}) = \frac{1}{k} \sum_{\beta=1}^{k} crossEn(\mathbf{X}_\beta^{(\tau)}, \mathbf{Y}_\beta^{(\tau)}), \tag{35}$$

where $\mathbf{X}^{(\tau)}$ and $\mathbf{Y}^{(\tau)}$ are computed with a multiscale procedure (see Section 3), k is the number of time series that are generated by the multiscale procedure ($k = 1$ for the coarse-graining procedure and $k = \tau$ for the time-shift and the composite coarse-graining procedures), and crossEn is the cross-entropy method used (see Section 2). Table 2 shows the multiscale cross-entropy methods that can be generalized with Equation (35). Before the computation of MCE, a pre-treatment can be performed. For example, the asymetric multiscale cross-SampEn (AMCSE) method [33] decomposes each signal into two, one for the positive trends and the other for the negative trends, before applying a coarse-graining procedure and cross-SampEn.

Table 2. Multiscale cross-entropy methods, in chronological order, that can be generalized with Equation (35). For each method, the multiscale procedure and the cross-entropy measure used and the reference are mentioned.

Method	Multiscale Procedure	Cross-entropy Measure	Reference
Multiscale cross-SampEn	Coarse-grained	cross-SampEn	Yan et al., 2009 [15]
Multiscale cross-ApEn	Coarse-grained	cross-ApEn	Wu et al., 2013 [31]
Asymetric multiscale cross-SampEn	Coarse-grained	cross-SampEn	Yin and Shang, 2015 [33]
Composite multiscale cross-SampEn	Composite coarse-grained	cross-SampEn	Yin et al., 2016 [16]
Multiscale cross-DistEn	Coarse-grained	cross-DistEn	Wang and Shang, 2018 [9]
Modified multiscale cross-SampEn based on symbolic representation and similarity	Coarse-grained	MCSEBSS	Wu et al., 2018 [6]
Modified multiscale cross-SampEn	Coarse-grained	mCSE	Castiglioni et al., 2019 [20]
Time-shift multiscale cross-SampEn	Time-shift	cross-SampEn	Jamin et al., 2019 [17]
Time-shift multiscale cross-DistEn	Time-shift	cross-DistEn	Jamin et al., 2019 [17]
Multiscale cross-trend SampEn	Coarse-grained	CTSE	Wang et al., 2019 [23]
Multiscale joint permutation entropy	Coarse-grained	JPE	Yin et al., 2019 [24]

4.2. Particular Cases

Some multiscale cross-entropy methods cannot follow the generalization previously introduced. In this part we detail three particular methods, in chronological order: The adaptive multiscale cross-SampEn, the refined composite multiscale cross-SampEn, and the percussion entropy.

4.2.1. Adaptive Multiscale Cross-Sample Entropy

The adaptive multiscale cross-sample entropy (AMCSE), introduced by Hu and Liang in 2011, assesses the nonlinear interdependency between different visual cortical areas [41]. The method uses the multivariate empirical mode decomposition (MEMD), introduced by Rehman and Mandic [42], to decompose two time series into intrinsic mode functions (IMFs) that represent the oscillation mode embedded in the data. For two time series **u** and **v**, AMCSE is calculated with the following three steps algorithm:

1. The MEMD on **u** and **v** is performed to obtain N IMFs;
2. The scales of data are computed in two directions, fine-to-coarse S^{τ}_{f2c} and coarse-to-fine S^{τ}_{c2f}, with the following two equations:

$$S^{\tau}_{f2c} = \sum_{i=\tau}^{N} \mathrm{IMF}_i, \quad (\tau \leqslant N), \tag{36}$$

$$S^{\tau}_{c2f} = \sum_{i=1}^{N+1-\tau} \mathrm{IMF}_i, \quad (\tau \leqslant N). \tag{37}$$

   The two directions can be used separately or used in tandem to reveal the underlying dynamics of complex time series;
3. For each scale factor τ, the cross-SampEn (see Section 2.2.1) is applied between the two scales of data (S^{τ}_{f2c} and S^{τ}_{c2f}) extracted from vectors **u** and **v**.

4.2.2. Refined Composite Multiscale Cross-Sample Entropy

Yin et al. introduced in 2016 the composite multiscale cross-sample entropy (CMCSE) that follows the generalization (see Section 4.1), where the composite coarse-graining procedure and cross-SampEn are used [16]. The main disadvantage of this method is that cross-SampEn generates some undefined values when the number of matched sample is zero. To overcome this limitation, Yin et al. introduced the refined CMCSE (RCMCSE). This method leads to better results with short time series. For two times series **u** and **v** of length N, RCMSE is computed with the following three steps algorithm:

1. Coarse-grained time series are obtained with the composite coarse-graining procedure detailed in Section 3.3;
2. For a scale factor τ, the number of matched vector pairs, $n^m_{k,\tau}$ and $n^{m+1}_{k,\tau}$, are calculated for all coarse-grained vectorsl
3. For each scale factor τ, RCMCSE is computed as:

$$\mathrm{RCMCSE}(\mathbf{u},\mathbf{v},\tau,m,r) = -\ln\frac{\sum_{k=1}^{\tau} n^{m+1}_{k,\tau}}{\sum_{k=1}^{\tau} n^{m}_{k,\tau}}, \tag{38}$$

where m is the dimension and of the matched vector pairs and r is the distance tolerance for the matched vector pairs.

4.2.3. Percussion Entropy

Wu et al. introduced, in 2013, the multiscale small-scale entropy index (MEI_{SS}) that is obtained by summing the values of entropy for the first five scale factors [43]. Percussion entropy, introduced by Wei et al. in 2019, allows one to quantify a percussion entropy index (PEI) [18]. The method has been introduced to assess baroreflex sensitivity. PEI compares the similarity in tendency of change between two time-series. This index has been compared to MEI_{SS}. For two time series **u** and **v** of length N, PEI is computed with the following three steps algorithm:

1. A binary transformation of **u** and **v** is used to obtain $\mathbf{x} = \{x_1, x_2, ..., x_{N-1}\}$ and $\mathbf{y} = \{y_1, y_2, ..., y_{N-1}\}$:

$$x_i = \begin{cases} 0 & u(i+1) \leqslant u(i) \\ 1 & u(i+1) > u(i) \end{cases} \quad y_i = \begin{cases} 0 & v(i+1) \leqslant v(i) \\ 1 & v(i+1) > v(i) \end{cases}; \tag{39}$$

2. The percussion rate for each scale factor τ is computed as:

$$P^m_\tau = \frac{1}{n-m-\tau+1}\sum_{i=1}^{n-m-\tau+1} \mathrm{count}(i), \tag{40}$$

where m is the embedded dimension vectors and $\mathrm{count}(i)$ represents the match number between $\mathbf{A}(i) = \{x_i, x_{i+1}, ..., x_{i+m-1}\}$ and $\mathbf{B}(i+\tau) = \{y_{i+\tau}, y_{i+\tau+1}, ..., y_{i+\tau+m-1}\}$;
3. PEI is calculated as:

$$\mathrm{PEI}(m, n_\tau) = \phi^m - \phi^{m+1}, \tag{41}$$

where $\phi^m = \ln\sum_{\tau=1}^{n_\tau} P^m_\tau$ and n_τ is the number of scales to consider. Wei et al. [18] have chosen $n_\tau = 5$ in accordance with MEI_{SS}.

This algorithm is a generalization of the method developed by Wei et al. [18] for a specific time series, amplitudes of successive digital volume pulse signals and changes in R-R intervals of successive cardiac cycles. At the moment, it has not been used to process other kinds of signals.

5. Conclusions

In this review we proposed a state-of-the-art of cross-entropy measures, multiscale procedures, and multiscale cross-entropy methods. Multiscale cross-entropy methods offer other interesting perspectives for time series analysis. Furthermore, all the cross-entropy methods, detailed in this review, can be translated into multiscale cross-entropy methods with the multiscale procedures presented in this review.

Supplementary Materials: The following are available online at http://www.mdpi.com/1099-4300/22/1/45/s1, Video S1: Permutation entropy–An exemple to obtain permutation vectors.

Author Contributions: Investigation, A.J. and A.H.-H.; supervision, A.H.-H.; writing–original draft, A.J.; writing–review and editing, A.H.-H. All authors have read and agreed to the published version of the manuscript.

Funding: A CIFRE grant N°2017/1165 was awarded by ANRT to the company COTTOS Médical to support the work of graduate student A.J.

Conflicts of Interest: The authors declare no conflict of interest.

References

1. Pincus, S.; Singer, B.H. Randomness and degrees of irregularity. *Proc. Natl. Acad. Sci. USA* **1996**, *93*, 2083–2088. [CrossRef] [PubMed]
2. Richman, J.S.; Moorman, J.R. Physiological time-series analysis using approximate entropy and sample entropy. *Am. J. Physiol. Heart Circ. Physiol.* **2000**, *278*, H2039–H2049. [CrossRef] [PubMed]
3. Xie, H.B.; Zheng, Y.P.; Guo, J.Y.; Chen, X. Cross-fuzzy entropy: A new method to test pattern synchrony of bivariate time series. *Inf. Sci.* **2010**, *180*, 1715–1724. [CrossRef]
4. Yin, Y.; Shang, P. Modified cross sample entropy and surrogate data analysis method for financial time series. *Phys. A Stat. Mech. Appl.* **2015**, *433*, 17–25. [CrossRef]
5. Škorić, T.; Mohamoud, O.; Milovanovic, B.; Japundzic-Zigon, N.; Bajic, D. Binarized cross-approximate entropy in crowdsensing environment. *Comput. Biol. Med.* **2017**, *80*, 137–147. [CrossRef]
6. Wu, Y.; Shang, P.; Li, Y. Multiscale sample entropy and cross-sample entropy based on symbolic representation and similarity of stock markets. *Commun. Nonlinear Sci. Numer. Simul.* **2018**, *56*, 49–61. [CrossRef]
7. He, J.; Shang, P.; Xiong, H. Multidimensional scaling analysis of financial time series based on modified cross-sample entropy methods. *Phys. A Stat. Mech. Appl.* **2018**, *500*, 210–221. [CrossRef]
8. Porta, A.; Baselli, G.; Lombardi, F.; Montano, N.; Malliani, A.; Cerutti, S. Conditional entropy approach for the evaluation of the coupling strength. *Biol. Cybern.* **1999**, *81*, 119–129. [CrossRef]
9. Wang, Y.; Shang, P. Analysis of financial stock markets through the multiscale cross-distribution entropy based on the Tsallis entropy. *Nonlinear Dyn.* **2018**, *94*, 1361–1376. [CrossRef]
10. He, J.; Shang, P.; Zhang, Y. PID: A PDF-induced distance based on permutation cross-distribution entropy. *Nonlinear Dyn.* **2019**, *97*, 1329–1342. [CrossRef]
11. Škorić, T. Automatic Determination of Cross-Approximate Entropy Parameters for Cardiovascular Time Series. In Proceedings of the 2018 26th Telecommunications Forum (TELFOR), Belgrade, Serbia, 20–21 November 2018; pp. 1–8. [CrossRef]
12. Wang, R.; Li, D.; Wang, J.; Cai, L.; Shi, L. Synchrony analysis using different cross-entropy measures of the electroencephalograph activity in Alzheimer's disease. In Proceedings of the 2016 9th International Congress on Image and Signal Processing, BioMedical Engineering and Informatics (CISP-BMEI), Datong, China, 15–17 October 2016; pp. 1541–1545. [CrossRef]
13. Zhu, K.; Song, X. Cross-fuzzy entropy-based approach for performance degradation assessment of rolling element bearings. *Proc. Inst. Mech. Eng. E J. Process Mech. Eng.* **2018**, *232*, 173–185. [CrossRef]
14. Costa, M.; Goldberger, A.L.; Peng, C.K. Multiscale Entropy Analysis of Complex Physiologic Time Series. *Phys. Rev. Lett.* **2002**, *89*, 068102. [CrossRef] [PubMed]
15. Yan, R.; Yang, Z.; Zhang, T. Multiscale Cross Entropy: A Novel Algorithm for Analyzing Two Time Series. In Proceedings of the 2009 Fifth International Conference on Natural Computation, Tianjin, China, 14–16 August 2009; Volume 1, pp. 411–413, ISSN 2157-9555, 2157-9563. [CrossRef]
16. Yin, Y.; Shang, P.; Feng, G. Modified multiscale cross-sample entropy for complex time series. *Appl. Math. Comput.* **2016**, *289*, 98–110. [CrossRef]
17. Jamin, A.; Duval, G.; Annweiler, C.; Abraham, P.; Humeau-Heurtier, A. A Novel Multiscale Cross-Entropy Method Applied to Navigation Data Acquired with a Bike Simulator. In Proceedings of the 2019 41st Annual International Conference of the IEEE Engineering in Medicine and Biology Society (EMBC), Berlin, Germany, 23–27 July 2019; pp. 733–736, ISSN 1558-4615, 1557-170X. [CrossRef]
18. Wei, H.C.; Xiao, M.X.; Ta, N.; Wu, H.T.; Sun, C.K. Assessment of Diabetic Autonomic Nervous Dysfunction with a Novel Percussion Entropy Approach. *Complexity* **2019**, *2019*, 11. [CrossRef]

19. Xiao, M.X.; Lu, C.H.; Ta, N.; Jiang, W.W.; Tang, X.J.; Wu, H.T. Application of a Speedy Modified Entropy Method in Assessing the Complexity of Baroreflex Sensitivity for Age-Controlled Healthy and Diabetic Subjects. *Entropy* **2019**, *21*. [CrossRef]
20. Castiglioni, P.; Parati, G.; Faini, A. Information-Domain Analysis of Cardiovascular Complexity: Night and Day Modulations of Entropy and the Effects of Hypertension. *Entropy* **2019**, *21*, 550. [CrossRef]
21. Xiao, M.X.; Wei, H.C.; Xu, Y.J.; Wu, H.T.; Sun, C.K. Combination of R-R Interval and Crest Time in Assessing Complexity Using Multiscale Cross-Approximate Entropy in Normal and Diabetic Subjects. *Entropy* **2018**, *20*. [CrossRef]
22. Lin, T.K.; Chien, Y.H. Performance Evaluation of an Entropy-Based Structural Health Monitoring System Utilizing Composite Multiscale Cross-Sample Entropy. *Entropy* **2019**, *21*. [CrossRef]
23. Wang, F.; Zhao, W.; Jiang, S. Detecting asynchrony of two series using multiscale cross-trend sample entropy. *Nonlinear Dyn.* **2019**. [CrossRef]
24. Yin, Y.; Shang, P.; Ahn, A.C.; Peng, C.K. Multiscale joint permutation entropy for complex time series. *Phys. A Stat. Mech. Appl.* **2019**, *515*, 388–402. [CrossRef]
25. Silva, L.E.V.; Murta, L.O., Jr. Evaluation of physiologic complexity in time series using generalized sample entropy and surrogate data analysis. *Chaos* **2012**, *22*, 043105. [CrossRef] [PubMed]
26. Bandt, C.; Pompe, B. Permutation Entropy: A Natural Complexity Measure for Time Series. *Phys. Rev. Lett.* **2002**, *88*, 174102. [CrossRef] [PubMed]
27. Porta, A.; Baselli, G.; Liberati, D.; Montano, N.; Cogliati, C.; Gnecchi-Ruscone, T.; Malliani, A.; Cerutti, S. Measuring regularity by means of a corrected conditional entropy in sympathetic outflow. *Biol. Cybern.* **1998**, *78*, 71–78. [CrossRef] [PubMed]
28. Chen, W.; Wang, Z.; Xie, H.; Yu, W. Characterization of Surface EMG Signal Based on Fuzzy Entropy. *IEEE Trans. Neural Syst. Reh. Eng.* **2007**, *15*, 266–272. [CrossRef] [PubMed]
29. Yin, Y.; Peng, C.K.; Hou, F.; Gao, H.; Shang, P.; Li, Q.; Ma, Y. The application of multiscale joint permutation entropy on multichannel sleep electroencephalography. *AIP Adv.* **2019**, *9*, 125214. [CrossRef]
30. Zhang, Y.C. Complexity and 1/f noise. A phase space approach. *J. Phys. I France* **1991**, *1*, 971–977. [CrossRef]
31. Wu, H.T.; Lee, C.Y.; Liu, C.C.; Liu, A.B. Multiscale Cross-Approximate Entropy Analysis as a Measurement of Complexity between ECG R-R Interval and PPG Pulse Amplitude Series among the Normal and Diabetic Subjects. *Comput. Math. Methods Med.* **2013**, *2013*, 7. [CrossRef]
32. Humeau-Heurtier, A. The multiscale entropy algorithm and its variants: a review. *Entropy* **2015**, *17*. [CrossRef]
33. Yin, Y.; Shang, P. Asymmetric asynchrony of financial time series based on asymmetric multiscale cross-sample entropy. *Chaos* **2015**, *25*, 032101. [CrossRef]
34. Costa, M.; Goldberger, A.L.; Peng, C.K. Multiscale entropy to distinguish physiologic and synthetic RR time series. In Proceedings of the Computers in Cardiology, Memphis, TN, USA, 22–25 September 2002; pp. 137–140. [CrossRef]
35. Wu, S.D.; Wu, C.W.; Lin, S.G.; Wang, C.C.; Lee, K.Y. Time Series Analysis Using Composite Multiscale Entropy. *Entropy* **2013**, *15*, 1069–1084. [CrossRef]
36. Higuchi, T. Approach to an irregular time series on the basis of the fractal theory. *Phys. D Nonlinear Phenom.* **1988**, *31*, 277–283. [CrossRef]
37. Pham, T.D. Time-Shift Multiscale Entropy Analysis of Physiological Signals. *Entropy* **2017**, *19*, 257. [CrossRef]
38. Dong, Z.; Zheng, J.; Huang, S.; Pan, H.; Liu, Q. Time-Shift Multi-scale Weighted Permutation Entropy and GWO-SVM Based Fault Diagnosis Approach for Rolling Bearing. *Entropy* **2019**, *21*. [CrossRef]
39. Zhu, X.; Zheng, J.; Pan, H.; Bao, J.; Zhang, Y. Time-Shift Multiscale Fuzzy Entropy and Laplacian Support Vector Machine Based Rolling Bearing Fault Diagnosis. *Entropy* **2018**, *20*. [CrossRef]
40. Kesić, S.; Spasić, S.Z. Application of Higuchi's fractal dimension from basic to clinical neurophysiology: A review. *Comput. Methods Prog. Biomed.* **2016**, *133*, 55–70. [CrossRef]
41. Hu, M.; Liang, H. Uncovering perceptual awareness of visual stimulus with adaptive multiscale entropy. In Proceedings of the 2011 3rd International Conference on Awareness Science and Technology (iCAST), Dalian, China, 27–30 September 2011; pp. 485–490, ISSN 2325-5986, 2325-5994. [CrossRef]

42. Rehman, N.; Mandic, D.P. Multivariate empirical mode decomposition. *Proc. Royal Soc. A Math. Phys. Eng. Sci.* **2010**, *466*, 1291–1302. [CrossRef]
43. Wu, H.T.; Lo, M.T.; Chen, G.H.; Sun, C.K.; Chen, J.J. Novel Application of a Multiscale Entropy Index as a Sensitive Tool for Detecting Subtle Vascular Abnormalities in the Aged and Diabetic. *Comput. Math. Methods Med.* **2013**, *2013*, 8. [CrossRef]

Article

Multiscale Distribution Entropy Analysis of Short-Term Heart Rate Variability

Dae-Young Lee and Young-Seok Choi *

Department of Electronics and Communications Engineering, Kwangwoon University, Seoul 01897, Korea; simv1350@kw.ac.kr

* Correspondence: yschoi@kw.ac.kr; Tel.: +82-2-940-5186

Received: 12 November 2018; Accepted: 9 December 2018; Published: 11 December 2018

Abstract: Electrocardiogram (ECG) signal has been commonly used to analyze the complexity of heart rate variability (HRV). For this, various entropy methods have been considerably of interest. The multiscale entropy (MSE) method, which makes use of the sample entropy (SampEn) calculation of coarse-grained time series, has attracted attention for analysis of HRV. However, the SampEn computation may fail to be defined when the length of a time series is not enough long. Recently, distribution entropy (DistEn) with improved stability for a short-term time series has been proposed. Here, we propose a novel multiscale DistEn (MDE) for analysis of the complexity of short-term HRV by utilizing a moving-averaging multiscale process and the DistEn computation of each moving-averaged time series. Thus, it provides an improved stability of entropy evaluation for short-term HRV extracted from ECG. To verify the performance of MDE, we employ the analysis of synthetic signals and confirm the superiority of MDE over MSE. Then, we evaluate the complexity of short-term HRV extracted from ECG signals of congestive heart failure (CHF) patients and healthy subjects. The experimental results exhibit that MDE is capable of quantifying the decreased complexity of HRV with aging and CHF disease with short-term HRV time series.

Keywords: electrocardiogram; heart rate variability; multiscale distribution entropy; RR interval; short-term inter-beat interval

---

1. Introduction

An electrocardiogram (ECG) is a record of electrical activity caused by the heart. ECG is a non-invasive tool that is effective for a variety of biomedical applications such as heart rate measurement, diagnosis of heart failure, emotion recognition, and so on [1]. One of the main areas of need for ECG analysis is the diagnosis of heart diseases. Since ECG is closely related to cardiac activity, it can play an important role in the diagnosis of heart diseases. Among the causes of many heart diseases, congestive heart failure (CHF) is a collective term for heart disease that causes congestion in the systemic venous system due to heart pumping dysfunction [2]. Since heart failure diseases may cause death to many people all over the world every year, the diagnosis of CHF is of great interest and remains challenging issue.

Among the features that can be extracted from the ECG recordings, variabilities in heart beat-to-beat intervals controlled by the autonomic nervous system (ANS) is usually used, which is referred to as heart rate variability (HRV) [3]. The HRV analysis helps us to represent CHF symptoms and is widely used to identify CHF patients [4]. In practice, the HRV analysis using short-term inter-beat (RR) interval is of greatly important because of its suitability for short-term patient monitoring and the need for almost immediate reception of test results. Therefore, the complexity analysis of HRV is mainly utilized for the distinction between healthy people and those with heart disease, such as CHF patients. Recently, it is known that HRV of a healthy person exhibits dynamic fluctuations and it is characterized by a decrease in the incidence of CHF heart disease and aging [5,6].

Quantitative analysis of the complexity of a time series is promising in analyzing physical, mechanical, and biological systems that exhibit non-static, nonlinear, and complex behaviors [7,8]. A quantitative measure of physiological signal plays an important role for computer-aided diagnosis in clinical applications [9]. In this regard, various entropy approaches have attracted attention in the complexity analysis [10]. Conventional entropy measures such as sample entropy (SampEn) [11], fuzzy entropy (FuzzyEn) [12], and permutation entropy (PE) [13] have been utilized for the complexity analysis of HRV [14–17]. However, since these methods measure the irregularity of the time series, the resultant quantifications may fail to characterize the complexity of the underlying time series. For example, the SampEn value of white Gaussian noise is assigned to be higher than that of $1/f$ noise, which is not consistent with the complexity analysis in the sense that $1/f$ noise has higher complexity owing to its long-range correlations [18]. Along this line, though the complexity of HRV of healthy person is higher compared to that of patients, the conventional entropy approaches may fail to reflect the higher complexity of HRV of healthy over diseased status.

To address this issue, Costa et al. [5,19] have proposed a multiscale entropy (MSE) method that consists of a coarse-graining process and SampEn computation to measure the complexity of a time series at different temporal scales. It is generally effective in identifying characteristics over multiple temporal scales because the biological system possesses distinct properties over several spatial and temporal scales [19]. Therefore, various studies using this MSE method have been performed to analyze the complexity of HRV on various temporal scales [20,21]. Subsequently, the coarse-graining process has been applied to the FuzzEn and PE methods, yielding the multiscale fuzzy entropy (MFE) [22] and multiscale permutation entropy (MPE) [23], respectively. However, the coarse-graining process reduces the length of the coarse-grained time series as scale increases, thus resulting in inaccurate or undefined entropy computation. This behavior of MSE makes it unsuitable for computing entropy of a short-term time series. Wu et al. [24] have proposed a modified multiscale procedure that uses a moving-averaging process instead of a coarse-graining process. The authors have shown that the use of a moving-averaging process leads to better capability to reflect long-range correlations of a short-term time series than a coarse-graining one. Thus, it can provide more reliable computation of entropy values in situations in which a short-term time series is given.

Moreover, the MSE and MFE methods have drawbacks of high dependency on predetermined parameters because they do not fully make use of the distance information between vectors in the state space during computation. Recently, the distribution entropy (DistEn) proposed by Li et al. [25] has been developed from the fact that the inherent information of the distances between vectors in the state space is maximized through the probability density estimation, leading to relatively lower sensitivity to predetermined parameters and data length. While SampEn makes use of uses only a fraction of the distance vector information, DistEn is capable of quantifying full distance information. It gives DistEn improved sensitivity and consistency. However, DistEn only considers the complexity computation at single scale.

Here, we proposed an effective way to quantify HRV using the short-term RR interval of ECG signals. The proposed method is based on a computation of DistEn over multiple scales by a moving-averaging process, which is referred to as the multiscale distribution entropy (MDE). The computation of the MDE, which inherits the merits of the DistEn, is able to address the shortcoming of the conventional MSE which may fail to capture the long-range correlation of the short-term time series. We compare the performance of MDE to the conventional MSE using several synthetic data by evaluating the stability and characteristics over multiple temporal scales for the short-term time-series. Then, the capability of the proposed MDE is examined for the RR intervals with various lengths extracted from actual ECG signals of the healthy subjects and the CHF patients.

The remainder of this paper is organized as follows: In Section 2, we describe the conventional entropies and the proposed MDE. In Section 3, the results on synthetic data and real ECG data are presented. Section 4 presents the conclusions of this work.

2. Materials and Methods

2.1. Sample Entropy

The SampEn method is a modified entropy computation from the approximate entropy (ApEn) method [8]. SampEn computes the conditional probability that quantifies that the similarity of two sequences of different length m and $m+1$ is maintained. Here, m denotes the length of sequences that are compared to each other. More specifically, the SampEn method consists of four steps: reconstruction, definition of distance, definition of the criterion for similarity, and entropy calculation.

First, for a N points time series $x_N = \{x_1, x_2, \ldots, x_N\}$, it is to reconstruct x_N into multidimensional vectors as follows:

$$X_m^\tau(i) = \left\{x_i,\ x_{i+\tau},\ \ldots,\ x_{i+(m-1)\tau}\right\} \tag{1}$$

where m denotes the embedding dimension and τ denotes the time delay factor.

Next, define the distances between two different vectors as the maximum difference of their corresponding components as follows:

$$d[X_m^\tau(i), X_m^\tau(j)] = \max\left\{\left|x_{i+k\tau} - x_{j+k\tau}\right| : 0 \le k \le m-1\right\} \tag{2}$$

where i and j are not equal and the distance $d[X_m^\tau(i), X_m^\tau(j)]$ is referred to as the Chebyshev distance.

Third, if the distance $d[X_m^\tau(i), X_m^\tau(j)]$ is less than a threshold parameter r, a match occurs and we count the number of vector pairs that satisfy this condition. This process proceeds when the embedding dimension is m and $m+1$, which are called B_i^m and B_i^{m+1}, respectively.

$$B^m = \frac{1}{N - m\tau}\sum_{i=1}^{N-m\tau} B_i^m,\ B^{m+1} = \frac{1}{N - m\tau}\sum_{i=1}^{N-m\tau} B_i^{m+1} \tag{3}$$

Finally, SampEn is defined by

$$\mathrm{SampEn}(x_N, m, r, \tau) = -\ln\left[\frac{B^{m+1}}{B^m}\right] \tag{4}$$

In general, r is selected in the range of $[0.1\sigma, 0.25\sigma]$, where σ is the standard deviation of original time series x_N [11].

2.2. Distribution Entropy

The DistEn method quantifies the amount of information in the state space of the univariate time series by estimating the distribution characteristic of the distances between vectors [25]. The computation of DistEn consists of four steps: reconstruction, construction of a distance matrix, probability density estimation, and entropy calculation.

First, for N points of a time series x_n, we reconstruct multidimensional vector $X_m^\tau(i) = \left\{x_i,\ x_{i+\tau},\ \ldots,\ x_{i+(m-1)\tau}\right\}$, where m is the embedding dimension.

Next, construct the distance matrix $D = \{d[X_m^\tau(i), X_m^\tau(j)]\}$, where the $d[X_m^\tau(i), X_m^\tau(j)]$ is the Chebyshev distance.

Third, the distribution characteristics of $d[X_m^\tau(i), X_m^\tau(j)]$ should completely quantify the information reflecting the distances matrix D. To do this, we estimate the empirical probability density function (ePDF) of the matrix D by using the histogram approach. Since i and j are different, the diagonal components of the matrix D are excluded. In addition, since the $d[X_m^\tau(i), X_m^\tau(j)]$ and the $d[X_m^\tau(j), X_m^\tau(i)]$ are the same, only the upper or lower triangular part of the matrix D needs to be considered. If the histogram has B bins, the probability p_t of each bin is obtained, where $t = 1, 2, \ldots, B$. The value of B is usually selected as an integer value in a range of [512, 1024].

Finally, DistEn is calculated as

$$\text{DistEn}(\boldsymbol{x}_N, m, B, \tau) = -\frac{1}{\log_2 B} \sum_{t=1}^{B} p_t \log_2(p_t) \tag{5}$$

Figure 1 shows the ePDF of the distance matrix D for the white noise ($N = 1000$) for exploiting the difference between the SampEn and DistEn methods in terms of distance information. The SampEn method uses only a fraction of the distance information (less than the threshold parameter r), and it corresponds to the left area of the red dotted line in Figure 1. On the other hand, since the DistEn method takes full advantage of the distance information, it is able of reflecting the complexity that the SampEn method can't measure.

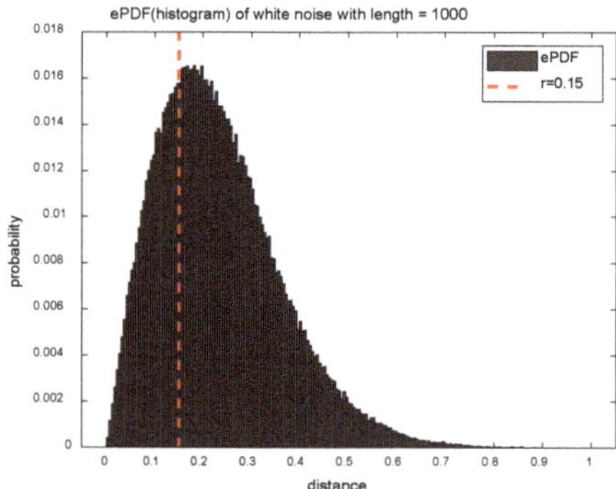

Figure 1. Empirical probability density function (ePDF) of white noise with length $N = 1000$ ($m = 2$, $r = 0.15\sigma$).

2.3. Multiscale Distribution Entropy

The coarse-graining procedure generates a number of sets of time series on a time scale s by considering different starting points of the time series. Therefore, the coarse-graining multiscale process can lead to inaccurate entropy values by reducing the length of the time series. To alleviate this drawback, we used the moving-averaging multiscale process, which has a better effect on short-term time series analysis [24]. Figure 2 shows the progress of two multiscale processes. It can be seen that the moving-averaging process (Figure 2b) leads to longer multiscale processed time series compared to the coarse-graining process (Figure 2a).

The moving-averaging multiscale process is composed of two procedures. First, for a N point time series $\boldsymbol{x}_N$ and a given scale factor s, we divide the original time series into several smaller time series overlapped of length scale factor s. Then, the continuous moving-averaged time series are constructed by averaging the number of data points on the scale s as follow:

$$y_i^s = \frac{1}{s} \sum_{j=i}^{i+s-1} x_j, 1 \le i \le N - s + 1 \tag{6}$$

The moving-averaging process generates multiple sets of time series on the time scale factor s. At the scale factor $s = 1$, the moving-averaged time series $\boldsymbol{y}^s$ is equal to the original time series.

Second, set the time delay factor τ of the DistEn to the scale factor s, and calculate the entropy value of MDE. In other words, the moving-averaged time series on each scale is used as an input signal for entropy calculation of DistEn as follows:

$$\mathrm{MDE}(x_N, m, B, s) = \mathrm{DistEn}(y^s, m, B,\ \tau = s) \tag{7}$$

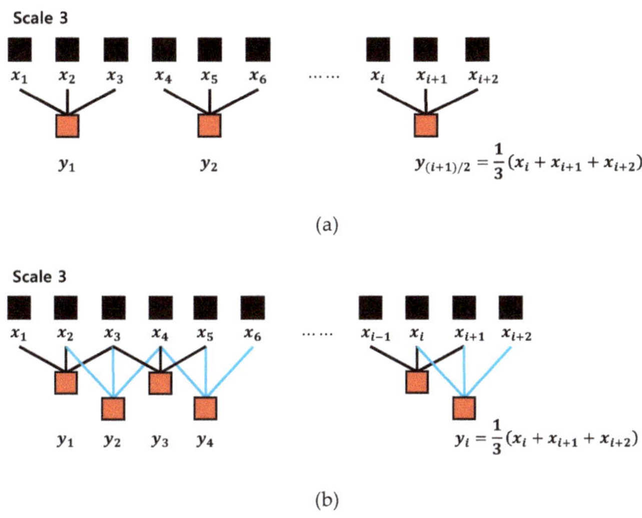

Figure 2. Progress of two multiscale processes: (**a**) coarse-graining; (**b**) moving-averaging.

2.4. Evaluation Data

2.4.1. Synthetic Data

To verify the performance of the MSE and MDE methods with respect to the length of data, we first employed entropy calculation on synthetic data. The synthetic data used in this work are the chaotic series, white Gaussian noise (or simply white noise), periodic signals, and MIX(p) processes. The chaotic series and periodic series are generated from the Logistic attractor $x(n+1) = \omega \times x(n) \times (1 - x(n))$ with $\omega = 4.0$ and $\omega = 3.5$, respectively. The MIX process is a kind of stochastic signal that is superimposed on a deterministic component, and randomly selected points of $N \times p$ are replaced by independent identically distributed random noise in a sinusoidal signal of length N [26]. Finally, white noise is that the values at any pair of times are identically distributed and statistically independent, and it is the case of uncorrelated noise.

For each signal, 100 realizations were randomly generated with data length of $N = 100,\ 300,$ and 1000, and used for the evaluation of the MSE and MDE.

2.4.2. Real ECG Data

Two real ECG datasets in *PhysioNet* are used [27]. Dataset I includes ECG data from *Fantasia* and *BIDMC CHF*. In addition, Dataset II includes other *CHF RR Interval* and *Normal Sinus Rhythm RR Interval* data. *BIDMC CHF* data includes ECG records from 15 patients with CHF (NYHA classes III, IV) consisting of 11 men aged 22–71 years and 4 women aged 54–63 years. Each record was measured for approximately 20 hours and contains two ECG signals sampled at 250 Hz. *Fantasia* data were measured from 20 healthy people aged 21–34 years and 20 elderly people aged 68–85 years. Each record was measured for approximately 2 hours, and the sampling frequency was 250 Hz. *CHF RR Interval* consists of beat annotation for 29 ECG signals (sampled at 128 Hz) of CHF (NYHA classes I, II, III) patients aged 34–79. In addition, *Normal Sinus Rhythm RR Interval* data consists of beat annotation for 54 ECG signals (sampled at 128 Hz) of subjects in normal sinus rhythm (NSR).

To find the R peak points from the ECG signals of Dataset I, we used a Pan–Tompkins algorithm [28]. Then, RR interval time series are constructed from the distances between two consecutive R peak points and can be seen in Figure 3. Figure 3 shows the representative RR intervals of CHF patient, healthy elderly, and healthy young subjects, respectively.

In this work, we used the RR interval time series of lengths of 100, 300, and 1000 extracted from ECG signals, respectively. Each time series was used for evaluation of the MSE, MPE, and MDE. Firstly, CHF patients (*BIDMC CHF*), healthy elderly, and healthy young groups' data (*Fantasia*) were analyzed. We then analyzed other CHF patients (*CHF RR Interval*) and NSR subject data (*NSR RR Interval*) to further evaluate the performance of discrimination between CHF patients and normal subjects. The parameters of the MSE and MPE were set to $r = 0.2\sigma$ and $m = 2$, and $m = 4$ and $t = 1$, respectively. The parameters of the MDE were set to $m = 2$ and $B = 512$.

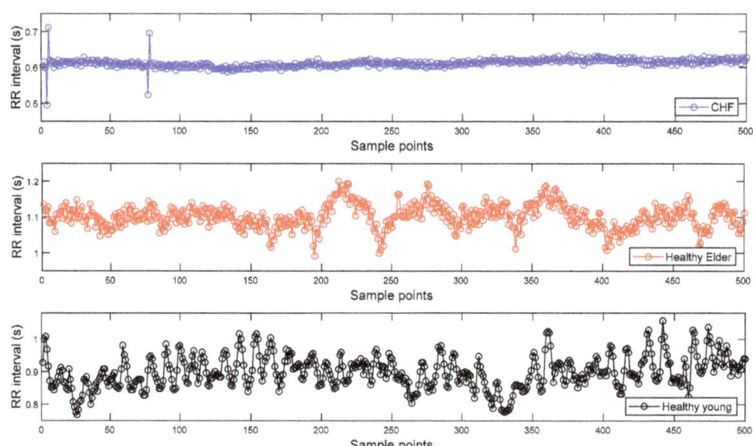

Figure 3. Examples of inter-beat (RR) interval time series extracted from the electrocardiogram (ECG) signals of Dataset I: Congestive heart failure (CHF) patient (**top panel**); Healthy elderly subject (**middle panel**); Healthy young subject (**bottom panel**).

3. Results

3.1. Simulation Result Using Synthetic Data

The results of the MDE and MSE evaluation of synthetic data are shown in Figure 4. Figure 4a,b shows the entropy values of the MSE and MDE for time series of length $N = 100$, respectively. Figure 4a shows that chaotic series has the highest values of MSE in the scales less than 3 and the periodic signal has entropy values of almost zero. Notably, the MSE values are mostly only defined on small scales. For instance, the white noise and the chaotic series are defined only on scale 1 and 2, with a large standard deviation. Compared with the MSE results, in Figure 4b, it is evident that the entropy values of MDE become smaller in order of chaotic series, Gaussian white noise, MIX (0.2), MIX (0.1), and periodic series. As the scale increases, the entropy values of MDE for white noise and chaotic series are comparable and higher than those of other synthetic data. The MDE values of MIX (0.1) and MIX (0.2) rise as scale factor increases and remain constant, while the periodic signal keeps the lowest entropy values. In the case of the periodic signal, due to the use of a period of 4 samples, lowest entropy values are repeated on scales of the multiples of 4.

Next, Figure 4c,d shows the MSE and MDE results for time series of length $N = 300$, respectively. In Figure 4c, white noise has the highest MSE values in a range of small scales. In the range of small scales, the entropy values of chaotic series, MIX (0.2), MIX (0.1), and periodic series follow in order. However, the MSE values of those synthetic data are still undefined over large scale factors, especially for white noise and chaotic series. As can be seen in Figure 4d, the similar results of Figure 4b are observed in Figure 4d in the sense that MDE computation leads to same order of entropy values on small scale factors and stable evaluation over all scales. In addition, it exhibits that the signals, except for the periodic signal, converge to almost similar entropy values on large scales, unlike the result of the data length $N = 100$ in Figure 4b.

Figure 4e,f shows the results of the MSE and MDE calculation for time series of length $N = 1000$, respectively. Here, since all values are defined in range of all scale, it is available to compare the entropy evaluation of the MDE and MSE. Both the MDE and MSE results have a smaller standard deviation at each scale than the results of length $N = 100$ and 300. For the MSE result of Figure 4e, as the scale becomes larger than scale 5, the entropy values of chaotic series and white noise gradually decrease and are comparable on large scales. Compared to Figure 4c, it results in reduced standard deviation at each scale.

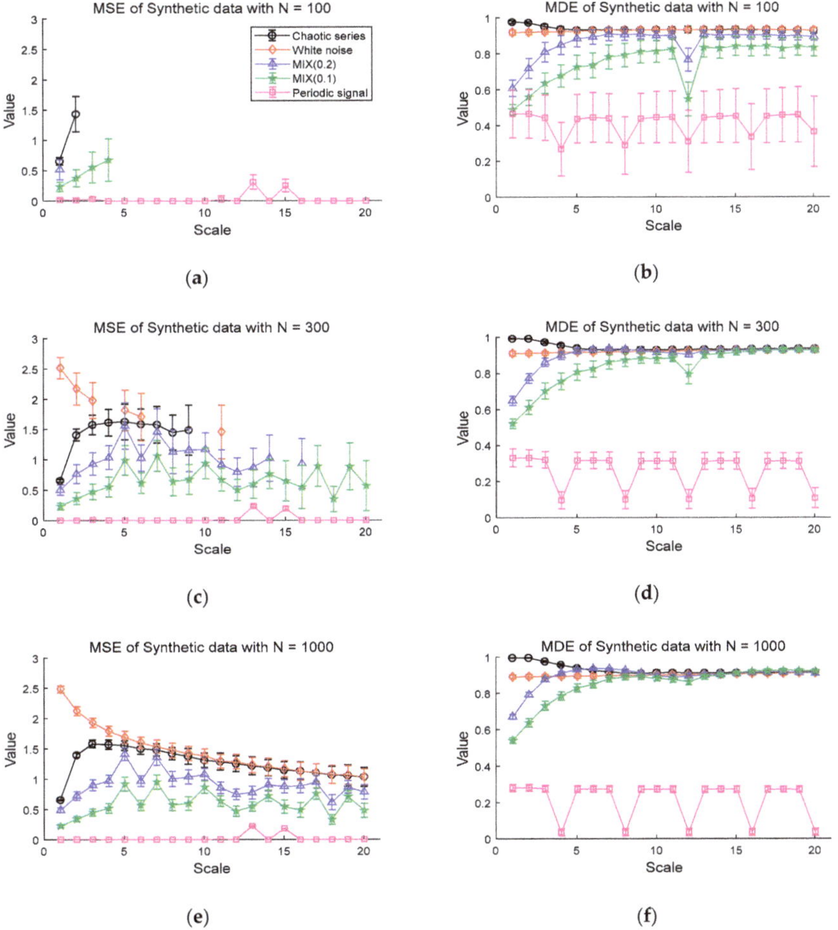

Figure 4. Entropy values for synthetic data: (**a**) multiscale entropy (MSE) value for $N = 100$; (**b**) multiscale distribution entropy (MDE) value for $N = 100$; (**c**) MSE value for $N = 300$; (**d**) MDE value for $N = 300$; (**e**) MSE value for $N = 1000$; (**f**) MDE value for $N = 1000$. Scales between 1 and 20 are used and the value at each scale represents a mean ± standard deviation.

The MDE results in Figure 4f indicate similar behaviors in the results in Figure 4d with the decreased standard deviation. Notably, the results of Figure 4b,d,f show that the MDE method leads to similar results in the complexity analysis for short-term and long-term time series, implying its insensitivity to the length of time series. In addition, MDE shows the smaller standard deviation than that of MSE and defined over all scales, indicating its superior stability and reliability over MSE.

3.2. Experiment Results Using Real Data

3.2.1. ECG Dataset I

We show the experimental results using RR interval time series extracted from the ECG signal database measured for CHF patients, healthy elderly and healthy young groups in Figure 5. Figure 5a–c shows the MSE, MPE, and MDE results for time series of length $N = 100$, respectively. In Figure 5a, the MSE values are not defined on most part of scales, indicating the shortcoming of MSE in analyzing a short-term RR interval time series. As for the results of MPE analysis in Figure 5b, the MPE values are present at all scales, but the entropy values decrease since the length of the time series get shorter as the multiscale process progresses. The distinction between CHF patients and healthy subjects seems to be difficult. On the other hand, the results of the MDE analysis in Figure 5c shows that the entropy values are defined over all scales for RR interval time series of length $N = 100$. In addition, MDE is capable of reflecting the difference of the complexity of RR interval time series not only between CHF patients and healthy subjects, but also between healthy subject groups, i.e., between the elderly and the young groups. As the scale increases, the MDE values of three groups get higher.

Next, Figure 5d–f shows the MSE, MPE, and MDE results for RR interval time series of length $N = 1000$, respectively. In Figure 5d, in a situation where the length of the time series is long enough, the result of the MSE values are defined over most scales except for healthy elderly group on scales 19 and 20. The distinction between CHF patients and healthy subject groups appears to be available over most scales. However, the entropy values of the healthy young group are higher than those of the healthy elder one until the scale 7, but after that the distinction between the two groups is difficult. As for the results of MPE analysis in Figure 5e, the MPE values show a slight decrease, which is less than the result ($N = 100$) in Figure 5b, and the distinction between the three groups seems possible. However, since the complexity of healthy subjects must be greater than the complexity of a CHF patients, it is possible to discriminate between CHF patients and healthy elderly group after the scale 5, and between CHF patients and healthy young group after the scale 2. In addition, the mean entropy values of the healthy young group are higher than those of the healthy elder group only on scales between 2 and 9, and the distinction is difficult on other scales. On the other hand, the MDE result in Figure 5f shows a similar evaluation results to Figure 5c in a situation where the length $N = 100$. In addition, as the scale gets larger, the entropy values reached is larger than those for the short-term time series with reduced variance. The MDE behaviors shown in Figure 5 are closely consistent with the previous finding of the decreased complexity with aging and pathological status, whereas the MSE and MPE results do not agree with known behaviors of physiological complexity [6].

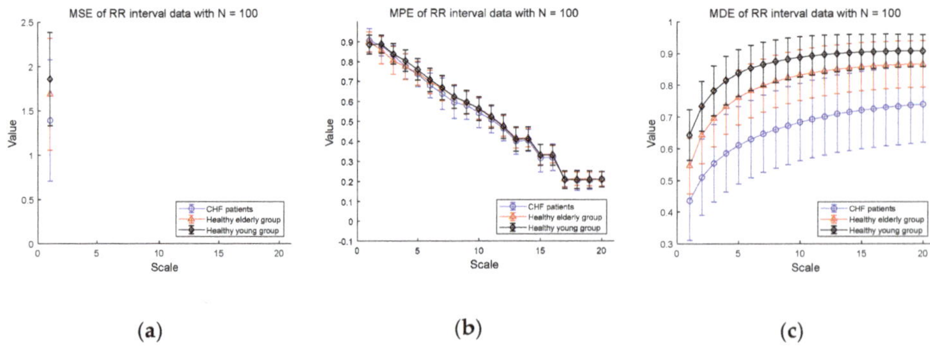

(a) (b) (c)

Figure 5. *Cont.*

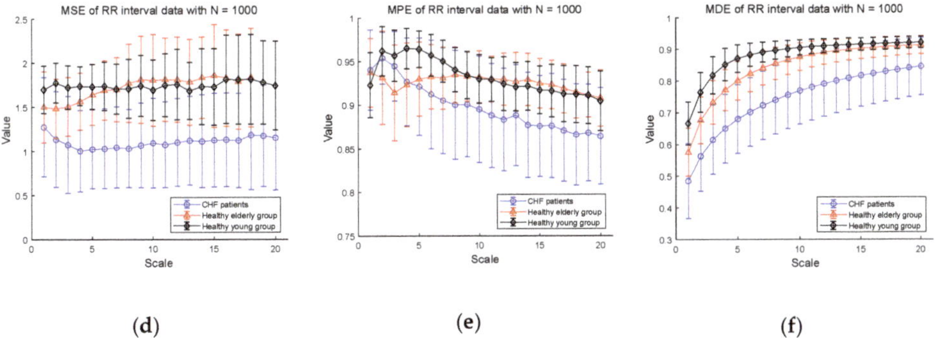

Figure 5. MSE, multiscale permutation entropy (MPE), and MDE results using real data (RR interval) for CHF patients, healthy elderly and healthy young groups: (**a**) MSE for $N = 100$; (**b**) MPE for $N = 100$; (**c**) MDE for $N = 100$; (**d**) MSE for $N = 1000$; (**e**) MPE for $N = 1000$; (**f**) MDE for $N = 1000$. Scale range of 1–20 were used and the value at each scale represents a mean ± standard deviation.

3.2.2. ECG Dataset II

Figure 6 shows the results of entropy computation for RR interval time series obtained from CHF patients and NSR subject datasets to further evaluate the distinction performance between CHF patients and normal individuals. Figure 6a–c shows the MSE, MPE, and MDE results for time series of length $N = 100$, respectively. MSE is not defined on most scales except scale 1 and MPE results show a dramatic decrease in entropy value as the scale becomes larger, indicating weak sensitivity of MPE on the length of a time series. Compared to conventional measures, MDE is able to reflect differences in the complexity of the RR interval time series between CHF patients and NSR group. Note that the MDE values increase as the scale gets higher, indicating robustness to the length of a time series. Next, in Figure 6d–f, the results of the MSE, MPE, and MDE are defined on most scales in situations where the length of the time series is long enough ($N = 1000$).

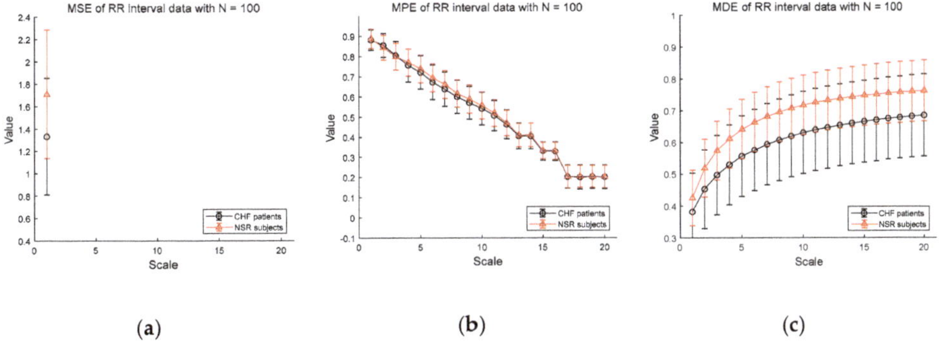

Figure 6. *Cont.*

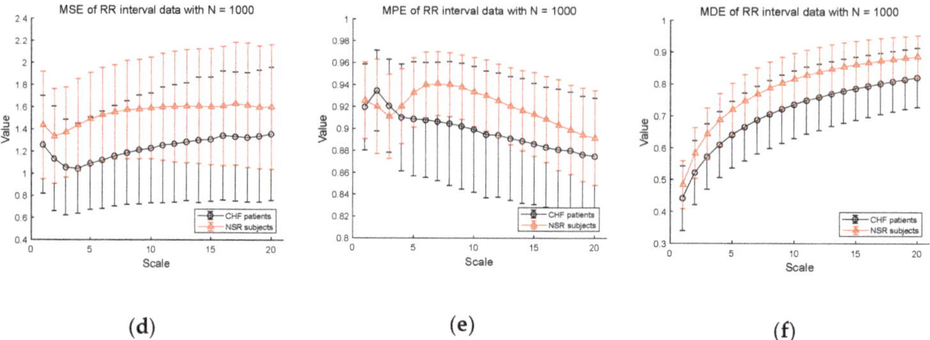

(d) (e) (f)

Figure 6. MSE, MPE, and MDE results using real data (RR interval) for other CHF patients and normal sinus rhythm (NSR) subjects: (**a**) MSE for $N = 100$; (**b**) MPE for $N = 100$; (**c**) MDE for $N = 100$; (**d**) MSE for $N = 1000$; (**e**) MPE for $N = 1000$; (**f**) MDE for $N = 1000$. Scale range of 1–20 were used and the value at each scale represents a mean ± standard deviation.

3.2.3. Statistical Analysis for CHF Patients, Healthy Elderly, and Healthy Young Groups

To verify the distinction capability of entropy results between CHF patients, healthy elderly, and healthy young groups in Dataset I, statistical analysis was conducted. First, the Kolmogorov–Smirnov test is used to check whether the MDE and MSE results satisfy the normal distribution. If they follow a normal distribution, the t-test method was conducted to test the statistical difference between three datasets, and if not, the Mann–Whitney U test method was performed. Generally, if the p-value is less than 0.05, statistical significance is accepted. The analysis results are shown in Tables 1–3.

Table 1 shows the p-value of the MSE comparison result of RR interval time series when the length N is 100, 300, and 1000, respectively. As can be seen, for $N = 100$ and 300, p-value computation is not available over most scales. For $N = 1000$, the distinction using MSE values between CHF patients and healthy elderly group in scales from 1 to 18 and CHF patients and healthy young group over all scales are statistically significant. However, it fails to differentiate healthy elderly group from healthy young one over most scales (p-value > 0.05). The shadows represent the cases which the distinction is statistically insignificant.

In Table 2, the MPE results show that the shorter the length of the time series ($N = 100$), the more difficult it is to distinguish between subject groups. As the length of RR interval increases, the discrimination performance gets better. However, even in $N = 1000$, the distinction between groups fails in several cases. In addition, similar to the MSE results, the distinction between CHF patients and healthy subjects for $N = 1000$ is statistically significant at most scales, but the distinction between healthy young and healthy elderly groups is available in limited scales.

The statistical results of MDE in Table 3 exhibit that the distinctions using the MDE values between three groups are statistically significant for all three lengths of RR interval, i.e., $N = 100$, 300, and 1000. Compared to the results of MSE and MPE, the MDE is capable of discriminating the complexities of different physiological groups for all scale values, implying its superiority over the conventional methods and insensitivity to the length of a time series.

Table 1. Statistical analysis results of MSE for RR interval of the CHF patients, healthy young, and healthy elder groups. The shadows mean that the distinction between the groups is unclear. *C*, *E*, and *Y* represent CHF, Elder, and Young, respectively. *s* denotes scale factor and N/A denotes 'Not Available'.

MSE Statistical Results for RR Interval Time Series									
	p-Value								
s	*N* = 100			*N* = 300			*N* = 1000		
	C - *E*	*C* - *Y*	*E* - *Y*	*C* - *E*	*C* - *Y*	*E* - *Y*	*C* - *E*	*C* - *Y*	*E* - *Y*
1	0.098	1.1×10^{-6}	8.7×10^{-5}	0.0014	2.5×10^{-8}	3.6×10^{-4}	1.3×10^{-5}	4.8×10^{-14}	8.5×10^{-7}
2	N/A	N/A	N/A	4.7×10^{-7}	5.5×10^{-13}	1.0×10^{-5}	6.7×10^{-10}	1.5×10^{-21}	8.0×10^{-15}
3	N/A	N/A	N/A	2.4×10^{-10}	1.8×10^{-17}	1.9×10^{-4}	8.4×10^{-14}	3.2×10^{-25}	5.3×10^{-10}
4	N/A	N/A	N/A	3.9×10^{-14}	3.1×10^{-19}	0.019	2.3×10^{-17}	2.2×10^{-26}	2.7×10^{-07}
5	N/A	N/A	N/A	1.3×10^{-11}	2.8×10^{-15}	0.311	1.3×10^{-20}	1.9×10^{-28}	0.005
6	N/A	N/A	N/A	N/A	N/A	N/A	1.9×10^{-21}	8.8×10^{-27}	0.089
7	N/A	N/A	N/A	N/A	N/A	N/A	3.7×10^{-19}	1.2×10^{-22}	0.512
8	N/A	N/A	N/A	N/A	N/A	N/A	1.6×10^{-20}	1.1×10^{-22}	0.847
9	N/A	N/A	N/A	N/A	N/A	N/A	3.5×10^{-21}	2.2×10^{-23}	0.353
10	N/A	N/A	N/A	N/A	N/A	N/A	1.4×10^{-19}	5.1×10^{-20}	0.175
11	N/A	N/A	N/A	N/A	N/A	N/A	5.2×10^{-18}	6.8×10^{-20}	0.673
12	N/A	N/A	N/A	N/A	N/A	N/A	1.6×10^{-19}	2.3×10^{-21}	0.359
13	N/A	N/A	N/A	N/A	N/A	N/A	1.3×10^{-17}	1.0×10^{-17}	0.086
14	N/A	N/A	N/A	N/A	N/A	N/A	3.7×10^{-18}	5.0×10^{-20}	0.853
15	N/A	N/A	N/A	N/A	N/A	N/A	1.9×10^{-13}	2.2×10^{-14}	0.448
16	N/A	N/A	N/A	N/A	N/A	N/A	1.6×10^{-13}	4.7×10^{-16}	0.945
17	N/A	N/A	N/A	N/A	N/A	N/A	1.0×10^{-11}	1.6×10^{-14}	0.695
18	N/A	N/A	N/A	N/A	N/A	N/A	1.6×10^{-15}	2.2×10^{-19}	0.936
19	N/A	N/A	N/A	N/A	N/A	N/A	N/A	3.2×10^{-16}	N/A
20	N/A	N/A	N/A	N/A	N/A	N/A	N/A	4.1×10^{-13}	N/A

Table 2. Statistical analysis results of MPE for RR interval of the CHF patients, healthy young, and healthy elder groups. The shadows mean that the distinction between the groups is unclear. *C*, *E*, and *Y* represent CHF, Elder, and Young, respectively. *s* denotes scale factor.

MPE Statistical Results for RR Interval Time Series									
	p-Value								
s	*N* = 100			*N* = 300			*N* = 1000		
	C - *E*	*C* - *Y*	*E* - *Y*	*C* - *E*	*C* - *Y*	*E* - *Y*	*C* - *E*	*C* - *Y*	*E* - *Y*
1	0.389	6.6×10^{-7}	6.5×10^{-5}	0.255	1.2×10^{-7}	8.9×10^{-6}	0.157	8.6×10^{-8}	5.6×10^{-4}
2	0.018	0.101	5.8×10^{-5}	0.006	0.006	1.1×10^{-7}	0.005	4.1×10^{-4}	7.9×10^{-7}
3	3.0×10^{-4}	0.519	2.9×10^{-4}	4.6×10^{-6}	0.005	1.6×10^{-14}	9.2×10^{-9}	0.012	5.9×10^{-13}
4	0.363	$1.9 \times 10{-}^{5}$	1.8×10^{-4}	0.674	8.6×10^{-13}	1.5×10^{-16}	0.030	5.2×10^{-12}	4.8×10^{-19}
5	0.139	6.3×10^{-8}	0.001	0.935	1.6×10^{-9}	7.3×10^{-13}	0.776	1.8×10^{-15}	1.2×10^{-16}
6	0.059	0.004	0.241	0.033	4.2×10^{-9}	1.1×10^{-5}	0.017	7.5×10^{-16}	9.5×10^{-10}
7	0.102	1.2×10^{-8}	0.655	0.002	3.7×10^{-7}	0.007	0.007	8.8×10^{-13}	3.2×10^{-6}
8	0.292	5.8×10^{-7}	0.619	0.004	4.9×10^{-4}	0.428	1.1×10^{-5}	3.5×10^{-8}	0.233
9	0.450	4.3×10^{-6}	0.129	9.7×10^{-4}	1.6×10^{-4}	0.567	5.3×10^{-5}	1.6×10^{-5}	0.720
10	0.433	4.3×10^{-4}	0.454	6.3×10^{-4}	7.0×10^{-6}	0.193	1.4×10^{-6}	1.4×10^{-5}	0.387
11	0.104	1.7×10^{-5}	0.965	2.8×10^{-5}	9.0×10^{-5}	0.656	1.8×10^{-8}	4.3×10^{-8}	0.701
12	0.034	0.050	0.873	1.6×10^{-4}	7.2×10^{-8}	0.041	8.6×10^{-10}	1.7×10^{-7}	0.159
13	0.025	0.062	0.686	9.1×10^{-7}	7.9×10^{-6}	0.701	9.6×10^{-9}	1.7×10^{-5}	0.009
14	0.702	0.914	0.780	1.2×10^{-7}	1.5×10^{-4}	0.056	4.8×10^{-13}	5.8×10^{-9}	0.020
15	0.912	0.779	0.854	3.7×10^{-5}	3.0×10^{-4}	0.505	3.4×10^{-11}	2.9×10^{-7}	0.020
16	0.074	0.411	0.296	1.1×10^{-4}	2.7×10^{-4}	0.745	2.6×10^{-13}	5.9×10^{-10}	0.070
17	0.821	0.520	0.648	3.8×10^{-5}	2.3×10^{-4}	0.672	5.4×10^{-10}	6.1×10^{-8}	0.309
18	0.171	0.531	0.431	1.1×10^{-5}	5.9×10^{-6}	0.880	2.6×10^{-9}	1.4×10^{-8}	0.729
19	0.448	0.437	0.103	3.7×10^{-5}	1.4×10^{-1}	0.758	2.0×10^{-11}	6.1×10^{-11}	0.947
20	0.272	0.556	0.588	0.002	0.005	0.671	1.0×10^{-9}	3.9×10^{-8}	0.496

Table 3. Statistical analysis results of MDE for RR interval of the CHF patients, healthy young, and healthy elder groups. *C*, *E*, and *Y* represent CHF, Elder, and Young, respectively. *s* denotes scale factor.

MDE Statistical Results for RR Interval Time Series									
	***p*-Value**								
s	$N = 100$			$N = 300$			$N = 1000$		
	C - *E*	*C* - *Y*	*E* - *Y*	*C* - *E*	*C* - *Y*	*E* - *Y*	*C* - *E*	*C* - *Y*	*E* - *Y*
1	2.8×10^{-15}	3.6×10^{-26}	6.9×10^{-15}	2.1×10^{-11}	9.2×10^{-26}	5.5×10^{-16}	1.8×10^{-10}	3.1×10^{-26}	1.1×10^{-16}
2	1.7×10^{-18}	7.2×10^{-31}	1.3×10^{-13}	2.7×10^{-15}	7.1×10^{-32}	1.1×10^{-14}	1.7×10^{-14}	1.6×10^{-31}	1.0×10^{-16}
3	4.9×10^{-21}	5.5×10^{-33}	2.0×10^{-12}	6.5×10^{-18}	2.3×10^{-34}	2.7×10^{-14}	2.8×10^{-17}	5.6×10^{-34}	3.5×10^{-17}
4	1.1×10^{-21}	1.5×10^{-33}	4.3×10^{-12}	1.3×10^{-18}	1.6×10^{-34}	1.3×10^{-14}	3.9×10^{-35}	3.9×10^{-35}	4.7×10^{-18}
5	2.5×10^{-23}	3.3×10^{-34}	1.5×10^{-11}	1.2×10^{-13}	9.7×10^{-35}	1.2×10^{-14}	1.8×10^{-35}	1.8×10^{-35}	6.1×10^{-17}
6	2.1×10^{-23}	1.7×10^{-34}	4.8×10^{-11}	2.6×10^{-20}	5.7×10^{-35}	2.4×10^{-14}	5.4×10^{-35}	5.4×10^{-35}	2.0×10^{-15}
7	3.1×10^{-23}	5.8×10^{-34}	3.0×10^{-10}	4.9×10^{-20}	2.2×10^{-34}	2.4×10^{-13}	1.1×10^{-33}	1.1×10^{-33}	8.1×10^{-13}
8	2.1×10^{-23}	8.0×10^{-34}	2.0×10^{-9}	4.6×10^{-20}	3.4×10^{-34}	3.0×10^{-12}	1.6×10^{-32}	1.6×10^{-32}	1.4×10^{-10}
9	6.7×10^{-23}	2.1×10^{-33}	4.5×10^{-9}	1.9×10^{-20}	4.4×10^{-34}	9.1×10^{-11}	6.9×10^{-32}	6.9×10^{-32}	1.9×10^{-9}
10	6.9×10^{-23}	2.1×10^{-33}	9.6×10^{-9}	3.3×10^{-20}	1.4×10^{-33}	2.9×10^{-9}	9.6×10^{-31}	9.6×10^{-31}	8.9×10^{-8}
11	1.7×10^{-22}	3.3×10^{-33}	2.0×10^{-8}	3.8×10^{-20}	2.3×10^{-33}	2.5×10^{-8}	3.3×10^{-30}	3.3×10^{-30}	4.8×10^{-6}
12	2.4×10^{-22}	2.9×10^{-33}	2.5×10^{-8}	4.6×10^{-20}	1.5×10^{-32}	1.6×10^{-7}	5.9×10^{-29}	5.9×10^{-29}	1.8×10^{-5}
13	2.6×10^{-22}	1.1×10^{-32}	5.7×10^{-8}	1.9×10^{-20}	4.6×10^{-32}	7.4×10^{-7}	2.4×10^{-27}	2.4×10^{-27}	1.4×10^{-4}
14	8.0×10^{-22}	7.9×10^{-33}	1.0×10^{-7}	8.8×10^{-20}	4.4×10^{-31}	2.1×10^{-6}	2.8×10^{-25}	2.8×10^{-25}	6.0×10^{-4}
15	3.2×10^{-21}	4.5×10^{-32}	2.3×10^{-7}	3.5×10^{-19}	3.8×10^{-30}	4.8×10^{-6}	4.0×10^{-23}	4.0×10^{-23}	0.001
16	1.0×10^{-20}	1.1×10^{-31}	3.8×10^{-7}	4.7×10^{-18}	1.1×10^{-28}	4.8×10^{-6}	7.6×10^{-21}	7.6×10^{-21}	8.8×10^{-4}
17	3.5×10^{-20}	2.4×10^{-31}	4.4×10^{-7}	1.6×10^{-17}	2.7×10^{-27}	1.7×10^{-5}	2.2×10^{-18}	2.2×10^{-18}	0.004
18	7.9×10^{-20}	3.4×10^{-31}	9.6×10^{-7}	1.6×10^{-17}	4.0×10^{-27}	3.7×10^{-5}	1.1×10^{-16}	1.1×10^{-16}	0.005
19	2.1×10^{-19}	1.4×10^{-30}	1.4×10^{-6}	7.6×10^{-17}	1.8×10^{-26}	1.2×10^{-9}	5.8×10^{-16}	5.8×10^{-16}	0.006
20	5.3×10^{-19}	1.9×10^{-30}	4.0×10^{-6}	1.7×10^{-16}	7.0×10^{-26}	1.1×10^{-5}	2.4×10^{-14}	2.4×10^{-14}	0.015

4. Conclusions

We have presented an improved multiscale entropy, named MDE, for capturing the complexity of short-term HRV time series. For analysis of short-term HRV, the conventional MSE method suffers from unreliable computation due to its dependency to the length of time series. On the other hand, the proposed MDE method is capable of quantifying the complexity even for short-term time series by integrating DistEn with the moving-averaging process. The use of DistEn and moving-averaging multiscale approach leads to improved stability and reliability of entropy evaluation for short-term time series. Thus, the proposed MDE outperforms the conventional MSE and MPE in the sense that MDE is insensitive to the length of time series. Through synthetic data analysis, the proposed MDE is shown to be effective for describing the complexity of various time series such as chaotic series, white noise, MIX process, and periodic series. The experimental results using real ECG recordings from CHF patients and healthy subjects indicate that MDE is useful to reflect the degree of the decreased complexity of HRV accompanied by aging and disease. Through this work, the proposed MDE has shown its potential to be a promising solution for short-term HRV analysis.

Author Contributions: D.-Y.L. and Y.-S.C. conceived and designed the methodology. Both authors were responsible for analyzing and writing the paper. Both the authors have read and approved the final manuscript.

Funding: This work was supported by Institute for information & communications Technology Promotion (IITP) grant funded by the Korea government (MISP) (No. 2018-0-00735, Media Interaction Technology based on Human Reaction and Intention to Content in UHD Broadcasting Environment), and the present research has been conducted by the research Grant of Kwangwoon University in 2017.

Conflicts of Interest: The authors declare no conflict of interest.

References

1. Kaplan Berkaya, S.; Uysal, A.K.; Sora Gunal, E.; Ergin, S.; Gunal, S.; Gulmezoglu, M.B. A survey on ECG analysis. *Biomed. Signal Process. Control* **2018**, *43*, 216–235. [CrossRef]
2. İşler, Y.; Kuntalp, M. Combining classical HRV indices with wavelet entropy measures improves to performance in diagnosing congestive heart failure. *Comput. Biol. Med.* **2007**, *37*, 1502–1510. [CrossRef] [PubMed]
3. Rajendra Acharya, U.; Paul Joseph, K.; Kannathal, N.; Lim, C.M.; Suri, J.S. Heart rate variability: A review. *Med. Biol. Eng. Comput.* **2006**, *44*, 1031–1051. [CrossRef] [PubMed]
4. Yu, S.-N.; Lee, M.-Y. Bispectral analysis and genetic algorithm for congestive heart failure recognition based on heart rate variability. *Comput. Biol. Med.* **2012**, *42*, 816–825. [CrossRef] [PubMed]
5. Costa, M.; Goldberger, A.L.; Peng, C.-K. Multiscale Entropy Analysis of Complex Physiologic Time Series. *Phys. Rev. Lett.* **2002**, *89*, 068102. [CrossRef] [PubMed]
6. Goldberger, A.L.; Peng, C.-K.; Lipsitz, L.A. What is physiologic complexity and how does it change with aging and disease? *Neurobiol. Aging* **2002**, *23*, 23–26. [CrossRef]
7. Villecco, F.; Pellegrino, A. Entropic Measure of Epistemic Uncertainties in Multibody System Models by Axiomatic Design. *Entropy* **2017**, *19*, 291. [CrossRef]
8. Villecco, F.; Pellegrino, A. Evaluation of Uncertainties in the Design Process of Complex Mechanical Systems. *Entropy* **2017**, *19*, 475. [CrossRef]
9. Sena, P.; Attianese, P.; Pappalardo, M.; Villecco, F. FIDELITY: Fuzzy Inferential Diagnostic Engine for on-LIne supporT to phYsicians. In Proceedings of the 4th International Conference on the Development of Biomedical Engineering, Ho Chi Minh City, Vietnam, 8–10 January 2012; IFMBE Proceedings. Springer: Berlin, Germany, 2013; pp. 396–400.
10. Coifman, R.R.; Wickerhauser, M.V. Entropy-based algorithms for best basis selection. *IEEE Trans. Inf. Theory* **1992**, *38*, 713–718. [CrossRef]
11. Richman, J.S.; Moorman, J.R. Physiological time-series analysis using approximate entropy and sample entropy. *Am. J. Physiol.-Heart Circ. Physiol.* **2000**, *278*, H2039–H2049. [CrossRef]
12. Chen, W.; Wang, Z.; Xie, H.; Yu, W. Characterization of Surface EMG Signal Based on Fuzzy Entropy. *IEEE Trans. Neural Syst. Rehabil. Eng.* **2007**, *15*, 266–272. [CrossRef] [PubMed]

13. Bandt, C.; Pompe, B. Permutation Entropy: A Natural Complexity Measure for Time Series. *Phys. Rev. Lett.* **2002**, *88*, 174102. [CrossRef] [PubMed]
14. Al-Angari, H.M.; Sahakian, A.V. Use of Sample Entropy Approach to Study Heart Rate Variability in Obstructive Sleep Apnea Syndrome. *IEEE Trans. Biomed. Eng.* **2007**, *54*, 1900–1904. [CrossRef] [PubMed]
15. Liu, C.; Li, K.; Zhao, L.; Liu, F.; Zheng, D.; Liu, C.; Liu, S. Analysis of heart rate variability using fuzzy measure entropy. *Comput. Biol. Med.* **2013**, *43*, 100–108. [CrossRef] [PubMed]
16. Shi, B.; Zhang, Y.; Yuan, C.; Wang, S.; Li, P. Entropy Analysis of Short-Term Heartbeat Interval Time Series during Regular Walking. *Entropy* **2017**, *19*, 568. [CrossRef]
17. Pan, W.-Y.; Su, M.-C.; Wu, H.-T.; Lin, M.-C.; Tsai, I.-T.; Sun, C.-K. Multiscale Entropy Analysis of Heart Rate Variability for Assessing the Severity of Sleep Disordered Breathing. *Entropy* **2015**, *17*, 231–243. [CrossRef]
18. Zhang, Y.-C. Complexity and 1/f noise. A phase space approach. *J. Phys. I* **1991**, *1*, 971–977. [CrossRef]
19. Costa, M.; Goldberger, A.L.; Peng, C.-K. Multiscale entropy analysis of biological signals. *Phys. Rev. E* **2005**, *71*, 021906. [CrossRef] [PubMed]
20. Valenza, G.; Nardelli, M.; Bertschy, G.; Lanata, A.; Scilingo, E.P. Mood states modulate complexity in heartbeat dynamics: A multiscale entropy analysis. *EPL* **2014**, *107*, 18003. [CrossRef]
21. Watanabe, E.; Kiyono, K.; Hayano, J.; Yamamoto, Y.; Inamasu, J.; Yamamoto, M.; Ichikawa, T.; Sobue, Y.; Harada, M.; Ozaki, Y. Multiscale Entropy of the Heart Rate Variability for the Prediction of an Ischemic Stroke in Patients with Permanent Atrial Fibrillation. *PLoS ONE* **2015**, *10*, e0137144. [CrossRef] [PubMed]
22. Jinde, Z.; Chen, M.-J.; Junsheng, C.; Yang, Y. Multiscale fuzzy entropy and its application in rolling bearing fault diagnosis. *Zhendong Gongcheng Xuebao/J. Vib. Eng.* **2014**, *27*, 145–151.
23. Ouyang, G.; Li, J.; Liu, X.; Li, X. Dynamic characteristics of absence EEG recordings with multiscale permutation entropy analysis. *Epilepsy Res.* **2013**, *104*, 246–252. [CrossRef] [PubMed]
24. Wu, S.-D.; Wu, C.-W.; Lee, K.-Y.; Lin, S.-G. Modified multiscale entropy for short-term time series analysis. *Phys. A Stat. Mech. Appl.* **2013**, *392*, 5865–5873. [CrossRef]
25. Li, P.; Liu, C.; Li, K.; Zheng, D.; Liu, C.; Hou, Y. Assessing the complexity of short-term heartbeat interval series by distribution entropy. *Med. Biol. Eng. Comput.* **2015**, *53*, 77–87. [CrossRef] [PubMed]
26. Pincus, S.M. Approximate entropy as a measure of system complexity. *PNAS* **1991**, *88*, 2297–2301. [CrossRef] [PubMed]
27. Goldberger, A.L.; Amaral, L.A.; Glass, L.; Hausdorff, J.M.; Ivanov, P.C.; Mark, R.G.; Mietus, J.E.; Moody, G.B.; Peng, C.K.; Stanley, H.E. PhysioBank, PhysioToolkit, and PhysioNet: Components of a new research resource for complex physiologic signals. *Circulation* **2000**, *101*, E215–E220. [CrossRef] [PubMed]
28. Pan, J.; Tompkins, W.J. A Real-Time QRS Detection Algorithm. *IEEE Trans. Biomed. Eng.* **1985**, *BME-32*, 230–236. [CrossRef]

Article

Quantifying the Multiscale Predictability of Financial Time Series by an Information-Theoretic Approach

Xiaojun Zhao [1], Chenxu Liang [1], Na Zhang [1,*] and Pengjian Shang [2]

1 School of Economics and Management, Beijing Jiaotong University, Beijing 100044, China
2 School of Science, Beijing Jiaotong University, Beijing 100044, China
* Correspondence: nazhang256@126.com

Received: 10 June 2019; Accepted: 8 July 2019; Published: 12 July 2019

Abstract: Making predictions on the dynamics of time series of a system is a very interesting topic. A fundamental prerequisite of this work is to evaluate the predictability of the system over a wide range of time. In this paper, we propose an information-theoretic tool, multiscale entropy difference (MED), to evaluate the predictability of nonlinear financial time series on multiple time scales. We discuss the predictability of the isolated system and open systems, respectively. Evidence from the analysis of the logistic map, Hénon map, and the Lorenz system manifests that the MED method is accurate, robust, and has a wide range of applications. We apply the new method to five-minute high-frequency data and the daily data of Chinese stock markets. Results show that the logarithmic change of stock price (logarithmic return) has a lower possibility of being predicted than the volatility. The logarithmic change of trading volume contributes significantly to the prediction of the logarithmic change of stock price on multiple time scales. The daily data are found to have a larger possibility of being predicted than the five-minute high-frequency data. This indicates that the arbitrage opportunity exists in the Chinese stock markets, which thus cannot be approximated by the effective market hypothesis (EMH).

Keywords: predictability; multiscale analysis; entropy rate; memory effect; financial time series

---

1. Introduction

Making predictions on the dynamics of time series of a system is a very interesting topic. Up to now, over thousands of methods have been proposed for the prediction of the systems' evolution [1]. A fundamental prerequisite of these works is to evaluate the predictability of the system over a wide range of time. For an isolated system, which does not exchange information with other systems, the predictability of the output time series is only determined by the degree of memory from the past values. In such a case, the time series in unpredictable if it is purely random, like Gaussian white noise; whereas, information can be extracted for prediction by analyzing the temporal structure of a time series with memory. In another way, examples of irreversible processes include typically chaotic dissipative processes, nonlinear stochastic processes, and processes with memory, operating away from thermodynamic equilibrium. One should be able to make easier predictions on irreversible processes, where the arrow of time is playing a role, than on reversible ones [2,3]. For a real-world system that may exchange information with other systems, the past values of other systems can also be utilized for prediction, except the past values of the underlying system itself [4,5].

In time series analysis, the multiscale analysis of time series has been broadly studied, which relies on the fact that the time series of complex systems, associated with a hierarchy of interacting regulatory mechanisms, usually generate complex fluctuations over multiple time scales. Analyzing the financial time series by amplification in different proportions with a coarse-graining algorithm [6] makes it possible to reveal both small-scale information and large-scale information at multiple resolutions.

This paper contributes to evaluating the multiscale predictability of financial time series. Another piece of evidence of this consideration is that the multiscale complexity (a tool of time series analysis that is associated with factors of the degree of memory, the temporal structure, and auto-correlations) have been measured [6,7], and hence, the predictability of time series, which is also closely related to those factors, can be analyzed on multiple time scales as well.

Financial time series analyses have played an important role in developing some of the fundamental economic theories. Furthermore, the understanding and analysis of financial time series, especially the evolution of stock markets, has been attracting the close attention of economists, statisticians, and mathematicians for many decades [8–14]. Recent research mostly focuses on the long-term average behavior of a market, and thus sheds little light on the temporal changes of a market. This type of method for analyzing financial time series may lead to a lack of analysis on the short-term predictability of time series, thus ignoring the critical information that is contained in the financial data, which may be used for the portfolio selection and pursuing an arbitrage opportunity [15].

If the efficient market hypothesis (EMH) is of some relevance to reality, then a market would be very unpredictable due to the possibility for investors to digest any new information instantly [16]. When a market behaves as the EMH stipulates, the market will be purely random without memory, and the variation of price will be very unpredictable. For an extensive review of the EMH, please see [17]. However, new evidence challenges the EMH with many empirical facts from observations, e.g., the leptokurtosis and fat tail of the non-Gaussian distribution, especially the fractal market hypothesis (FMH) [18]. The FMH asserts that (i) a market consists of many investors with different investment horizons, and (ii) the information set that is important to each investment horizon is different. As long as the market maintains this fractal structure, with no characteristic time scale, the market remains stable. When the market's investment horizon becomes uniform, the market becomes unstable because everyone is trading based on the same information set. In addition, Beben and Orlowski [19] and Di Matteo et al. [20,21] found that emerging markets were likely to have a stronger degree of memory than developed markets, suggesting that the emerging markets had a larger possibility of being predicted.

In this paper, we incorporate the multiscale analysis with an information-theoretic approach for characterizing the degree of memory of time series, so as to evaluate the predication of financial time series. We make use of the entropy rate in order to test the predictability of some synthetic data and of the Chinese stock markets. It is an interesting alternative to regression models, which are often used in financial time series. One advantage is that the method proposed is mainly model independent; another is that it deals with nonlinear systems, as well as with linear ones. The remainder of the paper is organized as follows. In the Methodology Section, we introduce a new entropy difference (ED) and its multiscale case, multiscale entropy difference (MED). We then apply these new methods to the numerical analysis of artificial simulations, including the logistic map, the Hénon map, the Lorenz system, and most importantly, the financial time series analysis. Finally, we give a brief conclusion.

2. Methodology

2.1. Entropy Difference

(i) For an isolated system, which does not exchange information with other systems, the degree of predictability of the time series can only be explained by the memory effects of its past values.

As the output of the underlying system, a time series $\{x_t\}$, $t = 1, \cdots, T$ is considered. First, the uncertainty of the time series at time t can be quantified by the Shannon entropy:

$$H[x_t] = \sum_{x_t \in \Theta} p(x_t) log_2 p(x_t). \tag{1}$$

$p(x_t)$ represents the probability distribution of x_t; Θ is the space of samples; and $H[x_t]$ describes the information of x at time t in bits.

The entropy rate measures the net information generated by the system at time t, given by $H[x_t|x_1,x_2,\cdots,x_{t-1}]$. We assume that the underlying system can be approximated by a p-order Markov process. That is to say, the value of the output time series at time t is only related to its nearest p neighbors and is independent of further values. Therefore, we obtain $H[x_t|x_1,x_2,\cdots,x_{t-1}] = H[x_t|x_{t-p},x_{t-p+1},\cdots,x_{t-1}] \equiv H[x_t|x_{t-1}^{(p)}]$, where:

$$H[x_t|x_{t-1}^{(p)}] = \sum_{x_t,x_{t-1}^{(p)}\in\Theta} p(x_t,x_{t-1}^{(p)})log_2\frac{p(x_t,x_{t-1}^{(p)})}{p(x_{t-1}^{(p)})}. \tag{2}$$

The uncertainty of the time series at time t is non-increasing given the past values, and hence, the entropy rate is no larger than the entropy itself: $H[x_t|x_{t-1}^{(p)}] \leq H[x_t]$.

The difference between the Shannon entropy and the entropy rate represents the contributions of the past values to reducing the uncertainty (and improving the predictability) of the time series at time t. It is given by:

$$D = H[x_t] - H[x_t|x_{t-1}^{(p)}]. \tag{3}$$

We name D the entropy difference (ED). For any (nonlinear) time series, $D \geq 0$. For a random walk process, the contribution of past values is negligible; hence, $H[x_t|x_{t-1}^{(p)}] = H[x_t]$, and $H[x_t] - H[x_t|x_{t-1}^{(p)}] = 0$. D equal to zero indicates that the time series cannot be predicted at all, as no past information can be utilized; whereas, if there exist autocorrelations/memory effects within the time series, the past values can be used to reduce the uncertainty of time series at time t, so $D > 0$.

The entropy difference D is non-negative, while the upper bound of D is uncertain. Thus, we further normalize D to the range of $[0,1]$, divided by its maximum value $H[x_t]$:

$$\mathcal{D} = \frac{H[x_t] - H[x_t|x_{t-1}^{(p)}]}{H[x_t]} = 1 - \frac{H[x_t|x_{t-1}^{(p)}]}{H[x_t]}. \tag{4}$$

Here, $0 \leq \mathcal{D} \leq 1$. The normalized ED, $\mathcal{D}$, quantifies the degree of predictability of the time series. Similarly, when $\mathcal{D}$ is approximately 0, the time series is unpredictable. When $\mathcal{D}$ attains a value of one, $H[x_t|x_{t-1}^{(p)}]$ is approximately 0. Therefore, there exists no uncertainty of x_t in the presence of the past values $x_{t-1}^{(p)}$, and the time series is completely specified (well predicted) at time t.

(ii) Next, consider a real-world system that exchanges information with other systems. Except the past values of the underlying system itself, the past values of other systems can also be exploited. Revisit the Granger causality, which is a statistical concept of causality that is based on prediction [22,23]. If a signal y "Granger-causes" a signal x, then past values of y should contain information that helps predict y above and beyond the information contained in past values of x alone. In the Granger causality, the value of x_t is predicted by two equations, respectively,

$$\begin{aligned} x_t &= \sum_{i=1}^{p} \alpha_i x_{t-i} + \epsilon_{1t}. \\ x_t &= \sum_{i=1}^{p} \beta_i x_{t-i} + \sum_{j=1}^{q} \gamma_j y_{t-j} + \epsilon_{2t}. \end{aligned} \tag{5}$$

The Granger causality is normally tested in the context of linear regression models. If the second forecast is found to be more successful, according to standard cost functions, then the past of y appears to contain information helping in forecasting x_t that is not in past $x_{t-1}^{(p)}$. The Akaike information criterion (AIC) or Bayesian information criterion (BIC) can be adopted to determine the lagged ranks p and q. The residual terms ϵ_{1t} and ϵ_{2t}, as a matter of fact, contain the information generated by the system at time t. A nonlinear extension of the Granger causality is the information-theoretic tool of transfer entropy [24,25], which measures the information flow from y to x:

$$\begin{aligned} T_{y\to x} &= H[x_t|x_{t-1}^{(p)}] - H[x_t|x_{t-1}^{(p)}, y_{t-1}^{(q)}] \\ &= \sum_{\substack{x_t, x_{t-1}^{(q)} \in \Theta \\ y_{t-1}^{(q)} \in \Xi}} p(x_t, x_{t-1}^{(p)}, y_{t-1}^{(q)}) log_2 \frac{p(x_t|x_{t-1}^{(p)}, y_{t-1}^{(q)})}{p(x_t|x_{t-1}^{(p)})}. \end{aligned} \tag{6}$$

Both the Granger causality and the transfer entropy indicate that the past values of another related system can be used to infer the trajectory of the underlying system. Hence, the *ED* of the isolated system can be extended to the multiple systems case.

The entropy rate of one system in the presence of another coupled system is given by $H[x_t|x_{t-1}^{(t-1)}, y_{t-1}^{(t-1)}]$. We further assume that these two systems can be approximated by the generalized Markov processes [24], that is $H[x_t|x_{t-1}^{(t-1)}, y_{t-1}^{(t-1)}] = H[x_t|x_{t-1}^{(p)}, y_{t-1}^{(q)}]$, and:

$$\begin{aligned} H[x_t|x_{t-1}^{(p)}, y_{t-1}^{(q)}] &= H[x_t, x_{t-1}^{(p)}, y_{t-1}^{(q)}] - H[x_{t-1}^{(p)}, y_{t-1}^{(q)}] \\ &= \sum_{\substack{x_t, x_{t-1}^{(q)} \in \Theta \\ y_{t-1}^{(q)} \in \Xi}} p(x_t, x_{t-1}^{(p)}, y_{t-1}^{(q)}) log_2 \frac{p(x_t, x_{t-1}^{(p)}, y_{t-1}^{(q)})}{p(x_{t-1}^{(p)}, y_{t-1}^{(q)})}. \end{aligned} \tag{7}$$

The uncertainty of system x can be given by the conditional probability distribution $p(x_t|x_{t-1}^{(p)}, y_{t-1}^{(q)})$. The conditional probability distribution $p(x_t|x_{t-1}^{(p)}, y_{t-1}^{(q)})$ describes the data range and the occurrence probability of x_t by knowing the past values of $x_{t-1}^{(p)}, y_{t-1}^{(q)}$. Consider an extreme case. If $p(x_t \equiv c|x_{t-1}^{(p)}, y_{t-1}^{(q)})$, where c is a constant, then x_t is fixed at point c with no uncertainty. Further, when the conditional distribution is fixed within a narrow range, the system is more deterministic at time t by knowing $x_{t-1}^{(p)}, y_{t-1}^{(q)}$, which can thus be well predicted. If the conditional distribution is still wide in the range, the system is full of uncertainty at time t and has a low possibility of being predicted.

The reduced uncertainty by knowing the past values of both x and y is estimated by the *ED*:

$$D = H[x_t] - H[x_t|x_{t-1}^{(p)}, y_{t-1}^{(q)}]. \tag{8}$$

Further, the *ED* is normalized by:

$$\mathcal{D} = \frac{H[x_t] - H[x_t|x_{t-1}^{(p)}, y_{t-1}^{(q)}]}{H[x_t]} = 1 - \frac{H[x_t|x_{t-1}^{(p)}, y_{t-1}^{(q)}]}{H[x_t]}. \tag{9}$$

$\mathcal{D}$ ranges between 0 and 1. $\mathcal{D}$ being approximately 0 indicates a low degree of predictability of the time series, and $\mathcal{D}$ close to 1 indicates a large degree of predictability. In addition, to set the *ED* in a fixed range, the normalization of *ED* also has other merits. Below is the explanation.

The predictability of a system is mainly subjected to the contributions of two aspects:

(i) The degree of the memory of the underlying system, that the past information can be well utilized to infer the future evolution of the system;
(ii) Whether a system is more deterministic than other systems. This is related to the range of the fluctuations of the time series, which can be partly explained by the variance of the time series. A time series with large variance (entropy) tends to be more difficult to predict than a time series with much small variance. Both the variance and the entropy reflect the diversity of the system. A system with more diverse states is likely to have large variance and entropy, whereas

a system with few states tends to have small ones. Obviously, a system with fewer states is easier to predict than that with diverse states.

Therefore, the normalization of ED by dividing D by $H[x_t]$ makes it possible to compare the degree of predictability between different systems, even if they have different ranges of fluctuations. Moreover, regarding the estimation of entropy values from time series, there may exist biases for different estimators. The normalization can offset those biases caused by the estimation of entropy if the numerator and the denominator use the same estimator.

Further, for a more complicated case of multiple subsystems (larger than 2 subsystems), e.g., the Lorenz system, the predictability of the time series can be given by:

$$\begin{aligned} \mathcal{D} &= \frac{H[x_t] - H[x_t|x_{t-1}^{(p)}, y_{t-1}^{(q)}, z_{t-1}^{(l)}]}{H[x_t]} \\ &= 1 - \frac{H[x_t|x_{t-1}^{(p)}, y_{t-1}^{(q)}, z_{t-1}^{(l)}]}{H[x_t]}, \end{aligned} \tag{10}$$

when the past values of x, y, and z can be used to predict x_t. Here, $Z_{t-1}^{(l)}$ could be a vector of possible explanatory variables.

2.2. Multiscale Entropy Difference

The predictability of time series estimated by ED and the normalized version is given on a unique time scale, on which the data are sampled. Here, we further evaluate that the multiscale predictability of time series relies on the fact that the time series of complex systems, associated with a hierarchy of interacting regulatory mechanisms, usually generate complex fluctuations over multiple time scales. There exist many approaches for the multiscale analysis in the framework of fractal theory [26], e.g., the data segments of detrended fluctuation analysis (DFA) [27], coarse-graining [6], and the time delay of phase space reconstruction [28,29], where the coarse-graining is one of the simplest methods.

We coarse grained the original data onto multiple time scales with a scale parameter s [2,6,7]. By the non-overlapping coarse-graining, the original time series x (with length T) is rescaled to $X(s)$:

$$X_t(s) = \frac{1}{s} \sum_{k=(t-1)s+1}^{ts} x_k. \tag{11}$$

t ranges from 1 to T/s. $X_t(s)$ represents the moving average of the system x at time t on the temporal scale s. The coarse-graining process is a low-pass filter, where the high-frequency fluctuations are filtered out. At small time scales, the details of the time series can be reserved, while at large scales, the details are ignored and only the profile of the time series is retained.

The procedure of the multiscale entropy difference (MED) mainly includes 3 steps:

Step 1. Coarse grain the original time series $\{x_t\}$ ($t = 1, \cdots, T$) to the coarse-grained time series $\{X_t(s)\}$ ($t = 1, \cdots, T/s$), with a time scale s.

Step 2. Estimate the ED and the normalized ED for the coarse-grained time series $\{X_t(s)\}$ ($t = 1, \cdots, T/s$), respectively.

Step 3. Change the time scale s and observe the changes of ED, and the normalized ED, on different time scales.

When the scale s is equal to 1, the MED method retrieves back the ED method. For other scales, the MED can evaluate the multiscale predictability of the time series. To be noted, for a short time series of length T, the multiscale analysis may be affected by the finite size effects at large time scales, which can be solved by the refined entropy estimators during the coarse-graining process. For more details, please see [5,30,31].

3. Numerical Simulations

In this section, we consider three examples to test our new methods, including one isolated system and two open systems.

We first consider the logistic map. It is a polynomial mapping of degree two, which consists of only one nonlinear system: $x_t = \mu x_{t-1}(1 - x_{t-1})$. For $\forall t$, $x_t \in [0, 1]$ can be used to represent the ratio of existing population to the maximum possible population in ecology. The values of interest for the parameter μ are those in the interval [0, 4]. Complex, chaotic behavior can arise from this very simple non-linear dynamical equation. Most values of μ beyond 3.56995 exhibit chaotic behavior. Here, we set $\mu = 3.7$ and let the data length $T = 10^5$. The initial value of x_0 was set to 0.5.

As only one equation is described in the logistic map, x_t changes no information with other variables. We added Gaussian white noises on the original time series x_t with different strengths to obtain a composite time series: $y_t = x_t + \lambda \epsilon_t$. ϵ_t is the Gaussian white noise (with zero mean and unit variance). $\lambda \geq 0$ is a parameter that tunes the strength of noises. x_t is the real signal corrupted by the external noise ϵ_t, and λ determines the signal-noise ratio. The larger λ, the smaller the signal-noise ratio is.

We used k-means clustering [32] to discretize the original data into k symbols, so as to estimate the entropies. k is a pre-defined parameter that determines the number of clusterings. Here, the parameter k for the k-means clustering was 10, i.e., we symbolized the original continuous time series as 10 discrete symbols. In Figure 1, we show the values of normalized *ED* $\mathcal{D}$ on multiple time scales $s = 1, 2, \cdots, 10$, with the noise strength parameter λ from 0.01 to 0.1 with a step of 0.01, since the variance of the original time series x of the logistic map ($T = 10^5$) was only 0.0412, the original data length was $T = 10^5$, therefore, even at $s = 10$, this ensured that the coarse-grained data length was 10^4. For $s = 1$ and $\lambda = 0$, corresponding to the original time series x, the degree of predictability was larger than 0.7. This indicates that the logistic map had a large possibility of being predicted, which coincides well with what the equation describes. When the scale increased, the predictability of the coarse-grained time series decreased, since the relationship between $X_t(s)$ and $X_{t-1}(s)$ became weaker on large scales. Moreover, the predictability of the time series also decreased with increasing λ, as the signal-noise ratio became lower. $\mathcal{D}$ reached a value very close to zero when $\lambda = 0.1$, so the composite time series could not be predicted. We also tested other values of k, for which it turned out that the values of larger k gave more reliable results; however, this was limited by the original data length. We further generated several groups of Gaussian white noises to add on the original time series and obtained very similar results, which verified the robustness of our new methods.

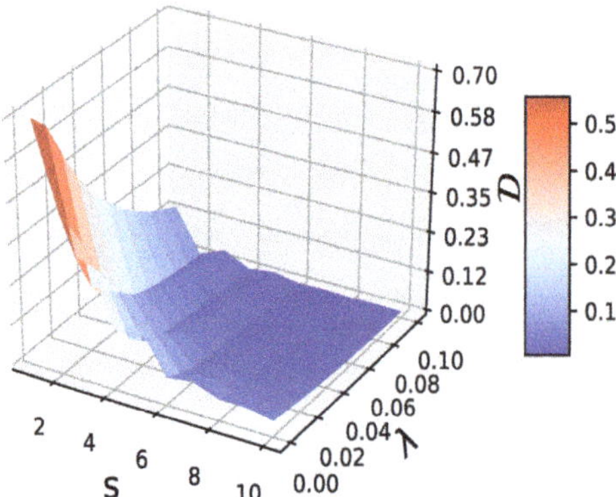

Figure 1. The values of $\mathcal{D}$ (Equation 4, lagged rank $p = 1$) on multiple time scales $s = 1, 2, \cdots, 10$, with the noise strength parameter $\lambda = 0.01, 0.02, \cdots, 0.1$ for the logistic map.

Next, we considered the Hénon map, which consists of two subsystems: $x_t = 1 - ax_{t-1}^2 + y_{t-1}$ and $y_t = bx_{t-1}$. The map depends on two parameters, a and b. For the classical Hénon map, it has values of $a = 1.4$ and $b = 0.3$. There exists nonlinear information flow from x_{t-1} and y_{t-1} to x_t, i.e., a one-step transition from the past data of one variable y to the current the data of another variable x. The initial values were set to $(1, 0)$.

We generated data with the classical Hénon map, with the data length $T = 10^5$. In Figure 2, we show the values of normalized ED $\mathcal{D}$ on multiple time scales $s = 1, 2, \cdots, 10$. For $s = 1$, which corresponds to the original time series x and y, the degree of predictability was 0.77. This indicates that x_t can be well predicted by using the past values of x and y. When the scale increased, the predictability of the coarse-grained time series decreased, as the relationship among $X_t(s)$, $X_{t-1}(s)$, and $Y_{t-1}(s)$ became weaker on large scales. We also compare $\mathcal{D}_{X_{t-1}(s),Y_{t-1}(s)\to X_t(s)} = 1 - H[X_t(s)|X_{t-1}(s), Y_{t-1}(s)]/H[X_t(s)]$ with $\mathcal{D}_{X_{t-1}(s)\to X_t(s)} = 1 - H[X_t(s)|X_{t-1}(s)]/H[X_t(s)]$ in Figure 2. Here, the lagged ranks p and q were both set to one. Obviously, if we only used the past values of x to predict x_t, the predictability of the time series would be much lower than if we incorporated both the past values of x and y. Therefore, we always obtained $\mathcal{D}_{X_{t-1}(s),Y_{t-1}(s)\to X_t(s)} \geq \mathcal{D}_{X_{t-1}(s)\to X_t(s)}$. Actually, $\mathcal{D}_{X_{t-1}(s),Y_{t-1}(s)\to X_t(s)} - \mathcal{D}_{X_{t-1}(s)\to X_t(s)}$ is just the normalized multiscale transfer entropy [5], and its unique scale case $\mathcal{D}_{x_{t-1},y_{t-1}\to x_t} - \mathcal{D}_{x_{t-1}\to x_t}$ is the normalized transfer entropy [24,33], from y to x.

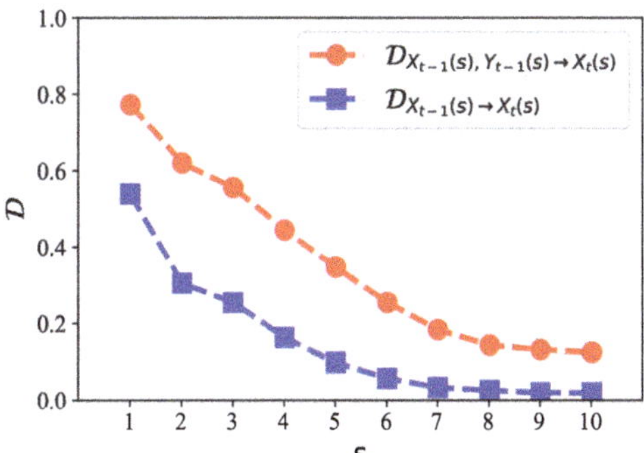

Figure 2. The values of $\mathcal{D}$ on multiple time scales $s = 1, 2, \cdots, 10$. The parameter k for the k-means clustering is 20. We compare $\mathcal{D}_{X_{t-1}(s),Y_{t-1}(s)\to X_t(s)}$ with $\mathcal{D}_{X_{t-1}(s)\to X_t(s)}$ and find that $\mathcal{D}_{X_{t-1}(s)\to X_t(s)}$ is always smaller than $\mathcal{D}_{X_{t-1}(s),Y_{t-1}(s)\to X_t(s)}$ on each time scale. This indicates that the predictability of x can be improved by incorporating the past values of y more than the past values of x alone. Therefore, the past values of y contain information for predicting x, which coincides well with the equations of the map.

Third, we studied the Lorenz system [34], which consists of three subsystems: $dx/dt = \sigma(y - x)$, $dy/dt = x(r - z) - y$, and $dz/dt = xy - bz$. Here, x, y, and z make up the system states, t time, and σ, r, and b the parameters: $\sigma = 10$, $r = 28$, $b = 8/3$. We integrated these equations numerically, applying a fourth-order Runge–Kutta method with the initial values of $(0.1, 0, 0)$.

We used the Lorenz system to generate data of length $T = 10^5$. In Figure 3, we give the values of normalized ED on multiple time scales $s = 1, 2, \cdots, 10$. For $s = 1$, which corresponds to the original time series x, y, and z, the degree of predictability of y_t reached 0.88. This indicates that y_t can be well predicted by using the past values of x, y, and z. When the scale increased, the predictability of the coarse-grained time series decreased, as the relationship among $Y_t(s)$, $X_{t-1}(s)$, $Y_{t-1}(s)$ and $Z_{t-1}(s)$ became weaker on large scales. We also compared $\mathcal{D}_{X_{t-1}(s),Y_{t-1}(s),Z_{t-1}(s)\to Y_t(s)}$ with $\mathcal{D}_{X_{t-1}(s),Y_{t-1}(s)\to Y_t(s)}$, $\mathcal{D}_{Y_{t-1}(s),Z_{t-1}(s)\to Y_t(s)}$, and $\mathcal{D}_{Y_{t-1}(s)\to Y_t(s)}$. Here, the lagged ranks p, q, and l were all set to one. We found that y_t can be well predicted giving the past values of x and y. Interestingly, the past values of z contributed much less to predicting y, although in the second equation of the Lorenz system, the change of y (dy/dt) is also explained by z. This can be explained as follows. In the x–y phase plane, x is closely related to y in the "diagonal" direction, as shown in Figure 3. However, in the y–z phase plane, no obvious relationship appears between y and z. Therefore, both the past values of x and y contribute to predicting y, rather than z. To predict other variables like x and z, we obtained very similar results.

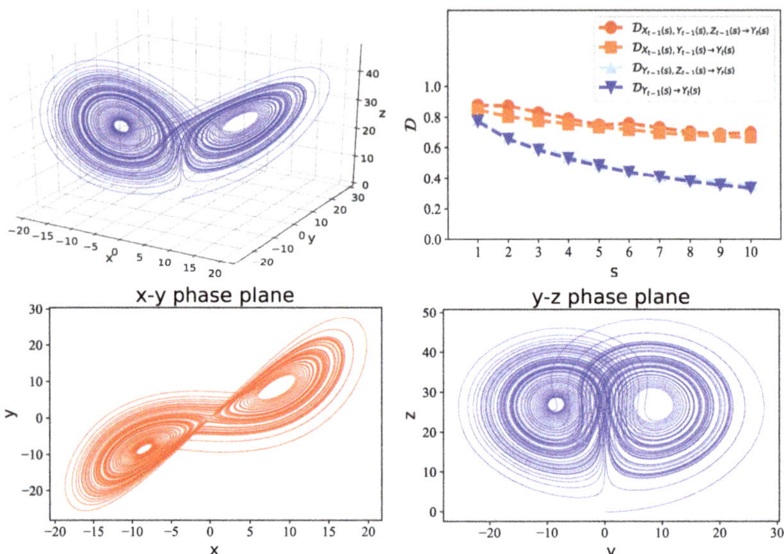

Figure 3. Upper left panel: A sample solution in the Lorenz system when $\sigma = 10$, $r = 28$, and $b = 8/3$, with initial values $(0.1, 0, 0)$. The data length is $T = 10^5$. Upper right panel: the values of $\mathcal{D}$ on multiple time scales $s = 1, 2, \cdots, 10$. The parameter k for the k-means clustering is 20. Lower left panel: the x–y phase plane. Lower right panel: the y–z phase plane.

4. Financial Time Series Analysis

The emerging stock markets have been found to have memory with the past values [35]; thus, the stock prices are not purely random. Past values can be used for the prediction of future stock prices. In this section, we study the predictability of the stock data of Shanghai and Shenzhen stock markets in China. We analyze the Shanghai Composite Index (SSE) and Shenzhen Composite Index (SZSE), both including the trading price and trading volume. At time t, the data related to trading price are given by x_t, and the data related to trading volume are given by y_t. Except the original data, we also analyzed the logarithmic change of stock price (i.e., logarithmic return): $log(x_t) - log(x_{t-1})$, the logarithmic change of trading volume: $log(y_t) - log(y_{t-1})$, the volatility (absolute return) of stock price: $|log(x_t) - log(x_{t-1})|$, and the volatility of trading volume: $|log(y_t) - log(y_{t-1})|$, respectively.

4.1. Five-Minute High-Frequency Data Analysis

We first analyzed the predictability of five-minute high-frequency data of SSE and SZSE. The data ranged from 3 March 2016 to 9 October 2018. In Figure 4, the left panels show the predictability of the stock price, logarithmic return, and price volatility for SSE, respectively. The right panels show the predictability of the stock price, logarithmic return, and price volatility for SZSE, respectively.

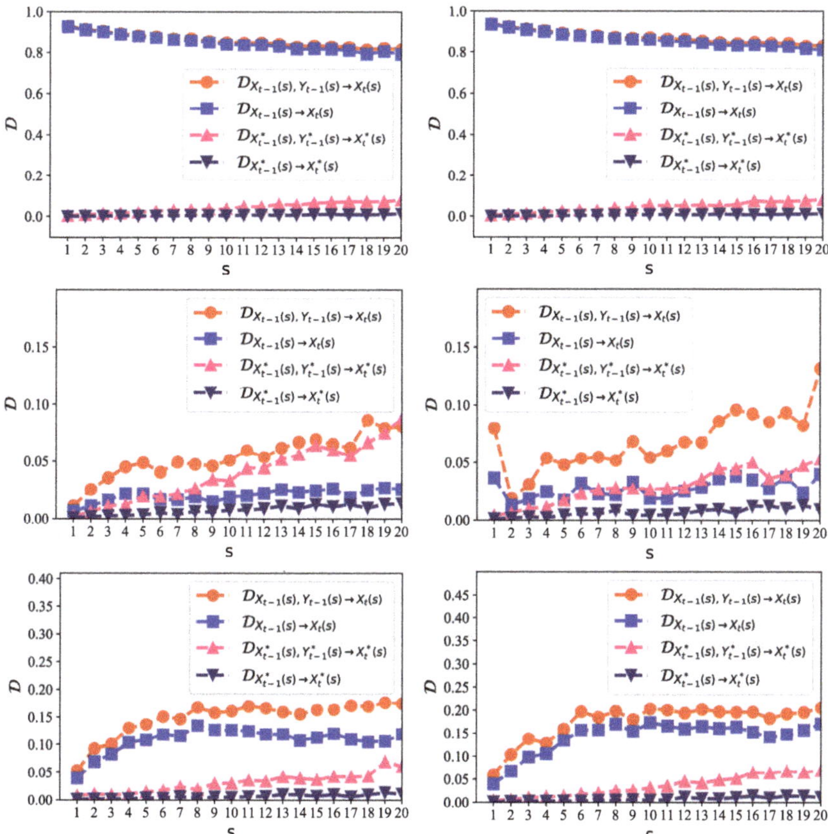

Figure 4. The multiscale entropy difference (MED) for the five-minute high-frequency stock data for the Shanghai and Shenzhen markets. Left panels show the results of the stock price (upper left), logarithmic return (middle left), and price volatility (lower left) for the Shanghai Composite Index (SSE), respectively. Right panels are those for the Shenzhen Composite Index (SZSE), respectively. The data related to the trading price are given by X, and the data related to trading volume are given by Y. X^* and Y^* represent the shuffled data.

For the original non-stationary stock prices, the predictability was very high on multiple time scales (as shown on the left panels of Figure 4), with $\mathcal{D}$ larger than 0.8, in the presence of either X_{t-1} alone or $X_{t-1}\&Y_{t-1}$. The reason is that we can just use X_{t-1} as the predicted value of X_t. The prediction error would be very small, because neighboring stock prices are very close. This explains why $\mathcal{D}$ was so large, but such a prediction is meaningless for the arbitrage. What makes investors more interested are the logarithmic return, which indicates the price going up or down, and the price volatility, which is the indicator of risk.

On the middle panels of Figure 4, the logarithmic return is more likely to be predicted given the past values of logarithmic return and the logarithmic change of trading volume than given the past values of logarithmic return alone, that is $\mathcal{D}_{X_{t-1}(s),Y_{t-1}(s)\to X_{t-1}(s)} > \mathcal{D}_{X_{t-1}(s)\to X_{t-1}(s)}$. This indicates that the trading volume contributes significantly to the prediction of the stock price. The close relationship between the stock price and trading volume was also found in previous studies, e.g., [36]. We shuffled the underlying data, represented by X^* and Y^*. The predictability for the shuffled data became much lower, because the shuffling process broke the memory among neighboring values for prediction, although it retained the distribution of the data.

We also show the results of price volatility on the lower panels of Figure 4. The degree of the predictability became larger than that of the logarithmic return. There existed long-range persistent correlations in the volatility series [37], so that the clustering of extreme volatilities emerged. A larger volatility was more likely to be followed by a large volatility, and vice versa [38,39]. The clustering of extreme volatilities made it possible to predict the volatility series from the neighboring past values. We found that the trading volume volatility can also help to predict the price volatility. The price volatility of SZSE was more likely to be inferred than SSE as $\mathcal{D}$ was larger. This is consistent with the previous findings [40]. The Shanghai market was relatively more stochastic than the Shenzhen market (i.e., the Shenzhen market was a little more structured and predictable). This reflects the fact that the Shenzhen market consists of most of the medium- to small-sized companies in China; they are relatively less stable than the large companies. Moreover, the predictability of the price volatility increased when the scale s increased.

4.2. Daily Data Analysis

We next analyze the predictability of the daily data of SSE and SZSE. The SSE data ranged from 19 December 1990 to 9 October 2018. The SZSE data ranged from 3 April 1991 to 8 October 2018. The correlations between Chinese stock markets and other major stock markets in the world were rather low most of the time. This indicates the fact that Chinese stock markets are relatively independent of the other stock markets, and therefore, we treated the Chinese stock market as an isolated system here. The left panels of Figure 5 show the predictability of the stock price, logarithmic return, and price volatility for SSE, respectively. The right panels show those for SZSE, respectively.

For the non-stationary daily stock prices, the predictability on the upper panels of Figure 5 is high, but meaningless, in the presence of either X_{t-1} alone or $X_{t-1}\&Y_{t-1}$. On the middle panels of Figure 5, the daily logarithmic return is more likely to be predicted given the past values of the logarithmic return and logarithmic change of trading volume than given the past values of the logarithmic return alone: $\mathcal{D}_{X_{t-1}(s),Y_{t-1}(s)\to X_{t-1}(s)} > \mathcal{D}_{X_{t-1}(s)\to X_{t-1}(s)}$. However, the shuffled data showed a bit confusing results, as $\mathcal{D}_{X_{t-1}(s),Y_{t-1}(s)\to X_{t-1}(s)}$ and $\mathcal{D}_{X^*_{t-1}(s),Y^*_{t-1}(s)\to X^*_{t-1}(s)}$ were very close to each other, especially for the Shenzhen market.

We show the results of daily price volatility on the lower panels of Figure 5. The degree of the predictability became larger than that of the daily logarithmic return. Moreover, the MED values were much larger for the daily data than the five-minute data. This indicates that the daily data were more deterministic and predictable than the high-frequency data. The Shanghai market and Shenzhen market showed very similar results.

Our daily data results showed an average degree of predictability. However, they involved times of both high and low volatilities, which implies a change in market behavior. During the times of high volatility (e.g., the 2008 world economic crisis), we found that the degree of predictability increased; while during the times of low volatility, the degree of predictability was much lower. This coincides well with previous studies [40] that the economic crisis can reduce the complexity of stock time series, making the volatility easier to predict.

To be noted, the reasons why we considered only one lag for each variable were two-fold: (i) Suppose that the sampling frequency of the original time series is f. In the multiscale analysis, the coarse-graining process, like a low-pass filter, can down sample the time series to f/s. Therefore, for the five-minute high-frequency data, although we used one lag for each variable, we still considered long-distance connections, which were much larger than five minutes. (ii) For most cases, we found that the low-frequency daily data could be approximated by one-order Markov processes. This means that major information could be be exposed by the current daily price and trading volume. Further, past daily data contributed little to predicting the market behavior of the following day, in the presence of the current data.

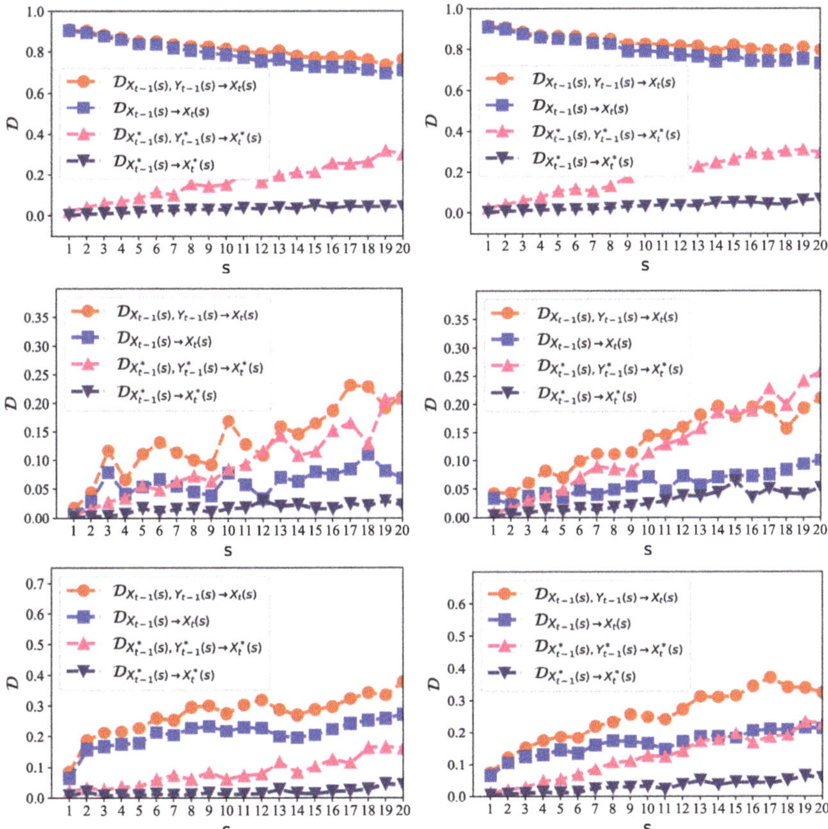

Figure 5. The MED for the daily stock data in the Shanghai and Shenzhen markets. Left panels show the results of the stock price (upper left), logarithmic return (middle left), and price volatility (lower left) for SSE, respectively. Right panels are those for SZSE, respectively. The data related to trading price are given by X, and the data related to trading volume are given by Y. X^* and Y^* represent the shuffled data.

5. Conclusions

In this paper, we introduced a new information-theoretic tool of MED to evaluate the degree of predictability for financial time series. The MED quantifies the contributions of the past values by reducing the uncertainty of the forthcoming values in time series on multiple time scales. For the isolated system, only the past values of the time series alone can be used. However, for the open systems, the past values of the time series and the past values of other time series (which have a close relationship with the underlying time series) can both be utilized. We performed several simulations based on the method, including the logistic map, the Hénon map, and the Lorenz system. All these simulations verified the accuracy and the robustness of our new method. We finally applied the MED method to the analysis of Chinese stock markets. The analysis on the five-minute high-frequency data and daily data of SSE and SZSE revealed that: (i) the logarithmic return had a lower possibility of being predicted than the price volatility; (ii) the trading volume volatility contributed significantly to the prediction of the stock price volatility on multiple time scales; (iii) the daily data were found to have a larger possibility of being predicted than the five-minute high-frequency data.

We note that our new evaluation methods of predictability were based on the assumption of the generalized Markov processes of the underlying time series. However, there still exist many other predicting methods that do not follow such a rule. For example, the k-nearest neighbors (KNN) prediction method [41] and the recurrence quantification analysis (RQA) tool [42] trace out more long-distance past values, so as to match them with the current states. In such a case, our methods would not be applicable any more.

Author Contributions: Conceptualization, X.Z. and P.S.; methodology, X.Z.; software, X.Z.; validation, N.Z.; formal analysis, X.Z.; investigation, C.L.; resources, N.Z. and P.S.; data curation, C.L.; writing–original draft preparation, X.Z.; writing–review and editing, N.Z. and P.S.; visualization, C.L. and P.S.; supervision, N.Z.; project administration, X.Z.; funding acquisition, X.Z.

Funding: National Natural Science Foundation of China (61703029, 61671048); Beijing Social Science (17YJC024).

Conflicts of Interest: The authors declare no conflict of interest.

References

1. Weigend, A.S. *Time Series Prediction: Forecasting the Future and Understanding the Past*; Routledge: Abingdon, UK, 2018.
2. Costa, M.D.; Peng, C.K.; Goldberger, A.L. Multiscale analysis of heart rate dynamics: Entropy and time irreversibility measures. *Cardiovasc. Eng.* **2008**, *8*, 88–93. [CrossRef]
3. Flanagan, R.; Lacasa, L. Irreversibility of financial time series: A graph-theoretical approach. *Phys. Lett. A* **2016**, *380*, 1689–1697. [CrossRef]
4. Zhao, X.; Shang, P.; Jing, W. Measuring the asymmetric contributions of individual subsystems. *Nonlinear Dyn.* **2014**, *78*, 1149–1158. [CrossRef]
5. Zhao, X.; Sun, Y.; Li, X.; Shang, P. Multiscale transfer entropy: Measuring information transfer on multiple time scales. *Commun. Nonlinear Sci. Numer. Simul.* **2018**, *62*, 202–212. [CrossRef]
6. Costa, M.D.; Goldberger, A.L.; Peng, C. Multiscale entropy analysis of complex physiologic time series. *Phys. Rev. Lett.* **2002**, *89*, 068102. [CrossRef]
7. Costa, M.; Goldberger, A.L.; Peng, C.K. Multiscale entropy analysis of biological signals. *Phys. Rev. E* **2005**, *71*, 021906. [CrossRef]
8. Borland, L.; Bouchaud, J.P. A non-Gaussian option pricing model with skew. *Quant. Finance* **2004**, *4*, 499–514. [CrossRef]
9. Cao, G.; Han, Y. Does the weather affect the Chinese stock markets? Evidence from the analysis of DCCA cross-correlation coefficient. *Int. J. Mod. Phys. B* **2014**, *29*, 1450236. [CrossRef]
10. Lai, K.K.; Yu, L.; Wang, S. Mean-variance-skewness-kurtosis-based portfolio optimization. *Int. Multi-Symp. Comput. Comput. Sci.* **2006**, *2*, 292–297.
11. Qiu, T.; Guo, L.; Chen, G. Scaling and memory effect in volatility return interval of the Chinese stock market. *Physics A* **2008**, *387*, 6812–6818. [CrossRef]
12. Yamasaki, K.; Muchnik, L.; Havlin, S.; Bunde, A.; Stanley, H.E. Scaling and memory in volatility return intervals in financial markets. *Proc. Natl. Acad. Sci. USA* **2005**, *102*, 9424–9428. [CrossRef]
13. Wang, F.; Yamasaki, K.; Havlin, S.; Stanley, H.E. Scaling and memory of intraday volatility return intervals in stock markets. *Phys. Rev. E* **2005**, *73*, 026117. [CrossRef]
14. Gabaix, X.; Gopikrishnan, P.; Plerou, V.; Stanley, H.E. Understanding the cubic and half-cubic laws of financial fluctuations. *Physics A* **2003**, *324*, 1–5. [CrossRef]
15. Zhao, X.; Shang, P.; Pang, Y. Power law and stretched exponential effects of extreme events in Chinese stock markets. *Fluct. Noise Lett.* **2012**, *9*, 203–217. [CrossRef]
16. Fama, E.F. Efficient capital markets: A review of theory and empirical work. *J. Finance* **1970**, *25*, 383–417. [CrossRef]
17. Sewell, M. History of the Efficient Market Hypothesis. Research Note. 2011. Available online: http://www.e-m-h.org/ (accessed on 5 July 2019).
18. Peters, E.E. *Fractal Market Analysis*; John Wiley & Sons: New York, NY, USA, 1994.
19. Beben, M.; Orłowski, A. Correlations in financial time series: Established versus emerging markets. *Eur. Phys. J. B* **2001**, *20*, 527–530. [CrossRef]

20. Matteo, T.D.; Aste, T.; Dacorogna, M.M. Scaling behaviors in differently developed markets. *Phys. A* **2003**, *324*, 183–188. [CrossRef]
21. Matteo, T.D.; Aste, T.; Dacorogna, M.M. Long-term memories of developed and emerging markets: Using the scaling analysis to characterize their stage of development. *J. Bank. Financ.* **2005**, *29*, 827–851. [CrossRef]
22. Wiener, N. The theory of prediction. *Mod. Math. Eng.* **1956**, *1*, 125–139.
23. Granger, C.W.J. Investigating causal relations by econometric models and cross-spectral methods. *Econometrica* **1969**, *37*, 424–438. [CrossRef]
24. Schreiber, T. Measuring information transfer. *Phys. Rev. Lett.* **2000**, *85*, 461–464. [CrossRef]
25. Barnett, L.; Barrett, A.B.; Seth, A.K. Granger causality and transfer entropy are equivalent for Gaussian variables. *Phys. Rev. Lett.* **2009**, *103*, 238701. [CrossRef]
26. Mandelbrot, B.B. *The Fractal Geometry of Nature*; Macmillan: London, UK, 1983; Volume 173.
27. Peng, C.K.; Buldyrev, S.V.; Havlin, S.; Simons, M.; Stanley, H.E.; Goldberger, A.L. Mosaic organization of DNA nucleotides. *Phys. Rev. E* **1994**, *49*, 1685. [CrossRef]
28. Takens, F. *Dynamical Systems and Turbulence*; Springer: Berlin, Germany, 1981.
29. Casali, K.R.; Casali, A.G.; Nicola, M.; Maria Claudia, I.; Fabricio, M.; Stefano, G.; Alberto, P. Multiple testing strategy for the detection of temporal irreversibility in stationary time series. *Phys. Rev. E* **2008**, *77*, 601–611. [CrossRef]
30. De Wu, S.; Wu, C.W.; Lin, S.G.; Wang, C.C.; Lee, K. Time series analysis using composite multiscale entropy. *Entropy* **2013**, *15*, 1069–1084.
31. Wu, S.D.; Wu, C.W.; Lin, S.G.; Lee, K.Y.; Peng, C.K. Analysis of complex time series using refined composite multiscale entropy. *Phys. Lett. A* **2014**, *378*, 1369–1374. [CrossRef]
32. Macqueen, J. Some methods for classification and analysis of multivariate observations. In *Proceedings of the Berkeley Symposium on Mathematical Statistics and Probability*; University of California Press: Berkeley, CA, USA, 1965.
33. Marschinski, R.; Kantz, H. Analysing the information flow between financial time series. *Eur. Phys. J. B* **2002**, *30*, 275–281. [CrossRef]
34. Lorenz, E.N. Deterministic nonperiodic flow. *J. Atmos. Sci.* **1963**, *20*, 130–141. [CrossRef]
35. Zhao, X.; Shang, P.; Shi, W. Multifractal cross-correlation spectra analysis on Chinese stock markets. *Physics A* **2014**, *402*, 84–92. [CrossRef]
36. Podobnik, B.; Horvatic, D.; Petersen, A.M.; Stanley, H.E. Cross-correlations between volume change and price change. *Proc. Natl. Acad. Sci. USA* **2009**, *106*, 22079–22084. [CrossRef]
37. Zhao, X.; Shang, P.; Lin, A. Universal and non-universal properties of recurrence intervals of rare events. *Physics A* **2016**, *448*, 132–143. [CrossRef]
38. Grau-Carles, P. Empirical evidence of long-range correlations in stock returns. *Physics A* **2000**, *287*, 396–404. [CrossRef]
39. Lin, A.; Shang, P.; Zhao, X. The cross-correlations of stock markets based on DCCA and time-delay DCCA. *Nonlinear Dyn.* **2011**, *67*, 425–435. [CrossRef]
40. Hou, Y.; Liu, F.; Gao, J.; Cheng, C.; Song, C. Characterizing complexity changes in Chinese stock markets by permutation entropy. *Entropy* **2017**, *19*, 514. [CrossRef]
41. Wang, J.; Shang, P.; Zhao, X. A new traffic speed forecasting method based on bi-pattern recognition. *Fluct. Noise Lett.* **2011**, *10*, 59–75. [CrossRef]
42. Marwan, N.; Romano, M.C.; Thiel, M.; Kurths, J. Recurrence plots for the analysis of complex systems. *Phys. Rep.* **2007**, *438*, 237–329. [CrossRef]

Article

Biometric Identification Method for Heart Sound Based on Multimodal Multiscale Dispersion Entropy

Xiefeng Cheng [1,2], Pengfei Wang [1,*] and Chenjun She [1]

1 College of Electronic and Optical Engineering & College of Microelectronics, Nanjing University of Posts and Telecommunications, Nanjing 210023, China; chengxf@njupt.edu.cn (X.C.); 1015020702@njupt.edu.cn (C.S.)

2 National and Local Joint Engineering Laboratory of RF Integration & Micro-Assembly Technology, Nanjing 210003, China

* Correspondence: 1017020702@njupt.edu.cn; Tel.: +86-18260031658

Received: 22 November 2019; Accepted: 17 February 2020; Published: 20 February 2020

Abstract: In this paper, a new method of biometric characterization of heart sounds based on multimodal multiscale dispersion entropy is proposed. Firstly, the heart sound is periodically segmented, and then each single-cycle heart sound is decomposed into a group of intrinsic mode functions (IMFs) by improved complete ensemble empirical mode decomposition with adaptive noise (ICEEMDAN). These IMFs are then segmented to a series of frames, which is used to calculate the refine composite multiscale dispersion entropy (RCMDE) as the characteristic representation of heart sound. In the simulation experiments I, carried out on the open heart sounds database Michigan, Washington and Littman, the feature representation method was combined with the heart sound segmentation method based on logistic regression (LR) and hidden semi-Markov models (HSMM), and feature selection was performed through the Fisher ratio (FR). Finally, the Euclidean distance (ED) and the close principle are used for matching and identification, and the recognition accuracy rate was 96.08%. To improve the practical application value of this method, the proposed method was applied to 80 heart sounds database constructed by 40 volunteer heart sounds to discuss the effect of single-cycle heart sounds with different starting positions on performance in experiment II. The experimental results show that the single-cycle heart sound with the starting position of the start of the first heart sound (S1) has the highest recognition rate of 97.5%. In summary, the proposed method is effective for heart sound biometric recognition.

Keywords: heart sound; ICEEMDAN; RCMDE; Fisher ratio; biometric characterization

1. Introduction

Heart sound is a complex, non-stationary and quasi-periodic signal that is consisted of multiple heartbeats or cardiac cycles, which mainly contain components such as the first heart sound S1, the second heart sound S2, systolic murmur and diastolic murmur. Heart sound originates from the opening and closing of the heart valve and the turbulence of blood, which contains physiological information, such as atria, ventricles, major vessels, cardiovascular vessels and functional status of various valves, and could reflect mechanical activity and structure status of heart. As the biometric characteristics, the biggest advantage of heart sound is universality, stability, uniqueness and collectability [1]. So far, there have been studies that have verified the feasibility of heart sound signals for biometric identification. The heart sound signal as an option for biometric identification was first introduced by Beritelli and Spadaccini [2]. Their method needs to locate and describe S1 and S2, then chirp-z transform (CZT) is performed to obtain the feature set, and finally, Euclidean distance (ED) is used as classifier. In another study, Phua et al. [3] introduced linear frequency band cepstrum (LFBC) for heart sound feature extraction and used two classifiers of vector quantization (VQ) and Gaussian mixture model (GMM) for classification and recognition. Beritelli and Spadaccini [4] continued improving the performance of phonocardiogram (PCG), building a human recognition system based on 13 MFCC extracted from S1 and S2 for feature extraction and First-to-Second ratio distance (FSR), achieving an equal error rate (EER) of 9% on 50 different people. Beritelli et al. [5] discussed that increasing the test set from 50 to 80 did not give a negative impact on EER. Fatemian et al. [6] Proposed a PCG signal identification and verification system. The system is based on wavelet preprocessing, feature extraction using short-time Fourier transform (STFT), feature dimension reduction using linear discriminant analysis (LDA) and majority voting using Euclidean distance for classification. Tran et al. [7] Used eight feature sets such as temporal shape, spectral shape, Mel-frequency cepstral coefficients (MFCC), linear frequency cepstral coefficients (LFCC), harmonic feature, rhythmic feature, cardiac feature, GMM-super vector as heart sound biometric recognition features, using two feature selection techniques and using support vector machine (SVM) for 52 users Classification recognition. Cheng et al. [8] introduced a human feature extraction method based on an improved circular convolution (ICC) slicing algorithm combined with independent subband function (ISF). The technology uses two recognition steps to obtain different human heart sound characteristics to ensure validity, and then uses similar distances for human heart sound pattern matching. Chen et al. [9] proposed a biometric recognition system based on heart sounds. The system uses wavelet for noise reduction, MFCC for feature extraction and Principal component analysis (PCA) for feature dimension reduction. Zhong et al. [10] proposed a biometric method based on cepstrum coefficients combined with GMM. These cepstrum coefficients are MFCC and LPCC, which are applied to 100 heart sounds of 50 people to test the algorithm. Zhao et al. [11] proposed a heart sound system based on marginal spectrum analysis and the classifier is based on VQ. Babiker et al. [12] present the design of a system for access control using a heart sound biometric signature based on energy percentage in each wavelet coefficients and MFCC feature. Db5 wavelet decomposition is used for noise reduction and ED is used for classification. The results show that the MFCC feature has better performance than the wavelet coefficient energy percentage. Akhter et al. [13] explored the possibility of using heart rate variability (HRV) in biometrics. They designed hardware and software for data collection. They also developed software for HRV analysis in Matlab, which uses various HRV Analysis techniques (such as statistics, spectrum, geometry, etc.) generate 101 HRV parameters (features), and use five different wrapper algorithms for feature selection, and obtain 10 reliable features from the 101 parameters, and finally use K Nearest Neighbor (KNN) classifies objects. The above introduces the common methods of heart sound biometrics. It can be found that the information entropy theory has not been applied in this field. At the same time, the information entropy theory has shown good results in the biological recognition of electrocardiogram (ECG) and electroencephalogram (EEG) signals [14–16]. In this paper, for the first time, multiscale entropy theory is introduced for the study of heart sound biometrics.

Ensemble empirical mode decomposition (EEMD) is a widely-used tool for the analysis of biomedical signals. It was proposed to overcome the deficiencies of ending effects and mode mixing in non-stationary signal decomposition when applying Empirical mode decomposition (EMD) to the time series. Recently, a new signal decomposition method based on the EEMD is presented, named as improved complete ensemble empirical mode decomposition with adaptive noise (ICEEMDAN), which provides a better spectral separation of the modes and a lesser number of sifting iterations is needed, reducing the computational cost. In this paper, ICEEMDAN is employed for heart sound signal decomposition to extract effective intrinsic mode functions (IMFs). To quantify the feature information of IMFs extracted from heart sound signals, dispersion entropy (DE), a new measure of uncertainty or irregularity, is introduced. The method tackles sample entropy (SE) and permutation entropy (PE) limitation. As a result of the relevance and the possible usefulness of DE in several signal analyses, it is important to understand the behavior of the technique for various kinds of classical signal concepts such as amplitude, frequency, noise power and signal bandwidth. In addition, the coarse-graining process is introduced to improve DE performance in estimating the complexity at the multiple time scales data, which is named as multiscale dispersion entropy (MDE). Recently, the refined composite

MDE (RCMDE) is proposed to improve the computing speed and stability, which is more applicable to process the short and noisy signals in biomedical applications.

To avoid the error, the testing data should be consistent with the length of the corresponding training data. Furthermore, the heart sound signal has pseudo-periodicity, and each cardiac cycle contains the dynamic acoustic characteristics of the heart structure. The cardiac cycle is different for each individual, which also reflects physiological characteristics between individuals. Therefore, this work takes single-cycle heart sound as the input of the proposed method, and RCMDE is combined with ICEEMDAN to quantify the important biometric information of the individual contained in every cardiac cycle. For the heart sound signal more than one cycle, it is firstly periodically segmented, and then the single cycle of heart sounds is decomposed into a group of IMFs by ICEEMDAN. These IMFs are then segmented to a series of frames, which is used to calculate the RCMDE as a characteristic representation of the heart sound. In addition, feature selection was performed to remove redundant features through the Fisher ratio (FR), and then ED is used to metric and match the features, and finally forming a new method based on ICEMDAN-RCMDE-FR-ED. At the same time, it can be considered that ICEEMDAN-RCMDE-FR has generated a kind of biometric characterization of heart sounds, which is named as the multimodal multiscale dispersion entropy. The feature generation flowchart of the multimodal multiscale dispersion entropy is as Figure 1.

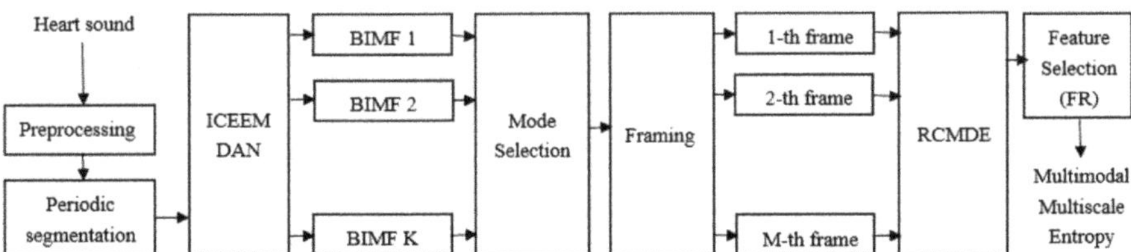

Figure 1. The feature generation flowchart of the multimodal multiscale dispersion entropy.

2. Materials and Methods

2.1. Mathematical Model of Heart Sound

In heart sound biometric recognition, heart sound is a non-stationary and quasi-periodic signal due to the rhythmicity of the heartbeat. Although the waveform of each cardiac cycle of heart sound has slight differences in time and magnitude, heart sound can be approximated to the periodic signals in the mathematics model. At the same time, since each cardiac cycle of heart sound contains four main components, including the first heart sound S1, the second heart sound S2, systolic murmur and diastolic murmur, cardiac cycles of heart sound are considered as the main objective of the biometric. The mathematical model of heart sound can be described as follows:

$$\begin{cases} x(i) = \sum_{n=0}^{N} x_T(i-nL) \\ x_T(i) = S_1(i) + Sysmur(i) + S_2(i) + Diasmur(i) \end{cases} \tag{1}$$

In the first step of Formula (1), $x_T(i), i = 1, 2, \ldots, L$, represents any period in the heart sound, the length of heart sound is $L = T * F_s$, where T represents the cardiac cycle, F_s represents the sampling frequency; $x(i), i = 1, 2, \ldots, N * L$, represents a heart sound signal containing N cardiac cycles. In the second step of Formula (1), $S_1(i)$ represents the first heart sound S1, $Sysmur(i)$ represents the systolic murmur, $S_2(i)$ represents the second heart sound S2 and $Diasmur(i)$ represents the diastolic murmur.

2.2. ICEEMDAN Method

Empirical Mode Decomposition (EMD) is an adaptive method for analyzing non-stationary signals originating from nonlinear systems. It decomposes the original signal into the sum of the intrinsic mode functions (IMFs), which can be described as follows [17]:

(1) Set $k = 0$ and find all extremums of $r_0 = x$.
(2) Interpolate between the minimum value (the maximum value) of r_k to obtain the lower (upper) envelope e_{min} (e_{max}).
(3) Calculate the average envelope: $me = (e_{min} + e_{max})/2$.
(4) Calculate the candidate IMF: $d_{k+1} = r_k - me$.
(5) Is d_{k+1} an IMF? Yes, save d_(k+1), calculate the residual $r_{k+1} = x - \sum_{i=1}^{k} d_i$, make $k = k+1$, and put r_k as input data in step (2). No, d_{k+1} is taken as input data in step (2).
(6) Continue to cycle until the final residual r_k meets some predefined stopping criteria.

The improved complete ensemble empirical mode decomposition with the adaptive noise (ICEEMDAN) method has been proved to be suitable for the processing of biomedical signals. The algorithm not only overcomes the mode mixing problem of EMD but also eliminates the spurious mode in CEEMDAN. Let $E_k(\cdot)$ denote the operator of the kth modal obtained by EMD, $\omega^{(i)}$ denotes the realization of white noise with zero mean and unit variance and $M(\cdot)$ denotes the operator for calculating the local mean of the signal. The realization steps of ICEEMDAN algorithm are as follows [18,19]:

(1) Calculate the local mean of the signal by I-times realization of EMD: $r_2 = r_1 + \beta_1 E_2(\omega^{(i)})$, get the first residual $r_1 = \left\langle M(x^{(i)}) \right\rangle$, where $\langle\cdot\rangle$ represents the average operator.
(2) The first modal is calculated from the residual r_1 obtained in the step (1): $\widetilde{d_1} = x - r_1$.
(3) The second residual is calculated by $x^{(i)} = x + \beta_0 E_1(\omega^{(i)})$, and defines the second mode: $\widetilde{d_1} = r_1 - r_2 = r_1 - \left\langle M(r_1 + \beta_1 E_2(\omega^{(i)})) \right\rangle$.
(4) For $k = 3, 4, \ldots, K$, calculate the kth residual: $r_k = \left\langle M(r_{k-1} + \beta_{k-1} E_k(\omega^{(i)})) \right\rangle$.
(5) Calculate the kth modal: $\widetilde{d_k} = r_{k-1} - r_k$.
(6) Go to the next k of step (4) until all modes are obtained.

The constant β_{k-1} is selected to adjust the signal-to-noise ratio (SNR) between the residual and the added noise. For $k = 1$, $\beta_0 = \varepsilon_0 std(x)/std(E_1(\omega^{(i)}))$, where $std(\cdot)$ represents the standard deviation, ε_0 is the reciprocal of the required SNR between the input signal x and the first added noise. For $k \geq 2$, $\beta_k = \varepsilon_0 std(r_k)$.

2.3. RCMDE Method

Multiscale dispersion entropy (MDE) is a combination of coarse-grained and dispersion entropy, and the refine composite multiscale dispersion entropy (RCMDE) improves MDE in that the different starting time of the coarse-grained time series corresponding to the different scale factors τ is adopted. Based on multiscale techniques, the main steps for calculating RCMDE are as follows [20–22]:

(1) The first is to construct a continuous coarse-grained time series. For a univariate signal $x(i)(i = 1, 2, \ldots, N)$, Its J-th coarse-grained time series $x_J^{(\tau)} = \left\{x_{J,1}^{(\tau)}, x_{J,2}^{(\tau)}, \ldots\right\}$ can be showed as follows:

$$x_{J,j}^{(\tau)} = \frac{1}{\tau} \sum_{i=(j-1)\tau+J}^{j\tau+J-1} x_i, 1 \leq j \leq N, 1 \leq J \leq \tau \tag{2}$$

where τ is the scale factor, and the original time series x is scaled by controlling the value of τ.

(2) Map $x_{J,j}^{(\tau)}$ into $y_{J,j}^{(\tau)}$ with the normal cumulative distribution function:

$$y_{J,j}^{(\tau)} = \frac{1}{\sigma\sqrt{2\pi}} \int_{-\infty}^{x_{J,j}^{(\tau)}} e^{\frac{-(\tau-\mu^2)}{2\sigma^2}} dt \tag{3}$$

where σ and μ represent the standard deviation and mean of $x_{J,j}^{(\tau)}$, respectively.

(3) Assign each $y_{J,j}^{(\tau)}$ to an integer from Label 1 to c using a linear algorithm. The mapped signal can be defined as follows:

$$z_{J,j}^{(\tau,c)} = round(c \cdot y_{J,j}^{(\tau)} + 0.5) \tag{4}$$

(4) Define embedding vector $z_{J,j}^{(\tau,c,m)}$ with embedding dimension m and time delay d as:

$$z_{J,j}^{(\tau,c,m)} = \left\{ z_{J,j}^{c}, z_{J,j+d}^{c}, \ldots, z_{J,j+(m-1)d}^{c} \right\} \tag{5}$$

Each time series $z_{J,j}^{(\tau,c,m)}$ is mapped to a dispersion pattern $\pi_{v_0 v_1 \ldots v_{m-1}}$, where:

$$z_{J,j}^{(\tau,c)} = v_0, z_{J,j+d}^{(\tau,c)} = v_1, \ldots, z_{J,j+(m-1)d}^{(\tau,c)} = v_{m-1}$$

(5) For each dispersion pattern, the relative frequency can be obtained as:

$$p(\pi_{v_0 v_1 \ldots v_{m-1}}) = \frac{Number\left\{ j \middle| j \leq N-(m-1)d, z_{J,j}^{(\tau,c,m)} has\_type\_v_0 v_1 \ldots v_{m-1} \right\}}{N-(m-1)d} \tag{6}$$

where $p(\pi_{v_0 v_1 \ldots v_{m-1}})$ represent the number of dispersion pattern which is assigned to $z_{J,j}^{(\tau,c,m)}$ divided by the total number of embedding signals with embedding dimension m.

(6) Based on Shannon's definition of entropy, multiscale dispersion entropy with embedding dimension m, time delay d, and the number of classes c can be defined as:

$$RCMDE(x, m, c, d, \tau) = -\sum_{\pi=1}^{c^m} p(\pi_{v_0 v_1 \ldots v_{m-1}}) Inp(\pi_{v_0 v_1 \ldots v_{m-1}}) \tag{7}$$

2.4. Feature Selection

Fisher Ratio (FR) is proposed on the basis of Fisher criterion. It is used to measure the classification and recognition ability of features and has been successfully used by Pruzansky and Mathews in the research of speech recognition [23]. In this paper, the Fisher ratio is used to select the optimal features and the steps are as follows:

(1) Calculate the inter-class dispersion σ_{between}, which is used to measure the degree of dispersion of the r-dimensional feature parameters between the heart sound signals of various categories. The calculation formula is:

$$\sigma_{\text{between}} = \sum_{i=1}^{M} \left(\mu_r^{(i)} - \mu_r\right)^2 \tag{8}$$

where M represents the total number of heart sound samples, $\mu_r^{(i)}$ is the mean value of the r-dimensional feature parameters of the i-th type heart sound signal, and μ_r is the mean value of the r-dimensional feature parameters in all heart sound signals.

(2) Calculate the intra-class dispersion σ_{within}, which is used to measure the degree of dispersion in the r-dimensional feature parameters of a certain type of heart sound signal. The calculation formula is:

$$\sigma_{\text{within}} = \sum_{i=1}^{M} [\frac{1}{n_i} \sum_{j=1}^{n_i} (x_r^{(j)} - \mu_r^{(i)})^2] \tag{9}$$

where n_i is the number of heart sound samples of the i-th type heart sound signal, $x_r^{(j)}$ is the r-dimensional feature parameter of the j-th heart sound sample of the i-th heart sound signal.

(3) To calculate the Fisher ratio, the calculation formula is:

$$F_r = \frac{\sigma_{between}}{\sigma_{within}} \tag{10}$$

where F_r is the Fisher ratio of the r-dimensional characteristic parameter.

(4) Sort the Fisher ratio of each dimension feature parameter in descending order:

$$F_1, F_2, \ldots, F_R \rightarrow F_{r_1} > F_{r_2} > \ldots > F_{r_R} \tag{11}$$

where $r_i \in \{1, 2, \ldots, R\}, i = 1, 2, \ldots, R$, R is the dimension of the characteristic parameter.

(5) The larger the Fisher ratio, the stronger the classification and recognition ability of the feature parameter of the dimension. According to this principle, the top N_r dimensional feature parameters ranked first in (4) is selected as the optimal features.

2.5. Matching Recognition

This paper adopts the Euclidean distance (ED) and the close principle to realize the pattern recognition of the user's heart sound. The idea of the algorithm is as follows: when the data and labels in the training set are known, compare the one-dimensional feature vector of the test data with the corresponding feature vector in the training set to find the one-dimensional feature vector most similar to it in the training set, then the category corresponding to the test feature vector is the category corresponding to the training feature vector. The algorithm steps are:

(1) Calculate the distance between the test data and each training data;

For the feature vector v of the test data and the feature vector v_A of the A-th training data in the heart sound database, the Euclidean distance d_A in the D dimension Euclidean space is calculated as follows:

$$d_A = \sqrt{\sum_{i=1}^{D} (v(i) - v_A(i))^2} \tag{12}$$

where $A = 1, 2, \ldots, C$, C is the number of training data, and D is the dimension of the feature vector.

(2) Sort in increasing order of distance:

$$d_1, d_2, \ldots, d_C \rightarrow d_{x_1} > d_{x_2} > \ldots > d_{x_C} \tag{13}$$

where $x_i \in \{1, 2, \ldots, C\}, i = 1, 2, \ldots, C$.

(3) According to the selection principle, the closer the distance, the higher the degree of matching between the two data. The category corresponding to the closest feature vector v_{x_1} is selected as the prediction classification of the test data.

2.6. Evaluation Methods

This paper uses the following three indicators to evaluate the proposed algorithm [24]:

(1) Average test accuracy: The $\overline{CRR}$ obtained by averaging the CRR of J experiments was used as the final experimental result, as shown in Equation (9). Considering the calculation amount and accuracy comprehensively, J = 200 is taken in the following experiment of parameter selection and algorithm comparison.

$$\begin{cases} CRR(j) = \frac{Number\_of\_correctly\_identified\_subjects\_in\_trial\_j}{Total\_number\_of\_subjects} \\ \overline{CRR} = \frac{1}{J}\sum_{j=1}^{J} CRR(j) \end{cases} \quad (14)$$

(2) Kappa coefficient: Kappa coefficient is an index to measure the accuracy of multi-classification. Its calculation formula is as follows:

$$Kappa = \frac{p_0 - p_e}{1 - p_e} \quad (15)$$

Among them, p_0 is the sum of the number of correctly classified samples of each class divided by the total number of samples, which is the overall classification accuracy. Suppose the number of true samples in each class is $a_1, a_2, \ldots, a_c$, and the number of predicted samples in each class is $b_1, b_2, \ldots, b_c$, and the total number of samples is num.

$$p_e = \frac{a_1 b_1 + a_2 b_2 + \ldots + a_c b_c}{num \cdot num} \quad (16)$$

The kappa coefficient is usually between 0 and 1 and can be divided into five groups to represent different levels of classification accuracy: 0.0 to 0.20 extremely low classification accuracy, 0.21 to 0.40 general classification accuracy, 0.41 to 0.60 high classification accuracy, 0.61–0.80 very high classification accuracy and 0.81–1 extremely high classification accuracy.

(3) t-test: t-test uses the t-distribution theory to infer the probability of a difference occurring, thereby comparing whether the difference between the two averages is significant. This paper is repeatedly training/testing by randomly dividing the training set/test set multiple times, therefore this will get multiple test accuracy rates. Therefore, the t test can be used to verify whether the $\overline{CRR}_{200}$ selected in this paper can be used as generalization Accuracy. Assuming the generalization accuracy $\mu_0 = \overline{CRR}_{200}$, we get n test accuracy rates, $CRR(i), i = 1, 2, \ldots, n$, then the average test accuracy μ and variance σ^2 are:

$$\mu = \frac{1}{n}\sum_{i=1}^{n} CRR(i) \quad (17)$$

$$\sigma^2 = \frac{1}{n-1}\sum_{i=1}^{n} (CRR(i) - \mu)^2 \quad (18)$$

Considering that the accuracy of these n tests can be regarded as independent sampling of the generalization accuracy μ_0, then the variable $t = \frac{\sqrt{n}(\mu - \mu_0)}{\sigma}$ follows a t-distribution with $n-1$ degrees of freedom. This paper uses the following t-test steps:

(1) First establish hypotheses and determine the test level α:

$H_0 : \mu = \mu_0$ (zero hypothesis), $H_1 : \mu \neq \mu_0$ (alternative hypothesis), using bilateral hypothesis, α commonly used values are 0.05 and 0.1, the test level is $\alpha = 0.05$ in this paper.

(2) Calculate the test statistics: $t = \frac{\sqrt{n}(\mu - \mu_0)}{\sigma}$.

(3) Check the corresponding critical value table to determine the critical value $t_{(\alpha,n)}$ and conclude: If the value of t is within the critical value range $[-t_{(\alpha,n)}, t_{(\alpha,n)}]$, you cannot reject the assumption that $H_0 : \mu = \mu_0$, you can think that the generalization accuracy is μ, the degree of confidence is $1 - \alpha$; otherwise, the hypothesis can be rejected, that is, under this significance degree, the generalization accuracy μ_0 can be considered to be significantly different from the average test accuracy μ.

3. Results and Discussion

3.1. Data Sources

To verify the effectiveness of the proposed method, the open databases of heart sound recordings and the heart sound database built by our research group are both analyzed. The open database used in this paper consists of 72 heart sounds from the three open heart sound databases Michigan, Washington and Littman, including 18 normal heart sounds and 54 abnormal heart sounds. Among them, 23 cases and 16 heart sounds were obtained from the Michigan and Washington heart sound database, and 33 heart sounds were selected from 3M's Littman heart sound database, because 3 heart sounds that did not meet the experimental conditions (i.e., not satisfied should contain at least two cardiac cycles) was abandoned. For the Michigan and Washington heart sound databases [25,26], the sampling frequency is 44.1 kHz, and the acquisition time is about 60 s, respectively including 23 and 16 heart sounds. For the Littman heart sound database [27], the sampling frequency is 11.025 kHz, and the acquisition time is about 3 s. The heart sound database built by our research group consisted of 80 cases of heart sound recordings from college student and teacher volunteers, which are collected by using the Ω shoulder-belt wireless heart sound sensor self-developed by our research group (patent number: 201310454575.6) with sampling frequency of 11,025 Hz. Every volunteer is recorded twice at least one-hour intervals, and every time keep approximately 5 s by properly contacting it with the skin of the front chest wall of the subject, as shown in Figure 2. The heart sound recording is from the apex located slightly inside the midline of the left intercostal bone of the fifth intercostal space obtained from the valve area. In addition, the heart sound recordings obtained from the subjects are collected in their calm state, and the recorded heart sound recordings are stored in a .wav format.

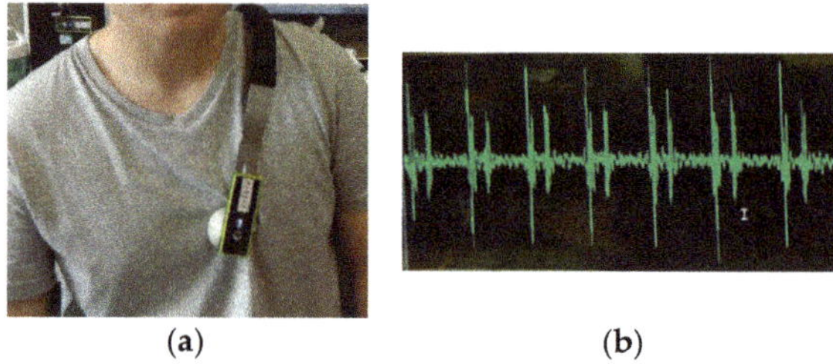

(**a**) (**b**)

Figure 2. Heart sound database collected by our group: (**a**) Ω shoulder-belt wireless heart sound sensor; (**b**) The processing of collecting heart sound.

3.2. Feature Extraction and Recognition

3.2.1. Pretreatment

The original signal is first preprocessed before performing feature extraction and matching recognition. The preprocessing module includes set the labels for heart sounds, downsampling, denoising, and cycle segmentation. Firstly, the 72 heart sound recordings of the three open heart sound databases are set the labels of 1–72 separately to distinguish the individual corresponding to each heart sound. Then the downsampling frequency is set to 2000 Hz, and the background noise when collecting heart sounds is eliminated by using the wavelet packet multi-threshold denoising method. Wavelet packet multi-threshold denoising is through setting a certain threshold value for each layer of wavelet packet coefficients to quantify and analyze each wavelet coefficient, retain useful data and eliminate unnecessary data. Different wavelets may cause different denoising effects, therefore, Biorthogonal HS wavelets developed for heart sound signals [28] is used here to filter in this work. The specific process is as follows:

(1) Performing four-layer HS wavelet packet transform on the noisy signal, and obtain a set of wavelet packet coefficients $wpt_i, i = 1, 2, \ldots, 16$;

(2) To quantify the threshold of wpt_i separately by selecting the Heursure function, and use the threshold to remove the useless data in wpt_i;

(3) To perform discrete wavelet reconstruction by using the denoised coefficient wpt_i, and the reconstructed signal is the denoised signal.

3.2.2. Periodic Segmentation

Since the proposed feature extraction method is based on single-period heart sounds, the logical regression (LR) and hidden semi-Markov model (HSMM) heart sound segmentation method proposed by Springer et al. is used in this work, which has proven in the 2016 PhysioNet/CinC Challenge to accurately segment heart sounds in noisy real heart sound recordings with good performance [29–31]. In this paper, the heart sound segmentation method is firstly used to assign four states such as S1 (the first heart sound), systole, S2 (the second heart sound) and diastole for the preprocessed heart sound recordings. The time point of the first jump from the initial state of the current heart sound recording to the next state is used as the initial split point. The following four situations may be obtained: (1) a series of cardiac cycles segmented from the beginning of S1 to the beginning of the next S1 of the current heart sound recording; (2) a series of cardiac cycles segmented from the beginning of the systole to the beginning of the next systole of the current heart sound recording; (3) a series of cardiac cycles segmented from the beginning of S2 to the beginning of the next S2 of the current heart sound recording; (4) a series of cardiac cycles segmented from the beginning of the diastole to the beginning of the next diastole of the current heart sound recording. The schematic diagram of the heart sound cycle segmentation corresponding to these four cases is shown in Figure 3.

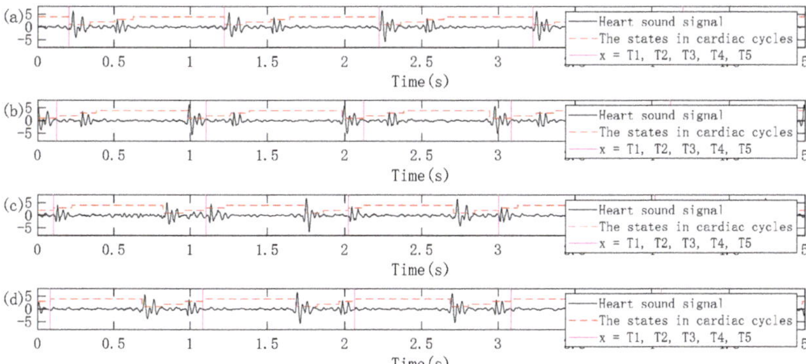

Figure 3. Four methods of heart sound cycle segmentation. (**a**) A series of cardiac cycles segmented from the beginning of S1 to the beginning of the next S1 of the current heart sound recording; (**b**) a series of cardiac cycles segmented from the beginning of the systole to the beginning of the next systole of the current heart sound recording; (**c**) a series of cardiac cycles segmented from the beginning of S2 to the beginning of the next S2 of the current heart sound recording; (**d**) a series of cardiac cycles segmented from the beginning of the diastole to the beginning of the next diastole of the current heart sound recording.

By the above segmentation method, 72 heart sound recordings in the open heart sound databases are divided into 2005 single-cycle heart sounds, where each heart sound recording is divided into 2–101 single-cycle heart sounds according to their length. In each of the following experiments, a single-period heart sound was randomly selected from the single-cycle heart sounds from the same heart sound recording as a test data, so that the test data contained 72 single-period heart sounds from different individuals, and the remaining 1933 single-cycle heart sounds were used as training data.

3.2.3. Framing and Windowing

Similar to the speech signal, heart sound is also a non-stationary and time-varying signal. Therefore, the heart sound signal is divided into a set of frames to analyze its characteristic parameters. For each frame, the length of the frame is called the frame length. The standard speech frame length is 20 ms to 25 ms, which is not suitable for heart sounds due to its pseudo-periodicity. Reference [32] thinks that the frame length of heart sounds should be longer than 20–25 ms, and it is best when the frame length equals to 256 ms. In this paper, the frame length of the heart sound should be related to the cardiac cycle, and frame lengths should be set different values according to the cardiac cycle. Further, the distance from the start of the frame to the start of the subsequent frame is called the frameshift. To smoothly change the feature parameters, a part of the overlap between adjacent frames is often provided in the case of framing. To prevent spectrum leakage, windowing is usually performed for each frame of heart sounds, usually a Hanning window or a Hamming window.

The single-cycle heart sound obtained after preprocessing and period segmentation is framed by overlap windowing, and then the RCMDE features of each frame are calculated. Then, the RCMDE features of each frame of the single-cycle heart sound are combined into a one-dimensional feature vector. When calculating RCMDE, four important parameters in RCMDE that may have a greater impact on the results, namely scale factor τ, categories c, embedding dimension r and delay time τ. In this experiment, a large number of experiments show when the scaling factor $\tau = 20$, the categories $c = 3$, the embedding dimension $m = 2$ and the delay time $d = 1$, the algorithm performance is the best. In Figure 4a, the RCMDE characteristics of two different single-cycle heart sounds of the same person after windowing and framing are compared. As can be seen from the figure, all feature points of the two single-cycle heart sounds are distributed near the 45° line. It shows that the two single-cycle heart sounds are close in their corresponding eigenvalues, and they are relatively matched. In Figure 4b, the RCMDE characteristics of two single-cycle heart sounds of different people after windowing and framing are compared. It can be seen that the two single-cycle heart sounds have more feature points distributed farther from the 45° line, which indicates that the two single-cycle heart sounds have relatively large differences in corresponding feature values, and are not well matched. In Figure 4, the frame length is taken as T/4, the frameshift is taken as T/8 and the Hanning window is used.

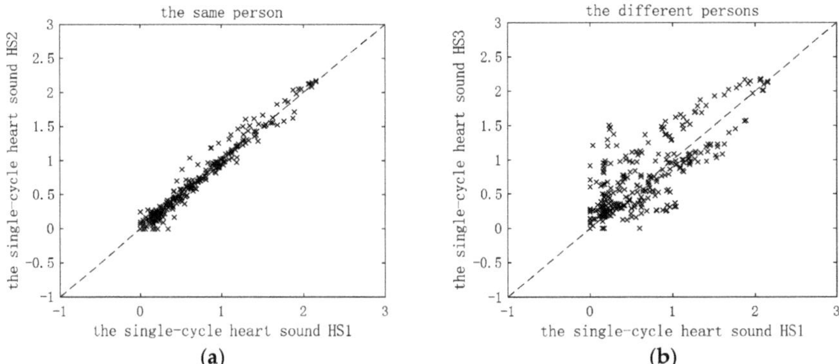

Figure 4. Comparison of refine composite multiscale dispersion entropy (RCMDE) characteristics of single-cycle heart sounds after windowing and framing: (**a**) Comparison of RCMDE characteristics of two single-cycle heart sounds of the same person; (**b**) comparison of RCMDE characteristics of two single-cycle heart sounds of different persons.

From the above analysis, it can be known that the RCMDE feature of single-cycle heart sounds after windowing and framing is feasible for the identification of different individuals. The effect of setting different frame lengths and frameshifts on the performance of the algorithm based on the cardiac cycle is discussed below. Here, adopting the control variable method, the above-mentioned parameters remain unchanged. It is discussed that the frame length takes win = T/i (i = 1, 2, . . . , 20) respectively in the condition of no frame overlap, and the corresponding $\overline{CRR}$ is as shown in the left half of Table 1. The result shows that the optimum frame length is T/4. Then, it is discussed that when the frame length takes T/4 unchanged, the frameshift takes inc = win/i (i = 1, 2, . . . , 10) respectively, and the corresponding $\overline{CRR}$ is as shown in the right half of Table 1. The result shows that when the frameshift is win/5, the best performance is achieved, and adding the frame overlap latter has not improved $\overline{CRR}$.

Table 1. Comparison of the recognition performance of setting different frame length and frameshift based on the cardiac cycle on the three open heart sound databases.

Heart Sound Database	Including 2005 Single-Cycle Heart Sounds from the Open Database Michigan, Washington, and Littman		
Algorithm	**RCMDE-ED**		
win (inc = win)	$\overline{CRR}$	**inc (win = T/4)**	$\overline{CRR}$
T	45.16%	win	84.55%
T/2	78.71%	win/2	84.82%
T/3	82.52%	win/3	88.88%
T/4	84.55%	win/4	87.11%
T/5	81.09%	win/5	90.08%
T/6	82.83%	win/6	88.68%
T/7	81.56%	win/7	88.20%
T/8	76.77%	win/8	88.64%
T/9	75.13%	win/9	89.66%
T/10	65.40%	win/10	89.47%

3.2.4. ICEEMDAN-RCMDE-FR-ED Algorithm

To achieve a higher $\overline{CRR}$, the ICEMDAN algorithm is used to decompose the training/test cycles into a group of IMFs, and then the hamming window with the window size of T/4 and the window shifting of T/20 is used to frame these IMFs. The training/test IMFs are segmented separately, and each of the obtained heart frames is subjected to RCMDE calculations. The result is sent to the ED algorithm, and the obtained $\overline{CRR}$ is shown in Table 2. Here, the parameters of the ICEEMDAN algorithm are selected as follows, the noise standard deviation is Nstd = 0.2, the number of EMD implementations is NR = 100, the maximum number of screening iterations allowed is MaxIter = 5000, and the SNRFlag = 1 indicates that the signal-to-noise ratio (SNR) is incremental with EMD implementation. Since ICEEMDAN is an adaptive decomposition algorithm, the number of modals obtained from different heart sound cycles may be different. For comparison, only the least number of IMFs obtained by the ICEEMDAN from the heart sound database is shown here.

Table 2. Comparison of the recognition performance of taking different intrinsic mode functions (IMFs) as the input of the algorithm on the three open heart sound databases.

Heart Sound Database	Including 2005 Single-Cycle Heart Sounds from the Open Database Michigan, Washington, and Littman		
Algorithm	**ICEEMDAN-RCMDE-ED**		
Input	$\overline{CRR}$	**Input**	$\overline{CRR}$
IMF 1	90.04%	IMF 5	41.08%
IMF 2	88.96%	IMF 6	24.28%
IMF 3	82.68%	IMF 7	14.94%
IMF 4	59.58%	IMF 8	12.15%

It is found through experiments that the first three IMFs of the heart sound cycle, respectively used as the algorithm input, can obtain a higher $\overline{CRR}$ compared with the others. It shows that the first three IMFs not only contain the majority of the information in the heart sound cycle but also dig deep the information in the entire heart sound cycle, which is expressed in a more detailed way. Therefore, adding the features from the above three IMFs to one feature vector as a new heart sound feature is considered. The original heart sound feature representation is shown in Figure 5a. The new heart sound feature representation is shown in Figure 5b. The red dot in the figure represents the single-cycle feature as the training, and the green dot represents the single-cycle feature as the testing. Since the single-cycle heart sounds as the training is much more than the single-cycle heart sounds as the testing, it is shown in the figure below that the green dot is wrapped by the red dot.

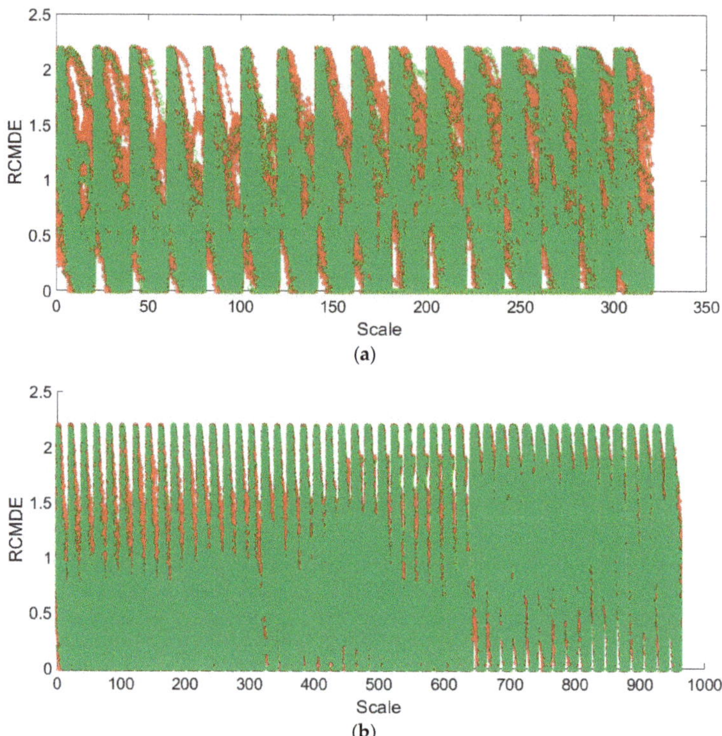

Figure 5. The feature characterization based on the different algorithms (**a**) the feature characterization based RCMDE; (**b**) the new feature characterization based on the combination of improved complete ensemble empirical mode decomposition with adaptive noise (ICEEMDAN) and RCMDE.

It can be found from Figure 5 that the merged features have twice as many feature dimensions as the original features and have great redundancy. Therefore, the Fisher ratio (FR) is used for feature selection. After the features are ranked according to the Fisher ratio, the features are selected. The first N_r feature dimensions are used as new heart sound features. After experimental verification, when $N_r = 300$, the recognition performance is optimal. The $\overline{CRR}_{200}$ and Kappa coefficients of respectively using the original heart sound feature and the new heart sound feature with ED and the close principle are shown in Table 3.

Table 3. Comparison of the recognition performance of RCMDE and ICEEMDAN-RCMDE-Fisher ratio (FR) algorithms on the three open heart sound databases.

Heart Sound Database	Including 2005 Single-Cycle Heart Sounds from the Open Database Michigan, Washington, and Littman		
Feature Extraction	**Numbers of Feature**	$\overline{CRR}$	**Kappa Coefficients**
RCMDE	320	90.08%	0.8994
ICEEMDAN-RCMDE-FR	300	96.08%	0.9602

It can be seen from Table 3 that the $\overline{CRR}_{200}$ and Kappa coefficients on the three public heart sound databases obtained from the feature extraction method based on ICEEMDAN-RCMDE-FR are higher than the feature extraction method based on RCMDE, and can achieve an average recognition rate of 96.08%. The Kappa coefficient is between 0.8 and 1, which indicates that the classification accuracy is extremely high. The following t-test is used to verify whether $\overline{CRR}_{200} = 96.08\%$ obtained in the above table can be regarded as the generalization accuracy. Here, n random experiments are performed, where n = 10, 20, 30, 50, 100, 200, 300, 400, 500, 600, respectively, the average test accuracy μ and standard deviation σ corresponding to n experiments are shown in the left half of Table 4. Here it is assumed that the generalization accuracy $\mu_0 = \overline{CRR}_{600}$, the test level α is 0.05 and then the t value of n experiments is obtained according to the t-test steps in Section 2.6, and the corresponding critical value range is also given.

Table 4. Comparison of average test accuracy μ, standard deviation σ and t value corresponding to n random trials.

n Random Trials	The Average Test Accuracy μ	Standard Deviation σ	*t* Value	The Critical Value Range $[-t_{(\alpha,n)}, t_{(\alpha,n)}]$	The Degree of Confidence $1-\alpha$
$n = 10$	0.9681	0.0147	1.570	[−2.262, 2.262]	0.95
$n = 20$	0.9653	0.0146	1.378	[−2.093, 2.093]	0.95
$n = 30$	0.9634	0.0144	0.989	[−2.045, 2.045]	0.95
$n = 50$	0.9639	0.0161	1.362	[−2.010, 2.010]	0.95
$n = 100$	0.9603	0.0162	−0.309	[−1.984, 1.984]	0.95
$n = 200$	0.9608	0.0162	0	[−1.972, 1.972]	0.95
$n = 300$	0.9606	0.0162	−0.214	[−1.968, 1.968]	0.95
$n = 400$	0.9608	0.0162	0	[−1.966, 1.966]	0.95
$n = 500$	0.9607	0.0162	−0.138	[−1.965, 1.965]	0.95
$n = 600$	0.9608	0.0162	0	[−1.964, 1.964]	0.95

It can be seen from Table 4 that the t values corresponding to n experiments are all within the critical value range $[-t_{(\alpha,n)}, t_{(\alpha,n)}]$, and the average test accuracy μ corresponding to n experiments can be considered Both can be regarded as generalization accuracy of μ_0 and the degree of confidence is 0.95. It can also be found from Table 4 that when the number of experiments n is greater than 200, the average test accuracy μ has basically stabilized at 96.08% and the t value has basically stabilized near 0. Therefore, considering the stability and calculation cost, the number of experiments is taken as $J = 200$, the best generalization accuracy is $\mu_0 = 96.08\%$.

In summary, the feature extraction method based on ICEEMDAN-RCMDE-FR proposed in this paper can achieve a generalization accuracy of 96.08% on three public heart sound databases with a confidence level of 0.95, which shows that the multimodal multiscale dispersion entropy generated by the ICEEMDAN-RCMDE-FR algorithm has good characterization of heart sounds, and it is suitable for the field of biometrics. Considering that the classifier currently used is rough, this may be one of the reasons why the $\overline{CRR}$ cannot be further improved. Therefore, different classifiers such as SVM and KNN are compared with ED, and the results are shown in Table 5. The SVM classifier used here is parameter-tuned. The two main penalty parameters c and the kernel function parameter g are 64 and 0.001, and the nearest neighbors of the KNN classifier are taken as k = 5, 3, 2, respectively. From the results in Table 5, the difference between the best performance of the three classifiers is within 1%. It can be found that the smaller the parameter k of KNN is, the higher the CRR is. When k = 1 or 2, KNN is the ED classifier. The heart sound recordings in the open database are different in length, therefore the data distribution in the single-cycle heart sound database generated by the segmentation is not balanced, which may be the reason that the classifier performance cannot be further improved. Since the ED classifier is relatively simpler, the matching recognition time is also faster. Considering the combination, the ED classifier is most suitable for the heart sound database.

Table 5. Comparison of recognition performance of SVM, KNN and ED classifier on the three open heart sound databases.

Heart Sound Database	**Including 2005 Single-Cycle Heart Sounds from the Open Database Michigan, Washington, and Littman**			
Feature Extraction	**ICEEMDAN-RCMDE-FR**			
Classifier	**Classifier Parameter**	**Speed**	$\overline{CRR}$	**Kappa Coefficients**
SVM	$c = 64, g = 0.001$	Slowest	95.91%	0.9585
KNN	$k = 5$	Medium	73.14%	0.7276
	$k = 3$		84.83%	0.8462
	$k = 2$		95.97%	0.9591
ED and the close principle	None	Fastest	96.08%	0.9602

3.3. Practical Application of ICEEMDAN-RCMDE-FR-ED Algorithm

Although it has been considered in the previous section that the single-cycle heart sound as the test data should be aligned with the corresponding single-period heart sound in the database, the position of the initial split point is not the same when the heart sound cycle is divided. In the practical application of the heart sound biometric identification system, when the single-cycle heart sound is to be segmented from the randomly collected heart sound signal, the position of the initial segmentation point must be fixed and kept consistent. Therefore, the heart sound segmentation method based LR-HSMM proposed by Springer et al. [31] is firstly used to assign the states of the heart sound recording of 40 volunteers collected in the natural environment, and then the following four initial dividing points are used to obtain four kinds of single-cycle heart sounds as training: (1) The starting position of the first S1 appearing in the heart sound recording is taken as the initial dividing point; (2) the starting position of the first systole appearing in the heart sound recording is taken as the initial dividing point; (3) the starting position of the first S2 appearing in the heart sound recording is taken as the initial dividing point; (4) the starting position of the first diastole appearing in the heart sound recording is taken as the initial dividing point. At least one hour later, the heart sound recordings of the 40 volunteers were collected again, and four single-cycle heart sounds as testing are respectively obtained in the same manner as the single-cycle heart sounds obtained as training. A schematic diagram of four heart sound cycle segmentation methods is shown in Figure 6.

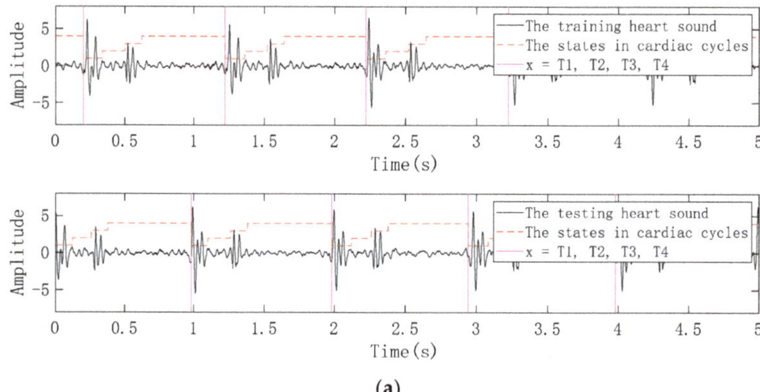

(a)

Figure 6. *Cont.*

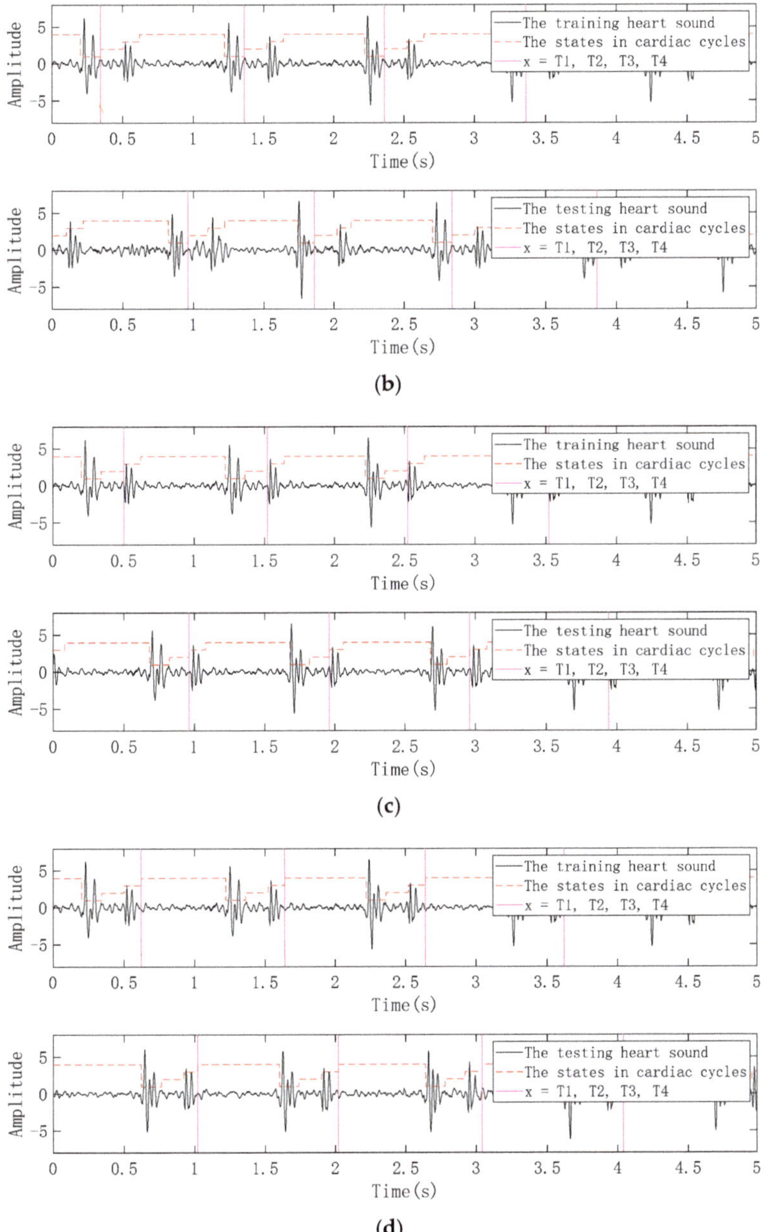

Figure 6. Four cardiac cycle segmentation methods based on different initial segmentation points. (**a**) The starting position of the first S1 appearing in the heart sound recording is taken as the initial dividing point; (**b**) the starting position of the first systole appearing in the heart sound recording is taken as the initial dividing point; (**c**) the starting position of the first S2 appearing in the heart sound recording is taken as the initial dividing point; (**d**) the starting position of the first diastole appearing in the heart sound recording is taken as the initial dividing point.

It is known from experiments that the four segments of (a), (b), (c) and (d) obtain the same number of single-cycle heart sounds: 209 single-cycle heart sounds as training and 190 single-cycle heart sounds as testing. Since the current data set is relatively balanced, the feature processing here adopts another method different from the previous one: that is, one-to-one method, the feature vectors of all single-period heart sounds as training\testing corresponding to each of the individual are averaged to obtain an average feature vector, so that each individual corresponds to only one average feature vector as the training\testing. Then, the average feature vector as training/testing obtained from (a), (b), (c) and (d) is used as the input of the ICEMDAN-RCMDE-FR-ED algorithm for verification experiments. The selected parameters of the experiment are still the selected parameters in the previous section. The experimental results in Table 6 show that the first segmentation method (a) achieves the highest CRR = 97.5%, which may be since the Springer algorithm is more accurate for S1 segmentation, so the segmentation (a) makes the training and testing features closer. The difference in CRR obtained by the four segmentation methods is between 0% and 5%, and the difference in Kappa coefficients obtained by the four segmentation methods is between 0 and 0.0513. The overall is very close, which may be due to the use of the average feature vector, which enhances the robustness of the algorithm and does not affect the result due to some bad single-cycle heart sounds. Therefore, the ICEEMDAN-RCMDE-FR-ED algorithm proposed in this paper combined with the heart sound segmentation method based on the logistic regression and hidden semi-Markov model (HSMM) has high practical application value in the field of biometric identification.

Table 6. The recognition effect of ICEEMDAN-RCMDE-FR-ED on the self-built heart sound database.

Heart Sound Database	**Including the 80 Heart Sound Recordings from the Self-Built Heart Sound Database**	
Algorithm	**ICEEMDAN-RCMDE-FR-ED**	
The Starting and Ending Position of the Input Single-Cycle Heart Sound	**CRR**	**Kappa Coefficients**
the starting position of S1—the starting position of next S1	97.5%	0.9744
the starting position of systole—the starting position of next systole	92.5%	0.9231
the starting position of S2—the starting position of next S2	95.0%	0.9487
the starting position of diastole—the starting position of next diastole	95.0%	0.9487

3.4. Comparison with Related Literature

Table 7 lists performance comparisons between the proposed study and other existing heart sound biometric work. Phua et al. [3] introduced linear frequency band cepstrum (LFBC) for heart sound feature extraction and used two classifiers of vector quantization (VQ) and Gaussian mixture model (GMM) for classification and recognition. The database used Composed of 10 users, the correct recognition rate is 96%. Fatemian et al. [6] proposed a PCG signal identification and verification system. The system is based on wavelet preprocessing, feature extraction using short-time Fourier transform (STFT), feature dimension reduction using linear discriminant analysis (LDA) and majority voting using Euclidean distance (ED) for classification. As a result, the recognition result for 21 subjects was 100%, and the equal error rate (EER) verification result was 33%. Tran et al. [7] used eight feature sets such as temporal shape, spectral shape, Mel-frequency cepstral coefficients (MFCC), linear frequency cepstral coefficients (LFCC), harmonic feature, rhythmic feature, cardiac feature, GMM-super vector as heart sound biometric recognition features, using two feature selection techniques, and using SVM for 52 users classification recognition, the first experiment achieved more than 80% accuracy and the second experiment achieved more than 90% accuracy. Jasper and Othman [32] applied wavelet transform (WT) to analyze the signals in the Time-Frequency representation, then selected Shannon energy envelogram (SSE) as the feature set, and tested the performance of the feature set in a database of 10 people with an accuracy of 98.67%. Cheng et al. [1] introduced a human feature extraction method based on an improved circular convolution (ICC) slicing algorithm combined with independent subband function (ISF). The technology uses two recognition steps to obtain different human heart sound characteristics to ensure validity, and then uses similar distances for human heart sound pattern matching. The method was verified using 10 recorded heart sounds. The results show that the two-step recognition accuracy is 85.7%. Cheng et al. [8] used heart sound linear band frequency cepstrum (HS-LBFC) for feature extraction and used similar distances for classification. The results were done on 12 heart sound signals, with a verification rate as high as 95%, false acceptance rate = 1% to 8%, and a false rejection rate of less than 3%. Zhao et al. [11] used the heart sound database of 280 samples constructed by 40 users to test their proposed marginal spectrum (MS) features and validated them using 80 samples randomly selected from the open database HSCT-11. Gautam and Deepesh [33] proposed a new method for heart sound recognition, which is based on preprocessing using a low-pass filter, using autocorrelation to detect the cardiac cycle, and segmenting S1 and S2 by windowing and thresholding. The method used WT for feature extraction and back propagation multilayer perceptron artificial neural network (BP-MLP-ANN) for classification and the accuracy rate on 10 volunteers reached 90.52%, the EER reached 9.48%. Tan et al. [34] demonstrated a new method for heart sound authentication. The pre-processing is based on low-pass filtering, and then the heart sounds are segmented using zero-crossing rate (ZCR) and short-term amplitude (STA) techniques to extract S1 and S2 sounds. Features are extracted using MFCC, and features are classified using a sparse representation classifier (SRC). Fifteen users were randomly selected, and the best effect of 85.45% can be achieved. Verma and Tanuja [35] proposed a heart sound-based biometric recognition system that uses MFCC for feature extraction and SVM for classification. They studied 30 topics with an accuracy rate of 96%. Abo Zahhad et al. [36] proposed a heart sound recognition system based on 17 subjects with an accuracy rate of 99%. Features were selected using MFCC, LFCC, bark frequency cepstral coefficients (BFCC) and discrete wavelet transform (DWT), and fused using Cone Correlation Analysis (CCA). GMM and Bayesian rules were used for classification. Abo Zahhad et al. [37] used HSCT-11 and BioSec. databases to compare the biometric performance of MFCC, LFCC, wavelet packet cepstral coefficient (WPCC) and non-linear frequency cepstral coefficients (NLFCC), and the conclusion is that WPCC and NLFCC have better biometric performance in high noise scenarios.

Compared with the above work using different feature extraction, classification methods and heart sound database, we can conclude that our method has the best effect in the same size heart sound database. The previous methods were performed on normal healthy subjects, without taking into account heart disease, and this paper conducted research on the open pathological heart sound library Michigan, Washington, and Littman. Different from the previous method, we first use the LR-HSMM-based heart sound segmentation method proposed by Springer [31] to segment the pre-processed heart sound record into a series of single-cycle heart sounds, and frame and window based on each cycle length to ensure that Each single cycle heart sound can get the same number of frames. Different from the previous method, we first introduced RCMDE features for heart sound biometrics, and selectively combined RCMDE with ICEEMDAN, FR and ED methods, and strived to improve the mixed Recognition rate of normal and pathological heart sounds with more fine characterization. The proposed method not only achieved a correct recognition rate of 96.08% on the open heart sound database, but also achieved a recognition rate of 97.5% on the 80 heart sound database composed of 40 healthy subjects constructed by the research group, and draw the conclusion that the single-cycle heart sound recognition rate from the first heart sound (S1) to the next S1 is the highest.

Table 7. Compared with the related literature.

Comparative Literature	Heart Sound Database	Feature Extraction	Classifier	Accuracy
Phua et al. [3]	10 people	LFBC	VQ GMM	CRR = 94% CRR = 96%
Fatemian et al. [6]	21 subjects	STFT	LDA and ED	CRR = 100% EER = 33%
Tran et al. [7]	52 users	temporal shape, spectral shape, MFCC, LFCC, harmonic feature, rhythmic feature, cardiac feature and GMM-super vector	RFE-SVM	CRR = 80% CRR = 90%
Jasper and Othman [32]	10 people	WT-SSE	Template matching	CRR = 98.67%
Cheng et al. [1]	12 people 300 HS	HS-LBFC	similar distances	CRR = 99%
Cheng et al. [8]	10 people	ICC-ISF	similar distances	CRR = 85.7%
Zhao et al. [11]	40 participants 280 samples HSCT-11 80 subjects	MS	VQ and ED	CRR = 94.16% CRR = 92%
Gautam and Deepesh [33]	10 subjects	segment S1 and S2 by windowing and thresholding +WT	BP-MLP-ANN	CRR = 90.52% EER = 9.48%
Tan et al. [34]	52 users	extract S1 and S2 by ZCR and STA techniques + MFCC	SRC	CRR = 85.45%
Verma and Tanuja [35]	30 people	MFCC	SVM	CRR = 96%
Abo Zahhad et al. [36]	17 subjects	MFCC, LFCC, BFCC and DWT+ CCA	GMM and Bayesian rules	CRR = 99%
Abo Zahhad et al. [37]	HSCT-11 206 subjects BioSec. 21 subjects	WPCC NLFCC WPCC NLFCC	LDA and Bayesian Decision Rules	CRR = 90.26% CRR = 92.94% CRR = 97.02% CRR = 98.1%
The proposed method	Michigan, Washington, and Littman 72 subjects 40 users 80 HS	segment cardiac cycle by LR-HSMM + framing and windowing + ICEEMDAN-RCMDE-FR	SVM KNN ED and the close principle ED and the close principle	CRR = 95.91% CRR = 95.97% CRR = 96.08% CRR = 97.5%

4. Conclusions

In the current research, based on the characteristics of the heart sound signal, the improved ensemble empirical mode decomposition (ICEEMDAN), fine composite multiscale dispersion entropy (RCMDE), Fisher ratio (FR) and Euclidean distance (ED) is used to study the mixed recognition of normal and pathological heart sounds, the following conclusions were reached:

(1) Given the quasi-periodic and non-steady-state characteristics of heart sound signals, this paper first uses LR-HSMM-based heart sound segmentation to divide heart sounds into a series of single-cycle heart sounds, and framing and windowing based on each cycle length to ensure each single cycle heart sound can get the same number of frames.

(2) To solve the problem of unified representation of heart sound frames with different lengths, this paper first introduces RCMDE for heart sound biometric identification and selectively combines RCMDE with ICEEMDAN, FR and ED methods for heart sound Biometric characterization.

(3) The recognition rate of this method on the open pathological heart sound database Michigan, Washington and Littman reached 96.08%, that is, the method can effectively recognize normal and pathological heart sounds.

(4) To enhance its practical application value, this paper applies the proposed method to a self-built heart sound database. Research shows that the single-cycle heart sound recognition rate from the first heart sound (S1) to the next S1 is the highest, which is 97.5%.

Although the features of this article have been proven to have a good effect on heart sound biometrics, we believe that each biometric has its limitations, and the future research direction is bound to integrate the outstanding performance features, and then use the latest powerful classifiers, such as deep learning methods, achieve optimal recognition. It is even possible to consider using a combination of feature extraction techniques for different signals, such as Abo-Zahhad et al. [38] proposed to use both ECG and PCG signals in a multimodal biometric authentication system, and Bugdol, M.D. et al. [39] proposed the multimodal biometric system combining ECG and sound signals. Of course, we also need to consider the impact of subject age, database size, race, gender and disease status on the performance of the biometric system. In the future, we can use features that are less affected by these factors to fuse or use only specific features to biometrically identify specific populations, such as using the method in this article to biometrically identify people with heart disease, so the research in this article can be used as a foundation for future biological identification research.

Author Contributions: X.C. designed the topic and the experiments and made suggestions for the paper organization; P.W. conducted the experiments and wrote the paper; C.S. help collect data and gave advice. All authors have read and agreed to the published version of the manuscript.

Funding: This work is supported by the Natural Science Foundation of China (Grant No. 61271334, No. 61373065) "Research and application on the heart sound features extraction for identification".

Conflicts of Interest: The authors declare no conflict of interest.

References

1. Cheng, X.F.; Ma, Y.; Liu, C.; Zhang, X.J.; Guo, Y.F. An introduction to heart sounds identification technology. *Sci. China-Inf. Sci.* **2012**, *42*, 237–251.
2. Beritelli, F.; Serrano, S. Biometric identification based on frequency analysis of cardiac sounds. *IEEE Trans. Inf. Forensics Secur.* **2007**, *2*, 596–604. [CrossRef]
3. Phua, K.; Chen, J.F.; Dat, T.H.; Shue, L. Heart sound as a biometric. *Pattern Recognit.* **2008**, *41*, 906–919. [CrossRef]
4. Beritelli, F.; Spadaccini, A. Human identity verification based on mel frequency analysis of digital heart sounds. In Proceedings of the 16th International Conference on Digital Signal Processing, Santorini, Greece, 5–7 July 2009.
5. Beritelli, F.; Spadaccini, A. An improved biometric identification system based on heart sound and gaussian mixture models. In Proceedings of the 2010 IEEE Workshop on Biometric Measurements and Systems for Security and Medical Applications, Taranto, Italy, 9 September 2010.
6. Fatemian, S.Z.; Agrafioti, F.; Hatzinakos, D. Heartid: Cardiac biometric recognition. In Proceedings of the Fourth IEEE International Conference on Biometrics: Theory, Applications and Systems (BTAS), Washington, DC, USA, 27–29 September 2010.
7. Tran, D.H.; Leng, Y.R.; Li, H. Feature integration for heart sound biometrics. International Conference on Acoustics Speech Signal Processing. In Proceedings of the 2010 IEEE International Conference on Acoustics, Speech and Signal Processing, Dallas, TX, USA, 14–19 March 2010; pp. 1714–1717.
8. Cheng, X.F.; Tao, Y.W.; Huang, Z.J. Cardiac sound recognition—A prospective candidate for biometric identification. *Adv. Mater. Res.* **2011**, *225*, 433–436. [CrossRef]
9. Chen, W.; Zhao, Y.; Lei, S.; Zhao, Z.; Pan, M. Study of biometric identification of Cardiac sound base on Mel-Frequency cepstrum coefficient. *J. Biomed. Eng.* **2012**, *29*, 1015–1020.
10. Zhong, L.; Wan, J.; Huang, Z.; Guo, X.; Duan, Y. Research on biometric method of Cardiac sound signal based on GMM. *Chin. J. Med. Instrum.* **2013**, *37*, 92–99.
11. Zhao, Z.D.; Shen, Q.Q.; Ren, F.Q. Heart sound biometric system based on marginal spectrum analysis. *Sensors* **2013**, *13*, 2530–2551. [CrossRef]
12. Babiker, A.; Hassan, A.; Mustafwa, H. Cardiac sounds biometric system. *J. Biomed. Eng. Med. Device* **2017**, *2*, 2–15. [CrossRef]

13. Akhter, N.; Tharewal, S.; Kale, V.; Bhalerao, A.; Kale, K.V. Heart-Based Biometrics and Possible Use of Heart Rate Variability in Biometric Recognition Systems. In Proceedings of the 2nd International Doctoral Symposium on Applied Computation and Security Systems (ACSS), Kolkata, India, 23–25 May 2015.
14. Bao, S.; Poon, C.C.Y.; Zhang, Y.; Shen, L. Using the Timing Information of Heartbeats as an Entity Identifier to Secure Body Sensor Network. *IEEE Trans. Inf. Technol. Biomed.* **2008**, *12*, 772–779.
15. Palaniappan, R. Two-stage biometric authentication method using thought activity brain waves. In Proceedings of the 7th International Conference on Intelligent Data Engineering and Automated Learning (IDEAL 2006), Burgos, Spain, 20–23 September 2006.
16. Mu, Z.; Hu, J.; Min, J. EEG-Based Person Authentication Using a Fuzzy Entropy-Related Approach with Two Electrodes. *Entropy* **2016**, *18*, 432. [CrossRef]
17. Huang, N.E.; Shen, Z.; Long, S.R.; Wu, M.C.; Shih, H.H.; Zheng, Q.; Yen, N.C.; Tung, C.C.; Liu, H.H. The empirical mode decomposition and the Hilbert spectrum for nonlinear and non-stationary time series analysis. *Proc. R. Soc. Lond. A* **1998**, *454*, 903–995. [CrossRef]
18. Colominas, M.A.; Schlotthauer, G.; Torres, M.E. Improve complete ensemble EMD: A suitable tool for biomedical signal processing. *Biomed. Signal Process. Control* **2014**, *14*, 19–29. [CrossRef]
19. Torres, M.E.; Colominas, M.A.; Schlotthauer, G.; Flandrin, P. A Complete Ensemble Empirical Mode Decomposition with Adaptive Noise. In Proceedings of the 36th IEEE International Conference on Acoustics, Speech and Signal Processing, Prague, Czech Republic, 22–27 May 2011.
20. Rostaghi, M.; Azami, H. Dispersion Entropy: A Measure for Time-Series Analysis. *IEEE Signal Process. Lett.* **2016**, *23*, 610–614. [CrossRef]
21. Azami, H.; Rostaghi, M.; Abasolo, D.; Escudero, J. Refined Composite Multiscale Dispersion Entropy and its Application to Biomedical Signals. *IEEE Trans. Biomed. Eng.* **2017**, *64*, 2872–2879.
22. Matlab Codes for Refined Composite Multiscale Dispersion Entropy and Its Application to Biomedical Signals. Available online: https://datashare.is.ed.ac.uk/handle/10283/2637 (accessed on 7 November 2019).
23. Pruzansky, S.; Mathews, M.V. Talker-Recognition Procedure Based on Analysis of Variance. *J. Acoust. Soc. Am.* **1964**, *36*, 2021–2026. [CrossRef]
24. Zhou, Z.H. *Machine Learning*, 3rd ed.; Tsinghua University Press: Beijing, China, 2016; pp. 24–47.
25. University of Michigan Department of Medicine. Michigan Heart Sound and Murmur Library. Available online: http://www.med.umich.edu/lrc/psb/heartsounds/ (accessed on 7 November 2019).
26. Washington Heart Sounds & Murmurs Library. Available online: https://depts.washington.edu/physdx/heart/tech5.html (accessed on 7 November 2019).
27. Littmann Heart and Lung Sounds Library. Available online: http://www.3m.com/healthcare/littmann/mmm-library.html (accessed on 7 November 2019).
28. Cheng, X.F.; Zhang, Z. A construction method of biorthogonal heart sound wavelet. *Acta Phys. Sin.* **2013**, *62*, 168701.
29. Gupta, C.N.; Palaniappan, R.; Swaminathan, S.; Krishnan, S.M. Neural network classification of homomorphic segmented heart sounds. *Appl. Soft Comput.* **2007**, *7*, 286–297. [CrossRef]
30. Liu, C.; Springer, D.; Li, Q.; Moody, B.; Juan, R.A.; Chorro, F.J.; Castells, F.; Roig, J.M.; Silva, I.; Johnson, A.E.; et al. An open access database for the evaluation of heart sound algorithms. *Physiol. Meas.* **2016**, *37*, 2181–2213. [CrossRef]
31. Springer, D.B.; Tarassenko, L.; Clifford, G.D. Logistic Regression-HSMM-Based Heart Sound Segmentation. *IEEE Trans. Biomed. Eng.* **2016**, *63*, 822–832. [CrossRef]
32. Jasper, J.; Othman, K.R. Feature extraction for human identification based on envelogram signal analysis of cardiac sounds in time-frequency domain. *Electron. Inf. Eng.* **2010**, *2*, 228–233.
33. Gautam, G.; Deepesh, K. Biometric system from Cardiac sound using wavelet based feature set. In Proceedings of the 2013 International Conference on Communication and Signal Processing, Melmaruvathur, India, 3–5 April 2013.
34. Tan, W.; Yeap, H.; Chee, K.; Ramli, D. Towards real time implementation of sparse representation classifier (SRC) based heartbeat biometric system. *Comput. Probl. Eng.* **2014**, *307*, 189–202.
35. Verma, S.; Tanuja, K. Analysis of Cardiac sound as biometric using mfcc and linear svm classifier. *IJAREEIE* **2014**, *3*, 6626–6633.

36. Abo-Zahhad, M.; Ahmed, S.M.; Abbas, S.N. PCG biometric identification system based on feature level fusion using canonical correlation analysis. In Proceedings of the 27th Canadian Conference on Electrical and Computer Engineering, Toronto, ON, Canada, 4–7 May 2014.
37. Abo-Zahhad, M.; Farrag, M.; Abbas, S.N.; Ahmed, S.M. A comparative approach between cepstral features for human authentication using heart sounds. *Signal Image Video Process.* **2016**, *10*, 843–851. [CrossRef]
38. Abo-Zahhad, M.; Ahmed, S.M.; Abbas, S.N. Biometric authentication based on PCG and ECG signals: Present status and future directions. *Signal Image Video Process.* **2014**, *8*, 739–751. [CrossRef]
39. Bugdol, M.D.; Mitas, A.W. Multimodal biometric system combining ECG and sound signals. *Pattern Recognit. Lett.* **2014**, *38*, 107–112. [CrossRef]

Article

An Improved Method of Handling Missing Values in the Analysis of Sample Entropy for Continuous Monitoring of Physiological Signals

Xinzheng Dong [1,2,†], Chang Chen [3,†], Qingshan Geng [4], Zhixin Cao [5], Xiaoyan Chen [6], Jinxiang Lin [6], Yu Jin [3], Zhaozhi Zhang [7], Yan Shi [5,8] and Xiaohua Douglas Zhang [3,*]

1 School of Software Engineering, South China University of Technology, Guangzhou 510006, China; xinzhengdong@163.com
2 Zhuhai Laboratory of Key Laboratory of Symbolic Computation and Knowledge Engineering of Ministry of Education, Zhuhai College of Jilin University, Zhuhai 519041, China
3 Faculty of Health Sciences, University of Macau, Taipa, Macau 999078, China; yb67646@connect.umac.mo (C.C.); yb67647@connect.umac.mo (Y.J.)
4 Guangdong General Hospital, Guangdong Academy of Medical Science, Guangzhou 510080, China; gengqs2010@163.com
5 Beijing Engineering Research Center of Diagnosis and Treatment of Respiratory and Critical Care Medicine, Beijing Chaoyang Hospital, Beijing 100043, China; 18301564184@163.com (Z.C.); shiyan@buaa.edu.cn (Y.S.)
6 Department of Endocrinology, First Affiliated Hospital of Guangzhou Medical University, Guangzhou 510120, China; gzscxy@126.com (X.C.); 13794353925@163.com (J.L.)
7 School of Law, Washington University, St. Louis, MO 63130, USA; zhazhang59@gmail.com
8 Department of Mechanical and Electronic Engineering, Beihang University, Beijing 100191, China
9 BARDS, Merck Research Laboratories, Upper Gwynedd, PA 19454, USA
* Correspondence: douglaszhang@umac.mo; Tel: +853-8822-4813
† These authors contributed equally.

Received: 11 January 2019; Accepted: 9 March 2019; Published: 12 March 2019

Abstract: Medical devices generate huge amounts of continuous time series data. However, missing values commonly found in these data can prevent us from directly using analytic methods such as sample entropy to reveal the information contained in these data. To minimize the influence of missing points on the calculation of sample entropy, we propose a new method to handle missing values in continuous time series data. We use both experimental and simulated datasets to compare the performance (in percentage error) of our proposed method with three currently used methods: skipping the missing values, linear interpolation, and bootstrapping. Unlike the methods that involve modifying the input data, our method modifies the calculation process. This keeps the data unchanged which is less intrusive to the structure of the data. The results demonstrate that our method has a consistent lower average percentage error than other three commonly used methods in multiple common physiological signals. For missing values in common physiological signal type, different data size and generating mechanism, our method can more accurately extract the information contained in continuously monitored data than traditional methods. So it may serve as an effective tool for handling missing values and may have broad utility in analyzing sample entropy for common physiological signals. This could help develop new tools for disease diagnosis and evaluation of treatment effects.

Keywords: sample entropy; missing values; physiological data; complexity; medical information

---

1. Introduction

The demand for more advanced, more personalized treatments, increased availability of healthcare and an aging population are pushing the market and expanding medical device technology, especially in the area of wearables for continuous monitoring of physiological signals. These advancements require better analytic methods to extract the useful information contained in these data more accurately due to the huge amount of data generated by these devices. Entropy, an indicator for the degree of irregularity in a dynamic system, has been applied to more and more disciplines since its definition was extended to information theory in 1950s [1]. As a nonlinear dynamic index, sample entropy (one type of entropy) is often used to measure the complexity of the physiological system in medical research for disease diagnosis and prognosis. Besides the routine detected indicators, this index can help doctors better confirm the diagnosis and prognosis, so as to provide better treatment and suggestions for patients' rehabilitation. One of the most well-known examples is the use of entropy as an indicator of heart rate variability to evaluate cardiac sympathetic and parasympathetic functions [2]. In addition to the widely used application in the diagnosis and prognosis of the cardiac autonomic nerve disease [3,4], entropy is also used in the diagnosis of diabetes [5–9], chronic obstructive pulmonary disease [10–12] and other diseases [13]. The change of entropy values (decrease or increase) have been shown to be a predictor of multiple diseases [3,10,11].

The data from signals measured by continuous monitoring devices, such as wearables [7,14], commonly have various degrees of missing values [15] because of the patient's unconscious movement, loose equipment and interference by other equipment. This issue of missing data is compounded by a study [16] which showed that sample entropy can be highly sensitive to missing values. Once the data has missing fragments, entropy fluctuations will be large. More worrisome, the abnormal fluctuations will increase as the percentage of missing values increases [17]. Handling missing values in the calculation of entropy is therefore imperative. Although there are a few studies to investigate the effect of missing values on the analytic results of nonlinearity including entropy [16,17], methods to deal with missing values have yet to be developed for the calculation of entropy.

Currently, there are two basic strategies to overcome this problem: ignoring/imputing missing values and modifying the method of calculating sample entropy [18]. The issue is that common methods for imputing missing values may not work effectively for entropy calculation [19]. An artificial effect will be introduced to the calculated value of entropy [15] if missing values cannot be imputed in the same way as the distribution of the original data. In this paper, we propose a new method to improve the algorithm of calculating sample entropy on data with missing values that does not involve imputing missing values before entropy calculation. This provides a less intrusive way of handling missing values in the analysis of sample entropy for continuously monitoring physiological data since the new method does not add new data points. Thus removing the danger that the original structure of the data is compromised.

To demonstrate the utility of our proposed method, the following key questions need to be answered: Can the new method be applied to common types of continuously monitoring physiological data? Is the new method robust to the data size represented by the length of a time series? Is it robust to the scheme of generating missing values? How does the new method perform compared with existing methods? In this paper, we designed simulations based on experimental data to address these questions.

Our article is organized as follows: in the Methods section, we first introduce the definition of sample entropy and four methods of handling missing values. Then we present the datasets and the method we used to construct a sequence with missing values from the original sequence without missing values. In the Results section, we investigate the utility and applicability of our method to most common physiological signal types as well as the robustness of our method to both data size and scheme for generating missing values, as compared to the currently, commonly used methods. We conclude with discussion on the applicability and robustness of our method.

2. Methods

2.1. Sample Entropy

Sample entropy is a measure of irregularity, which was first proposed by Richman and Moorman [20,21]. It is defined as the negative natural logarithm of the conditional probability that two sequences similar for m points remain similar at the next point with a tolerance r.

Let $X_i = \{x_1, \ldots, x_i, \ldots, x_N\}$ represent a time series of length N. We define the template vector of length m from X: $X_m(i) = \{x_i, x_{i+1}, x_{i+2}, \ldots, x_{i+m-1}\}$ and the distance function between two such vectors:

$$d(m,i,j) = d[X_m(i), X_m(j)] = \max_k\left\{\left|x_{i+k-1} - x_{j+k-1}\right|\right\}, \text{ where } k = 1, \ldots, m \quad (1)$$

Then given a distance threshold r, the number of (similar or matched) vectors that are within r of the distance between the vectors $X_m(i)$ and $X_{m+1}(i)$ can be defined respectively as:

$$C_i^m(r) = \sum\nolimits_{j=1, j\neq i}^{N-m} \Phi(m,i,j,r) \quad (2)$$

$$C_i^{m+1}(r) = \sum\nolimits_{j=1, j\neq i}^{N-m} \Phi(m+1,i,j,r) \quad (3)$$

where $\Phi(m,i,j,r)$ is defined as:

$$\Phi(m,i,j,r) = \begin{cases} 1, & d(m,i,j) \leq r \\ 0, & \text{otherwise} \end{cases} \quad (4)$$

Then the probability that two vectors of length m will match is defined as:

$$B^m(r) = \frac{1}{N-m}\sum\nolimits_{i=1}^{N-m} \frac{1}{N-m-1} C_i^m(r) \quad (5)$$

and the probability that two vectors of length m+1 will match as:

$$A^m(r) = \frac{1}{N-m}\sum\nolimits_{i=1}^{N-m} \frac{1}{N-m-1} C_i^{m+1}(r) \quad (6)$$

Finally, we define the sample entropy as:

$$SampEn(m,r,N) = -ln\frac{A^m(r)}{B^m(r)} \quad (7)$$

Pincus [22] has proved that when the parameter m = 2 and r is between $0.1 \times \sigma$ and $0.25 \times \sigma$, the sample entropy can retain enough information from time series and obtain effective statistical properties. Different r can give different conditional probability estimates. We have showed that there is no percentage error difference of four methods in four types data when r equal to $0.1 \times \sigma$, $0.15 \times \sigma$ and $0.2 \times \sigma$ (Appendix A, Figure A1). For different r, the trend of percentage error fluctuating with the percentage of missing values is consistent, so the parameters we used are: m = 2 and $r = 0.15 \times \sigma$, where σ is the standard deviation of a time series.

2.2. KeepSampEn, SkipSampEn, LinearSampEn and BootSampEn

The definition of sample entropy cannot be directly used on time series data containing missing values. In order to solve this problem, a common method [16,23] is to remove the missing values and connect the remaining points into a single time series, which we denote as SkipSampEn. Additionally, interpolation is commonly used in handling missing values [24]. Two interpolation methods, linear interpolation and bootstrapping [25,26] which are denoted as LinearSampEn and BootSampEn respectively, are also addressed in our article. For the bootstrapping method, as described in Keun's articles [25,26], ten reconstruction time series are generated from each time series containing missing values using bootstrapping method and the average value of entropy is obtained.

The major feature of our method, KeepSampEn, is that a new screening condition is added when $A^m(r)$ and $B^m(r)$ are calculated, which is that both $X_{m+1}(i)$ and $X_{m+1}(j)$ must contain only non-missing values. This condition not only ensures the existence of the distance function $d[X_{m+1}(i), X_{m+1}(j)]$, but also excludes the unmatched situation that $d[X_m(i), X_m(j)]$ exists but $d[X_{m+1}(i), X_{m+1}(j)]$ does not exist. So KeepSampEn is still the negative natural logarithm of the conditional probability $\frac{A^m(r)}{B^m(r)}$, but excludes the number of vector pairs of length m and $m + 1$ which contain missing values. This reduces the impact of missing values on the calculated value of sample entropy. In addition, in KeepSampEn, the standard deviation σ used to determine the tolerance value r ($r = 0.15 \times \sigma$) is computed using only non-missing values, thus eliminating the impact of imputed values on the tolerance value. KeepSampEn is implemented in C and extended to multiscale sample entropy on the basis of mse.c from Costa et al. [20,27], other methods and the overall analysis are implemented in R language.

2.3. Experimental Datasets

To evaluate the utility of our methods compared with existing methods, some common types of physiological signals are investigated. Physiological signal is divided into electrical signals and non-electrical signals. Electrical signals mainly include electrocardiogram (ECG), electroencephalogram (EEG) and electromyogram (EMG) signals. Sample entropy of electrical signals can be used to measure the complexity of the nervous system (ECG corresponds to cardiac nervous system, EEG corresponds to central nervous system, EMG corresponds to motor nerves). Since the purpose of measurement (complexity of nervous system) and signal type (electrical signals) are consistent, we selected the ECG and EEG signal as a representative for analysis and speculated that our method can obtain the similar results in EMG data too. In the non-electrical signal, we chose airflow data and blood glucose data for analysis. They can measure the complexity of the respiratory and glycemic metabolic systems respectively which correspond to respiratory diseases, diabetes and complications. Due to the high cost and low representativeness, other non-electrical signals weren't analyzed. In summary, these four types of data contain most of the physiological signals that can be used for continuous measurements, which can analyze the complexity of cardiovascular, diabetes, respiratory diseases and their complications.

For respiratory data, we used the retrospective data in a published study [28] for one regularly treated, chronic obstructive pulmonary disease patient in Beijing Chaoyang Hospital. Specifically we used the air flow data collected from ventilators over the course of 20 nights. In one night, the 5 Hz ventilator collects around 100,000 total points. However due to common issues in data collection, some measuring points are inevitably missing. Therefore, certain sections of respiratory sequences without missing points were selected for the analysis in this paper. All glucose data were obtained from the First Affiliated Hospital of Guangzhou Medical University. Twenty patients with type 2 diabetes (10 males and 10 females) used the Glutalor (Glutalor Medical Inc, Exton, PA, USA) for blood glucose measurement. That study was conducted according to the principles of the Helsinki Declaration and all participants gave their informed consent. The device measured blood glucose automatically every three minutes over the 7-day metrical period. The ECG data were obtained from the Fantasia Database in the PhysioNet (https://physionet.org/physiobank/database/fantasia/). Twenty young healthy subjects (21–34 years old) were selected and the original ECG sampling frequency was 250 Hz. They were measured while the study subjects were in a resting state while they were lying down. The RR interval is a time interval where two consecutive R waves crest in the ECG, and the sample entropy of it can reflect the complexity of ECG. After conversion, we obtained 20 original data series made up of RR intervals. The number of male and female subjects were equal. The EEG data were obtained from the CHB-MIT Scalp Database in the PhysioNet (https://www.physionet.org/pn6/chbmit/). Twenty pediatric epilepsy patients (1.5–22 years old) were selected and the original EEG sampling frequency was 256 Hz. We used the first column of each EEG data for analysis. Yentes et al. [29] found that sample entropy values stabilize around a length of 2000 points. Thus, except to verify the robustness to the impact of data size, we analyze the first 4000 points of each signals to rule out the effect of data length on the results (the length of glucose is only 2500, which is the maximum size in the dataset.). The choice of 4000 continuous points can make the shortest time series to reach 2000 points when the missing percentage reaches the maximum we set (i.e., 50%).

2.4. Scheme for Generating Missing Values

In actual practice, many situations lead to missing values. We consider two common situations: a single missing point and a group of missing points at a time. An example of the first situation is that outliers [20,30] must be identified and excluded in heartbeat signals before calculating sample entropy. A common example of the second situation is a disruption in the wireless or network transmission of data which may cause a group of missing values [31,32].

Given a continuous time series having N points without missing values, we calculate its sample entropy. For convenience, we denote it as original entropy value. The total number of points marked as missing values is determined by the percentage of marked missing points P. The P values adopted in our research are 0% (baseline), 10%, 20%, 30%, 40%, 50%, respectively. We have designed two schemes to simulate the distribution of missing values: random sample marking and group-based random marking (Figure 1), which correspond to the two common situations that create missing values in time series data respectively.

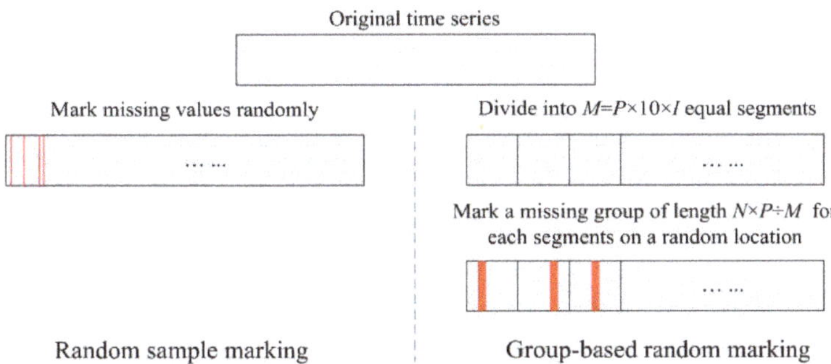

Figure 1. Two schemes for generating missing values.

Random sample marking: In this scheme, the missing points are randomly marked in the time series. First, the count of missing values to be marked C is calculated in accordance with the specified missing percentage P multiplied by the total number N ($C = P \times N$). Then, C positions are selected randomly from the original time series without replacement. Finally, each value on these positions is replaced by a special identifier (*NA*, for example, is commonly used in R) to indicate the missing values.

Group-based random marking: In this scheme, continuous missing points are seen as a group and the starting position of the group is randomly chosen. The process is displayed in Figure 1. First, the original time series is divided into M segments on average, where M ($M = P \times 10 \times I, I = 1, 2, 3...$) is determined by a tunable parameter factor I. Second, we mark a missing group of length $N \times P \div M$ for each segment, and the starting position of each group is selected randomly in the segment.

By marking some points as missing values, a time series containing missing values is generated. Then we calculate the entropy values of the generated data using the various methods and compare them with the original entropy value. The performance of a method is evaluated using percentage error defined as follows:

$$percentage\ error = \left|\frac{x - x_0}{x_0}\right| \times 100\%$$

where x is the entropy value of the modified time series with missing values; x_0 is the entropy value of the original time series without missing values.

The percentage error represents the percentage of absolute deviation of the experimental entropy value with missing values from the original entropy value based on the original dataset. Thus, it can be used to evaluate the performance of a method in handling missing values. The smaller the percentage error, the better the method.

3. Results

3.1. Robustness to the Impact of Physiological Types

To investigate whether our new KeepSampEn method is applicable to various types of physiological signals, we applied our method to four common types of physiological signals: air flow data, RR interval data (from ECG), EEG data and blood glucose levels. We first randomly selected one signal from each type of datasets, then used the random sample marking scheme to generate test data, finally repeated each method 10 times for a given proportion of missing values. The performance as measured by percentage error is shown in the top panels of Figure 2. The results clearly show that the KeepSampEn method had an average percentage error lower than 15% even in the situation with 50% of the values missing (top panels in Figure 2) in each of the four types of physiological signals. This demonstrates how our method has good performance and applicability over most common types of physiological signals.

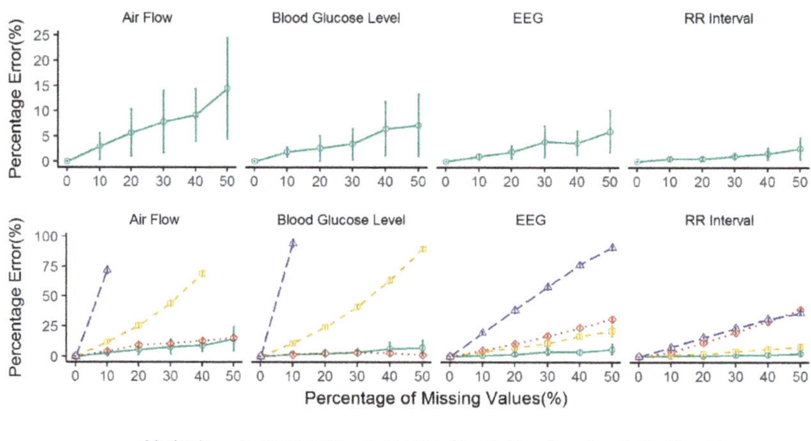

Figure 2. Performance of methods for handling missing values in four types of continuously monitoring physiological signals: air flow (**left**), blood glucose level (middle left-skewed), EEG (middle right-skewed), RR interval (**right**). Values are given as means $\pm$ standard deviation. The results for BootSampEn are not shown in the Figure 2 when the percentage error is too large and out of range of the figure.

Next, the performance of KeepSampEn is compared to the three other methods: SkipSampEn, LinearSampEn and BootSampEn. The results are shown in the bottom panels of Figure 2. KeepSampEn achieved the lowest percentage error in each percentage of missing values for each type of data except for blood glucose levels missing a high percentage (more than 40%) of values.

To verify the stability of the result, we selected randomly 20 records from each dataset, then used the random sample marking scheme to generate test data, finally calculated percentage errors and make pairwise comparison by paired T-test. The results are shown in Figure 3, which is same as Figure 2. In this scenario where a high percentage of values for blood glucose levels are missing, the percentage error of KeepSampEn is only slightly higher than LinearSampEn without any significant difference (p-value > 0.05).

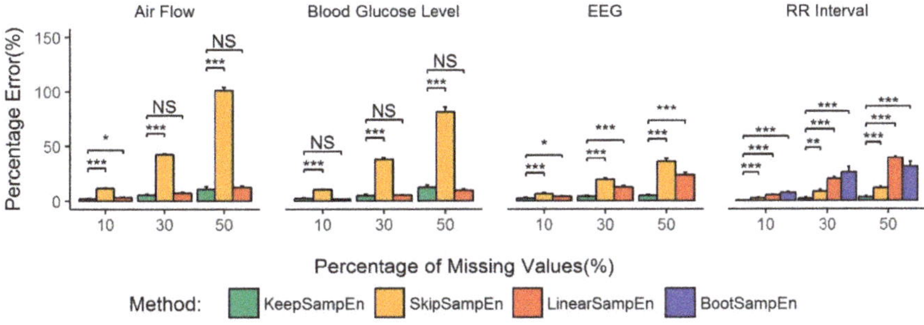

Figure 3. Average percentage errors for each method in four types of continuously monitoring physiological signals: air flow (**left**), blood glucose level (middle left-skewed), RR interval (**right**). The percentage errors by BootSampEn in the left and middle panels are higher than 120% and are not shown in the figure. Values are given as means ± standard error. NS means $p > 0.05$, * means $p < 0.05$, ** means $p < 0.01$, *** means $p < 0.001$.

BootSampEn performed poorly for all the four types of data. SkipSampEn has a good performance for RR interval data but a poor performance for rest three types of data. LinearSampEn has a good performance for air flow and blood glucose level but a poor performance for RR interval data and EEG data. By contrast, the performance of the new KeepSampEn method was robustly low for all four types of data (Figures 2 and 3). This demonstrates how our method is robust to data type. In addition, the percentage error increased rapidly when the percentage of missing values increased for BootSampEn for all four types of data, for SkipSampEn in air flow and blood glucose levels, and for LinearSampEn in RR interval data and EEG data (Bottom panels of Figure 2). By contrast, our method did not demonstrate such sensitivity for any of the four types of data considered, which indicates another aspect of how robust the method is.

3.2. Robustness to the Impact of Data Size

To investigate whether the KeepSampEn method is robust to data size (i.e., the length of the time series in a study), we first applied our method to nine datasets with a large size (4k–10k, 10k–20k, 20k–30k, 30k–40k, 40k–50k, 50k–60k, 60k–70k, 70k–80k, 80k+, where 1k represents 1000 data points) from the air flow dataset. For each data size, we selected three time series randomly for analysis. The performance is shown Figure 4. The result shows that, among the four methods of handling missing values, our method KeepSampEn obtains the smallest percentage error rate almost in any data size. The result further shows that KeepSampEn can effectively control the percentage error below 4.53% regardless of the amount of data when the percentage of missing value is less than 30%. When the percentage of missing value is above 30% but less than 50%, the percentage error can be greater than 5% but within 15%.

Considering the need of calculating sample entropy on a dataset with a small size in real clinical settings, we further explore the case of a dataset with a small size (i.e., 0.5k, 1k, 2k) from four types of physiological signals (i.e., airflow, glucose level, EEG and RR interval). Again, our proposed method achieves the smallest percentage error among the four methods in nearly all the settings except for airflow.

Based on the performance shown in Figure 5, our method can control the percentage error to be less than 5% for a small dataset with a small percentage (i.e., lower than 15%) of missing values for the glucose, RR interval and EEG data. For air flow data, linearSampEn performs better than our method for a small dataset. However, our method controls the percentage error to be less than 4.63% when percentage of missing value is 5% or less.

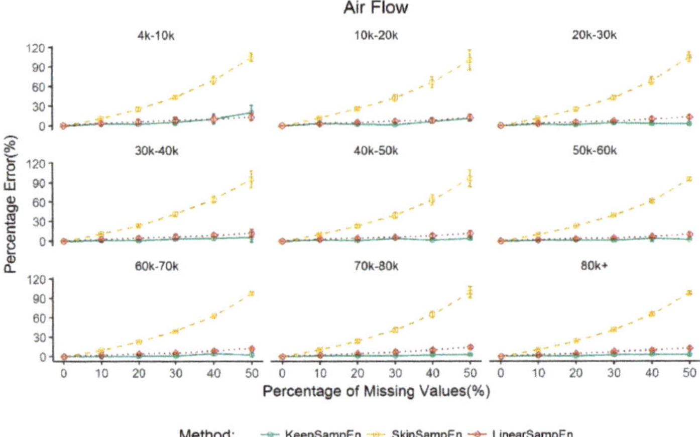

Figure 4. Performance of methods for handling missing values in air flow time series of nine large data sizes. Values are given as mean ± standard deviation. The percentage error for BootSampEn is out of the range and not shown in the figure.

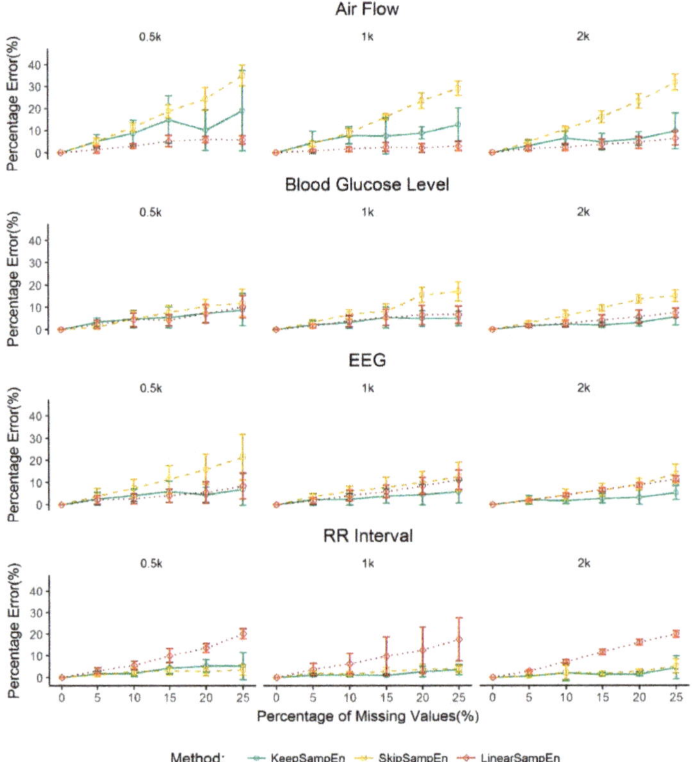

Figure 5. Performance of methods for handling missing values in a dataset with a small size (i.e., less than 2000 data points) for four types of physiological signals (i.e., air flow, blood glucose level, EEG and RR interval. Values are given as mean ± standard deviation. The percentage error for BootSampEn is out of the range and not shown in the figure.

3.3. Robustness to the Impact of Schemes for Generating Missing Values

To investigate whether our method is robust to the impact of schemes for generating missing values, we explore random sample marking and group-based random marking schemes with Factor $I = (1, 3, 5, 10, 15, 20, 30, 40, 50)$ as described in the Method Section. Note, the factor I indicates the degree of scatterings of missing values. The larger the value of I, the more scattered the missing values. We selected randomly 20 signals from the air flow dataset, used two kinds of scheme to generate test data, then calculated the average percentage error for each proportion of missing values. The performance is shown Figure 6. The percentage error of the proposed KeepSampEn method remains low for both schemes, and it is low regardless of the scatter of the missing values (Figure 6). Thus, the new method is robust when it comes to how the missing values are generated.

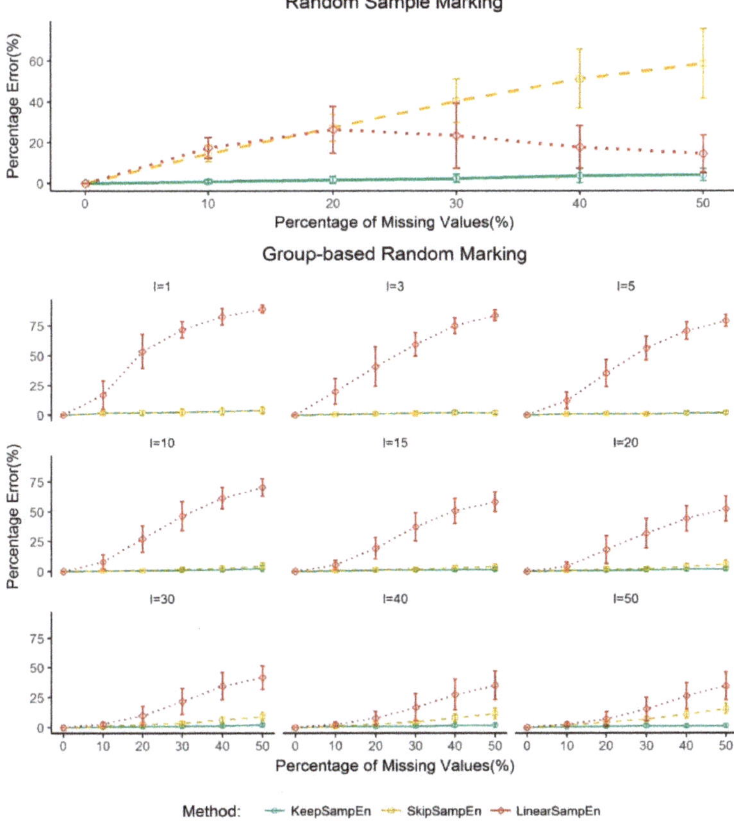

Figure 6. Percentage errors for each method using random sample marking and group-based random marking schemes to generate missing values, respectively. Values are given as mean±standard deviation. The percentage error for BootSampEn is out of the range and not shown in the figure.

It worth noting that LinearSampEn achieved a low percentage error, similar to KeepSampEn, when the missing values were generated randomly. However, it has a much higher percentage errors when the missing values are grouped together (red lines in Figure 6).

3.4. Exploration of the Computational Complexity of KeepSampEn

We evaluate the computational complexity by calculating running time of four methods on the data with the same length (time complexity). The time for calculating sample entropy based on the time series without any missing value has $O(n^2)$. For convenience, we denote this running time as the base so that the running time of all the four methods for handling missing values can be compared with it. SkipSampEn takes slightly less time than the base due to the reduction of the number of points. KeepSampEn takes almost the same time as the base. LinearSampEn takes slightly more time than the base due to additional interpolation operations. BootSampEn takes 10 times as much time as the base because it includes resampling and averaging 10 times. We further use a list of air flow data for verification. The result shown in Figure 7 confirms the theoretical derivation.

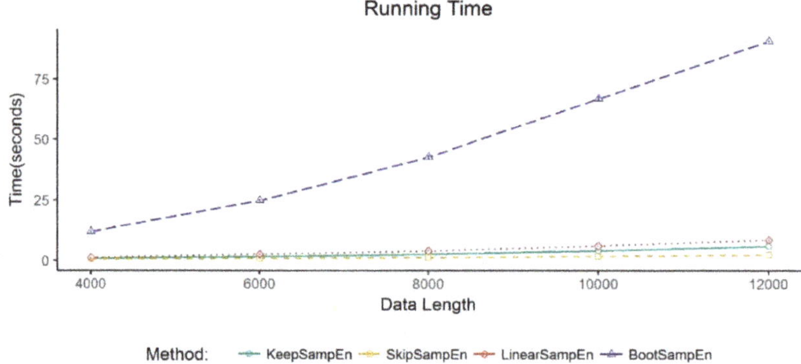

Figure 7. Evaluation of the running time of four methods for handling missing values. Each method is run on five values of percentage of missing values (i.e., 10%, 20%, 30%, 40% and 50%) on the air flow dataset. The running time for them are summed up to be the total running time. This process is repeated 10 times. The average total running time for the 10 repeats is shown in the *y*-axis. Values are given as mean ± standard deviation. Note, the standard deviation is tiny so that it can hardly been seen in the figure.

4. Discussion

Sample entropy is a widely used metric for assessing the irregularity of physiological signals. In practice, missing values or outliers commonly exist in the data of physiological signals. If we use conventional methods (neglect/interpolation) to calculate the signal with missing values, the resulting value will have some error from the original series. In this paper, we propose a new method, KeepSampEn, to minimize the error due to missing values in sample entropy calculation. This approach is less intrusive to the structure of the original data and provides a theoretical advantage in handling missing values in the analysis of sample entropy.

We further studied the utility, applicability and robustness of our method in practice by designing common different datasets with missing values from experimental data without missing values. This was done for most of types of physiological signals and corresponding diseases. In this paper, we use percentage error to judge the quality of various missing value processing methods. The results of our analysis demonstrate that for several common physiological data, our method can minimize the influence of missing values in the sample entropy calculation. In addition, the performance of our method is stable and robust, whereas other methods have either a poor or an unstable performance (Figure 3).

One limitation for the use of our method is data type. We only verified that our method is suitable for ECG data, EEG data, blood glucose data, airflow data rather than all physiological data. For other electrical physiological signals such as EMG signals, we speculate that conclusion is consistent with ECG and EEG signals due to the same signal types and measurement purposes. However, if researchers want to apply KeepSampEn to other types of data, simple verification is required.

We also investigated the performance of our method in various large sizes of data. The results show that the entropy value obtained using our method is always close to the true value (i.e., the entropy value from the original data without missing values), with a percentage error less than 4.53% when the percentage of missing values is less than 30% for data sizes varying from 4000 to 80,000. We further see that the percentage error can be greater than 5% but within 15% when the percentage of missing value is above 30% but less than 50%. For a dataset with a small size (i.e., from 500 to 2000 data points), our method can control the percentage error to be less than 5% for blood glucose level, EEG and RR interval when the percentage of missing values is less than 15%. This contrasts with the entropy values obtained using other methods which greatly deviated from the true value (Figures 4 and 5). These results may provide valuable information for data quality control in practice. That is, for the physiological signals, when a dataset has a percentage of missing values less than 30% for a large data size or 15% for a small data size of total series due to equipment or operational errors, our method can rescue the data through minimizing the impact of missing value (i.e., controlling the percentage error of the sample entropy to be less than 5%). On the other hand, if the percentage of missing values reaches 30% or more for a large data size or 15% or more for a small data size, the result of our research indicates that the calculated value of sample entropy is not reliable anymore even if we adopt the best method for handling missing values. In such a case, the data may be screened out from further analysis on sample entropy. It should be mentioned that, for air flow data, linearSampEn performs better than our method for a small dataset. However, it is easy and convenient to obtain a large dataset for air flow in real clinical settings and the size of airflow datasets is usually large in the clinical setting. Thus, we should usually not worry about the case of a small dataset of air flow. Moreover, even for air flow data, our method controls the percentage error to be less than 4.63% when percentage of missing value is 5% or less.

We further investigated whether our method will be affected by how the missing values are generated. The results indicate that our method is robust to how the missing values are generated in the continuous monitoring of physiological signals (Figure 6).

In conclusion, our proposed KeepSampEn method has the following merits for handling missing values in the analysis of entropy for continuously monitored physiological data. First, unlike the usual ways by modifying the input data, our method keeps the input data unchanged and modifies the calculation process, which is less intrusive to the structure of original data. Second, it is effective and applicable to most common types of physiological signal data for a variety of diseases. Third, it is robust in not only how long the data is but also how the missing values are generated. This is a marked improvement over the currently used methods for handling missing values in analysis of sample entropy. With these merits, our proposed method should have broad utility in handling missing values in the analysis of sample entropy for continuously monitored physiological signals.

Supplementary Materials: The supplementary materials are available online at http://www.mdpi.com/1099-4300/21/3/274/s1.

Author Contributions: All authors conceived of the idea. Data curation, X.D., Q.G., Z.C., X.C. and J.L.; Formal analysis, X.D. and C.C.; Funding acquisition, X.D.Z.; Methodology, C.C. and X.D.Z.; Resources, Q.G., Z.C., X.C. and J.L.; Software, Q.G. and Y.J.; Supervision, X.D.Z.; Writing—original draft, X.D. and C.C.; Writing—review & editing, Z.Z., Y.S. and X.D.Z. All authors reviewed and commented on the manuscript and approved the final draft.

Funding: This work was supported by University of Macau through Research Grants MYRG2018-00071-FHS, EF005/FHS-ZXH/2018/GSTIC and FHS-CRDA-029-002-2017.

Conflicts of Interest: The authors declare no potential conflicts of interest with respect to the research, authorship, and/or publication of this article.

Appendix A

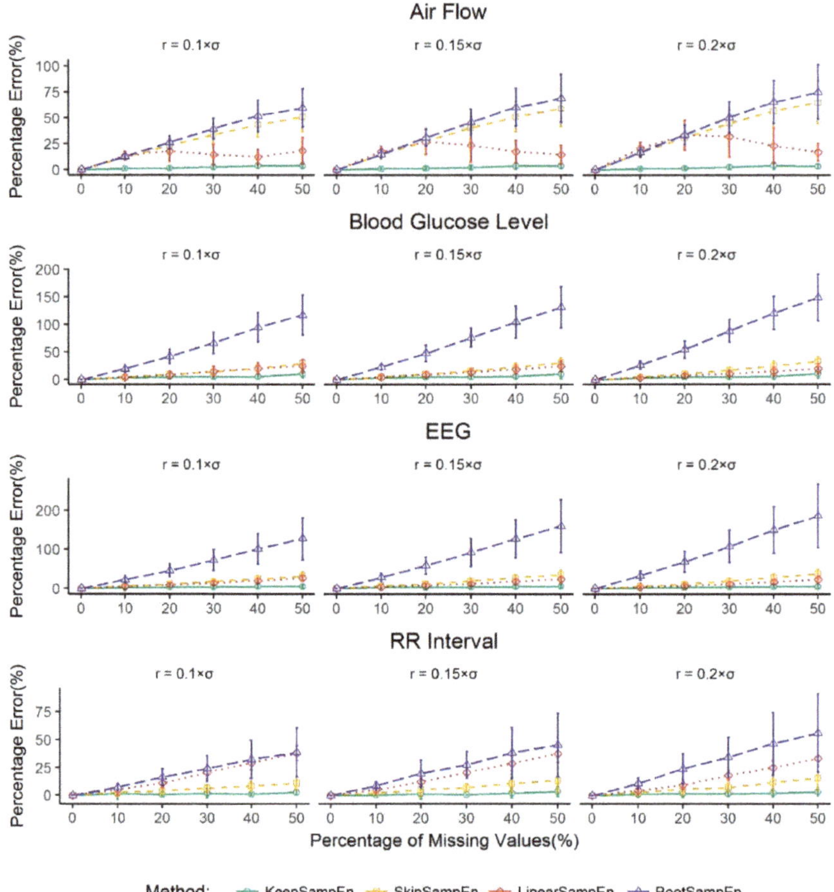

Figure A1. Percentage errors for each method in three types of continuously monitoring physiological signals: blood glucose level (top panel), RR interval (middle panel), air flow (bottom panel). Each panel includes three conditions: $r = 0.1 \times \sigma$ (**left**), $r = 0.15 \times \sigma$ (**middle**) and $r = 0.2 \times \sigma$ (**right**). Values are given as means ± standard deviation. The results for BootSampEn are not shown in the Figure 2 when the percentage error is too large and out of range of the figure.

References

1. Shannon, C.E.; Weaver, W. The Mathematical Theory of Information. *Math. Gazette* **1949**, *97*, 170–180.
2. Ivanov, P.C.; Amaral, L.A.; Goldberger, A.L.; Havlin, S.; Rosenblum, M.G.; Struzik, Z.R.; Stanley, H.E. Multifractality in Human Heartbeat Dynamics. *Nature* **1999**, *399*, 461. [CrossRef] [PubMed]
3. Chen, C.; Jin, Y.; Lo, I.L.; Zhao, H.; Sun, B.; Zhao, Q.; Zheng, J.; Zhang, X.D. Complexity Change in Cardiovascular Disease. *Int. J. Biol. Sci.* **2017**, *13*, 1320–1328. [CrossRef] [PubMed]
4. Henriques, T.S.; Costa, M.D.; Mathur, P.; Mathur, P.; Davis, R.B.; Mittleman, M.A.; Khabbaz, K.R.; Goldberger, A.L.; Subramaniam, B. Complexity of Preoperative Blood Pressure Dynamics: Possible Utility in Cardiac Surgical Risk Assessment. *J. Clin. Monit. Comput.* **2018**, *33*, 31–38. [CrossRef] [PubMed]

5. Kohnert, K.D.; Heinke, P.; Vogt, L.; Augstein, P.; Thomas, A.; Salzsieder, E. Associations of Blood Glucose Dynamics with Antihyperglycemic Treatment and Glycemic Variability in Type 1 and Type 2 Diabetes. *J. Endocrinol. Investig.* **2017**, *40*, 1201–1207. [CrossRef]
6. Investigators, F.S.T. Glucose Variability in a 26-Week Randomized Comparison of Mealtime Treatment with Rapid-Acting Insulin Versus GLP-1 Agonist in Participants with Type 2 Diabetes at High Cardiovascular Risk. *Diabetes Care* **2016**, *39*, 973–981.
7. Kumar, R.B.; Goren, N.D.; Stark, D.E.; Wall, D.P.; Longhurst, C.A. Automated Integration of Continuous Glucose Monitor Data in the Electronic Health Record Using Consumer Technology. *J. Am. Med. Inform. Assoc.* **2016**, *23*, 532–537. [CrossRef]
8. Wu, H.T.; Hsu, P.C.; Lin, C.F.; Wang, H.J.; Sun, C.K.; Liu, A.B.; Lo, M.T.; Tang, C.J. Multiscale Entropy Analysis of Pulse Wave Velocity for Assessing Atherosclerosis in the Aged and Diabetic. *IEEE Trans. Biomed. Eng.* **2011**, *58*, 2978–2981.
9. Watanabe, E.; Kiyono, K.; Hayano, J.; Yamamoto, Y.; Inamasu, J.; Yamamoto, M.; Ichikawa, T.; Sobue, Y.; Harada, M.; Ozaki, Y. Multiscale Entropy of the Heart Rate Variability for the Prediction of an Ischemic Stroke in Patients with Permanent Atrial Fibrillation. *PLoS ONE* **2015**, *10*, e0137144. [CrossRef]
10. Jin, Y.; Chen, C.; Cao, Z.; Sun, B.; Lo, I.L.; Liu, T.M.; Zheng, J.; Sun, S.; Shi, Y.; Zhang, X.D. Entropy Change of Biological Dynamics in COPD. *Int. J. Chronic Obstr. Pulm. Disease* **2017**, *12*, 2997. [CrossRef]
11. Xue, M.; Wang, D.; Zhang, Z.; Cao, Z.; Luo, Z.; Zheng, Y.; Lu, J.; Zhao, Q.; Zhang, X.D. Demonstrating the Potential of Using Transcutaneous Oxygen and Carbon Dioxide Tensions to Assess the Risk of Pressure Injuries. *Int. J. Biol. Sci.* **2018**, *14*, 1466–1471. [CrossRef]
12. Sun, S.; Jin, Y.; Chen, C.; Sun, B.; Cao, Z.; Lo, I.L.; Zhao, Q.; Zheng, J.; Shi, Y.; Zhang, X.D. Entropy Change of Biological Dynamics in Asthmatic Patients and Its Diagnostic Value in Individualized Treatment: A Systematic Review. *Entropy* **2018**, *20*, 402. [CrossRef]
13. Wang, D.; Zheng, P.; Zhang, T.; Li, C.; Chen, L.; Zhai, Y.; Leng, D.; Jin, Y.; Sun, B.; Zhang, X.D. Analysis of the Complexity Patterns in Respiratory Data. In Proceedings of the 2018 IEEE International Conference on Bioinformatics and Biomedicine, Madrid, Spain, 3–6 December 2018; pp. 2556–2560.
14. Sahoo, S.S.; Jayapandian, C.; Garg, G.; Kaffashi, F.; Chung, S.; Bozorgi, A.; Chen, C.H.; Loparo, K.; Lhatoo, S.D.; Zhang, G.Q. Heart Beats in the Cloud: Distributed Analysis of Electrophysiological 'Big Datac Using Cloud Computing for Epilepsy Clinical Research. *J. Am. Med. Inform. Assoc. Jamia* **2014**, *21*, 263–271. [CrossRef]
15. Zhang, X.D.; Zhang, Z.; Wang, D. CGManalyzer: an R package for Analyzing Continuous Glucose Monitoring Studies. *Bioinformatics* **2018**, *34*, 1609–1611. [CrossRef]
16. Cirugedaroldan, E.; Cuestafrau, D.; Miromartinez, P.; Oltracrespo, S. Comparative Study of Entropy Sensitivity to Missing Biosignal Data. *Entropy* **2014**, *16*, 5901–5918. [CrossRef]
17. Li, Y.; Wang, J.; Li, J.; Xu, W.; Zhang, P.; Zhang, X. Detecting Complexity Changes for Heart Rate Variability in the Case of Missing Data. *Sci. Sin. Inf.* **2015**, *45*, 1015. [CrossRef]
18. Zhang, X.D.; Pechter, D.; Yang, L.; Ping, X.; Yao, Z.; Zhang, R.; Shen, X.; Li, N.X.; Connick, J.; Nawrocki, A.R. Decreased Complexity of Glucose Dynamics Preceding the Onset of Diabetes in Mice and Rats. *PLoS ONE* **2017**, *12*, e0182810. [CrossRef]
19. Cho, K.; Miyano, T. Entropy Test for Complexity in Chaotic Time Series. *Nonlinear Theory Its Appl.* **2016**, *7*, 21–29. [CrossRef]
20. Costa, M.; Goldberger, A.L.; Peng, C.K. Multiscale entropy analysis of biological signals. *Phys. Rev. E* **2005**, *71*, 021906. [CrossRef]
21. Richman, J.S.; Moorman, J.R. Physiological time-series analysis using approximate entropy and sample entropy. *Am. J. Physiol. Heart Circ. Physiol.* **2000**, *278*, H2039–H2049. [CrossRef]
22. Pincus, S.M. Approximate Entropy as a Measure of System Complexity. *Proc. Natl. Acad. Sci. USA* **1991**, *88*, 2297–2301. [CrossRef]
23. Lake, D.E.; Richman, J.S.; Griffin, M.P.; Moorman, J.R. Sample Entropy Analysis of Neonatal Heart Rate Variability. *Am. J. Physiol. Regul. Integr. Comp. Physiol.* **2002**, *283*, R789–R797. [CrossRef]
24. Shara, N.M.; Umans, J.G.; Wang, W.; Howard, B.V.; Resnick, H.E. Assessing the Impact of Different Imputation Methods on Serial Measures of Renal Function: The Strong Heart Study. *Kidney Int.* **2007**, *71*, 701–705. [CrossRef]

25. Kim, K.K.; Baek, H.J.; Yong, G.L.; Park, K.S. Effect of Missing RR-interval Data on Nonlinear Heart Rate Variability Analysis. *Comput. Methods Programs Biomed.* **2012**, *106*, 210–218. [CrossRef]
26. Dai, X.; Zhang, D. Research of Interpolation Methods of Missing RR-interval Data. *Chin. J. Med. Phys.* **2013**, *30*, 3903–3905.
27. Costa, M.D.; Peng, C.K.; Goldberger, A.L. Multiscale Analysis of Heart Rate Dynamics: Entropy and Time Irreversibility Measures. *Cardiovasc. Eng.* **2008**, *8*, 88–93. [CrossRef]
28. Cao, Z.; Luo, Z.; Hou, A.; Nie, Q.; Xie, B.; An, X.; Wan, Z.; Ye, X.; Xu, Y.; Chen, X. Volume-Targeted Versus Pressure-Limited Noninvasive Ventilation in Subjects with Acute Hypercapnic Respiratory Failure: A Multicenter Randomized Controlled Trial. *Respir. Care* **2016**, *61*, 1440–1450. [CrossRef]
29. Yentes, J.M.; Hunt, N.; Schmid, K.K.; Kaipust, J.P.; McGrath, D.; Stergiou, N. The Appropriate Use of Approximate Entropy and Sample Entropy with Short Data Sets. *Ann. Biomed. Eng.* **2013**, *41*, 349–365. [CrossRef]
30. Voss, A.; Schulz, S.; Schroeder, R.; Baumert, M.; Caminal, P. Methods Derived from Nonlinear Dynamics for Analysing Heart Rate Variability. *Philos. Trans. A Math. Phys. Eng. Sci.* **2009**, *367*, 277–296. [CrossRef]
31. Bao, Y.; Li, H.; Sun, X.; Yu, Y.; Ou, J. Compressive Sampling-based Data Loss Recovery for Wireless Sensor Networks Used in Civil Structural Health Monitoring. *Struct. Health Monit.* **2013**, *12*, 78–95. [CrossRef]
32. Xu, Y.; Lee, W.C.; Xu, J. Analysis of a Loss-Resilient Proactive Data Transmission Protocol in Wireless Sensor Networks. In Proceedings of the IEEE International Conference on Computer Communications, Barcelona, Spain, 6–12 May 2007; pp. 1712–1720.

Article

Multi-Scale Heart Beat Entropy Measures for Mental Workload Assessment of Ambulant Users

Abhishek Tiwari [1,*], Isabela Albuquerque [1], Mark Parent [1], Jean-François Gagnon [2], Daniel Lafond [2], Sébastien Tremblay [3] and Tiago H. Falk [1]

1 Institut National de la Research Scientifique, Université du Québec, Montréal, QC H3A 0E7, Canada
2 Thales Research and Technology, Québec, QC G1P 4P5, Canada
3 School of Psychology, Université Laval, Québec, QC G1V 0A6, Canada
* Correspondence: abhishek.tiwari@emt.inrs.ca

Received: 26 June 2019; Accepted: 8 August 2019; Published: 10 August 2019

Abstract: Mental workload assessment is crucial in many real life applications which require constant attention and where imbalance of mental workload resources may cause safety hazards. As such, mental workload and its relationship with heart rate variability (HRV) have been well studied in the literature. However, the majority of the developed models have assumed individuals are not ambulant, thus bypassing the issue of movement-related electrocardiography (ECG) artifacts and changing heart beat dynamics due to physical activity. In this work, multi-scale features for mental workload assessment of ambulatory users is explored. ECG data was sampled from users while they performed different types and levels of physical activity while performing the multi-attribute test battery (MATB-II) task at varying difficulty levels. Proposed features are shown to outperform benchmark ones and further exhibit complementarity when used in combination. Indeed, results show gains over the benchmark HRV measures of 24.41% in accuracy and of 27.97% in F1 score can be achieved even at high activity levels.

Keywords: mental workload; motif; multi-scale entropy; permutation entropy; HRV; SVM

1. Introduction

Mental workload is defined as the part of a person's mental capacity that is needed to perform different demands brought on by a task [1]. Mental workload resources, if not used in a balanced way, can lead to a decrease in worker performance [2], as the resources are directly related to other mental concepts such as situational awareness, mental fatigue, and drowsiness, to name a few. In fact, in various jobs which require continuous attention for long durations of time, such as air traffic management, aircraft piloting, and emergency responders, overload of mental resources may lead to life-threatening outcomes. As such, mental workload assessment can be used to optimize operator performance by preventing the operators from becoming overwhelmed by the task at any point in time while also ensuring that they can perform the task efficiently. System design improvements can also be made to ensure balanced used of mental resources by proper assessment of mental workload. For example, car dashboards are designed in such a way that they provide the driver with all the necessary/critical information whilst minimizing mental workload. Similarly, aircraft cockpit design has been greatly simplified using mental workload considerations [3].

Mental workload is typically assessed using subjective, performance-related, and/or physiological methods. Subjective ratings are a direct way of mental workload assessment. Typically, such methods involve sampling the participant response to the amount of mental workload by using a questionnaire. The sampling can be done for every set amount of stimuli, or every set amount of time. Two of the most commonly used questionnaires are the (1) Subjective Workload Assessment Technique (SWAT) [4] and (2) NASA Task Load Index (NASA-TLX) [5]. The latter requires subjects to rate multiple dimensions that can then be aggregated into a single general workload index [6–8]. However, subjective ratings have some key limitations. First, they do not allow for continuous mental workload assessment and have poor temporal resolution. While increasing the sampling rate for operator feedback via the questionnaire could lead to improved temporal resolution, this may actually have negative impact of increasing the perceived workload due to the number of interruptions to the task being performed. Second, the ratings collected can be corrupted by subject bias, particularly if responses impact benefits received by the operators e.g., unpaid time off from work [6]. In turn, performance metrics related to the task can also be used to devise strategies for monitoring mental workload. Examples of such metrics include reaction time, number of errors, and task accuracy. However, one major barrier to applying these strategies to safety critical scenarios is the fact that the metrics can only be computed once the task is concluded.

In order to circumvent the limitations presented by these methods, operator physiological monitoring has emerged as a promising solution for mental workload assessment. Monitoring psycho-physiological signals allows for unbiased, continuous assessment of real-time mental workload while also being unobtrusive to the task at hand. In fact, with recent improvements in wearable technologies, remote monitoring of mental workload has been made possible even with operators in ambulatory conditions [9]. To this end, the electrocardiogram (ECG) has become an important modality to be measured. Heart rate variability (HRV), which is the variability in the inter-beat interval (RR) series derived from ECG, has shown to be an important correlate of psycho-social workload [10] (i.e., job stressors), mental workload and anxiety [11], as well as mental fatigue [12]. HRV is an indicator of the changes in the autonomic nervous system (ANS) and is controlled by both the sympathetic and para-sympathetic nervous system demands (increased activation of the sympathetic nervous system causes increases in heart rate, whereas an increased activation of the parasympathetic branch makes the heart rate slower). HRV has traditionally been quantified using time- and/or frequency-domain features computed from the RR time series [13]. The inter-beat interval series also exhibit complex fractal behavior with long-term correlations [14]. Over the last decade, these non-linear properties of the cardiac autonomic system have also been exploited using several complexity measures [15].

One such measure, called the multi-scale entropy (MSE) [16] has been proposed to characterize the complexity of physiologic time series at multiple scales. The algorithm is based on obtaining sample entropy at different time scales using a scaling algorithm. However, the originally proposed scaling algorithm (known as coarse graining) is sub-optimal and may lead to imprecise or undefined entropy values [17]. As a result, several variants of the multi-scale algorithm have been proposed in recent years [17], including permutation entropy (PE) [18,19] which has been shown to be robust to signal artifacts, as it deals with the shape of the time series, and not on magnitude values themselves [20,21]. Several variants of the coarse-graining method have also been proposed, including replacing it with moving average for short time series [22], a composite procedure that reduces the variance of entropy at higher scales [23], and the recently-proposed generalized multi-scale entropy measure [24], which quantifies the dynamics of the volatility (variance) of the time series over different scales [24,25].

Entropy-based measures have been used in the past to characterize aging and to diagnose different cardiac diseases [26–28]. Moreover, multi-scale analysis of the volatility series of the RR intervals [24] has also shown non-linear behavior and is able to successfully distinguish between healthy subjects and those with congestive heart failure. In [24], the authors argue that the coarse grained volatility series encapsulates additional information about the time series missed by the normal coarse graining procedure. Additionally the magnitude of the difference intervals of the RR series (i.e., $dRR_i = abs(RR_{i+1} - RR_i)$ has exhibited similar long-range correlations [29]. This property was used in [30] and showed better performance in distinguishing patients with congestive heart failure at lower scales compared to the RR series. Further, several improvements to the permutation entropy measure have been proposed. The modified permutation entropy (mPE), for example, takes into account cases in which instantaneous heart rate measures remain the same for two consecutive beats [31], while the weighted permutation entropy [32] tries to incorporate the amplitude information of the time series being analyzed.

While such multi-scale measures have been used in cardiac disease monitoring, they have received little attention for mental state monitoring. However, *PE* was recently used for emotion assessment [33] of stationary users. Here, we are particularly interested in assessing user mental states in an ambulatory setting, in which movement may not only introduce artifacts that play a detrimental role in signal quality, but also cause changes in cardiac dynamics that may alter HRV measurement. As such, we explore a number of existing multi-scale features and propose new ones for mental workload assessment as subjects performed two different physical activities at three different levels. We hypothesize that the noise robustness provided by the permutation entropy measure coupled with studying complexity at different scales would help better quantify heart rate changes due to mental workload at different levels of physical activity.

The remainder of this paper is organized as follows. Section 2 describes the materials and methods used, including the database considered, proposed and benchmark features, prediction method, and performance metrics used. Section 3 then presents and discusses the results obtained and conclusions are presented in Section 4.

2. Materials and Methods

Here, we describe the database used, benchmark features, the different multi-scale methods tested, as well as the feature selection scheme employed, classifiers, and analysis pipeline.

2.1. Data Collection

2.1.1. Participants

After screening, 47 participants were selected (23 female, 27.4 ± 6.6 years old). Screening was performed in order to prevent any potential risk to the participants during the experiments. Candidates with cardiovascular diseases, neurological disorders, history of feeling dizzy or fainting were excluded from the experiment. The participants were asked to wear comfortable sportswear. Twenty-two participants utilized a treadmill during the experiment and 26 a stationary bike. Participants consented to participating in the experiment and were remunerated for their time. The experimental protocol was approved by the Ethics Review Boards of INRS, Université Laval and the PERFORM Centre (Concordia University), the latter being the location in which data was collected.

2.1.2. Experimental Protocol

Before starting the data collection, a tutorial explaining the experimental procedure and task to be executed was shown to the participants. Next, various sensors were placed on the subject. These included a portable eight-channel wireless EEG headset (Enobio, Neuroelectrics), a portable chest strap (Bioharness 3, Zephyr) for ECG, respiration, and accelerometer signal recording, an E4 (Empatica) wrist watch which measures blood volume pulse, skin temperature, galvanic skin response and acceleration. Following this step, in the case of participants using the treadmill, a safety harness was placed on the participant's chest to avoid falls. For participants using the stationary bike, they were asked to adjust the seat according to their preference. The height of the screen was then adjusted to provide a more comfortable set-up. Three levels of physical activity were considered: no movement, medium (treadmill: 3 km/h, bike: 50 rpm), and high (treadmill: 5 km/h, bike: 70 rpm). Figure 1 shows the experimental setup for both bike and treadmill conditions.

In order to elicit high and low mental workload levels, the NASA multi-attribute task battery (MATB-II) was used. It is a computer-based task designed to evaluate operator performance and mental workload. Figure 2 depicts the MATB-II interface seen by the participant [34]. MATB-II encompass four tasks, all grouped under a single-window interface. These four tasks are the system monitoring (top-left), the tracking (top-center), the communications (bottom-left) and the resource management (bottom-center). In this study, the communication task was not used. The interface also includes a scheduler (top-right), which only showed the time remaining in each trials, as well as the pump status (bottom-right), which complemented information on the resource management task (see below). Since participants were simultaneously doing a physical task, a mouse could not be used to interact with MATB-II. Instead, participants were instructed to use an xBox One controller.

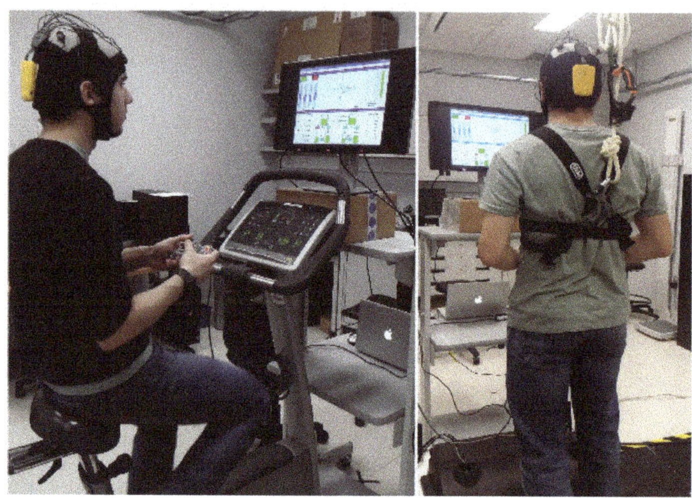

Figure 1. Experimental setup for both bike and treadmill conditions.

The system monitoring task requires the participant to monitor four sliders and report deviations from their normal state. The two warning lights (see F5 and F6 on Figure 2) were not used in this study. In their normal states, sliders were oscillating around the middle position. In their deviation state, sliders started oscillating around the top or the bottom of the sliders. Participant had to use the directional pad of the controller to report deviations. The tracking task requires the participant to keep a target (the circle) within a box (the square). As the trials went on, the target started to move randomly. Participants had to use the joystick of their controller to brink it back near the center of the square. The resource management task requires the participant to balance a network of fuel tanks. Participants were instructed to keep the level of tanks A and B as close as possible from 2500 units (this level is indicated by ticks on tanks A and B, see Figure 2). Fuel gradually depleted from tanks A and B. To keep the tanks at level, participants could use eight pumps (labeled 1–8, between tank) to move fuel between tanks. To activate pumps, participants had to use the other joystick of the controller to move the cursor and "click" on the pumps. Pumps were configured to fail from times to times. When a pump failed, it turned red and became unusable. Pumps were "repaired" automatically after a while and the participant could resume using it if needed.

Two levels of workload were used for the mental task. Compared to the low workload condition, the high workload condition had: more frequent sliders deviations (for system monitoring), faster random oscillations (for the tracking task) and more frequent pump failures (for resource management monitoring).

In total, six combinations of combined mental workload and physical activity were tested (two level of mental workload X three levels of physical activity). The experiment was then split into six sessions, each one corresponding to one of the six combinations previously described. The order in which the combinations were done was counterbalanced to avoid ordering effects. Before each session, two baseline sessions were performed. During the first session, there were neither physical or mental activity and the participants were asked to close their eyes and relax for 60 s. Then the subject was asked to open their eyes and start moving according to the corresponding sessions physical activity until reaching the desired level. After reaching a stable activity level, the second baseline was collected while the participant kept the pace for 2 min without any mental effort involved. This baseline has been added to ensure the stationarity of the heart rate dynamics for a given activity level prior to introducing the mental workload condition. This Lastly, the experimenter gave the joystick to the participant who then started the first experimental session for a duration of 10 min. After each session, a 5-min break was given. After each of the experimental sessions, participants were asked to fill the NASA-TLX questionnaire and report their perceived fatigue level based on the Borg scale. The NASA-TLX ratings were validated with respect to the ground truth in [9] Overall, the experimental protocol lasted roughly two hours.

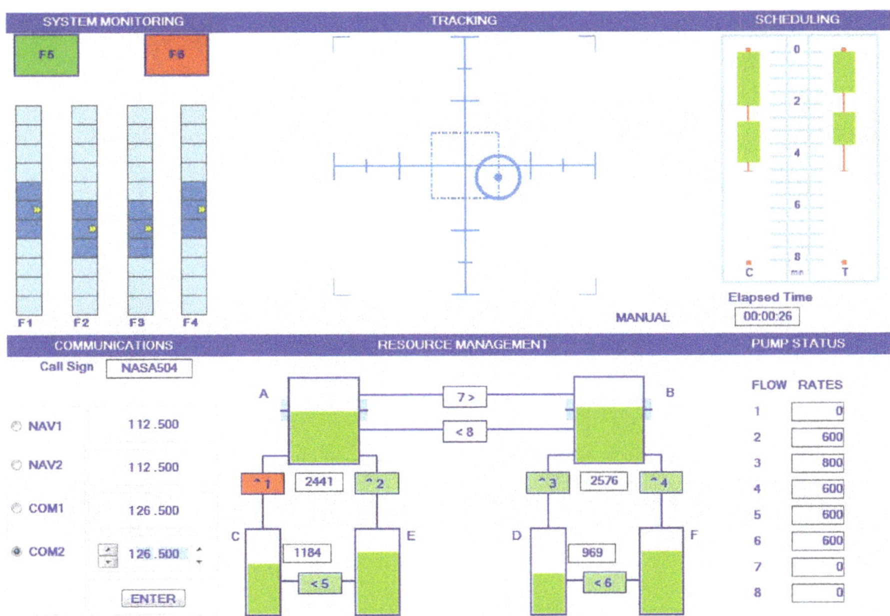

Figure 2. Multi-attribute task battery (MATB-II) game for eliciting different levels of mental workload.

2.2. Pre-Processing

In this study, we focused only on the ECG signal measured by the Bioharness Bh3 device, sampled at 250 Hz. This sampling rate allowed for continuous streaming throughout the experiment without the need to recharge the device. Such sampling rate has been successfully used in a number of different applications, including [35–37]. First, the ECG signal for all subjects was visually inspected and two subjects were removed as the data was corrupted due to sensor malfunction. For the remaining subjects the inter-beat interval series was extracted as follows. First, the ECG was filtered using a band-pass filter with a bandwidth 4–40 Hz to enhance the QRS complex. This was followed by an energy based QRS detection algorithm [38], which is an adaption of the popular Pan and Tompkins algorithm [39]. The RR series was further filtered to remove outliers using range based detection ($\geq$280 ms and $\leq$1500 ms), moving average outlier detection, and a filter based on percent change in consecutive RR values ($\leq$20%) as implemented in [40].

2.3. Benchmark HRV Features

Standard time- and frequency-domain HRV metrics were extracted and used as benchmark measures. A complete list of these conventional measures can be found in Table 1. The majority of these benchmark features have been shown in the literature to correlate with mental workload [41] and anxiety [11]. Complete details about these measures can be found in [13]. A total of 15 benchmark features were extracted over 5-min segments of RR series with a 4-min overlap, resulting in six RR series for each of the 10-min experimental sessions. The 5-min windows for HRV analysis follows recommendations from [13]. Notwithstanding, shorter time series may cause problems with multi-scale entropy estimation. In order to overcome this limitation, focus has been placed on multi-scale entropy methods designed specifically for short term analysis of HRV recordings [22,23].

Table 1. Benchmark heart rate variability (HRV) features extracted.

Time Domain Features
mean, standard deviation, coefficient of variation, rmsdd, pNN50, mean of 1st diff., standard deviation of absolute of 1st diff., normalized mean of absolute 1st diff
Frequency Domain Features
High frequency power (HF), normalized HF, Low frequency power (LF), normalized LF, very low frequency power, HF/LF

2.4. Multi-Scale Entropy Features

The multi-scale entropy methods rely on two steps: (i) scaling and (ii) entropy calculation over the different scales. Here, we explored different algorithms for both steps, as summarized in Table 2.

Table 2. Different scaling and entropy algorithms used.

Scaling Algorithms	Entropy Algorithms
Coarse graining (cg)	Sample Entropy
moving average ($mavg$)	Modified Permutation Entropy
second moment cg (mom)	Weighted Modified Permutation Entropy
moving average mom ($mavg\_mom$)	
composite coarse graining ($comp\_cg$)	

2.4.1. Scaling Algorithms

Several scaling algorithms have been proposed in the literature and attempt to convey fractal information at different scales. All of these methods take the original time series ($x(i)$) with an index i and produce the time series for a different scale. Details about the methods explored herein are given next.

- Coarse graining (cg): This is the original algorithm proposed to obtain different scales. A point j on the scaled series $y_s(j)$ for a scale s is given by:

$$y_s(j) = \frac{\sum_{i=(j-1)s+1}^{js} x(i)}{s}, \quad (1)$$

  where, $1 \leq j \leq N/s$. This method has been shown to be sub-optimal [17] and to lead to scaled series that decrease in size, which could increase the variance in the estimated entropy [23]. As a result, several variants have since been proposed, including the remainder listed below.
- Moving average ($mavg$): With moving average, a point j on the scaled series $y_s(j)$ for a scale s is given by:

$$y_s(j) = \frac{\sum_{i=j}^{i=j+s-1} x(i)}{s}, \quad (2)$$

  where $1 \leq j \leq N - s + 1$.

- Composite coarse graining (*comp_cg*): the composite method generates s different (from $k = 1, \ldots, s$) scaled series for a given scale s. The entropy estimates from the different series for the scale s are then averaged to get the entropy estimate. This helps reduce the error in the entropy estimation that occurs due to coarse graining produce. For a given scale s the point j of the kth coarse grained series ($y_{k,s}(j)$) is given by:

$$y_{k,s}(j) = \frac{\sum_{i=(j-1)s+k}^{js+k-1} x(i)}{s}, \tag{3}$$

where $1 \leq j \leq N/s$, and $1 \leq k \leq s$ gives the next index of the scaled series. The composite multi-scale entropy ($CMSE$) for a given scale is then given by:

$$CMSE(s) = \frac{\sum_{k=1}^{s} Ent(y_{k,s})}{s}, \tag{4}$$

where Ent is the entropy calculation algorithm. In the original $CMSE$ algorithm the sample entropy algorithm is used.
- The second moment coarse graining (*mom*): this method quantifies the standard deviation of the scaled series (also called the volatility series) rather than its mean as done by the coarse graining procedure. A point i on the scaled series $y_s(i)$ for a scale s is given by:

$$y_s(j) = \sigma|_{i=(j-1)s+1}^{js}(x(i)), \tag{5}$$

where $1 \leq j \leq N/s$ and σ represents the standard deviation.
- The second moment moving average (*mavg_mom*): we propose the moving average procedure for calculation of the second moment to adapt to short time series. This replaces the non-overlapping windows used in coarse graining to a sliding window, as in the case of the moving average algorithm.

Figures 3 and 4 show the RR series and RR volatility series (using *mavg_mom*) for scales $s = 1$ to $s = 3$ and scales $s = 2$ to $s = 4$, respectively, for a five minute ECG segment. As can be seen, scaling removes some high frequency information from the series, commonly associated with artifacts.

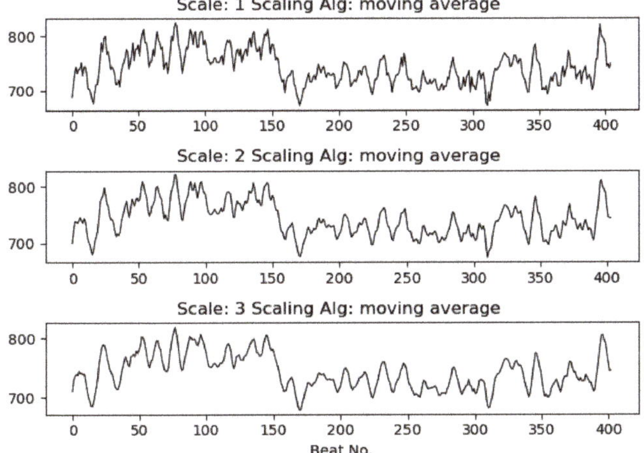

Figure 3. Scaled inter-beat interval (RR) time series with the moving average algorithm for scales $s = 1$ (original series) to $s = 3$.

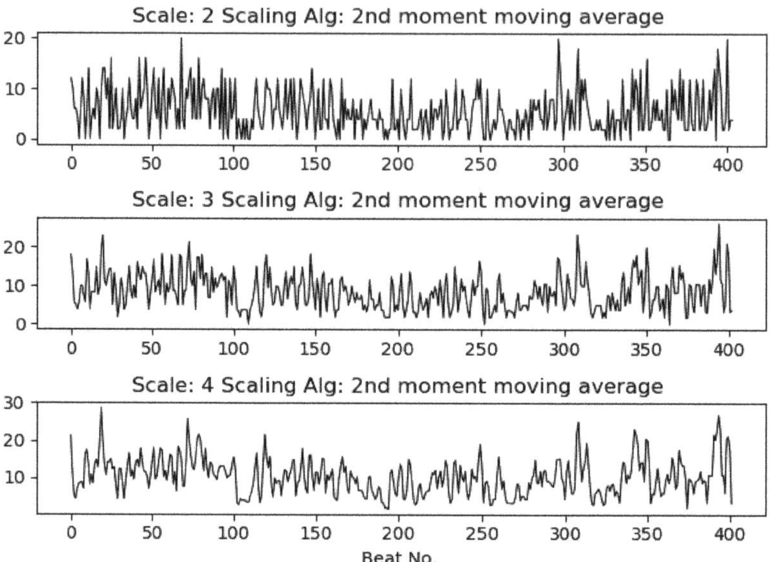

Figure 4. Scaled RR time series with the second moment moving average algorithm for scales $s = 2$ to $s = 4$.

2.4.2. Entropy Algorithms

- Sample entropy (*SampEn*): it is the negative natural logarithm of an estimate of the conditional probability that if two sets of vectors ($X_m(i)$ and $X_m(j)$) of length m have a distance $< r$, then two sets of vectors ($X_{m+1}(i)$ and $X_{m+1}(j)$) of length $m+1$ also have a distance $< r$, based on some distance metric $d_m(X,Y)$. It is formally defined as:

$$SampEn = -\log\frac{N_{m+1}}{N_m}, \tag{6}$$

where N_m is number of vector pairs with $d_m(X_m(i), X_m(j)) < r$ and N_{m+1} is number of vector pairs with $d_m(X_{m+1}(i), X_{m+1}(j)) < r$.
- Modified permutation entropy (*mPE*): the permutation entropy algorithm quantifies the occurrence of motifs in the series. Motifs are defined as recurring patterns in the time series with a degree m and lag λ. Based on the rank ordering of the motif pattern we assign it a specific symbol j. Representative motifs of degree ($m = 3$) and lag ($\lambda = 1$) are shown in Figure 5. For modified permutation entropy, to account for stationary consecutive beats, four additional symbolic representations have been added (from 7 to 10 as shown in the equation below). The time series ($X(t)$) is first converted to the ordinal series ($X^{m,\lambda}(j)$), where $1 \geq j \leq N - m$ where N is the size of the time series using the following relations:

$$X^{m,\lambda}(j) = \begin{cases} 1 & \text{if } X(i) < X(i+\lambda) \;\&\; X(i+\lambda) < X(i+2\lambda) \;\&\; X(i) < X(i+2\lambda) \\ 2 & \text{if } X(i) < X(i+\lambda) \;\&\; X(i+\lambda) > X(i+2\lambda) \;\&\; X(i) < X(i+2\lambda), \\ 3 & \text{if } X(i) > X(i+\lambda) \;\&\; X(i+\lambda) < X(i+2\lambda) \;\&\; X(i) < X(i+2\lambda), \\ 4 & \text{if } X(i) < X(i+\lambda) \;\&\; X(i+\lambda) > X(i+2\lambda) \;\&\; X(i) > X(i+2\lambda), \\ 5 & \text{if } X(i) > X(i+\lambda) \;\&\; X(i+\lambda) > X(i+2\lambda) \;\&\; X(i) > X(i+2\lambda), \\ 6 & \text{if } X(i) > X(i+\lambda) \;\&\; X(i+\lambda) < X(i+2\lambda) \;\&\; X(i) > X(i+2\lambda). \\ 7 & \text{if } X(i) == X(i+\lambda) \;\&\; X(i+\lambda) < X(i+2\lambda). \\ 8 & \text{if } X(i) == X(i+\lambda) \;\&\; X(i+\lambda) > X(i+2\lambda). \\ 9 & \text{if } X(i) > X(i+\lambda) \;\&\; X(i+\lambda) == X(i+2\lambda). \\ 10 & \text{if } X(i) < X(i+\lambda) \;\&\; X(i+\lambda) == X(i+2\lambda). \end{cases}$$

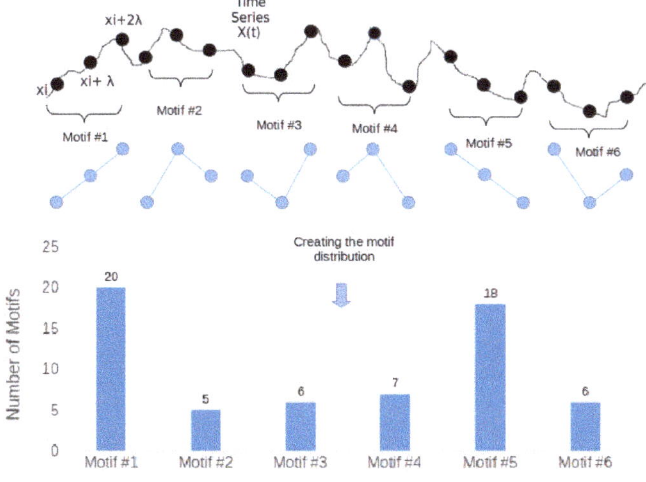

Figure 5. Original motifs of degree $m = 3$ appearing in a time series.

The modified permutation entropy (mPE) is then given by:

$$mPE = -\sum_{j}^{m!+n} p(\pi_j^{m,\lambda}) \cdot \log(p(\pi_j^{m,\lambda})), \tag{7}$$

where n is the number of additional motif patterns added for the modified permutation entropy and $p(\pi_j^{m,\lambda})$ is the relative frequency of the motif pattern represented by $\pi_j^{m,\lambda}$ and calculated as:

$$p(\pi_j^{m,\lambda}) = \frac{\sum_{j \le m!+n} \Bbbk_{u:type(u)=\pi_j}(X_j^{m,\lambda})}{\sum_{j \le m!+n} \Bbbk_{u:type(u)\in\Pi}(X_j^{m,\lambda})}, \tag{8}$$

where $\Bbbk_A(u)$ denotes the indicator function of set A defined as $\Bbbk_A(u) = 1$ if $u \in A$ and $\Bbbk_A(u) = 0$ otherwise and $type(.)$ denotes the map from pattern space to symbol space.

- Weighted modified permutation entropy (mPE_wt): weighted permutation entropy was proposed to incorporate the amplitude information into the permutation entropy algorithm. This is done by calculating the variances (referred as weights w_i) corresponding to each motif pattern in the time series. The relative frequency $p(\pi_i^{m,\lambda})$ of a given pattern is then calculated as:

$$p(\pi_j^{m,\lambda}) = \frac{\sum_{j\leq m!+n} \mathbb{1}_{u:type(u)=\pi_i}(X_j^{m,\lambda})w_j}{\sum_{j\leq m!+n} \mathbb{1}_{u:type(u)\in\Pi}(X_j^{m,\lambda})w_j}. \tag{9}$$

The permutation entropy for this adjusted relative frequency is then calculated using the standard permutation entropy equation in (8).

The entropy values were calculated for both the RR series and the dRR series for a scale of $s = 1, \ldots, 10$. Moreover, the mean and standard deviation of the entropy measures across all scales were calculated, thus resulting in 24 total features for each type of entropy measure.

2.4.3. Ordinal Distance Dissimilarity

Ordinal distance based dissimilarity [42] can be used to calculate the difference between the two ordinal series. The distance between two ordinal series X and Y is given by

$$D_m(X,Y) = \sqrt{\frac{m!}{m!-1}} \sqrt{\sum_j^{m!} (p_x(\pi_j^{m,\lambda}) - p_y(\pi_j^{m,\lambda}))^2}, \tag{10}$$

where $p_x(\pi^{m,\lambda})$ and $p_y(\pi^{m,\lambda})$ are the relative frequencies of the motif pattern represented by $\pi^{m,\lambda}$ in series X and Y, respectively, and m is the degree of the motif. As the RR series has statistical fractal properties (the statistical properties over different scales do not change), the ordinal distance has been calculated over the different scaled series (referred to as the inter-scale ordinal distance ($isod_{s_x,s_y}$)). To limit the feature space size, we have calculated the distances between scales $s_x = 1$ *to* $s_x = 3$ and $s_y = 1$ *to* $s_y = 10$ with $s_x \neq s_y$. This was done for both the RR and dRR series. Additionally, we calculate the statistics of *isod* for given $s_x = 1$ to $s_x = 3$ relative to all s_y, we calculate the mean, standard deviation and first difference of $isod_{s_x}$, thus resulting in a total number of ordinal distance features of 66. For simplicity, only the original permutation entropy based motif structures are considered and the scaling is done by the moving average scaling algorithm.

2.5. Feature Selection and Classification

For evaluation, a five-fold cross validation setup was used. Workload assessment was performed as a binary classification task, where the high and low mental workload ground truth labels are taken from the from the MATB-II task. A support vector machine (SVM) classifier with an RBF kernel was used. To explore the generalization performance, the above mentioned procedure is repeated 50 times with different random seeds. This leads to 250 (5-folds times 50 repetitions with different random seeds) training and test sets and classifications. To assess feature importance, we used feature selection and look at the frequency of features occurring in the top 20 sets for the 250 possible combinations.

To assess feature importance, recursive feature elimination was performed using the extra trees classifier [43]. Given an external estimator that assigns weights to features (an extra trees classifier in this case), the least important features are pruned from the current set of features. The procedure is recursively repeated on the pruned set until the desired number of features to be selected is reached. This technique considers the interaction of features with the learning algorithm to give the optimal subset of features. The feature selection is used to select the top 20 features for each fold of the cross-validation set. The implementation of the classifier and feature selection algorithms was done using sci-kit learn [44].

3. Results and Discussion

The performance of the different entropy and scaling methods, the inter-scale ordinal distance and benchmark features were compared for mental workload assessment. Comparison was done for different levels of physical workload, thus allowing for the robustness of the features to be assessed relative to increases in movement artifacts and changing dynamics of the heart rate brought on by physical activity. We first compared the performance of the different combinations of entropy and scaling approaches for different activity levels. The best performing algorithms were then compared to the performance of ordinal distance scale similarity measure and benchmark features. Additionally, we performed feature fusion where all the different feature sets were combined to test for feature set complementarity.

3.1. Comparing Different Multi-Scale Entropy Algorithms

We calculated the accuracy (Acc) for different activity levels with the combinations of the three entropy measures and five scaling algorithms. Figures 6–8 show the performance of the algorithms for no, medium and high physical activity levels, respectively. As can be seen, generally across all physical activity cases and for all entropy algorithms, the short time moving average based scaling (mov_avg and $mavg\_mom$) and composite based scaling ($comp\_cg$) methods outperform coarse graining based approaches (cg, mom). It can also be seen that the modified permutation entropy based on second moment based scaling methods (mom and $mavg\_mom$) (referred to as generalized permutation entropy in [24]) typically achieve higher predictive power across all physical workload cases, hence indicating the importance of volatility series of the RR series, as well as dRR series. Lastly, the modified permutation entropy based algorithms (mPE and mPE_wt) performs better than sample entropy based methods.

Specifically, looking at the performance of the features across different physical activity level conditions, it can be seen for the no physical activity condition that mPE with ($comp\_cg$ and $mavg_{m}om$) achieved significantly higher performance ($p < 0.01$) than most of the other methods. Moreover, by including the amplitude information into the mPE via the mPE_{wt} measure, a drop in performance is seen for the moving average scaled RR series, but the best performance for the volatility scaled series is achieved with our proposed moving average scaling of the second moment ($mavg\_mom$). These results suggest that amplitude information of the volatility series is important for mental workload assessment.

For the medium physical activity level condition, we observe similar performance trends to the no physical activity condition with mPE ($comp\_cg$ scaling) achieving significantly higher performance ($p < 0.01$) than the other methods. Interestingly, incorporating the amplitude information in this condition leads to a decrease in performance for both $comp\_cg$ and $mavg\_mom$. This could be due to the changing cardiac dynamics during physical activity.

Finally, for the high physical activity level condition, we see an overall drop in performance compared to the other two physical activity levels. We observe that the $SampEn$ performance is comparable to mPE and mPE_{wt} for certain scaling methods (cg, $mavg$ and $comp\_cg$). We also see a drop in performance for mPE when using $mavg$ compared to cg scaling, though this is not observed for $SampEn$ and mPE_wt, both of which show improvement on using the $mavg$ scaling. Overall, we achieve the best performance using the modified permutation entropy for proposed short time second moment calculation and $comp\_cg$, using the mPE, i.e., excluding the amplitude information. Higher performance without incorporating amplitude information (not using mPE_{wt}) in both medium and high physical activity level conditions could be due to higher noise in the RR series arising from misdetections in the QRS complex caused by movement artifacts. As mPE_{wt} is sensitive to such artifacts, ignoring the amplitude information is better in such cases. Overall, mPE with $comp\_cg$ and $mavg\_mom$ achieved the best results, with $mavg\_mom$ based scaling giving significantly higher results than all the other methods tested ($p < 0.01$).

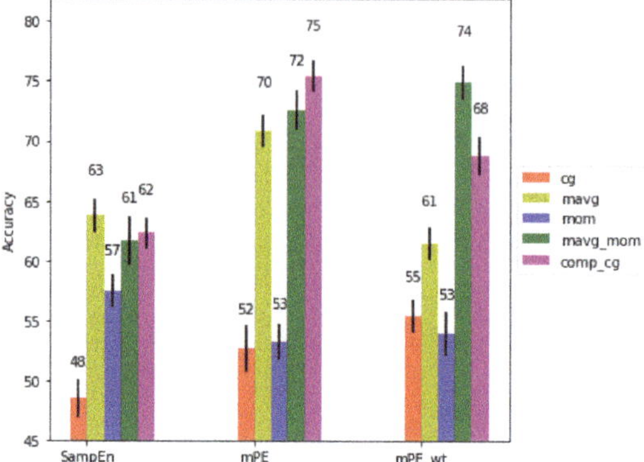

Figure 6. Performance comparison for the no physical activity condition.

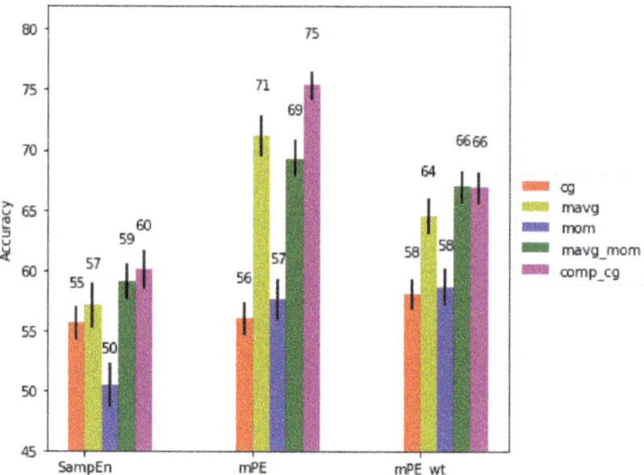

Figure 7. Performance comparison for the medium physical activity condition.

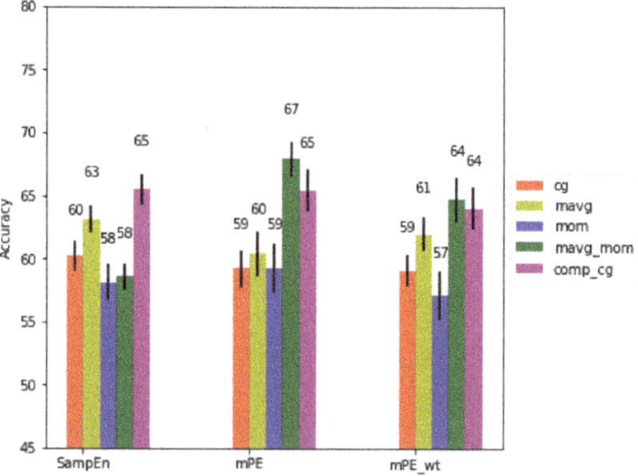

Figure 8. Performance comparison for the high physical activity condition.

Moreover, as [24] emphasizes the complementary nature of the volatility series, we further investigate the fusion of *comp_cg* and *mavg_mom* base scaling methods using the *mPE* algorithm, as these resulted in consistently better performance across all three physical activity conditions. Table 3 shows the results of fusion for the different physical activity levels. As can be seen, for the no and medium physical activity levels, fusion gave a significant ($p < 0.01$) improvement of 3.53% in accuracy and 3.30% in f1-score and 1.90% accuracy and 1.63% f1-score, respectively, over the best performing $comp\_cg$ + mPE algorithm. However, no improvement was seen for high physical activity level. Such findings corroborate those of [24].

Table 3. Performance of fused $comp\_cg$ and $mavg\_mom$ scaling with mPE algorithm for different physical workload levels (* represents cases which perform significantly better ($p < 0.01$) than chance).

Physical Activity Level	Acc	F1
No	0.7893 ± 0.0122 *	0.7886 ± 0.0131 *
Medium	0.7726 ± 0.0114 *	0.7701 ± 0.0111 *
High	0.6741 ± 0.0150 *	0.6698 ± 0.0164 *

3.2. Gauging Performance Against the Benchmark

Here, we compare the performance of the best performing algorithms from Section 3.1 with the benchmark features. We further perform feature fusion of the two sets to further explore their complementary. Tables 4–6 shows the benchmark, inter-scale ordinal distance, and best performing multi-scale entropy methods and their fusion for no, medium and high physical activity levels, respectively. In the tables, 'nof' indicates number of features used in each case.

Table 4. Benchmark performance comparison for the no physical activity condition (* represents cases which perform significantly better ($p < 0.01$) than chance).

Feature (Nof)	Acc	F1
benchmark (15)	0.5772 ± 0.0192 *	0.4991 ± 0.0206
isod (66)	0.7838 ± 0.0137 *	0.7882 ± 0.0137 *
multi-scale entropy (48)	0.7893 ± 0.0122 *	0.7886 ± 0.0131 *
fused (129)	0.8438 ± 0.0126 *	0.8428 ± 0.0132 *

Table 5. Benchmark performance comparison for the medium physical activity condition (* represents cases which perform significantly better ($p < 0.01$) than chance).

Feature (Nof)	Acc	F1
benchmark (15)	0.5318 ± 0.0169 *	0.6019 ± 0.0231 *
isod (66)	0.8189 ± 0.0133 *	0.8203 ± 0.0134 *
multi-scale entropy (48)	0.7726 ± 0.0114 *	0.7701 ± 0.0111 *
fused (129)	0.8401 ± 0.0128 *	0.8410 ± 0.0129 *

Table 6. Benchmark performance comparison for the high physical activity condition (* represents cases which perform significantly better ($p < 0.01$) than chance).

Feature (Nof)	Acc	F1
benchmark (15)	0.5751 ± 0.0160 *	0.5393 ± 0.0245 *
isod (66)	0.7825 ± 0.0128 *	0.7818 ± 0.0137 *
multi-scale entropy (48)	0.6741 ± 0.0150 *	0.6698 ± 0.0164 *
fused (129)	0.8015 ± 0.0152 *	0.7987 ± 0.0156 *

As can be seen, for the no physical activity condition, both multi-scale entropy and the inter-scale ordinal distance features perform significantly better than the benchmark with improvements of 21.21% in accuracy and 28.25% in f1-score and 20.66% in accuracy and 28.91% in f1-score, respectively. Additionally, fusion provides significant ($p < 0.01$) improvements of 5.45% accuracy and 5.42% f1-score over the multi-scale features alone.

Similarly for medium physical activity levels, both multi-scale entropy and the inter-scale ordinal distance features performed significantly better than the benchmark with improvements of 24.08% in accuracy and 16.82% in f1-score and 27.20% in accuracy and 21.84% in f1-score, respectively. In this case, the inter-scale ordinal distance features performed significantly better ($p < 0.01$) than the multi-scale entropy features. Fusion also improved performance significantly ($p < 0.01$) and results in gains of 2.12% accuracy and 2.07% f1-score over the inter-scale ordinal distance features. Lastly, for the high physical activity level condition, both multi-scale entropy and the inter-scale ordinal distance features performed significantly better than the benchmark with improvements of 9.90% in accuracy and 13.05% in f1-score and 20.74% in accuracy and 24.25% in f1-score, respectively. Again, the inter-scale ordinal distance features performed significantly better ($p < 0.01$) than the multi-scale entropy features. Additionally, fusion gave a significant ($p < 0.05$) improvement of 1.90% accuracy and 1.69% f1-score over the inter-scale ordinal distance features. The improvement in performance achieved with fusion for all three activity levels further corroborates the results of [24].

3.3. Feature Ranking

Feature importance was computed based on the outcomes of feature selection across the five cross validation steps, repeated 50 times. The top 20 features were selected for every fold. As such, the frequency of occurrence of a given feature in the top feature set was calculated over the 250 iterations. Features appearing more than 70% were further ranked according to their frequency of occurrence ($freq$) for no, medium and high physical activity levels. These values are reported in Tables 7–9, respectively, along with the feature names.

Table 7. Most frequently occurring features in the top-20 feature pool for the no physical activity condition.

Feature Name	$freq$
mean of RR	99.2
$isod_{s1,s4}$ dRR	98.8
$isod_{s3,s9}$ RR	98.4
Coefficient of variation	93.2
lf/hf	82.8
std. absolute first difference RR	81.2
mean $isod_{s3,:}$ RR	80
$(mPE + comp\_cg)_{s3}$ RR	70.4

Table 8. Most frequently occurring features in the top-20 feature pool for the medium physical activity condition.

Feature Name	$freq$
$isod_{s1,s2}$ RR	99.2
$isod_{s1,s7}$ RR	99.2
$isod_{s2,s10}$ RR	98
mean RR	95.2
$(mPE + mavg\_mom)_{s1}$ RR	91.2
$(mPE + comp\_cg)_{s9}$ RR	89.6
$(mPE + comp\_cg)_{s1}$ RR	88.8
$isod_{s1,s4}$ dRR	83.2
$(mPE + comp\_cg)_{s10}$ RR	81.6
$(mPE + mavg\_mom)_{s8}$ dRR	74.4
$isod_{s2,s5}$ RR	74.4
$(mPE + mavg\_mom)_{s8}$ RR	70.4

Table 9. Most frequently occurring features in the top-20 feature pool for the high physical activity condition.

Feature Name	*freq*
mean abs. first difference RR	100
$(mPE + comp\_cg)_{s8}$ RR	96
lfnu	95.2
$(mPE_{wt} + mavg\_mom)_{s3}$ dRR	93.6
hfnu	90.8
$(mPE + comp\_cg)_{s4}$ RR	90.8
$isod_{s3,s4}$ RR	84.8
$isod_{s2,s9}$ RR	79.6
$(mPE + comp\_cg)_{s6}$ dRR	79.6
$(mPE + comp\_cg)_{s2}$ dRR	70.8

As can be seen, for the no physical activity condition, we observe that three of the eight top-ranked features are from the inter-scale ordinal distance feature set with interaction of different scales with $s = 3$ being the case for two of the three *isod* features. Additionally, one multi-scale mPE features show up in the most frequent set with composite scaling based entropy of $s = 3$, Additionally, we see four benchmark features in the top feature set, with three statistical features as well as the ratio of low to high frequency (lf/hf). A consistent decrease in the mean of RR has been reported in the literature with increased mental stress, a similar trend in the standard deviation of RR intervals could explain the overall importance of the coefficient of variation which is a ratio of the two [41]. Similarly, an increase in the lf/hf ratio was reported for increased mental workload across various studies [41]. We also observed that one of the proposed feature was calculated over the dRR series (*isod* feature), reflecting the presence of long-term correlations and complexity in the magnitude difference of RR series as noted in [29]. The presence of the different feature sets along with the benchmark features further corroborates the complementary nature of the features.

Similarly, looking at medium physical activity level conditions, we observe that of the 12 most frequent features, five were from inter-scale ordinal distance features with interactions between $s = 1$ and $s = 2$ with other scales. Further, additionally six multi-scale mPE features showed up in the most frequent set with composite scaling based entropy of the original time series ($s = 1$) (same as for mov_mom with $s = 1$ as well), along with $s = 9$ and $s = 10$ as well as of second moment from $s = 5$ and $s = 9$. Additionally, scales $s = 8$ for the mPE of second moment also shows up in the top features for both the RR and dRR series. We only have one benchmark feature (mean RR) in this case among the top features, thus suggesting their sensitivity to movement artifacts. One of features were calculated on the dRR series.

Lastly, for the high physical activity level conditions, of the top 10 most occurring features, we observed that two features are from the inter-scale ordinal distance features, five features are from the multi-scale mPE entropy with composite scaling based entropy of the scales $s = 2$, $s = 4$, $s = 6$ and $s = 8$ as well as of second moment from $s = 3$. Interestingly, no entropy feature from the original time series ($s = 1$) was seen in the top features. We also observed three benchmark features in the top set, with both normalized low and high frequency along with mean of absolute first difference.

Mental workload has reported a drop in HRV features [45,46] attributed to sympathetic activation and/or para-sympathetic withdrawal [45–48]. Time and frequency domain HRV features were focused on characterizing the balance between these two systems. However, a lack of clear unbalance of the ANS due to mental workload has been reported in the literature [49]. This has shifted focus on the use of non-linear descriptors based on complex systems approach to better characterize the fractal RR time series [15]. These methods often characterize the complexity of the RR time series [50]. Recent studies indicate that this complexity is a result of both sympathetic and parasympathetic components of the ANS [51]. A recent study [49] has shown correlation dimension, which measures the fractal self-similarity of signal, decrease to comparable pathological values during mental workload inducing tasks, which indicates an suppression of the parasympathetic activity in the heart [52] and breakdown of long term correlations in the RR series [14] which can be quantified by complexity at higher scales [26].

A few studies have looked at the effects of exercise on HRV features. [53] reported an increase in overall complexity due to walking (4 km/hr) along with a significant increase in normalized low frequency power and a decrease in normalized high frequency power. A similar trend for low intensity exercise was reported in [54] with a contradictory increase in the high frequency component with increased exercise intensity on a bicycle. This increase has further been explained by the influence of breathing on heart rate (respiratory sinus arrhythmia (RSA)) which has a strong high frequency component during high intensity exercise [55]. Interestingly, when looking at the non-linear properties of the heart rate for high intensity exercise, entropy ($scale = 1$) decreases while complexity is still retained at different scales [56], something that can be exploited by multi-scale entropy measure.

The scaling process for the multi-scale entropy algorithm is equivalent to low pass filtered frequency bands with decreasing bandwidth with increasing scales [57]. This scaling can be achieved by different types of scaling operations. For our study we have focused on two methods, namely composite coarse graining and moving average scaling methods. Given the presence of two distinct frequency regions in the heart rate due to parasympathetic activity (corresponding to high frequency fluctuations in the RR series) and sympathetic activity (corresponding to lower frequency fluctuations) [13], the multiscale entropy algorithm represents the complexity of the overall series due to interaction of both sympathetic and parasympathetic systems at lower scales, while representing the complexity of lower frequency component (mostly due to sympathetic activity) at higher scales. Furthermore, the inter-scale ordinal distance feature tries to quantify the complex interaction between the different frequency regions.

In keeping with the above variations in ANS balance with mental workload and exercise we observe a scale $s = 3$ show importance for no physical workload case which captures more lower frequency information compared to original scale. With medium physical activity further contributing to the increase lower frequency components in mental workload we observe entropy of higher scale of $s = 8$ to $s = 10$ (capturing low frequency information) show up in the top feature sets. Finally for high physical activity where high frequency components show important contribution due to the influence of RSA to the heart rate we see both low ($s = 2$ and $s = 3$) and high ($s = 4$, $s = 6$ and $s = 8$) scales for entropy show significance in the top feature set. Additionally we see the normalized high and low frequency components among the top features which show significance during exercise [53]. We hypothesize that the RSA component which usually causes cardio-respiratory coherence is disrupted due to added mental workload [58] making these features important for distinguishing between the two states. The inter-scale ordinal distance feature also shows significance for all three physical activity levels hinting at non-linear interaction between the different frequency regions. The presence of features from the dRR series show the importance of complementary non-linear information present in the series which should be further investigated. Finally, the importance of generalized entropy features calculated on the volatility series hints at the multifractal characteristics holding vital information regarding mental workload. The link between generalized entropy and multifractal heart rate characteristics has been hypothesized in [24].

4. Conclusions

Traditional time and frequency HRV methods don't account for processes occurring at multiple time scales leading to complex behavior of heart rate at multiple scales. In this study, we explored the use of several multi-scale features for mental workload assessment for ambulatory users. We show that the multi-scale entropy features are robust to changing heart rate dynamics due to movement and activity and do better than the benchmark statistical and frequency domain features while also providing complementary information. This hints at the physiological processes at various time scales being effected by mental workload.

A number of different multi-scale entropy methods and scaling algorithms were tested and we found that, generally, composite coarse graining via a new second moment moving average scaling method, combined with the modified permutation entropy method outperformed other combinations, thus suggesting that multi-scale entropy methods specifically designed for short term time series are important for short term HRV analysis. Overall, the results reported herein suggest that increased workload can result in changes in ECG signal complexity of higher scales of the fractal RR series. These findings suggest that the long-term processes involved in heart rate regulation may be affected by mental workload changes. These findings allow for sufficiently accurate assessment of mental workload in safety-critical applications where users are ambulant, especially at moderate levels of physical activity. For future work, context aware models which can distinguish between different physical activity levels followed by mental workload classification could be developed.

Author Contributions: conceptualization, A.T., J.-F.G., D.L., S.T. and T.H.F.; methodology, A.T., M.P. , J.-F.G., D.L., S.T. and T.H.F; Software, A.T.; data curation, A.T., I.A., M.P. ; validation, A.T, I.A.; formal analysis, A.T., I.A., M.P., J.-F.G., D.L., S.T. and T.H.F.; investigation, A.T., M.P. , J.-F.G., D.L., S.T. and T.H.F.; resources, J.-F.G., D.L., S.T. and T.H.F.; writing—original draft preparation, A.T.; writing—review and editing, M.P., I.A., J.-F.G., D.L., S.T. and T.H.F..; visualization, A.T.; supervision, T.H.F.; project administration, J.-F.G., D.L., S.T. and T.H.F.; funding acquisition, J.-F.G., D.L., S.T. and T.H.F.

Funding: This research was funded by NSERC, Thales Canada, PROMPT, and MITACS.

Acknowledgments: We would like to acknowledge the Perform Center, University of Concordia (Loyola Campus) for providing lab space and certain equipment for carrying out the data collection.

Conflicts of Interest: The authors declare no conflict of interest. The funders had no role in the design of the study; in the collection, analyses, or interpretation of data; in the writing of the manuscript, or in the decision to publish the results.

Abbreviations

The following abbreviations are used in this manuscript:

RSA	Respiratory sinus arrhythmia
RR	inter-beat interval series
dRR	Absolute difference inter-beat interval series
ANS	Autonomic nervous system
NASA-TLX	NASA task load index
SWAT	Subjective workload assessment technique
ECG	Electrocardiogram
MSE	Multi-scale entropy
PE	Permutation entropy
mPE	Modified permutation entropy
HRV	Heart rate variability
MATB	Multi-attribute task battery

References

1. Boff, K.; Kaufman, L.; Thomas, J. *Handbook of Perception and Human Performance;* Wiley-Interscience: Hoboken, NJ, USA, 1986; Volume 2; pp. 42–45.
2. Hancock, P.; Caird, J.K. Experimental evaluation of a model of mental workload. *Hum. Factors* **1993**, *35*, 413–429. [CrossRef] [PubMed]
3. Sheridan, T.B.; Simpson, R. *Toward the Definition and Measurement of the Mental Workload of Transport Pilots;* Technical Report; Massachusetts Institute of Technology, Department of Aeronautics and Astronautics, Flight Transportation Laboratory: Cambridge, MA, USA, 1979.
4. Reid, G.B.; Eggemeier, F.T.; Shingledecker, C.A. *Subjective Workload Assessment Technique;* Technical Report; Air Force Flight Test Center: Edwards AFB, CA, USA, 1982.
5. Hart, S.G.; Staveland, L.E. Development of NASA-TLX (Task Load Index): Results of empirical and theoretical research. In *Advances in Psychology;* Elsevier: Amsterdam, The Netherlands, 1988; Volume 52, pp. 139–183.

6. Cain, B. *A Review of the Mental Workload Literature*; Technical Report; Defence Research And Development: Toronto, ON, Canada, 2007.
7. Cassenti, D.N.; Kelley, T.D.; Carlson, R.A. Modeling the workload-performance relationship. In Proceedings of the Human Factors and Ergonomics Society Annual Meeting, Austin, TX, USA, 4–7 October 2010; SAGE Publications: Sage/Los Angeles, CA, USA, 2010; Volume 54, pp. 1684–1688.
8. Matthews, G.; Reinerman-Jones, L.E.; Barber, D.J.; Abich, J., IV. The psychometrics of mental workload: Multiple measures are sensitive but divergent. *Hum. Factors* **2015**, *57*, 125–143. [CrossRef] [PubMed]
9. Albuquerque, I.; Tiwari, A.; Gagnon, J.F.; Lafond, D.; Parent, M.; Tremblay, S.; Falk, T. On the Analysis of EEG Features for Mental Workload Assessment During Physical Activity. In Proceedings of the 2018 IEEE International Conference on Systems, Man, and Cybernetics (SMC), Miyazaki, Japan, 7–10 October 2018; pp. 538–543.
10. Togo, F.; Takahashi, M. Heart Rate Variability in Occupational Health-A Systematic Review. *Ind. Health* **2009**, *47*, 589–602. [CrossRef] [PubMed]
11. Wen, W.H.; Liu, G.Y.; Mao, Z.H.; Huang, W.J.; Zhang, X.; Hu, H.;Yang, J.; Jia, W. Toward Constructing a Real-time Social Anxiety Evaluation System: Exploring Effective Heart Rate Features. *IEEE Trans. Affect. Comput.* **2018**. [CrossRef]
12. Patel, M.; Lal, S.K.; Kavanagh, D.; Rossiter, P. Applying neural network analysis on heart rate variability data to assess driver fatigue. *Expert Syst. Appl.* **2011**, *38*, 7235–7242. [CrossRef]
13. Camm, A.J.; Malik, M.; Bigger, J.T.; Breithardt, G.; Cerutti, S.; Cohen, R.J.; Coumel, P.; Fallen, E.L.; Kennedy, H.L.; Kleiger, R.E.; et al. Heart rate variability: Standards of measurement, physiological interpretation and clinical use. Task Force of the European Society of Cardiology and the North American Society of Pacing and Electrophysiology. *Circulation* **1996**, *93*, 1043–1065.
14. Goldberger, A.L.; Amaral, L.A.; Hausdorff, J.M.; Ivanov, P.C.; Peng, C.K.; Stanley, H.E. Fractal dynamics in physiology: Alterations with disease and aging. *Proc. Natl. Acad. Sci. USA* **2002**, *99*, 2466–2472. [CrossRef]
15. Sassi, R.; Cerutti, S.; Lombardi, F.; Malik, M.; Huikuri, H.V.; Peng, C.K.; Schmidt, G.; Yamamoto, Y.; Reviewers:, D.; Gorenek, B.; et al. Advances in heart rate variability signal analysis: Joint position statement by the e-Cardiology ESC Working Group and the European Heart Rhythm Association co-endorsed by the Asia Pacific Heart Rhythm Society. *Ep Eur.* **2015**, *17*, 1341–1353. [CrossRef]
16. Costa, M.; Goldberger, A.L.; Peng, C.K. Multiscale entropy analysis of complex physiologic time series. *Phys. Rev. Lett.* **2002**, *89*, 068102. [CrossRef]
17. Humeau-Heurtier, A. The multiscale entropy algorithm and its variants: A review. *Entropy* **2015**, *17*, 3110–3123. [CrossRef]
18. Wu, S.D.; Wu, P.H.; Wu, C.W.; Ding, J.J.; Wang, C.C. Bearing fault diagnosis based on multiscale permutation entropy and support vector machine. *Entropy* **2012**, *14*, 1343–1356. [CrossRef]
19. Li, Q.; Zuntao, F. Permutation entropy and statistical complexity quantifier of nonstationarity effect in the vertical velocity records. *Phys. Rev. E* **2014**, *89*, 012905. [CrossRef] [PubMed]
20. Bandt, C.; Pompe, B. Permutation entropy: A natural complexity measure for time series. *Phys. Rev. Lett.* **2002**, *88*, 174102. [CrossRef] [PubMed]
21. Zanin, M.; Zunino, L.; Rosso, O.A.; Papo, D. Permutation entropy and its main biomedical and econophysics applications: A review. *Entropy* **2012**, *14*, 1553–1577. [CrossRef]
22. Wu, S.D.; Wu, C.W.; Lee, K.Y.; Lin, S.G. Modified multiscale entropy for short-term time series analysis. *Phys. A Stat. Mech. Its Appl.* **2013**, *392*, 5865–5873. [CrossRef]
23. Wu, S.D.; Wu, C.W.; Lin, S.G.; Wang, C.C.; Lee, K.Y. Time series analysis using composite multiscale entropy. *Entropy* **2013**, *15*, 1069–1084. [CrossRef]
24. Costa, M.; Goldberger, A. Generalized multiscale entropy analysis: Application to quantifying the complex volatility of human heartbeat time series. *Entropy* **2015**, *17*, 1197–1203. [CrossRef] [PubMed]
25. Kalisky, T.; Ashkenazy, Y.; Havlin, S. Volatility of linear and nonlinear time series. *Phys. Rev. E* **2005**, *72*, 011913. [CrossRef] [PubMed]
26. Costa, M.D.; Peng, C.K.; Goldberger, A.L. Multiscale analysis of heart rate dynamics: Entropy and time irreversibility measures. *Cardiovasc. Eng.* **2008**, *8*, 88–93. [CrossRef]
27. Costa, M.; Healey, J. Multiscale entropy analysis of complex heart rate dynamics: discrimination of age and heart failure effects. In Proceedings of the Computers in Cardiology, Thessaloniki Chalkidiki, Greece, 21–24 September 2003; pp. 705–708.

28. Ho, Y.L.; Lin, C.; Lin, Y.H.; Lo, M.T. The prognostic value of non-linear analysis of heart rate variability in patients with congestive heart failure—A pilot study of multiscale entropy. *PLoS ONE* **2011**, *6*, e18699. [CrossRef]
29. Ashkenazy, Y.; Ivanov, P.C.; Havlin, S.; Peng, C.K.; Goldberger, A.L.; Stanley, H.E. Magnitude and sign correlations in heartbeat fluctuations. *Phys. Rev. Lett.* **2001**, *86*, 1900. [CrossRef] [PubMed]
30. Liu, C.; Gao, R. Multiscale entropy analysis of the differential RR interval time series signal and its application in detecting congestive heart failure. *Entropy* **2017**, *19*, 251.
31. Bian, C.; Qin, C.; Ma, Q.D.; Shen, Q. Modified permutation-entropy analysis of heartbeat dynamics. *Phys. Rev. E* **2012**, *85*, 021906. [CrossRef] [PubMed]
32. Fadlallah, B.; Chen, B.; Keil, A.; Príncipe, J. Weighted-permutation entropy: A complexity measure for time series incorporating amplitude information. *Phys. Rev. E* **2013**, *87*, 022911. [CrossRef] [PubMed]
33. Xia, Y.; Yang, L.; Zunino, L.; Shi, H.; Zhuang, Y.; Liu, C. Application of Permutation Entropy and Permutation Min-Entropy in Multiple Emotional States Analysis of RRI Time Series. *Entropy* **2018**, *20*, 148. [CrossRef]
34. Santiago-Espada, Y.; Myer, R.R.; Latorella, K.A.; Comstock, J.R., Jr. *The Multi-Attribute Task Battery II (MATB-II) Software for Human Performance and Workload Research: A User's Guide*; NASA: Washington, DC, USA, 2011.
35. Mahdiani, S.; Jeyhani, V.; Peltokangas, M.; Vehkaoja, A. Is 50 Hz high enough ECG sampling frequency for accurate HRV analysis? In Proceedings of the 2015 37th Annual International Conference of the IEEE Engineering in Medicine and Biology Society (EMBC), Milano, Italy, 25–29 August 2015; pp. 5948–5951.
36. Merri, M.; Farden, D.C.; Mottley, J.G.; Titlebaum, E.L. Sampling frequency of the electrocardiogram for spectral analysis of the heart rate variability. *IEEE Trans. Biomed. Eng.* **1990**, *37*, 99–106. [CrossRef] [PubMed]
37. El-Yaagoubi, M.; Goya-Esteban, R.; Jabrane, Y.; Muñoz-Romero, S.; García-Alberola, A.; Rojo-Álvarez, J.L. On the Robustness of Multiscale Indices for Long-Term Monitoring in Cardiac Signals. *Entropy* **2019**, *21*, 594. [CrossRef]
38. Behar, J.; Johnson, A.; Clifford, G.D.; Oster, J. A comparison of single channel fetal ECG extraction methods. *Ann. Biomed. Eng.* **2014**, *42*, 1340–1353. [CrossRef]
39. Pan, J.; Tompkins, W.J. A real-time QRS detection algorithm. *IEEE Trans. Biomed. Eng* **1985**, *32*, 230–236. [CrossRef]
40. Behar, J.A.; Rosenberg, A.A.; Weiser-Bitoun, I.; Shemla, O.; Alexandrovich, A.; Konyukhov, E.; Yaniv, Y. PhysioZoo: A novel open access platform for heart rate variability analysis of mammalian electrocardiographic data. *Front. Physiol.* **2018**, *9*, 1390. [CrossRef]
41. Castaldo, R.; Melillo, P.; Bracale, U.; Caserta, M.; Triassi, M.; Pecchia, L. Acute mental stress assessment via short term HRV analysis in healthy adults: A systematic review with meta-analysis. *Biomed. Signal Process. Control* **2015**, *18*, 370–377. [CrossRef]
42. Ouyang, G.; Dang, C.; Richards, D.A.; Li, X. Ordinal pattern based similarity analysis for EEG recordings. *Clin. Neurophysiol.* **2010**, *121*, 694–703. [CrossRef] [PubMed]
43. Geurts, P.; Ernst, D.; Wehenkel, L. Extremely randomized trees. *Mach. Learn.* **2006**, *63*, 3–42. [CrossRef]
44. Pedregosa, F.; Varoquaux, G.; Gramfort, A.; Michel, V.; Thirion, B.; Grisel, O.; Blondel, M.; Prettenhofer, P.; Weiss, R.; Dubourg, V.; et al. Scikit-learn: Machine learning in Python. *J. Mach. Learn. Res.* **2011**, *12*, 2825–2830.
45. Wang, Z.; Yang, L.; Ding, J. Application of heart rate variability in evaluation of mental workload. *Zhonghua Lao Dong Wei Sheng Zhi Ye Bing Za Zhi = Zhonghua Laodong Weisheng Zhiyebing Zazhi = Chin. J. Ind. Hyg. Occup. Dis.* **2005**, *23*, 182–184.
46. Hjortskov, N.; Rissén, D.; Blangsted, A.K.; Fallentin, N.; Lundberg, U.; Søgaard, K. The effect of mental stress on heart rate variability and blood pressure during computer work. *Eur. J. Appl. Physiol.* **2004**, *92*, 84–89. [CrossRef]
47. Taelman, J.; Vandeput, S.; Vlemincx, E.; Spaepen, A.; Van Huffel, S. Instantaneous changes in heart rate regulation due to mental load in simulated office work. *Eur. J. Appl. Physiol.* **2011**, *111*, 1497–1505. [CrossRef]
48. Collins, S.M.; Karasek, R.A.; Costas, K. Job strain and autonomic indices of cardiovascular disease risk. *Am. J. Ind. Med.* **2005**, *48*, 182–193. [CrossRef]
49. Chaumet, G.; Delaforge, A.; Delliaux, S. Mental workload alters heart rate variability lowering non-linear dynamics. *Front. Physiol.* **2019**, *10*, 565.
50. Melillo, P.; Bracale, M.; Pecchia, L. Nonlinear Heart Rate Variability features for real-life stress detection. Case study: Students under stress due to university examination. *Biomed. Eng. Online* **2011**, *10*, 96. [CrossRef]

51. Weippert, M.; Behrens, M.; Rieger, A.; Behrens, K. Sample entropy and traditional measures of heart rate dynamics reveal different modes of cardiovascular control during low intensity exercise. *Entropy* **2014**, *16*, 5698–5711. [CrossRef]
52. Osaka, M.; Saitoh, H.; Atarashi, H.; Hayakawa, H. Correlation dimension of heart rate variability: A new index of human autonomic function. *Front. Med Biol. Eng. Int. J. Jpn. Soc. Med Electron. Biol. Eng.* **1993**, *5*, 289–300.
53. Tulppo, M.P.; Hughson, R.L.; Mä, T.H.; Airaksinen, K.J.; Seppä nen, T.; Huikuri, H.V. Effects of exercise and passive head-up tilt on fractal and complexity properties of heart rate dynamics. *Am. J. Physiol. Heart Circ. Physiol.* **2001**, *280*, H1081–H1087. [CrossRef] [PubMed]
54. Cottin, F.; Médigue, C.; Leprêtre, P.M.; Papelier, Y.; Koralsztein, J.P.; Billat, V. Heart rate variability during exercise performed below and above ventilatory threshold. *Med. Sci. Sport. Exerc.* **2004**, *36*, 594–600. [CrossRef]
55. Blain, G.; Meste, O.; Bermon, S. Influences of breathing patterns on respiratory sinus arrhythmia in humans during exercise. *Am. J. Physiol. Heart Circ. Physiol.* **2005**, *288*, H887–H895. [CrossRef]
56. Goya-Esteban, R.; Barquero-Pérez, O.; Sarabia-Cachadina, E.; de la Cruz-Torres, B.; Naranjo-Orellana, J.; Rojo-Alvarez, J.L. Heart rate variability non linear dynamics in intense exercise. In Proceedings of the 2012 Computing in Cardiology, Krakow, Poland, 9–12 September 2012; pp. 177–180.
57. Valencia, J.F.; Porta, A.; Vallverdu, M.; Claria, F.; Baranowski, R.; Orlowska-Baranowska, E.; Caminal, P. Refined multiscale entropy: Application to 24-h holter recordings of heart period variability in healthy and aortic stenosis subjects. *IEEE Trans. Biomed. Eng.* **2009**, *56*, 2202–2213. [CrossRef]
58. Jerath, R.; Crawford, M.W. How does the body affect the mind? Role of cardiorespiratory coherence in spectrum of emotions. *Adv. Mind Body Med.* **2015**, *29*, 4–16.

Article

On the Statistical Properties of Multiscale Permutation Entropy: Characterization of the Estimator's Variance

Antonio Dávalos *, Meryem Jabloun, Philippe Ravier and Olivier Buttelli

Laboratoire Pluridisciplinaire de Recherche en Ingénierie des Systèmes, Mécanique, Énergétique (PRISME), University of Orléans, 45100 Orléans, INSA-CVL, France; meryem.jabloun@univ-orleans.fr (M.J.); philippe.ravier@univ-orleans.fr (P.R.); olivier.buttelli@univ-orleans.fr (O.B.)

* Correspondence: antonio.davalos-trevino@etu.univ-orleans.fr

Received: 30 March 2019; Accepted: 26 April 2019; Published: 30 April 2019

Abstract: Permutation Entropy (PE) and Multiscale Permutation Entropy (MPE) have been extensively used in the analysis of time series searching for regularities. Although PE has been explored and characterized, there is still a lack of theoretical background regarding MPE. Therefore, we expand the available MPE theory by developing an explicit expression for the estimator's variance as a function of time scale and ordinal pattern distribution. We derived the MPE Cramér–Rao Lower Bound (CRLB) to test the efficiency of our theoretical result. We also tested our formulation against MPE variance measurements from simulated surrogate signals. We found the MPE variance symmetric around the point of equally probable patterns, showing clear maxima and minima. This implies that the MPE variance is directly linked to the MPE measurement itself, and there is a region where the variance is maximum. This effect arises directly from the pattern distribution, and it is unrelated to the time scale or the signal length. The MPE variance also increases linearly with time scale, except when the MPE measurement is close to its maximum, where the variance presents quadratic growth. The expression approaches the CRLB asymptotically, with fast convergence. The theoretical variance is close to the results from simulations, and appears consistently below the actual measurements. By knowing the MPE variance, it is possible to have a clear precision criterion for statistical comparison in real-life applications.

Keywords: Multiscale Permutation Entropy; ordinal patterns; estimator variance; Cramér–Rao Lower Bound; finite-length signals

1. Introduction

Information entropy, originally defined by Shannon [1], has been used as a measure of information content in the field of communications. Several other applications of entropy measurements have been proposed, such as the analysis of physiological electrical signals [2], where a reduction in entropy has been linked to aging [3] and various motor diseases [4]. Another application is the characterization of electrical load behavior, which can be used to perform non-intrusive load disaggregation and to design smart grid applications [5].

Multiple types of entropy measures [6–8] have been proposed in recent years to assess the information content of time series. One notable approach is the Permutation Entropy (PE) [9], used to measure the recurrence of ordinal patterns within a discrete signal. PE is fast to compute and robust even in the presence of outliers [9]. To better measure the information content at different time scales, Multiscale Permutation Entropy (MPE) [10] was formulated as an extension of PE, by using the multiscale approach proposed in [11]. Multiscaling is particularly useful in measuring the information content in long range trends. The main disadvantage of PE is the necessity of a large data set for it to be reliable [12]. This is especially important in MPE, where the signal length is reduced geometrically at each time scale. Signal-length limitations have been addressed and improved with Composite MPE and Refined Composite MPE [13].

PE theory and properties have been extensively explored [14–16]. However, there is still a lack of understanding regarding the statistical properties of MPE. In our previous work [17], we already derived the expected value of the MPE measurement, taking into consideration the time scale and the finite-length constraints. We found the MPE expected value to be biased. Nonetheless, this bias depends only on the MPE parameters, and it is constant with respect to the pattern probability distribution. In practice, the MPE bias does not depend on the signal [17], and thus, can be easily corrected.

In the present article, our goal is to continue this statistical analysis by obtaining the variance of the MPE estimator by means of Taylor series expansion. We develop an explicit equation for the MPE variance. We also obtain the Cramér–Rao Lower Bound (CRLB) of MPE, and compare it to our obtained expression to assess its theoretical efficiency. Lastly, we compare these results with simulated data with known parameters. This gives a better understanding of the MPE statistic, which is helpful in the interpretation of this measurement in real-life applications. By knowing the precision of the MPE, it is possible to take informed decisions regarding experimental design, sampling, and hypothesis testing.

The reminder of the article is organized as follows: In Section 2, we lay the necessary background of PE and MPE, the main derivation of the MPE variance, and the CRLB. We also develop the statistical model to generate the surrogate data simulations for later testing. In Section 3, we show and discuss the results obtained, including the properties of the MPE variance, its theoretical efficiency, and similarities with our simulations. Finally, in Section 4, we add some general remarks regarding the results obtained.

2. Materials and Methods

In this section, we establish the concepts and tools necessary for the derivation of the MPE model. In Section 2.1 we review the formulation of PE and MPE in detail. In Section 2.2, we show the derivation of the MPE variance. In Section 2.3, we derive the expression for the CRLB of the MPE, and we compare it to the variance. Finally, in Section 2.4, we explain the statistical model used to generate surrogate signals, which are used to test the MPE model.

2.1. Theoretical Background

2.1.1. Permutation Entropy

PE [9] measures the information content by counting the ordinal patters present within a signal. An ordinal pattern is defined as the comparison between the values of adjacent data points in a segment of size d, known as the embedded dimension. For example, for a discrete signal of length N, $x_1, \ldots, x_n, \ldots, x_N$, and $d = 2$, only two possible patterns can be found within the series, $x_n < x_{n+1}$ and $x_n > x_{n+1}$. For $d = 3$, there are six possible patterns present, as shown in Figure 1. In general, for any integer value of d, there exists d factorial ($d!$) possible patterns within any given signal segment. To calculate PE, we must first obtain the sample probabilities within the signal, by counting the number of times a certain pattern $i = 1, \ldots, d!$ occurs. This is formally expressed as follows:

$$P(\pi_i) = \frac{\sharp\{n | n \leq N - (d-1), (x_n, \ldots, x_{n+d-1})\ type\ \pi_i\}}{N - (d-1)} = \hat{p}_i. \tag{1}$$

where π_i is the label of a particular ordinal pattern i, and $\hat{p}_i$ is the estimated pattern probability. Some authors [15] also introduce a downsampling parameter in Equation (1), to address the analysis of an oversampled signal. For the purposes of this article, we assume a properly sampled signal, and thus, we do not include this parameter.

Using these estimations, we can apply the entropy definition [1] to obtain the PE of the signal

$$\hat{\mathcal{H}} = \frac{-1}{\ln(d!)} \sum_{i=1}^{d!} \hat{p}_i \ln \hat{p}_i. \tag{2}$$

where $\hat{\mathcal{H}}$ is the estimated normalized PE from the data.

PE is a very simple and fast estimator to calculate, and it is invariant to non-linear monotonous transformations [15]. It is also convenient to note that we need no prior assumptions on the probability distribution of the patterns, which makes PE a very robust estimator. The major disadvantage in PE involves the length of the signal, where we need $N \gg d!$ for PE to be meaningful. This imposes a practical constraint in the use of PE for short signals, or in conditions where the data length is reduced.

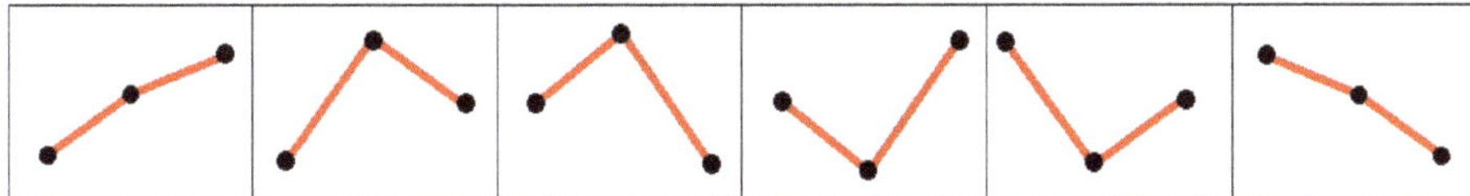

Figure 1. All possible patterns for embedded dimension $d = 3$, from a three-point sequence $\{x_n, x_{n+1}, x_{n+2}\}$. The patterns represented, from left to right, are $\pi_1 : x_n < x_{n+1} < x_{n+2}$, $\pi_2 : x_n < x_{n+2} < x_{n+1}$, $\pi_3 : x_{n+2} < x_n < x_{n+1}$, $\pi_4 : x_{n+1} < x_n < x_{n+2}$, $\pi_5 : x_{n+1} < x_{n+2} < x_n$, and $\pi_6 : x_{n+2} < x_{n+1} < x_n$. The difference in amplitude between data points does not affect the pattern, as long as the order is preserved.

2.1.2. Multiscale Permutation Entropy

The MPE consists of applying the classical PE analysis on coarse-grained signals [10]. First, we define m, as the time scale parameter for the MPE analysis. The coarse-graining procedure consists in dividing the original signal into adjacent, non-overlapping segments of size m, where m is a positive integer less than N. The data points within each segment are averaged to produce a single data point, which represents the segment at the given time scale [11].

$$x_k^{(m)} = \frac{1}{m} \sum_{j=m(k-1)+1}^{km} x_j. \tag{3}$$

MPE consists in applying the PE procedure in Equation (2) on the coarse-grained signals in Equation (3) for different time scales m. This technique allows us to measure the information inside longer trends and time scales, which is not possible using PE. Nonetheless, the resulting coarse-grained signals have a length of N/m, which limits the analysis. Moreover, for a sufficiently large m, the condition $N/m \gg d!$ is eventually not satisfied.

At this point, it is important to discuss some constraints regarding the interaction between time scale m and the signal length N. First, strictly speaking, N/m must be an integer. In practice, the coarse-grained signal length will be the integer number immediately below N/m. Second, the proper domain of m is $(0, N)$, as the segments size, at most, can be the same length as the signal itself. It is handy to use a normalized scale $\frac{m}{N}$ with domain $(0, 1)$ which does not change between signals. This normalized scale is the inverse of the coarse-grained signal length. Taking the length constraints from the previous section, this means that $\frac{m}{N} \ll \frac{1}{d!}$. For $d = 2$, a normalized scale of $\frac{m}{N} = \frac{1}{2}$ will result in a coarse-grained signal of 2 data points, which is not meaningful for this analysis. For $d = 3$, the normalized scale must be significantly less than $\frac{1}{6}$, which corresponds to a coarse-grained signal of six elements. Therefore, for practical applications, we restrict our analysis for values of $\frac{m}{N}$ close to zero, by selecting a small time scale m, a large signal length N, or both.

2.2. Variance of MPE Statistic

The calculated MPE can be interpreted as a sample statistic that estimates the true entropy value, with an associated expected value, variance, and bias, for each time scale. We have previously developed the calculation of the expected value of the MPE in [17], where the bias has been found to be independent from the pattern probabilities. We expand on this result by first proposing an explicit equation to the MPE statistic, and then we formulate the variance of MPE, as a function of the true pattern probabilities and time scale.

For the following development, we use H, the non-normalized version of Equation (2), for convenience.

$$\hat{H} = \hat{\mathcal{H}} \ln(d!) = -\sum_{i=1}^{d!} \hat{p}_i \ln \hat{p}_i. \tag{4}$$

By taking the Taylor expansion of Equation (4) on a coarse-grained signal in Equation (3), we get

$$\hat{H} = H - \frac{m}{N}\sum_{i=1}^{d!}(1 + ln(p_i))\Delta Y_i - \sum_{k=1}^{\infty}\frac{(-1)^{k+1}}{k(k+1)}\left(\frac{m}{N}\right)^{k+1}\sum_{i=1}^{d!}\frac{\Delta {Y_i}^{k+1}}{p_i^k} \tag{5}$$

where $\hat{H}$ is the MPE estimator, H is the true unknown MPE value, m is the time scale, N is the signal length, and d the embedded dimension. The probabilities p_i correspond to the true probabilities (unknown parameters) of each pattern, and ΔY_i correspond to the random part in the multinomially (Mu) distributed frequency of each pattern,

$$\begin{aligned} Y_i &= \frac{N}{m}\hat{p}_i = \frac{N}{m}p_i + \Delta Y_i, \\ \{Y_1, \ldots, Y_{d!}\} &\sim Mu\left(\frac{N}{m}, p_1, \ldots, p_{d!}\right) \end{aligned} \tag{6}$$

Y_i being the number of pattern counts of type i in the signal.

If we take into consideration the length constraints discussed in Section 2.1.2, we know that the normalized time scale $\frac{m}{N}$ is very close to zero. This implies that the high-order terms of Equation (5) will quickly vanish for increasing values of k. Therefore, we propose to make the simplest approximation of Equation (5) by taking only the term $k = 1$. By doing this, we get the following expression:

$$\hat{H} \approx H - \frac{m}{N}\sum_{i=1}^{d!}(1 + ln(p_i))\Delta Y_i - \frac{1}{2}\left(\frac{m}{N}\right)^2\sum_{i=1}^{d!}\frac{\Delta {Y_i}^2}{p_i}. \tag{7}$$

Using our previous results involving MPE [17], we know the expected value of Equation (7) is

$$E[\hat{H}] \approx H - \frac{d! - 1}{2}\left(\frac{m}{N}\right). \tag{8}$$

The statistic presents a bias that does not depend on the pattern probabilities p_i, and thus, can be easily corrected for any signal.

Now, we move further and calculate the variance of the MPE estimator. First, it is convenient to express Equation (4) in vectorial form.

$$\hat{H} = -\hat{\boldsymbol{l}}^T\hat{\boldsymbol{p}} \tag{9}$$

where

$$\begin{aligned} \hat{\boldsymbol{p}} &= [\hat{p}_1, \ldots, \hat{p}_{d!}]^T \\ \hat{\boldsymbol{l}} &= [\ln(\hat{p}_1), \ldots, \ln(\hat{p}_{d!})]^T. \end{aligned} \tag{10}$$

$\hat{p}$ being the column vector of pattern probability estimators, $\hat{l}$ the column vector of the logarithm of each pattern probability estimator, and T is the transpose symbol. We can now rewrite Equation (7) as

$$\hat{H} \approx H - \frac{m}{N}(\mathbf{1}+\boldsymbol{l})^T \Delta \boldsymbol{Y} - \frac{1}{2}\left(\frac{m}{N}\right)^2 \left(\boldsymbol{p}^{\circ -1}\right)^T \Delta \boldsymbol{Y}^{\circ 2} \tag{11}$$

where

$$\begin{aligned} \boldsymbol{p} &= [p_1, \ldots, p_{d!}]^T & \boldsymbol{p}^{\circ -1} &= \left[p_1^{-1}, \ldots, p_{d!}^{-1}\right]^T \\ \mathbf{1} &= [1, \ldots, 1]^T & \boldsymbol{l} &= [\ln(p_1), \ldots, \ln(p_{d!})]^T \\ \boldsymbol{p}^{\circ 2} &= \left[p_1^2, \ldots, p_{d!}^2\right]^T & \Delta \boldsymbol{Y}^{\circ 2} &= \left[\Delta Y_1^2, \ldots, \Delta Y p_{d!}^2\right]^T. \end{aligned} \tag{12}$$

The circle notation $\circ$ represents the Hadamard power (element-wise).

Now, we can obtain the variance of the MPE estimator (11),

$$\begin{aligned} var(\hat{H}) &= E[\hat{H}^2] - (E[\hat{H}])^2 \\ &\approx H^2 + (\tfrac{m}{N})^2(\mathbf{1}+\boldsymbol{l})^T E\left[\Delta\boldsymbol{Y}\Delta\boldsymbol{Y}^T\right](\mathbf{1}+\boldsymbol{l}) - (\tfrac{m}{N})^2(\boldsymbol{p}^{\circ -1})^T E\left[\Delta\boldsymbol{Y}^{\circ 2}\right] H \\ &\quad + (\tfrac{m}{N})^3(\mathbf{1}+\boldsymbol{l})^T E\left[\Delta\boldsymbol{Y}(\Delta\boldsymbol{Y}^{\circ 2})^T\right](\boldsymbol{p}^{\circ -1}) + \tfrac{1}{4}(\tfrac{m}{N})^4(\boldsymbol{p}^{\circ -1})^T E\left[\Delta\boldsymbol{Y}^{\circ 2}(\Delta\boldsymbol{Y}^{\circ 2})^T\right](\boldsymbol{p}^{\circ -1}) \\ &\quad + (\tfrac{m}{N})(d!-1)H - \tfrac{1}{4}(\tfrac{m}{N})^2(d!-1)^2 - H^2. \end{aligned} \tag{13}$$

Now, we know that $E\left[\Delta\boldsymbol{Y}\Delta\boldsymbol{Y}^T\right]$ is the Covariance matrix of $\Delta\boldsymbol{Y}$, which is the multinomial random variable defined in Equation (6). The matrix $E\left[\Delta\boldsymbol{Y}(\Delta\boldsymbol{Y}^{\circ 2})^T\right]$ is the Coskewness matrix, and $E\left[\Delta\boldsymbol{Y}^{\circ 2}(\Delta\boldsymbol{Y}^{\circ 2})^T\right]$ is the Cokurtosis. We know that (see Appendix A),

$$E\left[\Delta\boldsymbol{Y}\Delta\boldsymbol{Y}^T\right] = \tfrac{N}{m}(diag(\boldsymbol{p}) - \boldsymbol{p}\boldsymbol{p}^T) \tag{14}$$

$$E\left[\Delta\boldsymbol{Y}(\Delta\boldsymbol{Y}^{\circ 2})^T\right] = 2\tfrac{N}{m}\left(\boldsymbol{p}^{\circ 2}\boldsymbol{p}^T - diag(\boldsymbol{p}^{\circ 2})\right) + \tfrac{N}{m}(diag(\boldsymbol{p}) - \boldsymbol{p}\boldsymbol{p}^T) \tag{15}$$

$$\begin{aligned} E\left[\Delta\boldsymbol{Y}^{\circ 2}(\Delta\boldsymbol{Y}^{\circ 2})^T\right] &= 3\tfrac{N}{m}(\tfrac{N}{m}-2)\boldsymbol{p}^{\circ 2}(\boldsymbol{p}^{\circ 2})^T - \tfrac{N}{m}(\tfrac{N}{m}-2)(\boldsymbol{p}^{\circ 2}\boldsymbol{p}^T + \boldsymbol{p}(\boldsymbol{p}^{\circ 2})^T) \\ &\quad + (\tfrac{N}{m})^2\boldsymbol{p}\boldsymbol{p}^T - 4\tfrac{N}{m}(\tfrac{N}{m}-2)diag(\boldsymbol{p}^{\circ 3}) + 2\tfrac{N}{m}(\tfrac{N}{m}-3)diag(\boldsymbol{p}^{\circ 2}) \\ &\quad + \tfrac{N}{m}(diag(\boldsymbol{p}) - \boldsymbol{p}\boldsymbol{p}^T) \end{aligned} \tag{16}$$

where $diag(\boldsymbol{p})$ is a diagonal matrix, where the diagonal elements are the elements of $\boldsymbol{p}$.

We can further summarize the covariance matrix (14) as follows:

$$\tfrac{N}{m}(diag(\boldsymbol{p}) - \boldsymbol{p}\boldsymbol{p}^T) = \tfrac{N}{m}\boldsymbol{\Sigma}_p. \tag{17}$$

We also rewrite the following term in Equation (13),

$$(\boldsymbol{p}^{\circ -1})^T E\left[\Delta\boldsymbol{Y}^{\circ 2}\right] = (\tfrac{N}{m})(\boldsymbol{p}^{\circ -1})^T(\boldsymbol{p} - \boldsymbol{p}^{\circ 2}) = (\tfrac{N}{m})(d!-1). \tag{18}$$

By combining Equations (14) to (18) explicitly in Equation (13), we get the expression,

$$var(\hat{H}) \approx (\tfrac{m}{N})\boldsymbol{l}^T\boldsymbol{\Sigma}_p\boldsymbol{l} + (\tfrac{m}{N})^2\left(\mathbf{1}^T\boldsymbol{l} + d!H + \tfrac{1}{2}(d!-1)\right) + \tfrac{1}{4}(\tfrac{m}{N})^3\left(\mathbf{1}^T\boldsymbol{p}^{\circ -1} - (d!^2 + 2d! - 2)\right). \tag{19}$$

We note that Equation (19) is a cubic polynomial respect to the normalized scale m/N. Since we are still working in the region where m/N is close to (but not including) zero, we propose to further simplify this expression using only the linear term. This means that we can approximate Equation (19) as follows:

$$var(\hat{H}) \approx \left(\tfrac{m}{N}\right) \boldsymbol{l}^T \boldsymbol{\Sigma}_p \boldsymbol{l} = \tfrac{m}{N}\left(\sum_{i=1}^{d!} p_i \ln^2(p_i) - \sum_{i=1}^{d!}\sum_{j=1}^{d!} p_i p_j \ln(p_i)\ln(p_j)\right) = \tfrac{m}{N}\left(\sum_{i=1}^{d!} p_i \ln^2(p_i) - H^2\right). \quad (20)$$

We note that Equation (20) will be equal to zero for a perfectly uniform pattern distribution (which yields a maximum PE). In this particular case, Equation (20) will not be a good approximation for the MPE variance, and we will need, at least, the quadratic term in Equation (19). Except for extremely high or low values of MPE, we expect the variance linear approximation to be accurate. We discuss more properties of this statistic in the Section 3.2.

2.3. MPE Cramér–Rao Lower Bound

In this section, we compare the MPE variance (20) to the CRLB, to test the efficiency of our estimator. The CRLB is defined as

$$CRLB(H) = \frac{[1 - B'(H)]^2}{I(H)} \leq var(\hat{H}) \quad (21)$$

where B is the bias of the MPE expected value from Equation (8) and

$$\begin{aligned} B'(H) &= \frac{dB}{dH} = -\frac{d}{dH}\left(\frac{d! - 1}{2}\tfrac{m}{N}\right) = 0 \\ I(H) &= -E\left[\frac{\partial^2 \ln(f_H(\boldsymbol{y};\boldsymbol{p}))}{\partial H^2}\right] \end{aligned} \quad (22)$$

where $I(H)$ is the Fisher Information, and $f_H(\boldsymbol{y};\boldsymbol{p})$ is the probability distribution function of H.

Although we do not have an explicit expression for f_H, we can express $CRLB(H)$ as a function of $CRLB(\boldsymbol{p})$ as follows [18]:

$$CRLB(H) = \left(\frac{\partial H}{\partial \boldsymbol{p}}\right)^T CRLB(\boldsymbol{p}) \frac{\partial H}{\partial \boldsymbol{p}} \quad (23)$$

where

$$\frac{\partial H}{\partial \boldsymbol{p}} = \left[\frac{\partial H}{\partial p_1}, \cdots \frac{\partial H}{\partial p_{d!}}\right]^T \quad (24)$$

$$CRLB(\boldsymbol{p}) = \boldsymbol{I}(\boldsymbol{p})^{-1} \leq cov_{\boldsymbol{p}}(\hat{\boldsymbol{p}}). \quad (25)$$

$\boldsymbol{I}(\boldsymbol{p})$ is the Fisher information matrix for $\boldsymbol{p}$, which we know has a multinomial distribution related to Equation (6). Each element of $\boldsymbol{I}(\boldsymbol{p})$ is defined as

$$I_{j,k}(\boldsymbol{p}) = -E\left[\frac{\partial^2}{\partial p_j \partial p_k} \ln(f_{\hat{\boldsymbol{p}}}(\hat{\boldsymbol{p}};\boldsymbol{p}))\right]. \quad (26)$$

where $f_{\hat{\boldsymbol{p}}}(\hat{\boldsymbol{p}};\boldsymbol{p})$ is the probability distribution of $\hat{\boldsymbol{p}}$, which is identically distributed to Equation (6) (for the full calculation of $CRLB(H)$, see Appendix B). Thus, by obtaining the inverse of $\boldsymbol{I}(\boldsymbol{p})$ and all partial derivatives of H with respect to each element of $\boldsymbol{p}$, we obtain the $CRLB(H)$.

$$CRLB(H) = \tfrac{m}{N}\left(\sum_{i=1}^{d!} p_i \ln^2 p_i - H^2\right) = \tfrac{m}{N}\boldsymbol{l}^T \boldsymbol{\Sigma}_p \boldsymbol{l}. \quad (27)$$

As we recall from our results in Equations (19) and (20), the CRLB corresponds to the first term of the Taylor series expansion of the MPE variance. The high-order terms in Equation (19) increase the MPE variance above the CRLB. For small values of $\frac{m}{N}$, the higher-order terms in Equations (20) become neglectable, which make the MPE variance approximation converge to $CRLB(H)$.

2.4. Simulations

To test the MPE variance, we need an appropriate benchmark. We need to design a proper signal model with the following goals in mind: First, the model must preserve the pattern probabilities across all the signal generated; second, the function must have the pattern probability as an explicit parameter, easily modifiable for testing. The following equation satisfies these criteria for dimension $d = 2$:

$$x_n = x_{n-1} + \epsilon_n - \delta(p) \tag{28}$$

where

$$p = \frac{1}{2}\left(1 - erf\left(\frac{\delta(p)}{\sqrt{2}}\right)\right) \tag{29}$$

$$\delta(p) = \sqrt{2}\, erf^{-1}(1-2p). \tag{30}$$

This is a non-stationary process, with a trend function $\delta(p)$. ϵ is a Gaussian noise term with variance $\sigma^2 = 1$, without loss of generality. Although different values of σ^2 will indeed modify the trend function, it will not affect the pattern probabilities in the simulation, as PE is invariant to non-linear monotonous transformations [15].

For dimension $d = 2$, $p = p_1 = P(x_n < x_{n+1})$ and $1 - p = p_2 = P(x_n > x_{n+1})$, which are the probabilities of each of the two possible patterns. The formulation of $\delta(p)$ comes from the Gaussian Complementary Cumulative Distribution function, taking p as a parameter. This guarantees that we can directly modify the pattern probabilities p and $1 - p$ for simulation.

Although p is not invariant at different time scales, it is consistent within each coarse-grained signal, which suffices for our purposes. We chose to restrict our simulation analysis to the embedded dimension $d = 2$. Although our theoretical work holds for any value of d (see Section 2.2), it is difficult to visualize the results at higher dimensions.

This surrogate model (28) was implemented in the Matlab environment. For the test, we generated 1000 signals each, for 99 different values for $p = 0.01, 0.02, \ldots 0.99$. The signal length was set to $N = 1000$. Some sample paths are shown in Figure 2. These signals were then subject to the coarse-graining procedure (3) for time scale $m = 1, \ldots, 10$. The MPE measurement was taken for each coarse-grained simulated signal using Equation (2). Finally, we obtained the mean and variance of the resulting MPE's for each scale. This simulation results are discussed in Section 3.2.

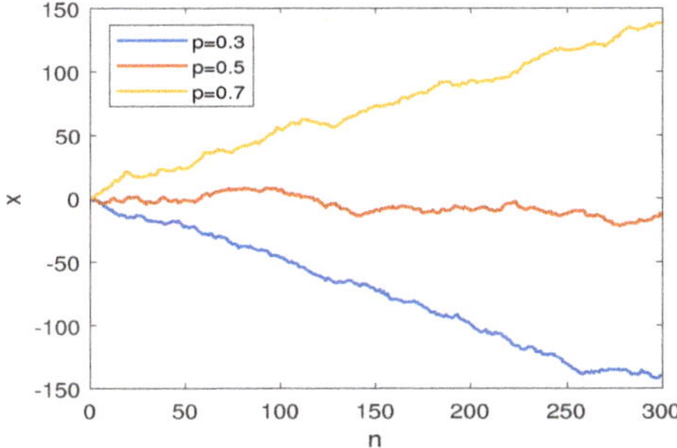

Figure 2. Simulated paths for MPE testing, where p is the probability of $x_n < x_{n+1}$ for dimension $d = 2$. The graph shows sample paths for $p = 0.3$, $p = 0.5$, $p = 0.7$

3. Results and Discussion

In this section, we explore the results from the theoretical MPE variance. In Section 3.1, we contrast the results from the model against the MPE variance measured from the simulations. In Section 3.2, we discuss these findings in detail.

3.1. Results

Here, we compare the theoretical results with the surrogate data obtained by means of the procedure described in Section 2.4. We use the cubic model from Equation (19) instead of the linear approximation in Equation (20), to take in consideration non-linear effects that could arise from simulations. Figure 3 shows the theoretical predictions in dotted lines, and simulation measurements in solid lines.

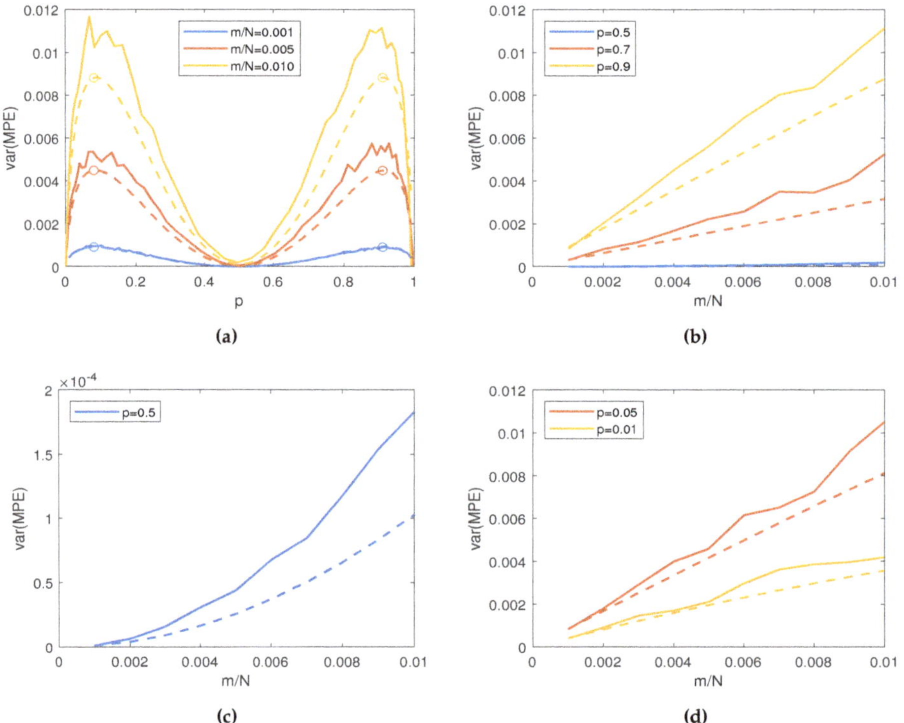

Figure 3. $var(\hat{H})$ for embedded dimension $d = 2$ from theory (dotted lines) and simulations (solid lines). (**a**) $var(\hat{H})$ vs. pattern probability p for different normalized scales m/N. (**b**) $var(\hat{H})$ vs. m/N for different values of p. (**c**) $var(\hat{H})$ vs. m/N at $p = 0.5$, which corresponds to maximum entropy. (**d**) $var(\hat{H})$ vs. m/N at with small p, which approaches minimum entropy.

Figure 3a shows the variance of the MPE (19) as a function of p for $d = 2$. The lines correspond to a normalized time scale of $m/N = 0.001, 0.005$, and 0.010. The variance presents symmetry with respect to the middle value of $p = 0.5$. This is to be expected, as for $d = 2$, the variance has only one degree of freedom. The structure is preserved, albeit scaled, for different values of m.

Figure 3b–d show the MPE variance versus the normalized time scale, for different values of pattern probability p. As we can see in Figure 3b, the variance increases linearly with respect to the normalized scale m/N at the positive vicinity of zero, as predicted in Equation (20). This is not the case for when $p = 0.5$ (maximum entropy), as shown in Figure 3c, where both the theory and simulations show a clear non-linear tendency. Finally, Figure 3d shows the case where we have extreme pattern probability distributions, with entropy close to zero. Although we use the complete cubic model (19), the predicted curves are almost linear.

In general, we can observe that the simulation results have greater variance than the prediction of the model. The real values from the simulations correspond to the sample variance from the signals, calculated from $\hat{\boldsymbol{p}}$ instead of $\boldsymbol{p}$. Nonetheless, the simulations and theoretical graphs have the same shape. It is interesting to note that the discrepancies increase with the scale. This effect is addressed in Section 3.2.

3.2. Discussion

It is interesting to explore the particular structure of the variance. As we can observe from Figure 3b, the MPE variance increases linearly with respect to the time scale for a wide range of pattern distributions, as described in Equation (20). This is even true with highly uneven distributions, where the entropy is very close to zero, as shown in Figure 3d. Nonetheless, when we observe the expression (20), the equation collapses to zero when all pattern probabilities p_i are the same. For any embedded dimension d, all probabilities $p_i = \frac{1}{d!}$. This particular pattern probability distribution is the discrete uniform distribution, which yield to the maximum possible entropy in Equation (2). As we can observe from Figure 3c, the linear approximation in Equation (19) is not enough to estimate the variance in this case. Nonetheless, the results suggest a quadratic increase. This agrees with previous results by Little and Kane [16], where they characterized the classical normalized PE variance for white noise under finite-length constraints.

$$var(\hat{\mathcal{H}}) \approx \frac{d!-1}{2(\ln d!)^2}\frac{1}{N^2} \tag{31}$$

This result matches our model in Equation (19) (taking the quadratic term) for the specific case of uniform pattern distribution and time scale $m = 1$. This suggest that, when we approach the maximum entropy, the quadratic approximation is necessary.

Contrary to the bias in the expected value of MPE [17], the variance is strongly dependent on the pattern probabilities present in the signal, as shown in Figure 3a. For the MPE with embedded dimension $d = 2$, the variance of MPE has a symmetric shape around equally probable patterns. We observe, for a fixed time scale, that the variance increases the further we deviate from the center, and sharply decreases for extreme probabilities. The variance presents its lowest points at the center and the extremes of the pattern probability distributions, which corresponds to the points where the entropy is maximum and minimum, respectively. For $d = 2$, we can calculate the variance (20) more explicitly,

$$var(\hat{H}) \approx \left(\tfrac{m}{N}\right) \boldsymbol{l}^T \boldsymbol{\Sigma}_p \boldsymbol{l}|_{d=2} = \left(\tfrac{m}{N}\right) p(1-p)\ln^2\left(\frac{p}{1-p}\right). \tag{32}$$

It is evident that the zeros of this equation correspond to $p = 0$, $p = 1$ (points of minimum entropy), and $p = 0.5$ (maximum entropy). We can get the critical points by calculating the first derivative of Equation (32) with respect to p

$$\left(\tfrac{m}{N}\right) \tfrac{d}{dp}\left(\boldsymbol{l}^T \boldsymbol{\Sigma}_p \boldsymbol{l}|_{d=2}\right) = \left(\tfrac{m}{N}\right) \ln\left(\frac{p}{1-p}\right)\left((1-2p)\ln\left(\frac{p}{1-p}\right)+2\right) = 0 \tag{33}$$

which is equal to zero to get the extreme points. From the first term of Equation (33), we obtain the critical point $p = 0.5$, which is a minimum. If the second term of the equation is equal to zero, we need to solve the transcendental function

$$\ln\left(\frac{p}{1-p}\right) = \frac{2}{2p-1.} \tag{34}$$

Numerically, we found the maximum points to be $p = 0.083$ and $p = 0.917$, as shown in Figure 3a. Both these values yield to a normalized PE value of $\hat{\mathcal{H}} = 0.413$. This implies that, when we obtain values of MPE close to this value, we will have a region with maximum variance. This effect arises directly from the pattern probability distribution. Hence, in the region around $\hat{\mathcal{H}} = 0.413$ for $d = 2$, we will have maximum estimation uncertainty, even before factoring the finite length of the signal, or the time scale. Therefore, Equation (20) implies an uneven variance across all possible values of the entropy measurement, regardless of time scale and embedded dimension. It also implies a region where this variance is maximum.

Lastly, as noted in Section 2.3, the MPE variance (20) approaches the CRLB for small values of m. This is further supported by the simulation variance MPE measurements shown in Figure 3, which are consistently above the theoretical prediction. We can attribute this effect to the number of iterations of the testing model in Equation (28), where an increasing number of repetitions will yield to a more precise MPE estimation, with a reduced variance. Since we already include the effect of the time scale in Equations (19) and (20), the difference between the theoretical results and the simulations does not come directly from the signal length or the pattern distribution.

4. Conclusions

By following on from our previous work [17], we further explored and characterized the MPE statistic. By using a Taylor series expansion, we were able to obtain an explicit expression of the MPE variance. We also derived the Cramér–Rao Lower bound of the MPE, and compared it to our obtained expression. Finally, we proposed a suitable signal model for testing our results against simulations.

By analyzing the properties of the MPE variance graph, we found the estimator to be symmetric around the point of equally probable patterns. Moreover, the estimator is minimum in both the points of maximum and minimum entropy. This implies, first, that the variance of the MPE is dependent on the MPE estimation itself. In regions where the entropy measurement is near the maximum or minimum, the estimation will have a minimum variance. On the other hand, there is an MPE measurement where the variance will have a maximum. This effect comes solely from the pattern distribution, and not from the signal length or the MPE time scale. We should take in account this effect when comparing real entropy measurements, as it could affect the interpretation of statistically significant difference.

Regarding the time scale, the MPE variance linear approximation is sufficiently accurate for almost all pattern distributions, provided that the time scale is small compared to the signal length. An important exception to this is the case where the pattern probability distribution is almost uniform. For equally probable patterns, the linear term of the MPE variance vanishes, regardless of time scale. In this case, we need to increase the order of the approximation to the quadratic term. Hence, for MPE values close to the maximum, the variance presents a quadratic growth respect to scale.

We found the MPE variance estimator obtained in this article to be almost efficient. When the time scale is small compared to the signal length, the MPE variance resembles the MPE CRLB closely. Since the CRLB is equal to the first term on the Taylor series approximation for the variance, this implies that the efficiency limit also changes with the MPE measurement itself. Since this effect also comes purely from the pattern distribution, we cannot correct it with the established improvements of the MPE algorithm, like Composite MPE or Refined Composite MPE [13].

By knowing the variance of the MPE, we can improve the interpretation of this estimator in real-life applications. This is important because researchers can impose a precision criterion over the MPE measurements, given the characteristics of the data to analyze. For example, the electrical load behavior analysis can be achieved using short-term measurements where the time scale is a limiting factor for MPE. By knowing the variance, we can compute the maximum time scale until which the MPE is still significant.

By better understanding the advantages and disadvantages of the MPE technique, it is possible to have a clear benchmark for statistically significant comparisons between signals at any time scale.

Author Contributions: Conceptualization, D.A. and J.M.; methodology, D.A.; software, D.A.; validation, J.M., R.P. and B.O.; formal analysis, D.A.; resources, D.A.; writing—original draft preparation, D.A.; writing—review and editing, J.M. and R.P.; visualization, D.A.; supervision, B.O.; project administration, B.O..

Funding: The work is part of the ECCO project supported by the french Centre Val-de-Loire region under the contract #15088PRI.

Acknowledgments: This work was supported by CONACyT, Mexico, with scholarship 439625.

Conflicts of Interest: The authors declare no conflict of interest.

Abbreviations

The following abbreviations are used in this manuscript:

PE	Permutation Entropy
MSE	Multiscale Entropy
MPE	Multiscale Permutation Entropy
CRLB	Cramér–Rao Lower Bound

Appendix A. Multinomial Moment Matrices

In this section we will briefly derive the expressions for the Covariance, Coskewness, and Cokurtosis matrices for a multinomial distribution. These will be necessary in the calculation of the MPE variance in Section 2.2.

First, we start with a multinomial random variable with the following characteristics,

$$\begin{aligned} \boldsymbol{Y} &= \begin{bmatrix} Y_1 \\ \vdots \\ Y_{d!} \end{bmatrix} = \begin{bmatrix} np_1 + \Delta Y_1 \\ \vdots \\ np_{d!} + \Delta Y_{d!} \end{bmatrix} = n\boldsymbol{p} + \Delta \boldsymbol{Y} \sim Mu\left(n, p_1, \ldots, p_{d!}\right) \\ \hat{\boldsymbol{p}} &= \tfrac{1}{n}\boldsymbol{Y} = \boldsymbol{p} + \tfrac{1}{n}\Delta \boldsymbol{Y} \end{aligned} \tag{A1}$$

where we use the same definitions as in Equations (10) and (12), where $\boldsymbol{p}$ is the vector composed of all pattern probabilities p_i, and $\hat{\boldsymbol{p}}$ is the estimation of $\boldsymbol{p}$. We will also use the definitions in Section 2.2, where $d!$ is the number of possible patterns (number of events in the sample space), m is the MPE time scale, and N is the length of the signal. For the remainder of this section, we will assume m and N are such that $n = \frac{N}{m}$ is integer.

We note that Y is composed of a deterministic part np_i, and a random part ΔY_i. It should be evident that $E[\Delta Y_i] = 0$ and ΔY_i is identically distributed to Y.

We proceed to calculate the Covariance matrix of the binomial random variable $\Delta \boldsymbol{Y}$. We know that

$$E\left[\Delta Y_i{}^2\right] = np_i(1 - p_i) \tag{A2}$$

$$E\left[\Delta Y_i \Delta y_j\right] = -np_i p_j \tag{A3}$$

for $i = 1, \ldots, d!$ and $j = 1, \ldots, d!$. Thus, if we gather all possible combinations of i and j in the Covariance matrix, we get

$$E\left[\Delta \boldsymbol{Y} \Delta \boldsymbol{Y}^T\right] = n(diag(\boldsymbol{p}) - \boldsymbol{p}\boldsymbol{p}^T) \tag{A4}$$

Similarly, the skewness and coskewness can be expressed as,

$$E\left[\Delta Y_i^3\right] = 2np_i^3 - 3np_i^2 + np_i \tag{A5}$$

$$E\left[\Delta Y_i^2 \Delta Y_j\right] = 2np_i^2 p_j - np_i p_j \tag{A6}$$

which yields to the Coskewness matrix

$$E\left[\Delta\boldsymbol{Y}(\Delta\boldsymbol{Y}^{\circ 2})^T\right] = 2n\left(\boldsymbol{p}^{\circ 2}\boldsymbol{p}^T - diag(\boldsymbol{p}^{\circ 2})\right) + n(diag(\boldsymbol{p}) - \boldsymbol{p}\boldsymbol{p}^T) \tag{A7}$$

where we use again the vector definitions in Equation (12).

Lastly, we follow the same procedure to obtain the Cokurtosis matrix, by obtaining the values

$$E\left[\Delta Y_i^4\right] = 3n(n-2)p_i^4 - 6n(n-2)p_i^3 + n(3n-7)p_i^3 + np_i \tag{A8}$$

$$E\left[\Delta Y_i^2 \Delta Y_j^2\right] = 3n(n-2)p_i^2 p_j^2 - n(n-2)(p_i^2 p_j + p_i p_j^2) + n(n-1)p_i p_j \tag{A9}$$

which combines in the matrix as follows

$$\begin{aligned} E[\Delta\boldsymbol{Y}^{\circ 2}(\Delta\boldsymbol{Y}^{\circ 2})^T] &= 3n(n-2)\boldsymbol{p}^{\circ 2}(\boldsymbol{p}^{\circ 2})^T - n(n-2)(\boldsymbol{p}^{\circ 2}\boldsymbol{p}^T + \boldsymbol{p}(\boldsymbol{p}^{\circ 2})^T) + n^2\boldsymbol{p}\boldsymbol{p}^T \\ &- 4n(n-2)diag(\boldsymbol{p}^{\circ 3}) + 2n(n-3)diag(\boldsymbol{p}^{\circ 2}) + n(diag(\boldsymbol{p}) - \boldsymbol{p}\boldsymbol{p}^T). \end{aligned} \tag{A10}$$

By taking advantage of this expressions, we are able to calculate the MPE variance in Section 2.2.

Appendix B. Cramér–Rao Lower Bound of MPE

In this section we will explicitly develop the calculations of $CRLB(H)$. Before calculating the elements of the Information matrix, we should note that the parameter vector $\boldsymbol{p}$ in Equation (12) has $d! - 1$ degrees of freedom. since the elements of $\boldsymbol{p}$ represent the probabilities of each possible pattern in the sample space, the sum of the elements of $\boldsymbol{p}$ must be one. We are subject to the restriction

$$\begin{aligned} \sum_{i=1}^{d!} p_i &= 1 \\ p_{d!} &= 1 - \sum_{i=1}^{d!-1} p_i \end{aligned} \tag{A11}$$

We should note that, as a consequence of this restriction, $p_{d!}$ is not an independent variable. We will define a new parameter vector $\boldsymbol{p}_*$, such that

$$\boldsymbol{p}_* = [p_1, \ldots, p_{d!-1}]^T \tag{A12}$$

which has the same degrees of freedom as $\boldsymbol{p}$. This will be necessary to use the lemma (A19) as we shall see later in this section.

Now, we calculate all the partial derivatives of H (4) respect to the pattern probabilities $\boldsymbol{p}$, such that, for $j = 1, \ldots, (d! - 1)$

$$\begin{aligned} H &= -\sum_{i=1}^{d!} p_i \ln p_i = -\sum_{i=1}^{d!-1} p_i \ln p_i - p_{d!} \ln p_{d!} \\ \frac{\partial H}{\partial p_j} &= \ln p_{d!} - \ln p_j \\ \frac{\partial H}{\partial \boldsymbol{p}_*} &= \ln p_{d!} \mathbf{1} - \boldsymbol{l}_* \end{aligned} \tag{A13}$$

where

$$\mathbf{1} = [1, \ldots, 1]^T \tag{A14}$$

$$\boldsymbol{l}_* = [\ln(p_1), \ldots, \ln(p_{d!-1})]^T \tag{A15}$$

following the same logic as Equation (A12).

The next step is to compute the Fisher Information matrix. We know from Equation (6) that $\hat{\boldsymbol{p}}$, the pattern probabilities estimated from a signal, are multinomially distributed. The number of trials in the multinomial variable is N/m.

$$f_{\hat{\boldsymbol{p}}}(\hat{\boldsymbol{p}};\boldsymbol{p}) = f_{\boldsymbol{Y}}(\boldsymbol{y};\boldsymbol{p}) = \tfrac{N}{m}!\prod_{i=1}^{d!}\frac{p_i^{y_i}}{y_i!} = \tfrac{N}{m}!\frac{p_{d!}^{y_{d!}}}{y_{d!}!}\prod_{i=1}^{d!-1}\frac{p_i^{y_i}}{y_i!} \tag{A16}$$

where

$$\boldsymbol{y} = [y_1,\ldots,y_{d!}]^T \tag{A17}$$

is the count of patterns in the signal, as described in Equations (1) and (6). Since $\hat{\boldsymbol{p}} = \frac{m}{N}Y$, they have the same probability distribution. Using (A16), $\boldsymbol{I}(\boldsymbol{p_*})$ is calculated as follows,

$$\begin{aligned}
\ln(f_{\boldsymbol{Y}}(\boldsymbol{y};\boldsymbol{p})) &= \ln(\tfrac{N}{m}!) + y_{d!}\ln(p_{d!}) + \sum_{i=1}^{d!-1} y_i\ln(p_i) - \ln(y_{d!}!) - \sum_{i=1}^{d!-1}\ln(y_i!)\\
\frac{\partial\ln(f_Y)}{\partial p_j} &= y_j p_j^{-1} - y_{d!}p_{d!}^{-1}\\
\frac{\partial^2\ln(f_Y)}{\partial p_j^2} &= -y_j p_j^{-2} - y_{d!}p_{d!}^{-2}\\
\frac{\partial^2\ln(f_Y)}{\partial p_j\partial p_k} &= -y_{d!}p_{d!}^{-2}\\
-E\left[\frac{\partial^2\ln(f_Y)}{\partial p_j^2}\right] &= \tfrac{N}{m}p_j^{-1} + \tfrac{N}{m}p_{d!}^{-1}\\
-E\left[\frac{\partial^2\ln(f_Y)}{\partial p_j\partial p_k}\right] &= \tfrac{N}{m}p_{d!}^{-1}\\
\boldsymbol{I}(\boldsymbol{p_*}) &= \tfrac{N}{m}\left(diag(\boldsymbol{p}_*^{\circ-1}) + p_{d!}^{-1}\mathbf{1}\cdot\mathbf{1}^T\right)
\end{aligned} \tag{A18}$$

where $\mathbf{1}\cdot\mathbf{1}^T$ is a square matrix of ones, with rank 1. We should note that $j = 1,\ldots,(d!-1)$ and $k = 1,\ldots,(d!-1)$, so $\boldsymbol{I}(\boldsymbol{p_*})$ is of size $(d!-1)x(d!-1)$. Because of the constraint in (A11), $p_{d!}$ offers no additional information. Moreover, we guarantee that $\boldsymbol{I}(\boldsymbol{p_*})$ is non-singular, and thus, invertible.

To get $CRLB(\boldsymbol{p_*})$, we will need the inverse of $\boldsymbol{I}(\boldsymbol{p_*})$: We will use the following lemma from [19]. If A and $A+B$ are non-singular matrices, and B has rank 1, then,

$$(A+B)^{-1} = A^{-1} - \frac{1}{1+tr(BA^{-1})}A^{-1}BA^{-1} \tag{A19}$$

therefore

$$\begin{aligned}
\boldsymbol{I}(\boldsymbol{p_*})^{-1} &= \tfrac{m}{N}(diag(\boldsymbol{p}_*^{\circ-1}) + p_{d!}^{-1}\mathbf{1}\cdot\mathbf{1}^T)^{-1}\\
&= \tfrac{m}{N}\left(diag(\boldsymbol{p}_*) - \frac{p_{d!}^{-1}}{1+p_{d!}^{-1}(1-p_{d!})}\boldsymbol{p}_*\boldsymbol{p}_*^T\right)\\
\boldsymbol{I}(\boldsymbol{p_*})^{-1} &= \tfrac{m}{N}\left(diag(\boldsymbol{p}_*) - \boldsymbol{p}_*\boldsymbol{p}_*^T\right) = CRLB(\boldsymbol{p}_*).
\end{aligned} \tag{A20}$$

Lastly, if we introduce Equations (A13) and (A20) into (23), we get

$$\begin{aligned}
CRLB(H) &= \tfrac{m}{N}\left(\ln(p_{d!})\mathbf{1} - \boldsymbol{l}_*\right)^T\left(diag(\boldsymbol{p}_*) - \boldsymbol{p}_*\boldsymbol{p}_*^T\right)\left(\ln(p_{d!})\mathbf{1} - \boldsymbol{l}_*\right)\\
&= \tfrac{m}{N}(\ln^2(p_{d!})\mathbf{1}^T diag(\boldsymbol{p}_*)\mathbf{1} - \ln^2(p_{d!})\mathbf{1}^T\boldsymbol{p}_*\boldsymbol{p}_*^T\mathbf{1} + \boldsymbol{l}_*^T diag(\boldsymbol{p}_*)\boldsymbol{l}_* - \boldsymbol{l}_*^T\boldsymbol{p}_*\boldsymbol{p}_*^T\boldsymbol{l}_*\\
&\quad - 2\ln(p_{d!})\mathbf{1}^T diag(\boldsymbol{p}_*)\boldsymbol{l}_* + 2\ln(p_{d!})\mathbf{1}^T\boldsymbol{p}_*\boldsymbol{p}_*^T\boldsymbol{l}_*).
\end{aligned} \tag{A21}$$

By noting the following relations,

$$
\begin{aligned}
\mathbf{1}^T diag(\boldsymbol{p}_*)\mathbf{1} &= \sum_{i=1}^{d!-1} p_i = 1 - p_{d!} \\
\mathbf{1}^T \boldsymbol{p}_* &= \sum_{i=1}^{d!-1} p_i = 1 - p_{d!} \\
\mathbf{1}^T \boldsymbol{p}_* \boldsymbol{p}_*^T \mathbf{1} &= \left(\mathbf{1}^T \boldsymbol{p}_*\right)^2 = (1 - p_{d!})^2 \\
\boldsymbol{l}_*^T diag(\boldsymbol{p}_*)\boldsymbol{l}_* &= \sum_{i=1}^{d!-1} p_i \ln^2 p_i \\
\boldsymbol{l}_*^T \boldsymbol{p}_* &= \sum_{i=1}^{D-1} p_i \ln p_i = -H - p_{d!} \ln p_{d!} \\
\boldsymbol{l}_*^T \boldsymbol{p}_* \boldsymbol{p}_*^T \boldsymbol{l}_* &= \left(\boldsymbol{l}_*^T \boldsymbol{p}_*\right)^2 = H^2 + 2Hp_{d!} \ln p_{d!} + p_{d!}^2 \ln^2 p_{d!} \\
\mathbf{1}^T diag(\boldsymbol{p}_*)\boldsymbol{l}_* &= \boldsymbol{p}_*^T \boldsymbol{l}_* = -H - p_{d!} \ln p_{d!} \\
\mathbf{1}^T \boldsymbol{p}_* \boldsymbol{p}_*^T \boldsymbol{l}_* &= -(1 - p_{d!})(H + p_{d!} \ln p_{d!}) \\
&= -H - p_{d!} \ln p_{d!} + p_{d!} H + p_{d!}^2 \ln p_{d!}
\end{aligned}
$$

we can simplify and rewrite Equation (A21) as

$$
CRLB(H) = \tfrac{m}{N} \left(\sum_{i=1}^{d!} p_i \ln^2(p_i) - H^2 \right) = \tfrac{m}{N} \boldsymbol{l}^T \boldsymbol{\Sigma}_p \boldsymbol{l}.
$$

which is what we obtained in Equation (27).

References

1. Shannon, C.E. A Mathematical Theory of Communication. *SIGMOBILE Mob. Comput. Commun. Rev.* **2001**, *5*, 3–55, [CrossRef]
2. Goldberger, A.L.; Peng, C.K.; Lipsitz, L.A. What is physiologic complexity and how does it change with aging and disease? *Neurobiol. Aging* **2002**, *23*, 23–26, [CrossRef]
3. Cashaback, J.G.A.; Cluff, T.; Potvin, J.R. Muscle fatigue and contraction intensity modulates the complexity of surface electromyography. *J. Electromyogr. Kinesiol.* **2013**, *23*, 78–83. [CrossRef] [PubMed]
4. Wu, Y.; Chen, P.; Luo, X.; Wu, M.; Liao, L.; Yang, S.; Rangayyan, R.M. Measuring signal fluctuations in gait rhythm time series of patients with Parkinson's disease using entropy parameters. *Biomed. Signal Process. Control* **2017**, *31*, 265–271. [CrossRef]
5. Aquino, A.L.L.; Ramos, H.S.; Frery, A.C.; Viana, L.P.; Cavalcante, T.S.G.; Rosso O.A. Characterization of electric load with Information Theory quantifiers. *PHYSICA A* **2017**, *465*, 277–284. [CrossRef]
6. Pincus, S.M. Approximate entropy as a measure of system complexity. *Proc. Natl. Acad. Sci. USA* **1991**, *88*, 2297–2301. [CrossRef] [PubMed]
7. Richman, J.S.; Moorman, J.R. Physiological time-series analysis using approximate entropy and sample entropy. *Am. J. Physiol. Heart Circ. Physiol.* **2000**, *278*, H2039–H2049. [CrossRef] [PubMed]
8. Liu, C.; Li, K.; Zhao, L.; Liu, F.; Zheng, D.; Liu, C.; Liu, S. Analysis of heart rate variability using fuzzy measure entropy. *Comput. Biol. Med.* **2013**, *43*, 100–108. [CrossRef] [PubMed]
9. Bandt, C.; Pompe, B. Permutation Entropy: A Natural Complexity Measure for Time Series. *Phys. Rev. Lett.* **2002**, *88*, 174102. [CrossRef] [PubMed]
10. Aziz, W.; Arif, M. Multiscale Permutation Entropy of Physiological Time Series. In Proceedings of the 2005 Pakistan Section Multitopic Conference, Karachi, Pakistan, 24–25 December 2005; pp. 1–6. [CrossRef]
11. Costa, M.; Peng, C.K.; Goldberger, A.L.; Hausdorff, J.M. Multiscale entropy analysis of human gait dynamics. *Physica A* **2003**, *330*, 53–60. [CrossRef]

12. Zanin, M.; Zunino, L.; Rosso, O.A.; Papo, D. Permutation Entropy and Its Main Biomedical and Econophysics Applications: A Review. *Entropy* **2012**, *14*, 1553–1577. [CrossRef]
13. Humeau-Heurtier, A.; Wu, C.W.; Wu, S.D. Refined Composite Multiscale Permutation Entropy to Overcome Multiscale Permutation Entropy Length Dependence. *IEEE Signal Process. Lett.* **2015**, *22*, 2364–2367. [CrossRef]
14. Bandt, C.; Shiha, F. Order Patterns in Time Series. *J. Time Ser. Anal.* **2007**, *28*, 646–665. [CrossRef]
15. Zunino, L.; Pérez, D.G.; Martín, M.T.; Garavaglia, M.; Plastino, A.; Rosso, O.A. Permutation entropy of fractional Brownian motion and fractional Gaussian noise. *Phys. Lett. A* **2008**, *372*, 4768–4774. [CrossRef]
16. Little, D.J.; Kane, D.M. Permutation entropy of finite-length white-noise time series. *Phys. Rev. E* **2016**, *94*, 022118. [CrossRef] [PubMed]
17. Dávalos, A.; Jabloun, M.; Ravier, P.; Buttelli, O. Theoretical Study of Multiscale Permutation Entropy on Finite-Length Fractional Gaussian Noise. In Proceedings of the 26th European Signal Processing Conference (EUSIPCO), Rome, Italy, 3–7 September 2018; pp. 1092–1096.
18. Friedlander, B.; Francos, J.M. Estimation of Amplitude and Phase Parameters of Multicomponent Signals. *IEEE Trans. Signal Process.* **1995**, *43*, 917–926. [CrossRef]
19. Miller, K.S. On the Inverse of the Sum of Matrices. *Math. Mag.* **1981**, *54*, 67–72. [CrossRef]

Article

Voronoi Decomposition of Cardiovascular Dependency Structures in Different Ambient Conditions: An Entropy Study

Dragana Bajic [1,*], Tamara Skoric [1], Sanja Milutinovic-Smiljanic [2] and Nina Japundzic-Zigon [3]

1 Faculty of Technical Sciences-DEET, University of Novi Sad, Trg Dositeja Obradovica 6, 21000 Novi Sad, Serbia; tamara.ceranic@gmail.com or ceranic@uns.ac.rs

2 Faculty of Dental Medicine, University of Belgrade, Dr. Subotica 1, 11000 Belgrade, Serbia; sanja.milutinovic@stomf.bg.ac.rs

3 School of Medicine, University of Belgrade, Dr. Subotica 1, 11000 Belgrade, Serbia; nina.japundzic@gmail.com or nzigon@med.bg.ac.rs

* Correspondence: dragana.bajic@gmail.com or draganab@uns.ac.rs; Tel.: +381-65-24-36-441

Received: 15 August 2019; Accepted: 9 November 2019; Published: 11 November 2019

Abstract: This paper proposes a method that maps the coupling strength of an arbitrary number of signals D, $D \geq 2$, into a single time series. It is motivated by the inability of multiscale entropy to jointly analyze more than two signals. The coupling strength is determined using the copula density defined over a $[0\ 1]^D$ copula domain. The copula domain is decomposed into the Voronoi regions, with volumes inversely proportional to the dependency level (coupling strength) of the observed joint signals. A stream of dependency levels, ordered in time, creates a new time series that shows the fluctuation of the signals' coupling strength along the time axis. The composite multiscale entropy (*CMSE*) is then applied to three signals, systolic blood pressure (*SBP*), pulse interval (*PI*), and body temperature (t_B), simultaneously recorded from rats exposed to different ambient temperatures (t_A). The obtained results are consistent with the results from the classical studies, and the method itself offers more levels of freedom than the classical analysis.

Keywords: copula density; dependency structures; Voronoi decomposition; multiscale entropy; ambient temperature; telemetry; systolic blood pressure; pulse interval; thermoregulation; vasopressin

1. Introduction

Approximate [1,2] and sample entropies [3], *ApEn* and *SampEn*, have been intensively implemented in a range of scientific fields to quantify the unpredictability of time series fluctuations. Contributions that apply *ApEn* and *SampEn* are measured by thousands [4], confirming their significance. The cross entropies—*XApEn* and *XSampEn*—are designed to measure a level of asynchrony of two parallel time series [3,5,6]. Descriptions of (cross) entropy concepts can be found in numerous articles, but a recent comprehensive review [7] provides an excellent tutorial with the guidelines aimed to help the research society to understand *ApEn* and *SampEn* and to apply them correctly [7].

Multiscale entropy (*MSE*) [8,9], based on *SampEn*, investigates the changes in complexity caused by a change of the time scale. Composite *MSE* (*CMSE*) performs an additional averaging, thus solving the problem of decreased reliability induced by temporal scaling [10,11]. A comprehensive study of fixed and variable thresholds at different scales also presents an excellent review of the *MSE* improvements [12].

The benefits offered by entropy are explored in cardiovascular data analysis. Entropy was implemented to determine the cardiac variability [13], the complexity changes in cardiovascular disease [14], a level of deterministic chaos of heart rate variability (*HRV*) [9], *HRV* complexity in diabetes patients [15], in heart failure [16], in stress, [17,18] or in different aging and gender groups [19,20], while multiscale cross-entropy was applied for health monitoring systems [11].

SampEn- or *ApEn*-based entropy estimates are designed for one signal, or at most for two signals (cross-entropy), but biomedical studies often require an analysis of three or more simultaneously recorded signals.

We propose a method that maps levels of interaction of two or more time series into a single signal. Levels of interaction are assessed using the copula density [21]. The transformation from the probabilistic copula domain to the beat-to-beat time domain is performed by Voronoi decomposition.

The method is applied to multivariate time series that comprises three simultaneously recorded signals: systolic blood pressure (*SBP*), pulse interval (*PI*), and body temperature (t_B) recorded at different ambient temperatures (t_A). It is well known that thermoregulation can affect cardiovascular homeostasis [22]. Analysis of heart rate (*HR*) and *SBP* in the spectral domain has shown that changes of ambient temperature modulate vasomotion in the skin blood vessels, reflected in the very-low-frequency range of *SBP* and reflex changes in *HR* spectra [23,24]. Thermoregulation is complex and involves autonomic, cardiovascular, respiratory as well as a metabolic adaptation [25–28]. The key corrector of blood pressure is the baroreceptor reflex (BRR). The disfunction of BRR is the hallmark of cardiovascular diseases with a bad clinical prognosis. Thus, evaluating its functioning is important not only for the diagnosis and prognosis of cardiovascular diseases but also for the evaluation of treatment.

The aims of this study are:

1. To propose a method that enables an application of multiscale entropy to an arbitrary number of signals and to analyze the outcome;
2. To compare the results of the classical multiscale method and the proposed method when applicable, i.e., in a case of two-dimensional signals;
3. To test whether the proposed method recognizes the changes of dependency level (coupling strength, level of interaction) of joint multivariate signals in different biomedical experiments.

The paper is organized as follows: the experimental setting for signal acquisition is explained in Section 2.1, together with surrogate signals and artificially generated control signals. The signal pre-processing that ensures the reliability of the results is explained in Section 2.2. Section 2.3. shows the mathematical tools assembled to create the proposed method: it gives an introduction to the copula theory, it outlines copula advantages and applications, and it discusses the various procedures for density estimation to justify the preference of Voronoi decomposition.

Section 3.1. shows the basic statistical analysis of the experimental *SBP*, *PI*, and t_B signals. For the sake of comparison, this section includes the outcomes of classical *(X)SampEn* and *CMSE* entropy analysis. Section 3.2. introduces the new signal, created by the proposed method, for a two-dimensional case (*SBP* and *PI* mapped into the new $D = 2$ signal) and a three-dimensional case (*SBP*, *PI*, and t_B mapped into the new $D = 3$ signal). In both cases, the *SBP-PI* offset (delay) is taken into account ranging from 0 to 5 beats [29]. The wide sense stationarity of the created signals is checked and the correction proposed. The signals' statistical properties, in terms of skewness and kurtosis, are estimated and discussed. In Section 3.3., the entropy parameters are analyzed and the proper ones that ensure the reliable estimates are selected. Then, the results of experiments performed to justify the consistency with the classical methods (in cases when the comparison is possible) are presented. The results showing that the method recognizes the changes in the level of signal interaction in various experimental environments are presented as well. The results are discussed in Section 4 with respect to the aims of this paper. The same section gives the conclusion and the possibilities for further method applications.

A brief description of well-known entropy concepts—*ApEn*, *SampEn*, *MSE*, and *CMSE*—is included in the Appendix A.

2. Materials and Methods

2.1. Experimental Setting and Signal Acquisition

All experimental procedures conformed to Directive 2010/63/EU National Animal Welfare Act 2009/6/RS and Rule Book 2010/RS. The protocol was approved by the University of Belgrade Ethics review board (license n°323-07-10519/2013-05/2).

Adult male Wistar outbred rats, weighing 300–350 g, housed under control laboratory conditions (temperature—22 ± 2 °C; relative humidity: 60–70%; lighting: 12:12 h light-dark cycle) with food (0.2% NaCl) and tap water ad libitum were used in experimentation. Vasopressin selective antagonists of V1a or V2 receptors were injected via cannula chronically positioned in the lateral cerebral ventricle of the rat. The concomitant measurement of blood pressure waveforms (*BP*) and body temperature was performed using TL11M2-C50-PXT (DSI, St. Paul, MN, USA) equipment implanted into the abdominal aorta. The measurements were performed at the neutral ambient temperature (NT), 27 rats at $t_A = 22 \pm 2$ °C, and the increased ambient temperature (HT), 28 rats at $t_A = 34 \pm 2$ °C. The four rats recorded at the low temperature (LT), $t_A = 12 \pm 2$ °C, were included as an illustrative example. There are five subgroups in NT and HT groups: control group, V1a-100 ng, V1a-500 ng, V2-100 ng, and V2-500 ng. The experimental timeline is shown in Figure 1.

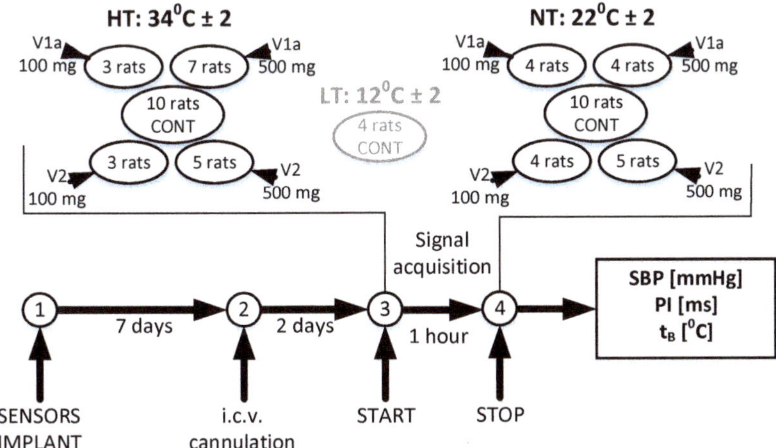

Figure 1. The experimental timeline and the signal subgroups. The high temperature (HT) experiment includes 28 animals exposed to 34 ± 2 °C ambient temperature; the neutral temperature (NT) experiment includes 27 animals exposed to 22 ± 2 °C ambient temperature; ten animals from each group were controls (CONT), the others got V1a and V2 antagonists, either 100 ng or 500 ng; the low temperature (LT) experiment contains four control animals exposed to 12 ± 2 °C ambient temperature; it is included as an illustration.

The experimental environment includes two types of control signals. The first controls are isodistributional surrogate data [30,31]. Surrogate data are derived from the experimental time series by randomizing the property that needs to be tested, keeping the other signal attributes intact. Thus, isodistributional surrogates randomly permute the signal to destroy the orderliness that is checked by entropy analysis. The signal distribution function remains unchanged. The second controls are artificially generated signals—a series of independent and identically distributed (i.i.d.) samples with Gaussian distribution and with exponential distribution. Gaussian signals possess a unique property in which linear independency implies statistical independency [32], and that it is an asymptotic distribution of the sum of i.i.d. samples (with some constraints) [32], an issue important for the multiscale entropy coarse-graining. Signals with exponential distribution are often implemented when there is a need to test the signals with large variance.

2.2. Signal Pre-Processing

Arterial blood pressure (*BP*) and body temperature signals were acquired using a sampling frequency of 1000 Hz. Systolic blood pressure (*SBP*) and pulse interval (*PI*) time series were derived from the *BP* waveforms as the local maxima and as the intervals between the successive maximal *BP* positive changes, respectively. The samples from the body temperature signals were taken simultaneously with *SBP* to create body temperature beat-to-beat time series t_B. Artifacts were detected semi-automatically using the filter [33] adjusted to the signals recorded from the laboratory rats. A visual examination was then performed to find the residual errors. A very low signal component (trend) was removed by a high-pass filter designed for biomedical time series [34], thus ensuring *SBP*, *PI*, and t_B signal stationarity. All the signals were cut to the length of the shortest time series, n = 14,405 samples. The time series $X_1 = SBP$, $X_2 = PI$ and $X_3 = t_B$ jointly create a single three-dimensional signal ($D = 3$). Its samples X_{1k}, X_{2k}, and X_{3k}, $k = 1, \ldots, N$ create points in the three-dimensional signal space.

2.3. Copula Density, Voronoi Regions and Dependency Time Series

A copula is a mathematical concept that provides a multidimensional probability density function, where density reflects the level of signal interaction (dependency, coupling). It is introduced in 1959 [21] as a multivariate distribution function with marginals uniformly distributed on $[0\ 1]^D$. If $X_1, \ldots, X_D$ are the source signals with joint distribution function H and univariate marginal distribution functions $F_1, \ldots, F_D$, then copula C is defined as [21]:

$$H(X_1, \ldots, X_D) = C(F_1(X_1), \ldots, F_D(X_D)) \tag{1}$$

and vice versa:

$$C(U_1, \ldots U_D) = H\left(F_1^{-1}(U_1), \ldots, F_D^{-1}(U_D)\right). \tag{2}$$

Sklar's theorem [21] states that any D-dimensional joint distribution H with arbitrary univariate marginals could be decomposed into D independent uniform marginal distributions, bound together by a new joint distribution function C, called copula.

The concept of the copula is based on the classical transformation of a random variable. Any continuous variable X_i with a distribution function $F_i\,(X_i)$ and density $f_i\,(X_i) = \frac{d\,F_i\,(X_i)}{dX_i}$, $i = 1, \ldots, D$ can be transformed using a monotone function $U_i = \varphi_i\,(X_i\,)$. The result is a variable U_i with a probability density function [32] $u_i(U_i\,) = \frac{f_i(X_i\,)}{|d(\varphi_i\,(X_i\,))/dX_i\,|}$, $X_i = \varphi_i\,^{-1}(U_i\,)$. The transformation function $\varphi_i\,(X_i)$ that creates a copula is the distribution function $F_i(X_i\,)$ of the signal X_i, i.e., $\varphi_i\,(X_i\,) = F_i(X_i)$. The new variable U_i is then defined in [0, 1], as the following holds: $0 \le F_i(X_i\,) \le 1$. It can be easily shown that the probability density function (PDF) of the new variable U_i is uniform:

$$u_i(U_i\,) = \frac{f_i(X_i\,)}{|d(F_i(X_i\,))/dX_i\,|} = \frac{f_i(X_i\,)}{f_i(X_i\,)} = 1. \tag{3}$$

The transformation of a random variable using its distribution function is known as probability integral transform, PI-transform, or PIT [35], and it is a core of the copula theory. It should be noted that the distribution function of a continuous variable, by definition, monotonically increases so the denominator in Equation (3) is positive, comprising just a single term.

The copula has been intensively used for the analysis and prediction of financial time series and the prediction of insurance risk [36–39], in hydrology and climate analysis [40–42] and communications [43]. Medical applications include aortic regurgitation study [44] and diagnostic classifiers for neuropsychiatric disorders [45]. A possibility to use a bivariate copula to analyze the cardiovascular dependency structures was introduced in [46] and pharmacologically validated by blocking the feedback paths using Scopolamine, Atenolol, Prazosin, and Hexamethonium. It was shown that Frank's copula is the most appropriate to quantify the level of dependency of cardiovascular signals.

Copula density $c(U) = \frac{\partial^N C(U_1,\cdots, U_D)}{\partial U_1 \cdots \partial U_D}$ is used to visualize the intensity of signal coupling. The regions of increased copula density indicate the regions where the dependency of the signal samples increases. The difference between a classical bivariate probability density function (PDF) and the corresponding copula density of *SBP* and *PI* signals is that PDF shows the distribution of amplitude levels, while copula density shows the distribution of coupling strength between these amplitudes, regardless of the absolute amplitude values. An illustration of this difference is shown in Figure 2. *SBP* and *PI* signals and their probability integral transformed (PIT) counterparts are separated in time, first by two heartbeats (SBP_k is coupled with PI_{k+2}, $k = 1, 2, \ldots, N{-}2$), and then by ten heartbeats (SBP_k is coupled with PI_{k+10}, $k = 1, 2, \ldots, N{-}10$). The copula density in panel b exhibits a distinct linear positive coupling structure that follows the known physiological relationships [47]. The copula density in panel d shows almost uniform distribution as the time offset between *SBP* and *PI* signals is sufficiently large to attenuate their mutual dependency. Contrary to copula density, the joint probability density functions are almost the same in both cases (panels a and c). The temporal separation of *SBP* and *PI* signals does not alter the mutual relationship of signal amplitudes, but it significantly alters the intensity of signal coupling.

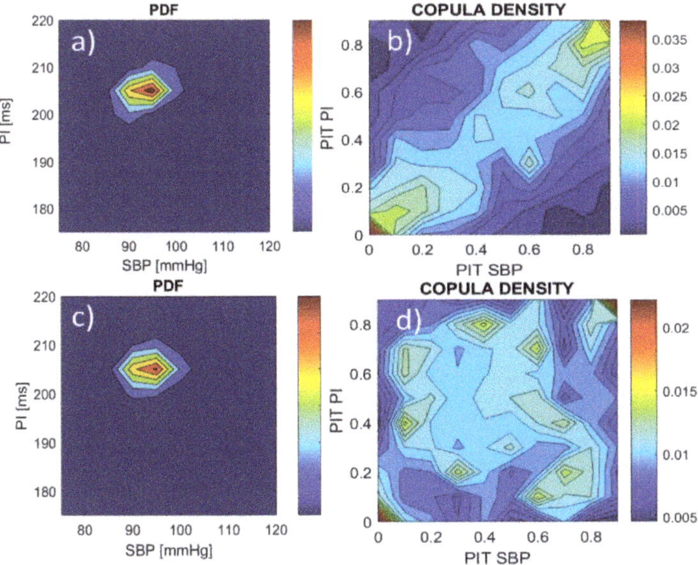

Figure 2. Bivariate probability density function (PDF) of *SBP* and *PI* signals and copula density of probability integral transformed (PIT) signals; (**a**,**b**) the offset between *PI* and *SBP* is equal to two beats; (**c**,**d**) the offset between *PI* and *SBP* is equal to ten beats; note that the PDFs in (**a**,**c**) are almost the same in spite of different *SBP-PI* offsets, while the copula density exhibits a strong positive dependency when offset is small (**b**), and a lack of dependency when offset is large (**d**).

The advantages of copula are numerous. Copula density visualizes the dependency structures of the observed signals, and it quantifies the signal coupling strength ("copula parameter"). It captures both linear and nonlinear relationships between the signals. It can quantify the intensity of signal coupling within the different regions of the copula domain, and, in particular, it can model the tail dependencies of the signals.

Such a visualization, in a case of *SBP-PI* signals, cannot be achieved by other methods: the Oxford method, the oldest and the referent procedure for the evaluation of the baroreceptor reflex, uses increasing doses of short-acting vasoconstrictors (e.g., phenylephrine) and vasodilators (e.g., nitroprusside) to trigger heart deceleration of acceleration. The *SBP* and *PI* relationship is plotted as a fitted sigmoid curve. It is an invasive method, and it does not show spontaneous BRR. The most acknowledged among the non-invasive approaches is the sequence method, with the visualization that shows the scatterplot of the signal points that are elements of BRR sequences (i.e., the scatterplot contains a subset of all signal points). The method quantifies the spontaneous BRR operating range and set point [48], but the visualization is similar to the classical probability density function.

The time offset (delay) between *SBP* and *PI* signals is important for the signal coupling, and it depends on species. It was shown [49] that the delay of 0, 1, and 2 beats is the most appropriate for humans, while the delays of 3, 4 and 5 beats are appropriate for rats [29] and mice [47]. In [50], it was shown that, in laboratory rats, the highest level of comonotonic behavior of pulse interval and systolic blood pressure is observed at time lags 0, 3, and 4 beats, while a strong counter-monotonic behavior occurs at time lags of 1 and 2 beats.

Copula density is a probabilistic quantity. To convert it into a time series, to each point in the time domain, an appropriate density (dependency level) DL_k should be assigned, thus creating a dependency signal $DL_k, k = 1, \ldots, N$.

A trivial way to estimate a copula density is to create a D-dimensional histogram. The obtained DL_k signal would be discrete, as the points within the same histogram bins would get the same value. An increased number of histogram bins would increase the number of discrete signal levels, but the estimation reliability would decrease.

Density estimation based on Markov chains [51] creates a stochastic matrix of "transition probabilities"—scaled distances—between the points, with the steady-state probabilities proportional to the required distribution. The method is computationally inefficient in multidimensional space, except for the short time series.

A D-dimensional sphere (or cube) around a particular signal point defines a local density according to the number of encircled neighbors. The procedure is efficient, but the neighboring spheres overlap inducing the bias, and the result depends on the sphere diameter (i.e., threshold) choice.

The chosen approach expresses the sample density proportionally to the non-overlapping free space surrounding the sample. Such a concept has long been known as the Voronoi region. It can be traced back to the scholars from the 17th and 18th centuries, but it was re-discovered, analyzed, and its applications outlined at the beginning of the 20th century [52].

The concept is simple: Let A be the set of all points in a $[0\ 1]^D$ copula space. Let $U_k = [U_{1k}, \ldots, U_{Dk}],\ k = 1, \ldots, N$ be a D-dimensional point from a PI-transformed multivariate time series. Then, the Voronoi region R_k^D around the point U_k comprises all the points from A that are closer to the particular point U_k then to any other point $U_j,\ j = 1, \ldots, N,\ j \neq k$. More formally,

$$R_k^D = \left\{a \in A \,\middle|\, d(a, U_k) \leq d\left(a, U_j\right),\ \forall\, j \neq k\right\}. \tag{4}$$

A classical Euclidean distance is typically chosen for the distance $d(a,\ U_k)$, but any other distance measure can be used as well, resulting in different Voronoi decompositions.

Figure 3 shows examples of the Voronoi regions in two and three dimensions. The line segments that separate particular Voronoi cells R_k^2 and R_j^2 in the left panel of Figure 3 are the sets of the points $a \in A$ that are equidistant to the points U_k and U_j, i.e., $d(a,\ U_k) = d\left(a, U_j\right)$. The Voronoi vertex $a \in A$ in the same panel is the point equidistant to three (or more) time series points, e.g., $d(a,\ U_k) = d\left(a,\ U_j\right) = d(a,\ U_l)$. The right panel ($D$ = 3) also shows Voronoi lines and vertices, but, in the $[0\ 1]^3$ domain, this is more difficult to visualize. Uncolored Voronoi regions are either unbounded, or the boundaries are outside the $[0\ 1]^D$ space. These regions are cut to fit the $[0\ 1]^D$ space.

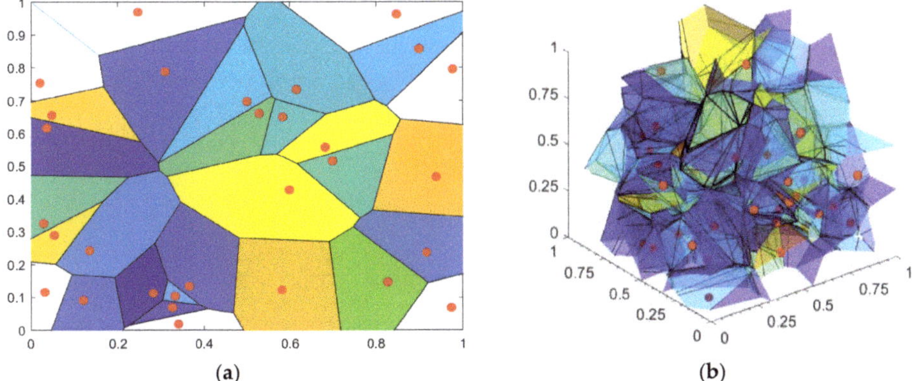

Figure 3. Voronoi region (polytope) and the corresponding signal points. (**a**) An example of Voronoi cells in a two-dimensional plane ($D = 2$); (**b**) An example of Voronoi polyhedrons in three dimensions ($D = 3$). The uncolored cells/polyhedrons both in (**a**) and (**b**) are cut to fit the $[0\ 1]^D$ space.

A series of surface areas in two-dimensional Voronoi regions and a series of volumes in three-dimensional Voronoi regions are a good foundation to quantify the dependency level and to form the time series $DL_k, k = 1, \ldots, N$, as:

(a) The surface/volume of R_k^D is inversely proportional to the dependency level of the point U_k. An increased density of dependency structures in $[0\ 1]^D$ space implies a decrease of available space between the points.

(b) The region R_k^D is shaped like the best distance separation of the point U_k, so its surface/volume is unambiguously calculated and unique, without a necessity to include any thresholds.

The drawback of the method is that a change of distance measure changes the shape of regions. We have opted for Euclidian distance as a classical approach for distance measurement, widely used in a wide range of applications.

3. Results

3.1. Source Signal Analysis

The total number of *SBP*-*PI*-t_B signal triplets is equal to 59. The basic statistical parameters, shown as a control, are presented in Table 1. Results in Table 1 show no significant changes in statistical parameters of *SBP*, *PI*, and t_B signals. An earlier study [26] revealed that V1a antagonists increase body temperature. The differences might be the outcome of different measurement procedures: in this study, the temperature is measured using a telemetric probe in the abdominal aorta, while, in [26], the temperature was measured rectally.

Figure 4 shows the results of the classical entropy analysis, performed for the sake of comparison. The left panels show the *CMSE* of the signals recorded from control animals at different ambient temperatures. The middle panels show the effect of drugs at the neutral temperature. The left panels show the effects of drugs at a high temperature. Each signal is accompanied by ten isodistributional surrogate signals, generated by a random temporal permutation of the signal samples [30,31].

The first three rows in Figure 4 show the classical composite multiscale entropy analysis of a single-dimensional time series, *SBP*, *PI*, and t_B, respectively. The last row shows multiscale *SBP*-*PI* cross-entropy that can be compared to the multiscale entropy of the new signals.

Table 1. Statistical parameters of the source data (mean ± standard deviation).

Ambient Temperature (°C)	Drug	SBP	(mmHg)	PI	(ms)	t_B	(°C)
NT 22 ± 2	Control	112.81	±19.54	179.22	±33.22	38.07	±0.29
	V1a, 100 mg	115.62	±12.17	173.74	±20.69	38.42	±0.10
	V1a, 500 mg	110.28	±15.35	184.77	±28.39	38.05	±0.10
	V2, 100 mg	119.98	±16.53	184.79	±38.30	38.54	±0.38
	V2, 500 mg	108.61	±14.79	176.16	±4.04	38.33	±0.41
HT 34 ± 2	Control	107.26	±4.19	188.63	±8.95	38.27	±0.34
	V1a, 100 mg	107.90	±10.52	197.08	±21.63	38.52	±0.26
	V1a, 500 mg	110.40	±10.07	177.21	±16.34	38.57	±0.57
	V2, 100 mg	113.26	±15.41	193.14	±30.65	38.01	±0.37
	V2, 500 mg	114.28	±6.14	184.23	±12.97	38.33	±0.47
LT 12 ± 2	Control	115.22	±5.23	164.54	±24.31	37.51	±0.43

Note: Results are presented as mean ± standard deviation; SBP: systolic blood pressure; PI: pulse interval; t_B: body temperature; NT: neutral temperature; HT: high temperature; LT: low temperature.

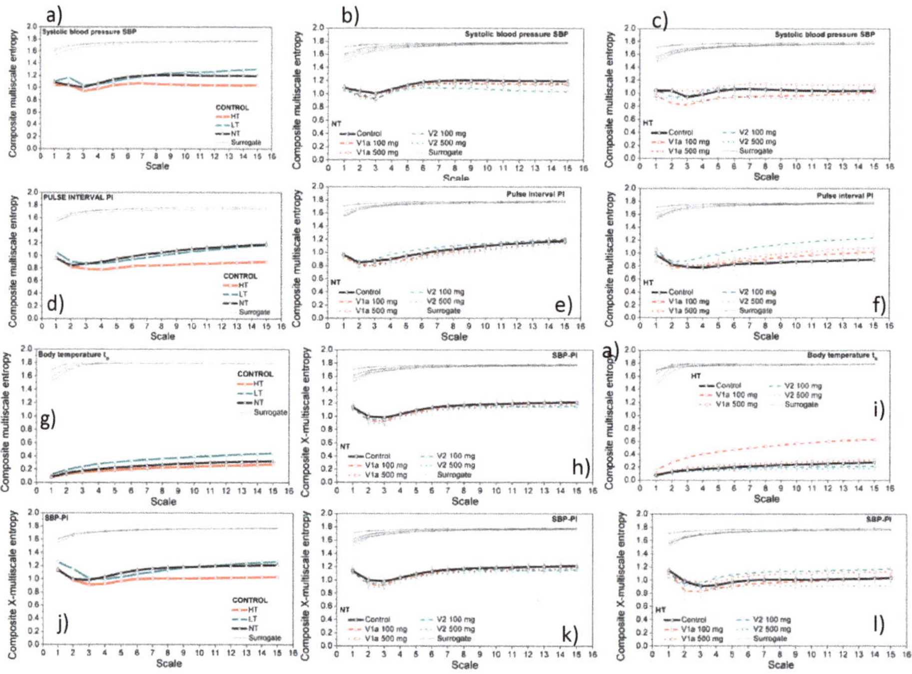

Figure 4. Composite multiscale entropy estimated from the source signals. From top to bottom, entropy was applied to *SBP, PI*, t_B signals, and *SBP-PI* signal pairs. Left panels: entropy of the control signals at different ambient temperatures; middle panels: effects of antagonists at neutral temperature; right panels: effects of antagonists at high temperature. Results are presented as a mean ± SE (standard error).

3.2. Properties of the Dependency Time Series

The created time series are new signals, so their statistical properties need to be checked before entropy analysis.

Mapping the signals into the dependency time series takes into account the delay (offset) between the *PI* and *SBP* signals. The time delay (offset) $DEL = 0, \cdots, 5$ [beats] applied to each pair of *SBP-PI* signals resulted in six two-dimensional (2D) time series (SBP_k, PI_{k+DEL}), and six three-dimensional (3D) time series $(SBP_k, PI_{k+DEL}, t_{Bk})$, $k = 1, \ldots, N - DEL$. The total of 354 *SBP-PI* pairs and 354 *SBP-PI-*t_B triplets were converted into two-dimensional and three-dimensional Voronoi cell time series. An average percentage of Voronoi cells that had to be cut to fit the $[0\ 1]^D$ space was 2.68% for two-dimensional, and 16.26% for three-dimensional signals (cf. Figure 3). Additionally, 11 signal points (0.0002%) were too close to the vertices of the $[0\ 1]^3$ cube to generate the three-dimensional polyhedrons, so they were managed manually.

Examples of Voronoi cell time series are shown in Figure 5.

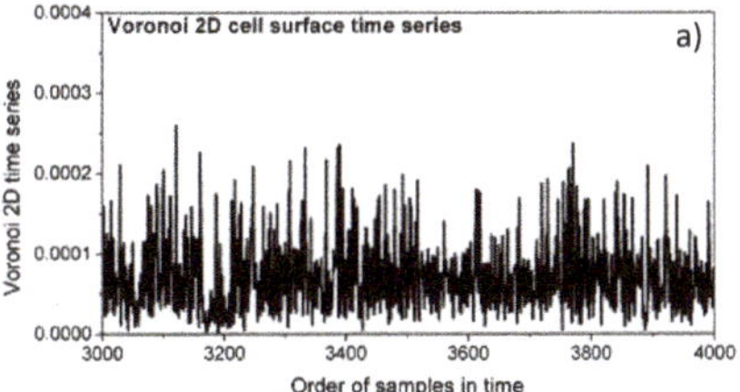

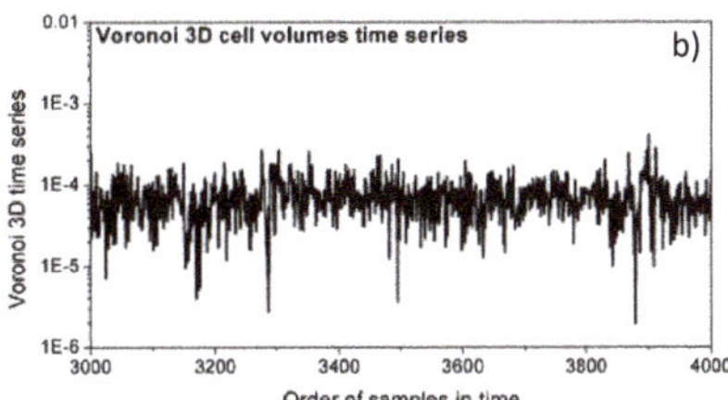

Figure 5. Samples of Voronoi cells time series. (**a**) two-dimensional signals (*SBP* and *PI* interaction, $D = 2$); (**b**) three-dimensional signals (*SBP*, *PI* and t_B interaction, $D = 3$).

A wide sense stationarity (WSS) test [53] is then applied, as the stationarity is an obligatory prerequisite for entropy estimation [7,38]. The test checks the stationarity of the first and the second statistical moments. The three-dimensional Voronoi cells time series failed the second-moment test.

Figure 6 shows the negative effects of non-stationarity: a three-dimensional non-stationary Voronoi cell time series is cut into 14 successive segments, each one comprising $n = 1000$ signal points. Then, mean, variance, and entropy were estimated from each segment and plotted in Figure 6a. Figure 6b shows the same parameters but estimated from the two-dimensional stationary Voronoi cells time series.

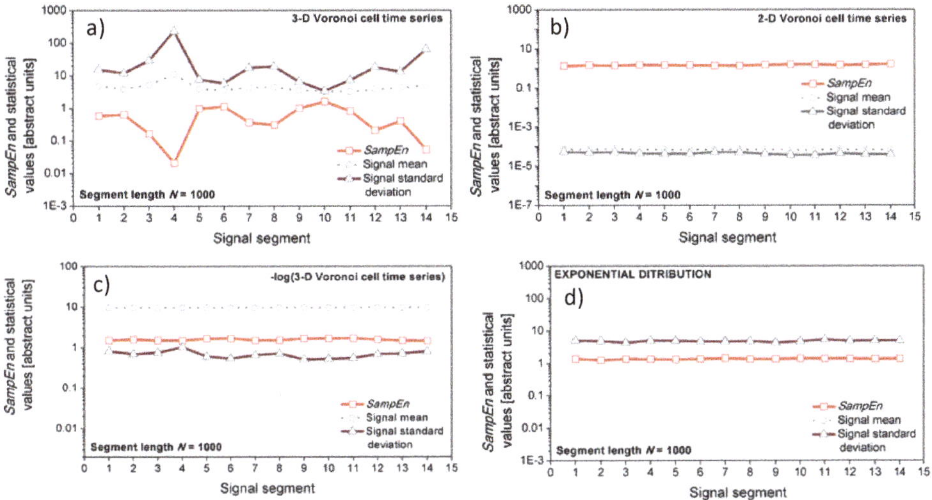

Figure 6. *SampEn*, mean and variance of a time series. Note the high variability of entropy estimated in the different segments of three-dimensional Voronoi cells time series (**a**) (*SBP*, *PI* and t_B interaction, $D = 3$), smoothed by logarithm; (**c**) two-dimensional Voronoi cells time series (*SBP* and *PI* interaction, $D = 2$) are stationary (**b**) as well as the signal created from the interaction of three exponentially distributed random signals, $D = 3$ (**d**).

The difference between two- and three- dimensional signals is a consequence of coverage. The number of signal points in two-dimensional space is sufficient to ensure good coverage. The same points are sparsely and unevenly scattered in three-dimensional space, so the estimation is unreliable, resulting in different values obtained from the different sections of the same signal.

Panel d in Figure 6 shows the results from artificially generated time series with exponentially distributed samples. It is a usual control example of a signal with large, but time-invariant, variance. It passed the stationarity test and the parameters estimated from it are constant in each segment.

Taking a logarithm is a procedure that ensures the stationarity of the second moment. A negative logarithm corresponds to the inverse of the Voronoi cell volume, and it is proportional to the local sample density. It is always positive as the inverse of any Voronoi cell volume in $[0\ 1]^D$ domain is greater than 1. Panel c of Figure 6 is a visual confirmation of a successful test outcome.

The dependency level time series, *DL*, is finally defined as the negative logarithm of the Voronoi cell time series. The number of signal points (N = 14,400) ensures the signal stationarity at least in the wide sense for two-dimensional (D = 2) and three-dimensional (D = 3) signals.

The statistical properties of the new signals—probability density function, skewness, and kurtosis—are shown in Figures 7 and 8.

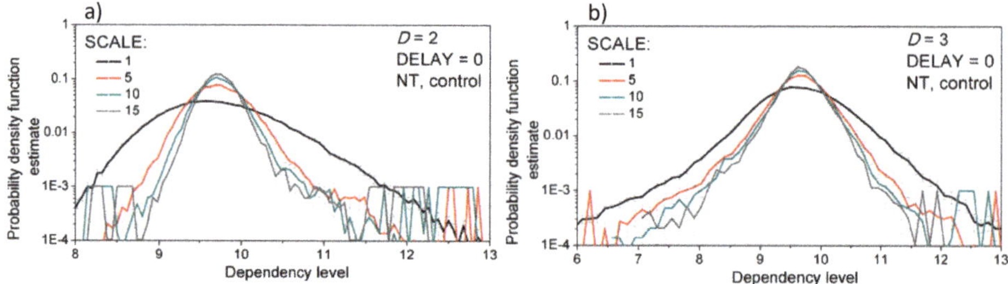

Figure 7. Empirical probability density function of the created signals, averaged over 10 control rats at neutral temperature (NT). (**a**) two-dimensional signals (*SBP* and *PI* interaction, D = 2); (**b**) three-dimensional signals (*SBP*, *PI* and t_B interaction, D = 3). The *SBP-PI* offset (DELAY) is equal to 0 beats. Results are presented as a mean ± SE (standard error).

Empirical probability density functions for different scaling levels are plotted in Figure 7. Signals are normalized and centered so the changes of mean and variance due to the convolution are not visible. There is no significant difference in the estimated probability density functions for the scaling levels greater than five.

Skewness is a third statistical moment that shows a level of signal asymmetry around the mean. The skewness of the two-dimensional dependency signal (*SBP* and *PI* interaction) is presented in Figure 8a). It is positive, with a right tail exhibited, indicating the existence of signals with a strong dependency level between systolic blood pressure and a pulse interval. The positive skewness increases with the increasing offset (delay) *DEL* between *SBP* in *PI*. It is in accordance with [29] that located the dominant *SBP-PI* relationships at offset of 3, 4, and 5 beats.

The skewness of the three-dimensional dependency signal (*SBP*, *PI*, and t_B interaction) is presented in Figure 8b). It is close to zero—slightly negative—so the level of the dependency between the three signals is almost symmetric. It may indicate that the inclusion of body temperature into the new signal attenuates the *SBP-PI* signal coupling. The increase of the *SBP-PI* offset (delay) *DEL* results in the increased skewness shifted closer to zero, towards the positive values, again in accordance with [29].

Kurtosis measures the intensity of probability density function "tails". It is shown in Figure 8b for two-dimensional signals (*SBP* and *PI* interaction) and in Figure 8d for three-dimensional signals (*SBP*, *PI*, and t_B interaction). The tails of dependency signals are heavy if compared to Gaussian distribution, indicating an increased number of signals with very high and very low dependency levels.

It is expected, due to the high variance of three-dimensional dependency signals. The intensity of tails increases with the increased *SBP-PI* offset *DEL*, and it also increases with an increase of scale. This is also expected, as scaling convolves the probability density functions of the components that are coarse-grained. The convolution emphasizes the tail parts of the distribution in spite of the normalization, as the convolved samples are not Gaussian.

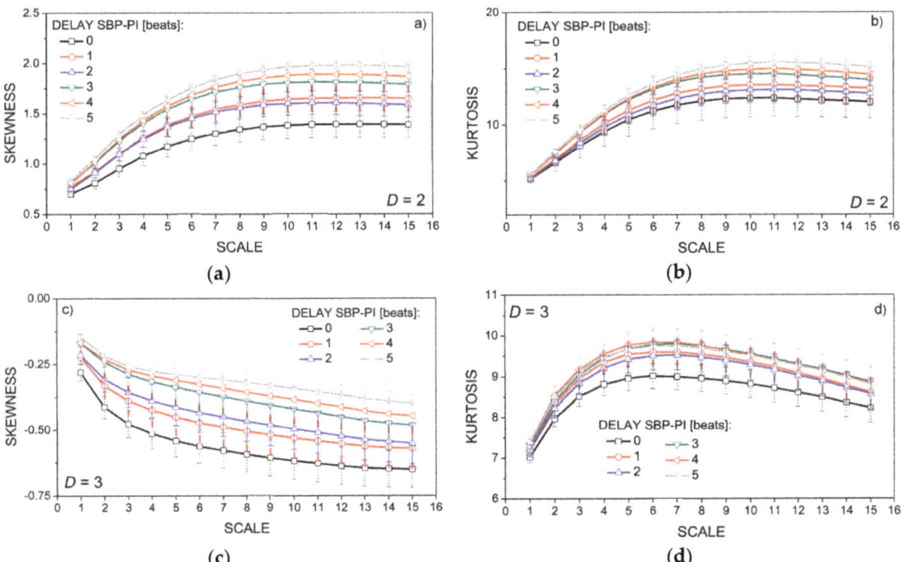

Figure 8. Skewness (panels (**a**,**c**)) and kurtosis (panels (**b**,**d**)) for different SBP-PI offset (DELAY), averaged over all 59 created signals; panels (**a**,**b**) two-dimensional signals (*SBP* and *PI* interaction, $D = 2$); panels (**c**,**d**) three-dimensional signals (*SBP*, *PI*, and t_B interaction, $D = 3$); results are presented as a mean ± SE.

3.3. Entropy Analysis of the Dependency Time Series

Entropy is a parametric method, with parameters determined to ensure its reliable estimation. The statistical analysis from Section 3.2, however, is insufficient to provide the guidelines for entropy parameter selection for the new dependency time series.

It has already been pointed out [18,54–59] that the threshold (filter) r (cf. Equation (A3)) is one of the major causes of inconsistency in entropy estimation and that its choice is related to the series length N. Thus, the threshold and length profiles of the dependency time series are plotted in Figure 9. Although the multiscale entropy is defined on the *SampEn* basis, the figure also includes *ApEn* as the worst-case example.

In a multiscale entropy approach, the time series length step-wise decreases with the increased scaling level. Maximal series length of our signals is equal to $n = 14{,}400$. If the scaling level is equal to 15, then the minimal series length would be equal to $n = 960$. Panels a and c of Figure 9 show the threshold profile of *ApEn* and *SampEn* for the maximal and for the minimal lengths. Stable results are achieved for threshold $r = 0.3$ [18].

The length profile is plotted in panels b and d for the typical threshold value $r = 0.15$ and the chosen threshold value $r = 0.3$. It can be seen that the results are not consistent for lengths below $n = 900$, so the choice of 15 scaling levels is justified.

Figures 10–14 show the main result of the composite multiscale entropy study of dependency level signals, with the scaling level set to 15, and the threshold level set to $r = 0.3$. These results are discussed in Section 4.

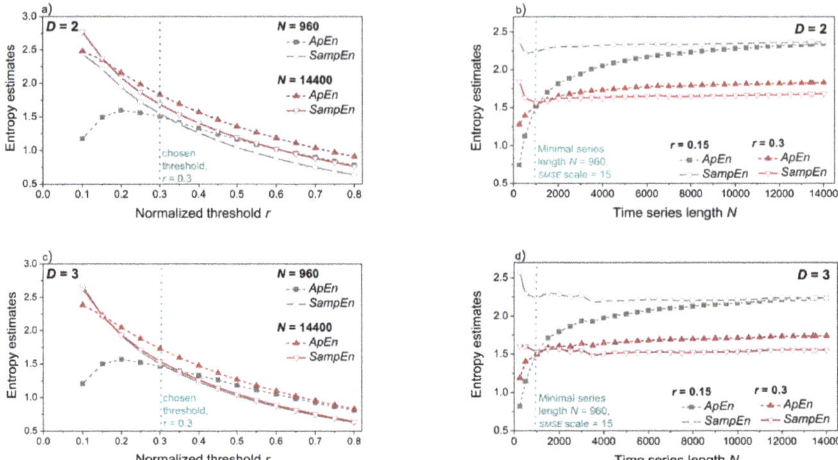

Figure 9. Threshold profile (panels (**a**,**c**)) and length profile (panels (**b**,**d**)) of a single subject; vertical dashed lines mark the chosen threshold $r = 0.3$ in panels (**a**,**c**) and minimal series length $n = 960$ (the highest scale) in panels (**b**,**d**); upper panels: two-dimensional signals (*SBP* and *PI* interaction, $D = 2$); lower panels: three-dimensional signals (*SBP*, *PI* and t_B interaction, $D = 3$.

Figure 10a presents *CSME* estimated from the dependency level signals of control rats at different ambient temperatures, for two-dimensional signals (*SBP* and *PI* interaction, $D = 2$). Each signal is accompanied by 10 isodistributional surrogates (the estimated entropy is averaged). These results can be compared to the results of the classical entropy analysis shown in Figure 6j. The figure also presents *CSME* estimated from the artificially generated two-dimensional signals with Gaussian distribution. Figure 10b shows the same entropies but estimated from the three-dimensional signals.

The aim of Figure 11 is to show whether the *CMSE* estimated from the signals of control rats at different ambient temperatures can distinguish different temporal offsets (DEL) between the *SBP* and *PI*. Such an analysis cannot be performed by classical entropy study.

Differences between experimental groups were analyzed by a Mann–Whitney U-test. Statistical significance was considered at $p < 0.05$.

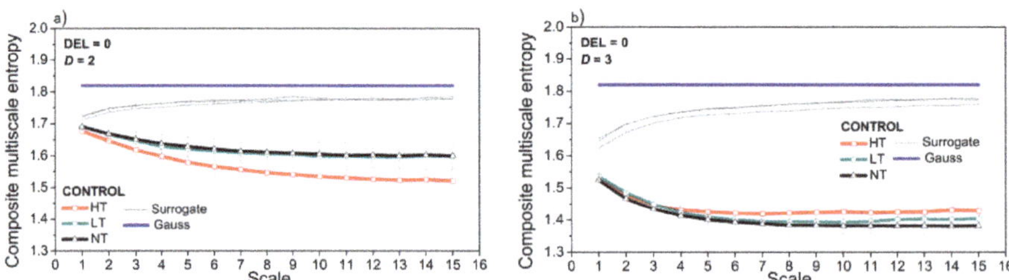

Figure 10. Comparison of CSME estimates for signals recorded from control animals at high ambient temperature (HT), low temperature (LT) and neutral temperature (NT). (**a**) two-dimensional signals (*SBP* and *PI* interaction, $D = 2$); (**b**) three-dimensional signals (*SBP*, *PI* and t_B interaction, $D = 3$). Delay (offset) of *SBP-PI* signals was set to DEL = 0 beats. Signals are accompanied by the control surrogate study and by the artificial two- and three- dimensional Gaussian signals. Results are presented as a mean ± SE.

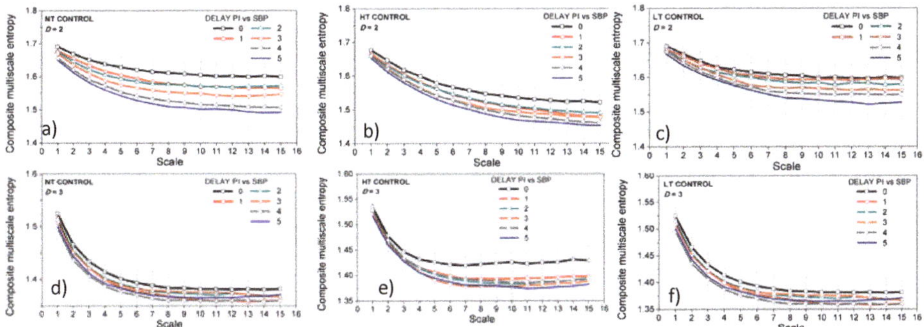

Figure 11. Composite multiscale entropy *CMSE* with *SBP-PI* offset (DELAY) as a parameter, estimated from control animals exposed to neutral temperature (NT, panels (**a**,**d**)), high temperature (HT, panels (**b**,**e**)), and low temperature (LT, panels (**c**,**f**)). Upper panels (**a**–**c**) two-dimensional signals (*SBP* and *PI* interaction, $D = 2$); lower panels (**d**–**f**) three-dimensional signals (*SBP*, *PI* and t_B interaction, $D = 3$). Results are presented as a mean ± SE. Statistically significant difference ($p < 0.05$) between the lowest and highest offsets, DEL = 0 and 5, are observed for the scale greater than 5 in panels (**a**–**c**,**e**).

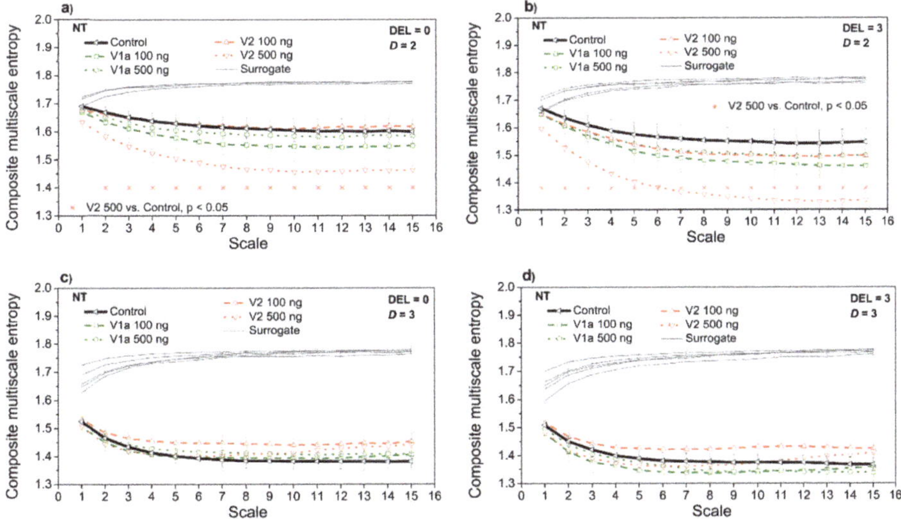

Figure 12. Composite multiscale entropy *CMSE* estimated from rats exposed to vasopressin antagonists at neutral temperature (NT). Panels (**a**,**c**) *SBP-PI* offset (delay) is set to DEL=0; panels (**b**,**d**) *SBP-PI* offset (delay) is set to DEL=3; Upper panels (**a**,**b**) two-dimensional signals (*SBP* and *PI* interaction, $D = 2$); lower panels (**c**,**d**) three-dimensional signals (*SBP*, *PI* and t_B interaction, $D = 3$). Results are presented as a mean ± SE.

Figure 12 presents the entropy estimates after the administration of vasopressin antagonists at neutral ambient temperature, for two different *SBP-PI* temporal offsets (delays), DEL = 0, and DEL = 3. Figure 13 presents the same entropy estimates but at the high ambient temperature. Both Figures 12 and 13 are accompanied by isodistributional surrogate data controls. The purpose of these two figures is to investigate whether CMSE can distinguish the V1a and V2 antagonist administration if compared to the control case (without the drugs). The two-dimensional cases can be compared to the classical entropy study shown in Figure 4.

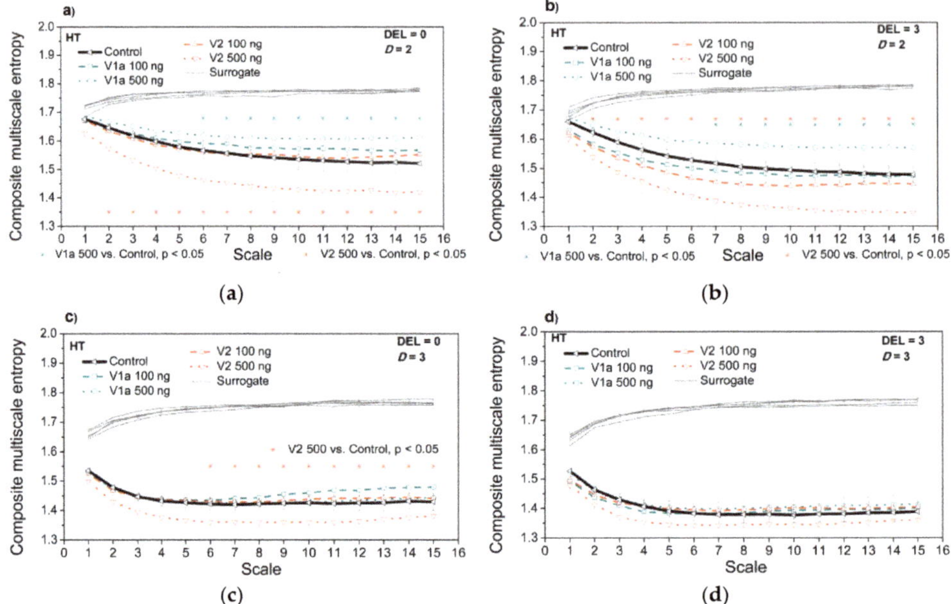

Figure 13. Composite multiscale entropy *CMSE* estimated from rats exposed to vasopressin antagonists at high temperature (HT). Panels (**a**,**c**) *SBP-PI* offset (delay) is set to DEL=0; panels (**b**,**d**) *SBP-PI* offset (delay) is set to DEL=3; Upper panels (**a**,**b**) two-dimensional signals (*SBP* and *PI* interaction, D = 2); lower panels (**c**,**d**) three-dimensional signals (*SBP*, *PI* and t_B interaction, D = 3). Results are presented as a mean ± SE.

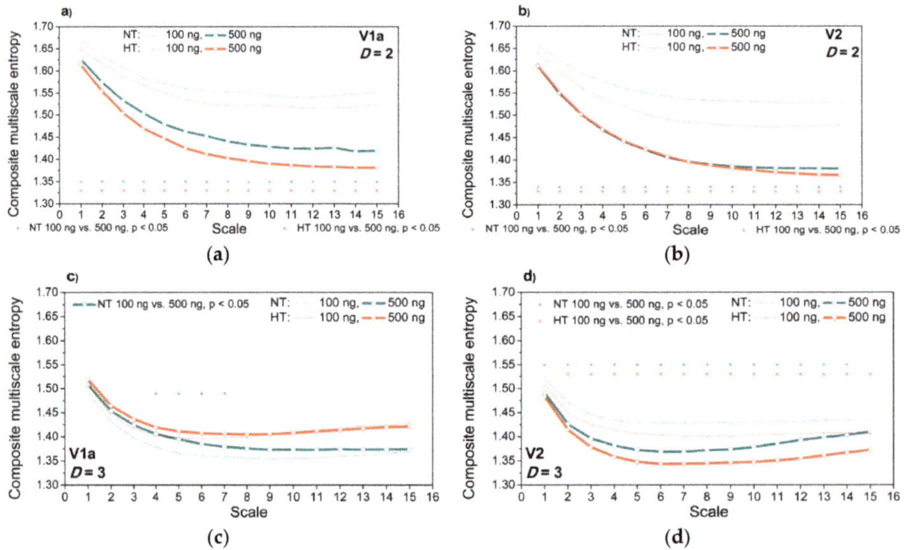

Figure 14. Effects of V1a and V2 antagonist dosage. Panels (**a**,**b**) two-dimensional signals (*SBP* and *PI* interaction, D = 2); panels (**c**,**d**) three-dimensional signals (*SBP*, *PI* and t_B interaction, D = 3). Panels (**a**,**c**) V1a antagonist; panels (**b**,**d**) V2 antagonist. Results are presented as a

The purpose of Figure 14 is to check whether the proposed entropy can distinguish the signals after administering the different doses of V1a and V2 antagonists. The same experiments are repeated separating the signals according to the *SBP-PI* offset (delay), and these results are presented as Supplementary data.

4. Discussion

The first aim of our study is to create a single time series that reflects the level of interaction between the arbitrary number of simultaneously recorded signals. It is accomplished by mathematical tools: copula density captures the level of signal interaction, and the Voronoi decomposition maps the interaction levels into the temporal signal.

Figure 10a shows that the *CMS* entropy of two-dimensional *SBP-PI* dependency signal decreases at high ambient temperature. The classical *CMSE* analysis provided the same result (Figure 4, panel j). The high ambient temperature in this experiment exceeds the boundary set to 29.5 °C [60], inducing heat dissipation in rats such as vasodilatation, evaporation, sweating, panting and affecting the blood vessel circulatory strain [60]. The existence of the dominant component reduces the influence of the other mechanism, the signals and their mutual interactions become less complex, and the entropy decreases. On the other hand, neutral and low ambient temperatures in our experiments are within the normal boundaries, and entropy estimates overlap, both in the proposed and classical entropy estimates—Figures 10a and 4j —respectively.

An administration of V2-500ng significantly decreases the multiscale entropy of a two-dimensional *SBP-PI* dependency signal, both at neutral (Figure 12a,b) and at high ambient temperature (Figure 13a,b). Additionally, V1a-500 significantly increases the entropy of a two-dimensional *SBP-PI* dependency signal, signals at high temperatures (Figure 13a,b). These analyses were performed at the particular offsets (delays) between *SBP* and *PI* signals. The results correspond to the spectral analysis of the signals after V1a and V2 antagonist administration: it was shown [27] that V2-500 ng administration increases the low-frequency signal component of the *SBP* signal; the signal becomes smoother, the number of repetitive patterns increases, and the signal becomes more predictive so the entropy decreases. On the other hand, V1a-500 ng increases the high-frequency signal component [27], the signal, and its interactions become more turbulent and the entropy increases.

While spectral analysis separates high and low signal components, a classical cross-entropy observes the signal as a whole. The template matching procedure (cf. Equation (A3)) averages all the temporal *SBP* and *PI* positions, so different offsets (delays) of *SBP-PI* signals cannot be distinguished. Panels k and l of Figure 4 show that classical *SBP-PI* cross-entropy does slightly decrease after V2-500 ng administration, but without the statistical significance, especially at the neutral temperature (Figure 4k).

It is consistent with the second aim of our contribution: the proposed analysis corresponds to the classical two-dimensional entropy results and with the results of spectral analysis. In cases when the consistency is not statistically significant, the possible explanation is that the classical entropy observes the complete signal, while the proposed method includes signals separated according to the SBP-PI signal offset.

When the body temperature time series is included to create three-dimensional *SBP-PI-*t_B dependency signals, the entropy decreases and differences caused by ambient temperature are attenuated (Figure 10b). It should be noted (Figure 4g,h,j) that signal t_B is a low-entropy signal. However, the significant entropy decrease induced by V2-500 ng is preserved in three-dimensional signals, but only at the high temperatures and DEL = 0 (Figure 13c).

The proposed method can make a distinction between the different *SBP-PI* offsets (delays), as shown in Figure 11. Regardless of the ambient temperature, the entropy of the two-dimensional dependency signals at DEL = 5 is significantly lower than at DEL = 0. It corresponds to [29], as the *SBP-PI* dependency in rats is the greatest for delays of 3, 4, and 5 beats. When three-dimensional signals are observed, the significant decrease of entropy for DEL = 5 occurs at high ambient temperatures only.

The proposed entropy can distinguish the signals after administering the different doses of V1a and V2 antagonists. It is shown in Figure 14. The distinction is statistically significant for the two-dimensional *SBP-PI* dependency signal, regardless of antagonist and ambient temperature, and for three-dimensional signals after V2 administration. V1a administration induces a statistically significant difference in three-dimensional *SBP-PI-*t_B dependency signal at the neutral temperature, and only for a couple of scaling levels.

Figure 14 presents the entropy estimates averaged over all the signals. The Supplementary data comprise Figures S1 and S2, where the dosages are separated according to the *SBP-PI* offset (delay).

The entropy of the artificial two- and three-dimensional Gaussian control signals is shown in Figure 10, but not repeated in the subsequent figures as it is always the same. The entropy of the surrogate data converges towards the Gaussian, but never reaches it: although the surrogate signals can be regarded as streams of i.i.d. random variables, their distribution remains equal to the distribution of the original dependency signals.

5. Conclusions

The signal framework created in this contribution provides a possibility for an easy analysis of signal dependency structures mapped into a single time series. The estimated entropies of two-dimensional dependency signals correspond to the classical cross-entropy and spectral analysis. The method can recognize entropy changes at different temperature levels, at different *SBP-PI* offsets, at different administered V1a and V2 dosages. The recognition of these differences is not random: the surrogate data analysis destroys the temporal coupling of the observed dependency signals, yielding in all cases entropy estimates that are almost identical and that approach (but do not reach) the entropy of Gaussian time series.

This method can be applied to any multivariate signals. It is necessary to conduct a deep analysis to find the minimal signal length that provides reliable results for the arbitrary number of signals and to check the possibility of applying other analytical tools besides the entropy. A comparative study of the various mapping procedure (Voronoi cells, eigenvalues of the transition matrix, classical multidimensional histograms) should be performed. Further analysis on coupling the cardiovascular data with a body and ambient temperature could reveal more adverse effects, an issue that could be extremely important regarding the climatic changes.

Supplementary Materials: The following are available online at http://www.mdpi.com/1099-4300/21/11/1103/s1, Figure S1: Effects of V1a antagonists at different time offsets between SBP and PI. Upper panels: two-dimensional signals (*SBP* and *PI* interaction, $D = 2$); lower panels: three-dimensional signals (*SBP*, *PI* and tB interaction, $D = 3$). Left panels: Neutral temperature. Right panels: High temperature. Figure S2 Effects of V2 antagonists at different time offsets between SBP and PI. Upper panels: two-dimensional signals (*SBP* and *PI* interaction, $D = 2$); lower panels: three-dimensional signals (*SBP*, *PI* and tB interaction, $D = 3$). Left panels: Neutral temperature. Right panels: High temperature.

Author Contributions: N.J.-Z. designed a medical experiment. S.M.-S. performed the surgery and did the measurements. D.B. found a way to convert a group of signals into a time series with samples that are proportional to the dependency density (copula density, dependency structures) and did the preliminary research. T.S. processed the results.

Funding: The research was partially funded by the Serbian Ministry of Education, Science and Technology Development under Grant Nos. TR32040 and III41013. The work acknowledges support from the EU COST-Action CA15104-Body Communications group.

Acknowledgments: The authors are grateful to the unknown Reviewers and Editor for their valuable comments that significantly improved the text.

Conflicts of Interest: The authors declare no conflict of interest.

Appendix A Entropy Concepts in Brief

Approximate entropy, *ApEn* [1], and sample entropy, *SampEn* [3], are tools that determine the regularity of a time series $x_{1k} \in X_{1k}, k = 1, \ldots, N$, based on the existence of similar patterns of increasing length [7]. Their cross (X) variants measure a level of asynchrony of two time series, $x_{1k} \in X_{1k}, k = 1, \ldots, N$ and $x_{2j} \in X_{2j}, j = 1, \ldots, N$ [7], with a notable difference that *XSampEn* is a symmetric measure, while *XApEn* is not [7]. Before analysis, the time series should be normalized and centered, a procedure known as z-normalization or standard scaling. Signal stationarity is an important prerequisite; otherwise, a threshold concept would be useless and the results would be unreliable [7,18,54].

The general X-entropy estimation starts with partitioning the time series into the overlapping vectors of length m:

$$\begin{aligned} \text{Template } X_{1k}^{(m)} &= \left[x_{1k}, x_{1,k+1}, \ldots, x_{1,k+m-1}\right], \; k = 1, 2, \ldots, N-m+z; \\ \text{Follower } X_{2j}^{(m)} &= \left[x_{2j}, x_{2,j+1}, \ldots, x_{2,j+m-1}\right], \; j = 1, 2, \ldots, N-m+z; \\ z &= \begin{cases} 1 \text{ for } (X)ApEn \\ 0 \text{ for } (X)SampEn \end{cases}. \end{aligned} \tag{A1}$$

Vector distance is defined as the maximal absolute difference of the signal samples:

$$d1\left(X_{1k}^{(m)}, X_{2j}^{(m)}\right) = \max_{i=0,\ldots,m-1} \left|x_{1,k+i} - x_{2,j+i}\right|, \; k, j = 1, 2, \ldots, N-m+z. \tag{A2}$$

The vectors are similar if the distance is less than or equal to the predefined threshold r. A probability that the time series X_2 is similar to the given template $X_{1k}^{(m)}$ is estimated as:

$$\hat{p}_k^{(m)}(r) = \Pr\left\{d1\left(X_1^{(m)}, X_2^{(m)}\right) \le r \middle| X_1^{(m)} = X_{1k}^{(m)}, X_2^{(m)} \in X_2, \; r > 0\right\} == \frac{1}{N-m+z} \cdot \sum_{k=1}^{N-m+z} \mathrm{I}\left\{d1\left(X_1^{(m)}, X_2^{(m)}\right) \le r\right\}. \tag{A3}$$

In Equation (A3) "^" denotes an estimate, while I{·} is an indicator function that enables a compact symbolic presentation of the counting process. It is equal to 1 if $d1\left(X_1^{(m)}, X_2^{(m)}\right) \le r$; otherwise, it is equal to zero. The value (1-z) eliminates the self-matching for *SampEn*.

To calculate *ApEn* and *SampEn*, it is sufficient to replace x_{2j}, X_{2j}, $X_2^{(m)}$, and $X_{2j}^{(m)}$ in Equations (A1), (A2), and (A3) with x_{1j}, X_{1j}, $X_1^{(m)}$ and $X_{1j}^{(m)}$. For *SampEn*, it is also necessary to subtract the self-matching in (A3). A negative logarithm of the estimated probability $\hat{p}_k^{(m)}(r)$ corresponds [61] to the information content stored in the similarity of the template $X_{1k}^{(m)}$ to the complete time series X_2. Arithmetic averaging over all the templates yields, for *(X)ApEn*:

$$\hat{\Phi}^{(m)}(r, N) = \frac{1}{N-m+1} \cdot \sum_{k=1}^{N-m+1} \log\left(\hat{p}_k^{(m)}(r)\right). \tag{A4}$$

Ithe *(X)SampEn* approach, the summation and logarithm change the order:

$$\hat{\Psi}^{(m)}(r, N) = \log\left(\sum_{k=1}^{N-m} \hat{p}_k^{(m)}(r)\right). \tag{A5}$$

The complete procedure is repeated for the vector length equal to m + 1, and with the quantity z in (A1), (A2), and (A3) fixed to zero. Then, *(X)ApEn* and *(X)SampEn* are calculated as:

$$\begin{aligned} XApEn(m, r, N) &= \hat{\Phi}^{(m)}(r, N) - \hat{\Phi}^{(m+1)}(r, N); \\ XSampEn(n, r, N) &= \hat{\Psi}^{(m)}(r, N) - \hat{\Psi}^{(m+1)}(r, N). \end{aligned} \tag{A6}$$

Multiscale entropy (*MSE*) [8,9] and its composite improvement *CMSE* [10–12] are based on *SampEn* estimation repeated over the time series with increasingly coarser time resolution. The coarse graining (or downsampling) of the time series $X = \{x_k\}$ is performed as follows:

$$\begin{array}{l} x_{CMS,l,j}^{(\,)} = \left(\sum\limits_{k\,=\,(j-1)\cdot\tau+l}^{j\cdot\tau+l-1} x_k \right) / \tau, \\ j = 1,\ldots,fix(N/\tau),\ l = 1,\ldots,\tau, \quad x_{MS,j}^{(\tau)} = \sum\limits_{k\,=\,(j-1)\cdot\tau+1}^{j\cdot\tau} x_k = x_{CMS,1,j}^{(\tau)} \end{array} \tag{A7}$$

CMSE and *MSE* are evaluated as follows:

$$\begin{array}{rcl} CMSE(\tau,m,r,N) & = & \left(\sum\limits_{l\,=\,1}^{\tau} SampEn(x_{CMS,l}^{(\tau)}, m, r, N \right) / \tau, \\ MSE(\tau,m,r,N) & = & SampEn\left(x_{MS}^{(\tau)}, m, r, N \right). \end{array} \tag{A8}$$

References

1. Pincus, S.M. Approximate entropy as a measure of system complexity. *Proc. Natl. Acad. Sci. USA* **1991**, *88*, 2297–2301. [CrossRef] [PubMed]
2. Pincus, S.M.; Huang, W.M. Approximate entropy: Statistical properties and applications. *Commun. Stat. Theory Methods* **1992**, *21*, 3061–3077. [CrossRef]
3. Richman, J.S.; Moorman, J.R. Physiological time-series analysis using approximate entropy and sample entropy. *Am. J. Physiol. Heart Circ. Physiol.* **2000**, *278*, 2039–2049. [CrossRef] [PubMed]
4. Yentes, J.M.; Hunt, N.; Schmid, K.K.; Kaipust, J.P.; McGrath, D.; Stergiou, N. The appropriate use of approximate entropy and sample entropy with short data sets. *Ann. Biomed. Eng.* **2013**, *41*, 349–365. [CrossRef]
5. Pincus, S.; Singer, B.H. Randomness and degrees of irregularity. *Proc. Natl. Acad. Sci. USA* **1996**, *93*, 2083–2088. [CrossRef]
6. Pincus, S.M.; Mulligan, T.; Iranmanesh, A.; Gheorghiu, S.; Godschalk, M.; Veldhuis, J.D. Older males secrete luteinizing hormone and testosterone more irregularly, and jointly more asynchronously than younger males. *Proc. Natl. Acad. Sci. USA* **1996**, *93*, 14100–14105. [CrossRef]
7. Delgado-Bonal, A.; Marshak, A. Approximate Entropy and Sample Entropy: A Comprehensive Tutorial. *Entropy* **2019**, *21*, 541. [CrossRef]
8. Costa, M.; Goldberger, A.L.; Peng, C.K. Multiscale entropy analysis of complex physiologic time series. *Phys. Rev. Lett.* **2002**, *89*, 068102. [CrossRef]
9. Costa, M.; Goldberger, A.L.; Peng, C.K. Multiscale entropy analysis of biological signals. *Phys. Rev. E* **2005**, *71*, 021906. [CrossRef]
10. Wu, S.D.; Wu, C.W.; Lin, S.G.; Wang, C.C.; Lee, K.Y. Time series analysis using composite multiscale entropy. *Entropy* **2013**, *15*, 1069–1084. [CrossRef]
11. Lin, T.K.; Chien, Y.H. Performance Evaluation of an Entropy-Based Structural Health Monitoring System Utilizing Composite Multiscale Cross-Sample Entropy. *Entropy* **2019**, *21*, 41. [CrossRef]
12. Castiglioni, P.; Parati, G.; Faini, A. Information-Domain Analysis of Cardiovascular Complexity: Night and Day Modulations of Entropy and the Effects of Hypertension. *Entropy* **2019**, *21*, 550. [CrossRef]
13. Marwaha, P.; Sunkaria, R.M. Cardiac variability time–series analysis by sample entropy and multiscale entropy. *Int. J. Med. Eng. Inform.* **2015**, *7*, 1–14. [CrossRef]
14. Chen, C.; Jin, Y.; Lo, I.L.; Zhao, H.; Sun, B.; Zhao, Q.; Zhang, X.D. Complexity Change in Cardiovascular Disease. *Int. J. Biol. Sci.* **2017**, *13*, 1320–1328. [CrossRef]
15. Li, X.; Yu, S.; Chen, H.; Lu, C.; Zhang, K.; Li, F. Cardiovascular autonomic function analysis using approximate entropy from 24-h heart rate variability and its frequency components in patients with type 2 diabetes. *J. Diabetes Investig.* **2015**, *6*, 227–235. [CrossRef]

16. Krstacic, G.; Gamberger, D.; Krstacic, A.; Smuc, T.; Milicic, D. The Chaos Theory and Non-linear Dynamics in Heart Rate Variability in Patients with Heart Failure. In Proceedings of the Computers in Cardiology, Bologna, Italy, 14–17 September 2008; pp. 957–959.
17. Storella, R.J.; Wood, H.W.; Mills, K.M.; Kanters, J.K.; Højgaard, M.V.; Holstein-Rathlou, N.H. Approximate entropy and point correlation dimension of heart rate variability in healthy subjects. *Integr. Physiol. Behav. Sci.* **1998**, *33*, 315–320. [CrossRef]
18. Boskovic, A.; Loncar-Turukalo, T.; Sarenac, O.; Japundžić-Žigon, N.; Bajić, D. Unbiased entropy estimates in stress: A parameter study. *Comput. Biol. Med.* **2012**, *42*, 667–679. [CrossRef]
19. Ryan, S.M.; Goldberger, A.L.; Pincus, S.M.; Mietus, J.; Lipsitz, L.A. Gender- and Age-Related Differences in Heart Rate Dynamics: Are Women More Complex Than Men? *J. Am. Coll. Cardiol.* **1994**, *24*, 1700–1707. [CrossRef]
20. Wang, S.Y.; Zhang, L.F.; Wang, X.B.; Cheng, J.H. Age dependency and correlation of heart rate variability, blood pressure variability, and baroreflex sensitivity. *J. Gravit. Physiol.* **2002**, *7*, 145–146.
21. Sklar, A. *Fonctions de Répartition à n Dimensions et Leurs Marges*; Institut de Statistique del Universit'e de Paris: Paris, France, 1959; Volume 8, pp. 229–231.
22. Claeys, M.J.; Rajagopalan, S.; Nawrot, T.S.; Brook, R.D. Climate and environmental triggers of acute myocardial infarction. *Eur. Heart J.* **2017**, *38*, 955–960. [CrossRef]
23. Akselrod, S.; Gordon, D.; Ubel, F.A.; Shannon, D.C.; Bezger, A.C.; Cohen, R.J. Power spectrum analysis of heart rate fluctuations: A quantitative probe of beat-to-beat cardiovascular control. *Science* **1981**, *213*, 220–222. [CrossRef] [PubMed]
24. Kinugasa, H.; Hirayanagi, K. Effects of skin surface cooling and heating on autonomic nervous activity and baroreflex sensitivity in humans. *Exp. Physiol.* **1999**, *84*, 369–377. [CrossRef] [PubMed]
25. Japundžić-Žigon, N. Effects of nonpeptide V1a and V2 antagonists on blood pressure fast oscillations in conscious rats. *Clin. Exp. Hypertens.* **2001**, *23*, 277–292. [CrossRef] [PubMed]
26. Japundžić-Žigon, N.; Milutinović, S.; Jovanović, A. Effects of nonpeptide and selective V1 and V2 antagonists on blood pressure short-term variability in spontaneously hypertensive rats. *J. Pharmacol. Sci.* **2004**, *95*, 47–55. [CrossRef] [PubMed]
27. Milutinović, S.; Murphy, D.; Japundžić-Žigon, N. The role of central vasopressin receptors in the modulation of autonomic cardiovascular controls: A spectral analysis study. *Am. J. Physiol. Regul. Integr. Comp. Physiol.* **2006**, *291*, 1579–1591. [CrossRef]
28. Milutinović-Smiljanić, S.; Šarenac, O.; Lozić-Djurić, M.; Murphy, D.; Japundžić-Žigon, N. Evidence for the involvement of central vasopressin V1b and V2 receptors in stress-induced baroreflex desensitization. *Br. J. Pharmacol.* **2013**, *169*, 900–908. [CrossRef]
29. Oosting, J.; Struijker-Boudier, H.A.; Janssen, B.J. Validation of a continuous baroreceptor reflex sensitivity index calculated from spontaneous fluctuations of blood pressure and pulse interval in rats. *J. Hypertens.* **1997**, *15*, 391–399. [CrossRef]
30. Schreiber, T.; Schmitz, A. Surrogate time series. *Phys. D Nonlinear Phenom.* **1999**, *142*, 346–382. [CrossRef]
31. Theiler, J.; Eubank, S.; Longtin, A.; Galdrikian, B.; Farmer, J.D. Testing for nonlinearity in time series: The method of surrogate data. *Phys. D Nonlinear Phenom.* **1992**, *58*, 77–94. [CrossRef]
32. Papoulis, A.; Pillai, S.U. *Probability, Random Variables, and Stochastic Processes*; McGraw-Hill: New York, NY, USA, 1984.
33. Wessel, N.; Voss, A.; Malberg, H.; Ziehmann, C.H.; Voss, H.U.; Schirdewan, A.; Meyerfeldt, U.; Kurths, J. Nonlinear analysis of complex phenomena in cardiological data. *Herzschrittmacherther. Elektrophysiol.* **2000**, *11*, 159–173. [CrossRef]
34. Tarvainen, M.P.; Ranta-aho, P.O.; Karjalainen, P.A. An advanced detrending approach with application to HRV analysis. *IEEE Trans. Biomed. Eng.* **2002**, *42*, 172–174. [CrossRef] [PubMed]
35. Angus, J.E. The Probability Integral Transform and Related Results. *SIAM Rev.* **1994**, *36*, 652–654. [CrossRef]
36. Cherubini, U.; Luciano, E.; Vecchiato, W. *Copula Methods in Finance*; Wiley Finance Series; Wiley: Chichester, UK, 2004.
37. Malevergne, Y.; Sornette, D. *Extreme Financial Risks*; Springer: Berlin/Heidelberg, Germany, 2006.

38. McNeil, A.J.; Frey, R.; Embrechts, P. *Quantitative Risk Management: Concepts, Techniques, and Tools*; Princeton Series in Finance; Princeton University Press: Princeton, NJ, USA, 2005.
39. Schönbucker, P. *Credit Derivatives Pricing Models: Models, Pricing, Implementation*; Wiley Finance Series; Wiley: Chichester, UK, 2003.
40. Genest, C.; Favre, A.C. Everything you always wanted to know about copula modeling but were afraid to ask. *J. Hydrol. Eng.* **2007**, *12*, 347–368. [CrossRef]
41. Salvadori, G.; DeMichele, C.; Kottegoda, N.T.; Rosso, R. *Extremes in Nature: An Approach Using Copulas*; Springer: Dordrecht, The Netherlands, 2007; Volume 56.
42. Vandenberghe, S.; Verhoest, N.E.C.; De Baets, B. Fitting bivariate copulas to the dependence structure between storm characteristics: A detailed analysis based on 105 years 10 min rainfall. *Water Resour. Res.* **2010**, *46*. [CrossRef]
43. Yan, X.; Clavier, L.; Peters, G.; Septier, F.; Nevat, I. Modeling dependence in impulsive interference and impact on receivers. In Proceedings of the 11th MCM COST IC1004, Krakow, Poland, 24–26 September 2014.
44. Kumar, P.; Shoukri, M.M. Copula based prediction models: An application to an aortic regurgitation study. *BMC Med. Res. Methodol.* **2007**, *7*, 21. [CrossRef] [PubMed]
45. Bansal, R.; Hao, X.J.; Liu, J.; Peterson, B.S. Using Copula distributions to support more accurate imaging-based diagnostic classifiers for neuropsychiatric disorders. *Magn. Reson. Imaging* **2014**, *32*, 1102–1113. [CrossRef]
46. Jovanovic, S.; Skoric, T.; Sarenac, O.; Milutinovic-Smiljanic, S.; Japundzic-Zigon, N.; Bajic, D. Copula as a dynamic measure of cardiovascular signal interactions. *Biomed. Signal Process. Control* **2018**, *43*, 250–264. [CrossRef]
47. Laude, D.; Baudrie, V.; Elghozi, J.L. Tuning of the sequence technique. *IEEE Eng. Med. Biol. Mag.* **2009**, *28*, 30–34.
48. Bajic, D.; Loncar-Turukalo, T.; Stojicic, S.; Sarenac, O.; Bojic, T.; Murphy, J.; Paton, R.; Japundzic-Zigon, N. Temporal analysis of the spontaneous baroreceptor reflex during mild emotional stress in the rat. *Stress Int. J. Biol. Stress* **2009**, *13*, 142–154. [CrossRef]
49. Laude, D.; Elghozi, J.L.; Girard, A.; Bellard, E.; Bouhaddi, M.; Castiglioni, P.; Cerutti, C.; Cividjian, A.; Di Rienzo, M.; Fortrat, J.O.; et al. Comparison of various techniques used to estimate spontaneous baroreflex sensitivity (the EUROBAVAR study). *Am. J. Physiol. Regul. Integr. Comp. Physiol.* **2004**, *286*, 226–231. [CrossRef]
50. Tasic, T.; Jovanovic, S.; Mohamoud, O.; Skoric, T.; Japundzic-Zigon, N.; Bajićc, D. Dependency structures in differentially coded cardiovascular time series. *Comput. Math. Methods Med.* **2017**, *2017*, 2082351. [CrossRef] [PubMed]
51. Bernard, W. *Silverman: Density Estimation for Statistics and Data Analysis*; CRC Press: Boca Raton, FL, USA, 1986.
52. Voronoi, G. Nouvelles applications des paramètres continuos à la théorie des formes quadratiques. *J. Reine Angew. Math.* **1908**, *133*, 198–287. [CrossRef]
53. Skoric, T.; Sarenac, O.; Milovanovic, B.; Japundzic-Zigon, N.; Bajic, D. On Consistency of Cross-Approximate Entropy in Cardiovascular and Artificial Environments. *Complexity* **2017**, *2017*, 8365685. [CrossRef]
54. Shannon, C.E. Communications in the presence of noise. *Proc. IRE* **1949**, *37*, 10–21. [CrossRef]
55. Bendat, J.S.; Piersol, A.G. *Random Data Analysis and Measurement Procedures*; Wiley Series in Probability and Statistics; Wiley: New York, NY, USA, 1986.
56. Lu, S.; Chen, X.; Kanters, J.K.; Solomon, I.C.; Chon, K.H. Automatic selection of the threshold value r for approximate entropy. *IEEE Trans. Biomed. Eng.* **2008**, *55*, 1966–1972.
57. Chon, K.H.; Scully, C.G.; Lu, S. Approximate entropy for all signals. *IEEE Eng. Med. Biol. Mag.* **2009**, *28*, 18–23. [CrossRef]
58. Castiglioni, P.; Di Rienzo, M. How the threshold "r" influences approximate entropy analysis of heart-rate variability. In Proceedings of the Conference Computers in Cardiology, Bologna, Italy, 14–17 September 2008; Volume 35, pp. 561–564.

59. Liu, C.; Shao, P.; Li, L.; Sun, X.; Wang, X.; Liu, F. Comparison of different threshold values r for approximate entropy: Application to investigate the heart rate variability between heart failure and healthy control groups. *Physiol. Meas.* **2011**, *32*, 167–180. [CrossRef]
60. Restrepo, J.F.; Schlotthauer, G.; Torres, M.E. Maximum approximate entropy and r threshold: A new approach for regularity changes detection. *Phys. A Stat. Mech. Appl.* **2014**, *409*, 97–109. [CrossRef]
61. Yang, Y.; Gordon, J. Ambient temperature limits and stability of temperature regulation in telemetered male and female rats. *J. Therm. Biol.* **1996**, *21*, 353–363. [CrossRef]

Article

Multivariate Multiscale Dispersion Entropy of Biomedical Times Series

Hamed Azami [1,2,*], Alberto Fernández [3,4] and Javier Escudero [1]

1 School of Engineering, Institute for Digital Communications, University of Edinburgh, King's Buildings, Edinburgh EH9 3FB, UK; javier.escudero@ed.ac.uk
2 Department of Neurology and Massachusetts General Hospital, Harvard Medical School, Charlestown, MA 02129, USA
3 Departamento de Psiquiatría y Psicología Médica, Universidad Complutense de Madrid, 28040 Madrid, Spain; aferlucas@med.ucm.es
4 Laboratorio de Neurociencia Cognitiva y Computacional, Centro de Tecnología Biomédica, Universidad Politecnica de Madrid and Universidad Complutense de Madrid, 28040 Madrid, Spain
* Correspondence: hazami@mgh.harvard.edu

Received: 22 July 2019; Accepted: 12 September 2019; Published:19 September 2019

Abstract: Due to the non-linearity of numerous physiological recordings, non-linear analysis of multi-channel signals has been extensively used in biomedical engineering and neuroscience. Multivariate multiscale sample entropy (MSE–mvMSE) is a popular non-linear metric to quantify the irregularity of multi-channel time series. However, mvMSE has two main drawbacks: (1) the entropy values obtained by the original algorithm of mvMSE are either undefined or unreliable for short signals (300 sample points); and (2) the computation of mvMSE for signals with a large number of channels requires the storage of a huge number of elements. To deal with these problems and improve the stability of mvMSE, we introduce multivariate multiscale dispersion entropy (MDE–mvMDE), as an extension of our recently developed MDE, to quantify the complexity of multivariate time series. We assess mvMDE, in comparison with the state-of-the-art and most widespread multivariate approaches, namely, mvMSE and multivariate multiscale fuzzy entropy (mvMFE), on multi-channel noise signals, bivariate autoregressive processes, and three biomedical datasets. The results show that mvMDE takes into account dependencies in patterns across both the time and spatial domains. The mvMDE, mvMSE, and mvMFE methods are consistent in that they lead to similar conclusions about the underlying physiological conditions. However, the proposed mvMDE discriminates various physiological states of the biomedical recordings better than mvMSE and mvMFE. In addition, for both the short and long time series, the mvMDE-based results are noticeably more stable than the mvMSE- and mvMFE-based ones. For short multivariate time series, mvMDE, unlike mvMSE, does not result in undefined values. Furthermore, mvMDE is faster than mvMFE and mvMSE and also needs to store a considerably smaller number of elements. Due to its ability to detect different kinds of dynamics of multivariate signals, mvMDE has great potential to analyse various signals.

Keywords: complexity; multivariate multiscale dispersion entropy; multivariate time series; electroencephalogram; magnetoencephalogram

1. Introduction

Multivariate techniques are needed to analyse data consisting of more than one time series [1–3]. The majority of physiological and pathophysiological activities, and even many non-physiological signals, include interactions between different kinds of single processes. Thus, we expect that parameters or measures with different origins are considered in a multivariate way [1,4]. Furthermore, recent developments in sensor technology enabling routine recordings of multi-channel signals have led to an increasing popularity of this kind of analysis on physiological data [1–3,5,6].

Advances on information theory and non-linear dynamical approaches have recently allowed the study of different kinds of multivariate time series [3,7–9]. Due to the intrinsic non-linearity of diverse physiological and non-physiological processes, non-linear analysis of multivariate time series has been broadly used in biomedical signal processing with the aim of studying the relationship between simultaneously recorded signals [3,7,8].

Multivariate multiscale entropy (mvMSE) as a powerful non-linear measure is based on a combination of multivariate sample entropy (SampEn–mvSE) and the coarse-graining process [8]. mvSE characterizes the likelihood that similar multi-channel embedded patterns, which consider both the time and spatial domains, within a time series will remain similar when the pattern length is increased [3]. mvMSE, by taking into account both the spatial and time domains, shows the complexity of multi-channel signals [8]. Complexity reflects the degree of structural richness of time series [8,10] and is different with that of irregularity or uncertainty defined from classical entropy methods such as SampEn [11], permutation entropy (PerEn) [12], and dispersion entropy (DisEn) [13]. That is to say, neither completely regular or certain nor completely irregular (uncorrelated random) time series are truly complex, since none of them is structurally rich at a global level [8,10,14–16].

The multivariate multiscale entropy-based analysis is interpreted based on: (1) the multivariate time series **X** is more complex than the multivariate time series **Y**, if for the most temporal scales, the mvSE measures for **X** are larger than those for **Y**; (2) a monotonic fall in the multivariate entropy values along the temporal scale factors shows that the signal only includes useful information at the smallest scale factors; and (3) a multivariate signal illustrating long-range correlations and complex creating dynamics is characterized by either a constant mvSE or this demonstrates a monotonic rise in mvSE with the temporal scale factor [8].

Although the mvMSE is a powerful and widely-used method, when applied to short signals, the results may be undefined or unreliable [17]. To alleviate this shortcoming, multivariate multiscale fuzzy entropy (mvMFE) based on multivariate fuzzy entropy (mvFE) and the coarse-graining process was suggested [18]. To decrease the running time of the mvMFE proposed in [18], we have recently proposed an mvMFE with a new fuzzy membership function [17]. Nevertheless, the mvMFE is still slow for real-time applications and may lead to unreliable results for short signals, as shown later.

To overcome the problem of unreliable values for mvMFE and mvMSE, multivariate multiscale PerEn (mvMPE) was proposed [19]. To have more information regarding the amplitude of multi-channel signals, multivariate weighted multiscale PerEn (mvWMPE) has recently been developed [20]. However, both the mvMPE and mvWMPE do not take into account the cross-statistical properties between multiple input channels and do not follow the concept of complexity for some signals such as white Gaussian noise (WGN) and $1/f$ noise [8,14,17].

mvMSE and mvMFE have growing appeal and broad use. They have been successfully used in a number of biomedical and mechanical engineering applications, such as, to characterise electroencephalogram (EEG) signals in Alzheimer's disease (AD) [21,22], to quantitatively distinguish different horizontal oil–water flow patterns [23], to analyze mechanical vibration noise to stimulate the patient's feet while wearing the shoes [24], to analyze the multivariate cardiovascular time series [25], to characterize focal and non-focal EEG time series [17], to analyze the complexity of interbeat interval and interbreath signals [8], and to analyze the postural fluctuations in fallers and non-fallers older adults [26].

However, mvMSE and mvMFE have the following shortcomings: (1) mvMSE and mvMFE values may be unreliable and unstable for short signals (300 sample points); (2) they are not quick enough for real-time applications; and (3) computation of mvMSE and mvMFE of a signal with a large number of channels needs to have large memory space, as shown later. To address these drawbacks and due to the advantages of multiscale dispersion entropy (DispEn-MDE) over the state-over-the-art multiscale entropy techniques in terms of distinguishing different kinds of dynamics of univariate synthetic and real time series and computation time [27–29], we propose four algorithms to extend our recently developed MDE to its multivariate forms, termed multivariate MDE (mvMDE). To evaluate the mvMDE methods, we use both synthetic and real multivariate datasets. Our results indicate that mvMDE is noticeably faster than the existing methods, leads to more stable results, better discriminates different kinds of biomedical time series, does not lead to undefined values for short multivariate time series, and needs to store a considerably smaller number of elements in comparison with mvMSE and mvMFE.

2. Multivariate Multiscale Dispersion Entropy (mvMDE)

In this study, we propose and explore three different alternative implementations of mvMDE until we arrive at a fourth and preferred one. All the mvMDE implementations include two main steps: (1) coarse-graining process for multivariate time series; and (2) multivariate DispEn (mvDE), as an extension of our recently developed DisEn [13]. It is worth noting that for all the mvMDE algorithms, the mapping based on the normal cumulative distribution function (NCDF) used in the calculation of mvDE for the first temporal scale factor is maintained fixed across all scales. In fact, in the mvMDE, μ and σ of the NCDF are respectively set at the average and standard deviation (SD) of the original time series and they remain constant for all temporal scale factors. This fact is similar to r in the mvMSE and mvMFE, setting at a certain percentage (usually 15%) of the SD of the original signal and remaining constant for all scales [8,17].

2.1. Coarse-Graining Process for Multivariate Signals

Assume we have a p-channel time series $\mathbf{U} = \{u_{k,b}\}_{k=1,2,\ldots,p}^{b=1,2,\ldots,L}$ of length L. In the mvMDE algorithms, for each channel, the original signal is first divided into non-overlapping segments of length τ, named scale factor. Next, for each channel, the average of each segment is calculated to derive the coarse-grained signals as follows [8,17]:

$$x_{k,i}{}^{(\tau)} = \frac{1}{\tau} \sum_{b=(i-1)\tau+1}^{i\tau} u_{k,b},\ 1 \le i \le \left\lfloor \frac{L}{\tau} \right\rfloor = N\,,\ 1 \le k \le p \tag{1}$$

where N denotes the length of the coarse-grained signal. The second step of mvMDE is calculating the mvDE of each coarse-grained signal.

2.2. Background Information for the mvDE

We build four diverse alternative implementations of mvDE (mvDE_I to III and mvDE) until we arrive at a preferred (or optimal) one, i.e., mvDE. However, here, we present all the simpler alternatives (mvDE_I to mvDE_III), since they can still be useful in some settings and allow for clearer comparisons with other current approaches.

2.2.1. mvDE_I

The mvDE_I of the multi-channel coarse-grained time series $\mathbf{X} = \{x_{k,i}\}_{k=1,2,\ldots,p}^{i=1,2,\ldots,N}$, which is based on the mvMPE algorithm [19], is calculated as follows:

(*a*) First, $\mathbf{X} = \{x_{k,i}\}_{k=1,2,\ldots,p}^{i=1,2,\ldots,N}$ are mapped to c classes with integer indices from 1 to c. To this aim, there are a number of linear and nonlinear mapping approaches [30]. The simple linear mapping technique may lead to the problem of assigning the majority of $x_{k,i}$ to limited classes when maximum or minimum values are noticeably larger or smaller than the mean/median value of the image [30]. The weak permanence of DispEn with linear mapping for the characterization of syntactic and real data was illustrated in [13].

A large number of natural processes illustrate a progression from small beginnings that accelerates and approaches a climax over time (e.g., a sigmoid function) [31,32]. When there is not detailed information, a sigmoid function is often used [30,32–34]. The choice of sigmoid function in the context of DispEn was discussed in [30]. We here use NCDF as a well-known sigmoid function like in [13]. Note that using NCDF for each channel also deals with the shortcoming of the amplitude values of each of series $\mathbf{x}_k$ $(k = 1, 2, \ldots, p)$ may be dominated by the components of vectors coming from the time series with the largest amplitudes. The NCDF maps $\mathbf{X}$ into $\mathbf{Y} = \{y_{k,i}\}_{k=1,2,\ldots,p}^{i=1,2,\ldots,N}$ from 0 to 1 as follows:

$$y_{k,i} = \frac{1}{\sigma_k\sqrt{2\pi}} \int_{-\infty}^{x_{k,i}} e^{\frac{-(t-\mu_k)^2}{2\sigma_k^2}} \, dt \tag{2}$$

where σ_k and μ_k are the SD and mean of time series $\mathbf{x}_k$, respectively. Then, we use a linear algorithm to assign each $y_{k,i}$ to an integer from 1 to c. To do so, for each member of the mapped signal, we use $z_{k,i}^c = \text{round}(c \cdot y_{k,i} + 0.5)$, where $z_{k,i}^c$ denotes the ith member of the classified signal in the kth channel and rounding involves either increasing or decreasing a number to the next digit. Note that, although this part is linear, the whole mapping approach is non-linear because of the use of NCDF.

(*b*) Time series $\mathbf{z}_{k,j}^{m,c}$ are made with embedding dimension m and time delay d according to $\mathbf{z}_{k,j}^{m,c} = \{z_{k,j}^c, z_{k,j+d}^c, + \cdots + z_{k,j+(m-1)d}^c\}, j = 1, 2, \ldots, N-(m-1)d$ [11–13]. Each time series $\mathbf{z}_{k,j}^{m,c}$ is mapped to a dispersion pattern $\pi_{v_0 v_1 \ldots v_{m-1}}$, where $z_{k,j}^c = v_0, z_{k,j+d}^c = v_1, \ldots, z_{k,j+(m-1)d}^c = v_{m-1}$. The number of possible dispersion patterns that can be assigned to each time series $\mathbf{z}_{k,j}^{m,c}$ is equal to c^m, since the signal has m members and each member can be one of the integers from 1 to c [13].

(*c*) For each channel $1 \le k \le p$ and for each of c^m potential dispersion patterns $\pi_{v_0 \ldots v_{m-1}}$, relative frequency is obtained as follows:

$$p(\pi_{v_0 \ldots v_{m-1}}) = \frac{\#\{j \,\big|\, j \le N-(m-1)d, \mathbf{z}_{k,j}^{m,c} \text{ has type } \pi_{v_0 \ldots v_{m-1}}\}}{(N-(m-1)d)p} \tag{3}$$

where # means cardinality. In fact, $p(\pi_{v_0 \ldots v_{m-1}})$ shows the number of dispersion patterns of $\pi_{v_0 \ldots v_{m-1}}$ that is assigned to $\mathbf{z}_{k,j}^{m,c}$, divided by the total number of embedded signals with embedding dimension m multiplied by the number of channels.

(*d*) Finally, based on the Shannon's definition of entropy, the mvDE$_\text{I}$ is calculated as follows:

$$mvDE_I(\mathbf{X}, m, c, d) = -\sum_{\pi=1}^{c^m} p(\pi_{v_0 \ldots v_{m-1}}) \cdot \ln\left(p(\pi_{v_0 \ldots v_{m-1}})\right) \tag{4}$$

In case all possible dispersion patterns have equal probability value, the highest value of mvDE$_\text{I}$ is obtained, which has a value of $ln(c^m)$. In contrast, if there is only one $p(\pi_{v_0 \ldots v_{m-1}})$ different from zero, which demonstrates a completely regular/certain signal, the smallest value of mvDE$_\text{I}$ is obtained. In the algorithm of mvDE$_\text{I}$, we compare Np dispersion patterns of a p-channel signal with c^m potential patterns. Thus, at least $c^m + Np$ elements are stored.

To work with reliable statistics to calculate MDE, it was recommended $c^m < \left\lfloor \frac{L}{\tau_{max}} \right\rfloor$ [27]. Since mvDE$_\text{I}$ counts the dispersion patterns for every channel of a multivariate time series, it is suggested $c^m < \left\lfloor \frac{pL}{\tau_{max}} \right\rfloor$. mvDE$_\text{I}$ extracts the dispersion patterns from each of channels regardless of their cross-channel information. Thus, mvDE$_\text{I}$ works appropriately when the components of a multivariate signal are statistically independent. However, the mvDE$_\text{I}$ algorithm, like mvPE [19], does not consider the spatial domain of time series. To overcome this problem, we propose mvDE$_\text{II}$ based on the Taken's theorem [17,35].

2.2.2. mvDE$_\text{II}$

The algorithm of mvDE$_\text{II}$ is as follows:

(*a*) First, like mvDE$_\text{I}$, $\mathbf{X} = \{x_{k,i}\}_{k=1,2,\ldots,p}^{i=1,2,\ldots,N}$ are mapped to $\mathbf{Z} = \{z_{k,i}\}_{k=1,2,\ldots,p}^{i=1,2,\ldots,N}$ based on the NCDF.

(*b*) To take into account both the spatial and time domains, multi-channel embedded vectors are generated according to the multivariate embedding theory [35]. The multivariate embedded reconstruction of **Z** is defined as:

$$\begin{aligned} Z_{\mathbf{m}}(j) = [&z_{1,j}, z_{1,j+d_1}, \dots, z_{1,j+(m_1-1)d_1}, \\ &z_{2,j}, z_{2,j+d_2}, \dots, z_{2,j+(m_2-1)d_2}, \dots, \\ &z_{p,j}, z_{p,j+d_p}, \dots, z_{p,j+(m_p-1)d_p}] \end{aligned} \tag{5}$$

where $\mathbf{m} = [m_1, m_2, \dots, m_p]$ and $\mathbf{d} = [d_1, d_2, \dots, d_p]$ denote the embedding dimension and the time lag vectors, respectively. Note that the length of $Z_{\mathbf{m}}(j)$ is $\sum_{k=1}^{p} m_k$. For simplicity, we assume $d_k = d$ and $m_k = m$, that is, all the embedding dimension values and all the delay values are equal.

(*c*) Each series $Z_{\mathbf{m}}(j)$ is mapped to a dispersion pattern $\pi_{v_0 v_1 \dots v_{mp-1}}$, where $z_{1,j}^c = v_0$, $z_{1,j+d}^c = v_1, \dots,$ $z_{p,j+(m-1)d} = v_{mp-1}$. The number of possible dispersion patterns that can be assigned to each time series $Z_{\mathbf{m}}(j)$ is equal to c^{mp}, since the signal has mp members and each member can be one of the integers from 1 to c.

(*d*) For each of c^{mp} potential dispersion patterns $\pi_{v_0 \dots v_{mp-1}}$, relative frequency is obtained based on the DisEn algorithm [13] as follows:

$$p(\pi_{v_0 \dots v_{mp-1}}) = \frac{\#\{j \,|\, j \leq N-(m-1)d, Z_{\mathbf{m}}(j) \text{ has type } \pi_{v_0 \dots v_{mp-1}}\}}{N-(m-1)d} \tag{6}$$

(*e*) Finally, based on the Shannon's definition of entropy, the mvDE$_{\text{II}}$ is calculated as follows:

$$mvDE_{II}(\mathbf{X}, m, c, d) = -\sum_{\pi=1}^{c^{mp}} p(\pi_{v_0 \dots v_{mp-1}}) \cdot \ln\left(p(\pi_{v_0 \dots v_{mp-1}})\right) \tag{7}$$

In the algorithm of mvDE$_{\text{II}}$, at least $c^{mp} + Np$ elements are stored. Thus, when p is large, the algorithm needs huge space of memory to store elements. To work with reliable statistics to calculate mvMDE$_{\text{II}}$, it is recommended $c^{mp} < \left\lfloor \frac{L}{\tau_{max}} \right\rfloor$. Thus, although mvDE$_{\text{II}}$ deals with both the spatial and time domains, the length of a signal and its number of channels should be very large and small, respectively, to reliably calculate mvDE$_{\text{II}}$ values. To alleviate the problem, we propose mvDE$_{\text{III}}$.

2.2.3. mvDE$_{\text{III}}$

The algorithm of mvDE$_{\text{III}}$ is as follows:

(*a*) First, like the mvDE$_{\text{I}}$ and mvDE$_{\text{II}}$ approaches, $\mathbf{X} = \{x_{k,i}\}_{k=1,2,\dots,p}^{i=1,2,\dots,N}$ are mapped to $\mathbf{Z} = \{z_{k,i}\}_{k=1,2,\dots,p}^{i=1,2,\dots,N}$.

(*b*) Multivariate embedded vectors $\mathbf{Z}_{k,\mathbf{m}}(j)$ with length $m + p - 1$ are generated according to the Taken's embedding theorem [35] with p embedding dimension vectors $\mathbf{m}_k = [1, 1, \dots, m_k, \dots, 1, 1]$ ($k = 1, \dots, p$), where m_k denotes the kth element of $\mathbf{m}$. For simplicity, we assume $m_k = m$ and $d_k = d$.

(*c*) Each series $\mathbf{Z}_{k,\mathbf{m}}(j)$ is mapped to a dispersion pattern $\pi_{v_0 v_1 \dots v_{m+p-2}}$. The number of possible dispersion patterns that can be assigned to each time series $\mathbf{Z}_{k,\mathbf{m}}(j)$ is equal to c^{m+p-1}, since the signal has $m + p - 1$ members and each member can be one of the integers from 1 to c [13]. Since we count the number of patterns for each of p different $\mathbf{m}_k$ leading to a considerable increase in the number of dispersion patterns, compared with mvDE$_{\text{II}}$, we have more reliable results for a signal with a small number of sampl than those fore points, as shown later.

(*d*) For each channel $1 \leq k \leq p$ and for each of c^{m+p-1} potential dispersion patterns $\pi_{v_0 \ldots v_{m+p-2}}$, relative frequency is obtained as follows:

$$p(\pi_{v_0 \ldots v_{m+p-2}}) = \frac{\#\{j \,\big|\, j \leq N-(m-1)d, \mathbf{Z}_{k,\mathbf{m}}(j) \text{ has type } \pi_{v_0 \ldots v_{m+p-2}}\}}{(N-(m-1)d)p} \tag{8}$$

(*e*) Finally, based on the Shannon's definition of entropy, the mvDE$_{\text{III}}$ is calculated as follows:

$$mvDE_{III}(\mathbf{X}, m, c, d) = -\sum_{\pi=1}^{c^{m+p-1}} p(\pi_{v_0 \ldots v_{m+p-2}}) \cdot \ln\left(p(\pi_{v_0 \ldots v_{m+p-2}})\right) \tag{9}$$

mvDE$_{\text{III}}$ assumes embedding dimension 1 for all signals except one, which might limit the potential to explore the dynamics. Moreover, in the algorithm of mvDE$_{\text{III}}$, at least $c^{m+p-1} + Np$ elements are stored. Although this number is noticeably smaller than that for mvDE$_{\text{II}}$, the algorithm still needs to have large memory space for a signal with a large number of channels. To work with reliable statistics to calculate mvMDE$_{\text{III}}$, it is recommended $c^{m+p-1} < \left\lfloor \frac{pL}{\tau_{max}} \right\rfloor$. Therefore, albeit mvDE$_{\text{III}}$ takes into account both the spatial and time domains and needs to smaller number of sample points in comparison with mvDE$_{\text{II}}$, there is a need to have a large enough number of samples and small number of channels. To alleviate these deficiencies, we propose mvDE.

2.3. Multivariate Dispersion Entropy (mvDE)

The mvDE algorithm is as follows:

(*a*) First, like mvDE$_{\text{I}}$ to III, the multivariate signal $\mathbf{X} = \{x_{k,i}\}_{k=1,2,\ldots,p}^{i=1,2,\ldots,N}$ is mapped to c classes with integer indices from 1 to c.

(*b*) Like mvDE$_{\text{II}}$, to consider both the spatial and time domains, multivariate embedded vectors $Z_{\mathbf{m}}(j), 1 \leq j \leq N-(m-1)d$ are created based on the Taken's embedding theorem [35]. For simplicity, we assume $d_k = d$ and $m_k = m$.

(*c*) For every $Z_{\mathbf{m}}(j)$, all combinations of the $\sum_{k=1}^{p} m_k$ elements in $Z_{\mathbf{m}}(j)$ taken m at a time, termed $\phi_q(j)$ $(q = 1, \ldots \binom{mp}{m})$, are created. The number of the combinations is equal to $\binom{mp}{m}$. Therefore, for all channels, we have $(N-(m-1)d)\binom{mp}{m}$ dispersion patterns.

(*d*) For each $1 \leq q \leq \binom{mp}{m}$ and for each of c^m potential dispersion patterns $\pi_{v_0 \ldots v_{m-1}}$, relative frequency is obtained as follows:

$$p(\pi_{v_0 \ldots v_{m-1}}) = \frac{\#\{j \,|\, j \leq N-(m-1)d, \phi_q(j) \text{ has type } \pi_{v_0 \ldots v_{m-1}}\}}{(N-(m-1)d)\binom{mp}{m}} \tag{10}$$

(*e*) Finally, based on the Shannon's definition of entropy, the mvDE is calculated as follows:

$$mvDE(\mathbf{X}, m, c, d) = -\sum_{\pi=1}^{c^m} p(\pi_{v_0 \ldots v_{m-1}}) \cdot \ln\left(p(\pi_{v_0 \ldots v_{m-1}})\right) \tag{11}$$

In fact, mvDE explores all combinations of patterns of length m within an mp-dimensional embedding vector. In the mvDE algorithm, at least $c^m + Np$ elements are stored. This number is noticeably smaller than those for mvDE$_{\text{II}}$ to III, leading to more stable results for signals with a short length and a large number of samples. As the number of patterns obtained by the mvMDE method is $(N-(m-1)d)\binom{mp}{m}$, it is suggested $c^m < \left\lfloor \frac{L\binom{mp}{m}}{\tau_{max}} \right\rfloor$ to work with reliable statistics. It is worth mentioning that if the order of channels in a multi-channel time series changes, although the assignment to each dispersion pattern obtained by the mvMDE-based methods may change, the entropy value will stay the same.

2.4. Parameters of the mvMDE, mvMSE, and mvMFE Methods

In addition to the maximum scale factor τ_{max} described before, there are three other parameters for the mvMDE methods, including the embedding dimension vector $\mathbf{m}$, number of classes c, and time delay vector $\mathbf{d}$. Although some information with regard to the frequency of signals may be ignored for $d_k > 1$, it is better to set $d_k > 1$ for oversampled time series. However, like previous studies about multivariate entropy methods [2,8], we set $d_k = 1$ for simplicity. Nevertheless, when the sampling frequency is considerably larger than the highest frequency component of a time series, the first minimum or zero crossing of the autocorrelation function or mutual information can be utilized for the selection of an appropriate time delay [36]. We need $1 < c$ to keep away the trivial case of having only one dispersion pattern. For simplicity, we use $c = 5$ and $m_k = 2$ for all signals used in this study, although the range $2 < c < 9$ leads to similar findings. For more information about c, m_k, and d_k, please refer to [13,30].

In this study, d_k, m_k, and r for the mvMSE and mvMFE were respectively set as 1, 2, and 0.15 of the SD of the original time series following recommendations in [8,17]. The maximum scale factor for mvMSE and mvMFE also follows [8,17]. In the algorithm of mvSE and mvFE, at least $\binom{Np}{2} + Np(pm+1)$ elements are stored (the mvSE code available at http://www.commsp.ee.ic.ac.uk/~mandic/research/Complexity_Stuff.htm). Matlab codes of mvMFE and mvMSE are available at http://dx.doi.org/10.7488/ds/1432. Overall, the characteristics and limitations of the mvSE, mvFE, and mvDE algorithms for a p-channel signal with length N are summarized in Table 1.

Table 1. Ability to deal with spatial domain and characterization of short signals (300 sample points), typical number of elements to be stored, and typical number of samples needed for each of the mvSE, mvFE, and mvDE algorithms for a p-channel signal with length N sample points.

Methods	Spatial Domain	Short Signals	Typical Number of Elements Stored	Typical Number of Samples
mvSE [3]	yes	undefined	$\binom{Np}{2} + Np(pm+1)$	$10^m < N$
mvFE [17]	yes	unreliable	$\binom{Np}{2} + Np(pm+1)$	$10^m < N$
mvPE [19] and mvWPE [20]	no	reliable	$m! + Np$	$m! < N$
mvDE$_{\text{I}}$	no	reliable	$c^m + Np$	$\frac{c^m}{p} < N$
mvDE$_{\text{II}}$	yes	unreliable	$c^{mp} + Np$	$c^{mp} < N$
mvDE$_{\text{III}}$	yes	unreliable	$c^{m+p-1} + Np$	$\frac{c^{m+p-1}}{p} < N$
mvDE	yes	reliable	$c^m + Np$	$\frac{c^m}{\binom{mp}{m}} < N$

3. Evaluation Signals

In this section, the descriptions of correlated and uncorrelated noise signals, bivariate autoregressive (BAR) process, and real time series used in this study are given.

3.1. Synthetic Signals

The irregularity of multivariate $1/f$ noise is lower than multivariate WGN, whereas the complexity of the former is higher than the latter [8,14,17]. Thus, $1/f$ noise and WGN signals have been commonly used to assess the multivariate multiscale entropy techniques [8,17,37]. For more information about the algorithms used for multivariate $1/f$ noise and WGN, please refer to [8,17].

To understand the behaviour of the mvMDE methods on uncorrelated WGN and $1/f$ noise, we first generated a trivariate time series, where originally all three data channels were realization of mutually independent $1/f$ noise. Then, we gradually decreased the number of data channels representing $1/f$ noise (from 3 to 0) and at the same time, increased the number of variates representing independent WGN (from 0 to 3) [37]. The number of channels was always three.

To create correlated bivariate noise time series, we first generated a bivariate uncorrelated random time series $\mathbf{H}$. Afterwards, $\mathbf{H}$ was multiplied with the standard deviation (hereafter, sigma) and then, the value of the mean (hereafter, mu) was added. Next, **H** was multiplied by the upper triangular matrix **L** obtained from the Cholesky decomposition of a defined correlation matrix **R** (which is positive and symmetric) to set the correlation. Here, we set $\mathbf{R} = \begin{bmatrix} 1 & 0.95 \\ 0.95 & 1 \end{bmatrix}$ according to [8,17]. An in-depth study on the effect of correlated and uncorrelated $1/f$ noise and WGN on multiscale entropy approaches can be found in [8,10].

Based on the fact that the larger the order of an autoregressive process, the more complex the AR process [8], we evaluate the mvMDE, mvMSE, and mvMFE methods on a BAR(α) process with the maximum lag α describing the evolution of a set of two variables as a linear function of their past values according to:

$$\mathbf{y}_n = \mathbf{e}_n + \sum_{\gamma=1}^{\alpha} \mathbf{y}_{n-\gamma}\mathbf{A}_\gamma \tag{12}$$

where $\mathbf{y}_n = \{y_n(1), y_n(2)\}$ is the nth sample of a bidimensional time series, $\mathbf{A}_\gamma$ denotes the 2×2 matrix of parameters corresponding to lag order γ, and $\mathbf{e}_n$ is the 2×1 vector of error terms assumed to be WGN [38].

3.2. Real Biomedical Datasets

(1) Dataset of Stride Interval Fluctuations: To investigate the ability of the proposed mvMDE methods to reveal the long-range correlations and dynamics of multivariate signals, the stride interval recordings are used [2,39]. The time series were recorded from ten young, healthy men. Mean age was 21.7 years, changing from 18 to 29 years. Height and weight were 1.77 ± 0.08 meters (mean $\pm$ SD) and 71.8 ± 10.7 kg (mean $\pm$ SD), respectively. All ten participants provided informed written consent walking for 1 hour at slow, 1 hour at normal, and 1 hour at fast paces and also walking a metronome set to each subject's mean stride interval. Three walking paces were considered as different variables from the same system. In this way, we expect to be able to discriminate between the metronomically-paced and self-spaced walking. For further information, please refer to [39].

(2) Dataset of Focal and Non-focal Brain Activity: The ability of the mvMDE methods, in comparison with mvMFE and mvMSE, to differentiate focal from non-focal recordings is evaluated using a publicly-available EEG dataset [40]. The dataset includes 5 patients and, for each patient, there are 750 focal and 750 non-focal bivariate signals. The length of each recording was 20 s with sampling frequency of 512 Hz (10,240 sample points). Further information can be found in [40]. Before computing the aforementioned methods, all recordings were digitally filtered employing an FIR band-pass filter with cut-off frequencies at 0.5 Hz and 40 Hz.

(3) Surface MEG Recordings in Alzheimer's Disease: We analysed resting state MEG time series recorded with a 148-channel whole-head magnetometer. All 62 participants agreed for the research, which was approved by the local ethics committee. To screen the cognitive status, a mini-mental state examination (MMSE) was done. There were 36 AD patients (age = 74.06 ± 6.95 years, all data given as mean $\pm$ SD, and MMSE score = 18.06 ± 3.36) and 26 controls (age = 71.77 ± 6.38 years, and MMSE score = 28.88 ± 1.18). The difference in age between two groups was not significant (p-value = 0.1911, Student's t-test) [41]. The distribution of MEG sensors is shown in Figure 2 in [41]. For each participant, five minutes of MEG resting state activity were recorded at a sampling frequency of 169.5 Hz. The signals were divided into 10 s segments (1695 samples) and visually inspected using an automated thresholding procedure to discard epochs noticeably contaminated with artifacts. All recordings were digitally band-pass filtered with a Hamming window FIR filter of order 200 and cut-off frequencies at 1.5 and 40 Hz. For more information, please see [41].

4. Results and Discussions

4.1. Synthetic Signals

4.1.1. Uncorrelated White Gaussian and $1/f$ Noises

We first apply the proposed and existing methods to 40 independent realizations of uncorrelated trivariate WGN and $1/f$ noise, described in Section 3. The number of sample points for each of the $1/f$ noise and WGN signals were 15,000. mvMSE and mvMFE are based on conditional entropy [2,8,17]. On the other hand, mvMDE is based on the Shannon's entropy definition applied to dispersion patterns. This means that the methods work on slightly different principles. However, the comparison of mvMDE with mvMSE and mvMFE is meaningful because the latter two are the most common multivariate entropy algorithms and MDE has been shown to have similar behaviour to MSE when analysing real and synthetic signals [27]. Thus, we compare the mvMDE methods with mvMSE and mvMFE. The average and SD of the results for mvMDE$_{\text{I}}$, mvMDE$_{\text{II}}$, mvMDE$_{\text{III}}$, mvMDE, mvMSE, and mvMFE are depicted in Figure 1a–f, respectively. Using all the existing and proposed methods, the entropy values of trivariate WGN signals are higher than those of the other trivariate time series at low scale factors. However, the entropy values for the coarse-grained trivariate $1/f$ noise signals stay almost constant or decrease slowly along the temporal scale factor, while the entropy values for the coarse-grained WGN signal monotonically decreases with the increase of scale factors. When the length of WGN signals, obtained by the coarse-graining process, decreases (i.e., the scale factor increases), the mean value of inside each signal converges to a constant value and the SD becomes smaller. Therefore, no new structures are revealed at higher temporal scales. This demonstrates a multivariate WGN time series has information only in small temporal scale factors. In contrast, for trivariate $1/f$ noise signals, the mean value of the fluctuations inside each signal does not converge to a constant value.

For all the methods, the higher the number of variates representing $1/f$ noise, the higher complexity the trivariate signal, in agreement with the fact that multivariate $1/f$ noise is structurally more complex than multivariate WGN [8,14,17]. Here, for multivariate $1/f$ noise and WGN, τ_{max} was 20 for mvMDE, according to Section 2.

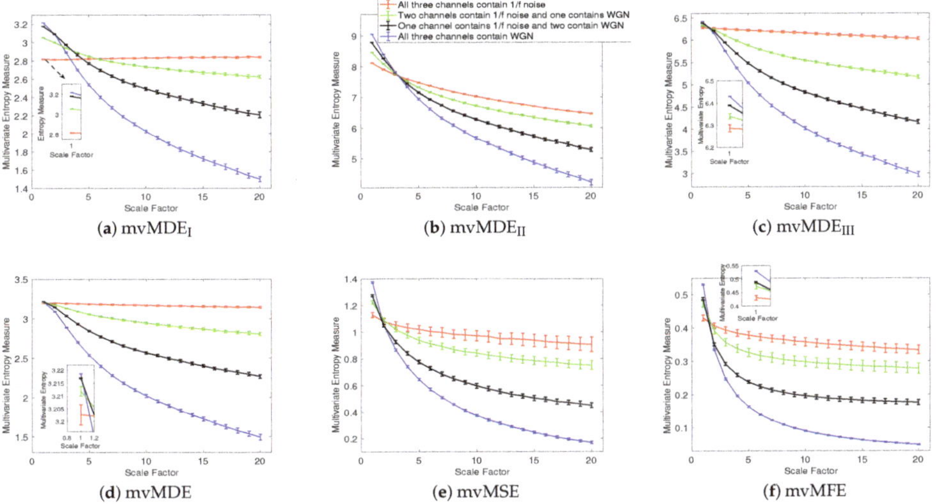

Figure 1. Mean value and SD of the results using (**a**) mvMDE$_{\text{I}}$, (**b**) mvMDE$_{\text{II}}$, (**c**) mvMDE$_{\text{III}}$, (**d**) mvMDE, (**e**) mvMSE, and (**f**) mvMFE computed from 40 different uncorrelated trivariate WGN and $1/f$ noise time series with length 15,000 sample points.

To compare the results obtained by the mvMDE, mvMSE, and mvMFE methods, we used the coefficient of variation (CV). CV, as a measure of relative variability, is defined as the SD divided by the mean of a time series. We use such a metric as the SDs of time series may increase or decrease proportionally to the mean. We investigate the results obtained by uncorrelated noise signals at scale factor 10, as a trade-off between short and long scale factors. As can be seen in Table 2, the smallest CV values for uncorrelated trivariate $1/f$ noise, an uncorrelated combination of bivariate $1/f$ noise and univariate WGN, an uncorrelated combination of bivariate WGN and univariate $1/f$ noise, and trivariate WGN are achieved by mvMDE, $mvMDE_{II}$, $mvMDE_{II}$, and $mvMDE_{I}$, respectively. Overall, the smallest CV values for trivariate $1/f$ noise and WGN profiles are reached by the mvMDE methods, showing the superiority of the mvMDE methods over mvMSE and mvMFE in terms of stability of results.

Table 2. CV values of the proposed and existing multivariate multiscale entropy-based analyses at scale factor 10 for the uncorrelated trivariate $1/f$ noise and WGN.

Time Series	$mvMDE_{I}$	$mvMDE_{II}$	$mvMDE_{III}$	mvMDE	mvMSE	mvMFE
All three channels contain $1/f$ noise	0.0028	0.0025	0.0037	0.0022	0.0405	0.0355
Two channels contain $1/f$ noise and one contains WGN	0.0042	0.0032	0.0036	0.0044	0.0283	0.0274
One channel contains $1/f$ noise and two contain WGN	0.0066	0.0052	0.0058	0.0061	0.0305	0.0292
All three channels contain WGN	0.0072	0.0080	0.0092	0.0101	0.0232	0.0211

To assess the ability of the mvMDE methods to characterize short signals in comparison with mvMFE and mvMSE, we use trivariate $1/f$ noise and WGN with length of 300 sample points. The results for the mvMDE, mvMSE, and mvMFE approaches at temporal scales 1 to 20 are depicted in Figure 2a–f, respectively. The results show that only $mvMDE_{I}$ is able to distinguish these four different kinds of noise signals at scale factor 1. For the higher temporal scale factors, $mvMDE_{I}$ and mvMDE distinguish these time series, showing a limitation of mvMDE for the discrimination of white from $1/f$ noise at lower scale factors and also the importance of considering higher temporal scales for the mvMDE technique. As can be seen in Figure 2a,d, the $mvMDE_{I}$ and mvMDE methods better discriminate different dynamics of the noise signals. However, the mvMSE values are undefined at higher scale factors. It is worth mentioning that we compared mvMDE with the original algorithms of mvMSE and mvMFE. However, more recent studies on entropy estimation of short physiological signals provided methods to deal with this issue [17,42].

Although the mvMFE- and $mvMDE_{II}$-based values are defined at all scale factors, they cannot distinguish the dynamics of different noise signals. The profiles obtained by $mvMDE_{III}$ are more distinguishable than $mvMDE_{II}$, as mentioned that $mvMDE_{III}$ needs a smaller number of sample points. Nevertheless, the profiles obtained by $mvMDE_{III}$ have overlaps at several scale factors. Overall, the results show the superiority of $mvMDE_{I}$ and mvMDE over $mvMDE_{II}$, $mvMDE_{III}$, mvMSE, and mvMFE for short uncorrelated signals.

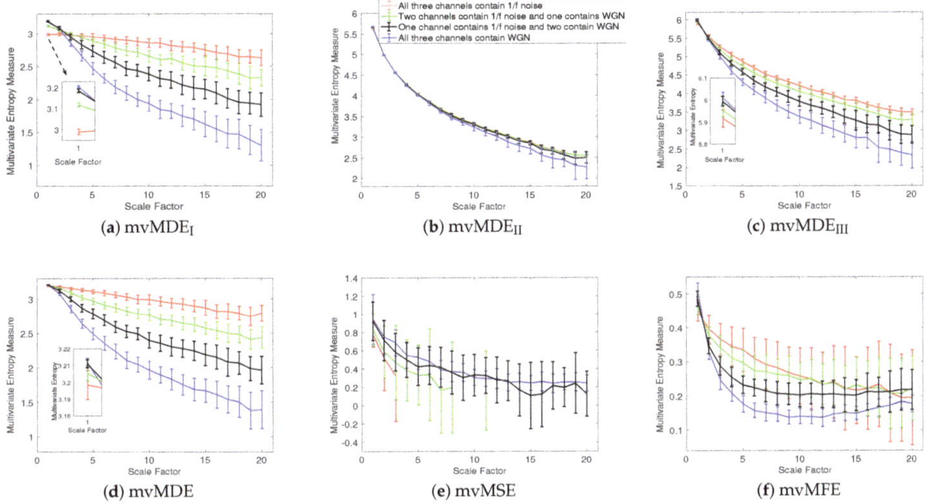

Figure 2. Mean value and SD of the results obtained by (**a**) mvMDE$_{\mathrm{I}}$, (**b**) mvMDE$_{\mathrm{II}}$, (**c**) mvMDE$_{\mathrm{III}}$, (**d**) mvMDE, (**e**) mvMSE, and (**f**) mvMFE computed from 40 different uncorrelated trivariate WGN and $1/f$ noise time series with length 300 sample points.

4.1.2. Computational Time

To evaluate the computational time of mvMSE, mvMFE, mvMDE$_{\mathrm{I}}$ to III, and mvMDE, we use uncorrelated multivariate WGN time series with different lengths, changing from 100 to 10,000 sample points, and different number of channels, changing from 2 to 8. The results are depicted in Table 3. The simulations have been carried out using a PC with Intel (R) Core (TM) i7-7820X CPU, 3.6 GHz and 16-GB RAM by MATLAB R2018b. The results show that the computation times for mvMSE and mvMFE are close. The slowest algorithm is mvMDE$_{\mathrm{II}}$, while the fastest ones are mvMDE$_{\mathrm{I}}$ and mvMDE, in that order. For an 8-channel signal with 10,000 samples, using mvMDE$_{\mathrm{II}}$, the array exceeded the memory available. Overall, in terms of computation time and memory space, mvMDE outperforms the other methods that take into account both the time and spatial domains. We used the mvMSE code provoided in [8] and the mvMDE, mvMSE, and mvMFE Matlab codes have not been optimized.

Table 3. Computational time of the mvMSE, mvMFE, and mvMDE algorithms with $\tau_{max} = 10$.

Number of Channels and Samples	mvMSE	mvMFE	mvMDE$_{\mathrm{I}}$	mvMDE$_{\mathrm{II}}$	mvMDE$_{\mathrm{III}}$	mvMDE
2 channels and 1000 samples	0.051 s	0.066 s	0.014 s	0.023 s	0.026 s	0.020 s
2 channels and 3000 samples	0.237 s	0.296 s	0.035 s	0.057 s	0.068 s	0.052 s
2 channels and 10,000 samples	1.821 s	2.016 s	0.111 s	0.190 s	0.223 s	0.181 s
5 channels and 1000 samples	0.209 s	0.223 s	0.028 s	43.096 s	0.490 s	0.050 s
5 channels and 3000 samples	1.129 s	1.204 s	0.080 s	82.246 s	1.137 s	0.137 s
5 channels and 10,000 samples	9.432 s	9.801 s	0.260 s	218.553 s	3.343 s	0.491 s
8 channels and 1000 samples	0.489 s	0.501 s	0.042 s	out of memory error	65.560 s	0.086 s
8 channels and 3000 samples	2.973 s	2.906 s	0.124 s	out of memory error	150.122 s	0.243 s
8 channels and 10,000 samples	27.993 s	25.951 s	0.398 s	out of memory error	363.752 s	0.824 s

4.1.3. Correlated white Gaussian and $1/f$ Noises

Univariate multiscale entropy approaches only consider every data channel separately and fail to take into account the cross-channel information of multivariate time series [8]. Uncorrelated multi-channel WGN has less structural complexity and more irregularity compared with multi-channel $1/f$ noise. To assess the ability of the existing and proposed multivariate entropy methods to reveal the dynamics across the channels, we created 40 independent realizations of different combinations of bivariate $1/f$ noise and WGN time series with length 20,000 (according to [8,17]), making the channels correlated. Figure 3a–d respectively show the results obtained using the mvMDE$_{I}$, mvMDE$_{II}$, mvMDE$_{III}$, and mvMDE to model both the within- and cross-channel properties in multivariate signals.

mvMDE$_{I}$ cannot discriminate the correlated from uncorrelated WGN or $1/f$ noise. This fact is revealed in Figure 3a. Therefore, mvMDE$_{I}$ should only be used when the components of a multi-channel time series are statistically independent. Multivariate multiscale entropy-based methods at scale factor 1 show the irregularity of multi-channel signals [8]. The mvMDE$_{II}$, mvMDE$_{III}$, and mvMDE values at scale 1 show that the uncorrelated WGN is the most irregular and unpredictable time series in agreement with [10], while the most irregular signals using mvMFE and mvMSE are the correlated WGN [8,17], in contrast with the fact that correlated multi-channel WGN signals are more predictable and regular than uncorrelated WGN ones [10,27]. Although mvMDE was able to distinguish all four different kinds of noises at the small scale factors, there are some overlaps between the results for the correlated and uncorrelated bivariate WGN time series at the high scale factors showing the importance both low and high temporal scale factors in mvMDE.

The correlated bivariate $1/f$ noise is the most complex signal using the mvMDE$_{II}$, mvMDE$_{III}$, and mvMDE. The second most complex signal is the uncorrelated bivariate $1/f$ noise, as can be seen in Figure 3. The decreases of the uncorrelated bivariate WGN profiles using mvMDE$_{II}$, mvMDE$_{III}$, and mvMDE are the largest, evidencing the fact that the uncorrelated WGN is the least complex time series. These facts are also in agreement with the previous studies [8,14,17]. Therefore, as desired, the mvMDE$_{II}$, mvMDE$_{III}$, and mvMDE deal with both the cross- and within-channel correlations.

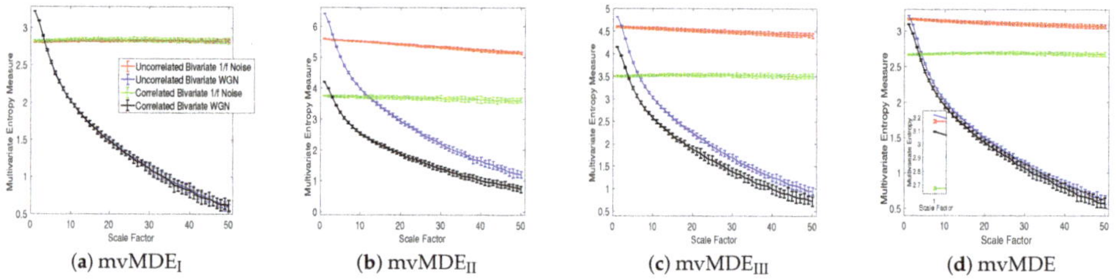

(**a**) mvMDE$_{I}$ (**b**) mvMDE$_{II}$ (**c**) mvMDE$_{III}$ (**d**) mvMDE

Figure 3. Mean value and SD of the results obtained by (**a**) mvMDE$_{I}$, (**b**) mvMDE$_{II}$, (**c**) mvMDE$_{III}$, and (**d**) mvMDE computed from 40 different correlated and uncorrelated bivariate WGN and $1/f$ noise time series with length 20,000 sample points.

4.1.4. Bivariate AR Processes

The ability of the mvMDE methods to characterize multivariate AR processes is further evaluated using combinations of BAR(1), BAR(3), and BAR(5) with $\mathbf{A}_{\gamma_1} = \begin{bmatrix} 0.05 & 0.05 \\ 0.05 & 0.05 \end{bmatrix}$, $\mathbf{A}_{\gamma_2} = \begin{bmatrix} 0.10 & 0.10 \\ 0.10 & 0.10 \end{bmatrix}$, and $\mathbf{A}_{\gamma_3} = \begin{bmatrix} 0.15 & 0.15 \\ 0.15 & 0.15 \end{bmatrix}$. The results obtained by the mvMDE$_{I}$, mvMDE$_{II}$, mvMDE$_{III}$, and mvMDE methods are shown in Figure 4. As expected, when the lag order increases, the complexity of the corresponding time series using the mvMDE approaches increases, in agreement with the fact that a larger lag order denotes a more complex time series [8]. As the elements of $\mathbf{A}_{\gamma_1}$ are smaller than those of $\mathbf{A}_{\gamma_2}$ and $\mathbf{A}_{\gamma_3}$, the behaviour of the profiles obtained by the mvMDE methods are more similar to the results for WGN (see Figure 1). In fact, the smaller the elements of $\mathbf{A}_{\gamma}$, the less complex the BAR, leading to lower entropy values at higher scale factors.

In order to investigate the dependence of the mvMDE methods on the sensitivity to changes in the signals, we generated BAR(3) with length of 10,000 sample points and sampling frequency of 150 Hz that $\mathbf{A}_{\gamma}$ linearly changes from $\begin{bmatrix} 0.17 & 0 \\ 0 & 0.17 \end{bmatrix}$ to $\begin{bmatrix} 0.17 & 0.17 \\ 0.17 & 0.17 \end{bmatrix}$. In fact, the elements of the diagonal of

A are constant and those of anti-diagonal linearly increase from 0 to 0.17, leading to more complex series. We moved a bivariate window—termed temporal window—with length 2000 samples and 20% overlap along this BAR(3) signal. The entropy of each bivariate temproal window is caculated. The results, depicted in Figure 5 show that when the time window is occupied at the beginning of the BAR(3) ($\mathbf{A} = \begin{bmatrix} 0.17 & 0 \\ 0 & 0.17 \end{bmatrix}$), the mvMDE$_{\text{I}}$, mvMDE$_{\text{II}}$, mvMDE$_{\text{III}}$, and mvMDE values at higher scale factors are the smallest, showing the least complexity of BAR(3) in lower temporal windows, while their corresponding entropy values in the end of BAR(3) process ($\mathbf{A} = \begin{bmatrix} 0.17 & 0.17 \\ 0.17 & 0.17 \end{bmatrix}$) are the largest. It is worth noting that as described before, mvMDE$_{\text{II}}$ needs a larger number of sample points to appropriately characterize the dynamics of signals. This fact can be observed in Figure 5, showing mvMDE$_{\text{II}}$ is the least able to distinguish such changes.

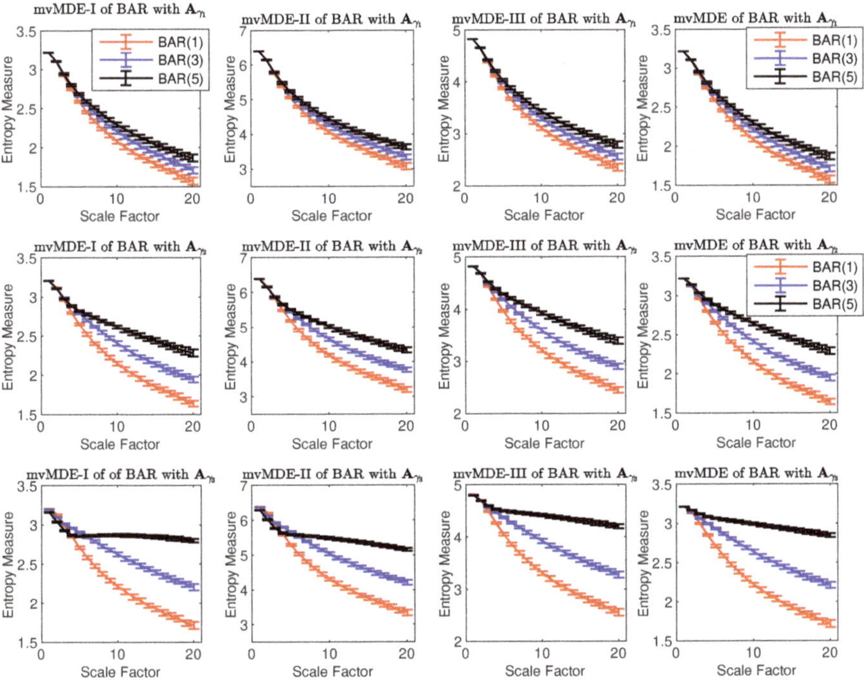

Figure 4. Mean and SD values of the results using mvMDE$_{\text{I}}$, mvMDE$_{\text{II}}$, mvMDE$_{\text{III}}$, and mvMDE computed from 40 different BAR(1), BAR(3), and BAR(5) time series with $\mathbf{A}_{\gamma_1}$ (first row), $\mathbf{A}_{\gamma_2}$ (second row), and $\mathbf{A}_{\gamma_3}$ (third row).

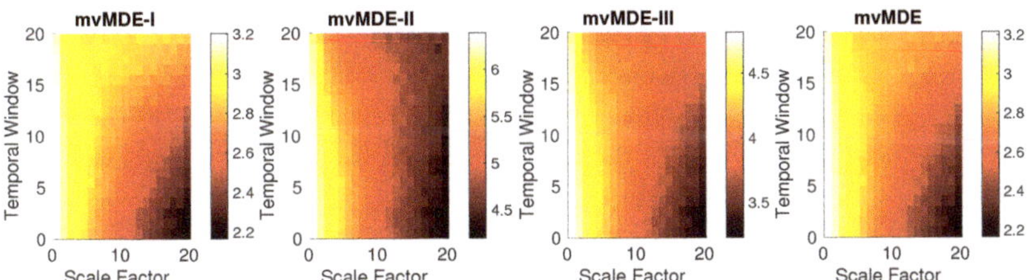

Figure 5. Results obtained by the mvMDE methods using a bivariate temporal window with length 2000 sample points moving along the BAR(3) signal, which the elements of anti-diagonal of the matrix **A** linearly increase from 0 to 0.17, leading to more complex series.

4.2. Real Biomedical Datasets

Discrimination of aged and diseased individuals' from control or healthy subjects' time series is a long-lasting challenge in the physiological complexity literature [8,10,17]. To this end, we use the mvMDE methods, in comparison with mvMFE as an improved version of mvMSE [17], to detect different types of dynamical variability of multivariate recordings of three physiological datasets. Of note is that we do not use the $mvMDE_I$ for biomedical signals, because it does not take into account both the spatial and time domains at the same time.

(1) Dataset of Stride Interval Fluctuations: For the self-paced versus metronomically-paced stride interval fluctuations, the results obtained by the $mvMDE_{III}$, mvMDE, and mvMFE, respectively depicted in Figure 6a–c, show that the self-paced unconstrained walk's fluctuations have more complexity and greater long-range correlations than the metronomically-paced walk's series, in agreement with those reportred in [2]. We did not use $mvMDE_{II}$, as the signals do not follow the typical number of samples required for $mvMDE_{II}$. To compare the results, the CV values for both the metronomically- and self-paced walk (MPW and SPW) at scale factor 4, as a trade-off between the long and short scales, are shown in Table 4. The CV values for the $mvMDE_{III}$- and mvMDE-based profiles are smaller than those for mvMFE, showing the superiority of the proposed methods over mvMFE in terms of the stability of results. The smallest CV values are achieved by the mvMDE.

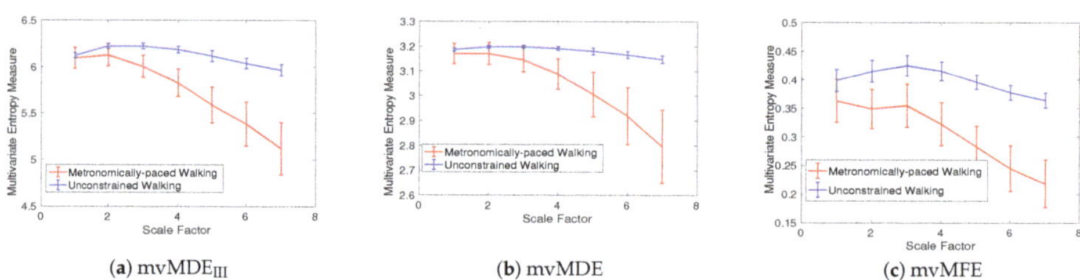

(a) $mvMDE_{III}$ (b) mvMDE (c) mvMFE

Figure 6. Mean value and SD of the results using (**a**) $mvMDE_{III}$, (**b**) mvMDE, and (**c**) mvMFE for self-paced vs. metronomically-paced stride interval fluctuations.

Table 4. CV values of the entropy results at scale factor 4 using $mvMDE_{III}$, mvMDE, and mvMFE for self-paced walk (SPW) vs. metronomically-paced walk (MPW).

Stride Interval Fluctuations	mvMFE	$mvMDE_{III}$	mvMDE
Self-paced walk	0.040	0.005	0.002
Metronomically-paced walk	0.116	0.025	0.019

(2) Dataset of Focal and Non-focal Brain Activity: For the focal and non-focal EEG recordings, the results obtained by $mvMDE_{II}$, $mvMDE_{III}$, mvMDE, and mvMFE, respectively depicted in Figure 7a–d, show that the focal time series are less complex than the non-focal ones, in agreement with previous studies [40,43]. The CV values for the focal- and non-focal-based results at scale 6 are shown in Table 5. All the mvMDE-based CV values are smaller than those using mvMFE, showing more stability of the results obtained by the proposed methods. Moreover, the CV values for mvMDE are smaller than those for $mvMDE_{III}$, and the latter ones are smaller than those for $mvMDE_{II}$, suggesting that the mvMDE leads to more stable profiles.

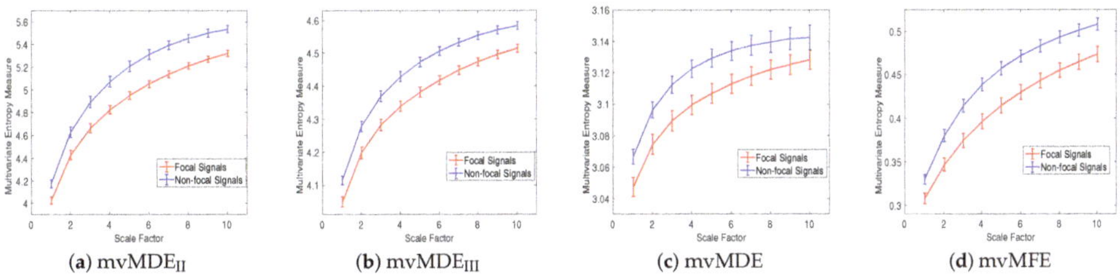

Figure 7. Mean value and SD of the results using (**a**) $mvMDE_{II}$, (**b**) $mvMDE_{III}$, (**c**) mvMDE, and (**d**) mvMFE for focal vs. non-focal time series.

Table 5. CV values of the entropy results at scale factor 6 using $mvMDE_{II}$, $mvMDE_{III}$, mvMDE, mvMSE, and mvMFE for focal vs. non-focal EEG recordings.

Signals	mvMSE	mvMFE	$mvMDE_{II}$	$mvMDE_{III}$	mvMDE
focal EEGs	0.019	0.019	0.006	0.003	0.002
Non-focal EEGs	0.021	0.015	0.008	0.003	0.002

(3) Surface MEG Recordings in Alzheimer's Disease: To assess the ability of mvMDE, in comparison with mvMFE, we applied the methods to the 148-channel MEG signals to discriminate AD patients from controls. Because mvMFE needs to store a huge number of elements for a signal with a large number of channels, mvMFE was not able to simultaneously analyse all 148 time series. However, the results using mvMDE are depicted in Figure 8. It represents an advantage of mvMDE over mvMFE for signals with a large number of channels. To compare the mvMFE and mvMDE, we applied the methods to five main scalp regions, namely, anterior (17 channels), right (34 channels) and left lateral (34 channels), central (29 channels), and posterior (34 channels) areas, leading to the smaller number of channels to noticeably decrease the number of elements stored by the use of the mvMFE algorithm.

The average and SD of mvMDE and mvMFE values for five regions are respectively shown in Figure 9a,b. As can be seen in Figures 8 and 9, the average mvMDE and mvMFE values for AD patients are smaller than those for controls at lower scale factors (short-time scale factors), while at higher scales, the AD subjects' recordings have larger entropy values (long-time scale factors) for both the mvMFE and mvMDE, in agreement with [21,44,45]. Because the larger the number of channels, the smaller the mvMSE and similarly mvMFE values [21], the entropy values for anterior region are larger than those for the other four regions. It is worth noting that we only use mvMDE, as the signals do not follow the typical number of samples required for $mvMDE_{II}$ and $mvMDE_{III}$.

The Mann–Whitney *U*-test was used to assess the differences between the mvMDE and mvMFE profiles at each temporal scale for AD patients versus controls, because the mvMDE and mvMFE values at each scale factor did not follow a normal distribution. The temporal scales with p-values smaller than 0.001 are shown with * in Figures 8 and 9. The p-values show that the mvMDE, compared with the mvMFE, significantly discriminated the controls from subjects with AD at a larger number of scale factors. Moreover, the smallest p-value was achieved by the mvMDE, evidencing the superiority of mvMDE over mvMFE.

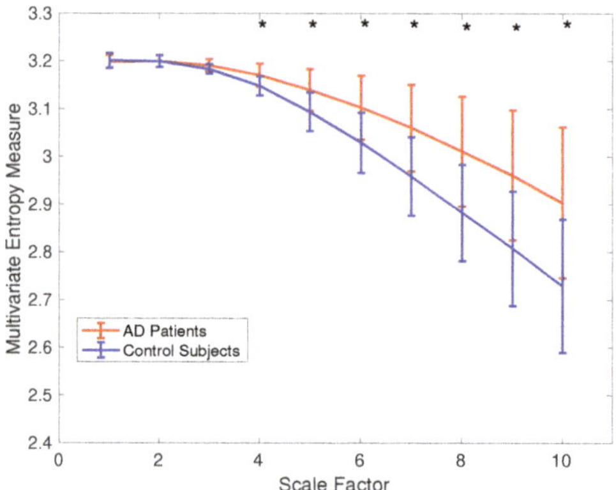

Figure 8. Mean value and SD of the results obtained by mvMDE computed from 36 AD patients versus 26 elderly controls for all the 148 channels. Red and blue respectively indicate AD patients and controls. The scales with p-values smaller than 0.001 are shown with *.

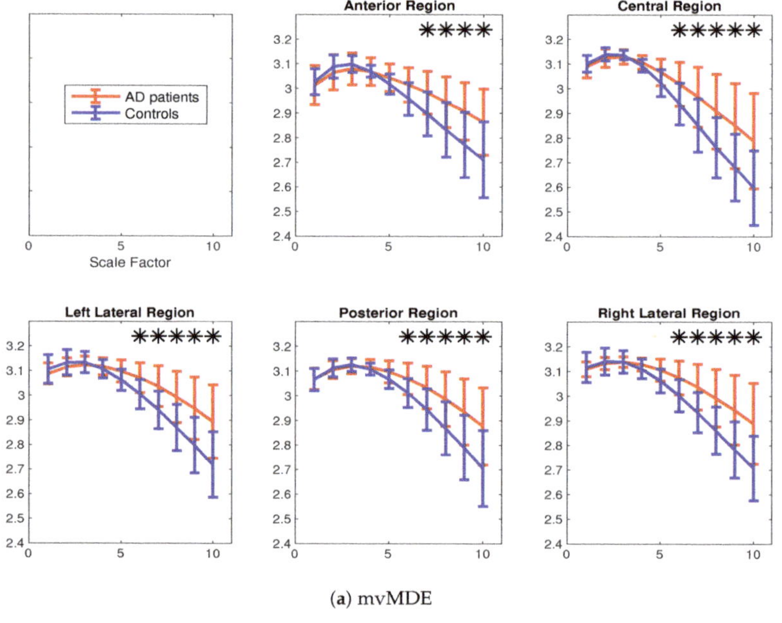

(**a**) mvMDE

Figure 9. *Cont.*

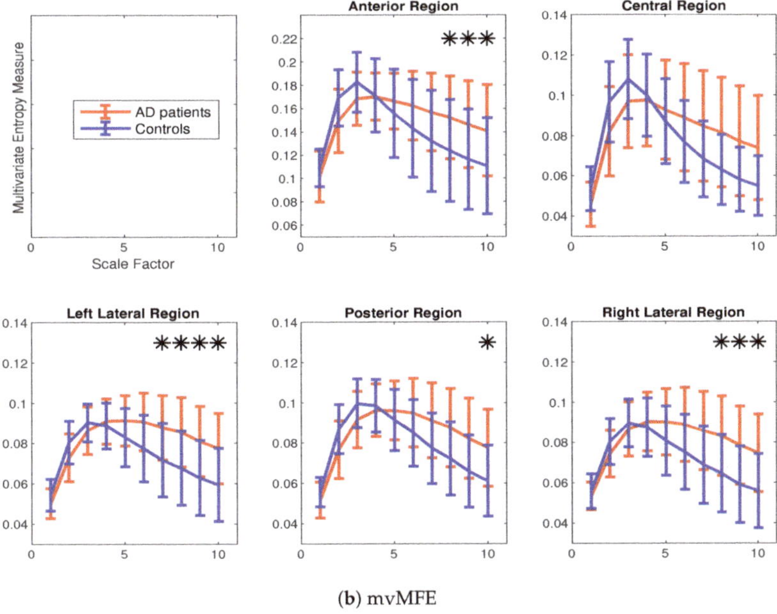

(b) mvMFE

Figure 9. Mean value and SD of the results obtained by (a) mvMDE and (b) mvMFE computed from 36 AD patients versus 26 elderly age-matched controls over five scalp regions. Red and blue indicate AD patients and controls, respectively. The scale factors with p-values smaller than 0.001 are shown with *.

The Hedges' g effect size [46] was also used to quantify the differences between the entropy values for the AD patients' vs. healthy controls' MEGs for the five main brain regions [47]. The Hedges' g test shows the difference between the means of two groups, divided by the weighted average of standard deviations for these two groups. The differences, illustrated in Table 6, show that the highest effect size is obtained by mvMDE, showing the advantage of this method over mvMFE.

Table 6. Differences between results for AD patients' vs. healthy controls' MEGs obtained by mvMFE and mvMDE for five main brain regions based on the Hedges' g effect size.

Region-Method	Scale 1	Scale 2	Scale 3	Scale 4	Scale 5	Scale 6	Scale 7	Scale 8	Scale 9	Scale 10
Anterior-mvMFE	0.36	0.73	0.57	0.04	0.33	0.53	0.63	0.70	0.72	0.73
Central-mvMFE	0.68	0.67	0.49	0.10	0.23	0.48	0.65	0.76	0.79	0.83
Left lateral-mvMFE	0.53	0.64	0.34	0.18	0.60	0.83	0.92	0.98	0.97	0.98
Posterior-mvMFE	0.46	0.72	0.58	0.16	0.30	0.57	0.73	0.78	0.82	0.85
Right lateral-mvMFE	0.30	0.50	0.22	0.18	0.53	0.71	0.84	0.92	0.97	0.95
Anterior-mvMDE	0.18	0.37	0.36	0.03	0.49	0.80	0.95	1.02	1.06	1.04
Central-mvMDE	0.29	0.45	0.29	0.48	0.78	0.88	0.97	1.01	1.03	1.04
Left lateral-mvMDE	0.37	0.40	0.24	0.24	0.77	1.07	1.17	1.20	1.19	1.19
Posterior-mvMDE	0.05	0.19	0.18	0.24	0.67	0.90	1.015	1.05	1.06	1.06
Right lateral-mvMDE	0.15	0.19	0.00	0.51	0.90	1.05	1.14	1.18	1.20	1.16

On the whole, the profiles for the real datasets show the advantage of $mvMDE_{II}$, $mvMDE_{III}$, and mvMDE over mvMFE to discriminate different types of dynamics of multi-channel signals as well as the superiority of mvMDE over mvMFE in terms of ability to discriminate various dynamics of time series, computational time, and memory cost. As mentioned before, mvMPE does not consider the spatial domain. We have also refined the mvMPE [19] on the basis of $mvMDE_{II}$, $mvMDE_{III}$, and mvMDE. These approaches have the following advantages over the first version of mvMPE [19]:

(1) they take into account both the spatial and time domains; (2) their results were more stable than the mvMPE-based ones; and (3) better distinguished different dynamics of multivariate signals. However, since the mvMDE methods are considerably faster, result in more stable profiles, and lead to larger differences between physiological conditions of recordings, for simplicity, we did not report the mvMPE-based results.

In this article, we proposed four implementations of the mvDE methods combined with the most commonly used coarse-graining process [3,8,17]. The key contribution of this study was introducing the mvDE methods. The alternative coarse-graining processes based on multivariate empirical mode decomposition [2,28,48–50], and FIR filters [28,51], though out of the scope of this paper, can be employed instead of the classical implementation of coarse-graining process used herein.

Our future study will aim at proposing the refined composite mvMDE (RCmvMDE) approaches according to [17]. Moreover, we will explore the mvMDE and RCmvMDE on other physiological and non-physiological time series. The similarity of two multi-channel signals based on mvMDE and cross-entropy [11] can also be developed as future work. An important step in making mvMDE a useful and stable metric is the mapping of the data to discrete set of integers via the normal cumulative distribution. Other mapping functions are available in [30]. The mvMDE method and its univariate form can also be generalized based on Renyi entropy [52].

5. Conclusions

To quantify the complexity of multivariate time series, we built four diverse alternative implementations of mvMDE as further developments of our recently introduced MDE [27]. These insights help towards a comprehensive understanding of four strategies to extend a univariate-based entropy method to its multivariate versions and therefore, provide invaluable information for future studies on multivariate time series. Although mvMDE was the best algorithm in terms of ability to discriminate dynamics of multivariate signals, computational time, and memory cost, the simpler alternatives ($mvDE_I$ to $mvDE_{III}$) may still be useful in some settings.

We assessed their performance on the correlated and uncorrelated multivariate noise signals, the bivariate AR time series, and three physiological datasets. The results showed the similar behavior of mvMSE-, mvMFE-, and mvMDE-based profiles. However, mvMDE had the following advantages over the existing methods: (1) it was faster than the existing methods; (2) mvMDE, in comparison with mvMSE and mvMFE, resulted in more stable profiles; (3) mvMDE better discriminated different kinds of biomedical signals; (4) for short multivariate time series (300 sample points), mvMDE did not result in undefined values; and (5) mvMDE, compared with mvMSE and mvMFE, needed to store a considerably smaller number of elements.

Overall, we expect the mvMDE approach to play a key role in the assessment of complexity in multivariate time series.

Author Contributions: H.A. and J.E. conceived and designed the methodology. H.A. was responsible for analysing and writing the paper. A.F. and J.E. contributed critically to revise the results and discussed them. All authors have read and approved the final manuscript.

Funding: This research received no external funding.

Acknowledgments: The MATLAB code of the mvMDE techniques will be made publicly-available upon publication. The MATLAB codes for mvMDE and its refined composite form are available at https://github.com/HamedAzami/mvMDE.

Conflicts of Interest: The authors declare no conflict of interest.

References

1. Cerutti, S.; Hoyer, D.; Voss, A. Multiscale, multiorgan and multivariate complexity analyses of cardiovascular regulation. *Philos. Trans. R. Soc. Lond. A Math. Phys. Eng. Sci.* **2009**, *367*, 1337–1358. [CrossRef] [PubMed]
2. Ahmed, M.; Rehman, N.; Looney, D.; Rutkowski, T.; Mandic, D. Dynamical complexity of human responses: A multivariate data-adaptive framework. *Bull. Pol. Acad. Sci. Tech. Sci.* **2012**, *60*, 433–445. [CrossRef]

3. Ahmed, M.U.; Mandic, D.P. Multivariate multiscale entropy analysis. *IEEE Signal Process. Lett.* **2012**, *19*, 91–94. [CrossRef]
4. Cerutti, S. Multivariate and multiscale analysis of biomedical signals: Towards a comprehensive approach to medical diagnosis. In Proceedings of the 2012 25th International Symposium on Computer-Based Medical Systems (CBMS), Rome, Italy, 20–22 June 2012; pp. 1–5.
5. Fernández-Sotos, A.; Martínez-Rodrigo, A.; Moncho-Bogani, J.; Latorre, J.M.; Fernández-Caballero, A. Neural Correlates of Phrase Quadrature Perception in Harmonic Rhythm: an Eeg Study (Using a Brain-Computer Interface). *Int. J. Neural Syst.* **2018**, *28*, 1750054. [CrossRef] [PubMed]
6. Spyrou, L.; Martín-Lopez, D.; Valentín, A.; Alarcón, G.; Sanei, S. Detection of intracranial signatures of interictal epileptiform discharges from concurrent scalp EEG. *Int. J. Neural Syst.* **2016**, *26*, 1650016. [CrossRef] [PubMed]
7. Pereda, E.; Quiroga, R.Q.; Bhattacharya, J. Nonlinear multivariate analysis of neurophysiological signals. *Prog. Neurobiol.* **2005**, *77*, 1–37. [CrossRef]
8. Ahmed, M.U.; Mandic, D.P. Multivariate multiscale entropy: A tool for complexity analysis of multichannel data. *Phys. Rev. E* **2011**, *84*, 061918. [CrossRef]
9. Mammone, N.; Bonanno, L.; Salvo, S.D.; Marino, S.; Bramanti, P.; Bramanti, A.; Morabito, F.C. Permutation disalignment index as an indirect, EEG-based, measure of brain connectivity in MCI and AD patients. *Int. J. Neural Syst.* **2017**, *27*, 1750020. [CrossRef]
10. Costa, M.; Goldberger, A.L.; Peng, C.K. Multiscale entropy analysis of biological signals. *Phys. Rev. E* **2005**, *71*, 021906. [CrossRef]
11. Richman, J.S.; Moorman, J.R. Physiological time-series analysis using approximate entropy and sample entropy. *Am. J. Physiol. Heart Circ. Physiol.* **2000**, *278*, H2039–H2049. [CrossRef]
12. Bandt, C.; Pompe, B. Permutation entropy: A natural complexity measure for time series. *Phys. Rev. Lett.* **2002**, *88*, 174102. [CrossRef] [PubMed]
13. Rostaghi, M.; Azami, H. Dispersion entropy: A measure for time series analysis. *IEEE Signal Process. Lett.* **2016**, *23*, 610–614. [CrossRef]
14. Fogedby, H.C. On the phase space approach to complexity. *J. Stat. Phys.* **1992**, *69*, 411–425. [CrossRef]
15. Silva, L.E.V.; Cabella, B.C.T.; da Costa Neves, U.P.; Junior, L.O.M. Multiscale entropy-based methods for heart rate variability complexity analysis. *Phys. A Stat. Mech. Its Appl.* **2015**, *422*, 143–152. [CrossRef]
16. Humeau-Heurtier, A. The multiscale entropy algorithm and its variants: A review. *Entropy* **2015**, *17*, 3110–3123. [CrossRef]
17. Azami, H.; Escudero, J. Refined composite multivariate generalized multiscale fuzzy entropy: A tool for complexity analysis of multichannel signals. *Phys. A Stat. Mech. Its Appl.* **2017**, *465*, 261–276. [CrossRef]
18. Li, P.; Ji, L.; Yan, C.; Li, K.; Liu, C.; Liu, C. Coupling between short-term heart rate and diastolic period is reduced in heart failure patients as indicated by multivariate entropy analysis. In Proceedings of the Computing in Cardiology Conference (CinC), Cambridge, MA, USA, 7–10 September 2014; pp. 97–100.
19. Morabito, F.C.; Labate, D.; La Foresta, F.; Bramanti, A.; Morabito, G.; Palamara, I. Multivariate multi-scale permutation entropy for complexity analysis of Alzheimer's disease EEG. *Entropy* **2012**, *14*, 1186–1202. [CrossRef]
20. Yin, Y.; Shang, P. Multivariate weighted multiscale permutation entropy for complex time series. *Nonlinear Dyn.* **2017**, *88*, 1707–1722. [CrossRef]
21. Azami, H.; Abásolo, D.; Simons, S.; Escudero, J. Univariate and Multivariate Generalized Multiscale Entropy to Characterise EEG Signals in Alzheimer's Disease. *Entropy* **2017**, *19*, 31. [CrossRef]
22. Labate, D.; La Foresta, F.; Morabito, G.; Palamara, I.; Morabito, F.C. Entropic measures of EEG complexity in alzheimer's disease through a multivariate multiscale approach. *Sens. J.* **2013**, *13*, 3284–3292. [CrossRef]
23. Gao, Z.K.; Ding, M.S.; Geng, H.; Jin, N.D. Multivariate multiscale entropy analysis of horizontal oil–Water two-phase flow. *Phys. A Stat. Mech. Its Appl.* **2015**, *417*, 7–17. [CrossRef]
24. Wei, Q.; Liu, D.H.; Wang, K.H.; Liu, Q.; Abbod, M.F.; Jiang, B.C.; Chen, K.P.; Wu, C.; Shieh, J.S. Multivariate multiscale entropy applied to center of pressure signals analysis: an effect of vibration stimulation of shoes. *Entropy* **2012**, *14*, 2157–2172. [CrossRef]
25. Zhao, L.; Wei, S.; Tang, H.; Liu, C. Multivariable Fuzzy Measure Entropy Analysis for Heart Rate Variability and Heart Sound Amplitude Variability. *Entropy* **2016**, *18*, 430. [CrossRef]

26. Ramdani, S.; Bonnet, V.; Tallon, G.; Lagarde, J.; Bernard, P.L.; Blain, H. Parameters Selection for Bivariate Multiscale Entropy Analysis of Postural Fluctuations in Fallers and Non-Fallers Older Adults. *IEEE Trans. Neural Syst. Rehabil. Eng.* **2016**, *24*, 859–871. [CrossRef] [PubMed]
27. Azami, H.; Rostaghi, M.; Abasolo, D.; Escudero, J. Refined Composite Multiscale Dispersion Entropy and its Application to Biomedical Signals. *IEEE Trans. Biomed. Eng.* **2017**, *64*, 2872–2879. [PubMed]
28. Azami, H.; Escudero, J. Coarse-Graining Approaches in Univariate Multiscale Sample and Dispersion Entropy. *Entropy* **2018**, *20*, 138. [CrossRef]
29. Azami, H.; Kinney-lang, E.; Ebied, A.; Fernández, A.; Escudero, J. Multiscale dispersion entropy for the regional analysis of resting-state magnetoencephalogram complexity in Alzheimer's disease. In Proceedings of the 39th Annual International Conference of the IEEE Engineering in Medicine and Biology Society (EMBC), Seogwipo, Korea, 11–15 July 2017; pp. 3182–3185.
30. Azami, H.; Escudero, J. Amplitude-and Fluctuation-Based Dispersion Entropy. *Entropy* **2018**, *20*, 210. [CrossRef]
31. Tufféry, S. *Data Mining and Statistics for Decision Making*; Wiley: Chichester, UK, 2011; Volume 2.
32. Baranwal, G.; Vidyarthi, D.P. Admission control in cloud computing using game theory. *J. Supercomput.* **2016**, *72*, 317–346. [CrossRef]
33. Gibbs, M.N.; MacKay, D.J. Variational Gaussian process classifiers. *IEEE Trans. Neural Netw.* **2000**, *11*, 1458–1464.
34. Duch, W. Uncertainty of data, fuzzy membership functions, and multilayer perceptrons. *IEEE Trans. Neural Netw.* **2005**, *16*, 10–23. [CrossRef]
35. Cao, L.; Mees, A.; Judd, K. Dynamics from multivariate time series. *Phys. D Nonlinear Phenom.* **1998**, *121*, 75–88. [CrossRef]
36. Kaffashi, F.; Foglyano, R.; Wilson, C.G.; Loparo, K.A. The effect of time delay on approximate & sample entropy calculations. *Phys. D Nonlinear Phenom.* **2008**, *237*, 3069–3074.
37. Humeau-Heurtier, A. Multivariate generalized multiscale entropy analysis. *Entropy* **2016**, *18*, 411. [CrossRef]
38. Penny, W.; Roberts, S. Bayesian multivariate autoregressive models with structured priors. *IEE Proc. Vis. Image Signal Process.* **2002**, *149*, 33–41. [CrossRef]
39. Hausdorff, J.M.; Purdon, P.L.; Peng, C.; Ladin, Z.; Wei, J.Y.; Goldberger, A.L. Fractal dynamics of human gait: Stability of long-range correlations in stride interval fluctuations. *J. Appl. Physiol.* **1996**, *80*, 1448–1457. [CrossRef] [PubMed]
40. Andrzejak, R.G.; Schindler, K.; Rummel, C. Nonrandomness, nonlinear dependence, and nonstationarity of electroencephalographic recordings from epilepsy patients. *Phys. Rev. E* **2012**, *86*, 046206. [CrossRef] [PubMed]
41. Escudero, J.; Sanei, S.; Jarchi, D.; Abásolo, D.; Hornero, R. Regional coherence evaluation in mild cognitive impairment and Alzheimer's disease based on adaptively extracted magnetoencephalogram rhythms. *Physiol. Meas.* **2011**, *32*, 1163. [CrossRef] [PubMed]
42. Lake, D.E.; Moorman, J.R. Accurate estimation of entropy in very short physiological time series: The problem of atrial fibrillation detection in implanted ventricular devices. *Am. J. Physiol. Heart Circ. Physiol.* **2010**, *300*, H319–H325. [CrossRef] [PubMed]
43. Sharma, R.; Pachori, R.B.; Acharya, U.R. Application of entropy measures on intrinsic mode functions for the automated identification of focal electroencephalogram signals. *Entropy* **2015**, *17*, 669–691. [CrossRef]
44. Yang, A.C.; Wang, S.J.; Lai, K.L.; Tsai, C.F.; Yang, C.H.; Hwang, J.P.; Lo, M.T.; Huang, N.E.; Peng, C.K.; Fuh, J.L. Cognitive and neuropsychiatric correlates of EEG dynamic complexity in patients with Alzheimer's disease. *Prog. Neuro Psychopharmacol. Biol. Psychiatry* **2013**, *47*, 52–61. [CrossRef] [PubMed]
45. Hornero, R.; Abásolo, D.; Escudero, J.; Gómez, C. Nonlinear analysis of electroencephalogram and magnetoencephalogram recordings in patients with Alzheimer's disease. *Philos. Trans. R. Soc. Lond. A Math. Phys. Eng. Sci.* **2009**, *367*, 317–336. [CrossRef]
46. Rosenthal, R.; Cooper, H.; Hedges, L. Parametric measures of effect size. *Handb. Res. Synth.* **1994**, *621*, 231–244.
47. Sullivan, G.M.; Feinn, R. Using effect size—Or why the P value is not enough. *J. Grad. Med. Educ.* **2012**, *4*, 279–282. [CrossRef]
48. Hu, M.; Liang, H. Adaptive multiscale entropy analysis of multivariate neural data. *IEEE Trans. Biomed. Eng.* **2012**, *59*, 12–15.

49. Hu, M.; Liang, H. Perceptual suppression revealed by adaptive multi-scale entropy analysis of local field potential in monkey visual cortex. *Int. J. Neural Syst.* **2013**, *23*, 1350005. [CrossRef]
50. Tonoyan, Y.; Looney, D.; Mandic, D.P.; Van Hulle, M.M. Discriminating multiple emotional states from EEG using a data-adaptive, multiscale information-theoretic approach. *Int. J. Neural Syst.* **2016**, *26*, 1650005. [CrossRef]
51. Valencia, J.F.; Porta, A.; Vallverdu, M.; Claria, F.; Baranowski, R.; Orlowska-Baranowska, E.; Caminal, P. Refined multiscale entropy: Application to 24-h holter recordings of heart period variability in healthy and aortic stenosis subjects. *IEEE Trans. Biomed. Eng.* **2009**, *56*, 2202–2213. [CrossRef]
52. Renner, R.; Wolf, S. Smooth Rényi entropy and applications. In Proceedings of the International Symposium onInformation Theory (ISIT 2004), Chicago, IL, USA, 27 June–2 July 2004; p. 233.

Article

Multivariate and Multiscale Complexity of Long-Range Correlated Cardiovascular and Respiratory Variability Series

Aurora Martins [1,2], Riccardo Pernice [3,*], Celestino Amado [1], Ana Paula Rocha [1,2], Maria Eduarda Silva [4,5], Michal Javorka [6,7] and Luca Faes [3]

1 Faculdade de Ciências, Universidade do Porto, Rua Campo Alegre, 4169-007 Porto, Portugal; aurora.ramalho@hotmail.com (A.M.); celestino.amado@gmail.com (C.A.); aprocha@fc.up.pt (A.P.R.)

2 Centro de Matemática da Universidade do Porto (CMUP), 4169-007 Porto, Portugal

3 Department of Engineering, University of Palermo, Viale delle Scienze, Bldg. 9, 90128 Palermo, Italy; luca.faes@unipa.it

4 Faculdade de Economia, Universidade do Porto, Rua Dr. Roberto Frias, 4169-007 Porto, Portugal; mesilva@fep.up.pt

5 Centro de Investigação e Desenvolvimento em Matemática e Aplicações (CIDMA)

6 Department of Physiology, Comenius University in Bratislava, Jessenius Faculty of Medicine, Mala Hora 4C, 03601 Martin, Slovakia; michal.javorka@uniba.sk

7 Biomedical Center Martin, Comenius University in Bratislava, Jessenius Faculty of Medicine, Mala Hora 4C, 03601 Martin, Slovakia

* Correspondence: riccardo.pernice@unipa.it

Received: 18 February 2020; Accepted: 09 March 2020 ; Published: 11 March 2020

Abstract: Assessing the dynamical complexity of biological time series represents an important topic with potential applications ranging from the characterization of physiological states and pathological conditions to the calculation of diagnostic parameters. In particular, cardiovascular time series exhibit a variability produced by different physiological control mechanisms coupled with each other, which take into account several variables and operate across multiple time scales that result in the coexistence of short term dynamics and long-range correlations. The most widely employed technique to evaluate the dynamical complexity of a time series at different time scales, the so-called multiscale entropy (MSE), has been proven to be unsuitable in the presence of short multivariate time series to be analyzed at long time scales. This work aims at overcoming these issues via the introduction of a new method for the assessment of the multiscale complexity of multivariate time series. The method first exploits vector autoregressive fractionally integrated (VARFI) models to yield a linear parametric representation of vector stochastic processes characterized by short- and long-range correlations. Then, it provides an analytical formulation, within the theory of state-space models, of how the VARFI parameters change when the processes are observed across multiple time scales, which is finally exploited to derive MSE measures relevant to the overall multivariate process or to one constituent scalar process. The proposed approach is applied on cardiovascular and respiratory time series to assess the complexity of the heart period, systolic arterial pressure and respiration variability measured in a group of healthy subjects during conditions of postural and mental stress. Our results document that the proposed methodology can detect physiologically meaningful multiscale patterns of complexity documented previously, but can also capture significant variations in complexity which cannot be observed using standard methods that do not take into account long-range correlations.

Keywords: multi-scale entropy (MSE); vector autoregressive fractionally integrated (VARFI) models; heart rate variability (HRV); systolic arterial pressure (SAP)

---

1. Introduction

Cardiovascular oscillations are influenced by the combined activity of different physiological regulation processes and, as a consequence, exhibit a rich dynamical structure [1]. Such different processes do not usually work at a single time-scale, but instead operate across different temporal scales, that for example reflect thermoregulatory or neural parasympathetic and sympathetic control. For this reason, different methods have been recently developed to evaluate the 'multiscale complexity' of cardiovascular oscillations. These methods allow both to characterize the physiological regulatory systems and to extract diagnostic parameters, and thus can have noteworthy clinical implications: in fact, a decrease of dynamical complexity can be related to an impaired capability of the subsystems composing the organism to interact among each other and it has been already proposed as a marker of pathological conditions [2,3]. Among the proposed methods, the one which is likely most popular is the so-called multiscale entropy (MSE), developed by Costa et al. [4]. This method calculates the conditional entropy (CE) of a single time series (usually the heart period, HP) as a function of the time scale of observation; the change of time scale is achieved through averaging consecutive segments of the time series via a procedure that has been lately recognized to correspond to a two-step process consisting of a low-pass filter followed by downsampling [5]. It is worth noting, however, that the initial formulation of multiscale entropy suffered from drawbacks related both to issues due to the filtering and downsampling steps, and to the unsuitability of CE analysis in conditions of data paucity caused by the availability of short time series and by the needs to explore multivariate time series at coarse time scales. Therefore, in the last years, the definition of MSE has been refined to take into account typical requirements of cardiovascular and cardiorespiratory signal analysis, and specifically: (i) to allow the joint calculation of complexity of multiple variables besides HP, for example systolic arterial pressure (SAP) and respiration (RESP) [6]; (ii) to allow the assessment of the complexity of shorter time series, usually few hundred beats long [7,8]. For short-term physiological time series, complexity has been related to the regularity of the temporal patterns observed in the signals, and thus is usually a measure of the unpredictability of the present sample given a limited number of past samples [9]. However, recent studies have recognized the importance of long-range correlations resulting in slowly varying dynamics also for the analysis of short-term complexity [10], and have started to account for these correlations in multiscale entopy-based analysis [8].

In this work, we introduce a novel method to compute multivariate and multiscale complexity of cardiovascular oscillations. This method fits a multivariate time series using a vector autoregressive fractionally integrated (VARFI) model and then provides the multiscale representation of the VARFI parameters using the theory of state space models, allowing to extract from such parameters multiscale and multivariate measures of complexity. This approach presents several advantages if compared to other previous works in the same field [4–8,10], and in particular: (i) the parametric formulation employed permits to work reliably on short time series; (ii) fractional integration allows to take into account not only short-term dynamics, but also the long-range correlations; (iii) the vector formulation permits to compute the overall complexity of multivariate time series, or the complexity of an individual time series when one or more other series are considered. In this work, such approach is applied to HP, SAP and RESP time series measured in healthy subjects monitored in a resting condition and during two types of physiological stress: postural stress provoked by head-up tilt and mental stress induced by mental arithmetics.

The algorithms for multivariate multiscale analysis of physiological time series presented in this work are collected in the MSE-VARFI MATLAB toolbox, which can be freely downloaded from http://www.lucafaes.net/LMSE-MSE_VARFI.html.

2. Methods

2.1. Measures of Complexity in Linear Multivariate Stochastic processes

Considering a dynamical system $\mathcal{X}$ whose activity is defined by the zero-mean stationary multivariate stochastic process $\mathbf{X} = [X_1 \cdots X_M]$, where each X_j, $j = 1, \ldots, M$, denotes a scalar stochastic process $X_j = [\cdots X_{j,1} \cdots X_{j,n} \cdots]$, let us define as $\mathbf{X}_n = [X_{1,n} \cdots X_{M,n}]$ the $M-$dimensional random variable describing the present state of the system, and as $\mathbf{X}_n^- = [\mathbf{X}_{n-1}\mathbf{X}_{n-2}...]$ the infinite-dimensional vector variable describing its past states. Then, a measure of complexity of the system, typically defined for univariate systems [10] and here extended to the multivariate case, is the entropy rate defined as

$$C_{\mathbf{X}} = H(\mathbf{X}_n|\mathbf{X}_n^-) = H(\mathbf{X}_{n+1}^-) - H(\mathbf{X}_n^-), \tag{1}$$

where $H(\mathbf{X}_n|\mathbf{X}_n^-)$ is the conditional entropy of the present given the past, with $H(\cdot)$ denoting the Shannon Entropy. If the observed process $\mathbf{X}$ is a jointly distributed Gaussian process, it can be described without loss of generality through a vector linear regression model fed by white and uncorrelated innovations $\mathbf{E}_n = [E_{1,n} \cdots E_{M,n}]$ such that, for each $j \in 1, \ldots, M$, $E_{j,n} = X_{j,n} - \mathbb{E}[X_{j,n}|\mathbf{X}_n^-]$ [11]. In such a case, the entropy of the present state of the process and the conditional entropy of the present given the past can be expressed analytically in terms of the covariance of the process, $\Sigma_{\mathbf{X}} = \mathbb{E}[\mathbf{X}_n^T\mathbf{X}_n]$, and the covariance of the innovations, $\Sigma_{\mathbf{E}} = \mathbb{E}[\mathbf{E}_n^T\mathbf{E}_n]$, as [11–13]:

$$H(\mathbf{X}_n) = \frac{1}{2}\ln((2\pi e)^M|\Sigma_{\mathbf{X}}|), \tag{2a}$$

$$H(\mathbf{X}_n|\mathbf{X}_n^-) = \frac{1}{2}\ln((2\pi e)^M|\Sigma_{\mathbf{E}}|), \tag{2b}$$

where $|\cdot|$ stands for matrix determinant. Then, it follows immediately that the complexity of a multivariate linear process with joint Gaussian distribution is given by Equation (2b). In this work we provide an alternative definition of complexity, which includes a normalization of the innovation covariance to the process covariance:

$$\overline{C}_{\mathbf{X}} = \frac{1}{2}\ln\left((2\pi e)^M\frac{|\Sigma_{\mathbf{E}}|}{|\Sigma_{\mathbf{X}}|}\right). \tag{3}$$

Such normalization, which is implicitly implemented in studies assessing the complexity of univariate time series where the series is normalized to unit variance before computing the complexity measure, is formalized here in Equation (3) for multivariate series, and will be fundamental in the definition of multiscale complexity where the process covariance intrinsically changes with the time scale.

The measure of global complexity defined above for a multivariate process can be particularized to the characterization of one of its constituent processes. Specifically, the complexity of a scalar process $X_j \in \mathbf{X}$, $j \in 1, \ldots M$, can be defined in an univariate context with respect to its own dynamics only, in a bivariate context with respect to its dynamics and to the dynamics of another scalar process $X_i \in \mathbf{X}, i \neq j$, or in a fully multivariate context with respect to the dynamics of the whole observed vector process $\mathbf{X}$ [14]. The derivations are based on the knowledge that, in Gaussian processes, a linear parametric representation captures all of the entropy differences that define the conditional entropies [11] and that these entropy differences are related to the partial variances of the present of the target given its past and the past of one or more other processes, intended as variances of the prediction errors resulting from linear regression [13,15]. Specifically, let us denote as $E_{j|j,n} = X_{j,n} - \mathbb{E}[X_{j,n}|X_{j,n}^-]$ and $E_{j|ij,n} = X_{j,n} - \mathbb{E}[X_{j,n}|X_{i,n}^-, X_{j,n}^-]$ the prediction error of a linear regression of $X_{j,n}$ performed respectively on $X_{j,n}^-$ and $(X_{j,n}^-, X_{i,n}^-)$, and consider that the prediction error of a linear regression of $X_{j,n}$ on $\mathbf{X}_n^-$ is the jth innovation process $E_{j,n}$ defined above. Then, denoting the variances of the three prediction errors as $\Sigma_{E_{j|j}} = \mathbb{E}[E^2_{j|j,n}]$, $\Sigma_{E_{j|ij}} = \mathbb{E}[E^2_{j|ij,n}]$ and $\Sigma_{E_j} = \mathbb{E}[E^2_{j,n}] = \Sigma_{\mathbf{E}}(j,j)$, the univariate, bivariate and multivariate normalized complexity measures relevant to the process X_j are defined as:

$$\overline{C}_{X_j|X_j} = \frac{1}{2}\ln 2\pi e \frac{\Sigma_{E_{j|j}}}{\Sigma_{X_j}}, \tag{4a}$$

$$\overline{C}_{X_j|X_i,X_j} = \frac{1}{2}\ln 2\pi e \frac{\Sigma_{E_{j|ij}}}{\Sigma_{X_j}}, \tag{4b}$$

$$\overline{C}_{X_j|\mathbf{X}} = \frac{1}{2}\ln 2\pi e \frac{\Sigma_{E_j}}{\Sigma_{X_j}}, \tag{4c}$$

where $\Sigma_{X_j} = \Sigma_{\mathbf{X}}(j,j)$ is the jth diagonal element of $\Sigma_{\mathbf{X}}$.

2.2. Linear Multivariate Stochastic Processes with Long Range Correlations

In this section we present the parametric approach to the description of linear multivariate Gaussian stochastic processes exhibiting both short-term dynamics and long-range correlations. The approach is based on representing the M-dimensional discrete-time, zero-mean and unit variance stochastic process $\mathbf{X}_n$ defined in Section 2.1 as a vector autoregressive fractionally integrated (VARFI) process fed by the uncorrelated Gaussian innovations $\mathbf{E}_n$. The VARFI process takes the form [16]:

$$\mathbf{A}(L)\mathrm{diag}(\nabla^{\mathbf{d}})\mathbf{X}_n = \mathbf{E}_n, \tag{5}$$

where L is the back-shift operator ($L^i\mathbf{X}_n = \mathbf{X}_{n-i}$), $\mathbf{A}(L) = \mathbf{I}_M - \sum_{i=1}^{p} \mathbf{A}_i L^i$ ($\mathbf{I}_M$ is the identity matrix of size M) is a vector autoregressive (VAR) polynomial of order p defined by the $M \times M$ coefficient matrices $\mathbf{A}_1, \ldots, \mathbf{A}_p$, and

$$\mathrm{diag}(\nabla^{\mathbf{d}}) = \begin{bmatrix} (1-L)^{d_1} & 0 & \ldots & 0 \\ 0 & (1-L)^{d_2} & \ldots & 0 \\ \vdots & \vdots & & \vdots \\ 0 & 0 & \ldots & (1-L)^{d_M} \end{bmatrix},$$

where $(1-L)^{d_i}, i = 1, \ldots, M$, is the fractional differencing operator defined by [17]:

$$(1-L)^{d_i} = \sum_{k=0}^{\infty} G_k^{(i)} L^k \quad , \quad G_k^{(i)} = \frac{\Gamma(k-d_i)}{\Gamma(-d_i)\Gamma(k+1)}, \tag{6}$$

with $\Gamma(\cdot)$ denoting the gamma (generalized factorial) function. The parameter $\mathbf{d} = (d_1, \ldots, d_M)$ in Equation (5) determines the long-term behavior of each individual process, while the coefficients of the polynomial $\mathbf{A}(L)$ allow description of the short-term dynamics. Note that the process defined in Equation (5) is a particular case of the broader class of VARFIMA($p, \mathbf{d}, l$) processes [16], which also contains the class of autoregressive processes VAR(p); here we restrict our analysis to the description of the VARFIMA($p, \mathbf{d}, 0$) process, which we denote as a VARFI($p, \mathbf{d}$) process.

The parameters of the VARFI model (5), namely the coefficients of $\mathbf{A}(L)$ and the variance of the innovations $\Sigma_{\mathbf{E}}$, are typically obtained from process realizations of finite length first estimating the differencing parameters d_i by means of the Whittle semi-parametric local estimator [17] separately for each individual process X_i, then defining the filtered data $X^{(f)}_{i,n} = (1-L)^{d_i} X_{i,n}$, and finally estimating the VAR parameters from the filtered data $\mathbf{X}^{(f)}_n$ using the ordinary least squares method to solve the VAR model $\mathbf{A}(L)\mathbf{X}^{(f)}_n = \mathbf{E}_n$, with model order p assessed through the Bayesian information criterion [18].

2.3. Multiscale Complexity of VARFI Processes

In this section we report how to compute across multiple temporal scales the complexity measures defined in Section 2.1, under the hypothesis that the analyzed multivariate process is properly described by the VARFI representation provided in Section 2.2. The procedure for multiscale complexity analysis, which extends the approach proposed in Reference [19] to multivariate processes incorporating long-range correlations, is presented here reporting the essential steps and described with more mathematical details in the Appendices. There are three appendices to the main text. Appendix A reports the procedure for computing the coefficients of a finite-order VAR process that approximates an assigned VARFI process. Appendix B recalls the derivations relevant to the identification of multiscale state space (SS) processes defined as filtered and downsampled versions of an assigned VAR process; the parameters of the SS process defined at an assigned time scale are exploited to compute the multivariate complexity measure at that time scale. Appendix C describes how to define restricted SS processes and to rearrange them in order to extract the partial variances needed for the computation of the univariate and bivariate multiscale complexity measures. The derivations described in Appendices B and C are taken from Refs. [7,8,19,20].

Before implementing the rescaling procedure, we approximate the VARFI process (5) with a finite order VAR process by truncating the fractional integration part at a finite lag q, that is, we perform the following approximation:

$$\mathrm{diag}(\nabla^{\mathbf{d}}) \approx \mathbf{G}(L) = \sum_{k=0}^{q} \mathbf{G}_k L^k, \tag{7a}$$

$$\mathbf{G}_k = \mathrm{diag}\left[G_k^{(1)}, \ldots, G_k^{(M)}\right]. \tag{7b}$$

This allows us to rewrite the VARFI($p, \mathbf{d}$) process as a VAR(m) process, where $m = p + q$:

$$\mathbf{B}(L)\mathbf{X}_n = \mathbf{E}_n, \tag{8}$$

where the new coefficients are $\mathbf{B}(L) = \mathbf{A}(L)\mathbf{G}(L)$, with $\mathbf{A}(L)$ as in Equation (5) and $\mathbf{G}(L)$ as in Equation (7a). The exact procedure to derive the coefficients of the VAR(m) process is explained in Appendix A.

In order to represent a scalar stochastic process at the temporal scale defined by the scale factor τ, a two step procedure is typically employed which consists first in filtering the process with a lowpass filter with cutoff frequency $f_\tau = 1/(2\tau)$, and then downsampling the filtered process using a decimation factor τ [5,19]. Extending this approach to the multivariate process $\mathbf{X}$, we first implement the following linear finite impulse response (FIR):

$$\mathbf{X}_n^{(r)} = \mathbf{D}(L)\mathbf{X}_n \quad , \tag{9}$$

where r denotes the filter order and $\mathbf{D}(L) = \sum_{k=0}^{r} \mathbf{I}_M D_k L^k$, where the coefficients of the polynomial D_k, $k = 1, \ldots, r$, are the same for all scalar processes $X_j \in \mathbf{X}$ and are chosen to set up a lowpass FIR configuration with cutoff frequency $1/(2\tau)$. The filtering step transforms the VAR(p+q) process (8) into a VARMA(p+q,r) process with moving average (MA) part determined by the filter coefficients:

$$\mathbf{B}(L)\mathbf{X}_n^{(r)} = \mathbf{D}(L)\mathbf{B}(L)\mathbf{X}_n = \mathbf{D}(L)\mathbf{E}_n \quad . \tag{10}$$

Then, we exploit the connection between VARMA processes and state space (SS) processes [21] to evidence that the VARMA process (10) can be expressed in SS form as:

$$\mathbf{Z}_{n+1}^{(r)} = \mathbf{B}^{(r)}\mathbf{Z}_n^{(r)} + \mathbf{K}^{(r)}\mathbf{E}_n^{(r)} \tag{11a}$$

$$\mathbf{X}_n^{(r)} = \mathbf{C}^{(r)}\mathbf{Z}_n^{(r)} + \mathbf{E}_n^{(r)}, \tag{11b}$$

where $\mathbf{Z}_n^{(r)} = [\mathbf{X}_{n-1}^{(r)} \cdots \mathbf{X}_{n-m}^{(r)}\mathbf{E}_{n-1} \cdots \mathbf{E}_{n-r}]^T$ is a $(m+r)$-dimensional state process, $\mathbf{E}_n^{(r)} = \mathbf{D}_0\mathbf{E}_n$ is the SS innovation process, and the vectors $\mathbf{K}^{(r)}$ and $\mathbf{C}^{(r)}$ and the matrix $\mathbf{B}^{(r)}$ can be obtained from $\mathbf{B}(L)$ and $\mathbf{D}(L)$ (see Appendix B).

The second step of the rescaling procedure is to downsample the filtered process in order to complete the multiscale representation. This is done exploiting theoretical findings [7,22,23] which allow to describe the filtered SS process after downsampling in the form:

$$\mathbf{Z}_{n+1}^{(\tau)} = \mathbf{B}^{(\tau)}\mathbf{Z}_n^{(\tau)} + \mathbf{K}^{(\tau)}\mathbf{E}_n^{(\tau)} \tag{12a}$$

$$\mathbf{X}_n^{(\tau)} = \mathbf{C}^{(\tau)}\mathbf{Z}_n^{(\tau)} + \mathbf{E}_n^{(\tau)}. \tag{12b}$$

Equation (12) provides the SS form of the filtered and downsampled version of the original VARFI(p, $\mathbf{d}$) process, and its parameters ($\mathbf{B}^{(\tau)}, \mathbf{C}^{(\tau)}, \mathbf{K}^{(\tau)}, \Sigma_{\mathbf{E}^{(\tau)}}$) can be obtained from the SS parameters before downsampling and from the downsampling factor τ; moreover, the variance of the downsampled process, $\Sigma_{\mathbf{X}^{(\tau)}}$, can be computed analytically from the parameters of the SS model (12) by solving a discrete-time Lyapunov equation (these steps are shown in the Appendix B). The parameters relevant to the computation of complexity at scale τ are the variance of the downsampled process, $\Sigma_{\mathbf{X}^{(\tau)}}$, and the variance of the corresponding innovations, $\Sigma_{\mathbf{E}^{(\tau)}}$. These variances can indeed be combined in a similar way to that of Equation (3) to yield the expression of the complexity of the original process $\mathbf{X}_n$ when it is observed at scale τ:

$$\overline{C}_{\mathbf{X}} = \frac{1}{2}\ln\left((2\pi\mathrm{e})^M \frac{|\Sigma_{\mathbf{E}^{(\tau)}}|}{|\Sigma_{\mathbf{X}^{(\tau)}}|}\right). \tag{13}$$

Note that this measure tends to the theoretical value $\frac{1}{2}\ln(2\pi e)^M$ when $\tau \to \infty$.

Finally, we show how to compute any partial variance appearing in Equation (4) from the parameters of an SS model in the form of (12), so that the three complexity measures relevant to the scalar process X_j can be computed at any assigned time scale τ. To do this, we exploit the formulations reported in Refs. [22,23], showing that the partial variance $\Sigma_{E_{j|a}^{(\tau)}}$, where the subscript a denotes any combination of indexes $\in 1,\ldots,M$, can be derived from the SS representation of the innovations of a submodel obtained removing the variables not indexed by a from the observation equation. Specifically, we need to consider the submodel with state Equation (12a) and observation equation:

$$\mathbf{X}_{a,n}^{(\tau)} = \mathbf{C}_a^{(\tau)}\mathbf{Z}_n^{(\tau)} + \mathbf{E}_{a,n}^{(\tau)}, \tag{14}$$

where the additional subscript $_a$ denotes the selection of the rows with indices a in a vector or a matrix. This restricted SS model can be rearranged to extract the partial variance $\Sigma_{E_{j|a}^{(\tau)}}$ from the covariance matrix of its innovations (see Appendix C), so that with this procedure the univariate, bivariate and multivariate normalized complexity measures can be computed inserting the partial variances $\Sigma_{E_{j|j}^{(\tau)}}$, $\Sigma_{E_{j|ij}^{(\tau)}}$ in Equation (4a) and Equation (4b), and using $\Sigma_{E_j}^{(\tau)} = \Sigma_{\mathbf{E}^{(\tau)}}(j,j)$ in Equation (4c), together with $\Sigma_{X_j}^{(\tau)} = \Sigma_{\mathbf{X}^{(\tau)}}(j,j)$ from Equation (A9b). These individual measures tend to the theoretical value $\frac{1}{2}\ln(2\pi e)$ when $\tau \to \infty$.

3. Application to Cardiovascular Variability Processes

In this section the proposed method is illustrated using cardiovascular and respiratory series: the heart period (HP), systolic arterial pressure (SAP), and respiration (RESP). Several studies report the interaction between the dynamics of these three time series [24–28], which motivates their use in a multivariate context. The variation of heart period, usually referred as heart rate variability (HRV), reflecting cardiovascular complexity and representing the capability of the organism to react to environmental and psychological stimuli, is the most studied variable and main target variable in cardiovascular and cardiorespiratory spontaneous variability [29–31]. For this reason, the conditional measures of a single scalar process will focus on the HP time series, as it has been shown that SAP and RESP have an effect (direct or indirectly) on this process [24–28].

3.1. Experimental Protocol

The HP, SAP and RESP time series were measured in a group of 62 healthy subjects (19.5 ± 3.3 years old, 37 females) monitored in the resting supine position (SU_1), in the upright position (UP) reached through passive head-up tilt, in the recovery supine position (SU_2) and during mental stress induced by mental arithmetics (MA) [32]. During the measurements, the subjects were free-breathing. For each subject and condition, the multivariate process $\mathbf{X}$ is defined as $\mathbf{X} = [X_{HP}, X_{SAP}, X_{RESP}]$. The acquired signals were the surface electrocardiogram (ECG), the finger arterial blood pressure recorded noninvasively by the photoplethysmographic method, and the respiration signal recorded through respiratory inductive plethysmography. For each subject and experimental condition, the values of HP, SAP and RESP were measured on a beat-to-beat basis respectively as the sequences of the temporal distances between consecutive R peaks of the ECG, the maximum values of the arterial pressure waveform taken within the consecutively detected heart periods, and the values of the respiratory signal sampled at the onset of the consecutively detected heart periods. The study was approved by Ethical Committee of the Jessenius Faculty of Medicine, Comenius University and all participants signed a written informed consent. A detailed description of the experimental protocol and signal measurement is reported in Ref. [32].

The analysis was performed on segments of at least 400 consecutive points, free of artifacts and deemed as weak-sense stationary through visual inspection, extracted from the time series for each subject and condition. Missing values and outliers were corrected through linear interpolation and, for HP and when possible, erroneous/missing intervals were substituted by pulse intervals measured as the difference in time between two consecutive SAP measurements ($\Delta_{t_{SAP}}(n) = t_{SAP}(n+1) - t_{SAP}(n)$). The three time series were normalized to zero mean and unit variance before multiscale analysis.

3.2. Data Analysis

Two different approaches were followed to compute multiscale complexity: (i) the "eVAR" approach, based on pure VAR model identification, that is, performing the whole procedure described in Section 2 after forcing $\mathbf{d} = [0, 0, 0]$ in Equation (5); (ii) the "eVARFI" approach, based on complete VARFI model identification, that is, following the whole procedure described in Section 2 with d_i, $i = 1, \ldots, 3$, estimated individually from the original time series and considered in the computations. Pursuing these approaches we compare, respectively, (i) the traditional complexity analysis where long-range correlations are neither removed nor modeled, and (ii) the complexity analysis performed by modeling the long-range correlations and considering them together with the short-term dynamics. Such a comparison is meant to infer the role of long-range correlations in contributing to the multiscale complexity and to its variation between conditions. The VARFI model fitting the time series $\mathbf{X}$ was identified first estimating the fractional differencing parameter d_i, $i = 1, \ldots, 3$, individually for each time series using the Whittle estimator, then filtering the time series with the fractional integration polynomial truncated at a lag $q = 50$, and finally estimating the parameters of the polynomial relevant to the short-term dynamics through least squares VAR identification. The value of q has to be selected to approximate the VARFI process, which is theoretically of infinite order, with a finite order VAR process. According to previous studies [8,33] $q = 50$ is an appropriate value for truncating the VARFI process. By increasing q, we can obtain a more precise approximation of the fractional integration part but with a higher computational cost, while a reduced value (and thus an excessive truncation) causes an underestimation of the complexity and the smoothing of the nonmonotonic trends with the timescale [8]. The VAR model order p was selected as the minimum of the Bayesian information criterion (BIC) figure of merit [34]. Then, multiscale complexity measures were computed implementing a FIR lowpass filter of order $r = 48$, for time scales τ in the range $(1,\ldots,30)$, which corresponds to lowpass cutoff normalized frequencies $f_\tau = (0.5,\ldots,0.01667)$. The order of the FIR filter determines its selectivity around the cutoff frequency. In this study, an order $r = 48$ was selected according to previous settings [8]. The changes in complexity related to the oscillatory rhythms are typically evaluated in cardiovascular variability in low-frequency (LF, from 0.04 Hz to 0.15 Hz) and high-frequency (HF, from 0.15 Hz to 0.4 Hz) spectral bands [29]. To account for the different mean heart rate for each subject and condition, multiscale analysis was performed determining the cutoff frequency of the rescaling filter based on the Nyquist frequency in Hz of the HP time series, which for the time scale τ is given by $f_{\tau_{\mathrm{Hz}}} = 1/(2\tau\overline{\mathrm{HP}})$, where $\overline{\mathrm{HP}}$ stands for the mean of the HP process. In this work, considering the physiological LF and HF frequency ranges, four cutoff frequencies f_{Hz} were chosen to perform the analysis: 0.4 Hz, 0.15 Hz, 0.1 Hz and 0.04 Hz. To obtain the complexity values at such frequencies, linear interpolation was performed on the profiles $f_{\tau_{\mathrm{Hz}}}$.

The differencing parameters d_i were estimated individually for each time series in the interval $[-0.5, 1]$ since the VARFI model is stationary for $-0.5 < d_i < 0.5$, while it is nonstationary but mean reverting for $0.5 \leq d_i < 1$. As such, the subjects with an estimation of $d_i \approx 1$ in at least one condition were removed given that the series is possibly non mean reverting; three subjects were removed, so that a total of 59 subjects was considered for the statistical analysis.

3.3. Statistical Analysis

Significant changes in complexity across the pairs of experimental conditions SU_1 vs. UP and SU_2 vs. MA were assessed via a linear mixed-effects model, that is, a linear regression model that incorporates both fixed and random effects [35]. In our case, the fixed-effects (or factors) were condition and scale, while the random-effect was the subject-dependent intercept that allows for the random variation between subjects. Additionally, the interaction between the factors is also considered. The significance of both the effects (fixed and random) and the interaction between fixed effects was assessed by the significance of the corresponding estimated coefficients at the level of $p < 0.05$. Residuals were checked for whiteness.

To evaluate the changes of interest, estimated marginal means (EMM) [36] were computed for each difference, or contrast, UP-SU_1 and MA-SU_2, at each frequency of interest (0.4 Hz, 0.15 Hz, 0.1 Hz and 0.04 Hz). A Z-test was applied to check the significance of these differences with a significance level of $p < 0.05$ and Tukey correction for multiple comparisons. The packages `lme4` [37] and `emmeans` [38] of the `R` software [39] were used to build the model and to compute EMM, respectively.

4. Results

This section presents the results of multiscale analysis performed for the multivariate complexity $\overline{C}_{\mathbf{X}}$ as defined in Equation (3), as well as for the complexity computed for the scalar process X_{HP} in four different ways: the univariate complexity of HP with respect to its own dynamics, $\overline{C}_{X_{\mathrm{HP}}|X_{\mathrm{HP}}}$ (Equation (4a)), the bivariate complexity of HP with respect to its own dynamics and to the dynamics of SAP, $\overline{C}_{X_{\mathrm{HP}}|X_{\mathrm{SAP}},X_{\mathrm{HP}}}$, or RESP, $\overline{C}_{X_{\mathrm{HP}}|X_{\mathrm{RESP}},X_{\mathrm{HP}}}$ (Equation (4b)), and the multivariate complexity of HP with respect to the dynamics of the whole trivariate process, $\overline{C}_{X_{\mathrm{HP}}|\mathbf{X}}$ (Equation (4c)).

Figure 1 presents the median and quartiles across subjects of the multivariate complexity $\overline{C}_{\mathbf{X}}$ computed for eVAR (first row) and eVARFI (second row) as a function of the time scale $\tau = 1,\ldots,30$, for SU_1 vs. UP (left column) and SU_2 vs. MA (right column). The measure always tends to its theoretical value $\frac{1}{2}\ln(2\pi e)^3$ when computed for eVAR at high values of τ. In fact, the complexity value is in general lower for eVARFI than for eVAR, and presents more variability across subjects. From a visual inspection of the multiscale patterns one can infer that for eVAR there is a decrease in complexity moving from SU_1 to UP, evident at short time scales ($\tau < 10$), and that the complexity seems not to change while moving from SU_2 to MA. For eVARFI, the decrease from SU_1 to UP is visible across the whole range of time scales and there seems to be a decrease from SU_2 to MA for scales larger than 4.

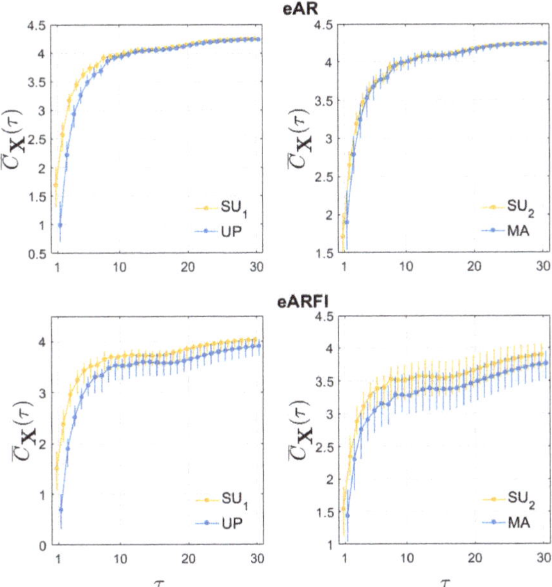

Figure 1. Distribution across subjects of the multivariate complexity measure $\overline{C}_{\mathbf{X}}$ as a function of the time scale τ for eVAR (first row) and eVARFI (second row), for SU_1 vs. UP (left column) and SU_2 vs. MA (right column).

Figure 2 presents the same as the previous figure but for the univariate (Figure 2a), bivariate (Figure 2b with SAP and Figure 2c with RESP) and multivariate (Figure 2d) complexity measures relevant to the HP time series. Again, eVARFI estimation presents more variability than eVAR, visible across all measures, and does not reach the theoretical value $\frac{1}{2}\ln(2\pi e)$ at long time scales. Although the four measures are similar with each other, slight changes occur when the complexity of HP is computed accounting for the dynamics of the other time series. When compared to the univariate measure $\overline{C}_{X_{HP}|X_{HP}}$, the median complexity value decreases for the bivariate measures at lower time scales, being lower with RESP ($\overline{C}_{X_{HP}|X_{RESP},X_{HP}}$) than with SAP ($\overline{C}_{X_{HP}|X_{SAP},X_{HP}}$); the multivariate measure $\overline{C}_{X_{HP}|\mathbf{X}}$ presents even lower median complexity values for lower time scales. For both eVAR and eVARFI, the complexity of HP decreases at short time scales for all four measures. The differences are more subtle during MA, where only a slight decrease in the median is noticeable at very short time scale for eVARFI.

Figure 3 reports the distribution of the multivariate complexity measure $\overline{C}_{\mathbf{X}}$ (blue dots) (values and boxplot) computed at the four predetermined cutoff frequencies of the multiscale filter (0.4 Hz, 0.15 Hz, 0.1 Hz and 0.04 Hz) for each experimental condition. Statistically significant changes ($p < 0.05$) in complexity across the pairs SU_1 vs UP or SU_2 vs MA, as assessed by the estimated marginal means based on the linear mixed-effects model, are marked with $*$. Comparing SU and UP, the multivariate complexity decreases significantly both for eVAR and eVARFI at 0.4 Hz, while no significant differences are observed at lower cutoff frequencies. Moving from SU_2 to MA, the measure increases significantly for eVAR at all frequencies except 0.04 Hz, but decreases for eVARFI at frequencies 0.1 Hz and 0.04 Hz.

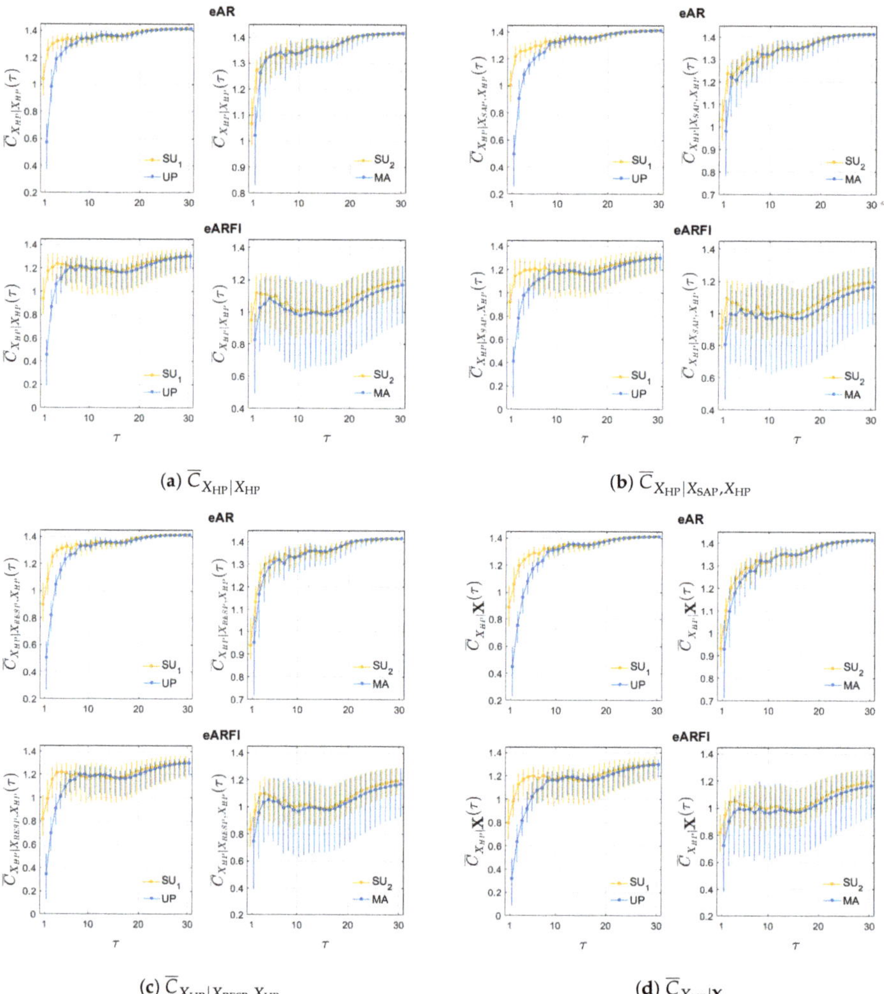

(a) $\overline{C}_{X_{HP}|X_{HP}}$ (b) $\overline{C}_{X_{HP}|X_{SAP},X_{HP}}$

(c) $\overline{C}_{X_{HP}|X_{RESP},X_{HP}}$ (d) $\overline{C}_{X_{HP}|\mathbf{X}}$

Figure 2. Distribution across subjects of the univariate (**a**), bivariate (**b**) with SAP and (**c**) with RESP, and multivariate (**d**) complexity measures as a function of the time scale τ for eVAR (first row) and eVARFI (second row), for SU_1 vs. UP (left column) and SU_2 vs. MA (right column).

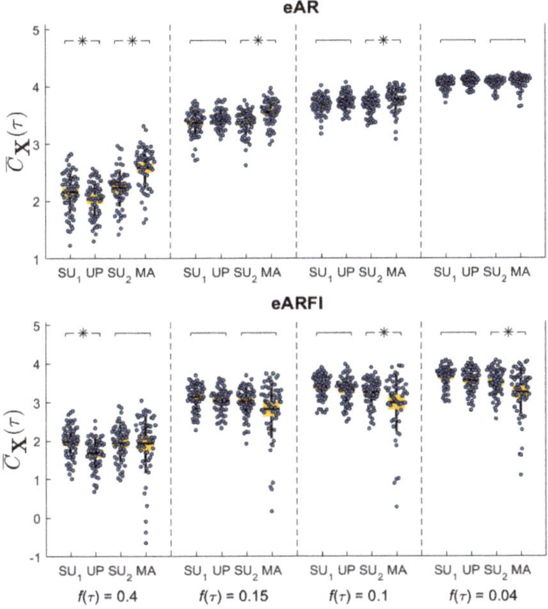

Figure 3. Distribution of the multivariate complexity measure $\overline{C}_{\mathbf{X}}$, depicted as boxplot (mean and confidence intervals, yellow filled box; standard deviation, black vertical line) and original values (dots) for selected frequencies ($f_{\tau_{\mathrm{Hz}}}$ = 0.4 Hz; 0.15 Hz; 0.1 Hz; 0.04 Hz), computed for the four experimental conditions using eVAR (first row) and eVARFI (second row) identification methods. Statistically significant differences between pairs of conditions are marked with an asterisk.

Figure 4 presents the same information as the previous figure but considering the HP only as target process and computing the corresponding univariate (Figure 4a), bivariate (Figure 4b with SAP and Figure 4c with RESP) and multivariate (Figure 4d) complexity measures. We find that, moving from SU to UP, the univariate measure $\overline{C}_{X_{\mathrm{HP}}|X_{\mathrm{HP}}}$ evaluated at a time scale corresponding to the cutoff frequency of 0.4 Hz decreases significantly for both eVAR and eVARFI; this decreased complexity of HP in the UP condition is observed identically when the dynamics of SAP, of RESP, and of both SAP and RESP are included in the conditional entropy measure. Moreover, the inclusion of one or both the other time series makes such a decrease statistically significant also with a cutoff of 0.15 Hz when eVAR analysis is performed. Moving from SU_2 to MA, we observe that only the eVARFI analysis detects statistically significant differences. Specifically, using eVARFI the univariate complexity of HP decreases during MA at time scales compatible with the cutoff frequency of 0.4 Hz, and also with cutoff 0.15 Hz when SAP dynamics (but not RESP dynamics) are considered. For both protocols and using both estimation methods, the analysis at longer time scales (cutoff 0.1 Hz and 0.04 Hz) does not lead to any significant variation in any of the HP complexity measures.

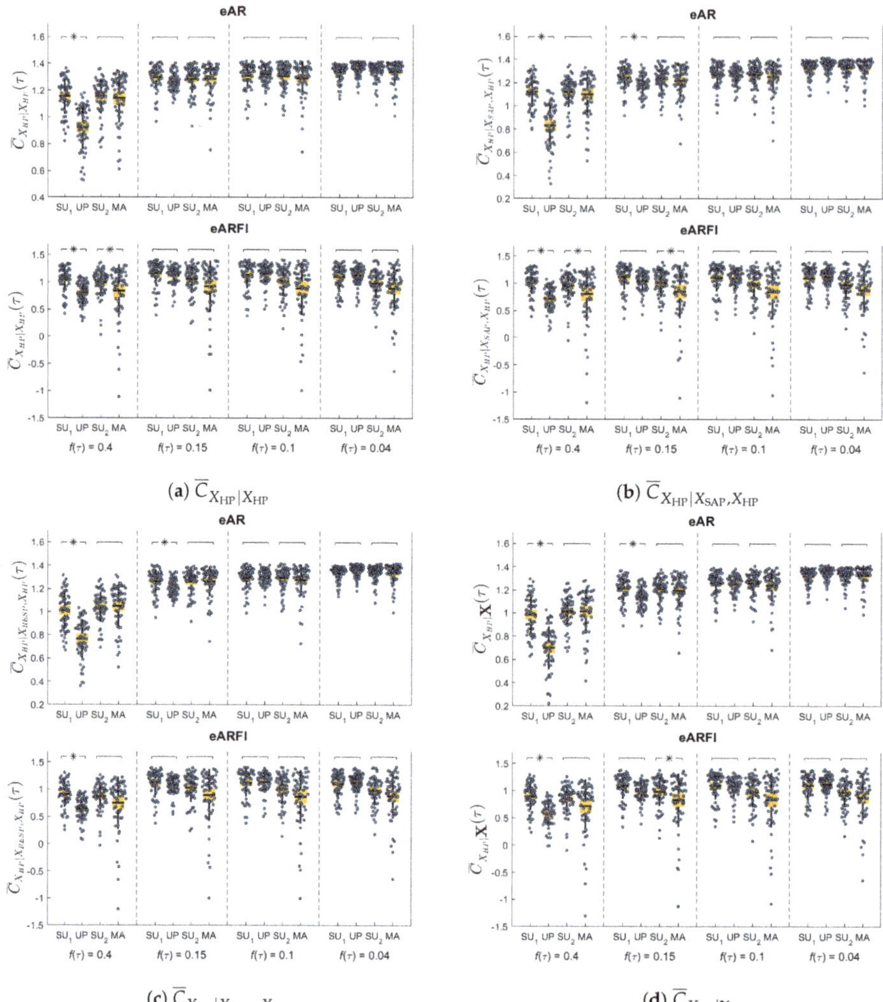

(**a**) $\overline{C}_{X_{\text{HP}}|X_{\text{HP}}}$ (**b**) $\overline{C}_{X_{\text{HP}}|X_{\text{SAP}},X_{\text{HP}}}$

(**c**) $\overline{C}_{X_{\text{HP}}|X_{\text{RESP}},X_{\text{HP}}}$ (**d**) $\overline{C}_{X_{\text{HP}}|\mathbf{X}}$

Figure 4. Distribution of the univariate (**a**), bivariate (**b**) with systolic arterial pressure (SAP) and (**c**) with respiration (RESP), and multivariate (**d**) complexity measures depicted as boxplot (mean and confidence intervals, yellow filled box; standard deviation, black vertical line) and original values (dots) for selected frequencies ($f_{\tau_{\text{Hz}}} = 0.4$ Hz; 0.15 Hz; 0.1 Hz; 0.04 Hz), computed for the four experimental conditions using eVAR (first row) and eVARFI (second row) identification methods. Statistically significant differences between pairs of conditions are marked with an asterisk.

Figure 5 depicts the distribution across subjects of the differencing parameter $\mathbf{d} = [d_{\text{HP}}, d_{\text{SAP}}, d_{\text{RESP}}]$ computed using the Whittle semiparametric estimator for the three time series and for each condition. The differencing parameter is positive for HP and SAP, while it is lower and centered around zero for RESP. The statistical significance of the variations of this parameter moving from SU_1 to UP, or from SU_2 to MA, was tested with the same linear mixed-effects model used for the complexity measures. Results indicate that there are statistically significant differences in the values of d only for SAP, documenting that conditions of stress evoked by UP or MA determine an increase of the differencing parameter estimated for this time series.

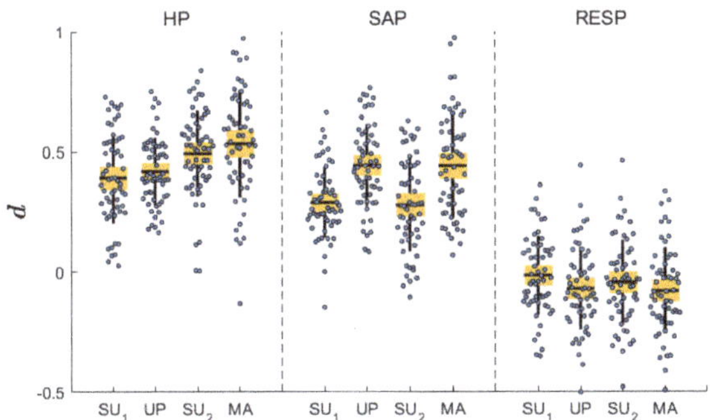

Figure 5. Distribution of the long-rang parameter d_i for each of the time series considered (heart period (HP), SAP and RESP) and for the four conditions.

5. Discussion

Building on recent developments regarding linear multiscale time series analysis [7,8,19,20], the present study introduces a novel approach to assess the dynamical complexity of multivariate processes accounting for the presence of short term dynamics and long-range correlations. The adopted linear parametric framework retains the advantage of previous formulations [7,19,20] related to the computational reliability of complexity estimated even at coarse time scales and over relatively short time series (few hundred points), and is here integrated with the incorporation of long-range dynamics (fundamental to a proper evaluation of complexity at coarse time scales [8]) and extended for the first time towards a fully multivariate implementation.

The usefulness of the proposed multivariate and multiscale measure of complexity was demonstrated in practice by the analysis of cardiovascular and cardiorespiratory interactions. In particular, the multivariate measure of complexity allowed us to evaluate the overall impact that the physiological mechanisms related to postural and mental stress have on the joint dynamics of HP, SAP and RESP. Simultaneously, the multiscale assessment of complexity led to elicit the contribution of mechanisms operating across multiple time scales (evaluation at fine time scales, with cutoff frequency of 0.4 Hz encompassing VLF, LF and HF oscillations) and that of mechanisms confined to slower oscillations (evaluation at coarser time scales, with cutoff at 0.15 Hz excluding HF oscillations and cutoffs at 0.1 Hz and 0.04 Hz excluding progressively also LF oscillations). In addition the evaluation of the complexity of HP, either considering its own dynamics only or the dynamics of SAP and/or RESP, was exploited to relate the overall complexity trends reflected in the multivariate measure to variations specifically related to physiologically well-known mechanisms of cardiovascular and cardiorespiratory interactions. To aid interpretation of the results and the discussion of the related physiological mechanisms, we report in Table 1 the statistically significant increase or decrease in complexity observed without (eVAR method) or with (eVARFI method) modeling long-range correlations during head-up tilt (comparison SU_1 vs. UP) or during mental arithmetics (comparison SU_2 vs. MA) at the time scales encompassing the whole spectral content (0.4 Hz), filtering the HF oscillations (0.15 Hz), and filtering in part (0.1 Hz) or completely (0.04 Hz) the LF oscillations. Complexity is linearly dependent on tilt inclination: a steeper inclination of tilt table produces higher degrees of sympathetic activation, with corresponding lower values of entropy-based complexity measures [40]. Studying the trends of the multivariate complexity measure, its decrease with tilt (Figure 3) documents a simplification of the overall dynamics of HP, SAP and RESP. This result can be mainly driven by the less complex dynamics of heart rate variability in the upright position, that we document calculating the complexity of HP (Figure 4a). Indeed it is well known that the dynamics of HP tend to become less complex as a consequence of the rise of LF oscillations and the weakening of HF oscillations related to the shift of the sympathovagal balance towards sympathetic activation and vagal deactivation during head-up tilt [15,40]. The fact that the decrease of multivariate complexity is not statistically significant for cutoff 0.15 Hz and lower suggests that it is related mainly to the weakening of HF oscillations; indeed, HF oscillations of both HP and SAP are known to be blunted with tilt [15,41,42]. In addition, the finding that the decrease in complexity is observed in the same way for VAR and VARFI analysis suggests that the impact of long-range correlations does not change substantially from rest to tilt. Overall, these results suggest that cardiovascular and respiratory oscillations in the LF and VLF band considered together do not alter significantly their complexity in the transition from the supine to the upright body position.

Table 1. Significant differences (*p-value* < 0.05) between pair of conditions for each measure and frequency. The arrows indicate if the measure increases or decreases from rest to stress.

Measure	Approach	$SU_1 \rightarrow UP$				$SU_2 \rightarrow MA$			
		0.4	0.15	0.1	0.04	0.4	0.15	0.1	0.04
$\overline{C}_{\mathbf{X}}$	eVAR	↘				↗	↗	↗	
	eVARFI	↘						↘	↘
$\overline{C}_{X_{HP}\|X_{HP}}$	eVAR	↘							
	eVARFI	↘				↘			
$\overline{C}_{X_{HP}\|X_{SAP},X_{HP}}$	eVAR	↘	↘						
	eVARFI	↘				↘	↘		
$\overline{C}_{X_{HP}\|X_{RESP},X_{HP}}$	eVAR	↘	↘						
	eVARFI	↘							
$\overline{C}_{X_{HP}\|\mathbf{X}}$	eVAR	↘	↘						
	eVARFI	↘					↘		

On the other hand, the increase of the global complexity with MA observed for eVAR modeling across multiple time scales (cutoffs 0.4, 0.15, 0.1) is in agreement with previous findings relevant to the HP time series [7]. The fact that such increase is not observed for eVARFI indicates that long range correlations have a different impact on the cardiovascular and respiratory dynamics during MA compared with the relaxed condition SU_2. In particular, eVARFI estimation reports values of the multivariate complexity which are unchanged at low time scales (cutoff 0.4 Hz), tend to decrease at cutoff 0.15 Hz, and decrease significantly at cutoffs 0.1 Hz and 0.04 Hz. We ascribe this lower complexity to a larger contribution of long-range correlations acting in LF and especially VLF bands, which is supported by the fact that the differencing parameter increases significantly, for HP and especially for SAP, during MA. This confirms the regularizing role of long range correlations on physiological dynamics [8,10].

The complexity of HP assessed at lower time scales (cutoff 0.4 Hz) always decreases in the UP position, for all measures (univariate, both bivariate, and multivariate), as shown in Figure 4. This documents the well-known simplification of heart rate variability induced by head-up tilt, which is known to evoke sympathetic activation and vagal withdrawal making the cardiac dynamics more regular [31,40,43,44]. The fact that this result is observed identically for the eVAR and eVARFI approaches indicates that long-range correlations do not impact significantly the evaluation of complexity performed at short time scales. On the other hand, focusing on intermediate time scales for which HF oscillations are cut off (f_τ = 0.15 Hz), we find that the decrease in complexity is lost when considering the HP dynamics individually, but is maintained when SAP and RESP, taken individually or together, are used to reduce the complexity of HP; moreover, this holds for eVAR but not for eVARFI. Taken together, these results suggest that slow oscillations of SAP and/or RESP reduce the complexity of HP in the transition from rest to tilt, and such regularizing action is related to long-range correlations. This finding is compatible with the known increase of cardiovascular interactions during tilt, documented in previous studies using a number of causality measures including information-theoretic ones [13,24,26,27], and also with the existence of slowly varying respiratory patterns that enhance their effect on the variability of HP during postural stress [24]. Here, the presence of cardiorespiratory and cardiovascular interactions is documented indirectly observing the lower values during tilt of the bivariate and multivariate complexity measures compared with the univariate one, through an approach similar to that proposed in Ref. [14].

The complexity of HP tends to decrease also with the mental stress induced by MA. Contrary to the postural stress induced by HUT, the decrease is observed only using the VARFI approach, and not using the VAR. The absence of significant changes in complexity using methods that do not model long-range correlations is in agreement with previous findings [7,14]. Our results document that the decrease in complexity is due to the different impact of long-range correlations, detected only modeling them through the VARFI approach, and again are confirmed by the higher values of the differencing parameter estimated for HP and for SAP (but indeed not for RESP) during MA compared with the relaxed condition (Figure 5). A differencing parameter $d = 0$ indicates that there are no long-range correlations. The rather small values of d observed for the RESP time series likely reflect the fact that, even if not controlled, the respiratory activity is confined in a narrow band of the frequency spectrum and thus it does not exhibit the slow trends typical of long-range correlated dynamics. Moreover, the changes of the differencing parameter d observed in the upright position for the systolic pressure (and only to a smaller extend for the heart period) document that the sympathetic activation produced by the tilt table inclination may modulate the impact that long range correlations have on the cardiovascular variability. We note also that, when SAP is considered, the significant changes extend to the second cutoff frequency (0.15 Hz), indicating the increasing contribution of SAP long-memory dynamics in reducing the complexity of HP. Therefore, we state the importance of accounting for long-range correlations in the assessment of the changes in the complexity of heart rate variability induced by mental stress. This may have relevance for the practical applications focused on the detection of mental workload or stress [45–48]).

Future extensions of the present work can be targeted first at investigating methodological aspects, such as the possibility of describing interactions between processes at the level of long-range correlations (possibly with the modeling of cointegration [49]) or the extension of the framework to the direct evaluation of Granger-causal interactions [20] in a multivariate context where long-range correlations are modeled [8]. Regarding applicative contexts, the analysis of multivariate multiscale dynamics is of particular interest in econometrics [50] and in the neurosciences [51,52], where dynamics spanning several temporal scales are commonly observed and multichannel data acquisition is ubiquitous.

Author Contributions: Conceptualization, A.P.R. and L.F.; methodology, L.F., R.P., A.P.R., A.M. and M.E.S.; software, L.F., R.P., A.P.R., A.M. and C.A.; validation, R.P., A.P.R., A.M. and M.E.S.; formal analysis, A.P.R., A.M. and M.E.S.; investigation, M.J.; resources, L.F. and M.J.; data curation, A.M. and M.J.; writing–original draft preparation, A.M.; writing—review and editing, L.F, R.P., A.P.R., C.A., A.M. and M.E.S.; visualization, A.P.R., A.M., M.E.S. and R.P.; supervision, L.F.; project administration, L.F. All authors have read and agreed to the published version of the manuscript.

Funding: This work was partially supported by UIDB/00144/2020 (CMUP) and UIDB/04106/2020 (CIDMA), which are funded by FCT with national (MEC) and European structural funds through the program FEDER, under PT2020, and project STRIDE-NORTE-01-0145-FEDER-000033 funded by ERDF-NORTE 2020), PRIN 2017 PRJ-0167, "Stochastic forecasting in complex systems", and PON R&I 2014-2020 AIM project no. AIM1851228-2, University of Palermo, Italy.

Conflicts of Interest: The authors declare no conflict of interest.

Appendix A

Since the computation of complexity works for finite-order VAR processes, while the VARFI process is equivalent to a VAR of infinite order, it is necessary to devise an approximation procedure. To do this, we truncate the fractional integration part of the VARFI process (5) at a finite lag q:

$$\operatorname{diag}(\nabla^{\mathbf{d}}) \approx \mathbf{G}(L) = \operatorname{diag}\left[\sum_{k=0}^{q} G_k^{(1)} L^k \quad \dots \quad \sum_{k=0}^{q} G_k^{(M)} L^k\right] = \sum_{k=0}^{q} \operatorname{diag}\left[G_k^{(1)}, \dots, G_k^{(M)}\right] L^k = \sum_{k=0}^{q} \mathbf{G}_k L^k, \tag{A1}$$

so that the VARFI process of order p and with differencing parameter $\mathbf{d}$ can be rewritten as a VAR process of order $m = p + q$ with parameters given by:

$$\mathbf{B}(L) = \mathbf{A}(L)\mathbf{G}(L) = \left(\mathbf{I}_M - \sum_{i=1}^{p} \mathbf{A}_i L^i\right)\left(\sum_{k=0}^{q} \mathbf{G}_k L^k\right). \tag{A2}$$

The coefficients of the VAR polynomial $\mathbf{B}(L) = \mathbf{I}_M - \sum_{k=0}^{p+q} \mathbf{B}_k L^k$ result from the multiplication of the two polynomials in (A2), which in the case $q \geq p$ yields:

$$\mathbf{B}_0 = \mathbf{I}_M \quad , \qquad \mathbf{B}_k = \begin{cases} -\mathbf{G}_k + \sum_{i=1}^{k} \mathbf{A}_i \mathbf{G}_{k-i}, & k = 1, \dots, p \\ -\mathbf{G}_k + \sum_{i=1}^{p} \mathbf{A}_i \mathbf{G}_{k-i}, & k = p+1, \dots, q \\ \sum_{i=0}^{p+q-k} \mathbf{A}_{i+k-q} \mathbf{G}_{q-i}, & k = q+1, \dots, q+p \end{cases}. \tag{A3}$$

Then, the Equation (A3) are exploited to identify the finite-order VAR process, finding its coefficients $\mathbf{B}_k$ from the differencing parameters d_i, which inserted in (6) yield the parameters $\mathbf{G}_k$, and from the VAR parameters $\mathbf{A}_k$.

Appendix B

The procedure that represents how the structure of a VAR process is altered when the process is represented at an assigned scale τ is based on the two steps of filtering the VAR process and of downsampling the filtered process [19,20]. The first step transforms the VAR process into a VARMA process (Equation (10)), which is equivalent of a SS process (eq. 11), for which the parameters $\mathbf{K}^{(r)}$, $\mathbf{C}^{(r)}$ and $\mathbf{B}^{(r)}$ are given by [19]:

$$\begin{aligned} \mathbf{C}^{(r)} &= \begin{bmatrix} \mathbf{B}_1 & \cdots & \mathbf{B}_m & \mathbf{D}_1 & \cdots & \mathbf{D}_r \end{bmatrix} \\ \mathbf{K}^{(r)} &= \begin{bmatrix} \mathbf{I}_M & \mathbf{0}_{M \times M(m-1)} & \mathbf{D}_0^{-T} & \mathbf{0}_{M \times M(r-1)} \end{bmatrix} \\ \mathbf{B}^{(r)} &= \begin{bmatrix} & \mathbf{C}^{(r)} & \\ \mathbf{I}_{M(m-1)} & \mathbf{0}_{M(m-1) \times M(r+1)} & \\ \mathbf{0}_{M \times M(m+r)} & & \\ \mathbf{0}_{M(r-1) \times Mm} & \mathbf{I}_{M(r-1)} & \mathbf{0}_{M(r-1) \times M} \end{bmatrix}. \end{aligned} \tag{A4}$$

The second step exploits theoretical findings showing that the downsampled version of an SS process has itself an SS representation [22,23]. Here, downsampling the SS process (11) with a factor τ yields the process $\mathbf{X}_n^{(\tau)} = \mathbf{X}_{n\tau}^{(r)}$, which has the SS representation:

$$\mathbf{Y}_{n+1} = \mathbf{B}^{(\tau)} \mathbf{Y}_n + \mathbf{W}_n \tag{A5a}$$

$$\mathbf{X}_n^{(\tau)} = \mathbf{C}^{(\tau)} \mathbf{Y}_n + \mathbf{V}_n, \tag{A5b}$$

where $\mathbf{V}_n$ and $\mathbf{W}_n$ are different white noise processes with variances $\Sigma_{\mathbf{W}}$ and $\Sigma_{\mathbf{V}}$ and covariance $\Sigma_{\mathbf{VW}}$, respectively serving as innovations for the downsampled process $\mathbf{X}_n^{(\tau)}$ and for the state process $\mathbf{Y}_n$. Thus, the process (A5) is an SS($\mathbf{B}^{(\tau)}, \mathbf{C}^{(\tau)}, \Sigma_{\mathbf{W}}, \Sigma_{\mathbf{V}}, \Sigma_{\mathbf{VW}}$) process whose parameters can be obtained as [22,23]:

$$\begin{aligned}
\mathbf{B}^{(\tau)} &= \left(\mathbf{B}^{(r)}\right)^{\tau} \\
\mathbf{C}^{(\tau)} &= \mathbf{C}^{(r)} \\
\Sigma_{\mathbf{V}} &= \Sigma_{\mathbf{E}^{(r)}} \\
\Sigma_{\mathbf{VW}} &= \left(\mathbf{B}^{(r)}\right)^{\tau-1} \mathbf{K}^{(r)} \Sigma_{\mathbf{E}^{(r)}} \\
\Sigma_{\mathbf{W}}(1) &= \Sigma_{\mathbf{E}^{(r)}} \left(\mathbf{K}^{(r)}\right)^{T} \mathbf{K}^{(r)}, \quad \tau = 1 \\
\Sigma_{\mathbf{W}}(\tau) &= \mathbf{B}^{(r)} \Sigma_{\mathbf{W}}(\tau-1) \left(\mathbf{B}^{(r)}\right)^{T} \\
&\quad + \Sigma_{\mathbf{E}^{(r)}} \left(\mathbf{K}^{(r)}\right)^{T} \mathbf{K}^{(r)}, \quad \tau >= 2
\end{aligned} \tag{A6}$$

Then, the SS($\mathbf{B}^{(\tau)}, \mathbf{C}^{(\tau)}, \Sigma_{\mathbf{W}}, \Sigma_{\mathbf{V}}, \Sigma_{\mathbf{VW}}$) process can be converted in the form of Equation (12), which evidences the innovations and has parameters ($\mathbf{B}^{(\tau)}, \mathbf{C}^{(\tau)}, \mathbf{K}^{(\tau)}, \Sigma_{\mathbf{E}^{(\tau)}}$). To move to this representation from (A5) it is necessary to solve the discrete algebraic Ricatti equation [22,23]:

$$\begin{aligned}
\mathbf{P} = \mathbf{B}^{(\tau)}&\mathbf{P}(\mathbf{B}^{(\tau)})^{T} + \Sigma_{\mathbf{W}} - (\mathbf{B}^{(\tau)}\mathbf{P}\mathbf{C}^{(\tau)} + \Sigma_{\mathbf{VW}}) \cdot \\
&\cdot (\mathbf{C}^{(\tau)}\mathbf{P}(\mathbf{C}^{(\tau)})^{T} + \Sigma_{\mathbf{V}})^{-1}(\mathbf{C}^{(\tau)}\mathbf{P}(\mathbf{B}^{(\tau)})^{T} + \\
&+ (\Sigma_{\mathbf{VW}})^{T}),
\end{aligned} \tag{A7}$$

which leads to the derivation of the two unknown parameters of the downsampled process:

$$\Sigma_{\mathbf{E}^{(\tau)}} = \mathbf{C}^{(\tau)}\mathbf{P}(\mathbf{C}^{(\tau)})^{T} + \Sigma_{\mathbf{V}} \tag{A8a}$$

$$\mathbf{K}^{(\tau)} = \frac{\mathbf{B}^{(\tau)}\mathbf{P}(\mathbf{C}^{(\tau)})^{T} + \Sigma_{\mathbf{VW}}}{\Sigma_{\mathbf{V}}}. \tag{A8b}$$

Finally, the covariance of the downsampled process can be computed from the process parameters by solving the following discrete-time Lyapunov equation:

$$\mathbf{\Omega} = \mathbf{B}^{(\tau)}\mathbf{\Omega}(\mathbf{B}^{(\tau)})^{T} + \Sigma_{\mathbf{E}^{(\tau)}}(\mathbf{K}^{(\tau)})^{T}\mathbf{K}^{(\tau)} \tag{A9a}$$

$$\Sigma_{\mathbf{X}^{(\tau)}} = \mathbf{C}^{(\tau)}\mathbf{\Omega}(\mathbf{C}^{(\tau)})^{T} + \Sigma_{\mathbf{E}^{(\tau)}}. \tag{A9b}$$

The two covariance matrices $\Sigma_{\mathbf{X}^{(\tau)}}$ and $\Sigma_{\mathbf{X}^{(\tau)}}$ are used in Equation (14) to derive the multivariate complexity of the process $\mathbf{X}$ observed at scale τ.

Appendix C

In order to derive the partial variances used in Equation (4a) and Equation (4b) for the computation of the complexity of scalar processes, SS submodels need to be formed in a way such that the observation equation (Equation (14)) contains only some of the scalar processes. It is important to note that these submodels are not in innovations form, but are rather SS models as in (A5) with parameters $(\mathbf{B}, \mathbf{C}^{(a)}, \mathbf{K}\Sigma_{\mathbf{X}}\mathbf{K}^{T}, \Sigma_{\mathbf{X}}(a,a), \mathbf{K}\Sigma_{\mathbf{X}}(:,a))$ [19]. Such SS models can be converted to the SS form as in (12), with innovation covariance $\Sigma_{\mathbf{E}_a^{(\tau)}}$, solving the discrete algebraic Ricatti Equations (A7) and (A8), so that the partial variance $\Sigma_{E_{j|a}^{(\tau)}}$ is derived as the diagonal element of $\Sigma_{\mathbf{E}_a^{(\tau)}}$ corresponding to the position of the target $X_{j,n}$.

References

1. Cohen, M.A.; Taylor, J.A. Short-term cardiovascular oscillations in man: Measuring and modelling the physiologies. *J. Phys.* **2002**, *542*, 669–683. [CrossRef] [PubMed]
2. Goldberger, A.L.; Peng, C.K.; Lipsitz, L.A. What is physiologic complexity and how does it change with aging and disease? *Neurobiol. Aging* **2002**, *23*, 23–26. doi:10.1016/S0197-4580(01)00266-4. [CrossRef]
3. Pincus, S.M. Greater signal regularity may indicate increased system isolation. *Math. Biosci.* **1994**, *122*, 161–181. doi:10.1016/0025-5564(94)90056-6. [CrossRef]
4. Costa, M.; Goldberger, A.L.; Peng, C.K. Multiscale entropy analysis of complex physiologic time series. *Phys. Rev. Lett.* **2002**, *89*, 068102. [CrossRef] [PubMed]
5. Valencia, J.; Porta, A.; Vallverdú, M.; Clariá, F.; Baranowski, R.; Orłowska-Baranowska, E.; Caminal, P. Refined multiscale entropy: Application to 24-h holter recordings of heart period variability in healthy and aortic stenosis subjects. *IEEE Trans. Biomed. Eng.* **2009**, *56*, 2202–2213. doi:10.1109/TBME.2009.2021986. [CrossRef] [PubMed]
6. Ahmed, M.U.; Mandic, D.P. Multivariate multiscale entropy analysis. *IEEE Signal Process. Lett.* **2011**, *19*, 91–94. [CrossRef]
7. Faes, L.; Porta, A.; Javorka, M.; Nollo, G. Efficient Computation of Multiscale Entropy over Short Biomedical Time Series Based on Linear State-Space Models. *Complexity* **2017**, *2017*. [CrossRef]
8. Faes, L.; Pereira, M.A.; Silva, M.E.; Pernice, R.; Busacca, A.; Javorka, M.; Rocha, A.P. Multiscale information storage of linear long-range correlated stochastic processes. *Phys. Rev. E* **2019**, *99*, 032115. doi:10.1103/PhysRevE.99.032115. [CrossRef]
9. Porta, A.; Guzzetti, S.; Furlan, R.; Gnecchi-Ruscone, T.; Montano, N.; Malliani, A. Complexity and nonlinearity in short-term heart period variability: Comparison of methods based on local nonlinear prediction. *IEEE Trans. Biomed. Eng.* **2007**, *54*, 94–106. [CrossRef]
10. Xiong, W.; Faes, L.; Ivanov, P.C. Entropy measures, entropy estimators, and their performance in quantifying complex dynamics: Effects of artifacts, nonstationarity, and long-range correlations. *Phys. Rev. E* **2017**, *95*, 062114. [CrossRef]
11. Barrett, A.B.; Barnett, L.; Seth, A.K. Multivariate Granger causality and generalized variance. *Phys. Rev. E* **2010**, *81*, 041907. [CrossRef] [PubMed]
12. Barnett, L.; Barrett, A.B.; Seth, A.K. Granger causality and transfer entropy are equivalent for Gaussian variables. *Phys. Rev. Lett.* **2009**, *103*, 238701. [CrossRef] [PubMed]
13. Faes, L.; Porta, A.; Nollo, G. Information decomposition in bivariate systems: Theory and application to cardiorespiratory dynamics. *Entropy* **2015**, *17*, 277–303. [CrossRef]
14. Valente, M.; Javorka, M.; Porta, A.; Bari, V.; Krohova, J.; Czippelova, B.; Turianikova, Z.; Nollo, G.; Faes, L. Univariate and multivariate conditional entropy measures for the characterization of short-term cardiovascular complexity under physiological stress. *Physiol. Meas.* **2018**, *39*, 014002. [CrossRef] [PubMed]
15. Faes, L.; Porta, A.; Nollo, G.; Javorka, M. Information decomposition in multivariate systems: Definitions, implementation and application to cardiovascular networks. *Entropy* **2016**, *19*, 5. [CrossRef]
16. Tsay, W.J. Maximum likelihood estimation of stationary multivariate ARFIMA processes. *J. Stat. Comput. Simul.* **2010**, *80*, 729–745. doi:10.1080/00949650902773536. [CrossRef]
17. Beran, J.; Feng, Y.; Ghosh, S.; Kulik, R. *Statistics for Long-Memory Processes: Probabilistic Properties and Statistical Methods*, 2012 ed.; Springer: New York, NY, USA, 2012.
18. Faes, L.; Erla, S.; Nollo, G. Measuring connectivity in linear multivariate processes: Definitions, interpretation, and practical analysis. *Comput. Math. Methods Med.* **2012**, *2012*. [CrossRef]
19. Faes, L.; Marinazzo, D.; Stramaglia, S. Multiscale information decomposition: Exact computation for multivariate Gaussian processes. *Entropy* **2017**, *19*, 408. [CrossRef]
20. Faes, L.; Nollo, G.; Stramaglia, S.; Marinazzo, D. Multiscale Granger causality. *Phys. Rev. E* **2017**, *96*, 042150. [CrossRef]
21. Aoki, M.; Havenner, A. State space modeling of multiple time series. *Econ. Rev.* **1991**, *10*, 1–59. doi:10.1080/07474939108800194. [CrossRef]
22. Solo, V. State-space analysis of Granger-Geweke causality measures with application to fMRI. *Neur. Comput.* **2016**, *28*, 914–949. [CrossRef] [PubMed]

23. Barnett, L.; Seth, A.K. Granger causality for state-space models. *Phys. Rev. E* **2015**, *91*, 040101. [CrossRef] [PubMed]
24. Krohova, J.; Faes, L.; Czippelova, B.; Turianikova, Z.; Mazgutova, N.; Pernice, R.; Busacca, A.; Marinazzo, D.; Stramaglia, S.; Javorka, M. Multiscale information decomposition dissects control mechanisms of heart rate variability at rest and during physiological stress. *Entropy* **2019**, *21*, 526. [CrossRef]
25. Faes, L.; Nollo, G.; Porta, A. Information-based detection of nonlinear Granger causality in multivariate processes via a nonuniform embedding technique. *Phys. Rev. E* **2011**, *83*, 051112. doi:10.1103/PhysRevE.83.051112. [CrossRef] [PubMed]
26. Porta, A.; Bassani, T.; Bari, V.; Tobaldini, E.; Takahashi, A.C.; Catai, A.M.; Montano, N. Model-based assessment of baroreflex and cardiopulmonary couplings during graded head-up tilt. *Comput. Biol. Med.* **2012**, *42*, 298–305. [CrossRef] [PubMed]
27. Faes, L.; Nollo, G.; Porta, A. Information domain approach to the investigation of cardio-vascular, cardio-pulmonary, and vasculo-pulmonary causal couplings. *Front. Phys.* **2011**, *2*, 80. [CrossRef] [PubMed]
28. Faes, L.; Porta, A.; Cucino, R.; Cerutti, S.; Antolini, R.; Nollo, G. Causal transfer function analysis to describe closed loop interactions between cardiovascular and cardiorespiratory variability signals. *Biol. Cybern.* **2004**, *90*, 390–399. [CrossRef]
29. Camm, A.; Malik, M.; Bigger, J.; Breithardt, G.; Cerutti, S.; Cohen, R.; Coumel, P.; Fallen, E.; Kennedy, H.; Kleiger, R.; et al. Heart rate variability: Standards of measurement, physiological interpretation and clinical use. Task Force of the European Society of Cardiology and the North American Society of Pacing and Electrophysiology. *Circulation* **1996**, *93*, 1043–1065.
30. Shaffer, F.; Ginsberg, J. An overview of heart rate variability metrics and norms. *Front. Public Health* **2017**, *5*, 258. [CrossRef]
31. Pernice, R.; Javorka, M.; Krohova, J.; Czippelova, B.; Turianikova, Z.; Busacca, A.; Faes, L. Comparison of short-term heart rate variability indexes evaluated through electrocardiographic and continuous blood pressure monitoring. *Med. Biol. Eng. Comput.* **2019**, *57*, 1247–1263. [CrossRef]
32. Javorka, M.; Czippelova, B.; Turianikova, Z.; Lazarova, Z.; Tonhajzerova, I.; Faes, L. Causal analysis of short-term cardiovascular variability: State-dependent contribution of feedback and feedforward mechanisms. *Med. Biol. Eng. Comput.* **2017**, *55*, 179–190. [CrossRef] [PubMed]
33. Bardet, J.M.; Lang, G.; Oppenheim, G.; Philippe, A.; Taqqu, M.S. Generators of long-range dependent processes: A survey. *Theory Appl. Long-Range Depend.* **2003**, 579–623.
34. Stoica, P.; Selen, Y. Model-order selection: a review of information criterion rules. *IEEE Signal Process. Mag.* **2004**, *21*, 36–47. [CrossRef]
35. Pinheiro, J.; Bates, D. *Mixed-Effects Models in S and S-PLUS*, 2000 ed.; Springer: New York, NY, USA, 2000.
36. Searle, S.R.; Speed, F.M.; Milliken, G.A. Population Marginal Means in the Linear Model: An Alternative to Least Squares Means. *Am. Stat.* **1980**, *34*, 216–221.
37. Bates, D.; Mächler, M.; Bolker, B.; Walker, S. Fitting Linear Mixed-Effects Models Using lme4. *J. Stat. Softw.* **2015**, *67*, 1–48. doi:10.18637/jss.v067.i01. [CrossRef]
38. Lenth, R. *emmeans: Estimated Marginal Means, aka Least-Squares Means*; R package version 1.3.3; R Foundation for Statistical Computing: Vienna, Austria, 2019.
39. R Core Team. *R: A Language and Environment for Statistical Computing*; R Foundation for Statistical Computing: Vienna, Austria, 2016.
40. Porta, A.; Gnecchi-Ruscone, T.; Tobaldini, E.; Guzzetti, S.; Furlan, R.; Montano, N. Progressive decrease of heart period variability entropy-based complexity during graded head-up tilt. *J. Appl. Physiol.* **2007**, *103*, 1143–1149. [CrossRef]
41. Porta, A.; Guzzetti, S.; Montano, N.; Furlan, R.; Pagani, M.; Malliani, A.; Cerutti, S. Entropy, entropy rate, and pattern classification as tools to typify complexity in short heart period variability series. *IEEE Trans. Biomed. Eng.* **2001**, *48*, 1282–1291. [CrossRef]
42. Mukai, S.; Hayano, J. Heart rate and blood pressure variabilities during graded head-up tilt. *J. Appl. Physiol.* **1995**, *78*, 212–216. [CrossRef]
43. El-Hamad, F.; Javorka, M.; Czippelova, B.; Krohova, J.; Turianikova, Z.; Porta, A.; Baumert, M. Repolarization variability independent of heart rate during sympathetic activation elicited by head-up tilt. *Med. Biol. Eng. Comput.* **2019**, *57*, 1753–1762. doi:10.1007/s11517-019-01998-9. [CrossRef]

44. Faes, L.; Gómez-Extremera, M.; Pernice, R.; Carpena, P.; Nollo, G.; Porta, A.; Bernaola-Galván, P. Comparison of methods for the assessment of nonlinearity in short-term heart rate variability under different physiopathological states. *Chaos Interdiscip. J. Nonlinear Sci.* **2019**, *29*, 123114. doi:10.1063/1.5115506. [CrossRef]
45. Zanetti, M.; Faes, L.; Nollo, G.; De Cecco, M.; Pernice, R.; Maule, L.; Pertile, M.; Fornaser, A. Information dynamics of the brain, cardiovascular and respiratory network during different levels of mental stress. *Entropy* **2019**, *21*, 275. [CrossRef]
46. Castaldo, R.; Montesinos, L.; Melillo, P.; James, C.; Pecchia, L. Ultra-short term HRV features as surrogates of short term HRV: A case study on mental stress detection in real life. *BMC Med. Inf. Dec. Mak.* **2019**, *19*, 1–13. [CrossRef] [PubMed]
47. Can, Y.S.; Arnrich, B.; Ersoy, C. Stress detection in daily life scenarios using smart phones and wearable sensors: A survey. *J. Biomed. Inf.* **2019**, 103139. [CrossRef] [PubMed]
48. Pernice, R.; Nollo, G.; Zanetti, M.; De Cecco, M.; Busacca, A.; Faes, L. Minimally Invasive Assessment of Mental Stress based on Wearable Wireless Physiological Sensors and Multivariate Biosignal Processing. In Proceedings of the IEEE EUROCON 2019-18th International Conference on Smart Technologies, Novi Sad, Serbia, 1–4 July 2019.
49. Reimers, H.E. Comparisons of tests for multivariate cointegration. *Stat. Pap.* **1992**, *33*, 335–359. [CrossRef]
50. Zhang, Y.; Shang, P.; Xiong, H. Multivariate generalized information entropy of financial time series. *Phys. A Stat. Mech. Appl.* **2019**, *525*, 1212–1223. [CrossRef]
51. Ahmed, M.U.; Mandic, D.P. Multivariate multiscale entropy: A tool for complexity analysis of multichannel data. *Phys. Rev. E* **2011**, *84*, 061918. [CrossRef]
52. Courtiol, J.; Perdikis, D.; Petkoski, S.; Müller, V.; Huys, R.; Sleimen-Malkoun, R.; Jirsa, V.K. The multiscale entropy: Guidelines for use and interpretation in brain signal analysis. *J. Neurosci. Methods* **2016**, *273*, 175–190. [CrossRef]

Article

Application of Time-Scale Decomposition of Entropy for Eye Movement Analysis

Katarzyna Harezlak * and Pawel Kasprowski

Silesian University of Technology, Akademicka 16, 44-100 Gliwice, Poland; pawel.kasprowski@polsl.pl
* Correspondence: katarzyna.harezlak@polsl.pl; Tel.: +48-32-237-1339

Received: 23 December 2019; Accepted: 30 January 2020; Published: 1 February 2020

Abstract: The methods for nonlinear time series analysis were used in the presented research to reveal eye movement signal characteristics. Three measures were used: approximate entropy, fuzzy entropy, and the Largest Lyapunov Exponent, for which the multilevel maps (MMs), being their time-scale decomposition, were defined. To check whether the estimated characteristics might be useful in eye movement events detection, these structures were applied in the classification process conducted with the usage of the kNN method. The elements of three MMs were used to define feature vectors for this process. They consisted of differently combined MM segments, belonging either to one or several selected levels, as well as included values either of one or all the analysed measures. Such a classification produced an improvement in the accuracy for saccadic latency and saccade, when compared with the previously conducted studies using eye movement dynamics.

Keywords: eye movement events detection; nonlinear analysis time series analysis; approximate entropy; fuzzy entropy; multilevel entropy map; time-scale decomposition

---

1. Introduction

Biological signals representing the electrical, chemical, and mechanical activities that occur during biological events attracted the attention of many researchers. These interests are aimed at discovering patterns which may prove useful in understanding the underlying physiological mechanisms or systems. The range of biological signals explored in various studies include, inter alia: electroencephalogram (EEG), electrocardiogram (ECG), surface electromyogram (sEMG), galvanic skin response (GSR), and arterial blood pressure (ABP). Obtaining these characteristics is of great importance in medicine, as this it may enable the differentiation of typical and atypical behaviors, supporting in this way diagnosing and treatment. However, analysis of some of them may also be applied, for example, in cognitive or psychological studies.

Acquiring bio-signal recordings requires various biomedical instruments to be applied. For example, eye movement signals, which are of interest in this article, can be measured by means of the electrical potential between electrodes placed at points close to the eye (electro-oculography (EOG)) or with the usage of less intrusive video-oculography (VOD), using video cameras and image processing algorithms for eye movement tracking.

Biological signals are naturally analogous; however, during measurement process conducted with a specific sampling rate, they are converted to a discrete-time form and constitute a biological time series. The distinct difficulty in the analysis of bio-signals is their nonlinear nature, therefore in order to extract useful features and components of the recorded signal, it is important to use appropriate processing methods. Frequently, methods of nonlinear time series analysis, which can be deployed for many types of bio-signals, are chosen for this purpose. Some of these are described in the following section.

1.1. Methods Used in Biological Signal Analysis

The approaches used for quantifying biological signal characteristics use fractal and dynamic feature analysis. In the former group of traits, fractal power spectra and the fractal dimension are investigated, usually, through detrended fluctuation analysis (DFA) [1]. This approach enables the assessment of the existence of self-similarity and long-range correlations. The main goal of these kinds of studies is revealing the long-term memory of an explored system and predicting the system's future states. This method was successfully applied to such biological processes as ECG, EEG, EMG [2–5].

Nonlinear system dynamics may also be represented by various entropy measures [6], and the Largest Lyapunov Exponent (LLE) [7]. The entropy-based algorithms ascertain the degree of disorder and uncertainty in the underlying system described by an observed variable. Two primary purposes of these studies can be mentioned. The first of them is the discovery of the dynamic characteristics of biological signals of healthy people to use them as reference data in revealing anomalous cases. The second one is to determine events occurring within these signals.

For example, Chen et al. in [8] provided a review of entropy measures used in cardiovascular diseases. Furthermore, *approximate entropy* (ApEn) was used in [9] to obtain a better understanding of abnormal dynamics in the brain in the case of Alzheimer's disease (AD). Yents et al. [10] used spatio-temporal gait data from young and elderly subjects to investigate the performance of ApEn and *sample entropy* (SampEn). Cao et al. [11] presented the *fuzzy entropy* (FuzEn) metrics comparison and pointed the FuzEn out as a better tool than ApEn and SampEn, for distinguishing EEG signals of people with AD from those obtained for non-AD. Additionally, the comparison of various fuzzy entropy measures was conducted in the work [12]. Some synthetic datasets, together with EEG and ECG clinical datasets, as well as the gait maturation database, were used for comparison purposes.

EEG, ECG, and gait characteristics were also explored with the usage of the LLE. The works conducted in [13,14] indicated that this measure can be used to identify sleep stages, whereas in [15] the LLE was applied for the analysis of nonlinear changes of neural dynamics in the processing of stressful anxiety-related memories. Furthermore, automated diagnosis of electrocardiographic changes was successfully explored in [16] by using the Lyapunov Exponents for the detection of four types of ECG beats. Additionally, some works were devoted to investigating the application of the LLE in human locomotion for gait stability assessment. For example, Josinski et al., in their research [17] focused on finding LLE-based, easy-to-measure, objective biomarkers that could classify PD (Parkinson's disease) patients in early (preclinical) stages of the disease.

Nonlinear analysis was also conducted in regard to eye movement signal. However, its dynamics in this scope was not so widely explored as in other biological signals. Nonetheless, in several studies, it was discovered, based on the LLE [18–20] and entropy measures: ApEn and SampEn ([21,22]), that various phases of eye movement signal reveal different dynamic characteristics.

1.2. Dynamic Eye Movement Signal Characteristics

Eye movements are triggered by brain activity, which is a response to visual stimuli or an intention to gain knowledge of the surrounding world. The biological signal evoked this way consists of two main components: fixations, occurring when eyes are focused on part of a scene, and saccades taking place when eyes move to another scene area. During both events, eyes are in constant motion, even during fixations, generating micro-movements such as tremors, microsaccades, and drifts [23,24]. Additionally, in the reaction to the observed stimulus movement, the brain needs some time to commit a saccade. This occurrence called *saccadic latency* [25] is shown, together with the preceding fixation and the following saccade, in Figure 1.

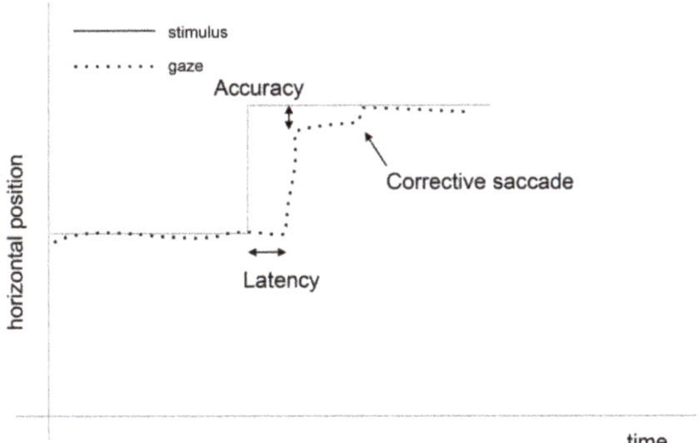

Figure 1. An example of eye movement events. At the beginning the eye is focused on the stimulus. When the stimulus moves to another location, the eye moves as well, yet after a period of saccadic latency [22].

Fixation and saccade differ in their characteristics. The first element is described by slow motions, small spatial dispersion, and relatively long duration (200–300 ms up to several sec.), while during the second one eyes move very quickly, reaching up to even 500°/s, with a short duration (30–80 ms) and a higher amplitude of movement [26]. These differences are of prominent importance when event detection algorithms are considered. The most common approaches for extracting fixations and saccades from registered signal use dispersion and velocity thresholds [27]. One of the major drawbacks of these solutions is the dependency of the obtained results on user-defined thresholds and the lack of their commonly accepted values. The alternative approach presented in [28] assumes the use of machine learning and 14-features vector constructed based on the 100–200-ms surroundings of each sample. This vector consists of features describing the data in terms of dispersion, velocity, sampling frequency, and precision. The results of these studies showed that the machine learning technique is a promising solution. It was taken into consideration when a new approach was being developed using the findings of the previously described nonlinear dynamic systems analysis. For eye movement events detection, in [22], a multilevel map of ApEn was defined, and entropy values calculated at each map level were used for the feature vectors construction. Determining particular eye movement segments was conducted using the kNN classifier for various values of the k parameter. The classification outcomes were promising as well, especially in the scope of saccadic period detection, confirming the usefulness of ApEn measure in such an application. However, because one measure is only a single index describing the behavior of a dynamic time series, further investigations are still required. The currently presented research is the answer to this need.

1.3. Contribution

These studies introduce an extended application of the multilevel map by adding two additional measures: FuzEn and the LLE. According to the authors' knowledge, the fuzzy entropy has not been applied in exploring eye movement dynamics so far. The idea of the application of the multilevel map, built based on FuzEn and the LLE, for eye movement time series analysis is also a novel approach, as well as combining three multilevel maps in order to define feature vectors for machine learning eye movement event detection.

2. Materials and Methods

2.1. The Dataset Description

The presented research was conducted with the usage of the same dataset, which was applied in the studies described in [22]. It was collected during the"jumping point" experiment, using 29 dark points (N_{pp}), distributed over a white 1280 × 1024 (370 mm × 295 mm) screen. The points' layout was designed in such a way to ensure both covering a screen area evenly and to obtain varying lengths of saccadic movements. Although the point was displayed in each location for 3 sec, only the first N_{EMr} = 1024 ms were taken into account during further analysis. Eye movement signals were acquired employing the head-mounted JAZZ-novo eye tracker [29], with a sampling rate equal to 1000 Hz. It uses Direct Infra Red Oculography (IROG), which is embedded in the Cyclop ODS sensor measuring the resultant rotations of the left and the right eye. During registration, eyes are illuminated with a low intensity infrared (IR) light. The difference between the amounts of IR reflected back from the eye surfaces, carries information pertaining to the eye position changes.

24 participants (N_p) aged between 22 and 24, with normal vision, took part in the experiment, which consisted of two sessions, separated by three weeks (N_{spp}). As a result,N_{EMs} participants' sessions were gathered, $N_{ps} = N_p * N_{spp}$, each of which comprised subsequent eye positions for all points' locations. Such a dataset was divided into N_{EMs} eye movement series ($N_{EMs} = N_p * N_{spp} * N_{pp}$), which are later referred to as *EM series*. Each of them included one participant's eye positions registered between the moment of the appearance of the stimulus and the time of its position change. Due to some registration and processing problems four participant session were removed from further analysis. Thus, the N_{ps} value decreased by 4, and N_{EMs} by 4 × N_{pp}. Finally, N_{EMs}—the number of eye movement time series—was 1276 (44 $participants\_sessions$ × 29 $points$)

Subsequently, by applying the standard procedure of the two-point signal differentiation, the first derivative of horizontal coordinates for each EM series was obtained. Thus, each series corresponds to the velocity of eye movement in a period between two appearances of the stimulus. This quantity was used to make research independent to the point positions.

2.2. The Method

The first step of the nonlinear time series analysis was the evaluation of the three chosen measures: ApEn, FuzEn and the LLE, for each EM series. These calculations were conducted taking into account the structure of the multilevel map (MM) introduced in [22]. It represents the concept of a time-scale decomposition of the particular measure, whose values are calculated for various segments of EM series following the schema presented in Figure 2.

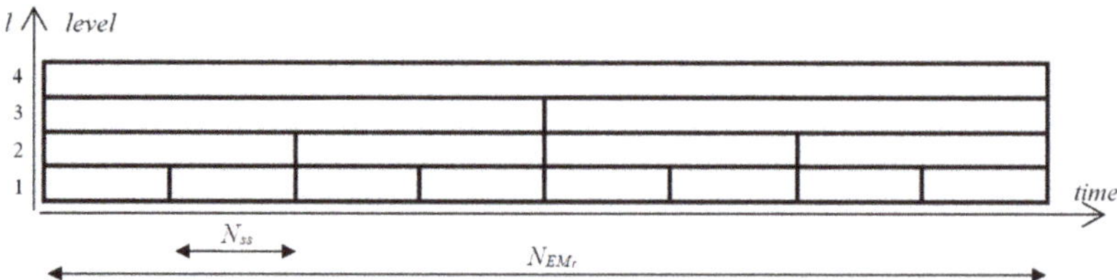

Figure 2. Segments in the multilevel time-scale structure.

A signal at the $(l+1)$-th level of the map is divided into two shorter segments at the l-th level. The smallest size of segments was used at the 1st level of the map, and for this research, it was set to N_{ss}=64 ms. As mentioned, the time series length N_{EMr} was set to 1024 ms, which resulted in $l = 5$ map levels. This means that at the 2nd level the length of a segment equaled 128 ms, at the 3rd: 256 ms, at the 4th: 512 ms and finally at the 5th: 1024 ms. For each map cell, all three measures were estimated independently; thus, three MMs were obtained for each EM series (Figure 4).

2.3. Approximate Entropy

The approximate entropy method was proposed in [30] as a measure of regularity for quantifying complexity within a time series. For a given integer parameter m and a positive real one r, for a time series $u(1), \ldots, u(N)$ in $\mathbf{R}$, the sequence of vectors $x(1), \ldots, x(N-m+1)$ in $\mathbf{R}^m$ can be defined, where $x(i) = [u(i), \ldots, u(i+m-1)]$.
For $1 \leq i \leq N-m+1$, using probability that any two vectors of length m are similar within a tolerance given by r:

$$C_i^m(r) = \frac{\text{number of } j \leq N-m+1 : d[x(i), x(j)] \leq r}{N-m+1} \tag{1}$$

where the distance d is defined as follows:

$$d[x(i), x(j)] = \max_{k \in (1,m)} (|u(i+k-1) - u(j+k-1)|) \tag{2}$$

and the average logarithmic probability over all m-patterns:

$$\Phi^m(r) = (N-m+1)^{-1} \sum_{i=1}^{N-m+1} \log C_i^m(r) \tag{3}$$

approximate entropy can be obtained:

$$\text{ApEn}(m,r) = \lim_{N \to \infty} [\Phi^m(r) - \Phi^{m+1}(r)]. \tag{4}$$

The asymptotic formula in Equation (4) cannot be used directly, and the estimator of approximate entropy (for sufficiently large values of N) is defined as:

$$\text{ApEn}(m,r,N) = \Phi^m(r) - \Phi^{m+1}(r). \tag{5}$$

There is usually significant variation in $ApEn(m,r,N)$ over the range of m and r, when a particular system is taken into consideration [31]. Thus, based on the analysis conducted in [22], in this research m was set to 2 and r as $0.2 \times SD$ (the standard deviation of the entire time series) in order to define the first MM.

2.4. Fuzzy Entropy

The second MM was prepared based on the fuzzy entropy values calculated according to the method introduced in [32], which consists of several steps.

For a given time series $u(1), \ldots, u(N)$ and integer parameter m, the sequence of vectors $X_1^m, \ldots, x_{(N-m+1)}^m$ is defined as:

$$X_i^m = u(i), \ldots, u(i+m-1) - u0(i) \tag{6}$$

where X_i^m represents m consecutive u values, commencing with the $i-th$ point and generalized by removing a baseline:

$$u0(i) = \frac{1}{m}\sum_{j=0}^{m-1} u(i+j) \tag{7}$$

Subsequently, the distance between two vectors is defined as:

$$d_{ij}^m = d[X_i^m, X_j^m] = \max_{k\in(0,m-1)} |(u(i+k) - u0(i)) - (u(j+k) - u0(j))| \tag{8}$$

and given n and r the similarity degree D_{ij}^m of X_j^m to X_i^m through a fuzzy function $\mu(d_{ij}^m, n, r)$ is calculated:

$$D_{ij}^m = \mu(d_{ij}^m, n, r) \tag{9}$$

where:

$$\mu(d_{ij}^m, n, r) = exp\left(-\frac{(d_{ij}^m)^n}{r}\right) \tag{10}$$

Finally, the fuzzy entropy for finite sets is estimated as:

$$FuzEn(m,n,r,N) = \ln\phi^m(n,r) - \ln\phi^{m+1}(n,r) \tag{11}$$

where

$$\phi^m = \frac{1}{N-m}\sum_{i=1}^{N-m}\left(\frac{1}{N-m-1}\sum_{j=1,j\neq i}^{N-m} D_{ij}^m\right) \tag{12}$$

N corresponds to the size of the time series, which depends on the MM level under consideration. m is the length of sequences to be compared and was set to the same value as for evaluating approximate entropy. The parameters n and r, determining the width and the gradient of the boundary of the exponential function, respectively, were chosen based on experimental data, following suggestions provided by the method's authors. Fuzzy entropy was evaluated for 20 time series with different r and n values. Subsequently, standard deviation (SD) of the results was calculated. The parameter values for which a slow decrease in SD was observed were chosen for the purpose of these studies. They were 2 for n and $0.075 \times SD$ (of the particular EM series) for r.

2.5. The Largest Lyapunov Exponent

The Largest Lyapunov Exponent is a measure which estimates the amount of chaos in dynamic systems. Chaos is observed when neighbouring paths followed by the system, starting from very close initial conditions, rapidly—exponentially fast—move to different states. The reconstruction of the system states is feasible based on measurements of a single observed property when Takens' embedding theorem is applied [33]. A time delay embedded series—with time lag denoted by τ and embedding dimension by m—can be defined as a transformation of the original time series u, such that:

$$x(i) = [u(i), u(i+\tau), \cdots, u(i+(m-1)\tau)], i = \{1, 2\cdots, M\} \tag{13}$$

where $M = N - (m-1)\times\tau$.

A pair $[x(i), x(j)]$ of nearest neighbours, starting close to one another in a chaotic system, diverges approximately at a rate given by the Largest Lyapunov Exponent λ [34]:

$$d_j(i) \approx d_{j0}e^{\lambda(i\Delta t)} \tag{14}$$

where $d_j(i)$ is the Euclidean distance after i time steps, Δt is the sampling rate of the time series and d_{j0} is the initial pair separation. Solving the Equation (14) by taking the logarithm of both sides, the Largest Lyapunov Exponent can be calculated as follows:

$$\lambda \approx \frac{1}{i\Delta t} ln(\frac{d_j(i)}{d_{j0}}) \tag{15}$$

Equation (15) provides the way for evaluating λ for two specific neighbouring points over a specific interval of time. Thus, to approximate the Lyapunov Exponent for a whole dynamic system, it has to be averaged for all $j = 1, 2, \ldots, M$, where $M = N - (m-1) \times \tau$. Negative Lyapunov Exponent values indicate convergence, while positive demonstrate divergence and chaos. The parameters m and τ are calculated with the usage of the False Nearest Neighbours (FNN) [35] and Mutual Information Factor [36], respectively. The above-described methods were used in the current research to define the third MM.

2.6. Detection of Eye Movement Events

Following the studies presented in [22], the elements of three MMs were used to define feature vectors for the classification process. Various vector types were defined. They consisted of differently combined MM segments, belonging either to one or several selected levels, as well as included values either of one or all the analysed measures (Figure 3). However, before the creation of the vector, data from all MMs levels were rescaled using min-max normalization:

$$x_{new} = \frac{x - x_{min}}{x_{max} - x_{min}} \tag{16}$$

Finally, the following sets were prepared (see Figure 3 for examples):

1. features based on one level (X={64, 128, 256, 512})
   - setX_ApEn, setX_FuzEn, setX_LLE, including one feature
   - setX_ApEn_FuzEn_LLE, including three features
2. features based on two levels (X_Y={64_128, 128_256, 256_512})
   - setX_Y_ApEn, setX_Y_FuzEn, setX_Y_LLE, including two features
   - setX_Y_ApEn_FuzEn_LLE, including six features
3. features based on three levels (X_Y_Z={64_128_256, 128_256_512})
   - setX_Y_Z_ApEn, setX_Y_Z_FuzEn, setX_Y_Z_LLE, including three features
   - setX_Y_Z_ApEn_FuzEn_LLE, including nine features

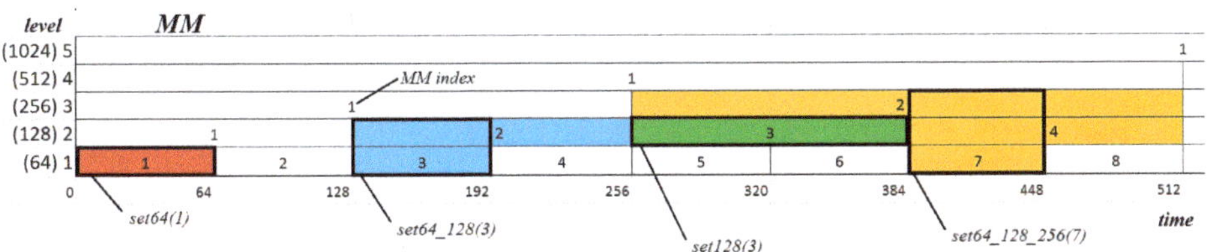

Figure 3. Sample one-, two-, and three-dimensional feature vectors embedded in the MM structure [22].

The number of classes (N_{cl})—the number of segments at the lowest level in the given feature vector —differed in particular feature vectors, depending on the segment used ([22]):

$$N_{cl}(setX\_Y\_Z) = \frac{N_{EMr}}{N_{ss} \times 2^{l_X - 1}} \tag{17}$$

where l_X is the map level and $X = \min(X, Y, Z)$. For example, if the set64_128_256 is considered: $X = 64$ and $l_X = 1$. This means that for l_X equal to:

- 1 – 16 classes were defined,
- 2 – 8,
- 3 – 4,
- 4 – 2.

These sets were separately used for feeding the *kNN* classifier, run with several values for the k parameter: $k = \{3, 7, 15, 31, 63, 127, 255\}$. Each feature vector set was divided into training and test sets, with the usage of the *leave-one-out cross-validation* approach: in each classifier run, N_{pp} maps—calculated for one participant and for one session—were always left for defining a test set.

3. Results

The multilevel maps were estimated for each EM series and each analysed measure separately. Subsequently, they were averaged over all N_{EMs} series, and finally, three resulting maps were created, as shown in Figure 4.

Figure 4. The multilevel maps obtained for the three measures: ApEN, FuzEn, the LLE, with values averaged for all EM series. The table at the bottom presents the maps' structure.

The maps cells were coloured according to their contents. Green indicates the lowest values, while red the highest ones. A repeating pattern can be noticed in each MM, in the 3rd and the 4th segments at the first level and the 2nd segment at the second one. These cells contain the lowest values calculated for each measure and relate to 128 ms of eye movement signal commencing with 128 ms. This dependency is easily noticeable in Figure 5, where the results obtained for the first two MMs levels are juxtaposed in the charts. However, to make this comparison feasible, the results had to be rescaled with the min-max normalization (Equation (16)).

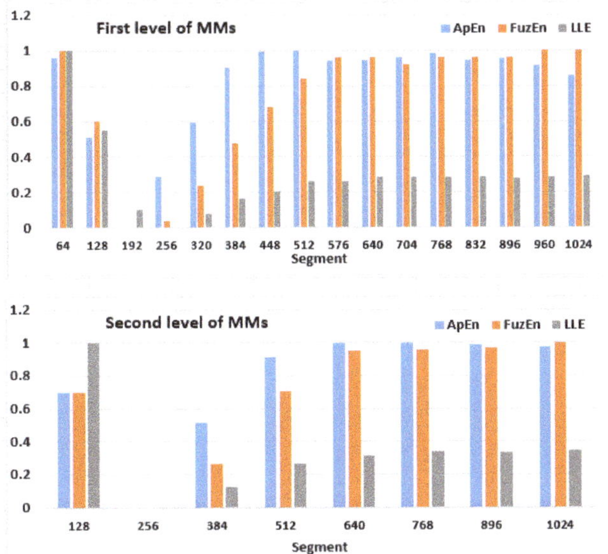

Figure 5. The charts presenting values from the two levels of the ApEn, FuzEn and LLE maps.The first level at the top and the second at the bottom.

Statistical significance of the outcomes was checked by means of ANOVA and Tukey's Honest Significance Difference tests. The two result sets—for ApEn and the LLE measures—verified by the Shapiro–Wilk test, turned out to be normally distributed, yet for the third one—FuzEn—the null hypothesis was rejected. In this case, the Kruskal-Wallis test was additionally run. All differences in the results obtained for the 3rd and the 4th times scopes at the first MM level, compared to the remaining set of segments turned out to be significant. However, there was one exception for the LLE—the difference between the 3rd and 5th segments—and a few for FuzEn—the differences between the 3rd, 4th, and 5th time scopes—which occurred not to be significant. While analysing the results for the second level of the maps, it was revealed that all measures provided significant differences for the 2nd segment when collated with all other time scopes. On the contrary, the outcomes from the second part of the EM series—starting from the 6th (ApEn) and the 8th (FuzEn and the LLE) segments at the first MM level, as well as from the 5th segment at the second level—disclosed the lack of significant differences.

In the second stage of the tests, the usefulness of the three MMs in recognizing eye movement segments was verified with the usage of the kNN method. During the first classification runs, feature vectors consisting of one, two, or three elements calculated only for one measure, were used. Because the attention was focused on the highest possible signal resolution, only features based on segments 64 and 128, and their combinations with values from other levels were taken into account. As can be seen in Figure 6, the performance of the segment recognition is rather low, independently of $k - value$ used, especially when time scopes later than 256 ms are considered. Thus, in further analysis, only the first four segments were addressed.

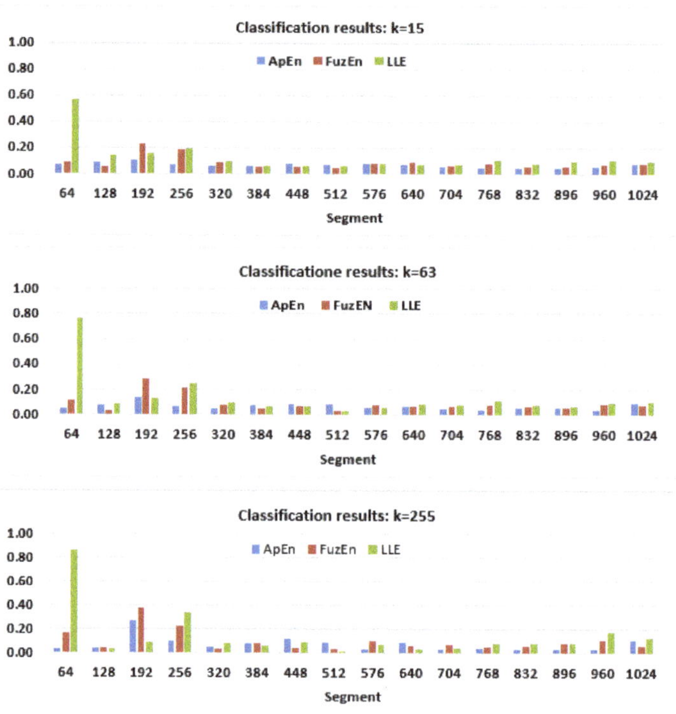

Figure 6. The classification accuracy based on the first level of the ApEn, FuzEn and LLE multilevel maps, and for different the k–parameter values: k = 15 (top), k = 63 (middle), k = 255 (bottom).

The exploration of the presented charts revealed that a higher k parameter did not always provide better results. For this reason, the attention was focused on $k - value$ equal to 63, for which the classification accuracy was similar to that, disclosed for k=255, and which ensures better performance of the execution.

The effect of the classification with the usage of feature vectors defined based on one measure, yet using several levels of the MMs is presented in Table 1. There are nine sets taken into account: three consisting of one feature: *set64_[ApEN, FuzEN, LLE]*, three with two elements: *set64_128_[ApEN, FuzEN, LLE]*, and three including three components: *set64_128_256_[ApEN, FuzEN, LLE]*.

Table 1. The percentage of the correctly classified samples for various feature vectors defined with the usage of one measure. The smallest segment size = 64 ms.

Vectors	set64_			set64_128_			set64_128_256_		
Segment	ApEn	FuzEn	LLE	ApEn	FuzEn	LLE	ApEn	FuzEn	LLE
1–64	0.05	0.12	0.76	0.07	0.13	0.63	0.25	0.34	0.71
64–128	0.08	0.04	0.09	0.07	0.05	0.53	0.34	0.29	0.60
128–192	0.14	0.29	0.13	0.39	0.29	0.26	0.38	0.30	0.40
192–256	0.07	0.21	0.25	0.42	0.25	0.27	0.42	0.28	0.40

The table content shows that the best results were obtained for the LLE measure in the first segment of eye movement signal, yet less than that got for k=255 (86% see Figure 6). However, its classification accuracy decreases significantly when other time scopes are considered. The opposite situation takes place for the two remaining groups of sets. As can be observed, both entropy measures at the beginning of the EM series revealed a low accuracy for vectors defined with only one element. The performance increases for later eye movement scopes when features are merged into two- or three-elements vectors (*set64_128_*, *set64_128_256_*).

The individual measure outcomes can be further collated with those presented in Table 2, obtained for the mixed feature vectors.

Table 2. The percentage of the correctly classified samples for the feature vectors defined with the usage of three measures: ApEN_FuzEn_LLE. The smallest segment size = 64 ms.

Segment	set64_	set64_128_	set64_128_256_
1–64	0.74	0.61	0.66
64–128	0.27	0.38	0.57
128–192	0.10	0.52	0.60
192–256	0.23	0.38	0.51

Similarly, the best performance, slightly worse than previously, was achieved for the first 64 ms of the signals, classified with samples coming from the first level of the MMs. Subsequently, the accuracy decreases, especially for the *set64_ApEN_FuzEn_LLE*. Thus, *set64_128_256_ApEN_FuzEn_LLE* seems to be the proper choice for the classification purpose, despite lower than the best accuracy produced for the first segment. It provides the best or close to the best efficiency for all analysed segments.

In the next two Tables 3 and 4, outcomes for the subsequent segment size (128 ms) were shown. They, in terms of the result pattern, are similar to those discussed above. The best detection was delivered by the LLE measure in the first 128 ms. In the further time scope, the efficiency of this measure dropped, and ApEn and FuzEn disclosed better performance; the best for vectors with three features *(set64_128_256_ApEn*, *set64_128_256_FuzEn*). While comparing the classification accuracy to the previous results (for 64ms segment), it turned out to be generally higher. However, it should not be surprising, because of the less by half the number of classes in a feature set, which facilitates the sample recognition. Additionally, once again, the *set128_256_512_ApEn_FuzEn_LLE* despite worse efficiency in the first time scope, provides, on average, the best results.

Table 3. The results–the percentage of the correctly classified samples for various feature vectors defined with the usage of one measure. The smallest segment size = 128 ms.

Vectors	**set128_**			**set128_256_**			**set128_256_512_**		
Segment	**ApEn**	**FuzEn**	**LLE**	**ApEn**	**FuzEn**	**LLE**	**ApEn**	**FuzEn**	**LLE**
1–128	0.08	0.05	0.84	0.52	0.48	0.80	0.60	0.49	0.82
128–256	0.72	0.64	0.48	0.67	0.61	0.57	0.80	0.71	0.59
256–384	0.07	0.16	0.23	0.30	0.32	0.45	0.60	0.42	0.49
384–512	0.17	0.08	0.14	0.32	0.41	0.47	0.42	0.45	0.54

Table 4. The results–the percentage of the correctly classified samples for the feature vectors defined with the usage of three measures: ApEn_FuzEn_LLE. The smallest segment size = 128 ms.

Segment	**Set128_**	**Set128_256_**	**Set128_256_512_**
1–128	0.78	0.81	0.81
128–256	0.69	0.75	0.79
256–384	0.20	0.52	0.61
384–512	0.20	0.53	0.66

The last step in the data exploration was the juxtaposition (Table 5) of the best results obtained in [22], using only ApEn (the third column), with the corresponding to them, achieved in these studies with the usage of three measures (the fourth column). The first column in the table represents the feature vector type, while in the second one, the eye movement segments were included. The last column in this table shows the best classification accuracy received in these studies for a feature set displayed in the first column and for the segment from the second one. All values presented in this column were estimated for the LLE measure in the first EM series scope.

In almost all cases, *Current accuracy* exposes better or similar performance; however, in one case (*set*64_128), the worse accuracy was achieved.

Table 5. The best accuracy obtained in [22] and in the current research for the same feature vector types and the same EM series segment.

Set	**Segment**	**Accuracy [22] (ApEn)**	**Current accuracy (ApEn_FuzEn_LLE)**	**Current the best (LLE)**
set64	128–192	0.19	0.27	0.76
set64_128	192–256	0.46	0.38	0.63
set64_128_256	128–192	0.43	0.57	0.71
set128	128–256	0.72	0.69	0.84
set128_256	128–256	0.71	0.75	0.80
set128_256_512	128–256	0.83	0.80	0.82

4. Discussion

Eye movement analysis has been conducted over many years, but research in this domain intensified over the last ten years. It resulted from the fact that acquiring knowledge concerning visual patterns has proved useful in many areas, because they provide essential data regarding people's behaviour, intentions, and interests. Proper reasoning in these fields requires an appropriate interpretation of collected recordings, which corresponds to correctly recognised eye movement events. Elements that constitute visual patterns are fixations and saccades, and extracting them adequately from the eye movement signal is crucial. For many years, approaches based on spatial recordings dispersion and velocity have been used. However, as mentioned, they rely on user-defined thresholds, which is their disadvantage as most of the biological patterns vary between individuals (Figure 7). Therefore, some steps were taken to search for other algorithms, ensuring more adjusted event detection.

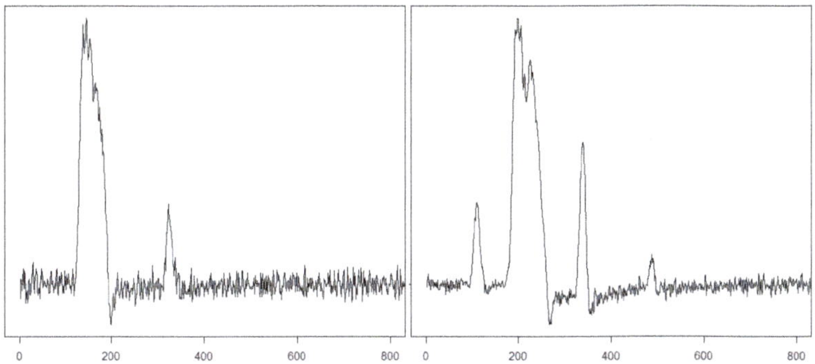

Figure 7. Example signal courses for eye movement velocity. For two more, refer to [22].

The studies presented here base their research on the dynamical aspects of eye movement signal. The main purpose of these investigations was to find characteristics of the particular signal parts, short enough to enable searching for the smallest eye movement events, saccades.

4.1. Multilevel Maps Analysis

The results presented in Figure 4 showed that measures calculated for one EM series, including three eye movement events—saccadic latency, saccade and fixation—have different values in various time scopes of the EM series. The previous studies [18,19,21,22] already provided some evidence in this regard for ApEn, SampEn, and the LLE; however, only during independent analyses. One of the advantages of the current studies is a comparison of these measures and introducing the additional signal description through the FuzEn usage.

When analysing the three multilevel maps from Figure 4, it may be noticed that two segments at the first level (128–192 ms and 192–256 ms) and one segment at the second level (128–256 ms) include the lowest values among all those calculated for the particular map. This means that all three measures detected, within the same period, an activity different than in the remaining EM series parts. This period is characterized by lower entropy values and a lack of chaotic behaviour. Collating MMs segments with signal recordings, for example, in Figure 7 shows that this period corresponds to the saccade. Further exploration of this figure unveils another distinction visible in the LLE map, in its first two segments at the first two levels. The positive values in this period show that neighboring points of EM series, which start close to one another, diverge with time approximately at a rate given by the LLE, as shown in Figure 8.

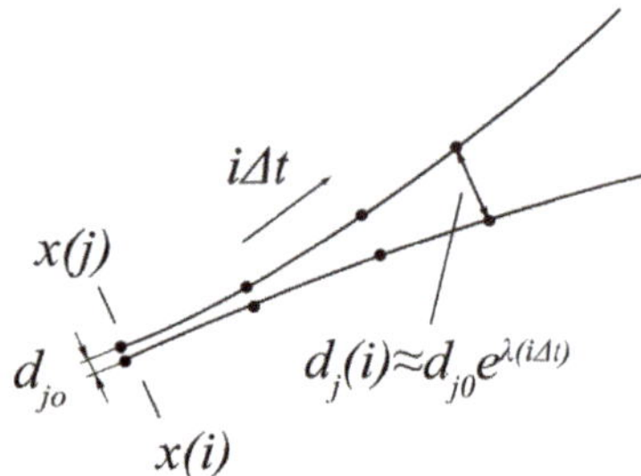

Figure 8. Example of divergent paths.

Such behaviour characterises chaotic dynamics, which, as it was revealed in [18,19], is related to saccadic latency. Fixations provide more homogeneous results than the saccadic latency and saccade, which is visible in the second part of the time series (512–1024 ms). The segments representing this component contain comparable values in the case of each of the studied metrics.

The above analysis demonstrate that given the presented MMs, it is feasible to indicate time series scopes within which three eye movement events take place.

4.2. Classification Performance

Promising visual inspection of the MMs found also a partial confirmation in the classification results, although overall accuracy was rather low. This outcome is a consequence of the fact that at least half of the samples—corresponding to a fixation—have very similar characteristics. It is especially visible in the charts presented in Figure 6. However, this is not the only reason, as for saccadic latency and fixation the similar entropy values were estimated (see Figure 4). Therefore, entropy metrics, at the lowest level of the MMs, do not introduce the distinction among these two events. The situation changes at the second level, where the fixation parts differ more noticeably from the first signal scopes.

Analysis of the results presented in Tables 1 and 3 shows that if saccadic latency is being searched for, the best approach is the usage of the LLE measure, for which accuracy at levels 76% and 84%, respectively, were achieved. Taking this event duration into account (approximately 120-200 ms), the segment of size equal to 128 ms appears to be sufficient. The advantage of the LLE over ApEn and FuzEn results from the fact that, while all three metrics quantifies irregularity in a time series, the first one is also able to describe how a system evolves in a particular period of its evolution. For saccadic latency segments, it indicates divergence, whereas during saccade, convergence. In the case of fixation, behaviour changes from convergent to chaotic and conversely. Therefore, this quantity is less efficient in fixation scopes, in which both entropy types yielded interchangeably better results. Yet, changing the measure during the classification process, depending on the segment under consideration seems not to be a proper solution. Thus, it is more reasonable to apply a feature vector consisting of all measures. Confirmation for such an approach provides the analysis of the third columns in Tables 2 and 4. The accuracy achieved for the combination of ApEn, FuzEn and the LLE was better than for any quantity except for the LLE. The decrease in its efficiency was approximately 3% for the longer segment (128 ms) and 10% for the shorter one (64 ms).

Generally, more satisfying classification outcomes were obtained for feature vectors with samples based on 128 ms-segments (Tables 3 and 4). As was previously noted, one of the reasons is the lower number of classes in the training and testing sets. However, it may also stem from the fact that for longer segments, entropy values for different eye movement events are more distinguishable.

Finally, the comparison between findings presented in [22] and provided here indicates that the usage of more than one measure may advance the events detection efficiency (Table 5). It especially refers to the shorter segments at the beginning of the eye movement signal. Thus, by applying the presented approach, better classification has become possible for the higher resolution of signal segments, even when using fewer than ten features. The high efficacy of the saccadic latency detection facilitates recognizing the following saccade, while finding saccade period opens the possibilities for more detailed exploration in this scope.

However, the results of this method still need further improvement, which can be achieved by, for example, extending the feature vector and usage of other classifiers. Additionally, decreasing the number of fixation segments is worth considering. In this research, the studied fixation period was approximately 700 ms and could be shortened by 200–300 ms. Furthermore, within the explored eye movement data, there was a lack of movement called *smooth pursuit*, when the eyes follow a moving object. The dynamics of such a motion should also be explored.

The suggestions mentioned above are plans for future work. Moreover, because the proposed method was verified only for one dataset gathered with the usage of a particular eye tracker type, other signal recordings are intended to be analysed.

5. Conclusions

The methods for nonlinear time series analysis were used in the presented research in order to reveal eye movement signal characteristics. The primary motivation for these investigations was to find patterns, which can characterise the main components of eye movements: fixation, saccade and saccadic latency. Their proper recognition in registered eye movement signals supports studies in many areas using eye tracking technology. Fixations uncover the gazed areas of observed scenes, while saccades reveal changes in gaze positions. Sometimes, an eye movement is caused by the motion of an object belonging to the scene. Such a situation introduces an additional period in the signal called saccadic latency: the time needed for the brain to react to the motion of the observed stimulus. Discovering this period is the indication that the subsequent signal segment will represent a saccade in the direction of a new stimulus position. Recognizing this element is, in turn, equivalent to revealing the subsequent fixation. Although there are some commonly used methods for eye movement events detection, new solutions, which will provide more accurate event detection, are still being sought.

In the presented approach, three measures were used: approximate entropy, fuzzy entropy and the Largest Lyapunov Exponent, for which multilevel maps, being their time-scale decomposition, were defined. To check if the estimated characteristics may be useful in distinguishing eye movement components, these structures were applied in the classification process. It produced an improvement in the accuracy, for saccadic latency and saccade, when compared to previously conducted studies using eye movement dynamics. Nonetheless, further investigations towards more precise results are planned in the future.

Author Contributions: Conceptualization, K.H.; Data curation, P.K.; Formal analysis, K.H.; Methodology, K.H.; Software, K.H. and P.K.; Validation, K.H. and P.K.; Writing – original draft, K.H.; Writing – review & editing, K.H. and P.K. All authors have read and agreed to the published version of the manuscript.

Funding: This research was funded by the Statutory Research funds of Faculty of Automatic Control, Electronics and Computer Science, Grant Number BK/2020.

Conflicts of Interest: The authors declare no conflict of interest.

Abbreviations

The following abbreviations are used in this manuscript:

Abbreviation	Description	Value
N_{pp}	Number of point positions	29
N_p	Number of participants	24
N_{spp}	Number of sessions per participant	2
N_{ps}	Number of participant sessions	44
N_{EMs}	Number of eye movement series	1276
N_{EMr}	Number of eye movement recordings	1024
N_{ss}	Number of samples in the smallest segment of eye movement recordings	64

References

1. Peng, C.K.; Buldyrev, S.V.; Havlin, S.; Simons, M.; Stanley, H.E.; Goldberger, A.L. Mosaic organization of DNA nucleotides. *Phys. Rev. E* **1994**, *49*, 1685. [CrossRef]
2. Golinska, A.K. Detrended fluctuation analysis (DFA) in biomedical signal processing: Selected examples. *Stud. Logic Grammar Rhetoric* **2012**, *29*, 107–115.
3. Márton, L.; Brassai, S.; Bakó, L.; Losonczi, L. Detrended Fluctuation Analysis of EEG Signals. *Procedia Technol.* **2014**, *12*, 125–132. [CrossRef]
4. Vargas-Luna, M.; Huerta-Franco, M.R.; Montes, J.B. Evaluation of the Cardiac Response to Psychological Stress by Short-Term ECG Recordings: Heart Rate Variability and Detrended Fluctuation Analysis. In Proceedings of the World Congress on Medical Physics and Biomedical Engineering, Beijing, China, 26–31 May 2012; Long, M., Ed.; Springer: Berlin, Germany, 2013; pp. 333–335.
5. Jiao, D.; Wang, Z.; Li, J.; Feng, F.; Hou, F. The chaotic characteristics detection based on multifractal detrended fluctuation analysis of the elderly 12-lead ECG signals. *Physica A* **2020**, *540*, 123234. doi:10.1016/j.physa.2019.123234. [CrossRef]
6. Borowska, M. Entropy-Based Algorithms in the Analysis of Biomedical Signals. *Stud. Logic Grammar Rhetoric* **2015**, *43*, 21–32. [CrossRef]
7. Kantz, H.; Schreiber, T. Human ECG - nonlinear deterministic versus stochastic aspects. *Sci. Measur. Technol. IEE Proc.* **1998**, *145*, 279–284. [CrossRef]
8. Chen, C.; Jin, Y.; Lo, I.L.; Zhao, H.; Sun, B.; Zhao, Q.; Zheng, J.; Zhang, X.D. Complexity Change in Cardiovascular Disease. *Int. J. Biol. Sci.* **2017**, *13*, 1320–1328. doi:10.7150/ijbs.19462. [CrossRef]
9. Abásolo, D.; Hornero, R.; Espino, P. Approximate Entropy of EEG Background Activity in Alzheimer's Disease Patients. *Intell. Auto Soft Comput.* **2009**, *15*, 591–603. doi:10.1080/10798587.2009.10643051. [CrossRef]
10. Yentes, J.; Hunt, N.; Schmid, K.; Kaipust, J.; McGrath, D.; Stergiou, N. The appropriate use of approximate entropy and sample entropy with short data sets. *Ann. Biomed. Eng.* **2013**, *41*, 349–365. doi:10.1007/s10439-012-0668-3. [CrossRef]
11. Cao, Z.; Lin, C.T. Inherent fuzzy entropy for the improvement of EEG complexity evaluation. *IEEE Trans. Fuzzy Syst.* **2018**, *26*, 1032–1035. [CrossRef]
12. Azami, H.; Li, P.; Arnold, S.E.; Escudero, J.; Humeau-Heurtier, A. Fuzzy Entropy Metrics for the Analysis of Biomedical Signals: Assessment and Comparison. *IEEE Access* **2019**, *7*, 104833–1048. [CrossRef]
13. Fell, J.; Röschke, J.; Beckmann, P. Deterministic chaos and the first positive Lyapunov exponent: A nonlinear analysis of the human electroencephalogram during sleep. *Biol. Cyber.* **1993**, *69*, 139–146. doi:10.1007/BF00226197. [CrossRef]
14. Lee, C.; Jo, H.; Yoo, S. Non-linear Analysis of Single Electroencephalography (EEG) for Sleep-Related Healthcare Applications. *Healthcare Inf. Res.* **2010**, *16*, 46–51. doi:10.4258/hir.2010.16.1.46. [CrossRef]
15. Bob, P.; Kukleta, M.; Riecansky, I.; Susta, M.; Kukumberg, P.; Jagla, F. Chaotic EEG patterns during recall of stressful memory related to panic attack. *Physiol. Res.* **2006**, *55*, S113–S119.

16. Übeyli, E.D. Detecting variabilities of ECG signals by Lyapunov exponents. *Neural Comput. Appl.* **2009**, *18*, 653–662. doi:10.1007/s00521-008-0229-8. [CrossRef]
17. Josiński, H.; Świtoński, A.; Michalczuk, A.; Grabiec, P.; Pawlyta, M.; Wojciechowski, K. Assessment of Local Dynamic Stability in Gait Based on Univariate and Multivariate Time Series. *Comput. Math. Methods Med.* **2019**, *2019*, 1–13. doi:10.1155/2019/6917658. [CrossRef]
18. Harezlak, K. Eye movement dynamics during imposed fixations. *Inf. Sci.* **2017**, *384*, 249–262. doi:http://dx.doi.org/10.1016/j.ins.2016.07.074. [CrossRef]
19. Harezlak, K.; Kasprowski, P. Understanding Eye Movement Signal Characteristics Based on Their Dynamical and Fractal Features. *Sensors* **2019**, *19*, 626. doi:10.3390/s19030626. [CrossRef]
20. Astefanoaei, C.; Creanga, D.; Pretegiani, E.; Optican, L.; Rufa, A. Dynamical Complexity Analysis of Saccadic Eye Movements In Two Different Psychological Conditions. *Rom. Rep. Phys.* **2014**, *66*, 1038–1055.
21. Harezlak, K.; Kasprowski, P. Searching for Chaos Evidence in Eye Movement Signals. *Entropy* **2018**, *20*, 32. doi:10.3390/e20010032. [CrossRef]
22. Katarzyna Harezlak, Dariusz R. Augustyn, P.K. An Analysis of Entropy-Based Eye Movement Events Detection. *Entropy* **2019**, *21*, 107. doi:10.3390/e21020107. [CrossRef]
23. Otero-Millan, J.; Troncoso, X.G.; Macknik, S.L.; Serrano-Pedraza, I.; Martinez-Conde, S. Saccades and microsaccades during visual fixation, exploration, and search: Foundations for a common saccadic generator. *J. Vision* **2008**, *8*, 21. [CrossRef]
24. Martinez-Conde, S.; Otero-Millan, J.; Macknik, S.L. The impact of microsaccades on vision: Towards a unified theory of saccadic function. *Nat. Rev. Neurosci.* **2013**, *14*, 83–96. [CrossRef]
25. Darrien, J.H.; Herd, K.; Starling, L.J.; Rosenberg, J.R.; Morrison, J.D. An analysis of the dependence of saccadic latency on target position and target characteristics in human subjects. *BMC Neurosci.* **2001**, *2*, 1–8. [CrossRef]
26. Holmqvist, K.; Nyström, M.; Andersson, R.; Dewhurst, R.; Jarodzka, H.; Van de Weijer, J. *Eye Tracking: A Comprehensive Guide to Methods and Measures*; Oxford University Press: Oxford, UK, 2011.
27. Salvucci, D.D.; Goldberg, J.H. Identifying Fixations and Saccades in Eye-tracking Protocols. In Proceedings of the 2000 Symposium on Eye Tracking Research & Applications, Gardens, FL, USA, 6–8 November 2000.
28. Zemblys, R.; Niehorster, D.C.; Komogortsev, O.; Holmqvist, K. Using machine learning to detect events in eye-tracking data. *Behav. Res. Methods* **2018**, *50*, 160–181. doi:10.3758/s13428-017-0860-3. [CrossRef]
29. Jazz Novo. Ober Consulting. Available online: http://www.ober-consulting.com/9/lang/1/ (accessed on 30 December 2019).
30. Pincus, S.M. Approximate entropy as a measure of system complexity. *PNAS* **1991**, *88*, 2297–2301, doi:10.1073/pnas.88.6.2297. [CrossRef]
31. Pincus, S.M.; Huang, W.M. Approximate entropy: Statistical properties and applications. *Commun. Stat. Theory Methods* **1992**, *21*, 3061–3077. [CrossRef]
32. Chen, W.; Wang, Z.; Xie, H.; Yu, W. Characterization of Surface EMG Signal Based on Fuzzy Entropy. *IEEE Trans. Neural Syst. Rehabil. Eng.* **2007**, *15*, 266–272. doi:10.1109/TNSRE.2007.897025. [CrossRef]
33. Takens, F. Detecting strange attractors in turbulence. In *Dynamical Systems and Turbulence, Warwick 1980*; Springer: Berlin, Germany, 1981, pp. 366–381.
34. Rosenstein, M.T.; Collins, J.J.; De Luca, C.J. A Practical Method for Calculating Largest Lyapunov Exponents from Small Data Sets. *Physica D* **1993**, *65*, 117–134. [CrossRef]
35. Kennel, M.B.; Brown, R.; Abarbanel, H.D.I. Determining embedding dimension for phase-space reconstruction using a geometrical construction. *Phys. Rev. A* **1992**, *45*, 3403–3411. [CrossRef]
36. Fraser, A.M.; Swinney, H.L. Independent coordinates for strange attractors from mutual information. *Phys. Rev. A* **1986**, *33*, 1134–1140. [CrossRef]

Article

Complexity-Based Measures of Postural Sway during Walking at Different Speeds and Durations Using Multiscale Entropy

Ben-Yi Liau [1], Fu-Lien Wu [2], Chi-Wen Lung [2,3], Xueyan Zhang [2], Xiaoling Wang [2] and Yih-Kuen Jan [2,4,*]

[1] Department of Biomedical Engineering, Hung Kuang University, Taichung 443, Taiwan; byliau@sunrise.hk.edu.tw
[2] Rehabilitation Engineering Lab, Department of Kinesiology and Community Health, University of Illinois at Urbana-Champaign, Champaign, IL 61820, USA; fulienwu@illinois.edu (F.-L.W.); lung@illinois.edu (C.-W.L.); mrkx88@126.com (X.Z.); xw39@illinois.edu (X.W.)
[3] Department of Creative Product Design, Asia University, Taichung 443, Taiwan
[4] Beijing Advanced Innovation Center for Biomedical Engineering, Beihang University, Beijing 10000, China
* Correspondence: yjan@illinois.edu; Tel.: +217-300-7253; Fax: +217-333-2766

Received: 15 October 2019; Accepted: 13 November 2019; Published: 16 November 2019

Abstract: Participation in various physical activities requires successful postural control in response to the changes in position of our body. It is important to assess postural control for early detection of falls and foot injuries. Walking at various speeds and for various durations is essential in daily physical activities. The purpose of this study was to evaluate the changes in complexity of the center of pressure (COP) during walking at different speeds and for different durations. In this study, a total of 12 participants were recruited for walking at two speeds (slow at 3 km/h and moderate at 6 km/h) for two durations (10 and 20 min). An insole-type plantar pressure measurement system was used to measure and calculate COP as participants walked on a treadmill. Multiscale entropy (MSE) was used to quantify the complexity of COP. Our results showed that the complexity of COP significantly decreased ($p < 0.05$) after 20 min of walking (complexity index, CI = −3.51) compared to 10 min of walking (CI = −3.20) while walking at 3 km/h, but not at 6 km/h. Our results also showed that the complexity index of COP indicated a significant difference ($p < 0.05$) between walking at speeds of 3 km/h (CI = −3.2) and 6 km/h (CI = −3.6) at the walking duration of 10 min, but not at 20 min. This study demonstrated an interaction between walking speeds and walking durations on the complexity of COP.

Keywords: center of pressure; complexity; falls; multiscale entropy; postural control

1. Introduction

Postural control is a complex process based on continuous and interactive information between our sensorimotor system and the environment [1–4]. Participation in various physical activities requires successful postural control in response to changes in the position of our body [5,6]. People with sensorimotor impairments due to disease or aging lose their postural control, which results in falls and foot injury. It is important for clinicians to assess the postural control function changes in at-risk populations to prevent fall-related injuries.

Various measures have been used to assess postural control functions, including center of mass, center of gravity, and center of pressure (COP) [1–4]. Among these measures, the analysis of COP has been widely used for quantitative assessments of postural control due to its easy measurement. COP is the point of the ground reaction force (GRF) vector acting on the plantar foot, which starts from heel strike, moves forward, and ends near the toes at toe-off. COP is a signal that reflects ankle torque and COP trajectory progression is time-varying and represents the muscle force for body stabilization [7]. Common COP-based indexes include the COP trajectory, the front/back maximum offset, and the left/right maximum offset. Research studies show that the shorter the trajectory and the smaller the offset, the better the balance [8–10].

Recently, researchers have shown that these traditional measurements may not fully characterize the changes of posture control associated with aging or pathological conditions [11,12]. This new evidence shows that traditional linear characteristics of COP trajectories may not be sensitive to detecting balance changes and the use of nonlinear analysis (e.g., complexity) may be more sensitive for detecting pathological changes of postural control and balance [11,12]. The complexity in a physiological system reflects the adaptability of the system to various stimuli [13–16]. Such complexity assessments have been widely introduced in various pathophysiological assessments for better diagnoses and assessments. In postural control, complexity of COP refers to a person's ability to adapt to various postural needs and a higher complexity of COP may imply a better capability for postural adaptability. Researchers have used entropy to study the complexity changes of COP in pathological conditions [17–21]. Purkayastha et al. suggested that the nonlinear complexity of COP might be more sensitive to detect insufficient postural control abilities, compared to linear variables [22]. Moreover, complexity-related variables of COP showed a better reliability than the COP velocity [23]. However, the COP data of most studies were derived from static double leg standing or single leg standing. The tasks adopted by these studies might not reflect the characteristics of postural control in daily life. Only two studies, conducted by Mei et al., compared the complexity of COP displacement, velocity, and acceleration, extracted from walking with sample entropy, concluding that sample entropy can be a possible evaluation method to identify different foot types [24,25]. Yet, they collected the plantar pressure data from a pressure plate system rather than an in-shoe plantar pressure system, which required participants to perform many walking trials to record enough data points.

Multiscale entropy (MSE) is a method to quantify complexity (i.e., regularity) of a time series (e.g., COP time series) at multiple scales [26–28]. MSE has been proposed to fully quantify complexity in the COP trajectories [12]. Although the complexity index of COP has shown some promising effects in evaluating static postural control or in identifying potential pathologies, the entropy used in these studies may not reveal different scales of complexity [29]. This might be the reason that the use of entropy in assessing COP is still not consistent and inconclusive. Understanding comprehensive postural control mechanism during walking by analyzing the MSE of COP may provide researchers and clinicians with valuable information to develop an appropriate postural control assessment or to design an optimal walking training program.

Participation in physical activities and activities of daily living involve various walking speeds and durations. However, it is unclear how engaging in activities with various walking speeds and durations affect postural sway and stability. This is particularly important in populations at risk for falls and foot injuries [30]. Although traditional COP analysis provides the information of COP displacement and velocity, the traditional analysis may lose the information of dynamical complexity due to the non-stationary nature of COP [31]. Therefore, we performed a nonlinear analysis of COP by multiscale entropy to observe changes in complexity during walking at different speeds and for different durations. The purpose of this study was to assess the complexity of COP during walking at different speeds and for different durations using MSE. To the best of our knowledge, this is the first study to explore the changes of complexity of COP during walking at different speeds and for different durations.

2. Methods

2.1. Subjects

Healthy subjects between 18 and 45 years of age were recruited from the university and nearby community. Exclusion criteria were active foot ulcers, diabetes, vascular diseases, hypertension, the inability to walk for 20 min independently, the inability to walking on the treadmill independently, or the use of vasoactive medications. Each subject signed the informed consent approved by the University of Illinois at Urbana-Champaign Institutional Review Board (#19225) before the screening and experimental procedures.

2.2. Experimental Procedures

All examinations were performed in the Rehabilitation Engineering Laboratory at the University of Illinois at Urbana-Champaign. Room temperature was fixed at 24 ± 2 °C. All subjects relaxed in the supine position for at least 30 min prior to testing to stabilize the baseline blood flow level and acclimate themselves to the room temperature. Two speeds (slow walking at 3 km/h and moderate walking at 6 km/h) [32], and two durations (10 min and 20 min) were tested in this study. All participants were asked to walk with an appropriate pair of shoes at a speed of 3 km/h on a treadmill at the first week, and 6 km/h at the second week. For each week, participants were randomly assigned into the 10 min or 20 min walking duration first, and the other duration later, with a balanced crossover design. For avoiding carryover effects and muscle fatigue, participants were allowed to rest for at least 20 min between 10 min and 20 min walking trials.

2.3. Center of Pressure Measurements

An F-scan system (Tekscan, South Boston, MA) was used to measure the plantar pressure data of the right foot in standardized shoes [33] during walking on the treadmill. Each F-scan in-shoe sensor contains 960 sensing pixels (sensels). The sensor was placed between the subject's sock and the insole of the shoe. A subject wore the sensors inside the shoes for 3–5 min of walking before the walking experiment. Each time, the sensor was calibrated according to the manufacturer's instructions. Data were sampled at 200 Hz. Center of pressure data were extracted from the Tekscan software.

2.4. Multiscale Entropy Analysis

The definition of "entropy" in thermodynamics is a measure of unavailability in a closed system, which describes the degree of the disorder state in a system [34]. Several algorithms based on the concept of entropy have been applied to measure the complexity of physiological signals. Multiscale entropy (MSE) is one of the methods developed by Costa et al. [13,28,34]. MSE uses the algorithm of "sample entropy" to estimate the regularity in different time scales, based on the approximate entropy, to assess the complexity degree [13,28,34].

First, a one-dimensional discrete time series $\{x_1,x_2, \ldots .x_n\}$ is reconstructed by the scale factor "τ" to be coarse-grained time series with different time scales. Each element of $y_j^{(\tau)}$ is according to Equation (1), as follows:

$$y_j^{(\tau)} = \frac{1}{\tau}\sum\nolimits_{i\,=\,(j-1)\tau+1}^{j\tau} x_i, 1 \le j \le \frac{N}{\tau}. \tag{1}$$

By the coarse-grained procedure, sample entropy (SE) can be estimated by the scale factor "τ". SE is defined by Equation (2), as follows:

$$\mathrm{SE}(m,\gamma,\tau) = -\log\frac{A_\tau}{B_\tau}, \tag{2}$$

where m is the a template vector of length, A is the number of template vector pairs having $d[x_{m+1}(i),x_{m+1}(j)] < \gamma$, and B is the number of template vector pairs having $d[x_m(i),x_m(j)] < \gamma$.

The complexity index (CI) can be estimated from SE by Equation (3), which is the summation of SE from scale factor 1 to the maximum.

$$CI = \sum_{i=1}^{\tau} SE(i). \quad (3)$$

Regarding the required data length for MSE (sensitivity), it has been suggested that 200 data points per window are needed to elicit consistent SE values, though some studies used 600 data points at the longest time scale [29]. Previous research showed that the shortest coarse-grained time series should be 300 data points [30]. In this study, we have almost 1,000 data points that are sufficient for MSE analyses. Two-way repeated measures ANOVA was used to examine the interaction between the walking speeds and walking durations on the complexity of COP. Paired t tests were used to examine the statistical significance. Cohen's effective size was also calculated to estimate the effect size [35]. The significance level was set as 0.05. All analyses were performed using MATLAB R2017b (MathWorks, Inc., Natick, MA, USA).

3. Results

Twelve healthy subjects (5 men, 7 women) were recruited in this study. The demographic data were as follows (mean ± standard deviation): Age, 25.7 ± 5.5 years (range 21–42 years); height, 171.3 ± 8.5 cm (range 157–188 cm); weight, 64.2 ± 13.5 kg (range 50–93 kg); and BMI, 22.04 ± 2.93 kg/m^2 (range 19.05–28.39 kg/m^2). Regarding BMI, 10 subjects were in the healthy range and 2 subjects were in the overweight range; none were in the obese range.

Figure 1 shows that time scale 10 of MSE and the complexity index of COP significantly decreased ($p < 0.05$) after 20 min of walking (complexity index, CI= −3.51) compared to 10 min of walking (CI = −3.20) at the waking speed of 3 km/h, but not at the walking speed of 6 km/h.

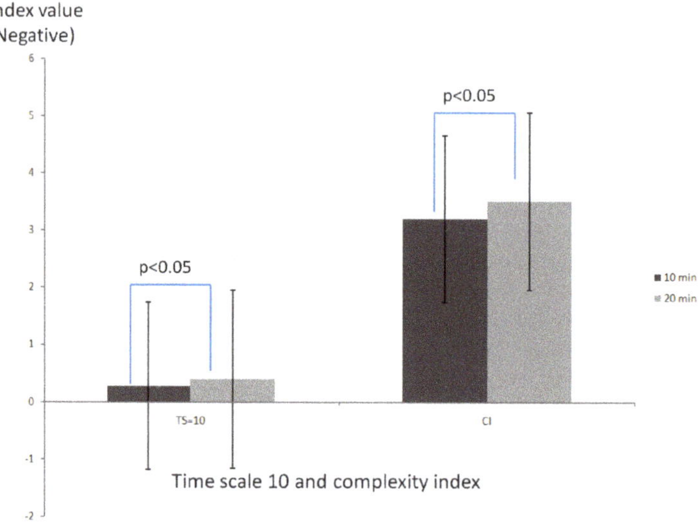

Figure 1. Comparison of complexity of center of pressure (COP) between 10 min and 20 min walking durations while walking at a speed of 3 km/h.

Figure 2 shows the complexity index of COP indicated a significant difference ($p < 0.05$) between walking at speeds of 3 km/h (CI= −3.2) and 6 km/h (CI = −3.6) at the walking duration of 10 min, but not at the walking duration of 20 min. The x-axis is the time scale and complexity index.

The y-axis is the index value. In this comparison, several time scales and complexity indexes revealed significant differences.

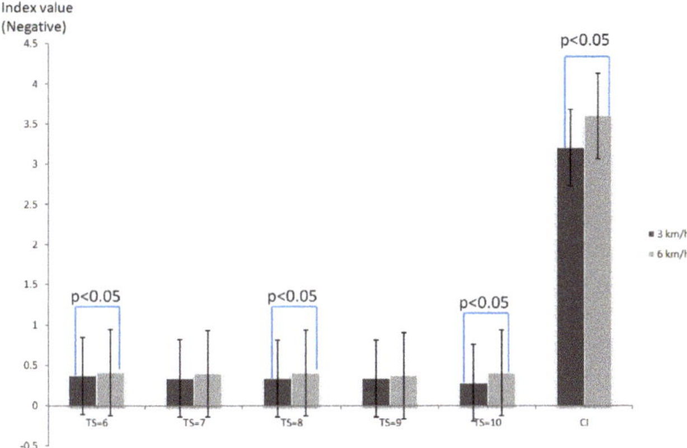

Figure 2. Comparison of complexity of COP between 3 km/h and 6 km/h walking speeds while walking for 10 min.

Two-way repeated measures ANOVA was performed to assess the interaction effect between the walking speed and walking duration on the CI of COP. The results showed the interaction effect (p-value < 0.01).

The size effect was calculated using Cohen's d algorithm. It showed that the value of Cohen's d estimated by the different walking durations while walking at 3 km/h was larger than 0.8 (large effect size). In addition, the value of Cohen's d, estimated by the different walking speeds for the 10 min walking duration, was larger than 0.9 (large effect size).

Figure 3 shows the trend of complexity of the COP between 3 km/h and 6 km/h walking speeds while walking for 10 and 20 min. It can be found that the trend of complexity of COP while walking for 10 min is separated more obviously with increasing the time scale until time scale 10, even in the contrasting trend, which is significantly different. However, while walking for 20 min, the effect of walking speed on the complexity of COP is mild due to there being no significant differences in each time scale.

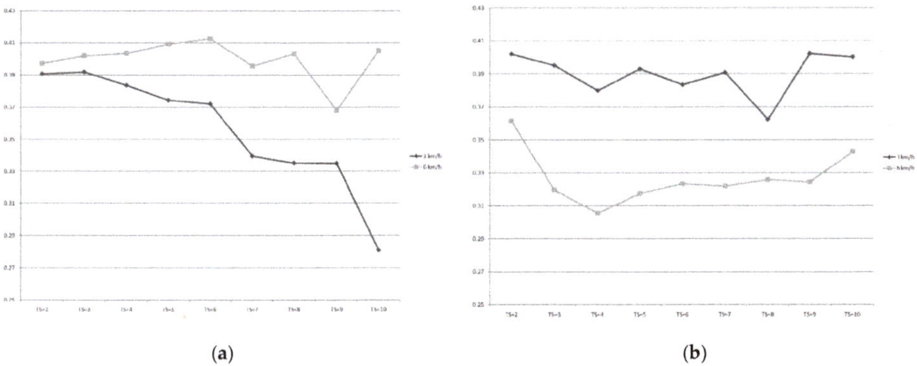

Figure 3. (**a**) Trend of complexity of COP between 3 km/h and 6 km/h in each time scale of walking speeds while walking for 10 min. (**b**) Trend of complexity of the COP between 3 km/h and 6 km/h in each time scale while walking for 20 min.

4. Discussion

The main finding of this study is that we demonstrated that both walking speed and walking duration factors may significantly affect the complexity of COP. Our results also showed that complexity of COP significantly decreased after 20 min of walking compared to 10 min of walking, and while walking at 3 km/h, but not at 6 km/h. Our results also showed that the complexity of COP indicated a significant difference between walking at speeds of 3 km/h and 6 km/h at the walking duration of 10 min, but not at 20 min.

Our study demonstrated that there is no significant difference between CI during 10 and 20 min walking at 6 km/h, which is significantly different from walking at 3 km/h. Few studies have investigated how people change their gait patterns or adjust their posture after various durations of walking [36,37]. Thomas and colleagues studied static postural sway every 5-min interval over 35 min of walking at three different speeds, suggesting that physically active young adults demonstrated the ability of static postural adaptation at the beginning of the fast walking and then maintained the ability until the end of the task [37]. Stolwijk et al. measured plantar pressure data from people who completed a 4-day marching event, indicating that after marching people tend to walk with less roll-off movement and more heal loading. The authors argued that altered gait patterns after prolonged walking might be from lower limb muscle fatigue [36]. In this study, the non-significant difference at 6 km/h between 10 min and 20 min walking durations could be due to an easier adaptation to the walking speed at 6 km/h.

In contrast to 3 km/h, our results showed that slow walking at 3 km/h for 20 min significantly decreased the complexity compared to 10 min. This could be due to lower leg muscle fatigue. The finding indicates that walking at a slower speed (3 km/h) may easily decrease the complexity of COP compared to walking at a higher speed (6 km/h). This may imply that slower walking speed may not improve postural control. Our finding is supported by the literature, which showed that higher walking speed and shorter step length may improve stability [38,39]. Older people, usually considered to have poor postural control, tend to adopt the strategy of a slower speed (slower than preferred walking speed) and a shorter step length than young people during walking [40–42]. Short step length and low walking speed, which were exhibited by the elder people, however, may cause opposite effects on postural stability during walking. Future studies need to investigate the effects of walking speeds on postural control by different variables in order to to broaden the understanding about the differences between fallers and non-fallers.

A study conducted by McClymont et al. showed that the mean square error, one of the variability measures, of plantar pressures significantly increased when walking faster, while the coefficient of variation, another standard variability measure, was not related to walking speed [43]. Lu et al. performed a series of studies to determine the effects of different gait speeds controlled by treadmill or participants on the inclination angle of center of mass and the COP, suggesting that the minimum value of the range of the frontal inclination angle was detected at the preferred walking speed, compared to that at lower or higher speeds, either in over-ground walking or in treadmill walking [44,45]. These findings indicate that people might control their posture with the least effort during walking at their preferred speed. Thus, that older people choose a non-preferred, slower walking speed may not help them improve balance, but rather it may cause them to lose balance and postural control ability. Moreover, people could adjust their posture quickly within 5 min of fast walking; thus, this may be the reason that we could not observe any difference between CI during 10 and 20 min of walking at 6 km/h.

The use of MSE on COP may provide a new window to assess postural control. The MSE approach could evaluate how disorders or diseases affect postural oscillations via time series scales that reflect the time domain dynamics of the complex network of our postural control system. Our results are consistent with previous reports of complexity studies. With the same time scale factor and complexity index, a longer walking duration would decrease the complexity index (CI) significantly ($p < 0.05$). This is consistent with clinical observations that walking for a longer time could decrease balance capacity; in our case, 20 min compared to 10 min of walking. With the walking velocity factor, only the results of 3 km/h vs. 6 km/h after 10 min of walking showed a significant difference ($p < 0.05$). As walking faster could reduce the CI value of COP, it can be speculated that walking faster would also reduce the walking balance. Traditional measures may be useful in patients with major impairments in postural control, while complexity analysis may be more sensitive for early detection of postural deficits in various pathological conditions, as well as in the elderly.

Various methods have been developed to identify people at risks for falls and foot injuries. Among them, COP analysis is an easy to use method in various settings. COP, or parameters associated with plantar pressures, has been used as an evaluation tool to quantify the postural control function. Previous research showed that peak plantar pressures of most plantar regions generally increased with increased walking speed [46–50]. Elevated plantar pressures may result in abnormally cumulated stress over the plantar soft tissue. COP, an important index calculated from plantar pressure data, is able to assess postural control and to predict the risk of falls [51–55]. Previous researchers have examined the trajectories, velocity, or variability of COP to evaluate the dynamic postural control during walking at different speeds. Chiu et al. found that the COP velocity significantly increased with increased walking speed, whereas there was no significant difference in the COP progression angle between different walking speeds [56]. In this study, we used MSE to analyze COP while waking at different speeds and for different durations and demonstrated that this method could be useful to complement current methods for assessing postural control by providing information about the complexity status of postural control.

In this study, the age of participants ranged from 21 to 42 years. According to the literature [57,58], people aged above 60 years demonstrate significant changes in COP progression and variability from the younger population. Age between 18 and 60 years is not a significant factor affecting COP. Thus, the age factor may not significantly affect our results. Regarding the range of body mass index in this study, 10 subjects were in the healthy range and 2 subjects were in the overweight range; none were in the obese range. Obesity (BMI above 30 kg/m^2) may increase postural sway during quiet standing [59–62] or gait initiation [63]. We believe that the body weight and BMI factors may not significantly affect the current results.

Our findings have several clinical implications. Patients with impaired sensorimotor function tend to walk at a slower speed to improve postural control. In this study, we demonstrated that walking duration may be a more significant factor for postural control, compared to walking speeds. This finding suggests that people at risk for falls and foot injuries may need to avoid walking for a longer duration, for which it may be the time (walking for a long time) that causes them to fall. Investigating the COP alteration when individuals perform walking at different speeds and for different durations for various activities is essential to understand the postural control mechanism so that clinical practitioners can design an appropriate exercise program for those who have balance control problems.

There are limitations of this study. First, this study only recruited healthy subjects, which limits its generalization to patients at risks for falls and foot injury. However, our study demonstrated that, even in healthy subjects, the MSE of COP revealed significant changes between walking for 10 and 20 min. Our method could be applied to study pathological changes in patients. Second, a study design with longer walking durations and faster walking speeds should be investigated to further understand how walking speeds and durations affect the complexity of COP. Third, the COP data were calculated using the in-shoe pressure measurement system. The accuracy of the in-shoe pressure measurement system is not as accurate as the force plate system. However, our in-shoe system provides the advantage of less restricted data collection requirements for activities.

5. Conclusions

Using MSE analysis of COP during walking at different speeds and durations, we found that both walking speed and walking duration factors significantly affect the complexity of COP. The MSE analysis of COP may provide new information to evaluate postural control in people at risk for falls and foot injuries.

Author Contributions: Conceptualization, B.-Y.L., W.-F.W., and Y.-K.J.; methodology, B.-Y.L., W.-F.W., and Y.-K.J.; formal analysis, B.-Y.L., F.-L.W., and C.-W.L.; resources, B.-Y.L. and Y.-K.J.; data curation, W.-F.W. and Y.-K.J.; writing—original draft preparation, B.-Y.L., F.-L.W., C.-W.L., X.Z., X.W., and Y.-K.J.; writing—review and editing, B.-Y.L., F.-L.W., C.-W.L., X.Z., X.W., and Y.-K.J.; supervision, Y.-K.J.; project administration, Y.-K.J.

Conflicts of Interest: The authors declare no conflict of interest.

References

1. Gow, B.J.; Peng, C.K.; Wayne, P.M.; Ahn, A.C. Multiscale Entropy Analysis of Center-of-Pressure Dynamics in Human Postural Control: Methodological Considerations. *Entropy* **2015**, *17*, 7926–7947. [CrossRef]
2. Hu, X.Y.; Zhao, J.; Peng, D.S.; Sun, Z.L.; Qu, X.D. Estimation of Foot Plantar Center of Pressure Trajectories with Low-Cost Instrumented Insoles Using an Individual-Specific Nonlinear Model. *Sensors* **2018**, *18*, 421. [CrossRef] [PubMed]
3. Huang, C.W.; Sue, P.D.; Abbod, M.F.; Jiang, B.C.; Shieh, J.S. Measuring Center of Pressure Signals to Quantify Human Balance Using Multivariate Multiscale Entropy by Designing a Force Platform. *Sensors* **2013**, *13*, 10151–10166. [CrossRef] [PubMed]
4. Wei, Q.; Liu, D.H.; Wang, K.H.; Liu, Q.; Abbod, M.F.; Jiang, B.C.; Chen, K.P.; Wu, C.; Shieh, J.S. Multivariate Multiscale Entropy Applied to Center of Pressure Signals Analysis: An Effect of Vibration Stimulation of Shoes. *Entropy* **2012**, *14*, 2157–2172. [CrossRef]
5. Grant, P.M.; Dall, P.M.; Kerr, A. Daily and hourly frequency of the sit to stand movement in older adults: A comparison of day hospital, rehabilitation ward and community living groups. *Aging Clin. Exp. Res.* **2011**, *23*, 437–444. [CrossRef]
6. Bestaven, E.; Petit, J.; Robert, B.; Dehail, P. Center of pressure path during Sit-to-walk tasks in young and elderly humans. *Ann. Phys. Rehabi. Med.* **2013**, *56*, 644–651. [CrossRef]
7. Baratto, L.; Morasso, P.G.; Re, C.; Spada, G. A new look at posturographic analysis in the clinical context: Sway-density versus other parameterization techniques. *Motor. Control.* **2002**, *6*, 246–270. [CrossRef]
8. Bauer, C.; Groger, I.; Rupprecht, R.; Gassmann, K.G. Intrasession Reliability of Force Platform Parameters in Community-Dwelling Older Adults. *Arch. Phys. Med. Rehabil.* **2008**, *89*, 1977–1982. [CrossRef]
9. Corriveau, H.; Hebert, R.; Raiche, M.; Prince, F. Evaluation of postural stability in the elderly with stroke. *Arch. Phys. Med. Rehabil.* **2004**, *85*, 1095–1101. [CrossRef]
10. Paillard, T.; Noe, F. Techniques and Methods for Testing the Postural Function in Healthy and Pathological Subjects. *BioMed Res. Int.* **2015**. [CrossRef]
11. Fino, P.C.; Mojdehi, A.R.; Adjerid, K.; Habibi, M.; Lockhart, T.E.; Ross, S.D. Comparing Postural Stability Entropy Analyses to Differentiate Fallers and Non-fallers. *Ann. Biomed. Eng.* **2016**, *44*, 1636–1645. [CrossRef] [PubMed]
12. Busa, M.A.; Jones, S.L.; Hamill, J.; van Emmerik, R.E.A. Multiscale entropy identifies differences in complexity in postural control in women with multiple sclerosis. *Gait Posture* **2016**, *45*, 7–11. [CrossRef] [PubMed]
13. Liao, F.Y.; Yang, T.D.; Wu, F.L.; Cao, C.M.; Mohamed, A.; Jan, Y.K. Using Multiscale Entropy to Assess the Efficacy of Local Cooling on Reactive Hyperemia in People with a Spinal Cord Injury. *Entropy* **2019**, *21*, 90. [CrossRef]
14. Liao, F.Y.; Struck, B.D.; MacRobert, M.; Jan, Y.K. Multifractal analysis of nonlinear complexity of sacral skin blood flow oscillations in older adults. *Med. Biol. Eng. Comput.* **2011**, *49*, 925–934. [CrossRef] [PubMed]
15. Liao, F.Y.; Jan, Y.K. Using multifractal detrended fluctuation analysis to assess sacral skin blood flow oscillations in people with spinal cord injury. *J. Rehabil. Res. Dev.* **2011**, *48*, 787–799. [CrossRef] [PubMed]
16. Jan, Y.K.; Liao, F.Y.; Cheing, G.L.Y.; Pu, F.; Ren, W.Y.; Choi, H.M.C. Differences in skin blood flow oscillations between the plantar and dorsal foot in people with diabetes mellitus and peripheral neuropathy. *Microvasc. Res.* **2019**, *122*, 45–51. [CrossRef] [PubMed]
17. Isableu, B.; Hlavackova, P.; Diot, B.; Vuillerme, N. Regularity of Center of Pressure Trajectories in Expert Gymnasts during Bipedal Closed-Eyes Quiet Standing. *Front. Hum. Neurosci.* **2017**, *11*, 317. [CrossRef]
18. Jelinek, H.F.; Khalaf, K.; Poilvet, J.; Khandoker, A.H.; Heale, L.; Donnan, L. The Effect of Ankle Support on Lower Limb Kinematics During the Y-Balance Test Using Non-linear Dynamic Measures. *Front. Physiol.* **2019**, *10*, 935. [CrossRef]

19. Raffalt, P.C.; Chrysanthou, M.; Duda, G.N.; Agres, A.N. Dynamics of postural control in individuals with ankle instability: Effect of visual input and orthotic use. *Comput. Biol. Med.* **2019**, *110*, 120–126. [CrossRef]
20. Mirahmadi, M.; Karimi, M.T.; Esrafilian, A. An Evaluation of the Effect of Vision on Standing Stability in the Early Stage of Parkinson's Disease. *Eur. Neurol.* **2018**, *80*, 261–267. [CrossRef]
21. Li, Y.; Mache, M.A.; Todd, T.A. Complexity of Center of Pressure in Postural Control for Children with Autism Spectrum Disorders was Partially Compromised. *J. Appl. Biomech.* **2019**, *35*, 190–195. [CrossRef] [PubMed]
22. Purkayastha, S.; Adair, H.; Woodruff, A.; Ryan, L.J.; Williams, B.; James, E.; Bell, K.R. Balance Testing Following Concussion: Postural Sway versus Complexity Index. *PM&R* **2019**. [CrossRef]
23. Bizovska, L.; Janura, M.; Svoboda, Z.; Cerny, M.; Krohova, J.; Smondrk, M. Intra- and inter-session reliability of traditional and entropy-based variables describing stance on a wobble board. *Med. Eng. Phys.* **2017**, *50*, 29–34. [CrossRef] [PubMed]
24. Mei, Z.; Ivanov, K.; Zhao, G.; Li, H.; Wang, L. An explorative investigation of functional differences in plantar center of pressure of four foot types using sample entropy method. *Med. Biol. Eng. Comput.* **2017**, *55*, 537–548. [CrossRef]
25. Mei, Z.; Zhao, G.; Ivanov, K.; Guo, Y.; Zhu, Q.; Zhou, Y.; Wang, L. Sample entropy characteristics of movement for four foot types based on plantar centre of pressure during stance phase. *Biomed. Eng. Online* **2013**, *12*, 101. [CrossRef]
26. Costa, M.; Goldberger, A.L.; Peng, C.K. Multiscale entropy analysis of complex physiologic time series. *Phys. Rev. Lett.* **2002**, *89*, 068102. [CrossRef]
27. Humeau-Heurtier, A. The Multiscale Entropy Algorithm and Its Variants: A Review. *Entropy* **2015**, *17*, 3110–3123. [CrossRef]
28. Liao, F.Y.; Cheing, G.L.Y.; Ren, W.Y.; Jain, S.; Jan, Y.K. Application of Multiscale Entropy in Assessing Plantar Skin Blood Flow Dynamics in Diabetics with Peripheral Neuropathy. *Entropy* **2018**, *20*, 127. [CrossRef]
29. Busa, M.A.; van Emmerik, R.E.A. Multiscale entropy: A tool for understanding the complexity of postural control. *J. Sport Health Sci.* **2016**, *5*, 44–51. [CrossRef]
30. Costa, M.; Priplata, A.A.; Lipsitz, L.A.; Wu, Z.; Huang, N.E.; Goldberger, A.L.; Peng, C.K. Noise and poise: Enhancement of postural complexity in the elderly with a stochastic-resonance-based therapy. *EPL* **2007**, *77*, 68008. [CrossRef]
31. Pajala, S.; Era, P.; Koskenvuo, M.; Kaprio, J.; Tormakangas, T.; Rantanen, T. Force platform balance measures as predictors of indoor and outdoor falls in community-dwelling women aged 63-76 years. *J. Gerontol. A* **2008**, *63*, 171–178. [CrossRef] [PubMed]
32. Bohannon, R.W.; Andrews, A.W. Normal walking speed: A descriptive meta-analysis. *Physiotherapy* **2011**, *97*, 182–189. [CrossRef] [PubMed]
33. Lung, C.W.; Hsiao-Wecksler, E.T.; Bums, S.; Lin, F.; Jan, Y.K. Quantifying Dynamic Changes in Plantar Pressure Gradient in Diabetics with Peripheral Neuropathy. *Front. Bioeng. Biotech.* **2016**, *4*, 54. [CrossRef] [PubMed]
34. Costa, M.; Goldberger, A.L.; Peng, C.K. Multiscale entropy analysis of biological signals. *Phys. Rev. E* **2005**, *71*, 021906. [CrossRef]
35. Cohen, J. *Statistical Power Analysis for the Behavioral Sciences*, 2nd ed.; Lawrence Earlbaum Associates: Hillsdale, NJ, USA, 1988.
36. Stolwijk, N.M.; Duysens, J.; Louwerens, J.W.; Keijsers, N.L. Plantar pressure changes after long-distance walking. *Med. Sci. Sports Exerc.* **2010**, *42*, 2264–2272. [CrossRef]
37. Thomas, K.S.; VanLunen, B.L.; Morrison, S. Changes in postural sway as a function of prolonged walking. *Eur. J. Appl. Physiol.* **2013**, *113*, 497–508. [CrossRef]
38. Bhatt, T.; Wening, J.D.; Pai, Y.C. Influence of gait speed on stability: Recovery from anterior slips and compensatory stepping. *Gait Posture* **2005**, *21*, 146–156. [CrossRef]
39. McAndrew Young, P.M.; Dingwell, J.B. Voluntarily changing step length or step width affects dynamic stability of human walking. *Gait Posture* **2012**, *35*, 472–477. [CrossRef]
40. Menz, H.B.; Lord, S.R.; Fitzpatrick, R.C. Age-related differences in walking stability. *Age Ageing* **2003**, *32*, 137–142. [CrossRef]

41. Stief, F.; Schäfer, A.; Vogt, L.; Kirchner, M.; Hubscher, M.; Thiel, C.; Banzer, W.; Meurer, A. Differences in Gait Performance, Quadriceps Strength, and Physical Activity Between Fallers and Nonfallers in Women with Osteoporosis. *J. Aging Phys. Act.* **2016**, *24*, 430–434. [CrossRef]
42. DeVita, P.; Hortobagyi, T. Age causes a redistribution of joint torques and powers during gait. *J. Appl. Physiol.* **2000**, *88*, 1804–1811. [CrossRef] [PubMed]
43. McClymont, J.; Pataky, T.C.; Crompton, R.H.; Savage, R.; Bates, K.T. The nature of functional variability in plantar pressure during a range of controlled walking speeds. *R. Soc. Open Sci.* **2016**, *3*, 160369. [CrossRef] [PubMed]
44. Lu, H.L.; Kuo, M.Y.; Chang, C.F.; Lu, T.W.; Hong, S.W. Effects of gait speed on the body's center of mass motion relative to the center of pressure during over-ground walking. *Hum. Mov. Sci.* **2017**, *54*, 354–362. [CrossRef] [PubMed]
45. Lu, H.L.; Lu, T.W.; Lin, H.C.; Hsieh, H.J.; Chan, W.P. Effects of belt speed on the body's center of mass motion relative to the center of pressure during treadmill walking. *Gait Posture* **2017**, *51*, 109–115. [CrossRef]
46. Rosenbaum, D.; Hautmann, S.; Gold, M.; Claes, L. Effects of walking speed on plantar pressure patterns and hindfoot angular motion. *Gait Posture* **1994**, *2*, 191–197. [CrossRef]
47. Segal, A.; Rohr, E.; Orendurff, M.; Shofer, J.; O'Brien, M.; Sangeorzan, B. The effect of walking speed on peak plantar pressure. *Foot Ankle Int.* **2004**, *25*, 926–933. [CrossRef]
48. Chung, M.J.; Wang, M.J. Gender and walking speed effects on plantar pressure distribution for adults aged 20–60 years. *Ergonomics* **2012**, *55*, 194–200. [CrossRef]
49. Burnfield, J.M.; Few, C.D.; Mohamed, O.S.; Perry, J. The influence of walking speed and footwear on plantar pressures in older adults. *Clin. Biomech.* **2004**, *19*, 78–84. [CrossRef]
50. Warren, G.L.; Maher, R.M.; Higbie, E.J. Temporal patterns of plantar pressures and lower-leg muscle activity during walking: Effect of speed. *Gait Posture* **2004**, *19*, 91–100. [CrossRef]
51. Melzer, I.; Benjuya, N.; Kaplanski, J. Postural stability in the elderly: A comparison between fallers and non-fallers. *Age Ageing* **2004**, *33*, 602–607. [CrossRef]
52. Merlo, A.; Zemp, D.; Zanda, E.; Rocchi, S.; Meroni, F.; Tettamanti, M.; Recchia, A.; Lucca, U.; Quadri, P. Postural stability and history of falls in cognitively able older adults: The Canton Ticino study. *Gait Posture* **2012**, *36*, 662–666. [CrossRef] [PubMed]
53. Laughton, C.A.; Slavin, M.; Katdare, K.; Nolan, L.; Bean, J.F.; Kerrigan, D.C.; Phillips, E.; Lipsitz, L.A.; Collins, J.J. Aging, muscle activity, and balance control: Physiologic changes associated with balance impairment. *Gait Posture* **2003**, *18*, 101–108. [CrossRef]
54. Prieto, T.E.; Myklebust, J.B.; Hoffmann, R.G.; Lovett, E.G.; Myklebust, B.M. Measures of postural steadiness: Differences between healthy young and elderly adults. *IEEE Trans. Biomed. Eng.* **1996**, *43*, 956–966. [CrossRef] [PubMed]
55. Qiu, H.; Xiong, S. Center-of-pressure based postural sway measures: Reliability and ability to distinguish between age, fear of falling and fall history. *Int. J. Ind. Ergon.* **2015**, *47*, 37–44. [CrossRef]
56. Chiu, M.C.; Wu, H.C.; Chang, L.Y. Gait speed and gender effects on center of pressure progression during normal walking. *Gait Posture* **2013**, *37*, 43–48. [CrossRef] [PubMed]
57. Sole, G.; Pataky, T.; Sole, C.C.; Hale, L.; Milosavljevic, S. Age-related plantar centre of pressure trajectory changes during barefoot walking. *Gait Posture* **2017**, *57*, 188–192. [CrossRef]
58. Chiu, M.C.; Wu, H.C.; Chang, L.Y.; Wu, M.H. Center of pressure progression characteristics under the plantar region for elderly adults. *Gait Posture* **2013**, *37*, 408–412. [CrossRef]
59. Hita-Contreras, F.; Martinez-Amat, A.; Lomas-Vega, R.; Alvarez, P.; Mendoza, N.; Romero-Franco, N.; Aranega, A. Relationship of body mass index and body fat distribution with postural balance and risk of falls in Spanish postmenopausal women. *Menopause* **2013**, *20*, 202–208. [CrossRef]
60. Menegoni, F.; Galli, M.; Tacchini, E.; Vismara, L.; Cavigioli, M.; Capodaglio, P. Gender-specific effect of obesity on balance. *Obesity* **2009**, *17*, 1951–1956. [CrossRef]
61. Hue, O.; Simoneau, M.; Marcotte, J.; Berrigan, F.; Dore, J.; Marceau, P.; Marceau, S.; Tremblay, A.; Teasdale, N. Body weight is a strong predictor of postural stability. *Gait Posture* **2007**, *26*, 32–38. [CrossRef]

62. Handrigan, G.; Hue, O.; Simoneau, M.; Corbeil, P.; Marceau, P.; Marceau, S.; Tremblay, A.; Teasdale, N. Weight loss and muscular strength affect static balance control. *Int. J. Obes.* **2010**, *34*, 936–942. [CrossRef] [PubMed]
63. Cau, N.; Cimolin, V.; Galli, M.; Precilios, H.; Tacchini, E.; Santovito, C.; Capodaglio, P. Center of pressure displacements during gait initiation in individuals with obesity. *J. Neuroeng. Rehabil.* **2014**, *11*, 82. [CrossRef] [PubMed]

Article

Development of Postural Stability Index to Distinguish Different Stability States

Nurul Retno Nurwulan [1,*], Bernard C. Jiang [1] and Vera Novak [2]

[1] Department of Industrial Management, National Taiwan University of Science and Technology, 43 Keelung Road Section 4, Daan District, Taipei City 10607, Taiwan; bcjiang@mail.ntust.edu.tw

[2] Department of Neurology, Beth Israel Deaconess Medical Center, Harvard Medical School, Boston, MA 02215, USA; vnovak@bidmc.harvard.edu

* Correspondence: nurul.retno@gmail.com

Received: 22 February 2019; Accepted: 20 March 2019; Published: 22 March 2019

Abstract: A key factor for fall prevention involves understanding the pathophysiology of stability. This study proposes the postural stability index (PSI), which is a novel measure to quantify different stability states on healthy subjects. The results of the x-, y-, and z-axes of the acceleration signals were analyzed from 10 healthy young adults and 10 healthy older adults under three conditions as follows: Normal walking, walking with obstacles, and fall-like motions. The ensemble empirical mode decomposition (EEMD) was used to reconstruct the acceleration signal data. Wearable accelerometers were located on the ankles and knees of the subjects. The PSI indicated a decreasing trend of its values from normal walking to the fall-like motions. Free-walking data were used to determine the stability based on the PSI. The segmented free-walking data indicated changes in the stability states that suggested that the PSI is potentially helpful in quantifying gait stability.

Keywords: postural stability index; stability states; ensemble empirical mode decomposition; gait

1. Introduction

Each year, approximately one third of older adults aged >65 years experience falls [1]. Falls can cause physical injuries that may lower the quality of life and health or even lead to death in older adults. Additionally, falls are a common cause of psychological stress and extended hospitalization for older adults [2]. Falls are potentially related to the difficulty in maintaining walking stability. Therefore, quantifying walking stability is potentially key to preventing falls.

Previous studies focused on the fall detection method [3–5]. However, the indexes can only be used to count the number of falls. They detect falls based on the sudden changes in a series of data. Despite the ability to detect falls, the aforementioned methods merely count the number of falls that occur in a given time. Although falls are caused by poor postural stability, it is difficult to determine the stability of the movement if the fall does not occur.

A previous study used dynamic stability to determine postural stability [6]. The acceleration root mean square (RMS), step and stride regularity, and sample entropy (SampEn) are parameters used to measure dynamic stability. Dynamic stability can be used to identify asymmetrical stability patterns related to ageing and illness [7–9]. However, the approach may not be as sensitive in discriminating stability patterns in healthy subjects.

In 2014, Cui et al. [10] used ensemble empirical mode decomposition (EEMD) to construct the step stability index (SSI) to discriminate between the walking patterns of fallers from non-fallers. When compared to fall detection methods, the SSI is a more promising approach to evaluate human postural stability because it evaluates the characteristics of movements irrespective of whether or not subjects fall during the evaluation. The SSI can be used to quantify gait dynamics and discriminate non-fallers from fallers although the real question concerns the extent of the stability of the movement of non-fallers. The evaluation of the postural stability of non-fallers may not be as simple as comparing fallers and non-fallers. EEMD was developed to overcome the empirical mode decomposition (EMD) mode mixing issues [11]. EMD decomposes signal data into a set of zero-mean underlying components called intrinsic mode functions (IMF). The main advantage of EMD is that its algorithm depends only on the signal under analysis. However, performance evaluation of EMD compared to discrete wavelet transform (DWT) and wavelet packet decomposition (WPD) showed that the EMD performed the worst in detecting seizures in electroencephalogram (EEG) signals. Machine learning methods, such as random forest (RF), k-Nearest neighbor (k-NN), artificial neural network (ANN), and support vector machines (SVM), were used to predict the accuracy of the EMD, DWT, and WPD [12].

The current study proposed a measure that can distinguish different stability states in healthy subjects using an accelerometer. This study used EEMD and multiscale entropy (MSE) to develop the measure. The performance of EEMD was also evaluated by comparing its performance with the wavelet transform method. The organization of this paper is prepared as follows. The methods section provides the experiment protocol and a brief description of previous postural stability measures as well as the theoretical background behind EEMD, MSE, and wavelet transform methods. Results are then presented in Section 3 and the discussion is in Section 4. Finally, a conclusion section is presented in Section 5.

2. Methods

2.1. Subjects

Ten young adults (24 $\pm$ 0.94 years) and 10 older adults (69 $\pm$ 6.77 years) were recruited. All subjects were free of any postural stability-related disorder based on self-reports. The study was approved by the Institutional Review Board and informed consent forms were obtained from all subjects before their participation. The subjects were asked to perform walking, walking with obstacles, free walking, and fall-like motions (Figure 1). Young adults performed free-falling and fall-like motions while older adults only performed fall-like motions. The subjects were instructed to walk at their own pace regardless of the distance for the duration of 60 s. For the fall-like motions, the subjects were asked to sit on a mattress from the standing-still position. The fall-like motion was repeated 10 times. The subjects were allowed to hold on to the pole in front of them while performing the fall-like motions. The results of the pilot study indicated that the subjects did not feel comfortable wearing safety harnesses while performing fall-like motions. The safety harness made the subjects bounce back during the task. For the free-fall task, subjects were asked to jump down from the standing position on the chair to imitate the free-fall. This free-fall task was repeated five times. Wearable accelerometers were attached to the ankle and knee of each subject with the assumption that the ankle and knee corresponded to the most relevant body parts during the experiment. The acceleration data were acquired at 30 Hz and were imported into Matlab R2016a [13] for feature computation.

2.2. Subject Characteristics

Twenty healthy subjects with no history of falling in the past two years were grouped by age. Table 1 presents the descriptive statistics of the subjects. The Mann Whitney test was used to compare the two groups. The differences between the two groups were all statistically insignificant, except for age.

Table 1. Descriptive statistics of the subjects.

	Young Adults	Older Adults	p-Value
Age (years)	24 (0.88)	70 (5.27)	<0.001 *
Gender (% Female)	50%	60%	0.66
Height (cm)	164 (6.84)	160 (9.93)	0.36
Weight (kg)	59 (10.67)	60 (9.12)	0.79

* p-values were obtained using the Mann Whitney test. For the SSI and the new PSI, higher scores indicate better performance.

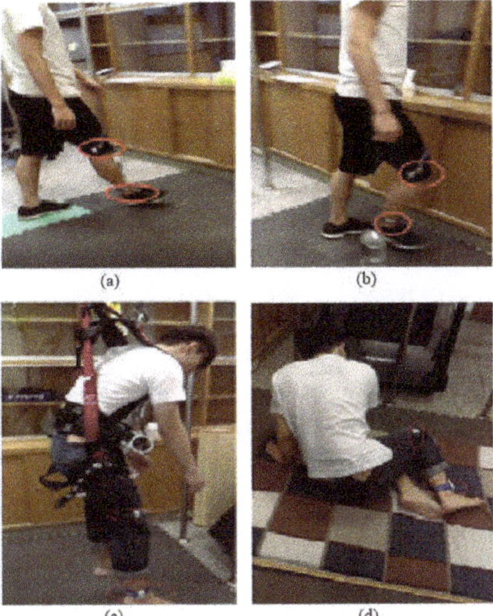

Figure 1. Subject performs the activities: (a) normal walking, (b) walking with obstacles, (c) free-falling, and (d) fall-like motions. The older adults did not perform the free-falling task.

2.3. Ensemble Empirical Mode Decomposition

Ensemble empirical mode decomposition (EEMD) was introduced to overcome the mode mixing phenomenon in empirical mode decomposition (EMD). Given signal intermittence, mode mixing situations typically occur during the EMD decomposition process. Mode mixing occurs when a single intrinsic mode function (IMF) either consists of oscillations with different frequencies or a signal with a similar frequency is in different IMF components. Mode mixing can change the physical meaning of each IMF component by replacing the part of the IMF and driving it to the next IMF, thereby falsely suggesting that different physical processes may exist in an IMF [11].

White noises of the same length were added to the original signal to form a mixture signal, xi(t), and subsequently the EMD decomposition was performed to obtain n layers of IMFs (IMF11, IMF12, . . . , IMF1n). We assumed that the process was conducted for m times, and this is defined as the number of ensemble members. The original signals were subsequently added by m white noises of the same energies to form m mixture signals. In the current study, white noises of an amplitude that was about 0.2 standard deviation of the signals were added [11]. Each mixture signal, xi(t), was decomposed into n layers of IMF components. Thus, there were m x n IMF components, where m denotes the number of ensemble members and n denotes the nth layer of the IMF. The decomposed results were averaged and subsequently each layer of the IMF was calculated as follows:

$$IMF_n = \frac{1}{m}\sum_{i=1}^{m} IMF_n \quad (1)$$

The purpose of adding the white noise of the finite amplitude to the signal was to populate the whole time-frequency space uniformly with the constituting components of different scales. The white noise cancels each other out in the time-space ensemble mean. Thus, only the true and physically meaningful signal can survive in the EEMD [11].

2.4. Wavelet Transform

Wavelet transform decomposes a signal into a set of basic functions by scaling and shifting the mother wavelet function. These basic functions are called wavelets because they wave up and down across the axis, and integrate to zero. Wavelets are highly effective in analyzing non-stationary signals, such as for noise reduction [12]. The discrete wavelet transform (DWT) transforms a discrete time signal to a discrete wavelet representation. It converts an input series, $x_0, x_1, \ldots x_m$, into one high-pass wavelet coefficient series and one low-pass coefficient series.

Wavelet packet decomposition (WPD) is a continuous wavelet transform. The difference between DWT and WPD is how the scale parameter is discretized. The WPD discretizes the scale more finely than DWT. Thus, it gives a better frequency resolution for the decomposed signal [12]. However, the WPD is less stable for signal reconstruction. Whereas the DWT is able to provide perfect reconstruction of the signal upon inversion, and its coefficients can be used to reproduce an exact signal within numerical precision. In the current study, the number of decomposition levels in DWT was selected to be 6, whereas the daubechies4 (db4) mother wavelet function with 4 levels of decomposition was used for WPD [12].

2.5. Multiscale Entropy

Following its introduction by Costa et al. [14], multiscale entropy (MSE) was successfully applied to quantify the complexity of signals in different research fields, such as biomedical [15–19], electroseismic [20], and the vibration of rotary machines [21].

MSE proposed a method to measure complexity by constructing a consecutive coarse-grained time series by averaging a successively increasing number of data points in non-overlapping windows as follows:

$$y_j^{(\tau)} = \frac{1}{\tau}\sum_{i=(j-1)\tau+1}^{j\tau} x_i,\ 1 \le j \le \frac{N}{\tau} \tag{2}$$

where τ denotes the scale factor, and the length of each coarse-grained time series is N/τ. For scale 1, the time series, $\left\{y^{(1)}\right\} = \{x_1, x_2, x_3, \ldots x_N\}$, simply corresponds to the original time series. The length of each coarse-grained time series is equal to the length of the original time series divided by the scale factor, τ. Subsequently, sample entropy (SampEn) is calculated for each of the coarse-grained time series plotted as a function of the scale factor as follows:

$$\boldsymbol{SampEn(m,r,N) = -\ln(A/B)} \tag{3}$$

where A denotes the total number of forward matches of a length, m + 1, and B denotes the total number of template matches of a length, m [22].

The complexity index (CI) is obtained from the total value of the SampEn as a function of the scale factor as follows:

$$CI = \sum_{i=1}^{N} SampEn(i) \tag{4}$$

2.6. Step Stability Index

The step stability index (SSI) was proposed to distinguish the walking patterns of fallers from non-fallers [10]. The magnitude of the acceleration signals is decomposed using EEMD with 8-modes of IMFs. Subsequently, the standard deviations of the IMF1 to IMF4 are used to develop the index as follows:

$$SSI = \frac{SD\ IMF4}{SD\ IMF1 + SD\ IMF2 + SD\ IMF3} \tag{5}$$

where SSI denotes the step stability index, and SD denotes the standard deviation.

The non-fallers exhibited a higher SSI value that indicated a more stable gait pattern because the energy (as measured by the standard deviation) of the component at the step frequency exceeds the energy of higher frequency components that are potentially related to subtle unsteadiness of stepping.

The SSI is used to quantify gait dynamics and discriminates between subjects with and without a history of falls. However, it is potentially not sufficiently sensitive to distinguish the postural stability of healthy subjects. The difference of the postural stability in non-fallers is potentially not as evident. Thus, the SSI may not be adequate to measure the likelihood of falls or to provide a better understanding as to humans' maintenance of stability given that it distinguishes non-fallers from fallers without the ability to understand how postural stability corresponds to the movement of non-fallers.

2.7. Dynamic Stability

Dynamic stability parameters include the ratio of root mean square (RMS), step and stride regularity, and sample entropy (SampEn) [6]. The ratio of RMS was used to capture the variability as obtained from the ratio of each axis RMS relative to the resultant vector RMS. The step and stride regularity were used to capture the consistency of the gait. Step regularity was obtained from the primary dominant unbiased autocorrelation coefficient while stride regularity is the second dominant coefficient [23]. The SampEn was used to capture the periodicity of the gait with a higher value indicating less periodicity [22].

2.8. Machine Learning

Machine learning can be used to evaluate the accuracy of measurement methods. This current study used Fisher's linear discriminant analysis (LDA), artificial neural network (ANN), support vector machines (SVM), and random forest (RF). The LDA is the oldest classifier, and it is used to find a linear combination, which characterizes two or more classes of objects or events. The ANN has been used extensively in classification problems, and it can be understood as a parallel-distributed processing system. This study used multilayer perception (MLP) as it is the most used and powerful neural network [24]. The SVM is a discriminative classifier that can efficiently perform both linear and non-linear classification. The last classifier used in the current study is RF, which is one of the decision-tree's classifiers that improves the classification performance of a single-tree classifier by combining the bootstrap aggregating method and randomization in the selection of segmenting data nodes in the construction of a decision tree [25]. The majority vote of the different decisions provided by each tree constituting the forest is used to assign the new observation vector to a class. In the comparative studies, RF outperformed the other classifiers [26]. However, RF requires large amounts of labeled data to achieve high performance.

3. Results

3.1. Evaluation Using Dynamic Stability

Dynamic stability can distinguish the different stability patterns due to ageing and illness. However, the approach cannot detect the difference between normal walking and walking with obstacles in healthy subjects, as shown in Table 2. For young adults, the significant differences were observed in fall-like motions when compared to normal walking and walking with obstacles for the ratio of RMS, step regularity, and stride regularity in all axes for the ankle data. Conversely, for the knee data, significant differences were observed in fall-like motions when compared to normal walking and walking with obstacles for the ratio of RMS, step regularity in the vertical (VT) and anteroposterior (AP) axes, and stride regularity in the vertical axis. For older adults, significant differences were observed in fall-like motions when compared to normal walking and walking with obstacles for the ratio of RMS in the VT and AP axes, step and stride regularity in all axes, and sample entropy in the VT and mediolateral (ML) axes for the ankle data. Significant differences of the knee data for older adults were only observed in fall-like motions when compared to normal walking and walking with obstacles for the ratio of RMS in the VT and AP axes and sample entropy in the VT axis.

Table 2. Activity differences of the dynamic stability in young and older adults. Data are reported as mean (SD). (**a**) Activity differences of the dynamic stability in young adults. (**b**) Activity differences of the dynamic stability in older adults.

(**a**)

Measures	Axis	NW	WO	FLM	FF	*p*-Value		
						NW-WO	NW-FLM	WO-FLM
Ankle Data								
Ratio of RMS	VT	0.50 (0.05)	0.48 (0.03)	0.64 (0.06)	0.64 (0.06)	0.16	<0.001	<0.001
	ML	0.64 (0.03)	0.62 (0.05)	0.57 (0.13)	0.57 (0.13)	0.51	0.03	0.03
	AP	0.59 (0.05)	0.62 (0.03)	0.48 (0.05)	0.48 (0.05)	0.19	0.04	0.009
Step regularity	VT	0.61 (0.09)	0.58 (0.08)	0.11 (0.06)	0.46 (0.23)	0.53	<0.001	<0.001
	ML	0.61 (0.07)	0.53 (0.09)	0.16 (0.09)	0.48 (0.11)	0.04	<0.001	<0.001
	AP	0.54 (0.10)	0.51 (0.09)	0.20 (0.11)	0.36 (0.09)	0.32	<0.001	<0.001
Stride regularity	VT	0.34 (0.11)	0.27 (0.08)	0.06 (0.05)	0.29 (0.16)	0.18	<0.001	<0.001
	ML	0.44 (0.09)	0.39 (0.07)	0.06 (0.04)	0.27 (0.13)	0.07	<0.001	<0.001
	AP	0.39 (0.10)	0.35 (0.08)	0.05 (0.03)	0.17 (0.10)	0.10	<0.001	<0.001
Sample entropy	VT	0.15 (0.11)	0.10 (0.03)	0.13 (0.06)	0.04 (0.01)	0.19	0.77	0.16
	ML	0.15 (0.13)	0.10 (0.03)	0.13 (0.07)	0.05 (0.03)	0.26	0.76	0.42
	AP	0.18 (0.27)	0.10 (0.03)	0.13 (0.06)	0.06 (0.03)	0.33	0.62	0.14
Knee Data								
Ratio of RMS	VT	0.43 (0.06)	0.45 (0.05)	0.69 (0.04)	0.55 (0.08)	0.20	<0.001	<0.001
	ML	0.66 (0.04)	0.64 (0.03)	0.56 (0.03)	0.65 (0.08)	0.25	<0.001	<0.001
	AP	0.61 (0.04)	0.46 (0.05)	0.51 (0.08)	0.52 (0.04)	0.45	<0.001	<0.001
Step regularity	VT	0.34 (0.14)	0.36 (0.14)	0.22 (0.12)	0.60 (0.12)	0.33	0.07	0.02
	ML	0.57 (0.11)	0.55 (0.10)	0.34 (0.05)	0.52 (0.14)	0.60	<0.001	<0.001
	AP	0.51 (0.10)	0.54 (0.09)	0.33 (0.10)	0.47 (0.15)	0.16	<0.001	<0.001
Stride regularity	VT	0.09 (0.06)	0.13 (0.07)	0.08 (0.04)	0.31 (0.14)	0.05	0.64	0.09
	ML	0.29 (0.11)	0.29 (0.12)	0.11 (0.05)	0.25 (0.13)	0.93	<0.001	<0.001
	AP	0.24 (0.09)	0.27 (0.08)	0.14 (0.08)	0.24 (0.11)	0.11	0.01	0.005
Sample entropy	VT	0.09 (0.05)	0.20 (0.11)	0.11 (0.05)	0.04 (0.03)	0.02	0.30	0.02
	ML	0.07 (0.03)	0.11 (0.05)	0.12 (0.04)	0.06 (0.04)	0.01	0.003	0.58
	AP	0.06 (0.02)	0.14 (0.08)	0.11 (0.04)	0.56 (0.03)	0.02	0.02	0.33

(b)

Measures	Axis	NW	WO	FLM	p-Value		
					NW-WO	NW-FLM	WO-FLM
Ankle Data							
Ratio of RMS	VT	0.52 (0.05)	0.48 (0.05)	0.60 (0.06)	0.08	0.002	<0.001
	ML	0.56 (0.03)	0.58 (0.05)	0.56 (0.06)	0.19	0.39	0.19
	AP	0.64 (0.03)	0.65 (0.04)	0.56 (0.09)	0.30	0.007	0.005
Step regularity	VT	0.48 (0.21)	0.46 (0.22)	0.28 (0.15)	0.41	0.012	0.02
	ML	0.55 (0.14)	0.54 (0.16)	0.41 (0.09)	0.46	0.008	0.02
	AP	0.55 (0.11)	0.53 (0.16)	0.39 (0.15)	0.40	0.01	0.03
Stride regularity	VT	0.21 (0.14)	0.22 (0.15)	0.12 (0.08)	0.46	0.04	0.04
	ML	0.35 (0.17)	0.35 (0.17)	0.15 (0.11)	0.49	0.003	0.003
	AP	0.36 (0.15)	0.35 (0.14)	0.14 (0.14)	0.47	0.002	0.001
Sample entropy	VT	0.09 (0.04)	0.11 (0.04)	0.06 (0.03)	0.05	0.03	<0.001
	ML	0.08 (0.04)	0.10 (0.04)	0.06 (0.03)	0.06	0.10	0.001
	AP	0.11 (0.09)	0.12 (0.06)	0.12 (0.17)	0.38	0.48	0.45
Knee Data							
Ratio of RMS	VT	0.47 (0.06)	0.47 (0.10)	0.60 (0.08)	0.47	<0.001	0.001
	ML	0.58 (0.03)	0.61 (0.04)	0.56 (0.06)	0.08	0.12	0.02
	AP	0.66 (0.03)	0.63 (0.07)	0.56 (0.06)	0.16	<0.001	0.01
Step regularity	VT	0.38 (0.18)	0.43 (0.22)	0.34 (0.18)	0.32	0.31	0.18
	ML	0.56 (0.12)	0.61 (0.10)	0.52 (0.12)	0.13	0.25	0.04
	AP	0.55 (0.14)	0.59 (0.11)	0.53 (0.11)	0.27	0.36	0.14
Stride regularity	VT	0.14 (0.10)	0.21 (0.21)	0.15 (0.12)	0.16	0.38	0.22
	ML	0.30 (0.09)	0.36 (0.12)	0.21 (0.18)	0.14	0.08	0.02
	AP	0.29 (0.12)	0.31 (0.09)	0.30 (0.19)	0.37	0.48	0.42
Sample entropy	VT	0.13 (0.05)	0.14 (0.04)	0.08 (0.04)	0.28	0.02	0.001
	ML	0.09 (0.05)	0.08 (0.02)	0.07 (0.05)	0.24	0.24	0.40
	AP	0.09 (0.05)	0.07 (0.02)	0.08 (0.06)	0.25	0.41	0.37

VT: vertical (y-axis), ML: mediolateral (x-axis), AP: anteroposterior (z-axis), NW: normal walking, WO: walking with obstacles, FLM: fall-like motions, FF: free-falling.

As shown in Table 2, the parameters of dynamic stability evidently did not distinguish the difference between normal walking and walking with obstacles. This was potentially caused by the slight difference in the stability pattern between normal walking and walking with obstacles in healthy subjects.

3.2. Development of Postural Stability Index

Each resultant of the x-, y-, and z-axes of the acceleration signal was decomposed using the 8-modes EEMD. This study used the resultant as opposed to a single axis to avoid information loss. Body sway indicates poor postural stability-ability and can occur in any axis of the signal. The complexity index of each IMF was calculated and subsequently normalized by dividing the complexity index of each IMF by the total complexity index of all IMFs to minimize the influence of individual differences.

The postural stability index (PSI) is defined as follows:

$$PSI = \frac{CI\ of\ IMF3}{CI\ of\ IMF1 + CI\ of\ IMF2 + \ldots + CI\ of\ IMF6} \tag{6}$$

where *CI* denotes the complexity index obtained from the MSE.

The IMF3 was selected as the dominant IMF considering its frequency is closely related to the frequency of walking. Based on past studies, the range of walking frequency was 1.4–2.5 Hz while less stable movement exhibited a lower frequency [27,28]. Table 3 shows the frequency of each IMF component of the walking data. The frequency of IMFs was obtained from the instantaneous frequency of the Hilbert-Huang transform.

Table 3. The average frequencies of intrinsic mode functions (IMFs) of walking data decomposed by ensemble empirical mode decomposition (EEMD).

IMF	Average Frequency
1	6.67 Hz
2	2.49 Hz
3	1.48 Hz
4	0.82 Hz
5	0.36 Hz
6	0.22 Hz

The results of the Pearson correlation showed that the IMF3 was correlated with the gait variability parameters of dynamic stability, as shown in Table 4. Step and stride variability in all axes were selected to represent gait variability. As shown in Table 4, based on the ankle data, IMF3 indicates more correlations than the other IMFs. However, the correlations between IMF3 and dynamic stability parameters were mostly negative. This implies that the IMF3 decreased with increases in gait variability. Conversely, based on the knee data, IMF1 to IMF3 were positively correlated in most dynamic stability parameters except for step and stride regularity in the vertical and anteroposterior axes. Although all three IMFs were correlated with the gait variability parameters, the correlation between dynamic stability and IMF3 exceeded those of the other IMFs.

Table 4. Correlations between the step stability index (SSI), the postural stability index (PSI), the intrinsic mode functions (IMFs), and dynamic stability.

		IMF1	IMF2	IMF3	IMF4	IMF5	IMF6
		r (*p*-value)	r (*p*-value)	r (*p*-value)	r (*p*-value)	r (*p*-value)	r (*p*-value)
Ankle Step Reg-ML	All Subjects	NS	0.28 (<0.001)	0.29 (<0.001)	0.2 (<0.001)	0.23 (<0.001)	0.28 (<0.001)
	Young Adults	NS	−0.26 (<0.001)	−0.21 (<0.001)	−0.24 (<0.001)	NS	NS
	Older Adults	NS	−0.21 (<0.001)	−0.42 (<0.001)	NS	−0.24 (<0.001)	NS
Step Reg-VT	All Subjects	NS	0.21 (<0.001)	NS	NS	NS	NS
	Young Adults	NS	NS	NS	−0.23 (<0.001)	NS	−0.22 (<0.001)
	Older Adults	NS	−0.29 (<0.001)	−0.25 (<0.001)	NS	NS	NS
Step Reg-AP	All Subjects	NS	NS	0.22 (<0.001)	NS	NS	NS
	Young Adults	NS	NS	NS	NS	NS	−0.22 (<0.001)
	Older Adults	NS	NS	−0.37 (<0.001)	NS	−0.28 (<0.001)	NS
Stride Reg-ML	All Subjects	NS	NS	0.27 (<0.001)	NS	NS	NS
	Young Adults	NS	NS	−0.22 (<0.001)	NS	NS	NS
	Older Adults	−0.20 (<0.001)	NS	−0.46 (<0.001)	NS	−0.28 (<0.001)	NS
Stride Reg-VT	All Subjects	NS	0.22 (<0.001)	0.24 (<0.001)	NS	NS	NS
	Young Adults	NS	−0.21 (<0.001)	−0.24 (<0.001)	−0.27 (<0.001)	NS	−0.24 (<0.001)
	Older Adults		−0.36 (<0.001)	−0.28 (<0.001)	NS	NS	NS
Stride Reg-AP	All Subjects	NS	NS	0.25 (<0.001)	NS	NS	NS
	Young Adults	NS	NS	NS	−0.21 (<0.001)	NS	−0.22 (<0.001)
	Older Adults	NS	NS	−0.41 (<0.001)	NS	−0.34 (<0.001)	NS
Knee Step Reg-ML	All Subjects	0.45 (<0.001)	0.36 (<0.001)	0.48 (<0.001)	NS	NS	NS
	Young Adults	0.40 (<0.001)	0.45 (<0.001)	0.41 (<0.001)	0.28 (<0.001)	0.34 (<0.001)	0.33 (<0.001)
	Older Adults	0.39 (<0.001)	0.24 (<0.001)	0.57 (<0.001)	NS	NS	−0.25 (<0.001)
Step Reg-VT	All Subjects	0.24 (<0.001)	0.26 (<0.001)	0.26 (<0.001)	NS	NS	NS
	Young Adults	NS	0.33 (<0.001)	NS	NS	0.26 (<0.001)	NS
	Older Adults	0.20 (<0.001)	0.21 (<0.001)	0.49 (<0.001)	NS	NS	NS
Step Reg-AP	All Subjects	0.38 (<0.001)	0.30 (<0.001)	0.42 (<0.001)	NS	NS	NS
	Young Adults	NS	0.35 (<0.001)	0.44 (<0.001)	0.39 (<0.001)	0.27 (<0.001)	0.20 (<0.001)
	Older Adults	0.43 (<0.001)	0.26 (<0.001)	0.29 (<0.001)	NS	NS	−0.46 (<0.001)
Stride Reg-ML	All Subjects	0.36 (<0.001)	0.28 (<0.001)	0.35 (<0.001)	NS	NS	NS
	Young Adults	0.39 (<0.001)	0.37 (<0.001)	0.29 (<0.001)	0.24 (<0.001)	0.29 (<0.001)	0.25 (<0.001)
	Older Adults	0.23 (<0.001)	0.20 (<0.001)	0.43 (<0.001)	NS	NS	NS
Stride Reg-VT	All Subjects	NS	NS	NS	NS	NS	NS
	Young Adults	NS	NS	NS	NS	NS	−0.20 (<0.001)
	Older Adults	NS	NS	0.32 (<0.001)	NS	0.36 (<0.001)	NS
Stride Reg-AP	All Subjects	0.27 (<0.001)	NS	0.25 (<0.001)	NS	NS	NS
	Young Adults	NS	0.26	0.27	0.25	NS	NS
	Older Adults	0.26 (<0.001)	NS	NS	NS	NS	−0.23 (<0.001)

Reg = regularity, NS = not significant.

4. Discussion

The differences of the SSI and PSI values for both young and older adults are shown in Table 5. The Mann Whitney test was used to compare the differences between the two groups. The significant difference between young and older adults were only observed in normal walking and walking with obstacles with the sensor location corresponding to the ankle. The results between the groups were similar, and this is explained by the fact that both groups were healthy. Unintentional falls were absent during the experiment. The Wilcoxon test was used, and significant differences were observed in all activities for the PSI ($p = 0.005$). Conversely, there was no significant difference for the SSI ($p > 0.05$).

Table 5. Group differences of the SSI and the new PSI measure. Data are reported as mean (SD).

	Young Adults	Older Adults	p-Value
Ankle Data			
SSI Normal walking	0.13 (0.02)	0.12 (0.03)	0.50
SSI Walking + obstacles	0.13 (0.03)	0.12 (0.02)	0.33
SSI Fall-like motions	0.11 (0.02)	0.11 (0.02)	0.50
SSI Free-fall	0.29 (0.06)	N/A	N/A
PSI Normal walking	0.33 (0.04)	0.25 (0.04)	<0.001
PSI Walking + obstacles	0.27 (0.04)	0.19 (0.03)	0.001
PSI Fall-like motions	0.16 (0.03)	0.14 (0.02)	0.16
PSI Free-fall	0.12 (0.03)	N/A	N/A
Knee Data			
SSI Normal walking	0.09 (0.01)	0.10 (0.02)	0.14
SSI Walking + obstacles	0.09 (0.01)	0.10 (0.02)	0.13
SSI Fall-like motions	0.12 (0.02)	0.11 (0.03)	0.60
SSI Free-fall	0.28 (0.04)	N/A	N/A
PSI Normal walking	0.28 (0.06)	0.27 (0.03)	0.60
PSI Walking + obstacles	0.21 (0.06)	0.19 (0.02)	0.41
PSI Fall-like motions	0.14 (0.04)	0.14 (0.02)	0.51
PSI Free-fall	0.10 (0.04)	N/A	N/A

The Wilcoxon test for the SSI on the ankle and knee did not indicate a significant difference between different activities ($p > 0.05$). Conversely, the dynamic stability did not distinguish between normal walking and walking with obstacles (Table 2). These findings are in agreement with the comparisons of the postural stability indexes shown in Table 6.

The aim of this study was to develop a measure to quantify walking stability in healthy subjects. The dynamic stability and the SSI were used as comparisons to evaluate the sensitivity of the PSI. The differences between normal walking and walking with obstacles in healthy subjects were not as evident as the differences between fallers and non-fallers. However, the PSI and step regularity in the ML axis distinguished between normal walking and walking with obstacles.

As shown in Table 6, the PSI using EEMD outperformed the other methods using wavelets decomposition. Step regularity in the ML axis was selected to represent the dynamic stability approach because it is the only parameter that can differentiate between normal walking, walking with obstacles, and fall-like motions (Table 2a). There was no decomposition process for dynamic stability since it used the original signal data.

To determine the location of the sensor that is more sensitive, the results of the PSI for both groups of subjects were grouped based on the ankle and knee. The comparison indicated that the knee group performed better than the ankle group. Thus, the knee corresponds to a better sensor location than the ankle with an accuracy of 82.22% based on the ANN.

Table 6. Accuracy for activity recognition on the ankle and knee.

	LDA			ANN			SVM			RF		
	DS	**SSI**	**PSI**	**DS**	**SSI**	**PSI**	**DS**	**SSI**	**PSI**	**DS**	**SSI**	**PSI**
Ensemble Empirical Mode Decomposition (EEMD)												
Young—Ankle	86.67	66.67	86.67	82.22	60.00	88.89	82.22	57.78	82.22	73.33	53.33	82.22
Young—Knee	68.89	64.44	68.89	68.89	73.33	72.22	68.89	64.44	64.44	74.44	71.11	68.89
Elder—Ankle	66.67	51.11	77.78	60.00	51.11	80.00	60.00	42.22	71.11	48.89	46.67	75.56
Elder—Knee	57.78	53.33	93.33	64.44	53.33	93.33	60.00	53.33	71.11	53.33	42.22	**93.33**
All—Ankle	71.11	54.44	74.44	71.11	53.33	80.00	68.89	57.78	74.44	62.22	51.11	74.44
All—Knee	63.33	60.00	82.22	63.33	60.00	82.22	58.89	62.22	74.44	61.11	50.00	81.11
Discrete Wavelet Transform (DWT)												
Young—Ankle	86.67	80.00	68.89	82.22	80.00	70.00	82.22	71.11	66.67	73.33	75.56	57.78
Young—Knee	68.89	55.56	44.44	68.89	60.00	55.56	68.89	55.56	48.89	74.44	55.56	66.67
Elder—Ankle	66.67	71.11	75.56	60.00	71.11	68.89	60.00	55.56	68.89	48.89	55.56	71.11
Elder—Knee	57.78	62.22	71.11	64.44	57.78	68.89	60.00	51.11	68.89	53.33	44.44	64.44
All—Ankle	71.11	76.67	71.11	71.11	75.56	71.11	68.89	66.67	71.11	62.22	66.67	63.33
All—Knee	63.33	57.78	51.11	63.33	57.78	56.67	58.89	57.78	51.11	61.11	48.89	63.33
Wavelet Packet Decomposition (WPD)												
Young—Ankle	86.67	70.00	56.67	82.22	73.33	61.11	82.22	71.11	60.00	73.33	64.44	60.00
Young—Knee	68.89	46.67	46.67	68.89	48.89	53.33	68.89	60.00	48.89	74.44	51.11	51.11
Elder—Ankle	66.67	60.00	50.00	60.00	55.56	51.11	60.00	60.00	46.67	48.89	51.11	60.00
Elder—Knee	57.78	53.33	51.11	64.44	57.78	53.33	60.00	57.78	55.56	53.33	53.33	55.56
All—Ankle	71.11	63.33	63.33	71.11	61.11	63.33	68.89	63.33	55.56	62.22	57.78	65.56
All—Knee	63.33	57.78	44.44	63.33	58.89	48.89	58.89	55.56	50.00	61.11	56.67	52.22

LDA = Fisher's linear discriminant analysis, ANN = artificial neural network, SVM = support vector machines, RF = random forest, DS = dynamic stability, All = all subjects.

The original SSI (using EEMD) performed the worst in both age groups. The poor performance of the SSI happened because the SSI only considers the vertical axis data. Conversely, the PSI and dynamic stability consider the data in all axes since instability can occur in any axis. However, other than the ankle data of young adults, the accuracies for dynamic stability were generally low. The poor performance of the SSI and dynamic stability for older adults potentially occurred because those methods could not distinguish between the different activities of healthy subjects. Thus, the activities were not classified correctly.

Interestingly, the performance of the PSI dropped when using DWT and WPD as the decomposition method. Conversely, the performance of the SSI increased although it was not as good as the performance of the PSI with EEMD. Other than that, the performance of the ankle improved while the performance of the knee decreased. The poor performance of wavelet decomposition compared to EEMD might be because of the short data records (60 s). Other than that, EEMD was able to estimate the subtle changes that were obtained via the first temporal derivative of the phase angle time series, scaled by the sampling rate. The decomposition by wavelet was not optimal because of frequency smoothing and it assumed frequency stationarity during the time span of the wavelet. The results in the current study are different from a previous study evaluating the performance of EMD and wavelets [12], because the EMD in the previous study could not determine the number of IMFs in the signal and there was mode mixing issues in the EMD.

In contrast to the previous study [12], the decompositions by WPD resulted in the lowest accuracy compared to EEMD and DWT. The DWT performed better than WPD in the current study, and this may be a result of differences in the sampling frequency; 30 Hz in this study and 256 Hz in the previous study [12]. Further, the nature of gait and brain signals are different. Another study by Barralon et al. [29] showed that decomposition using a discrete wavelet was more efficient than a continuous wavelet for gait signals with a 20 Hz sampling rate.

Walking Stability Determination

To better quantify the different stability states, the PSI values of the normal walking, walking with obstacles, and fall-like motions were evaluated. The PSI of the normal walking was used to determine the postural stability limit of each subject with the assumption that normal walking was at least 80% of the upper limit. Subsequently, the scales were developed by normalizing each PSI of the less stable movement to the upper limit value. There were small differences between young and older adults. This indicated that the MSE can eliminate the range variations between young and older adults. The normalization was calculated individually, and thus the similarity in the ranges was not equal to the same index values for all subject groups. The values of the MSE in young subjects were generally higher than those in older adults although the percentage of each movement when compared to the upper limit of postural stability for both young and older groups, were significantly similar. Therefore, the stability scales for both young and older adults can be unified as follows.

80–100%: Stable.

70–79%: Fairly stable, requires minor attention.

45–69%: Unstable, requires high attention.

<45%: Danger, may cause fall.

To determine the walking stability of the subjects, the free-walking data were evaluated using the stability index. The evaluation was divided into two parts, namely general stability determination and segmented stability state determination. In the general stability determination, the stability was determined from the data across the entire 60 s period. Conversely, in the segmented determination, stability was evaluated every 10 s.

Twenty subjects (10 young adults, 10 older adults) were asked to perform a free-walking task to determine the stability of their walk in normal circumstances. The subjects were instructed to move as they liked without any intervention. As shown in Table 6, there was one unstable movement for the ankle and five unstable movements for the knee. During the experiment, the young adults were in a fit condition and did not exhibit any problems during the free-walking performance, and they tended to walk normally in almost a straight line. However, the older adults tended to move around the circle. However, only subject E3 admitted feeling dizzy while performing the free-walking task. The results for the subject, E3, with a sensor on the knee, confirmed the real condition of the subject during the experiment.

With respect to the subject, E7, complaints related to the free-walking task were absent. However, this subject walked slower when compared to that in the normal walking and walking with obstacles tasks. This subject also paused several times during his free-walking task. This potentially occurred because the instruction involved walking freely as per the subjects' wishes, and thus the subject, E7, was potentially not sufficiently motivated to perform his best in the task.

Other unstable movements were detected on the knee for subjects, Y3, Y5, E1, and E3. With respect to the situations during the experiment, the results did not indicate those subjects performed poorly in the free-walking task. With respect to Table 6, the knee data performed better than the ankle data. Thus, the knee is a more reliable sensor location since the accuracies on the knee exceeded those on the ankle. This is potentially the reason why the stability determination of the free-walking task indicated that more unstable movements are detected on knee. The acceleration values varied, although the same type of sensor was used to simultaneously detect similar movements on a subject. The placement of the sensor on the body of the subject significantly affected the performance of the accelerometer. The placement on the ankle was potentially less sensitive than that on the knee.

Although the stability of a movement can be considered as stable in general, it is possible that the postural stability quality of the movement is not the same all the time. Less stable movement can occur in a particular time interval. To evaluate the quality of the movement, the free-walking data of the subjects, Y3, Y5, E1, E3, and E7, were further analyzed. The sensor location on the knee was more accurate, and thus only the knee data were analyzed to determine the stability states. Thus, 60 s of walking data was divided into six segments, and postural stability was evaluated every 10 s, as shown in Table 7.

Table 7. General walking stability determination.

Subject	Upper Limit		Free Walking		Normalization		Category	
	Ankle	Knee	Ankle	Knee	Ankle	Knee	Ankle	Knee
Y1	0.44	0.33	0.39	0.28	88.46%	84.19%	S	S
Y2	0.33	0.39	0.25	0.30	76.39%	77.44%	F	F
Y3	0.35	0.49	0.26	0.25	73.76%	51.41%	F	U
Y4	0.41	0.40	0.34	0.35	83.16%	87.98%	S	S
Y5	0.46	0.35	0.34	0.24	73.50%	67.00%	F	U
Y6	0.40	0.27	0.35	0.23	88.50%	85.11%	S	S
Y7	0.45	0.28	0.35	0.24	78.66%	85.00%	F	S
Y8	0.39	0.38	0.31	0.32	80.78%	83.99%	S	S
Y9	0.44	0.22	0.35	0.21	79.05%	93.97%	F	S
Y10	0.44	0.32	0.37	0.26	84.07%	81.90%	S	S
E1	0.36	0.32	0.31	0.21	85.92%	65.40%	S	U
E2	0.35	0.30	0.29	0.27	82.42%	89.88%	S	S
E3	0.24	0.35	0.17	0.18	70.93%	51.53%	F	U
E4	0.28	0.43	0.24	0.34	84.16%	80.47%	S	S
E5	0.27	0.33	0.24	0.29	90.19%	89.20%	S	S
E6	0.26	0.37	0.23	0.29	88.71%	76.25%	S	F
E7	0.38	0.31	0.16	0.12	41.20%	39.34%	D	D
E8	0.29	0.30	0.25	0.25	85.35%	82.50%	S	S
E9	0.32	0.29	0.26	0.22	79.38%	77.60%	F	F
E10	0.32	0.33	0.28	0.27	87.47%	83.05%	S	S

Category = stability category, S = stable, F = fairly stable, U = unstable, D = danger, Y = young adults, E = older adults.

As shown in Table 8, changes occurred in the postural stability within 60 s of walking for all subjects. For example, the general stability state for the subject, Y3, was unstable. However, based on the segmented stability states, the movement was not always unstable. The first 10 s was stable and, subsequently, the stability state dropped to one in the dangerous category. Fortunately, the stability state improved to unstable then alternated back and forth between the danger and unstable category. The segmented stability state indicated that the subject, Y3, evidently attempted to maintain stability to avoid falling.

Table 8. Segmented walking stability determination.

Time Interval	Y3		Y5		E1		E3		E7	
	Norm	Cat	Norm	Cat	Norm	Cat	Norm	Cat	Norm	Cat
0–10s	84.92%	S	94.64%	S	85.21%	S	12.00%	D	55.50%	U
10–20 s	36.78%	D	56.15%	U	83.43%	S	44.00%	D	44.22%	D
20–30 s	46.30%	U	90.36%	S	85.90%	S	37.35%	D	54.63%	U
30–40 s	36.08%	D	71.90%	F	36.28%	D	80.34%	S	11.25%	D
40–50 s	66.52%	U	70.51%	F	78.99%	F	33.22%	D	13.77%	D
50–60 s	42.01%	D	44.28%	D	46.21%	U	68.95%	U	23.84%	D

Norm = normalization, Cat = stability category, S = stable, F = fairly stable, U = unstable, D = danger, Y3 = young subject#3, Y5 = young subject#5, E1 = older adult#1, E3 = older adult#3, E7 = older adult#7.

The PSI was able to evaluate the postural stability states of movement. However, given the individual differences in human movement, it was only relevant to individually evaluate the movement. The approach represented the characteristics of the stability of a particular subject. However, it was not possible to use the PSI to immediately analyze the movement. Stable state data are required to categorize the movement stability of the subject in question.

5. Conclusions

Quantification of human stability is required to understand the mechanism underlying balance control. The contribution of this study involves providing a novel measure of the postural stability index (PSI) to distinguish between different postural stability states in healthy individuals. The PSI discriminated between normal walking and walking with obstacles in healthy subjects while SSI and dynamic stability did not. A previous study indicated that the SSI differentiates non-fallers from fallers. However, the SSI did not capture the differences between two walking tasks in healthy subjects. This potentially occurred because the differences were excessively small, and the SSI algorithm only used vertical acceleration data in the evaluation. Conversely, the PSI used the resultant of the three-axes acceleration data by assuming that postural sway can occur in any axis.

The present study involved several limitations as follows: (1) The present study used normal walking data to determine the upper postural stability limit. Therefore, the PSI cannot be used to quantify stability without existing normal walking data; (2) the present study used six IMFs to develop the PSI, and thus future studies are required to examine the applicability of the PSI for other decompositions with a different number of IMFs; (3) the present study used wearable accelerometers, and thus future studies are necessary to investigate the effect of the sensitivity of the accelerometers on the PSI; and (4) with respect to older adults who did not perform free-falls due to safety concerns, a past study compared the intentional and unintentional falls and indicated that the difference between the fall-like motions and falling only corresponds to the inclination angles [30]. Therefore, if the evaluation does not include such angles, the fall-like motions can be used to represent the falls in general.

In conclusion, the results of the present study suggest that the EEMD and MSE algorithm can be utilized to quantify postural stability in healthy subjects. The PSI method adequately measures different postural stability states in healthy subjects.

Author Contributions: Conceptualization, N.R.N., B.C.J. and V.N.; Methodology, N.R.N. and B.C.J.; Formal Analysis, N.R.N. and B.C.J.; Writing—Review & Editing, N.R.N., B.C.J. and V.N.

Funding: This study was financially supported by the Ministry of Science and Technology (MOST) of Taiwan (MOST 105-2221-E-011-104-MY3).

Conflicts of Interest: The authors declare no conflict of interest.

References

1. Stevens, J.A.; Ballesteros, M.F.; Mack, K.A.; Rudd, R.A.; deCaro, E.; Adler, J. Gender differences in seeking care for falls in the aged medicare population. *Am. J. Prev. Med.* **2012**, *43*, 59–62. [CrossRef] [PubMed]
2. Fletcher, P.C.; Hirdes, J.P. Restriction in activity associated with fear of falling among community-based seniors using home care services. *Age Ageing* **2009**, *33*, 273–279. [CrossRef] [PubMed]
3. Yoshida, T.; Mizuno, F.; Hayasaka, T.; Tsubota, K.; Wada, S.; Yamaguchi, T.A. Wearable computer system for a detection and prevention of elderly users from falling. In Proceedings of the 12th International Conference on Biomedical Engineering, Singapore, 7–10 December 2005; Magjarevic, R.R., Nagel, J.H., Eds.; Springer: Berlin, Germany, 2005.
4. Sporaso, F.; Tyson, G. iFall: An Android application for fall monitoring and response. In Proceedings of the Annual International Conference at IEEE Engineering in Medicine and Biology Society, Minneapolis, MN, USA, 3–6 September 2009; He, B., Kim, Y., Eds.; IEEE: Piscataway, NJ, USA, 2009.
5. Dai, J.; Bai, X.; Yang, Z.; Shen, Z.; Xuan, D. Mobile phone-based pervasive fall detection. *Pers. Ubiquit. Comput.* **2010**, *14*, 633–643. [CrossRef]
6. Schutte, K.H.; Aeles, J.; Op De Beek, T.; van der Zwaard, B.C.; Venter, R.; Vanwanseele, B. Surface effects on dynamic stability and loading during outdoor running using wireless trunk accelerometry. *Gait Posture* **2016**, *48*, 220–225. [CrossRef] [PubMed]
7. Tura, A.; Raggi, M.; Rocchi, L.; Cutti, A.G.; Chiari, L. Gait symmetry and regularity in transfemoral amputees assessed by trunk accelerations. *J. Neuroeng. Rehabil.* **2010**, *7*, 4. [CrossRef]
8. Kobsar, D.; Olson, C.; Paranjape, R.; Hadjistavropoulos, T.; Barden, J.M. Evaluation of age-related differences in the stride-to-stride fluctuations, regularity and symmetry of gait using a waist-mounted tri-axial accelerometer. *Gait Posture* **2014**, *39*, 553–557. [CrossRef] [PubMed]
9. Saether, R.; Helbostad, J.L.; Adde, S.; Braendvik, S.; Lydersen, S.; Vik, T. Gait characteristics in children and adolescents with cerebral palsy assessed with a trunk-worn accelerometer. *Res. Dev. Disabil.* **2014**, *35*, 1773–1781. [CrossRef] [PubMed]
10. Cui, X.; Peng, C.K.; Costa, M.D.; Weiss, A.; Goldberger, A.L.; Hausdorff, J.M. Development of a new approach to quantifying stepping stability using ensemble empirical mode decomposition. *Gait Posture* **2014**, *39*, 495–500. [CrossRef] [PubMed]
11. Wu, Z.; Huang, N.E. Ensemble empirical mode decomposition: A noise-assisted data analysis method. *Adv. Adapt. Data Anal.* **2009**, *1*, 1–41. [CrossRef]
12. Alikovic, E.; Kevric, J.; Subasi, A. Performance evaluation of empirical mode decomposition, discretewavelet transform, and wavelet packed decomposition for automatedepileptic seizure detection and prediction. *Biomed. Signal Process. Control* **2018**, *39*, 94–102. [CrossRef]
13. *MATLAB R2016a*; The MathWorks Inc.: Natick, MA, USA, 2016.
14. Costa, M.; Goldberger, A.L.; Peng, C.K. Multiscale entropy analysis of complex physiologic time series. *Phys. Rev. Lett.* **2002**, *89*, 068102. [CrossRef] [PubMed]
15. Costa, M.; Goldberger, A.L.; Peng, C.K. Multiscale entropy analysis of biological signals. *Phys. Rev. E* **2005**, *71*, 021906. [CrossRef] [PubMed]
16. Costa, M.; Priplata, A.; Lipsitz, L.A.; Wu, Z.; Huang, N.E.; Goldberger, A.L.; Peng, C.K. Noise and poise: Enhancement of postural complexity in the elderly with a stochastic-resonance-based therapy. *Europhys. Lett.* **2007**, *77*, 68008. [CrossRef] [PubMed]
17. Humeau, A.; Mahe, G.; Durand, S.; Abraham, P. Multiscale entropy study of medical laser speckle contrast images. *IEEE Trans. Biomed. Eng.* **2013**, *60*, 872–879. [CrossRef] [PubMed]
18. Nurwulan, N.R.; Jiang, B.C.; Iridiastadi, H. Posture and texting: Effect on balance in young adults. *PLoS ONE* **2015**, *10*, e0134230. [CrossRef]

19. Nurwulan, N.R.; Jiang, B.C. Possibility of using entropy method to evaluate the distracting effect of mobile phones on pedestrians. *Entropy* **2016**, *18*, 390. [CrossRef]
20. Guzman-Vargas, L.; Ramirez-Rojas, A.; Angulo-Brown, F. Multiscale entropy analysis of electroseismic time series. *Nat. Hazards Earth Syst. Sci.* **2008**, *8*, 855–860. [CrossRef]
21. Wu, S.D.; Wu, P.H.; Wu, C.W.; Ding, J.J.; Wang, C.C. Bearing fault diagnosis based on multiscale permutation entropy and support vector machine. *Entropy* **2012**, *14*, 1343–1356. [CrossRef]
22. Richman, J.S.; Moorman, J.R. Physiological time-series analysis using approximate entropy and sample entropy. *Am. J. Physiol. Heart Circ. Physiol.* **2000**, *278*, H2039–H2049. [CrossRef]
23. Moe-Nilssen, R.; Helbostad, J.L. Estimation of gait cycle characteristics by trunk accelerometry. *J. Biomech.* **2004**, *37*, 121–126. [CrossRef]
24. Maroco, J.; Silva, D.; Rodrigues, A.; Guerreiro, M.; Santana, I.; de Mendonça, A. Data mining methods in the prediction of Dementia: A real-data comparison of the accuracy, sensitivity and specificity of linear discriminant analysis, logistic regression, neural networks, support vector machines, classification trees and random forests. *BMC Res. Notes* **2011**, *4*, 299. [CrossRef] [PubMed]
25. Breiman, L.; Friedman, J.; Stone, C.J.; Olshen, R.A. *Classification and Regression Trees*; CRC Press: Boca Raton, FL, USA, 1984.
26. Bedogni, L.; di Felice, M.; Bononi, L. By train or by car? Detecting the user's motion type through smartphone sensors data. In Proceedings of the 2012 IFIP Wireless Day (WD), Dublin, Ireland, 21–23 November 2012; pp. 1–6.
27. Pachi, A.; Ji, T. Frequency and velocity of people walking. *Struct. Eng.* **2005**, *83*, 36–40.
28. Curtze, C.; Hof, A.L.; Postema, K.; Queen, B. Over rough and smooth: Amputee gait on an irregular surface. *Gait Posture* **2010**, *33*, 292–296. [CrossRef] [PubMed]
29. Barralon, P.; Vuillerme, N.; Noury, N. Walk detection with a kinematic sensor: Frequency and wavelet comparison. In Proceedings of the 28th IEEE EMBS Annual International Conference, New York, NY, USA, 30 August–3 September 2006.
30. Li, Q.; Stankovic, J.A.; Hanson, M.A.; Barth, A.T.; Lach, J. Accurate, fast fall detection using gyroscopes and accelerometer-derived posture information. In Proceedings of the Sixth International Workshop on Wearable and Implantable Body Sensor Networks, Berkeley, CA, USA, 3–5 June 2009.

Article

Evidence for Maintained Post-Encoding Memory Consolidation Across the Adult Lifespan Revealed by Network Complexity

Ian M. McDonough *, Sarah K. Letang, Hillary B. Erwin and Rajesh K. Kana

Department of Psychology, The University of Alabama, Tuscaloosa, AL 35487, USA;
sletang@crimson.ua.edu (S.K.L.); hberwin@crimson.ua.edu (H.B.E.); rkkana@ua.edu (R.K.K.)
* Correspondence: immcdonough@ua.edu

Received: 15 October 2019; Accepted: 31 October 2019; Published: 1 November 2019

Abstract: Memory consolidation is well known to occur during sleep, but might start immediately after encoding new information while awake. While consolidation processes are important across the lifespan, they may be even more important to maintain memory functioning in old age. We tested whether a novel measure of information processing known as network complexity might be sensitive to post-encoding consolidation mechanisms in a sample of young, middle-aged, and older adults. Network complexity was calculated by assessing the irregularity of brain signals within a network over time using multiscale entropy. To capture post-encoding mechanisms, network complexity was estimated using functional magnetic resonance imaging (fMRI) during rest before and after encoding of picture pairs, and subtracted between the two rest periods. Participants received a five-alternative-choice memory test to assess associative memory performance. Results indicated that aging was associated with an increase in network complexity from pre- to post-encoding in the default mode network (DMN). Increases in network complexity in the DMN also were associated with better subsequent memory across all age groups. These findings suggest that network complexity is sensitive to post-encoding consolidation mechanisms that enhance memory performance. These post-encoding mechanisms may represent a pathway to support memory performance in the face of overall memory declines.

Keywords: aging; consolidation; default mode network; episodic memory; fMRI; multiscale entropy; network complexity; resting state

---

1. Introduction

As people age, experiencing memory decline is a common occurrence [1]. According to the associative deficit hypothesis [2], these age-related decreases in episodic memory are due to weakened abilities to encode simple, unrelated units of information together into a more complex unit (i.e., associating a picture of a scene with a picture of a face), and to retrieve that complex unit. The ability to associate and bind features together has been shown to be mediated by the hippocampus (for reviews, see [3,4]). Despite these clear links, a meta-analysis conducted on fMRI studies investigating age differences in successful and unsuccessful memory encoding revealed overall stability in hippocampal functioning in old age [5]. This finding suggests that, under some circumstances, the hippocampus can be successfully recruited to aid memory performance across the adult lifespan. In the present study, we investigated the degree to which a novel measure of information processing might be sensitive to key episodic memory mechanisms within the hippocampus and associated brain regions in a sample of young, middle-aged, and older adults.

One mechanism that might contribute to age-related memory declines is consolidation. The most well-characterized consolidation mechanisms occur during slow-wave sleep and rapid eye movement stages of sleep via neural replay and changes in synaptic strengths [6–8]. These processes occur between hippocampal and neocortical regions to help create enduring episodic contexts and promote the generalization of semantic representations [9]. Moreover, it has been well documented that sleep becomes more disrupted with increasing age, including disruptions in slow-wave sleep and rapid eye movement sleep [10,11]. These age differences have been shown to reduce the benefits of sleep for cognition normally found in young adults [12].

More recently, accumulating evidence in animals has suggested that similar neural replay processes related to consolidation occur while awake soon after new memories are encoded [13]. In humans, fMRI has been used to support post-encoding consolidation mechanisms in an awake state [14–16]. In some of these studies, resting-state scans were collected in young adults before and after memory encoding to assess changes in functional connectivity between the hippocampus and the neocortex. Across these studies, functional connectivity increased from pre-encoding to post-encoding, and this increase has been associated with better subsequent memory performance. Together, these findings have been interpreted as evidence for systems-level consolidation in an awake state, at least in young adults.

Only a few studies have investigated post-encoding consolidation in older adults [17–19]. Of these aging studies, only one directly tested whether changes in functional connectivity within the default mode network (DMN) before and after memory encoding differed with age [17]. Interestingly, they did not find age differences in pre-post change in DMN connectivity, but rather found greater age-related changes in connectivity within the salience network and reduced age-related changes in connectivity within the occipito-temporal network. In addition to finding these age-related differences in non-DMN networks, they found that the coupling between the DMN and executive function network decreased after encoding for older adults and increased for younger adults. Furthermore, the decreased coupling between the two networks was associated with better memory performance only in older adults. The authors interpreted this finding as the need for older adults to suppress interference from competing networks during post-encoding consolidation. While this interpretation remains a possibility, measures other than functional connectivity may capture unique aspects of neural replay mechanisms that unfold over time between the hippocampus and neocortical regions.

Multiscale entropy (MSE) is one such measure that might capture novel properties of the brain over time. This method of analysis estimates the complexity of a physiological time series using temporal coarse-graining procedures to evaluate signals at multiple temporal scales, in recognition of the likelihood that the dynamic complexity of biological signals might operate across a range of temporal scales [20,21]. Across these different temporal scales, the degree of randomness (i.e., entropy) is estimated by searching for repeated patterns of small temporal segments. To the extent that repeated patterns are found, then the physiological signal is quantified as having lower entropy and is interpreted as containing less unique information. In contrast, researchers have argued that higher entropy across temporal scales is associated with richer information [22,23] or more integrated information [24,25]. For example, research has found that brain entropy increases when retrieving new episodic information compared with known semantic information [26]. More recently, applying repetitive transcranial magnetic stimulation to create a "virtual lesion" in the frontal cortex reduced brain entropy [27], further suggesting that greater entropy is associated with more information processing. Notably, greater entropy is not simply interpreted as randomness that is equivalent to noise. When entropy is estimated across temporal scales, clear differences in patterns emerge [28,29]. For example, many types of noise decrease as temporal scales increase, whereas entropy showed dynamic patterns of increases from fine to mid temporal scales followed by slow declines as the temporal scales became coarser [28]. Thus, this analysis is often referred to as "complex" because of these dynamic changes in entropy and deviation from patterns of noise.

Other evidence for the usefulness of quantifying the complexity of brain signals comes from neuropsychiatric disorders. Studies have found that aberrant functional connectivity is a hallmark feature of many disorders, such as Alzheimer's disease [30], schizophrenia [31], autism spectrum disorder (ASD) [32,33], attention deficit hyperactivity disorder [34], and mood disorders [35]. Nevertheless, traditional functional connectivity analyses have fallen short of explaining these syndromes comprehensively, primarily due to their neural heterogeneity. To fill this gap, MSE analyses have been applied to investigate neuropsychiatric disorders and have found differences depending on both temporal scale and brain region. Using electroencephalography (EEG), neural complexity has been related to observation/imitation tasks in children with ASD [36], severity of ASD [37], and has been used to predict autism and risk for autism with relatively high accuracy [38]. Using fMRI, neural complexity has been shown to differ in patients with schizophrenia, bipolar disorder, and schizoaffective disorder compared to healthy control individuals [39–50].

From an aging perspective, complexity has been proposed to decrease across many physiological systems [41]. Specifically, EEG and magnetoencephalography (MEG) studies have found decreases in neural complexity in mild cognitive impairment and Alzheimer's disease [42]. Decreases in neural complexity in old age and in disorders might underlie the deficits in selective cognitive processing in these conditions. Using fMRI during rest, decreased neural complexity has been reliably found in various regions across the brain in older adults [29,43–46], and these decreases have sometimes been related to poorer cognition [46]. However, one study highlighted some of the limitations of using MSE analyses to estimate neural complexity in fMRI in comparison to MEG [47]. In this study, neural complexity decreased with age when using both fMRI and MEG. However, the temporal precision of MEG allowed for a greater number of temporal scales, revealing additional relationships with hypoperfusion surrounding neuronal damage due to stroke. To the extent that more time points are collected, however, more temporal scales can be estimated [28,48].

Participants

Here, we capitalized on the well-established finding that temporal patterns cluster across different sets of brain regions to form intrinsic connectivity networks [49–51]. Such similar temporal patterns suggest that the complexity of those patterns also would be similar within each brain network (at least at rest), allowing us to estimate network complexity [28,48] involved in memory consolidation. Specifically, we aimed to test the extent that network complexity in the DMN might be used as a novel measure of post-encoding consolidation mechanisms across the adult lifespan. To the extent that network complexity is sensitive to and a proxy for information processing within memory networks, we predicted that an increase in network complexity from pre-encoding to post-encoding would be associated with memory performance for the task administered in the fMRI session. To the extent that the well-known age-related deficits in episodic memory for recollected details are due to impaired consolidation mechanisms, then we also should see an age-related decline in such post-encoding processes as measured by network complexity. We focused on the changes in network complexity within the DMN because of the high degree of connectivity between the DMN and the hippocampus and the known relationships between the two for consolidation.

2. Materials and Methods

2.1. Participants

Participants were drawn from the Alabama Brain Study on Risk for Dementia. Details from the study can be found in our earlier publication assessing dementia risk and brain activity during memory retrieval [52]. All participants were recruited from the Tuscaloosa and Birmingham areas within Alabama through word of mouth, flyers, Facebook ads, and newsletters. Participants were excluded if they had contra-indicators for magnetic resonance imaging (MRI), were left-handed, had a prior diagnosis of any neurological condition, stroke, traumatic brain injury, claustrophobia, or history of substance abuse. Young adults aged 20–30 were recruited from the local community to serve as a baseline group. Middle-aged and older adults ranging in age from 50 to 74 were included if they were free of dementia as measured by the St. Louis University Mental Status (SLUMS) [53], spoke English fluently, were right-handed, and had at least one of the following self-reported risks for dementia: subjective memory complaints, less than a high school education, African American or Hispanic ethnoracial category, mild head trauma, family history of Alzheimer's disease, current diagnosis of hypertension or systolic blood pressure greater than 140 mmHg, current diagnosis or a family history of heart disease, current diagnosis of high total cholesterol, history or current use of smoking tobacco, current diagnosis or family history of diabetes, and body mass index greater than 30 kg/m^2. All participants gave informed consent using methods approved by the institutional review board at the University of Alabama. Participants' vision was normal or corrected to normal using MR-compatible glasses or contact lenses. Demographic characteristics can be found in Table 1.

Table 1. Participant Characteristics.

	Young Adults	Middle-Aged Adults	Older Adults	Group Differences
N	20	31	35	-
Mean Age	23.35 (3.25)	54.29 (2.84)	66.17 (4.02)	F (2,83) = 987.70, p < 0.001
Age Range	20–30	50–60	61–74	-
Sex (F/M)	11 (55%)/9 (45%)	17 (54%)/14 (46%)	23 (66%)/12 (34%)	χ^2 (2) = 1.01, p = 0.61
Ethnoracial Category				χ^2 (4) = 22.66, p < 0.001
Non-Hispanic White	12 (60%)	14 (45%)	28 (80%)	-
African American	1 (5%)	13 (42%)	7 (20%)	-
Other	7 (35%)	4 (13%)	0 (0%)	-
Years of Education	15.00 (2.20)	14.26 (2.67)	13.83 (2.93)	F (2,83) = 1.20, p = 0.30
SLUMS Score	-	26.48 (2.95) [1]	26.21 (2.91)	t (62.7) = 0.38, p = 0.70
Associative Memory Performance	0.57 (0.17)	0.36 (0.14)	0.32 (0.09)	F (2,83) = 24.55, p < 0.001
Premorbid IQ	106.65 (10.69)	95.16 (16.08)	107.55 (16.70)	F (2,83) = 6.20, p = 0.003
Dementia Risk	-	5.32 (1.85)	5.03 (2.08)	t (62.9) = 0.66, p = 0.51

[1] Missing score for one participant. SLUMS = St. Louis University Mental Status.

2.2. Procedures

Across two sessions, participants completed cognitive and MRI batteries. From the cognitive battery, we used the scaled word reading subtest from the Wide Range Achievement Test-4 to control for premorbid IQ in all analyses. Note that one participant's premorbid IQ score was missing and was imputed using a regression-based matching technique [52]. The MRI session included scans in the following order: resting-state, memory encoding, T1-structural scan, resting-state, memory retrieval, and a visual-motor checkerboard task. The analyses here focus on the resting-state scans that occurred before and after the memory encoding phase.

2.3. fMRI Scans

The two resting-state scans consisted of 175 volumes over 5 min each. Participants were told to close their eyes but not fall asleep. After the first resting-state scan, participants studied 64 pairs of pictures for the memory encoding task over two 8-minute sessions. The pictures consisted of either a face-object or face-scene pair for 3 s. After viewing each pair, participants were given 2.16 s to predict how likely they would remember the pair on a later memory test on a 3-point scale corresponding to likely, maybe, or unlikely. The inter-trial interval ranged from 1.72 to 17.20 s. The memory test consisted of a five-alternative-choice test in which participants were presented with a previously seen face and were asked to choose the object or scene that was previously paired with the face from five options: two objects, two scenes, and "never seen." Of the four possible picture choices, one was the target and three were lures. Because all options were previously seen, participants needed to rely on recollection processes to answer correctly. The "never seen" option specifically referred to not remembering the face, thus precluding participants from making a correct response without complete guessing. Each of the 64 memory trials was presented for 5.16 s, with intertrial intervals that ranged from 1.72 to 10.32 s. The memory test was divided into two runs lasting for 5 min.

2.4. *fMRI Acquisition and Preprocessing*

A 3T Siemens PRISMA scanner at the UAB Civitan International Neuroimaging Laboratory was used to collect MRI scans. High resolution T1-weighted structural MPRAGE scans were acquired using (parallel acquisition acceleration type = GRAPPA; acceleration factor = 3, TR = 5000 ms, TE = 2.93 ms, TI 1 = 700 ms, TI 2 = 2030 ms, flip angle 1 = 4°, flip angle 2 = 5°, FOV = 256 mm, matrix = 240 × 256 mm^2, in-plane resolution = 1.0 × 1.0 mm^2). All functional scans used T2*-weighted EPI sequences (56 interleaved axial slices, 2.5 mm thickness, TR = 1720 ms, TE = 35.8 ms, flip angle = 73°, FOV = 260 mm, matrix = 104 × 104 mm, in-plane resolution = 2.5 × 2.5 mm^2, multi-band acceleration factor = 4).

The functional data were unwarped, coregistered to the structural scan, and spatially smoothed (8 mm FWHM kernel) using Statistical Parametric Mapping 12 (SPM12). The blood oxygen level dependent (BOLD) signal was then denoised using Multivariate Exploratory Linear Optimized Decomposition into Independent Components (MELODIC) [54]. The resulting spatiotemporal components were flagged using an in-house script that applied machine learning to frequency and temporal elements indicative of potential artifacts. The flagged components were then regressed from the BOLD signal, also using MELODIC. The denoised data were then warped into a study template using Advanced Neuroimaging Tools (ANTs) [55].

2.5. *Resting-state fMRI Analysis*

Dual regression analyses using FMRIB Software Library (FSL)'s "dual_regression" function [56,57] were implemented to isolate the time series within 10 major resting-state networks (RSNs) [51] to estimate subject-specific functional connectivity patterns within each of the networks. This method was used to extract the single time series common across all voxels within each network, which can then be used to estimate network complexity [28]. The template for the networks was transformed into the space from our sample template using ANTS. To implement dual regression, each RSN template is used as a spatial predictor for each subject's denoised 4D BOLD data in the first general linear model (GLM) regression. This regression is used to find the best matching time course for a given RSN. In a second regression, the resulting time course from the first regression is used as a set of temporal regressors in a GLM to estimate individual regression weights in the spatial domain, which represents the degree to which a time series in each voxel matches the time series for that component. The output from these dual regressions is a subject and network-specific time series along with a spatial Z-scored map.

2.6. *MultiScale Entropy (MSE) Analysis*

Network complexity was calculated by computing MSE on each of the time series that were created from the dual regression analysis. MSE estimates sample entropy across multiple temporal scales [20,21]. Different temporal scales are created by averaging neighboring time points within non-overlapping windows. This process is repeated to create a new coarse-grained time series that captures neural dynamics at different levels. Sample entropy is separately estimated for each of the created time series. We estimated seven temporal scales due to the length of the resting-state scans. Previous work has shown that the variability of MSE increases as the temporal scale increases, leading to unreliable estimates in the BOLD signal [28,48]. We used the heuristic of dividing the number of time points (N = 175) by 25 for our upper bound number of temporal scales to use. Sample entropy is defined as the natural logarithm of the conditional probability that a given pattern of data of a specified length (m) repeats at the next time point for the entire time series at a given scale factor (of a dataset with a total length N). It considers subsequent patterns to be a repeat of the given pattern if they match within a certain tolerance (r) such that larger tolerance values increase the number of matches [58,59]. To the extent that a time series has a greater number of pattern matches, the time series is less random, and the entropy value is lower. In contrast, a smaller number of pattern matches is characterized as being more random, yielding a greater entropy value. We selected our parameters based on those used in prior studies investigating MSE using fMRI: $m = 2$ and $r = 0.5$ [28,29,48,60].

2.7. Multilevel Modeling Analyses

Given the nested nature of the temporal scales within a given network for each subject, multilevel modeling (MLM) was used to test the effects of MSE differences as a function of age and memory performance. The lme4 package in R was used for data analysis [61]. The MLM analyses used a random intercept and random slope for Timescale, modeled an auto-correlation structure of 1, and used maximum likelihood estimation. MSE was modeled at the first level and Timescale at the second level. Thus, each of the seven temporal scales were modeled simultaneously in the analyses. Adding an interaction term in the analyses with Timescale would indicate that the effects differed by Timescale. However, none of our primary analyses revealed significant interactions with Timescale, indicating that the relationships among MSE and the factors of interest (e.g., age and memory accuracy) did not depend on Timescale. For simplicity, the models with the Timescale interaction terms are not reported. All MLM analyses controlled for pre-encoding network complexity, sex (male = 0, female = 1), and premorbid IQ.

Note that our sample was recruited to enrich risk factors for dementia across middle-aged and older adults. Because the young adults were not assessed for dementia risk, a covariate across all participants could not be calculated in the primary analyses. Instead, we created a dementia risk score in the middle-aged and older adults only by summing the presence of any of the dementia risk inclusion criteria (see Section 2.1). We then reconducted the MLM analysis in the middle-aged and older adults with the dementia risk score in the model (Model 4) to determine (1) if dementia risk modifies the post-encoding network complexity in the DMN, and (2) if the inclusion of the dementia risk score modifies the effects of aging or memory accuracy on post-encoding complexity.

Although theories of consolidation propose that the hippocampus closely coordinates with regions within the DMN to successfully store information, networks that process object representations also might be involved in post-encoding consolidation [16]. Additionally, to the extent that the post-encoding network complexity is due to covert rehearsal (i.e., re-visualization) of the study information following encoding, then we might expect similar changes in network complexity in attention or cognitive control networks. To test these possibilities, we conducted additional MLM analyses investigating the change in network complexity in four additional networks [51]: lateral occipito-temporal network (Model 5), cingulo-opercular network (Model 6), left fronto-parietal network (Model 7), and right fronto-parietal network (Model 8).

3. Results

Table 1 summarizes the descriptive characteristics among the three age groups. The biggest differences between the groups were in relation to cognition. A one-way analysis of variance (ANOVA) indicated that associative memory performance in the fMRI task differed between groups, (F (2,83) = 24.55, $p < 0.001$) such that memory for young adults was significantly greater than that of middle-aged ($p < 0.001$) and older adults ($p < 0.001$). Memory performance did not differ between middle-aged and older adults ($p = 0.40$). In addition, premorbid IQ differed with age (F (2,83) = 6.20, $p = 0.003$) such that middle-aged adults scored significantly poorer than young adults ($p = 0.028$) and older adults ($p = 0.004$), who did not differ from one another ($p = 0.98$). The proportion of self-reported belonging to a particular ethnoracial category also differed between age groups (χ^2 (4) = 22.66, $p = 0.001$).

The dual regressions successfully captured the 10 RSNs [51]. Of particular interest was the DMN both pre and post encoding. Notably, the hippocampus also was strongly correlated with the DMN in our sample (Figure 1). Thus, the resulting time series spanned both the DMN and bilateral hippocampi. As shown in Figure 2, network complexity increased slightly across the seven temporal scales, but these increases were not significant at pre-encoding rest or post-encoding rest (p's > 0.20). Additionally, mean network complexity in the DMN did not differ between pre- and post-encoding ($p = 0.71$), but they were correlated with one another ($r = 0.13$, $p = 0.001$).

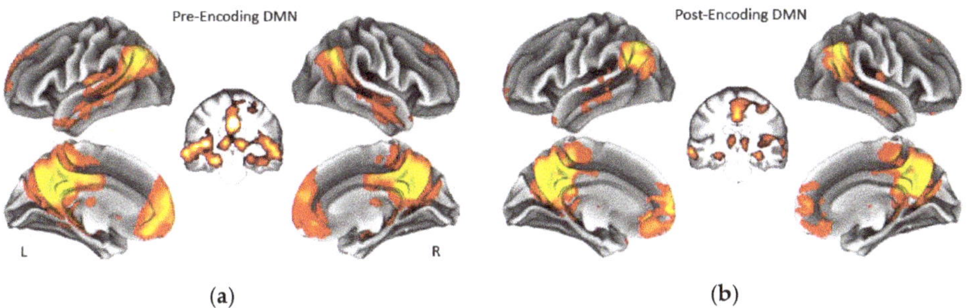

Figure 1. Group average of the default mode network during rest before memory encoding (**a**) and during rest after memory encoding (**b**). Coronal slices show that the default mode network in our sample also included bilateral hippocampus before and after encoding. Displayed maps were thresholded at a Z-score > 7. L = Left, R = Right. DMN = default mode network.

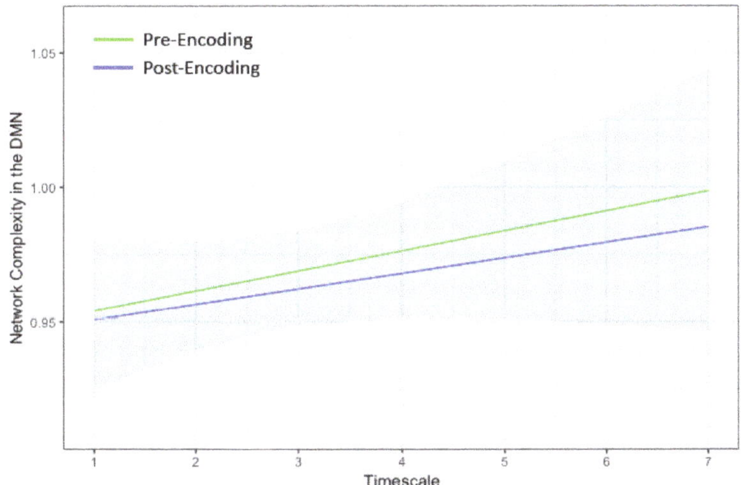

Figure 2. Marginal plots showing network complexity in the DMN estimated using multiscale entropy (MSE) on the y-axis at the seven temporal scales within the default mode network on the x-axis during rest before encoding (green) and after encoding (blue). The shaded regions represent 95% confidence intervals. Network complexity increased linearly with temporal scale and did not differ between pre- and post-encoding rest. DMN = default mode network.

In the MLM analysis, our primary predictors were Age and Memory Accuracy nested within Timescale, and our dependent variable was the difference in Network Complexity with the DMN between pre-encoding and post-encoding (Model 1). Both older age ($p = 0.0088$) and higher memory accuracy ($p = 0.043$) were associated with a greater increase in network complexity from pre- to post-encoding (Figure 3). The full results from the MLM analysis can be found in Table 2. An additional test for an Age × Memory Accuracy interaction was not significant ($p = 0.61$). Although we found evidence consistent with our hypothesis that network complexity increased following the encoding of information, difference scores can be driven by effects at pre-encoding only, post-encoding only, or both.

Thus, we conducted two additional tests to ensure that there were no pre-existing differences before memory encoding (Model 2) and to verify that the network complexity difference score was primarily due to differences at post-encoding (Model 3). We found no effect of Age or Memory Accuracy on Network Complexity during pre-encoding ($p > 0.59$). However, we did find significant effects of Age ($p = 0.010$) and Memory Accuracy ($p = 0.041$) on Network Complexity during post-encoding (see Table 2). Thus, the locus of the network complexity difference score appears to be isolated to changes in MSE at post-encoding.

Table 2. Fixed Effects for Network Complexity Multilevel Modeling Analyses.

	DMN Post–Pre Encoding (Model 1)	DMN Pre-Encoding (Model 2)	DMN Post-Encoding (Model 3)	DMN Post–Pre Encoding (Model 4)	OT Post–Pre Encoding (Model 5)	CO Post–Pre Encoding (Model 6)	LFP Post–Pre Encoding (Model 7)	RFP Post–Pre Encoding (Model 8)
Intercept	−0.0084 (0.013)	0.9764 (0.012) ***	0.9683 (0.013) ***	−0.0242 (0.024)	−0.0051 (0.016)	−0.0162 (0.017)	−0.0674 (0.015) ***	−0.0745 (0.017) ***
Timescale	0.0099 (0.013)	0.0148 (0.012)	0.0117 (0.013)	0.0065 (0.015)	−0.0010 (0.006)	0.0021 (0.007)	0.0175 (0.006) **	0.0156 (0.007) *
MSE Pre-Encoding	−0.1789 (0.009) ***	-	-	−0.1813 (0.010) ***	−0.1836 (0.009) ***	−0.1800 (0.009) ***	−0.1913 (0.009) ***	−0.1992 (0.009) ***
Age	0.0376 (0.014) **	−0.0002 (0.014)	0.0374 (0.014) *	0.0750 (0.036) *	0.0017 (0.016)	−0.0155 (0.015)	−0.0155 (0.014)	−0.0039 (0.015)
Sex (Ref. = Male)	−0.0036 (0.010)	0.0096 (0.010)	−0.0027 (0.011)	−0.0034 (0.012)	0.0014 (0.012)	0.0021 (0.011)	0.0104 (0.011)	0.0032 (0.011)
Premorbid IQ	−0.0090 (0.013)	0.0015 (0.013)	−0.0084 (0.013)	−0.0222 (0.015)	0.0123 (0.014)	−0.0055 (0.014)	−0.0020 (0.013)	−0.0074 (0.014)
Memory Accuracy	0.0321 (0.016) *	0.0083 (0.016)	0.0329 (0.016) *	0.0538 (0.021) **	0.0156 (0.018)	−0.0039 (0.017)	−0.0238 (0.016)	0.0144 (0.017)
Dementia Risk	-	-	-	0.0052 (0.007)	-	-	-	-

*** $p < 0.001$, ** $p < 0.01$, * $p < 0.05$; Values represent standardized beta coefficients and their standard errors in parentheses; MSE = Multiscale entropy; DMN = Default mode network; OT = Occipito-temporal Network; CO = Cingulo-opercular network; LFP = Left fronto-parietal network; RFP = Right fronto-parietal network.

To assess whether controlling for dementia risk would alter the findings, we conducted additional sensitivity analyses excluding the younger adults. We found that, even in this smaller sample, older age ($p = 0.040$) and memory accuracy ($p = 0.011$) were associated with a larger network complexity change in the DMN (see Table 2). In addition, the effect sizes for age and memory accuracy nearly doubled after young adults were removed from the analyses. No significant association was found for dementia risk ($p = 0.44$).

Finally, to test the extent that other networks also were involved in post-encoding consolidation [16,62] or involved covert rehearsal (i.e., re-visualization) of the study information following encoding, we conducted additional MLM analyses in four additional networks [51]: lateral occipito-temporal network (Model 5), cingulo-opercular network (Model 6), left fronto-parietal network (Model 7), and right fronto-parietal network (Model 8). The results from these analyses can be found in Table 2 In short, neither Age nor Memory Accuracy were significantly related to change in Network Complexity across any of the other RSNs (p's > 0.13).

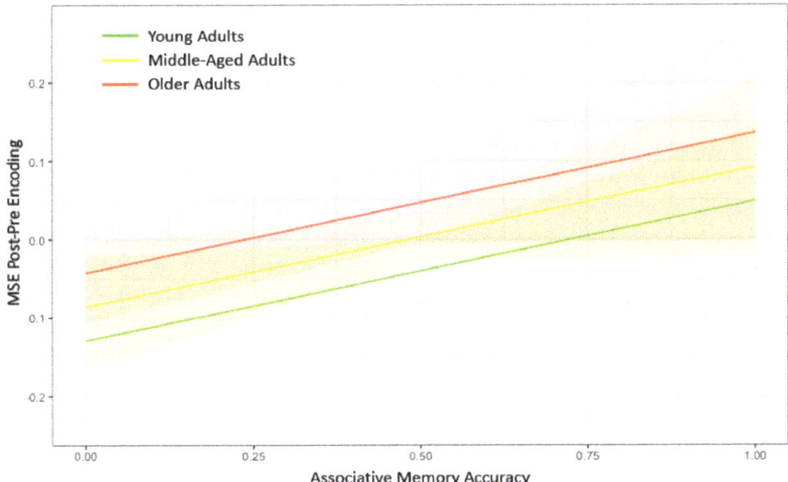

Figure 3. Marginal plots showing the relationship between post–pre encoding network complexity in the default mode network on the y-axis and associative memory performance on the x-axis for younger adults (green), middle-aged adults (yellow), and older adults (red). The shaded regions represent 95% confidence intervals. Greater associative memory performance was associated with greater increases in network complexity in the default mode network for each of the age groups. Older age was associated with greater increases in network complexity in the default mode network. No interactions were found between age and memory performance on change in network complexity. MSE = multiscale entropy.

4. Discussion

The present study aimed to test the degree to which network complexity would be sensitive to post-encoding consolidation mechanisms across the adult lifespan. This study is the first to our knowledge to assess post-encoding consolidation mechanisms using network complexity in fMRI. We predicted that network complexity within the DMN would increase following memory encoding, that this change in network complexity would be correlated with subsequent memory performance, and that changes in network complexity would partially explain age-related declines in episodic memory. Across participants, network complexity did not increase following memory encoding. However, substantial individual differences existed such that a greater increase in network complexity from pre- to post-encoding in the DMN was associated with better memory performance, as predicted. Although this change in network complexity was associated with age, the direction of the effect was contrary to our predictions. Older age was associated with greater increases in network complexity. We elaborate on these main findings below.

Greater increases in network complexity predicted better memory performance, suggesting that this measure successfully captured portions of post-encoding consolidation processes. This finding converges with work showing that the intensity of reactivation during neural replay at rest is associated with better memory performance in rats [63]. We also found that these effects were limited to the DMN. Importantly, the DMN captured by our dual regression analyses encompassed bilateral hippocampi, suggesting that the complexity in the BOLD signal may represent interactions between the hippocampus and portions of the neocortex that included the medial prefrontal cortex and the medial parietal cortex. Prior research in both human and animal models has established that consolidation-related stabilization of memories relies on the connections between the hippocampus and the neocortex, supporting the use of our network-based analyses [64,65]. Moreover, brain regions within the DMN specifically are well-known for supporting episodic memory (for review, see [66]). Indeed, studies using resting-state functional connectivity have found that reduced connectivity between the hippocampus and other parts of the DMN is associated with poorer memory in older adults [67,68].

We did not find the same evidence for other brain networks that might aid in suppressing competing information during post-encoding consolidation via cognitive control networks [17] or that might aid in consolidating perceptual information in the occipito-temporal network [14,17,69]. One reason we may not have found evidence for the involvement of these other networks is that specific brain regions involved in processing faces (e.g., fusiform gyrus) or places (e.g., parahippocampal gyrus) may have shown subtle alterations in MSE that were not captured by our relatively broader network approach. Another consideration is that our sample consisted of mostly middle-aged and older adults who show age-related declines in specialized processing of visual information [70–72]. Such reductions in visual-information processing may have reduced the ability of these regions to contribute substantially to post-encoding consolidation mechanisms. Supporting this idea, age-related decreases in pre- and post-encoding functional connectivity have been found in the occipito-temporal network [17].

Why might network complexity capture post-encoding consolidation? Current evidence in animal models suggests that neural replay occurs in a temporally reversed order from how the encoded events originally unfolded [13]. Therefore, the temporal patterns of brain activity, as assessed via network complexity, should provide insightful information into consolidation processes. Additionally, neural replay is thought to unfold more rapidly than the original event and repeated many times, suggesting that new and different information is coded within brain networks mediating consolidation processes [13]. To the extent that network complexity represents greater information processing offline and can capture neural dynamics at multiple time scales, then network complexity should be sensitive to consolidation processes.

In the context of the positive correlations we found between network complexity and memory performance, we were surprised to find that aging also was accompanied by increases in network complexity within the DMN. Older age was not associated with changes in network complexity in the other networks, suggesting that this aging effect was specific to the DMN and not a general physiological effect of the aging brain. One intriguing possibility for these unexpected age effects is that the increases in network complexity represent a novel compensatory mechanism to help ameliorate age-related memory declines. Some cognitive aging theories propose that much of the age-related deficits in memory occur because of reduced processing resources that limits their ability to initiate effortful strategies [73,74]. This deficit in controlled processing also is believed to be the origin of the enhanced reliance on automatic memory processes with age [75,76]. The present findings suggest that post-encoding consolidation processes might be considered an automatic process that is not only intact, but also helps maintain cognition in older age.

A slightly different perspective is that older adults may rely more on processes subserved by the DMN generally and, depending on the context, this over-reliance might aid or hinder memory [77]. Indeed, older adults often rely more on prior knowledge and schemas than younger adults [78]. Processing such schemas has been associated with connectivity between the medial temporal lobes and brain regions within the DMN, such as the medial prefrontal cortex in similar post-encoding paradigm, at least in young adults [16,79]. Thus, the present findings are consistent with the notion that middle-aged and older adults rely more on the DMN than younger adults [77].

In contrast to these optimistic perspectives, the present findings stand in contrast to two recent studies that assessed MSE using EEG in a similar rest–task–rest paradigm, but that did not measure episodic memory [80,81]. For example, MSE has been found to increase at fine temporal scales in anterior regions and decrease at coarse temporal scales in posterior brain regions from pre- to post-task in young adults [81]. In that study, no differences in MSE were found from pre- to post-task in middle-aged or older adults. Similar reductions in old age were found during a visual stimulation task [80]. Any differences between studies could easily be attributed to differences between EEG and fMRI. However, we also believe that their findings were sensitive to a different process altogether.

In [81], the positive correlations between MSE at rest with behavioral performance before and after the task (separately) were similar from fine to mid temporal scales. Because their study was not measuring episodic memory (but rather perception and working memory), the authors did not need to assess correlations between pre- to post-task changes in MSE and behavioral performance as we did in the current study. Thus, while any task can be considered an "encoding" task because information is constantly being encoded, an alternative hypothesis is that individual differences in resting MSE might provide a reliable individual difference estimate for cognitive health [81]. Our findings appear to be different from this perspective because individual differences in network complexity before encoding were not related to either age or memory performance as it was in their study. The specificity of our results suggests that our analyses are not capturing general individual differences in cognitive health, but specific post-encoding factors.

However, another possibility is that brain activity during rest after encoding represents the recovery and restoration of the brain following an effortful task, especially in the DMN [19,82–85]. On the one hand, the increased network complexity with age might be interpreted as an increase in "effort" with age to restore the brain to a pre-task state. This explanation would be broadly in line with theories proposing age-related deficits in neural efficiency [86]. Of course, such "effort" would have to be interpreted in a non-traditional manner, given that one does not have conscious awareness of brain processes occurring at rest. A related argument is that increases in network complexity relate to overt rehearsal of the encoding material in preparation for the later memory test. Indeed, such overt rehearsal would support later memory performance. Several reasons make this explanation unlikely. First, attempting to rehearse faces verbally can lead to later memory impairments known as verbal overshadowing [87,88]. Second, the encoding phase consisted of 64 face-picture pairs, which would be difficult to rehearse. Lastly, a structural scan was collected immediately after the last encoding scan. Therefore, any effects of rehearsal would likely have come to an end or at least would have been minimized by the time the second resting-state scan began.

5. Conclusions

Network complexity in the DMN during post-encoding awake states may represent a novel assessment of memory consolidation. We showed that pre- to post-change in network complexity related to better subsequent memory performance, supporting the idea that greater network complexity represents the degree of information processing in the brain [28,48]. This change in network complexity also increased with age, possibly representing a mechanism for older adults to maintain memory in the face of overall age-related declines in episodic memory. Network complexity and other measures of brain entropy have quickly become adopted as novel measures to assess optimal brain functioning, but more research is needed to understand further how these measures relate to more established brain measures such as task activations and how to interpret different temporal scales. Additionally, future work is needed to replicate and provide supporting evidence for this novel mechanism underlying post-encoding mechanisms of memory, including how network complexity might be related to neural replay.

Author Contributions: Conceptualization, I.M.M.; methodology, I.M.M.; formal analysis, I.M.M., and S.K.L.; investigation, I.M.M., S.K.L., and H.B.E.; writing—original draft preparation, I.M.M., S.K.L., H.B.E., and R.K.K.; writing—review and editing, I.M.M., S.K.L., and H.B.E.; visualization, I.M.M.; supervision, I.M.M.; project administration, I.M.M.; funding acquisition, I.M.M.

Funding: This research was funded by The University of Alabama through the College Academy of Research, Scholarship, and Creative Activity to I.M. and the University of Alabama, Birmingham (#A18-0284-001) to I.M. The APC was funded by I.M.

Acknowledgments: We would like to thank Lindsey Hessler, Tara Richardson, and William Miller for their recruitment and scheduling efforts for the Alabama Brain Study on Risk for Dementia. We also would like to thank Ashley Stinson for help collecting the data.

Conflicts of Interest: The authors declare no conflict of interest. The funders had no role in the design of the study; in the collection, analyses, or interpretation of data; in the writing of the manuscript, or in the decision to publish the results.

References

1. Luo, L.; Craik, F.I. Aging and memory: A cognitive approach. *Can. J. Psychiat.* **2008**, *53*, 346–353. [CrossRef]
2. Naveh-Benjamin, M. Adult age differences in memory performance: Tests of an associative deficit hypothesis. *J. Exp. Psychol. Learn. Mem. Cogn.* **2000**, *26*, 1170–1187. [CrossRef]
3. Bastin, C.; Besson, G.; Simon, J.; Delhaye, E.; Geurten, M.; Willems, S.; Salmon, E. An Integrative Memory model of recollection and familiarity to understand memory deficits. *Behav. Brain Sci.* **2019**, *1*, 1–66. [CrossRef]
4. Mitchell, K.J.; Johnson, M.K. Source monitoring 15 years later: What have we learned from fMRI about the neural mechanisms of source memory? *Psychol. Bull.* **2009**, *135*, 638–677. [CrossRef]
5. Maillet, D.; Rajah, M.N. Age-related differences in brain activity in the subsequent memory paradigm: A meta-analysis. *Neurosci. Biobehav. Rev.* **2014**, *45*, 246–257. [CrossRef] [PubMed]
6. Buzsáki, G. The hippocampo-neocortical dialogue. *Cereb. Cortex* **1996**, *6*, 81–92. [CrossRef] [PubMed]
7. Diekelmann, S.; Born, J. The memory function of sleep. *Nat. Rev. Neurosci.* **2010**, *11*, 114–126. [CrossRef] [PubMed]
8. Tononi, G.; Cirelli, C. Sleep function and synaptic homeostasis. *Sleep Med. Rev.* **2006**, *10*, 49–62. [CrossRef] [PubMed]
9. Nadel, L.; Hupbach, A.; Gomez, R.; Newman-Smith, K. Memory formation, consolidation and transformation. *Neurosci. Biobehav. Rev.* **2012**, *36*, 1640–1645. [CrossRef] [PubMed]
10. Bliwise, D.L. Sleep in normal aging and dementia. *Sleep* **1993**, *16*, 40–81. [CrossRef]
11. Ohayon, M.M.; Carskadon, M.A.; Guilleminault, C.; Vitiello, M.V. Meta-analysis of quantitative sleep parameters from childhood to old age in healthy individuals: Developing normative sleep values across the human lifespan. *Sleep* **2004**, *27*, 1255–1273. [CrossRef] [PubMed]
12. Scullin, M.K. Sleep, memory, and aging: The link between slow-wave sleep and episodic memory changes from younger to older adults. *Psychol. Aging* **2013**, *28*, 105–114. [CrossRef] [PubMed]
13. Carr, M.F.; Jadhav, S.P.; Frank, L.M. Hippocampal replay in the awake state: A potential substrate for memory consolidation and retrieval. *Nat. Neurosci.* **2011**, *14*, 147–153. [CrossRef] [PubMed]
14. Tambini, A.; Ketz, N.; Davachi, L. Enhanced brain correlations during rest are related to memory for recent experiences. *Neuron* **2010**, *65*, 280–290. [CrossRef]
15. Schlichting, M.L.; Preston, A.R. Hippocampal-medial prefrontal circuit supports memory updating during learning and post-encoding rest. *Neurobiol. Learn. Mem.* **2016**, *134*, 91–106. [CrossRef]
16. van Kesteren, M.T.R.; Fernández, G.; Norris, D.G.; Hermans, E.J. Persistent schema-dependent hippocampal-neocortical connectivity during memory encoding and postencoding rest in humans. *Proc. Natl. Acad. Sci. USA* **2010**, *107*, 7550–7555. [CrossRef]
17. Jacobs, H.I.; Dillen, K.N.; Risius, O.; Göreci, Y.; Onur, O.A.; Fink, G.R.; Kukolja, J. Consolidation in older adults depends upon competition between resting-state networks. *Front. Aging Neurosci.* **2015**, *6*, 344. [CrossRef]
18. Mary, A.; Wens, V.; Op de Beeck, M.; Leproult, R.; De Tiège, X.; Peigneux, P. Age-related differences in practice-dependent resting-state functional connectivity related to motor sequence learning. *Hum. Brain Mapp.* **2017**, *38*, 923–937. [CrossRef]
19. Oren, N.; Ash, E.; Shapira-Lichter, I.; Elkana, O.; Reichman-Eisikovits, O.; Chomsky, L.; Lerner, Y. Changes in resting-state functional connectivity of the hippocampus following cognitive effort predict memory decline at the older age—A longitudinal fMRI study. *Front. Aging Neurosci.* **2019**, *11*, 163. [CrossRef]
20. Costa, M.; Goldberger, A.L.; Peng, C.K. Multiscale entropy analysis of complex physiologic time series. *Phys. Rev. Lett.* **2002**, *89*, 68–102. [CrossRef]
21. Costa, M.; Goldberger, A.L.; Peng, C.K. Multiscale entropy analysis of biological signals. *Phys. Rev. E* **2005**, *71*, 021906. [CrossRef] [PubMed]
22. Tononi, G.; Sporns, O.; Edelman, G.M. A measure for brain complexity: Relating functional segregation and integration in the nervous system. *Proc. Natl. Acad. Sci. USA* **1994**, *91*, 5033–5037. [CrossRef] [PubMed]

23. Nakagawa, T.T.; Jirsa, V.K.; Spiegler, A.; McIntosh, A.R.; Deco, G. Bottom up modeling of the connectome: Linking structure and function in the resting brain and their changes in aging. *Neuroimage* **2013**, *80*, 318–329. [CrossRef] [PubMed]
24. Vakorin, V.A.; Lippé, S.; McIntosh, A.R. Variability of brain signals processed locally transforms into higher connectivity with brain development. *J. Neurosci.* **2011**, *31*, 6405–6413. [CrossRef] [PubMed]
25. McIntosh, A.R.; Vakorin, V.; Kovacevic, N.; Wang, H.; Diaconescu, A.; Protzner, A.B. Spatiotemporal dependency of age-related changes in brain signal variability. *Cereb. Cortex* **2014**, *24*, 1806–1817. [CrossRef]
26. Heisz, J.J.; Vakorin, V.; Ross, B.; Levine, B.; McIntosh, A.R. A trade-off between local and distributed information processing associated with remote episodic versus semantic memory. *J. Cogn. Neurosci.* **2013**, *26*, 41–53. [CrossRef]
27. Song, D.; Chang, D.; Zhang, J.; Peng, W.; Shang, Y.; Gao, X.; Wang, Z. Reduced brain entropy by repetitive transcranial magnetic stimulation on the left dorsolateral prefrontal cortex in healthy young adults. *Brain Imaging Behav.* **2019**, *13*, 421–429. [CrossRef] [PubMed]
28. McDonough, I.M.; Nashiro, K. Network complexity as a measure of information processing across resting-state networks: Evidence from the Human Connectome Project. *Front. Hum. Neurosci.* **2014**, *8*, 409. [CrossRef]
29. Smith, R.X.; Yan, L.; Wang, D.J. Multiple time scale complexity analysis of resting state FMRI. *Brain Imaging Behav.* **2014**, *8*, 284–291. [CrossRef]
30. Kavcic, V.; Ni, H.; Zhu, T.; Zhong, J.; Duffy, C.J. White matter integrity linked to functional impairments in aging and early Alzheimer's disease. *Alzheimers Dement.* **2008**, *4*, 381–389. [CrossRef]
31. Friston, K.J. The disconnection hypothesis. *Schizophr. Res.* **1998**, *30*, 115–125. [CrossRef]
32. Just, M.A.; Cherkassky, V.L.; Keller, T.A.; Minshew, N.J. Cortical activation and synchronization during sentence comprehension in high-functioning autism: Evidence of underconnectivity. *Brain* **2004**, *127*, 1811–1821. [CrossRef] [PubMed]
33. Kana, R.K.; Libero, L.E.; Moore, M.S. Disrupted cortical connectivity theory as an explanatory model for autism spectrum disorders. *Phys. Life Rev.* **2011**, *8*, 410–437. [CrossRef] [PubMed]
34. Murias, M.; Swanson, J.M.; Srinivasan, R. Functional connectivity of frontal cortex in healthy and ADHD children reflected in EEG coherence. *Cereb. Cortex* **2006**, *17*, 1788–1799. [CrossRef]
35. Carballedo, A.; Scheuerecker, J.; Meisenzahl, E.; Schoepf, V.; Bokde, A.; Möller, H.J.; Doyle, M.; Wiesmann, M.; Frodl, T. Functional connectivity of emotional processing in depression. *J. Affect. Disorders* **2011**, *134*, 272–279. [CrossRef]
36. Liu, T.; Chen, Y.; Chen, D.; Li, C.; Qiu, Y.; Wang, J. Altered electroencephalogram complexity in autistic children shown by the multiscale entropy approach. *Neuroreport* **2017**, *28*, 169. [CrossRef]
37. Hadoush, H.; Alafeef, M.; Abdulhay, E. Brain Complexity in Children with Mild and Severe Autism Spectrum Disorders: Analysis of Multiscale Entropy in EEG. *Brain Topogr.* **2019**, *32*, 914–921. [CrossRef]
38. Bosl, W.; Tierney, A.; Tager-Flusberg, H.; Nelson, C. EEG complexity as a biomarker for autism spectrum disorder risk. *BMC Med.* **2011**, *9*, 18. [CrossRef]
39. Hager, B.; Yang, A.C.; Brady, R.; Meda, S.; Clementz, B.; Pearlson, G.D.; Sweeney, J.A.; Tamminga, C.; Keshavan, M. Neural complexity as a potential translational biomarker for psychosis. *J. Affect. Disorders* **2017**, *216*, 89–99. [CrossRef]
40. Yang, A.C.; Hong, C.J.; Liou, Y.J.; Huang, K.L.; Huang, C.C.; Liu, M.E.; Lo, M.T.; Huang, N.E.; Peng, C.K.; Lin, C.P.; et al. Decreased resting-state brain activity complexity in schizophrenia characterized by both increased regularity and randomness. *Hum. Brain Mapp.* **2015**, *36*, 2174–2186. [CrossRef]
41. Lipsitz, L.A. Physiological complexity, aging, and the path to frailty. *Sci. Aging Knowl. Environ.* **2004**, *16*, pe16. [CrossRef] [PubMed]
42. Mcbride, J.C.; Zhao, X.; Munro, N.B.; Smith, C.D.; Jicha, G.A.; Hively, L.; Broster, L.S.; Schmitt, F.A.; Kryscio, R.J.; Jiang, Y. Spectral and complexity analysis of scalp EEG characteristics for mild cognitive impairment and early Alzheimer's disease. *Comput. Meth. Prog. Biomed.* **2014**, *114*, 153–163. [CrossRef] [PubMed]
43. Jia, Y.; Gu, H.; Luo, Q. Sample entropy reveals an age-related reduction in the complexity of dynamic brain. *Sci. Rep.* **2017**, *7*, 7990. [CrossRef] [PubMed]

44. Liu, C.Y.; Krishnan, A.P.; Yan, L.; Smith, R.X.; Kilroy, E.; Alger, J.R.; Ringman, J.M.; Wang, D.J. Complexity and synchronicity of resting state blood oxygenation level-dependent (BOLD) functional MRI in normal aging and cognitive decline. *J. Magn. Reson. Imaging* **2013**, *38*, 36–45. [CrossRef]
45. Sokunbi, M.O. Sample entropy reveals high discriminative power between young and elderly adults in short fMRI data sets. *Front. Neuroinform.* **2014**, *8*. [CrossRef]
46. Yang, A.C.; Huang, C.C.; Yeh, H.L.; Liu, M.E.; Hong, C.J.; Tu, P.C.; Chen, J.F.; Huang, N.E.; Peng, C.K.; Lin, C.P.; et al. Complexity of spontaneous BOLD activity in default mode network is correlated with cognitive function in normal male elderly: A multiscale entropy analysis. *Neurobiol. Aging* **2013**, *34*, 428–438. [CrossRef]
47. Kielar, A.; Deschamps, T.; Chu, R.K.O.; Jokel, R.; Khatamian, Y.B.; Chen, J.J.; Meltzer, J.A. Identifying dysfunctional cortex: Dissociable effects of stroke and aging on resting state dynamics in MEG and fMRI. *Front. Aging Neurosci.* **2016**, *8*, 40. [CrossRef]
48. McDonough, I.M.; Siegel, J.T. The Relation between White Matter Microstructure and Network Complexity: Implications for Processing Efficiency. *Front. Int. Neurosci.* **2018**, *12*, 43. [CrossRef]
49. Greicius, M.D.; Krasnow, B.; Reiss, A.L.; Menon, V. Functional connectivity in the resting brain: A network analysis of the default mode hypothesis. *Proc. Natl. Acad. Sci. USA* **2003**, *100*, 253–258. [CrossRef]
50. Laird, A.R.; Fox, P.M.; Eickhoff, S.B.; Turner, J.A.; Ray, K.L.; McKay, D.R.; Glahn, D.C.; Beckmann, C.F.; Smith, S.M.; Fox, P.T. Behavioral interpretations of intrinsic connectivity networks. *J. Cogn. Neurosci.* **2011**, *23*, 4022–4037. [CrossRef]
51. Smith, S.M.; Fox, P.T.; Miller, K.L.; Glahn, D.C.; Fox, P.M.; Mackay, C.E.; Filippini, N.; Watkins, K.E.; Toro, R.; Laird, A.R.; et al. Correspondence of the brain's functional architecture during activation and rest. *Proc. Natl. Acad. Sci. USA* **2009**, *106*, 13040–13045. [CrossRef] [PubMed]
52. McDonough, I.M.; Letang, S.K.; Stinson, E.A. Dementia Risk Elevates Brain Activity during Memory Retrieval: A Functional MRI Analysis of Middle Aged and Older Adults. *J. Alzheimers Dis.* **2019**, *70*, 1005–1023. [CrossRef] [PubMed]
53. Tariq, S.H.; Tumosa, N.; Chibnall, J.T.; Perry, M.H., III; Morley, J.E. Comparison of the Saint Louis University mental status examination and the mini-mental state examination for detecting dementia and mild neurocognitive disorder—A pilot study. *Am. J. Geriatr. Psychiatry* **2006**, *14*, 900–910. [CrossRef] [PubMed]
54. Beckmann, C.F.; Smith, S.M. Probabilistic independent component analysis for functional magnetic resonance imaging. *IEEE Trans. Med. Imaging* **2004**, *23*, 137–152. [CrossRef] [PubMed]
55. Avants, B.; Gee, J.C. Geodesic estimation for large deformation anatomical shape averaging and interpolation. *Neuroimage* **2004**, *23*, S139–S150. [CrossRef]
56. Beckmann, C.F.; Mackay, C.E.; Filippini, N.; Smith, S.M. Group comparison of resting-state FMRI data using multi-subject ICA and dual regression. *Neuroimage* **2009**, *47*, S148. [CrossRef]
57. Filippini, N.; MacIntosh, B.J.; Hough, M.G.; Goodwin, G.M.; Frisoni, G.B.; Smith, S.M.; Matthews, P.M.; Beckmann, C.F.; Mackay, C.E. Distinct patterns of brain activity in young carriers of the APOE-ε4 allele. *Proc. Natl. Acad. Sci. USA* **2009**, *106*, 7209–7214. [CrossRef]
58. Lake, D.E.; Richman, J.S.; Griffin, M.P.; Moorman, J.R. Sample entropy analysis of neonatal heart rate variability. *Am. J. Physiol. Reg.* **2002**, *283*, R789–R797. [CrossRef]
59. Richman, J.S.; Moorman, J.R. Physiological time-series analysis using approximate entropy and sample entropy. *Am. J. Physiol. Heart Circ. Physiol.* **2000**, *278*, H2039–H2049. [CrossRef]
60. Sokunbi, M.O.; Fung, W.; Sawlani, V.; Choppin, S.; Linden, D.E.; Thome, J. Resting state fMRI entropy probes complexity of brain activity in adults with ADHD. *Psychiatry Res. Neuroimaging* **2013**, *214*, 341–348. [CrossRef]
61. Bates, D.; Maechler, M.; Bolker, B.; Walker, S. Fitting Linear Mixed-Effects Models Using lme4. *J. Stat. Softw.* **2015**, *67*, 1–48. [CrossRef]
62. Murty, V.P.; Tompary, A.; Adcock, R.A.; Davachi, L. Selectivity in postencoding connectivity with high-level visual cortex is associated with reward-motivated memory. *J. Neurosci.* **2017**, *37*, 537–545. [CrossRef] [PubMed]
63. Dupret, D.; O'Neill, J.; Pleydell-Bouverie, B.; Csicsvari, J. The reorganization and reactivation of hippocampal maps predict spatial memory performance. *Nat. Neurosci.* **2010**, *13*, 995–1002. [CrossRef] [PubMed]

64. Alvarez, P.; Squire, L.R. Memory consolidation and the medial temporal lobe: A simple network model. *Proc. Natl. Acad. Sci. USA* **1994**, *91*, 7041–7045. [CrossRef]
65. Eichenbaum, H.; Cohen, N.J. *From Conditioning to Conscious Recollection*; Oxford Univ. Press: New York, NY, USA, 2001.
66. Rugg, M.D.; Vilberg, K.L. Brain networks underlying episodic memory retrieval. *Curr. Opin. Neurobiol.* **2013**, *23*, 255–260. [CrossRef]
67. He, J.; Carmichael, O.; Fletcher, E.; Singh, B.; Iosif, A.-M.; Martinez, O.; Reed, B.; Yonelinas, A.; DeCarli, C. Influence of functional connectivity and structural MRI measures on episodic memory. *Neurobiol. Aging* **2012**, *33*, 2612–2620. [CrossRef]
68. Wang, L.; LaViolette, P.; O'Keefe, K.; Putcha, D.; Bakkour, A.; Van Dijk, K.R.A.; Pihlajamaki, M.; Dickerson, B.C.; Sperling, R.A. Intrinsic connectivity between the hippocampus and posteromedial cortex predicts memory performance in cognitively intact older individuals. *Neuroimage* **2010**, *51*, 910–917. [CrossRef]
69. Tambini, A.; Davachi, L. Persistence of hippocampal multivoxel patterns into postencoding rest is related to memory. *Proc. Natl. Acad. Sci. USA* **2013**, *110*, 19591–19596. [CrossRef]
70. Park, D.C.; Polk, T.A.; Park, R.; Minear, M.; Savage, A.; Smith, M.R. Aging reduces neural specialization in ventral visual cortex. *Proc. Natl. Acad. Sci. USA* **2004**, *101*, 13091–13095. [CrossRef]
71. Park, J.; Carp, J.; Kennedy, K.M.; Rodrigue, K.M.; Bischof, G.N.; Huang, C.M.; Rieck, J.R.; Polk, T.A.; Park, D.C. Neural broadening or neural attenuation? Investigating age-related dedifferentiation in the face network in a large lifespan sample. *J. Neurosci.* **2012**, *32*, 2154–2158. [CrossRef] [PubMed]
72. Grady, C.L.; Haxby, J.V.; Horwitz, B.; Schapiro, M.B.; Rapoport, S.I.; Ungerleider, L.G.; Mishkin, M.; Carson, R.E.; Herscovitch, P. Dissociation of object and spatial vision in human extrastriate cortex: Age-related changes in activation of regional cerebral blood flow measured with [15 O] water and positron emission tomography. *J. Cogn. Neurosci.* **1992**, *4*, 23–34. [CrossRef] [PubMed]
73. Craik, F.I.M.; Byrd, M. Aging and cognitive deficits: The role of attentional resources. In *Aging and Cognitive Processes*; Craik, F.I.M., Trehub, S., Eds.; Plenum: New York, NY, USA, 1982; pp. 191–211.
74. Craik, F.I.M.; Jennings, J.M. Human memory. In *The Handbook of Aging and Cognition*; Craik, F.I.M., Salthouse, T.A., Eds.; Erlbaum: Hillsdale, MI, USA, 1992; pp. 51–110.
75. Dywan, J.; Jacoby, L. Effects of aging on source monitoring: Differences in susceptibility to false fame. *Psychol. Aging* **1990**, *5*, 379–387. [CrossRef] [PubMed]
76. Jennings, J.M.; Jacoby, L.L. Automatic versus intentional uses of memory: Aging, attention, and control. *Psychol. Aging* **1993**, *8*, 283–293. [CrossRef] [PubMed]
77. Maillet, D.; Schacter, D.L. Default network and aging: Beyond the task-negative perspective. *Trends Cogn. Sci.* **2016**, *20*, 646–648. [CrossRef] [PubMed]
78. Baltes, P.B.; Baltes, M.M. Psychological perspectives on successful aging: The model of selective optimization with compensation. In *Successful Aging: Perspectives from the Behavioral Sciences*, 1st ed.; University of Cambridge: Cambridge, UK, 1990; pp. 1–34.
79. van Kesteren, M.T.; Rijpkema, M.; Ruiter, D.J.; Morris, R.G.; Fernández, G. Building on prior knowledge: Schema-dependent encoding processes relate to academic performance. *J. Cogn. Neurosci.* **2014**, *26*, 2250–2261. [CrossRef] [PubMed]
80. Takahashi, T.; Cho, R.Y.; Murata, T.; Mizuno, T.; Kikuchi, M.; Mizukami, K.; Kosaka, H.; Takahashi, K.; Wada, Y. Age-related variation in EEG complexity to photic stimulation: A multiscale entropy analysis. *Clin. Neurophysiol.* **2009**, *120*, 476–483. [CrossRef] [PubMed]
81. Wang, H.; McIntosh, A.R.; Kovacevic, N.; Karachalios, M.; Protzner, A.B. Age-related multiscale changes in brain signal variability in pre-task versus post-task resting-state EEG. *J. Cogn. Neurosci.* **2016**, *28*, 971–984. [CrossRef]
82. Barnes, A.; Bullmore, E.T.; Suckling, J. Endogenous human brain dynamics recover slowly following cognitive effort. *PLoS ONE* **2009**, *4*, e6626. [CrossRef]
83. Grigg, O.; Grady, C.L. Task-related effects on the temporal and spatial dynamics of resting-state functional connectivity in the default network. *PLoS ONE* **2006**, *5*, e13311. [CrossRef]
84. Pyka, M.; Beckmann, C.F.; Schöning, S.; Hauke, S.; Heider, D.; Kugel, H.; Arolt, V.; Konrad, C. Impact of working memory load on fMRI resting state pattern in subsequent resting phases. *PLoS ONE* **2009**, *4*, e7198. [CrossRef]

85. Sala-Llonch, R.; Pena-Gomez, C.; Arenaza-Urquijo, E.M.; Vidal-Piñeiro, D.; Bargallo, N.; Junque, C.; Bartres-Faz, D. Brain connectivity during resting state and subsequent working memory task predicts behavioural performance. *Cortex* **2012**, *48*, 1187–1196. [CrossRef] [PubMed]
86. Reuter-Lorenz, P.A.; Park, D.C. How does it STAC up? Revisiting the scaffolding theory of aging and cognition. *Neuropsychol. Rev.* **2014**, *24*, 355–370. [CrossRef] [PubMed]
87. Schooler, J.W.; Engstler-Schooler, T.Y. Verbal overshadowing of visual memories: Some things are better left unsaid. *Cogn. Psychol.* **1990**, *22*, 36–71. [CrossRef]
88. Meissner, C.A.; Brigham, J.C. A meta-analysis of the verbal overshadowing effect in face identification. *Appl. Cogn. Psychol.* **2001**, *15*, 603–616. [CrossRef]

 *entropy*

Article

A Study of Brain Neuronal and Functional Complexities Estimated Using Multiscale Entropy in Healthy Young Adults

Sreevalsan S. Menon and K. Krishnamurthy *

Department of Mechanical and Aerospace Engineering, Missouri University of Science and Technology, Rolla, MO 65409, USA; sm2hm@mst.edu

* Correspondence: kkrishna@mst.edu

Received: 5 September 2019; Accepted: 10 October 2019; Published: 12 October 2019

Abstract: Brain complexity estimated using sample entropy and multiscale entropy (MSE) has recently gained much attention to compare brain function between diseased or neurologically impaired groups and healthy control groups. Using resting-state functional magnetic resonance imaging (rfMRI) blood oxygen-level dependent (BOLD) signals in a large cohort (n = 967) of healthy young adults, the present study maps neuronal and functional complexities estimated by using MSE of BOLD signals and BOLD phase coherence connectivity, respectively, at various levels of the brain's organization. The functional complexity explores patterns in a higher dimension than neuronal complexity and may better discern changes in brain functioning. The leave-one-subject-out cross-validation method is used to predict fluid intelligence using neuronal and functional complexity MSE values as features. While a wide range of scales was selected with neuronal complexity, only the first three scales were selected with functional complexity. Fewer scales are advantageous as they preclude the need for long BOLD signals to calculate good estimates of MSE. The presented results corroborate with previous findings and provide a baseline for other studies exploring the use of MSE to examine changes in brain function related to aging, diseases, and clinical disorders.

Keywords: brain complexity; dynamic functional connectivity; edge complexity; fluid intelligence; multiscale entropy; node complexity; resting-state functional magnetic resonance imaging; sample entropy

1. Introduction

Fluctuations in the resting-state functional magnetic resonance imaging (rfMRI) blood oxygen-level dependent (BOLD) signals have recently received considerable attention for their use in studies of functional brain networks and discovering "neuromarkers" for brain function in diseases and clinical disorders [1]. Functional connectivity (FC) describes the correlation of two time series from different regions of the brain, which are also referred to as nodes, and similarities among the time series from a set of nodes have been used to identify resting-state networks (RSNs). In addition, these functional networks have been associated with different cognitive functions [2]. The connections between the nodes are referred to as edges, and a matrix of all pair-wise edge strengths is used to denote the FC. This matrix is symmetric with the rows and columns representing the nodes and edges, respectively.

In the past, many studies made the simplifying assumption that the correlation or functional connection between different regions of the brain is static [3]. However, dynamic functional connectivity (dFC) more accurately reflects non-stationary brain activity and is, therefore, receiving increased attention [4,5]. A common approach to find the dFC is to use a sliding window, the length of which determines the final number of time points, of the BOLD time series data to determine repeated states by using a clustering algorithm [6]. Another approach is to use the BOLD phase coherence connectivity (PCC) [7–9]. Dynamic functional connectivity has been explored for its ability to identify intrinsic individual brain connectivity patterns [10,11], to study age-related cognitive changes [9], and to classify brain disorders [12], to cite a few examples.

The variability and complexity of the brain's neural signals are two quantitative measures that can be used to assess the health of a brain [13]. These measures indicate the ability of the brain to react to uncertainty and adapt to and function in a dynamic environment. Higher variability is associated with better behavioral performance, as it allows for the formation of functional networks and probes various functional configurations. Higher complexity, on the other hand, is associated with a higher information processing capacity. Healthy systems exhibit chaotic and complex behaviors; a loss of this complexity or transition to less complicated, predictable behaviors is an indication of disorders or impairments in fixed-point attractor systems and vice versa in oscillatory systems [14–16].

Several entropy measures have been used to study brain complexity using fMRI data. Approximate entropy [17] and sample entropy (SampEn) [18] are two such measures. Both measures estimate Kolmogorov entropy [19], which is the rate of generation of new information, and are attractive because of their immunity to low-level noise, robustness to occasional large or small artifacts, the ability to perform even with missing data, and preclude the requirement of a large number of data points. However, SampEn is advantageous because it is less dependent on the time series length being in the range 10^m–20^m (m being the pattern length) and displays relative consistency compared to approximate entropy. SampEn is the natural logarithm of the conditional probability that a pattern length of m points will repeat itself, without including self matches, for $m + 1$ points within a tolerance of r in a time series of length N [18]. Entropy is a measure of randomness and predictability of a stochastic process and, in general, increases with greater randomness. Therefore, higher SampEn values mean that the system has more complexity.

One of the issues in calculating SampEn is selecting the appropriate values for m and r. Choosing a large m and small r will result in fewer patterns, whereas a small m and large r will result in more pattern matches. These two extremes result in higher statistical variations in calculating SampEn and the reduced ability of SampEn to model temporal dynamics, respectively. It is also possible to fail in estimating SampEn for some combinations of m and r because no matching patterns can be found, and these combinations must be avoided. Adopting the strategy of [20], a systematic approach was presented in [21] to overcome limitations in previous studies wherein parameters are selected in an ad hoc fashion or those resulting in the maximum difference in population groups. Parameters are selected to keep the relative error of SampEn less than 0.1, which corresponds to about 10% of the coefficient of variation value in SampEn estimates. The relative error is estimated from the mean and standard deviation of SampEn of cerebrospinal fluid (CSF) signals, which contain minimal physiologic information (see Section 4.3 for details). Parameters that minimized this relative error are then used to calculate SampEn in BOLD signals in gray matter regions.

To more effectively model complex temporal fluctuations, multiscale entropy (MSE) was proposed to estimate dynamic complexity in a time series by considering different time scales [22]. In this method, multiple "coarse-grained" time series of length (N/l), where l is the scale factor, are first formed by averaging consecutive, non-overlapping data points of increasing length. Then the SampEn of each coarse-grained time series is calculated. Because complexity is evaluated over different time scales (high frequencies at fine scales (i.e., low scale factors) and low frequencies at coarse scales (i.e., high scale factors)), MSE can better identify frequency-dependent neuropathophysiological processes in different brain regions.

MSE has been used to study changes in brain complexity that are related to aging [16,23,24], Alzheimer's disease [25,26], and schizophrenia [27,28]. SampEn was used to measure complexity within dFC in [28,29], the only two studies where SampEn was calculated using dFC between nodes and are therefore different from the other studies. However, the entropy of the dFCs was averaged to find the SampEn of nodes, RSNs, and whole brain. Patients showed a significantly higher SampEn than the controls at the whole-brain level, in two RSNs (visual recognition and auditory networks), and three nodes (right middle occipital gyrus, right inferior occipital gyrus and left superior occipital gyrus) [28]. Although the whole brain and RSNs were not correlated to the clinical variables, the three nodes did show a strong positive correlation. The results in [28,29] motivate further studies on the use of MSE of dFC, particularly at the edge level without any averaging, to develop new approaches that more effectively utilize the spatiotemporal fluctuations in the brain activity.

Using publically available rfMRI data of about 1200 healthy young adults from the Human Connectome Project (HCP) S1200 release [30], brain complexity estimated using SampEn over multiple scales is studied here. One of the distinguishing features of the HCP is that rfMRI data were collected in four 15-min sessions over two consecutive days, providing a large time series data set (4800 volumes) to more accurately estimate complexity and study variations of SampEn over a wide range of scale factors. The long rfMRI time series data of a large cohort is exploited here to map the brain complexity across scales at four different levels:

1. Edge level—edge complexity is estimated by the MSE of each edge, calculated from its dFC time series data (eMSE) that is obtained using BOLD PCC.
2. Node level—node complexity is estimated by the MSE of each node, calculated from its BOLD time series data (nMSE) or calculated as the mean eMSE of all the edges connected to the node (edge-based nMSE).
3. Network level—network complexity is estimated by the mean nMSE or mean edge-based nMSE of all the nodes in the RSN.
4. Whole-brain, consisting of only the cortex and subcortical gray matter, level—whole-brain complexity is estimated by the mean nMSE of all the nodes or mean eMSE of all the edges in the brain.

By virtue of using the dFC edge time series, eMSE and edge-based nMSE are measures of functional complexity. On the other hand, nMSE is a measure of neuronal complexity because it is estimated from the BOLD time series. Past studies have focused mostly on neuronal complexity, exploring patterns at the node level and/or RSN level. Functional complexity explores patterns in a dimension higher than the node level, and the utility of functional complexity has not yet been fully explored.

Fluid intelligence [31,32], which is the capacity to identify patterns and solve problems independent of previously acquired knowledge, is known to be involved in individual differences [33]. The eMSE, nMSE, and edge-based nMSE values are used as features to predict fluid intelligence by using the linear-kernel support vector regression (SVR) leave-one-subject-out cross-validation scheme. The prediction accuracy results obtained using eMSE, nMSE, and edge-based nMSE features will help to better understand the use of brain complexity measures as a complementary tool to the traditional functional connectivity analysis. Moreover, the results of this study involving healthy young adults will serve as a baseline for better identification of changes in brain function due to age, diseases, and clinical disorders.

2. Results

2.1. Relative Error of SampEn

Figure 1 shows a color map of the median relative error of SampEn of CSF signals across all subjects. The median relative error was calculated for 48 different combinations of pattern lengths (m = 1, 2, and 3) and tolerance values (r = 0.05 to 0.8 in increments of 0.05). These 48 error values are mapped to a color and then shown in a 3 $\times$ 16 grid of pattern lengths and tolerance values. Several expected trends can be clearly seen when the MSE parameters are varied. At fine scales, the relative error was low, and this error increased with coarser scales. For pattern lengths $m = 2$ and $m = 3$, the relative error for low and high tolerance values was higher compared to other values, with a "skewed U-pattern" toward the lower end. An increase in the parameter length increased the relative error in

SampEn estimates. The combinations when there were failures in estimating SampEn, which occurred only when $m = 3$, are shown with an "X". The tolerance had to be increased with higher pattern lengths to avoid failures. In general, failures occurred when the tolerance was low and the pattern length was high, and the relative error increased with an increase in the scales.

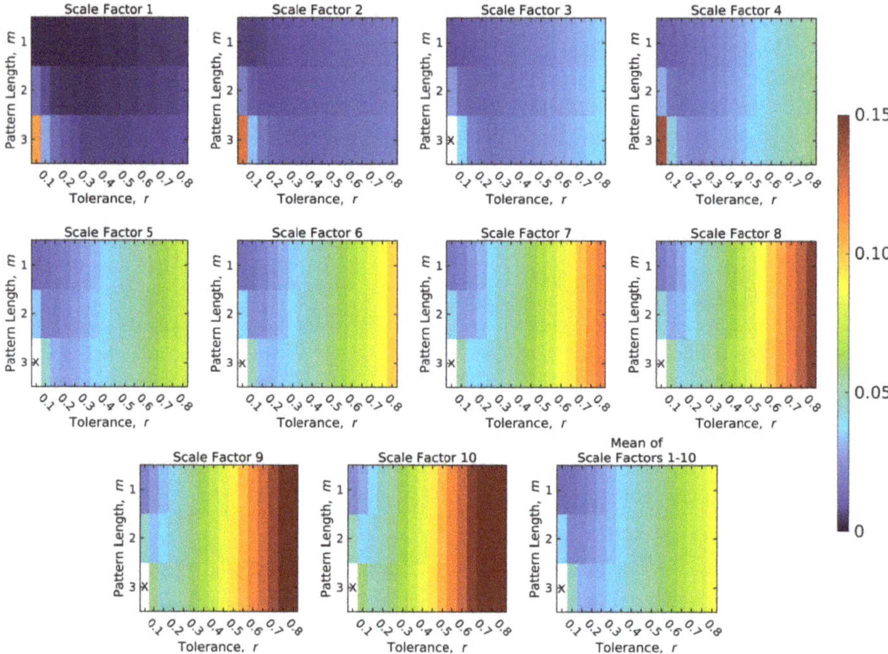

Figure 1. Median relative error of SampEn in CSF signals for scale factors 1–10. The color bar on the right displays the mapping of the median relative error to the color map. An "X" is used to denote failure to find a SampEn value.

The lowest mean relative errors across all scales for each parameter length were calculated as 0.016 ($m = 1$ and $r = 0.05$), 0.024 ($m = 2$ and $r = 0.1$), and 0.032 ($m = 3$ and $r = 0.2$), respectively. Instead of picking the case with the lowest relative error, the results are presented in Section 3 for the case $m = 3$ and $r = 0.2$, because $m = 3$ is a stronger test of repeating patterns compared to $m = 1$ and $m = 2$. As previously noted, relative errors of SampEn of CSF signals were used as estimates of the relative errors of BOLD signals and BOLD PCC.

2.2. Brain Complexity Across Different Levels

Figure 2 shows variations in the group mean (i.e., average over all subjects) SampEn across scales for the whole brain, considering only the cortex and subcortical gray matter, as previously defined. All three SampEn values decreased at coarse scales, but with different trends. The exponential decrease in CSF SampEn is similar to that of white noise [34]. There was an initial increase in eMSE, followed by a decreasing trend that also occurred in nMSE. In addition, the decrease in the slope of nMSE and eMSE was less than CSF SampEn, and the overall brain complexity estimated by nMSE was higher than eMSE for all scales. The two-tailed paired t-test for the worst case is $t_{966} = 40.9057$, $p < 5 \times 10^{-213}$ (uncorrected). The higher nMSE and eMSE values compared to CSF SampEn at coarse scales clearly delineate neuronal signals from non-neuronal signals.

Figures 3 and 4 show the network-level results. Figure 3 shows variations in the group mean SampEn in RSNs across the scales. Except for the precuneus, higher visual, and right executive control network (RECN), nMSE shows a decreasing trend similar to that seen in the whole brain. In the case of edge-based nMSE, all networks (with the exception of the precuneus and higher visual) show a trend similar to the whole brain, where a decrease follows an initial increase in SampEn. The nMSE values are higher than those of edge-based nMSE across all scales in every RSN. The two-tailed paired t-test for the worst case is $t_{966} = 13.7579$, $p < 2 \times 10^{-39}$ (uncorrected). The results also show that the networks have different levels of complexity when compared with the whole brain.

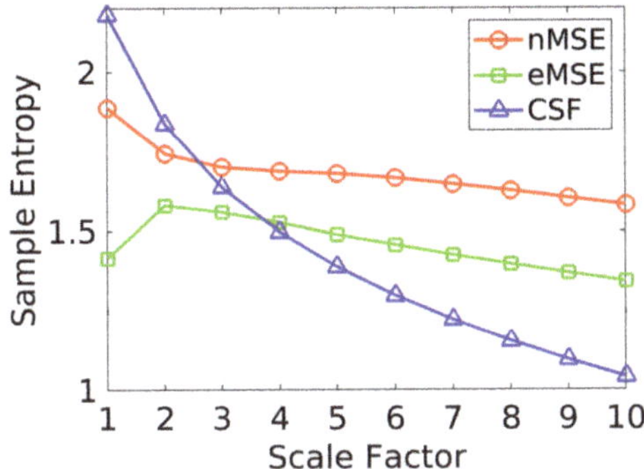

Figure 2. Group mean SampEn of whole brain (gray matter) and cerebrospinal fluid (CSF) signals across scale factors.

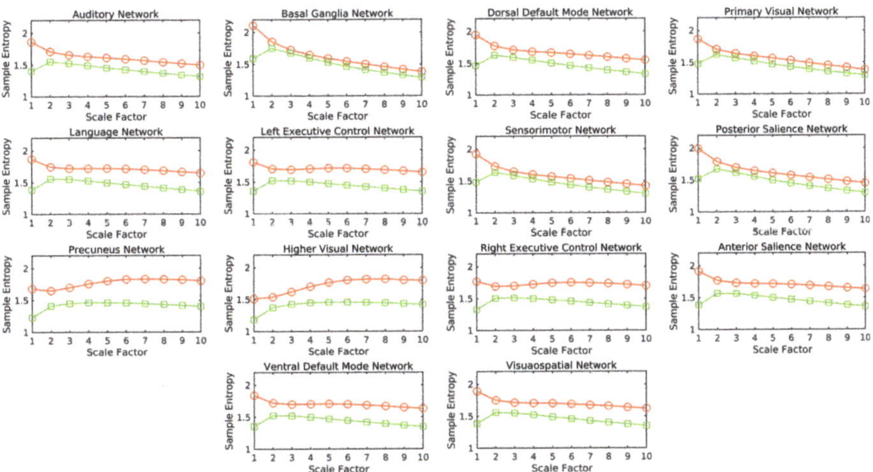

Figure 3. Group mean SampEn across scale factors of resting-state networks (RSNs). Red circles and line—nMSE, green squares and line—edge-based nMSE.

Figure 4 shows the variations of the mean (i.e., average of scale factors 1–10) MSE value of each RSN. There is a significant average difference between the mean nMSE and the mean edge-based nMSE values in every RSN. The two-tailed paired *t*-test for the worst case is $t_{966} = 34.2972$, $p < 10^{-169}$ (uncorrected), and the mean nMSE has more outliers than the mean edged-based nMSE. In addition, the mean nMSE has a smaller range, between the 25th and 75th percentile values. This suggests that the mean nMSE does not capture as much of the intrinsic patterns in subjects as the mean edge-based MSE does, but it captures much of the network differences. Another clear difference in most subjects is that the mean nMSE is higher than the mean edge-based nMSE in all RSNs. In short, the results from the network-level MSE analysis suggest that the inter-network differences are captured much better by the mean nMSE, while the intrinsic patterns in subjects are captured much better by the mean edge-based nMSE.

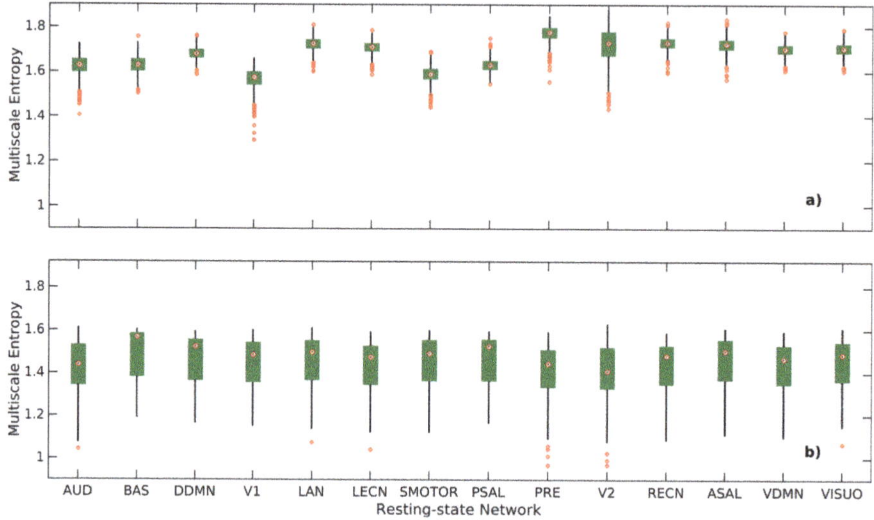

Figure 4. Box plots of the average MSE values of scale factors 1–10 for each RSN. (**a**) mean nMSE; and (**b**) mean edge-based nMSE. On each box, the central mark shows the median, bottom and top edges show the 25th and 75th percentiles, respectively, and the whiskers extend to the most extreme data points that are not considered as outliers, which are defined as 1.5 times the interquartile range away from the top or bottom of the box. AUD-auditory, BAS-basal ganglia, DDMN-dorsal default mode network, V1-primary visual, LAN-language, LECN-left executive control network, SMOTOR-sensorimotor, PSAL-posterior salience, PRE-precuneus, V2-higher visual, RECN-right executive control network, ASAL-anterior salience, VDMN-ventral default mode network, VISUO-visuospatial.

Figure 5 shows the node-level results (i.e., the variations of the mean MSE value of each node). Similar to the network-level case, there is a significant average difference between the mean nMSE and mean edge-based nMSE values in every node. The two-tailed paired t-test for the worst case is $t_{966} = 4.6786$, $p < 3 \times 10^{-6}$ (uncorrected), and the spread of the mean nMSE data was smaller compared to the mean edge-based nMSE. These results are consistent with the network-level MSE analysis (see Figure 4) in that the mean edge-based nMSE is capturing much more intrinsic information compared to the mean nMSE, which captures the difference among networks. In addition, Figure 5 shows that the mean MSE of some intra-network nodes has different median values.

Figure 6 shows the edge-level results (i.e., the group mean eMSE of edges across scales). The upper triangular part of each matrix shows the group mean eMSE of edges for the respective scale, and the lower triangular part shows the grand mean (averaged over all subjects and scales) eMSE of edges. Figure 7 shows the divergence color map for the edges obtained using the difference between SampEn at scale ten and scale two. Note that scale two was chosen as the reference because this scale had the highest whole-brain entropy. Therefore, a positive value in the divergence color map represents an increase in SampEn and vice versa. The major observations from the edge-level results are as follows:

1. The group mean eMSE of most edges was highest for scale 2, consistent with the whole-brain (see Figure 2) and network-level (see Figure 3) results.
2. The matrix for scale 5 shows that the group mean eMSE values are most similar to the grand mean eMSE values.
3. There is a large variation in complexity among all edges at fine scales, which decreases at coarse scales.
4. The complexity of intra-network edges is lower than inter-network edges in many RSNs across scales. An example of this is the ventral default mode network (VDMN).
5. There is a large variation in complexity of all edges in many nodes in various RSNs at fine scales. An example of this is the dorsal default mode network (DDMN), and these variations decrease at coarse scales.
6. About 70% of edges show a decrease in SampEn as the scale increases. The remaining 30% show a diverging pattern with the SampEn increasing, which shows that the complexity is not correlated to the variance of coarse-grained signals [35].

Thus, additional information is obtained from an edge-level analysis.

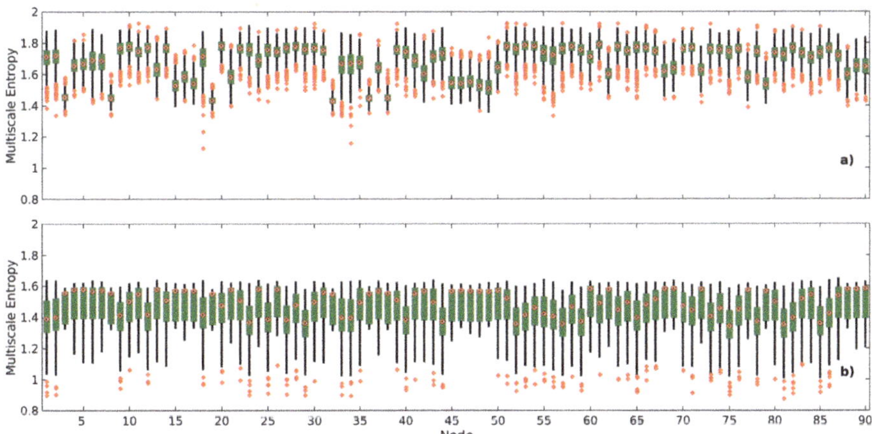

Figure 5. Box plots of the average MSE values of scale factors 1–10 for each node. (**a**) mean nMSE; and (**b**) mean edge-based nMSE. On each box, the central mark shows the median, bottom and top edges show the 25th and 75th percentiles, respectively, and the whiskers extend to the most extreme data points that are not considered as outliers, which are defined as 1.5 times the interquartile range away from the top or bottom of the box. Nodes belonging to the RSNs are shown in parentheses—auditory (1–3), basal ganglia (4–7), dorsal default mode network (8–17), primary visual (18–19), language (20–26), left executive control network (27–32), sensorimotor (33–38), posterior salience (39–50), precuneus (51–54), higher visual (55–56), right executive control network (57–62), anterior salience (63–69), ventral default mode network (70–79), visuospatial (80–90).

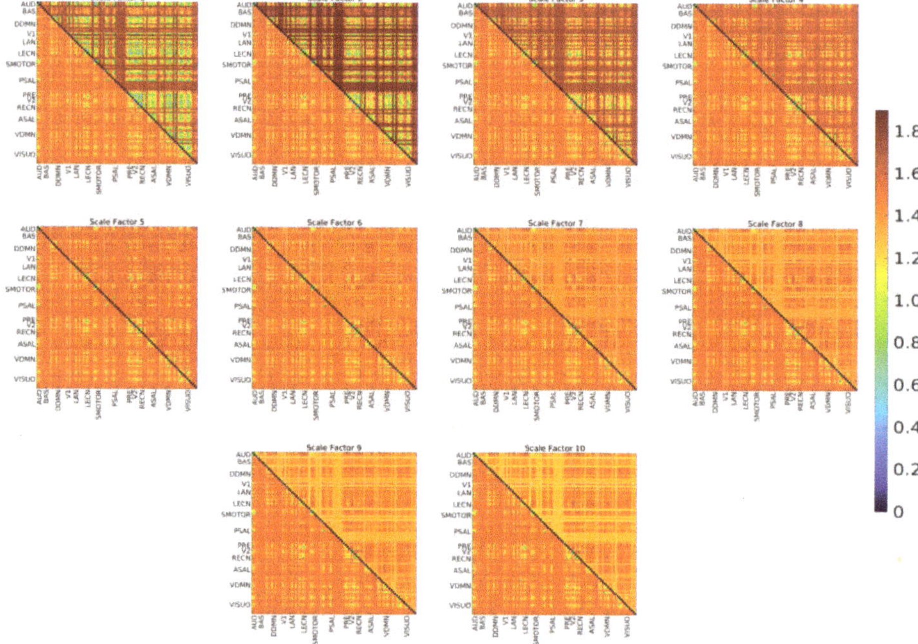

Figure 6. eMSE of edges. Upper triangular part—group mean eMSE for the noted scale factor, lower triangular part—grand mean eMSE (same in all matrices). The color bar on the right displays the mapping of the eMSE value to the color map. AUD-auditory, BAS-basal ganglia, DDMN-dorsal default mode network, V1-primary visual, LAN-language, LECN-left executive control network, SMOTOR-sensorimotor, PSAL-posterior salience, PRE-precuneus, V2-higher visual, RECN-right executive control network, ASAL-anterior salience, VDMN-ventral default mode network, VISUO-visuospatial.

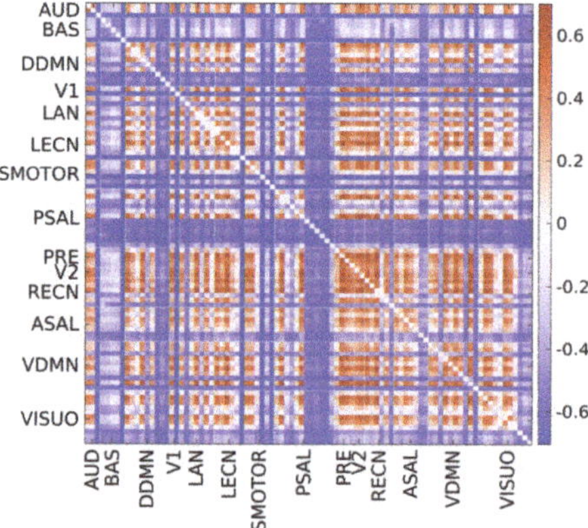

Figure 7. Divergence color map for the edges. The color bar on the right displays the mapping of the SampEn difference to the color map. AUD-auditory, BAS-basal ganglia, DDMN-dorsal default mode network, V1-primary visual, LAN-language, LECN-left executive control network, SMOTOR-sensorimotor, PSAL-posterior salience, PRE-precuneus, V2-higher visual, RECN-right executive control network, ASAL-anterior salience, VDMN-ventral default mode network, VISUO-visuospatial.

2.3. Cognitive Behavioral Prediction Correlation Scores and Selected Features

The leave-one-subject-out SVR models that used the MSE of nodes and edges as features were able to predict individual fluid intelligence scores. Table 1 shows the prediction results for several combinations of pattern length and tolerance and the number of selected features. With $m = 3$, $r = 0.2$, and 50 features selected in the SVR training, the correlation of predicted and actual fluid intelligence scores using nMSE, edge-based nMSE, and eMSE features were 0.249, 0.202, and 0.240, respectively. The permutation test resulted in a p-value < 0.001 with 1000 permutations; therefore, the correlation results are statistically significant. Note that when 900 nMSE and edge-based nMSE features were selected, all 90 nodes and 10 scale factors were used to predict fluid intelligence. However, 900 selected features using eMSE represented only 2.25% of the 4005 edges and 10 scale factors. Although no attempt was made to find an optimal number of features to include in the SVR models, it is clear that including a large number of features does not improve the prediction accuracy. Surprisingly, a small number of features was sufficient to obtain the best or second-best correlation value for the cases that were studied.

Further analysis was carried out to identify the most significant node and edge features involved in predicting fluid intelligence when 50 features were selected. Table A1 shows the nodes and the RSNs that the nodes were part of when nMSE and edge-based nMSE were used to predict fluid intelligence. While several nodes in different RSNs were involved, the most significant number of nodes came from the RECN and the left executive control network (LECN), followed by the language and precuneus networks. Although nodes selected with nMSE were mostly from scales 1 and 2, others were selected from scales 5, 7, 9, and 10. On the other hand, with edge-based nMSE, nodes were selected only from scales 1–3. Many nodes selected by both nMSE and edge-based nMSE were the same. Figure 8 shows the sagittal views of these common brain regions, as well as those that were specific to each complexity method.

Table 1. Correlation of predicted and actual fluid intelligence scores. The results for the case described in Section 2.3 are highlighted in bold italic font.

Complexity Level	Pattern Length, m	Tolerance, r	Number of Selected Features			
			25	50	100	900
nMSE	1	0.05	0.168	0.174	0.134	0.178
	2	0.10	0.231	0.179	0.200	0.150
	3	0.20	0.200	***0.249***	0.133	0.158
Edge-based nMSE	1	0.05	0.079	0.239	0.319	0.237
	2	0.10	0.159	0.152	0.246	0.181
	3	0.20	0.178	***0.202***	0.195	0.207
eMSE	1	0.05	0.180	0.127	0.171	0.165
	2	0.10	0.160	0.238	0.210	0.126
	3	0.20	0.237	***0.240***	0.234	0.165

Similar to the edge-based nMSE, features selected with eMSE came from scales 1–3. Figure 9 shows these edges for scales 1 and 2. The figure does not include scale 3 because there was only one intra-network edge between node #58 (right middle frontal gyrus) and node #59 (right inferior parietal gyrus, supramarginal gyrus and angular gyrus) in the RECN network. The edges were predominantly inter-network edges between the RECN and the VDMN, and a few intra-network edges were found in the RECN and precuneus.

Figure A1 shows the group mean correlation values of the selected nMSE features, which were common in all subjects to predict fluid intelligence in the leave-one-subject-out SVR models. The nMSE values were negatively correlated with fluid intelligence scores at fine scales and positively correlated at coarse scales. On the other hand, selected eMSE features had all negative correlations. However, it should be noted that the eMSE features had both negative and positive correlation values across scales; those selected to predict fluid intelligence were all negatively correlated.

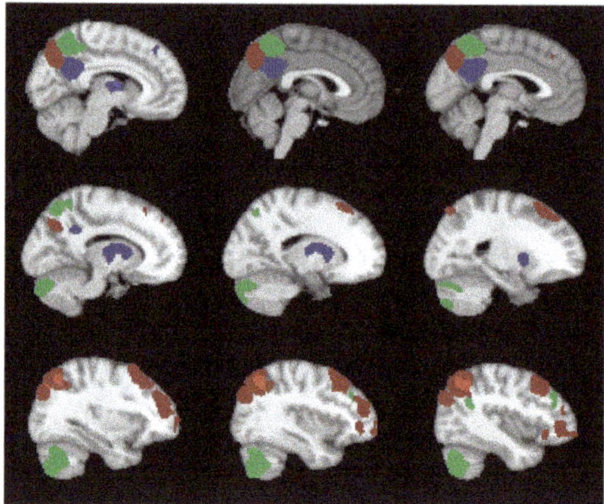

Figure 8. Sagittal views of brain regions selected in predicting fluid intelligence. Red—common regions selected when using nMSE and edge-based nMSE features, blue—regions selected with only nMSE features, green—regions selected with only edge-based nMSE features.

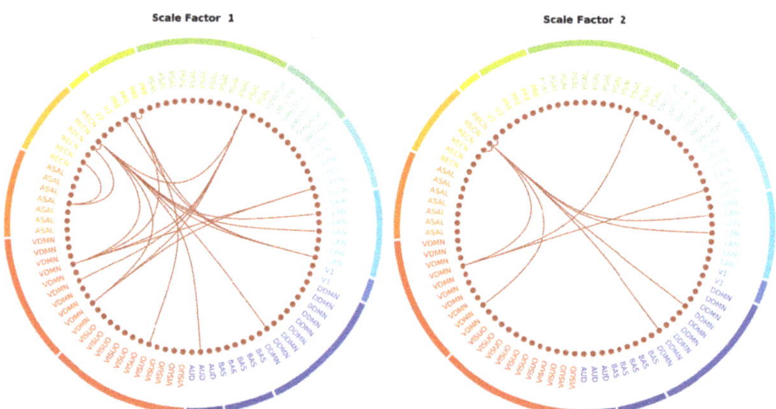

Figure 9. Significant edges in predicting fluid intelligence using eMSE features. Connectogram for scale factor 3 not shown because it had only one edge with an intra-node connection in the RECN. AUD-auditory, BAS-basal ganglia, DDMN-dorsal default mode network, V1-primary visual, LAN-language, LECN-left executive control network, SMOTOR-sensorimotor, PSAL-posterior salience, PRE-precuneus, V2-higher visual, RECN-right executive control network, ASAL-anterior salience, VDMN-ventral default mode network, VISUO-visuospatial.

3. Discussion

This study is the first to consider MSE of functional complexity using dFC obtained from the BOLD PCC, and expands brain complexity analysis to the edge level. In addition, neuronal and functional complexities estimated using MSE were compared at different levels of the brain's organization. Furthermore, neuronal and functional complexities were used to predict fluid intelligence. A large number of subjects (n = 967) with high temporal resolution data and long rfMRI data sets was studied here.

The brain complexity at the whole-brain level estimated using nMSE and eMSE was different because nMSE captures neuronal complexity from the BOLD time series, while eMSE captures functional complexity from the dFC edge time series. The decrease in neuronal complexity at higher scales was similar to some previous findings [24,36], but is different from [34], which may be due to differences in the number of subjects included in the study, the length of time series data, and the use of the parcellation scheme. However, the initial increase followed by a decrease in functional complexity when the scale is increased is similar to the "skewed inverted-U pattern" in [34]. The network-level complexity analysis reveals that the neuronal and functional complexities differ among networks. Except for the precuneus, higher visual, and RECN, all networks had trends similar to the whole-brain complexity. The results showed that some information is lost in averaging the SampEn due to mixing of complexity measures [34] because MSE trends were not the same in every RSN as in the whole brain. The edge-level results show that additional information is obtained compared to those from the node- and network-level analyses. A previous study with windowed dFC had shown that there exists a difference in SampEn measured throughout the brain at different organization levels [29].

The abbreviated version of Raven's Progressive Matrices would be considered as a "fair" quality measure to assess fluid intelligence, with the expected correlation in the range 0.50 to 0.71 [37]. The fair quality measure in itself may lead to a lower correlation between the fluid intelligence and the BOLD time series data. Moreover, a correlation of 0.2 is considered "typical" in research related to studying individual differences [38]. Considering the large number of subjects, the correlation results for predicting fluid intelligence were well within the range of some published results [10,39,40]. Although from an absolute perspective the correlation would be considered as small, valuable insights were obtained on the most significant features selected with different complexity levels (See Table A1 and Figure 9).

The brain regions in the RECN and LECN, such as frontal gyrus, angular gyrus, and parietal gyrus were selected as significant features in the leave-one-subject-out SVR models. These regions are known to be influential on fluid intelligence [33,39,41]. In addition, fluid intelligence is related to activity in the bilateral superior, inferior, and middle frontal gyri, as well as the anterior cingulate and paracingulate cortex [41]. The selected features corroborate with these past studies. The group mean correlation of nMSE with fluid intelligence was negative at fine scales and positive at coarse scales. Recent studies have shown the FC and SampEn have a similar correlation trend [34,36,42,43], and FC had a positive correlation with fluid intelligence [10,39,41,44]. In addition, studies have shown that fluid intelligence depends on synchrony in the brain's intrinsic network [45]. Higher intelligence was shown to be associated with the brain's capacity to adapt to a new state during the task with small changes in FC [44]. The synchronization and smaller changes in network dynamics are related to decreasing SampEn at fine scales [34,36], which is related to well-organized brain network dynamics.

Selecting appropriate MSE parameters is crucial to accurately estimate brain complexity. The relative error method [20,21] can be successfully implemented to find parameter combinations that satisfy some user-defined optimization criterion. Here, it can be seen that the relative error was well below the selected threshold of 0.1, and potential combinations with a relative error of less than 0.05 were present, which is the best value for the efficiency metric reported in [20]. The relative error was small compared to some recent studies [13,21], which may be due to the high quality of data used, high temporal resolution, and long rfMRI data sets.

Limitations and Future Direction

The results presented in this study, which considered a large number of healthy young adults, provide a baseline for other studies that explore the use of MSE to examine changes in brain function related to aging, diseases, and clinical disorders. Further studies are needed to ascertain if functional complexity provides a better differentiation ability than neuronal complexity by comparing between cohorts with diseases or neurological impairments and healthy control cohorts. Although many common regions were identified to be significantly correlated to fluid intelligence by considering neuronal and functional complexities, there were other areas specific to the two types of complexities.

With neuronal complexity, features were selected from a wide range of scales. On the other hand, features were selected from the first three scales with functional complexity, which is advantageous as it precludes the need for long BOLD signals to calculate good estimates of MSE. Future work is required to determine if only the first few scales are sufficient for estimating brain complexity. In addition, advanced nonlinear mathematical models are needed to predict cognitive performance with higher accuracy using the BOLD time series data.

Another limitation of this study is that MSE was estimated in a traditional manner by considering the tolerance r to be scale-invariant. A number of modifications and alternate coarse-graining procedures have been proposed to improve the accuracy of estimating MSE and mitigate the limitation of using scale-invariant tolerance values. See [46] for a review of these refined methods. More recently, a theoretical approach was presented to analytically calculate a new MSE measure using state-space models [47–49]. In addition, the importance of considering signal normalization and spectral content was shown using simulated and empirical data [50]. Recommendations were made for the steps to be followed to safeguard against biases in traditional MSE implementations. A comparison of the results presented in this study, including the diverging patterns shown in Figures 3 and 7, with recently published refined methods being of great interest; they are recommended for future studies.

4. Materials and Methods

4.1. *fMRI Data*

The data was acquired with multiband echo-planar imaging at a temporal resolution (TR) of 0.72 s per volume and 2-mm isotropic voxels for about 1200 young adults (ages 22–35) from families with twins and non-twins using a 3T imaging scanner at Washington University in St. Louis. The 3T rfMRI data was acquired in four runs of approximately 15 min each, two runs with right-to-left and left-to-right phase encoding protocols on day 1 and two similar runs on day 2 were used in this study. To remove the effects of structured artifacts, the data have been run through HCP FIX-ICA denoising. A detailed description of the minimal preprocessing steps applied to the data can be found in [1,51], and a description of the FIX approach in [52].

Some subjects in the HCP data set were excluded for the following reasons.

1. Missing rfMRI sessions—174 subjects were missing in one of the resting-state sessions.
2. High average framewise displacement [53]—16 subjects had an average framewise displacement greater than four standard deviations of the group, which introduces head motion artifacts.
3. Missing time series data—six subjects had less than 1200 points in one of the sessions.
4. Misalignment—one subject was misaligned in standard space and was missing some brain regions.
5. Low Mini-Mental Status Exam (MMSE) score [54]—33 subjects had an MMSE score of 26 or lower, which can be an indicator of cognitive impairment.
6. Missing fluid intelligence score—nine subjects were missing the fluid intelligence score.

The 967 subjects (451 males and 516 females) that remained after applying the exclusion criteria were included in this study.

Brain parcellation was performed using an atlas with 90 functional brain regions of interest, also known as nodes, defined across 14 major RSNs (https://findlab.stanford.edu/functional_ROIs.html). Time series of the 90 nodes, which were obtained using MATLAB based SPM (https://www.fil.ion.ucl.ac.uk/spm/) and REX toolbox (https://web.mit.edu/swg/rex/rex.m) from the four sessions, were detrended [MATLAB-detrend] and z-score normalized [MATLAB-zscore] before being concatenated to yield a total of 4800 data points. The time series were not filtered because studies have shown that useful neuronal related signals are present at higher frequencies up to 0.5 Hz [1,55,56]. Note that for clarity and completeness, the software package and commands, with any options, that were used to obtain the results are included within square brackets throughout the manuscript. MATLAB 2018a and SPM12 were used to obtain the results presented here.

4.2. Node and Edge Complexities

Brain complexity was estimated in two steps using MSE. First, the multiple coarse-grained time series of length (N/l) was formed by averaging consecutive, non-overlapping data points of increasing length. Second, the SampEn of each coarse-grained time series was calculated (WFDB Toolbox for MATLAB—https://physionet.org/physiotools/matlab/wfdb-app-matlab/). Three parameters must be defined: (i) pattern length m—the number of data points for pattern matching; (ii) tolerance (similarity factor) r—the fraction of standard deviation of the time series; and (iii) scale factor l—the scale factor of coarse-graining.

The time series of the 90 nodes, each having 4800, time points were used to calculate the nMSE values. On the other hand, the eMSE values were calculated using PCC [57], which yields dynamic connectivity matrices of size 90×90 at every time point. The phase of the time series data for the ith node, $\theta(i,t)$, was first calculated using Hilbert transform [MATLAB-hilbert,angle]. Next, the phase value was used to calculate the dFC matrix at time t using phase coherence between brain regions i and j as [57]:

$$dFC(i,j,t) = cos(\theta(i,t) - \theta(j,t)) \tag{1}$$

where $cos()$ is the cosine function. Because the dFC matrices are symmetric, only the upper triangular part was used to obtain the edge connectivity changes with time. This yielded the time series for 4005 edges with their connectivity variations along 4800 time points, and eMSE was calculated for these edges. For any given time series $\{x_i,\ i = 1,2,3,\ldots,N\}$, the course-grained time series, y^l, and SampEn are calculated as follows:

$$y_j^l = \frac{1}{l}\sum_{i=(j-1)l+1}^{jl} x_i,\ 1 \le j \le \frac{N}{l} \tag{2}$$

$$SampEn(m,r,N) = -ln\frac{P_{m+1}(r)}{P_m(r)} \tag{3}$$

where P is the probability that the time points are within tolerance r.

4.3. Optimal Parameter Selection for Calculating SampEn of BOLD Signals

Selecting the optimal combination of MSE parameters m, r and l is a challenging task. The relative error method presented in [21] was used in this study to find the acceptable MSE parameter space. Using the time series of all CSF signals, the relative error in SampEn estimation was calculated as:

$$RE = 1.96 \times \frac{\frac{\sigma_{SampEn}}{\mu_{SampEn}}}{2} \tag{4}$$

where μ_{SampEn} and σ_{SampEn} are the mean and standard deviation of the SampEn of CSF signals, respectively. The acceptable MSE parameter space of BOLD signals was determined by using a threshold value of 0.1 for RE [13,21] and excluding parameter combinations that failed to calculate SampEn for a subject using CSF signals.

4.4. Cognitive Behavioral Prediction

Fluid intelligence, which is related to the intrinsic cognitive ability that is correlated with reasoning and problem solving irrespective of acquired knowledge [39], was used to evaluate the significance of neuronal and functional complexities. A conservative leave-one-subject-out cross-validation method with support vector regression [MATLAB - fitrsvm] was used to predict individual fluid intelligence scores. These scores were measured in the HCP using Form A of an abbreviated version of Raven's

Progressive Matrices [32]. Three different cases were studied where features were selected to predict fluid intelligence using nMSE, edge-based nMSE, and eMSE values across multiple scales.

Figure 10 shows the four main steps for identifying the features and predicting fluid intelligence. First, the features (MSE values) across multiple scales were concatenated for each subject. For the nMSE and edge-based nMSE cases, the ($90 \times l$) matrix for each subject resulted in a column vector with dimension 90 l. In the higher dimensional eMSE case, the ($4005 \times l$) matrix for each subject resulted in a column vector with dimension 4005 l. Second, the correlation between the features and fluid intelligence was calculated. Third, the features were sorted from the lowest to the highest p-value of correlation, and the first n features were selected [39,58] for use in the SVR model. No correction for multiple comparisons was made since the feature selection method has a built-in guard against false positives, and the model fails to generalize for independent data when the false-positive proportion is higher in feature selection [39]. Finally, the selected features were then used to learn the SVR models, which were in turn used to predict the fluid intelligence scores of the left-out subjects. Because the choice of the number of features to be selected was arbitrary, a range of numbers was explored. After the leave-one-subject-out trials were completed, the prediction performance was calculated by correlating the predicted and actual fluid intelligence scores. The raw fluid intelligence scores were used here, and no control for confounding effects of age, gender, motion, and family was considered.

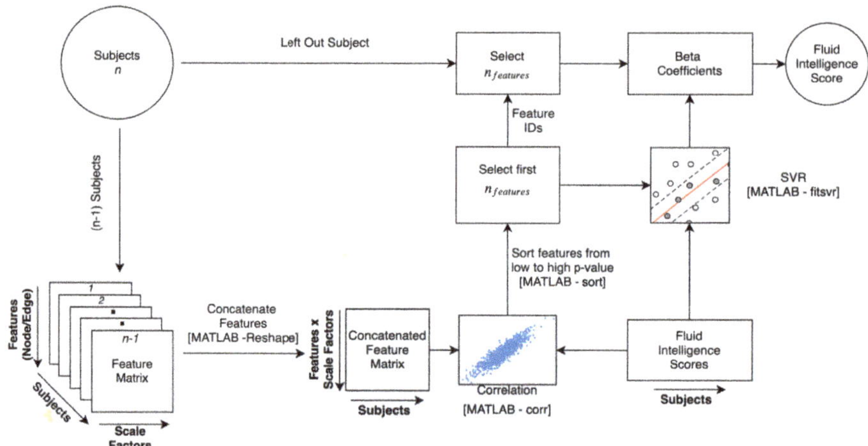

Figure 10. Schematic of feature selection and fluid intelligence prediction.

The statistical significance of the leave-one-subject-out method was assessed using permutation testing [10,58,59], which verifies whether the final prediction correlation was significantly better than the results expected by chance. Keeping feature selection and SVR training steps the same, the fluid intelligence scores of subjects were permuted 1000 times, and the correlation between predicted and actual fluid intelligence scores were re-calculated. The p-value then represents the probability of observing the reported accuracy by chance.

Author Contributions: Conceptualization, S.S.M. and K.K.; methodology, S.S.M. and K.K.; software, S.S.M; formal analysis, S.S.M. and K.K.; writing–original draft preparation, S.S.M. and K.K.; writing–review and editing, S.S.M. and K.K.

Funding: This research received no external funding.

Acknowledgments: Data were provided in part by the Human Connectome Project, WU-Minn Consortium (Principal Investigators: David Van Essen and Kamil Ugurbil; 1U54MH091657) funded by the 16 NIH Institutes and Centers that support the NIH Blueprint for Neuroscience Research; and by the McDonnell Center for Systems Neuroscience at Washington University.

Conflicts of Interest: The authors declare no conflict of interest.

Abbreviations

The following abbreviations are used in this manuscript:

ASAL	Anterior salience network
AUD	Auditory network
BAS	Basal ganglia network
BOLD	Blood oxygen-level dependent
CSF	Cerebrospinal fluid
DDMN	Dorsal default mode network
dFC	Dynamic functional connectivity
eMSE	Edge multiscale entropy
HCP	Human Connectome Project
LAN	Language network
LECN	Left executive control network
MSE	Multiscale entropy
nMSE	Node multiscale entropy
PCC	Phase coherence connectivity
PRE	Precuneus network
PSAL	Posterior salience network
rfMRI	Resting-state functional magnetic resonance imaging
RECN	Right executive control network
RSN	Resting-state networks
SampEn	Sample entropy
SMOTOR	Sensorimotor network
SVR	Support vector regression
V1	Primary visual network
V2	Higher visual network
VDMN	Ventral default mode network
VISUO	Visuospatial network

Appendix A. Predicting Fluid Intelligence

Table A1 shows the selected nodes that were common in all subjects when predicting fluid intelligence using nMSE and edge-based nMSE features. Although 50 nodes were selected for leave-one-subject-out SVR training, the total number of nodes listed in the table are less than 50 because a few nodes differed in the leave-one-subject-out trials. Figure A1 shows the group mean correlation of selected nMSE features with fluid intelligence across scales.

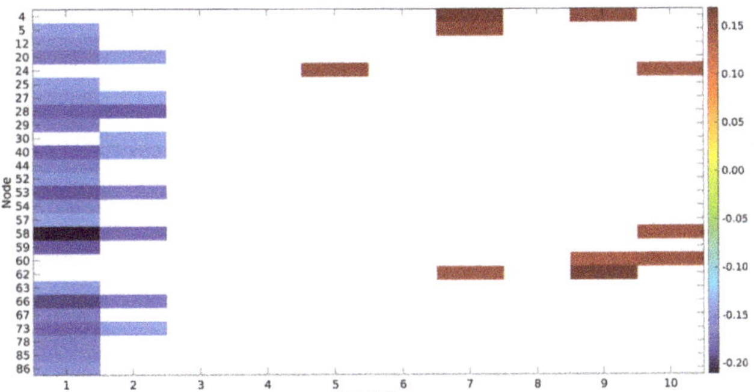

Figure A1. Group mean correlation of selected nMSE features with fluid intelligence across scale factors. The color bar on the right displays the mapping of the correlation value to the color map.

Table A1. Common nodes from all leave-one-subject-out trials for predicting fluid intelligence with $m = 3$, $r = 0.2$, and number of features = 50.

Network	Node Number	Brain Region	Scale Factor	
			nMSE	Edge-Based nMSE
BAS	4	Left Thalamus, Caudate	7, 9	
	5	Right Thalamus, Caudate, Putamen	1,7	
DDMN	12	Posterior Cingulate Cortex, Precuneus	1	
LAN	20	Inferior Frontal Gyrus	1, 2	1, 2
	22	Left Middle Temporal Gyrus, Angular Gyrus		1
	23	Left Middle Temporal Gyrus, Superior Temporal Gyrus, Supramarginal Gyrus, Angular Gyrus		1
	24	Right Inferior Frontal Gyrus	5, 10	
	25	Right Supramarginal Gyrus, Superior Temporal Gyrus, Middle Temporal Gyrus	1	1
LECN	27	Left Middle Frontal Gyrus, Superior Frontal Gyrus	1, 2	1, 2
	28	Left Inferior Frontal Gyrus, Orbitofrontal Gyrus	1, 2	1, 2, 3
	29	Left Superior Parietal Gyrus, Inferior Parietal Gyrus, Precuneus, Angular Gyrus	1	1, 2
	30	Left Inferior Temporal Gyrus, Middle Temporal Gyrus	2	1, 2
PSAL	39	Left Middle Frontal Gyrus		1, 2
	40	Left Supramarginal Gyrus, Inferior Parietal Gyrus	1, 2	1, 2
	44	Right Supramarginal Gyrus, Inferior Parietal Gyrus	1	1
PRE	52	Precuneus	1	1
	53	Left Angular Gyrus	1, 2	1, 2, 3
	54	Right Angular Gyrus	1	1, 2
RECN	57	Right Middle Frontal Gyrus, Right Superior Frontal Gyrus	1	1, 2
	58	Right Middle Frontal Gyrus	1, 2, 10	1, 2, 3
	59	Right Inferior Parietal Gyrus, Supramarginal Gyrus, Angular Gyrus	1	1, 2
	60	Right Superior Frontal Gyrus	9, 10	
	61	Left Crus 1, Crus II, Lobule VI		1, 2
	62	Right Caudate	7, 9	1
ASAL	63	Left Middle Frontal Gyrus	1	
	66	Right Middle Frontal Gyrus	1, 2	1, 2
	67	Right Insula	1	
VDMN	73	Left Middle Occipital Gyrus	1, 2	1, 2
	75	Precuneus		1
	78	Right Angular Gyrus, Middle Occipital Gyrus	1	1, 2
VISUO	85	Right Inferior Parietal Lobule	1	1
	86	Right Frontal Operculum. Inferior Frontal Gyrus	1	1

References

1. Smith, S.M.; Vidaurre, D.; Beckmann, C.F.; Glasser, M.F.; Jenkinson, M.; Miller, K.L.; Nichols, T.E.; Robinson, E.C.; Salimi-Khorshidi, G.; Woolrich, M.W.; et al. Functional connectomics from resting-state fMRI. *Trends Cogn. Sci.* **2013**, *17*, 666–682. [CrossRef] [PubMed]
2. Shirer, W.R.; Ryali, S.; Rykhlevskaia, E.; Menon, V.; Greicius, M.D. Decoding Subject-Driven Cognitive States with Whole-Brain Connectivity Patterns. *Cereb. Cortex* **2011**, *22*, 158–165. [CrossRef]
3. Lee, M.; Smyser, C.; Shimony, J. Resting-State fMRI: A Review of Methods and Clinical Applications. *Am. J. Neuroradiol.* **2013**, *34*, 1866–1872. [CrossRef] [PubMed]
4. Preti, M.G.; Bolton, T.A.; Van De Ville, D. The dynamic functional connectome: State-of-the-art and perspectives. *NeuroImage* **2017**, *160*, 41–54. [CrossRef] [PubMed]
5. Filippi, M.; Spinelli, E.G.; Cividini, C.; Agosta, F. Resting State Dynamic Functional Connectivity in Neurodegenerative Conditions: A Review of Magnetic Resonance Imaging Findings. *Front. Neurosci.* **2019**, *13*. [CrossRef] [PubMed]
6. Allen, E.A.; Damaraju, E.; Plis, S.M.; Erhardt, E.B.; Eichele, T.; Calhoun, V.D. Tracking whole-brain connectivity dynamics in the resting state. *Cereb. Cortex* **2014**, *24*, 663–676. [CrossRef] [PubMed]
7. Deco, G.; Cabral, J.; Woolrich, M.W.; Stevner, A.B.; van Hartevelt, T.J.; Kringelbach, M.L. Single or multiple frequency generators in on-going brain activity: A mechanistic whole-brain model of empirical MEG data. *NeuroImage* **2017**, *152*, 538–550. [CrossRef]
8. Glerean, E.; Salmi, J.; Lahnakoski, J.M.; Jääskeläinen, I.P.; Sams, M. Functional Magnetic Resonance Imaging Phase Synchronization as a Measure of Dynamic Functional Connectivity. *Brain Connect.* **2012**, *2*, 91–101. [CrossRef]
9. Cabral, J.; Kringelbach, M.L.; Deco, G. Functional connectivity dynamically evolves on multiple time-scales over a static structural connectome: Models and mechanisms. *NeuroImage* **2017**, *160*, 84–96. [CrossRef]
10. Liu, J.; Liao, X.; Xia, M.; He, Y. Chronnectome fingerprinting: Identifying individuals and predicting higher cognitive functions using dynamic brain connectivity patterns. *Hum. Brain Mapp.* **2018**, *39*, 902–915. [CrossRef]

11. Menon, S.S.; Krishnamurthy, K. A Comparison of Static and Dynamic Functional Connectivities for Identifying Subjects and Biological Sex Using Intrinsic Individual Brain Connectivity. *Sci. Rep.* **2019**, *9*, 5729. [CrossRef] [PubMed]
12. Du, Y.; Fu, Z.; Calhoun, V.D. Classification and Prediction of Brain Disorders Using Functional Connectivity: Promising but Challenging. *Front. Neurosci.* **2018**, *12*, 525. [CrossRef] [PubMed]
13. Easson, A.K.; McIntosh, A.R. BOLD signal variability and complexity in children and adolescents with and without autism spectrum disorder. *Dev. Cogn. Neurosci.* **2019**, *36*, 100630. [CrossRef] [PubMed]
14. Pool, R. Is it healthy to be chaotic? *Science* **1989**, *243*, 604–607. [CrossRef] [PubMed]
15. Vaillancourt, D.E.; Newell, K.M. Changing complexity in human behavior and physiology through aging and disease. *Neurobiol. Aging* **2002**, *23*, 1–11. [CrossRef]
16. Sokunbi, M.O. Sample entropy reveals high discriminative power between young and elderly adults in short fMRI data sets. *Front. Neuroinform.* **2014**, *8*, 69. [CrossRef] [PubMed]
17. Pincus, S.M. Approximate entropy as a measure of system complexity. *Proc. Natl. Acad. Sci. USA* **1991**. *88*, 2297–2301. [CrossRef] [PubMed]
18. Richman, J.S.; Moorman, J.R. Physiological time-series analysis using approximate entropy and sample entropy. *Am. J. Physiol.-Heart Circ. Physiol.* **2000**, *278*, H2039–H2049. [CrossRef] [PubMed]
19. Wang, Z.; Li, Y.; Childress, A.R.; Detre, J.A. Brain Entropy Mapping Using fMRI. *PLoS ONE* **2014**, *9*, e89948. [CrossRef]
20. Lake, D.E.; Richman, J.S.; Griffin, M.P.; Moorman, J.R. Sample entropy analysis of neonatal heart rate variability. *Am. J. Physiol.-Regul. Integr. Comp. Physiol.* **2002**, *283*, R789–R797, [CrossRef]
21. Yang, A.C.; Tsai, S.J.; Lin, C.P.; Peng, C.K. A strategy to reduce bias of entropy estimates in resting-state fMRI signals. *Front. Neurosci.* **2018**. [CrossRef] [PubMed]
22. Costa, M.; Goldberger, A.L.; Peng, C.K. Multiscale Entropy Analysis of Complex Physiologic Time Series. *Phys. Rev. Lett.* **2002**, *89*, 068102. [CrossRef] [PubMed]
23. Yang, A.C.; Huang, C.C.; Yeh, H.L.; Liu, M.E.; Hong, C.J.; Tu, P.C.; Chen, J.F.; Huang, N.E.; Peng, C.K.; Lin, C.P.; et al. Complexity of spontaneous BOLD activity in default mode network is correlated with cognitive function in normal male elderly: a multiscale entropy analysis. *Neurobiol. Aging* **2013**, *34*, 428–438. [CrossRef] [PubMed]
24. Smith, R.X.; Yan, L.; Wang, D.J. Multiple time scale complexity analysis of resting state FMRI. *Brain Imaging Behav.* **2014**. [CrossRef] [PubMed]
25. Niu, Y.; Wang, B.; Zhou, M.; Xue, J.; Shapour, H.; Cao, R.; Cui, X.; Wu, J.; Xiang, J. Dynamic Complexity of Spontaneous BOLD Activity in Alzheimer's Disease and Mild Cognitive Impairment Using Multiscale Entropy Analysis. *Front. Neurosci.* **2018**, *12*, 677. [CrossRef] [PubMed]
26. Grieder, M.; Wang, D.J.J.; Dierks, T.; Wahlund, L.O.; Jann, K. Default Mode Network Complexity and Cognitive Decmidrule in Mild Alzheimer's Disease. *Front. Neurosci.* **2018**, *12*, 770. [CrossRef]
27. Wang, X.; Zhang, Y.; Han, S.; Zhao, J.; Chen, H. Resting-State Brain Activity Complexity in Early-Onset Schizophrenia Characterized by a Multi-scale Entropy Method. In *Intelligence Science and Big Data Engineering*; Springer: Cham, Switzerland, 2017; pp. 580–588. [CrossRef]
28. Jia, Y.; Gu, H. Identifying nonlinear dynamics of brain functional networks of patients with schizophrenia by sample entropy. *Nonlinear Dyn.* **2019**, *96*, 2327–2340. [CrossRef]
29. Jia, Y.; Gu, H.; Luo, Q. Sample entropy reveals an age-related reduction in the complexity of dynamic brain. *Sci. Rep.* **2017**, *7*, 7990. [CrossRef]
30. Essen, D.C.V.; Smith, S.M.; Barch, D.M.; Behrens, T.E.J.; Yacoub, E.; Ugurbil, K. The WU-Minn Human Connectome Project: An overview. *NeuroImage* **2013**, *80*, 62–79. [CrossRef]
31. Gray, J.R.; Burgess, G.C.; Schaefer, A.; Yarkoni, T.; Larsen, R.J.; Braver, T.S. Affective personality differences in neural processing efficiency confirmed using fMRI. *Cogn. Affect. Behav. Neurosci.* **2005**, *5*, 182–190. [CrossRef]
32. Bilker, W.B.; Hansen, J.A.; Brensinger, C.M.; Richard, J.; Gur, R.E.; Gur, R.C. Development of Abbreviated Nine-Item Forms of the Raven's Standard Progressive Matrices Test. *Assessment* **2012**, *19*, 354–369. [CrossRef] [PubMed]
33. Gray, J.R.; Chabris, C.F.; Braver, T.S. Neural mechanisms of general fluid intelligence. *Nat. Neurosci.* **2003**, *6*, 316–322. [CrossRef] [PubMed]

34. McDonough, I.M.; Nashiro, K. Network complexity as a measure of information processing across resting-state networks: evidence from the Human Connectome Project. *Front. Hum. Neurosci.* **2014**, *8*. [CrossRef] [PubMed]
35. Costa, M.; Goldberger, A.L.; Peng, C.K. Costa, Goldberger, and Peng Reply. *Phys. Rev. Lett.* **2004**, *92*, 089804. [CrossRef]
36. Wang, D.J.J.; Jann, K.; Fan, C.; Qiao, Y.; Zang, Y.F.; Lu, H.; Yang, Y. Neurophysiological Basis of Multi-Scale Entropy of Brain Complexity and Its Relationship with Functional Connectivity. *Front. Neurosci.* **2018**, *12*. [CrossRef]
37. Gignac, G.E.; Bates, T.C. Brain volume and intelligence: The moderating role of intelligence measurement quality. *Intelligence* **2017**, *64*, 18–29. [CrossRef]
38. Gignac, G.E.; Szodorai, E.T. Effect size guidelines for individual differences researchers. *Personal. Individ. Differ.* **2016**, *102*, 74–78. [CrossRef]
39. Finn, E.S.; Shen, X.; Scheinost, D.; Rosenberg, M.D.; Huang, J.; Chun, M.M.; Papademetris, X.; Constable, R.T. Functional connectome fingerprinting: Identifying individuals using patterns of brain connectivity. *Nat. Neurosci.* **2015**, *18*, 1664–1671. [CrossRef]
40. Noble, S.; Spann, M.N.; Tokoglu, F.; Shen, X.; Constable, R.T.; Scheinost, D. Influences on the Test–Retest Reliability of Functional Connectivity MRI and its Relationship with Behavioral Utility. *Cereb. Cortex* **2017**, *27*, 5415–5429. [CrossRef]
41. Saxe, G.N.; Calderone, D.; Morales, L.J. Brain entropy and human intelligence: A resting-state fMRI study. *PLoS ONE* **2018**. [CrossRef]
42. Vakorin, V.A.; Lippe, S.; McIntosh, A.R. Variability of Brain Signals Processed Locally Transforms into Higher Connectivity with Brain Development. *J. Neurosci.* **2011**, *31*, 6405–6413. [CrossRef] [PubMed]
43. McIntosh, A.R.; Vakorin, V.; Kovacevic, N.; Wang, H.; Diaconescu, A.; Protzner, A.B. Spatiotemporal Dependency of Age-Related Changes in Brain Signal Variability. *Cereb. Cortex* **2014**, *24*, 1806–1817. [CrossRef] [PubMed]
44. Schultz, D.H.; Cole, M.W. Higher Intelligence Is Associated with Less Task-Related Brain Network Reconfiguration. *J. Neurosci.* **2016**, *36*, 8551–8561. [CrossRef] [PubMed]
45. Ferguson, M.A.; Anderson, J.S.; Spreng, R.N. Fluid and flexible minds: Intelligence reflects synchrony in the brain's intrinsic network architecture. *Netw. Neurosci.* **2017**, *1*, 192–207. [CrossRef] [PubMed]
46. Humeau-Heurtier, A. The Multiscale Entropy Algorithm and Its Variants: A Review. *Entropy* **2015**, *17*, 3110–3123. [CrossRef]
47. Faes, L.; Porta, A.; Javorka, M.; Nollo, G. Efficient Computation of Multiscale Entropy over Short Biomedical Time Series Based on Linear State-Space Models. *Complexity* **2017**, *2017*, 1768264. [CrossRef]
48. Faes, L.; Marinazzo, D.; Stramaglia, S. Multiscale Information Decomposition: Exact Computation for Multivariate Gaussian Processes. *Entropy* **2017**, *19*, 408. [CrossRef]
49. Faes, L.; Pereira, M.A.; Silva, M.E.; Pernice, R.; Busacca, A.; Javorka, M.; Rocha, A.P. Multiscale information storage of linear long-range correlated stochastic processes. *Phys. Rev. E* **2019**, *99*, 032115. [CrossRef]
50. Kosciessa, J.Q.; Kloosterman, N.A.; Garrett, D.D. Standard multiscale entropy reflects spectral power at mismatched temporal scales: What's signal irregularity got to do with it? *bioRxiv* **2019**. [CrossRef]
51. Glasser, M.F.; Sotiropoulos, S.N.; Wilson, J.A.; Coalson, T.S.; Fischl, B.; Andersson, J.L.; Xu, J.; Jbabdi, S.; Webster, M.; Polimeni, J.R.; et al. The minimal preprocessing pipelines for the Human Connectome Project. *NeuroImage* **2013**, *80*, 127. [CrossRef]
52. Salimi-Khorshidi, G.; Douaud, G.; Beckmann, C.F.; Glasser, M.F.; Griffanti, L.; Smith, S.M. Automatic denoising of functional MRI data: Combining independent component analysis and hierarchical fusion of classifiers. *NeuroImage* **2014**, *90*, 46. [CrossRef] [PubMed]
53. Siegel, J.S.; Mitra, A.; Laumann, T.O.; Seitzman, B.A.; Raichle, M.; Corbetta, M.; Snyder, A.Z. Data quality influences observed links between functional connectivity and behavior. *Cereb. Cortex* **2017**, *27*, 4492–4502. [CrossRef] [PubMed]
54. Dubois, J.; Galdi, P.; Paul, L.K.; Adolphs, R. A distributed brain network predicts general intelligence from resting-state human neuroimaging data. *Philos. Trans. R. Soc. Biol. Sci.* **2018**, *373*, 20170284. [CrossRef] [PubMed]

55. Feinberg, D.A.; Moeller, S.; Smith, S.M.; Auerbach, E.; Ramanna, S.; Glasser, M.F.; Miller, K.L.; Ugurbil, K.; Yacoub, E. Multiplexed Echo Planar Imaging for Sub-Second Whole Brain FMRI and Fast Diffusion Imaging. *PLoS ONE* **2010**, *5*, e15710. [CrossRef] [PubMed]
56. Bijsterbosch, J.; Smith, S.M.; Beckmann, C.F. *Introduction to Resting State FMRI Functional Connectivity*; Oxford University Press: Oxford, UK, 2017; p. 34.
57. Cabral, J.; Vidaurre, D.; Marques, P.; Magalhães, R.; Silva Moreira, P.; Miguel Soares, J.; Deco, G.; Sousa, N.; Kringelbach, M.L. Cognitive performance in healthy older adults relates to spontaneous switching between states of functional connectivity during rest. *Sci. Rep.* **2017**, *7*. [CrossRef] [PubMed]
58. Dosenbach, N.U.F.; Nardos, B.; Cohen, A.L.; Fair, D.A.; Power, J.D.; Church, J.A.; Nelson, S.M.; Wig, G.S.; Vogel, A.C.; Lessov-Schlaggar, C.N.; et al. Prediction of Individual Brain Maturity Using fMRI. *Science* **2010**, *329*, 1358–1361. [CrossRef]
59. Cui, Z.; Su, M.; Li, L.; Shu, H.; Gong, G. Individualized Prediction of Reading Comprehension Ability Using Gray Matter Volume. *Cereb. Cortex* **2018**, *28*, 1656–1672. [CrossRef]

Article

Decomposition of a Multiscale Entropy Tensor for Sleep Stage Identification in Preterm Infants

Ofelie De Wel [1,*], Mario Lavanga [1], Alexander Caicedo [2], Katrien Jansen [3,4], Gunnar Naulaers [3] and Sabine Van Huffel [1]

1 Department of Electrical Engineering (ESAT), STADIUS Center for Dynamical Systems, Signal Processing and Data Analytics, KU Leuven, 3001 Leuven, Belgium; mario.lavanga@kuleuven.be (M.L.); Sabine.VanHuffel@esat.kuleuven.be (S.V.H.)

2 Department of Applied Mathematics and Computer Science, Universidad del Rosario, Bogotá 111711, Colombia; alexander.caicedo@urosario.edu.co

3 Department of Development and Regeneration, Neonatal Intensive Care Unit, University Hospitals Leuven, 3000 Leuven, Belgium; katrien.jansen@uzleuven.be (K.J.); gunnar.naulaers@uzleuven.be (G.N.)

4 Department of Development and Regeneration, Child Neurology, University Hospitals Leuven, 3000 Leuven, Belgium

* Correspondence: ofelie.dewel@kuleuven.be

Received: 10 August 2019; Accepted: 23 September 2019; Published: 25 September 2019

Abstract: Established sleep cycling is one of the main hallmarks of early brain development in preterm infants, therefore, automated classification of the sleep stages in preterm infants can be used to assess the neonate's cerebral maturation. Tensor algebra is a powerful tool to analyze multidimensional data and has proven successful in many applications. In this paper, a novel unsupervised algorithm to identify neonatal sleep stages based on the decomposition of a multiscale entropy tensor is presented. The method relies on the difference in electroencephalography(EEG) complexity between the neonatal sleep stages and is evaluated on a dataset of 97 EEG recordings. An average sensitivity, specificity, accuracy and area under the receiver operating characteristic curve of 0.80, 0.79, 0.79 and 0.87 was obtained if the rank of the tensor decomposition is selected based on the age of the infant.

Keywords: CPD; EEG; multiscale entropy; sleep staging; tensor decomposition; preterm neonate

1. Introduction

In human infants, the emergence of sleep cycles occurs at approximately 26 to 28 weeks postmenstrual age (PMA) [1]. During maturation of the sleep architecture, the distribution and duration of specific sleep states change gradually. Very young preterm infants have an abundant amount of sleep, and active sleep is the dominant sleep stage. From then on, the proportion of time spent asleep decreases, while the relative amount of quiet sleep increases. Near term age, both active sleep and quiet sleep constitute approximately half of the total sleep time [2,3]. Existing research recognises the importance of sleep in early brain development [1,3,4]. Sleep and established sleep cycling play a vital role in normal neurosensory development, learning processes, memory consolidation and in the protection of the infant's brain plasticity [1]. Moreover, studies such as that conducted by Shellhaas et al [5] have shown that the presence of sleep cycling and the quantity and quality of each sleep state are associated with neurodevelopmental outcomes [6–8].

Most prematurely born infants stay in the neonatal intensive care unit (NICU) during the first critical weeks of rapid growth and development of the brain. In the NICU, neonates are exposed to a myriad of noxious environmental stimuli, such as high noise and light levels and painful procedures, which might disrupt their sleep state organization. In recent years, there has been an increasing interest in strategies to promote sleep in the NICU environment (e.g., kangaroo care, massage therapy, cycle lighting, etc.) [7,9].

Real-time automated identification of behavioural states can be used to optimize the planning of NICU caregiving in order to reduce disturbance of sleep-wake cyclicity [10]. Moreover, an automated sleep staging algorithm can be used to assess the sleep architecture and by that the functional brain maturation. In view of all that has been mentioned so far, one may suppose that there is a need to assess the sleep staging of neonates in order to provide developmentally appropriate care.

A number of algorithms for sleep stage classification in preterm neonates have been developed. The majority of these approaches are supervised and combine a set of electroencephalography (EEG) features (e.g., temporal features, spectral features, spatial features, complexity features) with a classification algorithm [11,12]. Recently, deep learning has also found its way to sleep staging in preterm infants [13]. Finally, Dereymaeker et al. [14] have proposed a cluster-based algorithm for quiet sleep detection.

This paper proposes a novel unsupervised method to discriminate quiet sleep from non-quiet sleep in preterm infants. In this study, a tensor-based method exploiting the differences in EEG complexity between different vigilance states will be used. Due to the increasing amount of data being collected, and the specific properties of tensor decompositions, multiway analysis has received increasing attention during recent decades. Tensor algebra has been used in a broad range of applications, such as image and video processing, machine learning and biomedical applications [15,16]. To the best of our knowledge, this is the first paper where tensor decompositions are used to discriminate sleep stages in preterm neonates. Therefore, this paper can serve as a proof of concept and illustrate how tensor decompositions can be used in biomedical applications, and more specifically in a classification problem based on neonatal EEG.

The remaining part of the paper proceeds as follows. The first section of this paper will describe the dataset. It will then go on to the explanation of the different steps of the proposed method. Afterwards the results of the algorithm will be reported and discussed.

2. Database

The proposed method is evaluated on a dataset consisting of EEG signals recorded at the neonatal intensive care unit of the University Hospitals of Leuven, Belgium. All neonates included in the study were born between 2012 and 2014 at a gestational age below 32 weeks (range: 24 weeks 4 days–32 weeks). In total 97 multichannel EEG signals were measured from 26 neonates at a postmenstrual age (PMA) between 27 and 42 weeks. So, each subject had at least two serial EEG recordings during their stay in the NICU. The study was approved by the Ethics committee of the University Hospitals of Leuven and parental consent was obtained for all recruited patients. Criteria for selecting the subjects were as follows: a normal neurodevelopmental outcome at 9 and 24 months corrected age (Bayley Scales of Infant Development-II, mental and motor score > 85), no severe brain lesions assessed by ultrasound and not taking any sedative or antiepileptic drugs during the EEG registration. EEG was recorded using nine electrodes: Fp1, Fp2, C3, C4, T3, T4, O1, O2 and Cz using the modified 10–20 system [17]. Electrode Cz served as a reference and was not taken into account during the analysis. The EEG data were acquired at a sampling frequency of 250 Hz using BrainRT equipment (OSG bvba, Rumst, Belgium). The analysis of the data was carried out in Matlab 2017b (The MathWorks, Inc., Natick, MA, USA), and the tensor decompositions were performed using Tensorlab [18].

Quiet sleep periods were annotated by two independent expert clinicians upon agreement. All other sleep states are merged and will be referred to as non-quiet sleep. The goal of the proposed algorithm is to automatically label EEG segments as either quiet sleep (QS) or non-quiet sleep (NQS). The most important feature in discriminating quiet sleep from non-quiet sleep is the continuity of the EEG trace. Generally, the EEG signal is relatively more discontinuous during quiet sleep compared to non-quiet sleep. Moreover, non-quiet sleep is characterized by a higher variability of the cardiorespiratory pattern and more body movements [4].

3. The Proposed Tensor-Based Sleep Stage Identification Method

The pipeline of the proposed algorithm consists of the following six steps: (1) Preprocessing of the EEG; (2) Assessment of the EEG complexity via computation of the multiscale entropy; (3) Tensorization of each EEG recording, (4) Decomposition of the multiscale entropy tensor; (5) Selecting the component of interest; and (6) Postprocessing and clustering of the temporal signature. Each step of the algorithm will be extensively described in the next paragraphs. Finally, the metrics to assess the classification performance will be explained.

3.1. EEG Preprocessing

In order to avoid distortion of the EEG time series by high or low frequency artifacts, a bandpass finite impulse response filter between 1 and 40 Hz was applied on each EEG channel. Moreover, an additional notch filter at 50 Hz is used to remove any remaining powerline interference. The filters were applied in both forward and reverse directions in order to avoid phase distortion. The EEG signal is then downsampled by a factor of two to reduce the computational complexity. No advanced artifact removal or visual preselection of data has been performed.

3.2. Multiscale Entropy Computation

After filtering the data, the EEG signal is segmented into nonoverlapping windows of 100 s [11]. To assess the complexity of each of these multichannel EEG segments, the multiscale entropy is computed. Multiscale entropy evaluates the complexity of a signal by measuring the regularity of the signal at multiple time scales [19,20]. So, the first step in its computation is to obtain signals at different scales. The coarse-grained signal y_j^τ at scale τ is obtained by taking the average of all data points within consecutive nonoverlapping windows of length τ. So the coarse-grained time series of the signal $\{x_1, x_2, ..., x_N\}$ at scale factor τ can then be written as:

$$y_j^\tau = \frac{1}{\tau} \sum_{i=(j-1)\tau+1}^{j\tau} x_i, \quad 1 \leqslant j \leqslant \frac{N}{\tau}. \tag{1}$$

This actually corresponds to a moving average operation with a window of length τ followed by a downsampling with factor τ. Hence, at scale 1 the signal is simply the original time series, while with increasing scale the coarse-grained signal length reduces progressively.

Following the coarse-graining of the EEG segment, the regularity of the signal at each scale τ is quantified using sample entropy. Sample entropy is a measure of the regularity or predictability of a time series. It is computed as the negative natural logarithm of the conditional probability that two sequences of m consecutive data points matching within a tolerance r, will also be similar when an additional data point is added to the sequence. For a discrete time series of length N, this can be expressed as [21]:

$$SampEn(m, r, N) = -ln\frac{A}{B} \tag{2}$$

where A and B denote the number of matches of length $m + 1$ and m within tolerance r, respectively. In this study, the template length m is set equal to 2. The multiscale entropy (MSE) is computed for scales ranging from 1 to 20. The tolerance r is defined as $0.2\times$ the standard deviation of the original time series [11,22]. Once the sample entropy is computed for each of the coarse-grained time series, a multiscale entropy (MSE) curve can be constructed. This curve shows the sample entropy in function of the scale factor τ. Hence, it reflects the regularity of the signal across multiple scales. The procedure to compute multiscale entropy of a time series is presented in Figure 1. The multiscale entropy curve of each 100 s multichannel EEG segment is computed.

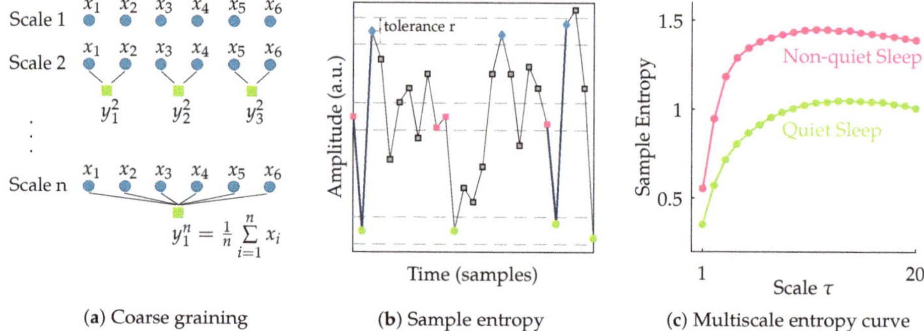

Figure 1. Illustration of the procedure to compute multiscale entropy (MSE). The first step (**a**) consists of coarse graining the time series into different scales; In the second step (**b**), the sample entropy of each of these coarse grained time series is computed; Finally, in (**c**), a multiscale entropy curve can be constructed by plotting the sample entropy in function of scale. The multiscale entropy curve is usually lower during quiet sleep compared to non-quiet sleep.

3.3. Tensorization

The entropy values of each EEG recording are then organized in a third order tensor with modes: channels, scales and time segments. So, the multiscale entropy curves of consecutive time segments are stacked in the third mode of the tensor $\mathcal{X} \in \mathbb{R}^{N \times S \times T}$. Using this tensorization the structural information among the leads is preserved. In this study, the number of EEG channels N is equal to 8, the number of scales S for which the sample entropy is computed is equal to 20 and the number of time segments T depends on the length of the EEG recording. Thus, the data of each EEG recording is transformed into a third order tensor $\mathcal{X} \in \mathbb{R}^{8 \times 20 \times T}$, where each row fiber (mode-2 fiber) of the tensor represents a multiscale entropy curve from a specific EEG channel and time segment. In the remainder of this paper, this tensor will be referred to as the multiscale entropy tensor.

3.4. Tensor Decomposition

Throughout this paper, a scalar, vector, matrix and tensor will be denoted by lowercase letters (a), boldface lowercase letters (**a**), boldface capitals (**A**) and letters in calligraphic script ($\mathcal{T}$), respectively. The outer product is denoted by $\circ$ and $\|.\|_F$ stands for the Frobenius norm.

The canonical polyadic decomposition (CPD) or parallel factor analysis (PARAFAC) of a rank-R tensor $\mathcal{T}$ factorizes the tensor in a sum of R rank-1 tensors [23]. This can be written as:

$$\mathcal{T} = \sum_{r=1}^{R} \mathbf{a}_r \circ \mathbf{b}_r \circ \mathbf{c}_r = [\![\mathbf{A}, \mathbf{B}, \mathbf{C}]\!] \tag{3}$$

where the factor vectors $\mathbf{a}_r$, $\mathbf{b}_r$ and $\mathbf{c}_r$ are the rth ($1 \leqslant r \leqslant R$) column of the factor matrices **A**, **B** and **C**, respectively. The factor matrices **A**, **B** and **C** are obtained by solving the least-squares optimization problem with objective function:

$$\min_{\mathbf{A},\mathbf{B},\mathbf{C}} \frac{1}{2} \|\mathcal{T} - [\![\mathbf{A}, \mathbf{B}, \mathbf{C}]\!]\|_F^2 . \tag{4}$$

The advantage of CPD compared to matrix factorizations is that the decomposition is unique (up to scaling and permutation ambiguity) under mild conditions [24]. Moreover, prior knowledge of the data properties can easily be taken into account by imposing constraints on the factor matrices (e.g., sparsity, smoothness, non-negativity, etc.) [15]. Because entropy values are positive, non-negativity is enforced during the decomposition of the multiscale entropy tensor. Moreover, the non-negativity constraint circumvents the occurrence of degenerate components.

The CPD of the third-order multiscale entropy tensor with rank R will result in three non-negative factor matrices: the spatial signatures will form the columns of $\mathbf{A} \in \mathbb{R}^{N \times R}$, the scale signatures the columns of $\mathbf{B} \in \mathbb{R}^{S \times R}$ and the temporal signatures the columns of $\mathbf{C} \in \mathbb{R}^{T \times R}$. An example of this decomposition is shown in Figure 2. The factor vectors in the first mode $\mathbf{a}_r$, will show the variation over the different EEG channels, the factor vectors in the second mode $\mathbf{b}_r$, contain information about the distribution over scales, while the factor vectors in the third mode $\mathbf{c}_r$, will capture the variation of the EEG complexity over the different time segments.

Starting from the assumption that the EEG complexity is different depending on the sleep stage [11], we expect that one of the temporal signatures will reflect the sleep cycling. Since the goal of the algorithm is to perform automated sleep staging, only the factor vectors in the third mode, the temporal signatures, are of interest.

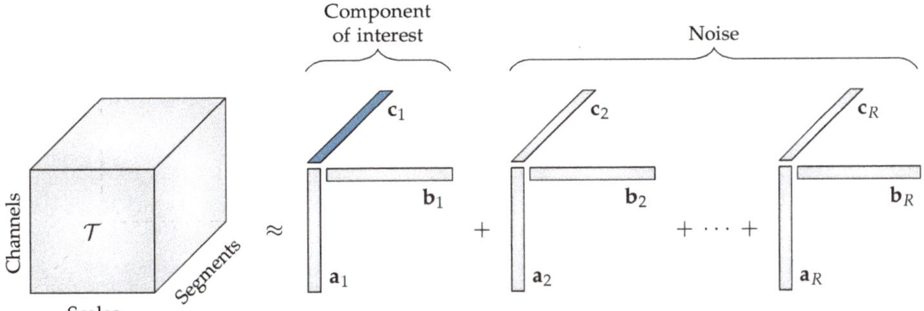

Figure 2. The rank-R canonical polyadic decomposition of a multiscale entropy tensor $\mathcal{T}$. When the rank R is greater than 1, the component of interest related to the neonatal sleep staging has to be selected. The temporal signature of interest $\mathbf{c}_1$, which is highlighted in blue, will then be used to discriminate the neonatal sleep stages. All other components will be discarded during further analysis.

3.4.1. Detection of Stable Solution

Due to the fact that (4) is a non-convex optimization problem, we may end up in a local minimum rather than a global minimum. Thus, different factor matrices can be obtained depending on their initialization [25]. To ensure that a reliable solution is found, the decomposition is repeated 50 times with different random initializations. Next, we want to detect and select the most occurring, stable solution among these 50 repetitions. To this end, the similarity between all possible combinations of components is assessed by means of cosine similarity. In this study we assume that each component should match in all its modes, so the triple cosine product or congruence is computed. To assess the similarity of two rank 1 tensors $\mathcal{X} = \mathbf{k} \circ \mathbf{l} \circ \mathbf{m}$ and $\mathcal{Y} = \mathbf{p} \circ \mathbf{q} \circ \mathbf{r}$ this can be written as [26–28]:

$$\text{cong}(\mathcal{X}, \mathcal{Y}) = \cos(\mathbf{k}, \mathbf{p}) \cos(\mathbf{l}, \mathbf{q}) \cos(\mathbf{m}, \mathbf{r}) = \frac{\mathbf{k}^T \mathbf{p}}{||\mathbf{k}||\ ||\mathbf{p}||} \frac{\mathbf{l}^T \mathbf{q}}{||\mathbf{l}||\ ||\mathbf{q}||} \frac{\mathbf{m}^T \mathbf{r}}{||\mathbf{m}||\ ||\mathbf{r}||}. \tag{5}$$

The R components of each repetition are sorted based on (5) to account for the permutation indeterminacy of the decomposition. Subsequently, the similarity between the corresponding components of different repetitions is investigated. On the assumption that all R components should match, the congruence of the R components is multiplied. This can be organized in a symmetric similarity matrix $\mathbf{S} \in \mathbb{R}^{it \times it}$, where it represents the number of iterations (in this case 50). So, s_{ij} represents how similar the factor matrices of the ith and jth iteration are. Finally, the solution of the iteration with the highest similarity to the other repetitions is selected as the stable and reproducible solution.

3.4.2. Number of Components

One of the principal challenges when applying tensor decompositions is selecting an appropriate number of components for the problem at hand [29]. A multitude of strategies to define the rank have been proposed. First of all, we investigated the multilinear singular spectrum in order to get an initial estimate of the rank. Then, we performed the decomposition for rank R going from 1 to 5. To compare the different ranks, we investigated the core consistency diagnostic (CORCONDIA), which assesses how close the core tensor is to being superdiagonal. In addition, a diagnostic based on the relative reduction of the fitting error for an increase of the rank, called the difference of fit (DIFFIT) [30], was computed to assist in determining the appropriate number of components.

3.5. Selection of the Component of Interest

When decomposing the multiscale entropy tensor, we expect that one component will reflect the sleep staging of the neonate. However, if the number of components R is greater than one, an automatic selection of the component of interest is required. So, the goal is to find the temporal signature of the component of interest. In Figure 2 this is the temporal signature of the first component, ($\mathbf{c}_1$) which is marked in blue.

We expect that the component related to sleep staging will have a more regular, cyclic pattern compared to the other (noise) components. Therefore, the autocorrelation functions of the temporal signatures are computed. Since the temporal signature reflecting the sleep staging has a stronger correlation in time, we expect that its area under the absolute value of the autocorrelation function (ACF) will be larger [31,32]. Therefore, the component whose area is the largest will be selected as the component of interest and used for further processing.

Figure 3 shows an example of this procedure for a rank-2 CPD of a multiscale entropy tensor. The autocorrelation of the two temporal signatures is plotted on the left. The area under the absolute value of the autocorrelation is equal to 31.85 and 13.82 for the temporal signature of the first (blue) and second (red) component, respectively. Therefore, the first component will be selected as the component of interest. The temporal signatures of the two components are plotted on the right with highlighted periods of quiet sleep in light grey. From these graphs we can see that the first component is indeed reduced during quiet sleep segments compared to non-quiet sleep segments, whereas the second component is not related to the sleep stages.

3.6. Postprocessing and Clustering

Once the non-negative polyadic decomposition of the multiscale entropy tensor is performed, the temporal signature of the selected component is used to define the neonate's sleep stage. This temporal signature shows a (slow) cyclic pattern reflecting the sleep staging of the infant. However, there are high frequency oscillations superimposed on this pattern, which could lead to incorrect sleep stage identification. Therefore, a postprocessing step consisting of smoothing the temporal signature using a moving average filter with length L equal to 5 is applied. This moving average filter is applied in both directions (to avoid phase distortion), so actually a weighted moving average filter is used with triangular shape and length 2 $\times$ L. This smoothing operation accounts for the fact that a sleep stage does not change instantly. More specifically, sleep periods were only visually labeled as quiet sleep if 3 consecutive minutes or 3 out 4 min were clinically detected as quiet sleep.

In order to divide the data then into two distinct clusters, k-means clustering (k = 2) is performed using the smoothed temporal loading. So, each data point of the smoothed temporal signature of interest $\mathbf{c} \in \mathbb{R}^{T\times 1}$ will be assigned to one of the two clusters. Thus, the result of the clustering is a vector of length T containing a cluster label for each of the 100 s EEG segments. Since k-means clustering is heavily dependent on its initialization, k-means clustering is repeated 100 times with different initial cluster centroids and the clustering with the lowest sum of within-cluster point-to-centroid distances is selected. The cluster with the lowest EEG complexity will be assigned the label of quiet sleep, while the other cluster is labeled as non-quiet sleep [11,33].

The effect of the smoothing and the result of the clustering is illustrated in Figure 4. In this example, the multiscale entropy tensor of an EEG recording is decomposed for rank R equal to 1 and 2. The raw temporal signatures of the rank-1 and rank-2 decomposition (the only temporal signature of interest) are shown on the left, while their smoothed versions are plotted on the right. The smoothing clearly removes the unwanted variations. The sleep labels obtained by the algorithm based on k-means clustering are marked by yellow squares (cluster 1 corresponding to quiet sleep) and red dots (cluster 2 corresponding to non-quiet sleep). The rank-1 CPD does not correlate well with the clinical annotations of quiet sleep highlighted in light grey, while the smoothed temporal signature of the rank-2 CPD is clearly reduced during quiet sleep.

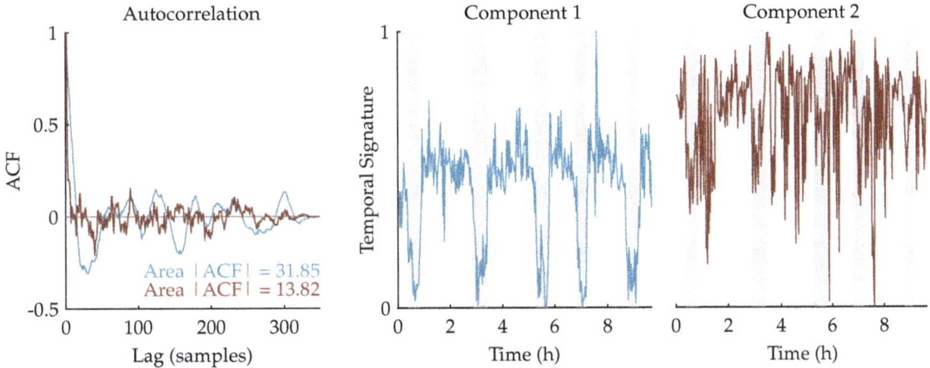

Figure 3. On the left, the autocorrelations of the temporal signatures of a rank-2 canonical polyadic decomposition (CPD) of a multiscale entropy tensor are shown. The area under the absolute value of the autocorrelation is equal to 31.85 and 13.82 for the first component (in blue) and the second component (in red), respectively. Hence, the first component will be selected as the component of interest. On the right, the corresponding temporal signatures are plotted with the annotated quiet sleep periods highlighted in light grey. The temporal signature of the first component shows a clear reduction during quiet sleep, while the second component is not correlated to the sleep cycling.

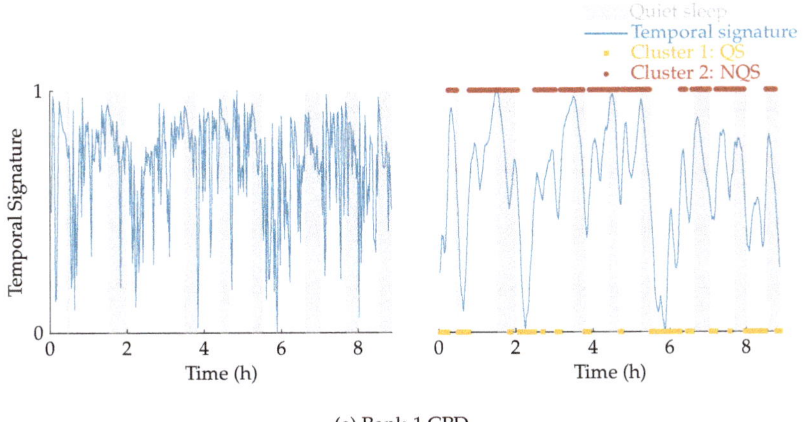

(a) Rank-1 CPD.

Figure 4. *Cont.*

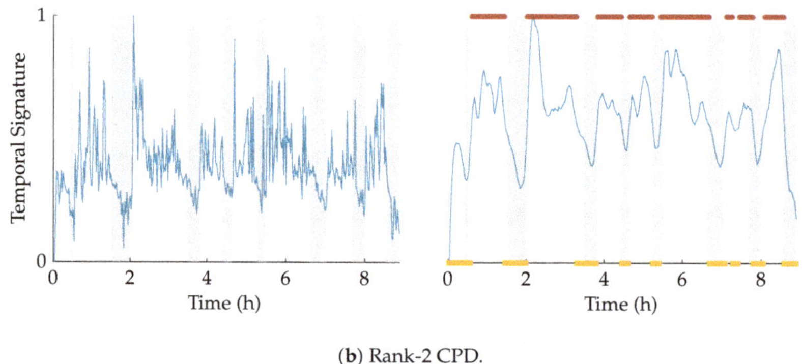

(**b**) Rank-2 CPD.

Figure 4. Illustration of the postprocessing and clustering of the temporal signature for a rank-1 (**a**) and rank-2 (**b**) decomposition of the multiscale entropy tensor (PMA = 40 weeks 5 days). On the left half of the figure, the temporal signature of interest is plotted in blue. The right half of the figure shows the smoothed temporal signature after applying the weighted moving average filter. Moreover, the quiet sleep and non-quiet sleep segments estimated by the algorithm are marked by the yellow squares and red dots, respectively. The clinically labelled quiet sleep periods are highlighted in light grey. The rank-1 CPD does not give a good indication of the sleep stages, while the (smoothed) temporal signature of the rank-2 CPD is clearly reduced during the quiet sleep periods.

3.7. Classification Performance

The performance of the algorithm is evaluated using the annotations by expert clinicians. The sensitivity, specificity and accuracy will be reported. Moreover, to investigate the performance without the k-means clustering, receiver operating characteristic (ROC) curves are constructed based on the smoothed temporal signatures and the clinical labels. The area under the ROC curve (AUC) will be reported. Finally, Cohen's kappa will be presented as well.

3.8. Statistical Analysis

Statistical analysis is performed to gain insight in the choice of the parameters. A statistical test is used to investigate whether the area under the absolute value of the autocorrelation function is a good feature to detect the component of interest. In order to do this, the agreement between each of the R temporal signatures obtained from a rank-R CPD and the clinical sleep labels are evaluated using the Kappa score. The temporal signatures are then sorted in descending order based on their Kappa statistic. As a result, the first component is the most related to sleep staging and should have the largest area under the absolute value of its autocorrelation function (ACF). A statistical test is then performed to assess whether the area under the ACF of the component of interest is significantly different from the other components. In addition, a statistical test is carried out to investigate the influence of the rank R on the performance. More specifically, we compared Cohen's Kappa between models with different values of the rank R. In all analyses, the Shapiro-Wilk test is used to test for normality. If the data is normally distributed, one-way ANOVA is used, otherwise a Kruskal-Wallis test is performed. The significance level is always set equal to 0.05.

4. Results

Comparison of the factor matrices obtained using the rank indicated by CORCONDIA or DIFFIT with the ones for other ranks revealed that the diagnostics were often not suitable to define the appropriate number of components in this application. Therefore, the rank has not been fixed for each recording, instead the results are reported for different values of the rank R. Table 1 presents the mean and standard deviation of the performance measures. The first five rows show the performance for a fixed rank R for all recordings with R going from 1 to 5. From these data, we can see that the average performance is slightly higher for rank 2 compared to rank 1 (higher AUC and kappa). Moreover, Table 1 also shows that the performance decreases gradually for an increasing rank beyond 2. However, a Kruskal-Wallis test with multiple comparisons revealed that this performance difference between the rank-1 and rank-2 model is not significant, while Cohen's kappa of the rank-1 and rank-2 model are both significantly better compared to the rank-5 model. In order to get more insight in the performance difference between a rank-1 and rank-2 CPD, the performance of each recording is plotted as a function of the postmenstrual age in Figure 5. In this figure each green circle and pink square represents the area under the ROC curve of an EEG recording for a rank-1 and rank-2 decomposition, respectively. The dashed line indicates 37 weeks postmenstrual age. The most interesting aspect of this graph is that when a rank-1 decomposition is used, a high performance is only obtained up to around 36 to 37 weeks PMA. From that age onwards a clear drop in the AUC can be observed in Figure 5. The rank-2 CPD on the other hand, has a high performance for most recordings at term equivalent age, but has a lower performance for some measurements recorded at a younger age.

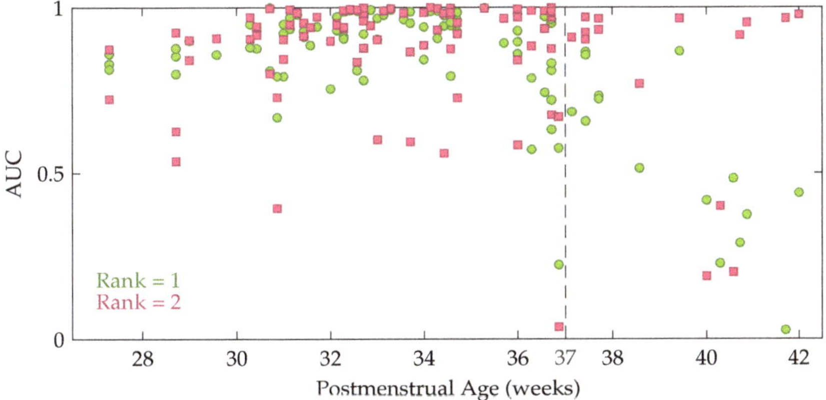

Figure 5. The area under the ROC curve as a function of the postmenstrual age at the moment of the recording. Each green circle shows the performance of one of the EEG recordings for a rank-1 CPD, whereas the pink squares represent the performance for a rank-2 CPD.

Based on these findings we decided to also report the results for an age-dependent rank, where the rank is chosen equal to 1 for preterm recordings and equal to 2 for EEGs recorded from 37 weeks PMA onwards. The performance of this approach is set out in the sixth row of Table 1. It is clear that the average performance of using an age-dependent rank is better compared to a fixed rank for all recordings, however this performance difference is not significant.

Finally, the last two rows of the table show the classification performance when the optimal rank (between 1 and 5) for each recording was selected based on Cohen's kappa computed with the ground truth annotations. Specifically, the last row corresponds to the highest attainable performance (optimal rank and best component), while in the penultimate row the optimal rank was used but the component of interest was automatically selected using the strategy based on the autocorrelation explained above. As a consequence, the difference between these two rows is caused by failures of the automatic component selection procedure.

As explained in Section 3.8, a statistical analysis is performed to examine whether the area under the ACF is a suitable feature to rely on for the automatic component selection. The boxplots in Figure 6 show the area under the absolute value of the ACF after sorting the temporal signatures based on their agreement with the clinical annotations. The component of interest, with the largest Kappa score, is the first component and is marked in blue. A Mann-Whitney U test and Kruskal-Wallis test with multiple comparisons determined that the first component is significantly different from all other components for the rank-2 and rank-3 CPD, respectively (Figure 6a,b). For the rank-4 and rank-5 CPD the area under the ACF of the sleep staging component is significantly different compared to all other components except the second one. This is also indicated on the boxplots in Figure 6c,d.

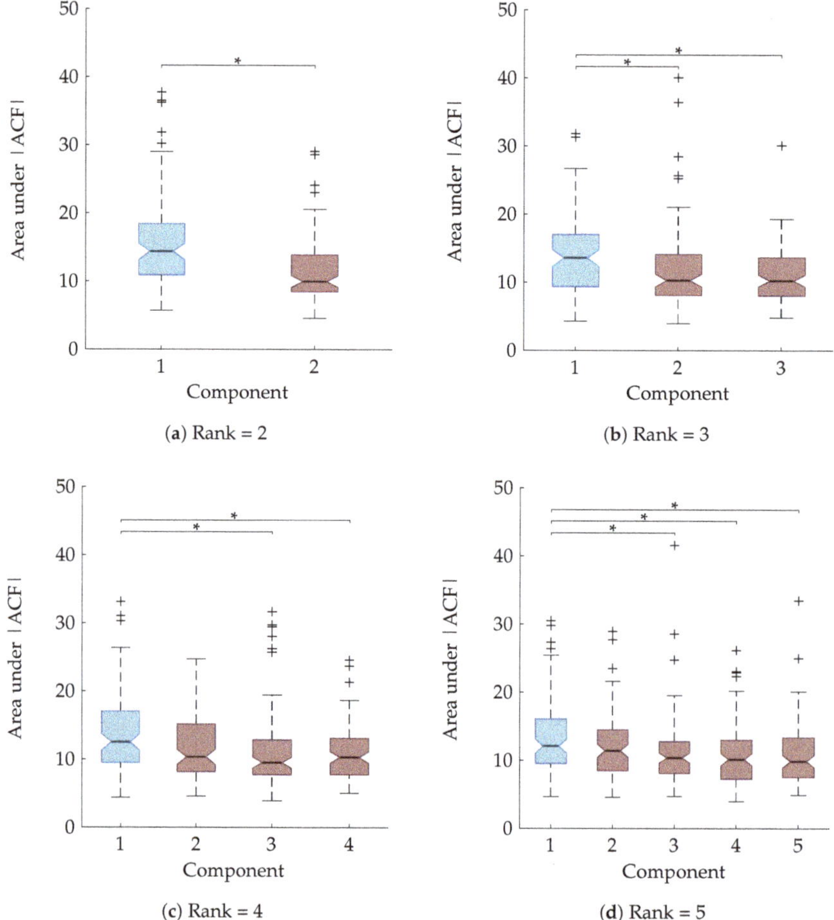

Figure 6. Boxplots of the area under absolute value of autocorrelation of the temporal signatures after sorting them in descending order according to their Kappa score. The component with the largest Kappa score and therefore most related to sleep staging is the first one and is marked in blue, while the other (noise) components are marked in red. "*" indicates a statistically significant difference with $p < 0.05$. For the rank-2 CPD (**a**) and the rank-3 CPD (**b**) there is a significant difference between the area under the ACF of the sleep staging component and the noise component(s); For a rank-4 (**c**) and rank-5 CPD (**d**) the component most related to sleep staging is significantly different from all other components except one.

Table 1. The classification performance of the proposed tensor-based method for different values of the rank R. The mean (standard deviation) of the sensitivity, specificity, accuracy, area under the ROC curve (AUC) and Cohen's kappa are presented.

	Sensitivity	Specificity	Accuracy	AUC	Kappa
Fixed rank					
R = 1	0.73(0.29)	0.79(0.16)	0.78(0.15)	0.84(0.19)	0.47(0.34)
R = 2	0.82(0.27)	0.74(0.23)	0.76(0.18)	0.87(0.19)	0.49(0.33)
R = 3	0.74 (0.34)	0.71(0.2)	0.72(0.17)	0.80(0.24)	0.38(0.36)
R = 4	0.75(0.33)	0.67(0.22)	0.70(0.18)	0.80(0.24)	0.35(0.36)
R = 5	0.73(0.35)	0.64(0.24)	0.67(0.21)	0.75(0.27)	0.31(0.38)
Age-dependent rank					
Preterm: R = 1, term: R = 2	0.80(0.23)	0.79(0.17)	0.79(0.14)	0.87(0.16)	0.53(0.28)
Optimal rank					
Automatic component selection	0.76(0.32)	0.78(0.18)	0.79(0.17)	0.85(0.21)	0.50(0.38)
Optimal component	0.86(0.32)	0.81(0.18)	0.82(0.17)	0.91(0.21)	0.60(0.38)

5. Discussion

This study aimed to discriminate quiet sleep (QS) from non-quiet sleep (NQS) using the nonlinear dynamics of the EEG signal in a data-driven way. The proposed algorithm relies on the fact that the EEG complexity is different depending on the sleep state. In order to quantify the complexity of the EEG signal, multiscale entropy is computed for consecutive segments of a multichannel EEG recording. These multiscale entropy values are then used to construct a third-order tensor with modes channels, scales and segments. Subsequently, a rank-R nonnegative CPD of the multiscale entropy tensor is computed. The temporal signature of interest is detected based on the area under the absolute value of its autocorrelation function. After smoothing, this temporal signature shows a cyclic pattern reflecting the neonatal sleep staging. Clustering is then performed to discriminate quiet sleep from non-quiet sleep.

The performance of the algorithm is reported for different ranks. The average Kappa is equal to 0.47 and 0.49, for a rank-1 and rank-2 CPD, respectively. However, if a rank-1 decomposition is used for EEGs recorded at a postmenstrual age below 37 weeks and a rank-2 decomposition is used for EEGs recorded at an older age, a Kappa of 0.53 is obtained. This indicates that a higher rank is preferred for neonates at term equivalent age, while rank 1 is suitable for most of the preterm recordings. The fact that for some recordings rank 1 is not sufficient, is demonstrated by the example in Figure 4. An even higher performance could be obtained if the optimal rank is used and the component of interest is detected correctly. However, as can be seen from Table 1, the performance when the optimal rank is used and the component of interest is selected automatically is slightly lower compared to an age-dependent rank. This reduction in performance is due to the fact that the procedure to select the component of interest is more likely to fail for a higher number of components. From the data presented in Table 1 we can see that the performance gradually decreases if a rank higher than two is used. This finding suggests that for the majority of the recordings in the current database a lower rank is more appropriate. Nevertheless, this performance reduction can also be partly attributed to the more difficult component selection for higher ranks.

There are various possible explanations for the need for a higher rank for recordings at term-equivalent age. First of all, it is likely that this is due to the increased amount of artifacts in near term EEG recordings. It is expected that an older neonate will move more, which could lead to more severe distortions and lower quality of the EEG signal. The higher performance for higher ranks in noisy preterm EEG recordings confirms this reasoning. Secondly, the emergence of four distinct sleep states around 36 weeks PMA might also play a role in this phenomenon [4].

The performance of the described algorithm is lower compared to state-of-the-art algorithms [13,14,34]. However, there are multiple advantages of the proposed method. The major advantage of the algorithm is that it is data-driven and unsupervised. As a consequence, the method can be easily applied on a completely new dataset and used in new centers, where there is little expertise about EEG sleep labelling. Moreover, in this study the algorithm is assessed on eight channel EEG recordings measured using the restricted 10–20 system for neonates. However, the same approach can also be applied on datasets with fewer or more electrodes without adjustments. Finally, the tensor decomposition can be updated in an efficient way whenever a new batch of EEG data is available [35]. This allows real time tracking of neonatal sleep states.

The main weakness of this study is the lack of a method to define the rank for each new EEG recording automatically in a data-driven way. This is a general issue in tensor applications, therefore it is suggested to combine different tools to asses the rank [36,37]. We tested several diagnostics, but none of them resulted in a reliable estimate of the number of components. More research is needed to determine the optimal number of components for this specific application. An additional drawback of the proposed algorithm is the procedure for selecting the component of interest. The temporal signature of interest does not always have the largest area under the absolute value of the autocorrelation function. For that reason more advanced strategies for component selection should be investigated in future studies. Another limitation of the study is that artifacts commonly lead to misclassification of the segment. More specifically, EEG segments distorted by artifacts are often classified as quiet sleep because they have a reduced EEG complexity. The cause of this is two-fold. First of all, artifacts are more predictable and less complex compared to noise-free EEG [38]. Secondly, a high amplitude artifact will drastically increase the standard deviation of the segment, resulting in a much higher tolerance r and thus more template matches and consequently lower entropy. Detecting artifacts as quiet sleep periods will lead to a reduced performance, since the majority of the (motion) artifacts occur during non-quiet sleep. The smoothing operation is implemented to reduce the influence of artifacts and can deal with short duration artifacts (sudden drop in EEG complexity). Yet, the postprocessing cannot avoid false detections caused by longer artifacts (on multiple channels). Further research might explore incorporating information about the noise in the algorithm to make it more robust to artifacts. Besides, an estimation of the amount of noise could be used to get an initial estimate of the proper number of components. Notwithstanding these limitations, the study suggests that a decomposition of a multiscale entropy tensor can be used to discriminate neonatal sleep stages.

This study is an exploratory analysis on the use of tensor decompositions for sleep stage identification in preterm infants. Only the differences in complexity between different sleep stages are used. However, in future work the temporal signature could be combined with other discriminating features (e.g., spectral edge frequency, power in specific EEG frequency bands, etc.) to boost the performance. Moreover, as explained before, the heart rate variability, respiration and body movements of the infant are also dependent on the sleep state. Hence, in future research the combination of these complementary modalities can be studied. Finally, other tensorization techniques can be examined in future investigations.

6. Conclusions

This study confirmed that the complexity of brain dynamics exhibits fundamental differences between vigilance states in preterm newborns. The EEG complexity is significantly lower during quiet sleep compared to non-quiet sleep. This property is exploited to develop an unsupervised algorithm that can detect quiet sleep in preterm infants based on the data-driven factorization of the multiscale entropy tensor.

Author Contributions: Conceptualization, methodology, O.D.W., M.L., A.C.; data curation, K.J.; investigation, formal analysis, software, visualization, writing-original draft preparation, O.D.W.; writing-review and editing, all; supervision, S.V.H., G.N. and A.C.

Funding: This research received no external funding.

Acknowledgments: Research supported by Bijzonder Onderzoeksfonds KU Leuven (BOF): The effect of perinatal stress on the later outcome in preterm babies #: C24/15/036. European Research Council: The research leading to these results has received funding from the European Research Council under the European Union's Seventh Framework Programme (FP7/2007–2013)/ERC Advanced Grant: BIOTENSORS (n°339804). This paper reflects only the authors' views and the Union is not liable for any use that may be made of the contained information. Mario Lavanga is a SB PhD fellow at Fonds voor Wetenschappelijk Onderzoek-Vlaanderen (FWO), supported by Flemish government.

Conflicts of Interest: The authors declare no conflict of interest.

Abbreviations

The following abbreviations are used in this manuscript:

ACF	Autocorrelation function
AUC	Area under the curve
CORCONDIA	Core consistency diagnostic
CPD	Canonical polyadic decomposition
DIFFIT	Difference of fit
EEG	Electroencephalogram
FIR	Finite impulse response
MSE	Multiscale entropy
NICU	Neonatal intensive care unit
NQS	Non-quiet sleep
PMA	Postmenstrual age
QS	Quiet sleep
ROC	Receiver operating characteristic

References

1. Graven, S.N.; Browne, J.V. Sleep and Brain Development The Critical Role of Sleep in Fetal and Early Neonatal Brain Development. *Newborn Infant Nurs. Rev.* **2008**, *8*, 173–179. [CrossRef]
2. Graven, S.N. Sleep and Brain Development. *Clin. Perinatol.* **2006**, *33*, 693–706. [CrossRef]
3. Calciolari, G.; Montirosso, R. The sleep protection in the preterm infants. *J. Mater.-Fetal Neonatal Med.* **2011**, *24*, 12–14. [CrossRef]
4. Dereymaeker, A.; Pillay, K.; Vervisch, J.; De Vos, M.; Van Huffel, S.; Jansen, K.; Naulaers, G. Review of sleep-EEG in preterm and term neonates. *Early Hum. Dev.* **2017**, *113*, 87–103. [CrossRef]
5. Shellhaas, R.A.; Burns, J.W.; Hassan, F.; Carlson, M.D.; Barks, J.D.E.; Chervin, R.D. Neonatal Sleep-Wake Analyses Predict 18-month Neurodevelopmental Outcomes. *Sleep* **2017**, *40*. [CrossRef]
6. Osredkar, D.; Toet, M.C.; van Rooij, L.G.M.; van Huffelen, A.C.; Groenendaal, F.; de Vries, L.S. Sleep-Wake Cycling on Amplitude-Integrated Electroencephalography in Term Newborns With Hypoxic-Ischemic Encephalopathy. *Pediactrics* **2005**, *115*, 327–332. [CrossRef]
7. Barbeau, D.Y.; Weiss, M.D. Sleep Disturbances in Newborns. *Children* **2017**, *4*, 90. [CrossRef]
8. Kidokoro, H.; Kubota, T.; Hayashi, N.; Hayakawa, M.; Takemoto, K.; Kato, Y.; Okumura, A. Absent cyclicity on aEEG within the first 24 h is associated with brain damage in preterm infants. *Neuropediatrics* **2010**, *41*, 241–245. [CrossRef]
9. van den Hoogen, A.; Teunis, C.J.; Shellhaas, R.A.; Pillen, S.; Benders, M.; Dudink, J. How to improve sleep in a neonatal intensive care unit: A systematic review. *Early Hum. Dev.* **2017**, *113*, 78–86. [CrossRef]
10. Werth, J.; Atallah, L.; Andriessen, P.; Long, X.; Zwartkruis-Pelgrim, E.; Aarts, R.M. Unobtrusive sleep state measurements in preterm infants—A review. *Sleep Med. Rev.* **2017**, *32*, 109–122. [CrossRef]
11. De Wel, O.; Lavanga, M.; Caicedo Dorado, A.; Jansen, K.; Dereymaeker, A.; Naulaers, G.; Huffel, S.V. Complexity Analysis of Neonatal EEG Using Multiscale Entropy : Applications in Brain Maturation and Sleep Stage Classification. *Entropy* **2017**, *19*, 516. [CrossRef]
12. Koolen, N.; Oberdorfer, L.; Rona, Z.; Giordano, V.; Werther, T.; Klebermass-Schrehof, K.; Stevenson, N.; Vanhatalo, S. Automated classification of neonatal sleep states using EEG. *Clin. Neurophysiol.* **2017**, *128*, 1100–1108. [CrossRef]

13. Ansari, A.H.; De Wel, O.; Lavanga, M.; Caicedo, A.; Dereymaeker, A.; Jansen, K.; Vervisch, J.; De Vos, M.; Naulaers, G.; Van Huffel, S. Quiet sleep detection in preterm infants using deep convolutional neural networks. *J. Neural Eng.* **2018**, *15*, 066006. [CrossRef]
14. Dereymaeker, A.; Pillay, K.; Vervisch, J.; Van Huffel, S.; Naulaers, G.; Jansen, K.; De Vos, M. An Automated Quiet Sleep Detection Approach in Preterm Infants as a Gateway to Assess Brain Maturation. *Int. J. Neural Syst.* **2017**, *27*, 1750023. [CrossRef]
15. Cichocki, A.; Mandic, D.; De Lathauwer, L.; Zhou, G.; Zhao, Q.; Caiafa, C.; Phan, H.A. Tensor decompositions for signal processing applications: From two-way to multiway component analysis. *IEEE Signal Process. Mag.* **2015**, *32*, 145–163. [CrossRef]
16. Zhou, G.; Zhao, Q.; Zhang, Y.; Adalı, T.; Xie, S.; Cichocki, A. Linked Component Analysis From Matrices to High-Order Tensors: Applications to Biomedical Data. *Proc. IEEE* **2016**, *104*, 310–331. [CrossRef]
17. Cherian, P.J.; Swarte, R.M.; Visser, G.H. Technical standards for recording and interpretation of neonatal electroencephalogram in clinical practice. *Ann. Indian Acad. Neurol.* **2009**, *12*, 58–70. [CrossRef]
18. Vervliet, N.; Debals, O.; Sorber, L.; Van Barel, M.; De Lathauwer, L. Tensorlab 3.0. 2016. Available online: https://www.tensorlab.net (accessed on 10 August 2019).
19. Costa, M.; Goldberger, A.L.; Peng, C.K. Multiscale Entropy Analysis of Complex Physiologic Time Series. *Phys. Rev. Lett.* **2002**, *89*, 068102. [CrossRef]
20. Costa, M.; Goldberger, A.L.; Peng, C.K. Multiscale entropy analysis of biological signals. *Phys. Rev. E* **2005**, *71*, 1–18. [CrossRef]
21. Humeau-Heurtier, A. The Multiscale Entropy Algorithm and Its Variants: A Review. *Entropy* **2015**, *17*, 3110–3123. [CrossRef]
22. Escudero, J.; Abásolo, D.; Hornero, R.; Espino, P.; López, M. Analysis of electroencephalograms in Alzheimer's disease patients with multiscale entropy. *Physiol. Meas.* **2006**, *27*, 1091–1106. [CrossRef]
23. Kolda, T.G.; Bader, B.W. Tensor decompositions and applications. *SIAM Rev.* **2009**, *51*, 455–500. [CrossRef]
24. Domanov, I.;De Lathauwer, L. On the Uniqueness of the Canonical Polyadic Decomposition of Third-Order Tensors—Part I: Basic results and uniquensess of one factor matrix and part II: Uniqueness of the overall decomposition. *SIAM J. Matrix Anal. Appl.* **2013**, *34*, 855–903. [CrossRef]
25. Van Eyndhoven, S.; Vervliet, N.; De Lathauwer, L.; Van Huffel, S. Identifying Stable Components of Matrix/Tensor Factorizations via Low-Rank Approximation of Inter-Factorization Similarity. In Proceedings of the 27th European Signal Processing Conference (EUSIPCO), A Coruna, Spain, 2–6 September 2019.
26. Bro, R. PARAFAC. Tutorial and applications. *Chemom. Intell. Lab. Syst.* **1997**, *38*, 149–171. [CrossRef]
27. Tomasi, G.; Bro, R. A comparison of algorithms for fitting the PARAFAC model. *Comput. Stat. Data Anal.* **2006**, *50*, 1700–1734. [CrossRef]
28. Lorenzo-Seva, U.; Ten Berge, J.M.F. Tucker's Congruence Coefficient as a Meaningful Index of Factor Similarity. *Hogrefe Huber Publ. Methodol.* **2006**, *2*, 57–64. [CrossRef]
29. Sidiropoulos, N.D.; De Lathauwer, L.; Fu, X.; Huang, K.; Papalexakis, E.E.; Faloutsos, C. Tensor Decomposition for Signal Processing and Machine Learning. *IEEE Trans. Signal Process.* **2017**, *65*, 3551–3582. [CrossRef]
30. Timmerman, M.E.; Kiers, H.A.L. Three-mode principal components analysis: Choosing the numbers of components and sensitivity to local optima. *Br. J. Math. Stat. Psychol.* **2000**, *53*, 1–16. [CrossRef]
31. Kobayashi, H.; Kakihana, W.; Kimura, T. Combined effects of age and gender on gait symmetry and regularity assessed by autocorrelation of trunk acceleration. *J. Neuroeng. Rehabil.* **2014**, *11*, 109. [CrossRef]
32. Deburchgraeve, W. Development of an Automated Neonatal EEG Seizure Monitor. Ph.D. Thesis, KU Leuven: Leuven, Belgium, 2010.
33. Janjarasjitt, S.; Scher, M.S.; Loparo, K.A. Nonlinear dynamical analysis of the neonatal EEG time series: The relationship between sleep state and complexity. *Clin. Neurophysiol.* **2008**, *119*, 1812–1823. [CrossRef]
34. Piryatinskaa, A.; Terdik, G.; Woyczynskic, W.A.; Loparo, K.A.; Scher, M.S.; Zlotnikf, A. Automated detection of neonate EEG sleep stages. *Comput. Methods Programs Biomed.* **2009**, *95*, 31–46. [CrossRef] [PubMed]
35. Vandecappelle, M.; Vervliet, N.; De Lathauwer, L. Nonlinear least squares updating of the canonical polyadic decomposition. In Proceedings of the 2017 25th European Signal Processing Conference (EUSIPCO), Kos, Greece, 28 August–2 September 2017; pp. 663–667. [CrossRef]
36. Acar, E.; Yener, B. Unsupervised Multiway Data Analysis: A Literature Survey. *IEEE Trans. Knowl. Data Eng.* **2009**, *21*, 6–20. [CrossRef]

37. Cong, F.; Lin, Q.H.; Kuang, L.D.; Gong, X.F.; Astikainen, P.; Ristaniemi, T. Tensor decomposition of EEG signals: A brief review. *J. Neurosci. Methods* **2015**, *248*, 59–69. [CrossRef] [PubMed]
38. Mariani, S.; Borges, A.F.T.; Henriques, T.; Goldberger, A.L.; Costa, M.D. Use of multiscale entropy to facilitate artifact detection in electroencephalographic signals. In Proceedings of the 2015 37th Annual International Conference of the IEEE Engineering in Medicine and Biology Society (EMBC), Milan, Italy, 25–29 August 2015; pp. 7869–7872. [CrossRef]

Article

Investigation of Linear and Nonlinear Properties of a Heartbeat Time Series Using Multiscale Rényi Entropy

Herbert F. Jelinek [1,2,*], David J. Cornforth [3], Mika P. Tarvainen [4,5] and Kinda Khalaf [6]

1 Australian School of Advanced Medicine, Macquarie University, Sydney 2109, Australia
2 School of Community Health, Charles Sturt University, Albury 2640, Australia
3 School of Design, Communication and IT, University of Newcastle, Newcastle 2308, Australia
4 Department of Applied Physics, University of Eastern Finland, 70210 Kuopio, Finland
5 Department of Clinical Physiology and Nuclear Medicine, Kuopio University Hospital, 70210 Kuopio, Finland
6 Department of Biomedical Engineering, Khalifa University, Abu Dhabi 127788, UAE
* Correspondence: hjelinek@csu.edu.au

Received: 27 May 2019; Accepted: 20 July 2019; Published: 25 July 2019

Abstract: The time series of interbeat intervals of the heart reveals much information about disease and disease progression. An area of intense research has been associated with cardiac autonomic neuropathy (CAN). In this work we have investigated the value of additional information derived from the magnitude, sign and acceleration of the *RR* intervals. When quantified using an entropy measure, these time series show statistically significant differences between disease classes of Normal, Early CAN and Definite CAN. In addition, pathophysiological characteristics of heartbeat dynamics provide information not only on the change in the system using the first difference but also the magnitude and direction of the change measured by the second difference (acceleration) with respect to sequence length. These additional measures provide disease categories to be discriminated and could prove useful for non-invasive diagnosis and understanding changes in heart rhythm associated with CAN.

Keywords: heart rate variability; entropy; nonlinear dynamics; cardiac autonomic neuropathy; diabetes

---

1. Introduction

Biological signals, including electrocardiograms (ECG) or the electrical activity of the heart, exhibit complex dynamics which are characterized by nonlinearity and nonstationarity and often include random noise due to movement artefacts [1]. Heartbeat time series associated with health and disease have been extensively investigated, where time and frequency domains, as well as nonlinear methods, are being proposed and summarized in a number of communications [2–11].

Physiological dynamics of the heartbeat time series change with healthy aging [12,13] and disease, but also during different activities such as sleeping [14,15] and exercise [16–21]. Other changes in dynamics can be attributed to pathology including cardiovascular disease and heart failure [22–24], diabetes [25], depression [26,27] and Parkinson's disease [24,28,29]. For all physiological and pathophysiological models of autonomic function, heart rate variability (HRV) is calculated from the cardiac interbeat intervals (IBI) of the time series. All models assume that the extrinsic modulation of the heartbeat by the autonomic nervous system (ANS) and the endocrine system affect HRV by either increasing the interbeat interval (parasympathetic influence), or decreasing the interbeat interval (sympathetic influence), or a combination of both. Disturbance of the ANS modulation by pathophysiological processes, such as oxidative stress, can then lead to the characteristic changes in sympathovagal input to the heart associated with cardiac autonomic neuropathy (CAN).

1.1. Heartbeat Interval Time Series

Non-invasive methods that are independent of patient cooperation are preferable in the diagnosis of CAN, but still require further research to confirm their sensitivity and specificity in stratification of CAN progression. The most common method currently used is heart rate variability analysis [30,31]. HRV is a useful indication of the health of the cardiovascular system, and is commonly used in assessing the regulation of cardiac autonomic function. HRV has been described by a variety of measures such as time domain, frequency domain, and non-linear dynamic (NLD) measures. However, time domain and power spectral density determination are not suitable for the analysis of non-linear and long-range correlated time signals [32]. Application of new signal processing techniques based on NLD, on the other hand, provides supplementary information (i.e., hidden underlying mechanisms) regarding physiological and pathophysiological processes involved in cardiovascular function and pathology. Two components of a time series in particular, being the sign and magnitude, allow further investigation into the characteristics of the time series and have been discussed previously [33–35].

1.2. Decomposition of the RR Interval Time Series

The current work is based on Ashkenazy [33], who introduced a decomposition algorithm of the RR interval time series by calculating the beat-to-beat increment or first difference ($\Delta RR = RR_{\mathrm{n}} - RR_{\mathrm{n-1}}$). The first difference series is then decomposed into the magnitude and sign of the increments ($|\Delta RRRR|$ and sign (signΔRR) respectively).

Here, we extend this work by introducing the acceleration, defined as the difference between two successive differences, i.e.,

$$\Delta^2 RR = (RR_n - RR_{n-1}) - (RR_{n-1} - RR_{n-2}) \tag{1}$$

$$\Delta^2 RR = RR_n - 2RR_{n-1} + RR_{n-2} \tag{2}$$

Velocity is defined as the rate of change in the RR interval length and therefore the first order difference (ΔRR). Acceleration is then the second order difference ($\Delta^2 RR$). The second difference or acceleration is a measure of the change of RR points with respect to time and indicates the instantaneous acceleration of the heart rate. We propose that acceleration represents an additional descriptive term for a time series. The scaling properties in sign, magnitude and acceleration can then be analyzed by HRV measures, which define the temporal organization of the original time series. Previously detrended fluctuation analysis (DFA) was applied to the sign and magnitude time series [33]. Here, we analyze, sign, magnitude and acceleration using the multiscale Rényi entropy [36–38].

1.3. The Rényi Entropy

Entropy measures can be used to quantify the diversity, uncertainty, or randomness of a system, and are hence considered as beneficial tools for analyzing nonlinear time series, including those of short duration, towards identifying underlying pathology [39–41]. Global entropy measures, such as approximate entropy (ApEn) [42] and sample entropy (SampEn) [43], were adapted from the correlation dimension [44,45] and Kolmogorov entropy [46]. RR time series, however, are multifactorial and display multiscale characteristics, and thus neither ApEn nor SampEn are ideal for such types of biosignal processing. Nonlinear, multiscale dynamic systems can however be described by scaling exponents [47], as well as several multiscale measures [48,49]. Rényi entropy has several advantages. The major advantage of Rényi entropy is that it is robust for short time series, nonlinearity and nonstationarity. The Rényi entropy introduced here also has the advantage of addressing how the probabilities are calculated by applying a density method rather than a histogram method, which is the standard for calculation of multiscale entropy [49].

In the current work we use the Rényi entropy, which generalizes the Shannon entropy [50] and is defined as:

$$H(\alpha) = \frac{1}{1-\alpha} log_2 \left(\sum_{i=1}^{n} p_i^{\alpha} \right) \quad (3)$$

where p_i is the probability that a random variable takes a given value out of n values, and α is the order of the entropy measure [50]. *H(0)* is simply the logarithm of n. As α increases, the measures become more sensitive to the values occurring at higher probability and less to those occurring at lower probability, which provides a picture of the *RR* length distribution within a signal. The probability, p, can be estimated for any sub-sample of *RR* intervals, by considering the sub-sample as a point embedded in a multi-dimensional space. The sub-sample is assigned a density measure by evaluating other sub-samples in its vicinity. This addresses the coarse-graining problem for the determination of scaling behavior in biosignal time series inherent in previous applications of the entropy measures by applying a Gaussian kernel [49,51]. The Gaussian kernel is calculated as the sum of all contributions from other *RR* sub-samples with index *j*:

$$\rho_i = \frac{1}{\sigma\sqrt{2\pi}} \sum_{j=1}^{n} e^{-\frac{dist_{ij}^2}{2\sigma^2}} \quad (4)$$

where σ is the dispersion of the function, and replaces the tolerance as suggested by Costa [48]. We designate the number of *RR* intervals in the sub-sample as π (not to be confused with the irrational number pi), and use the Euclidean distance measure in π dimensions:

$$dist_{ij} = \sum_{k=0}^{\pi} \left(x_{i+k} - x_{j+k} \right)^2. \quad (5)$$

Here, we investigate the efficacy of applying multiscale Rényi entropy as a measure of HRV with respect to the sign, magnitude and the rate of change (acceleration) of the biosignal over time.

2. Methods

2.1. Patient Selection

Heart rate tachograms were obtained from data collected at the Charles Sturt Diabetes Complications Screening Clinic (DiScRi), Australia [52] and were approved by the Charles Sturt University Human Ethics Committee. Written informed consent was obtained from all participants. A 20-min lead II ECG recording was taken from participants attending the clinic, using Powerlab hardware with Chart 7 software (ADInstruments, Sydney) during the morning in an ambient temperature room and after the participants were relaxed. Participants were comparable for age, gender, and heart rate, and after initial screening, those with heart disease, presence of a pacemaker, kidney disease or polypharmacy (including multiple anti-arrhythmic medications) were excluded from the study. The status of CAN was defined using the Cardiac Autonomic Reflex Test battery criteria [53]. Each participant was assigned as either without CAN (71 participants), early CAN (67 participants) or definite CAN (NN participants) [54,55].

2.2. ECG Recording and Obtaining the RR Intervals

From the 20-min *RR* tachogram, a 10 min segment was selected from the middle in order to remove transient start up artefacts and movement at the end of the recording. The *RR* intervals were then extracted from this shorter recording, and data were visually verified to not include any missed, extra or misaligned (including ectopic) beat detections. No other information was used in this study. The raw *RR* interval series for each participant was detrended based on smoothness priors formulation [56]. For the purposes of an initial examination of the *RR* interval recordings, we have selected one recording from each of these three classes as follows. For each participant class (Normal, Early and Definite), the Standard deviation of *RR* intervals in the time series were calculated. For each participant, we calculated the difference between this and the Median values of the Standard deviation obtained for all participants of the same class. We selected the *RR* time series closest to the median for that class. The 10 min *RR* time series for these representatives are shown in Figure 1. Horizontal scales are the same to allow comparison, but vertical scales are as indicated on each graph. Figure 1 shows that the participant from the Normal class manifested *RR* intervals with mostly low deviation from the mean, but some large excursions (standard deviation = 0.0357). In comparison, the participant from the Definite class showed fewer large excursions (SD = 0.0173), while the participant from the Early class was in between these (SD = 0.025926).

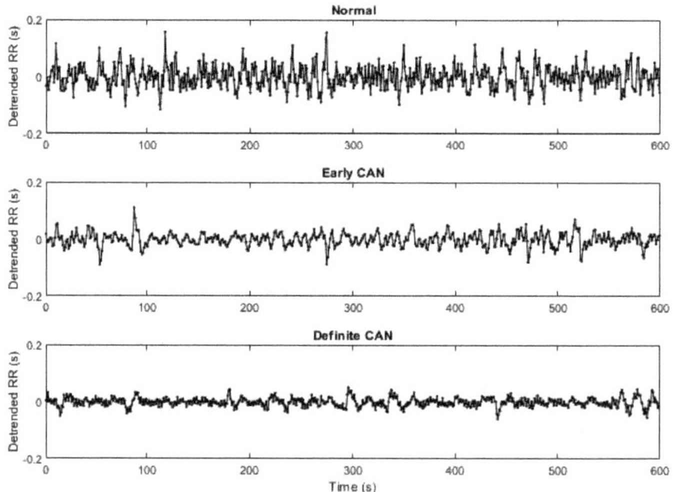

Figure 1. *RR* interval time series of normal, early cardiac autonomy neuropathy (eCAN) and definite CAN (dCAN).

2.3. Decomposition

The *RR* interval time series was decomposed into increment, magnitude, sign and acceleration, as discussed above. Raw *RR* intervals were filtered and the trend was removed. Increments were calculated as the difference between successive *RR* intervals. The magnitude, sign and acceleration of the increments were then determined. Finally, the Rényi entropy was calculated for the sign, magnitude and acceleration time series, using a variety of values for parameter sequence length π, exponent α, and width of the kernel function σ. This results in four different measures:

- Rényi entropy calculated from a sequence of the magnitude of the difference in *RR* intervals
- Rényi entropy calculated from a sequence of the sign of the difference in *RR* intervals
- Rényi entropy calculated from a sequence of the acceleration of *RR* intervals

2.4. Calculating the Multiscale Rényi (MSRen) Entropy

The Rényi entropy was calculated for scaling exponents α of integer values from -5 to $+5$. The entropy values were then normalized by dividing by $\log_2$ of the number of length of the *RR* interval time series. A range of sequence lengths, π, was also used, and the dispersion of the Gaussian function (σ) was varied in proportion. Sequence lengths of 1, 2, 4, 8 and 16 *RR* intervals were adopted, with corresponding values of σ as 0.01, 0.02, 0.04, 0.08 and 0.16, respectively. A Mann-Whitney test was performed to compare the Rényi value obtained for the Normal to that obtained for the Early CAN group, and a similar comparison between the Early CAN and the Definite CAN group, and between the Definite CAN and the Normal group.

3. Results

For each patient group, we calculated the median value of the standard deviation of the *RR* intervals and selected the patients with standard deviation of *RR* intervals closest to the median for the group. The resulting three representative patients are used to illustrate the differences in sign, magnitude and acceleration between Normal, Early CAN and Definite CAN. Figures 2–4 present a sample of 100 filtered *RR* intervals and their decomposition, using data from the three representative groups to illustrate the effect of working with the sign of the *RR* interval, first and second difference. All vertical axes are numbered in seconds. It can be observed, for example, that the increment ΔRR (b) varies between ±0.2 s with excursions up to 0.5 s, while the acceleration (e) varies between ±0.7 ms. There are frequent reversals of sign (d) with some periods of a continuation of the same sign.

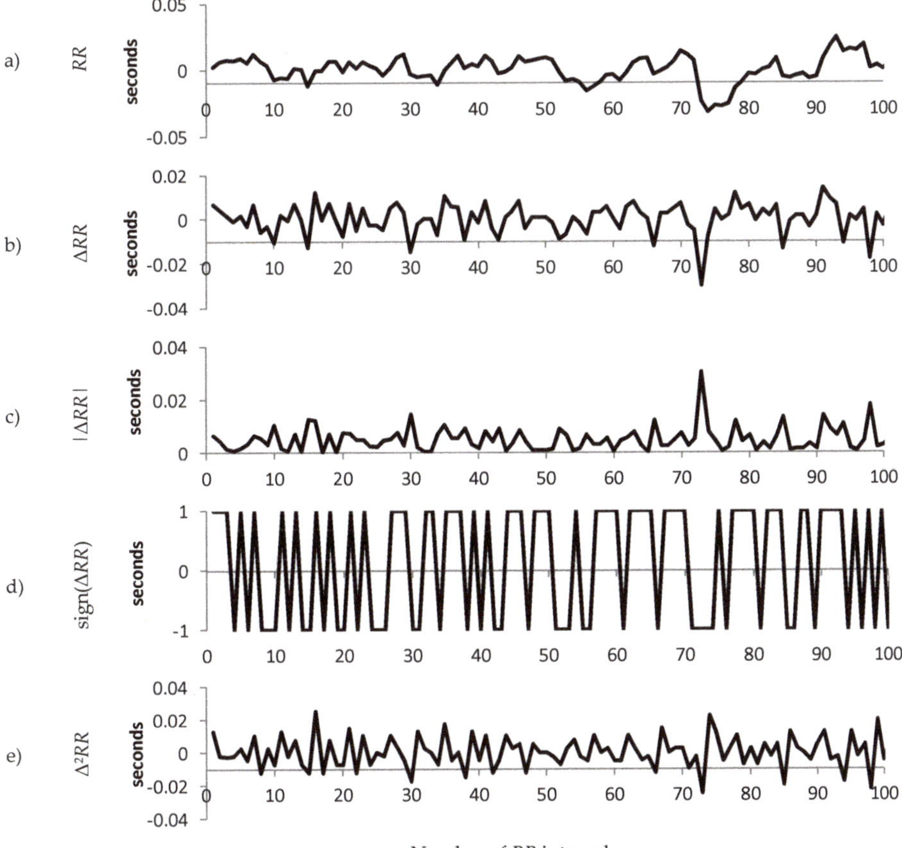

Figure 2. An illustration of the composition of the raw 100 *RR* tachogram for participants without CAN (Normal group). (**a**) The *RR* time series after filtering and pre-processing; (**b**) The increment ($\Delta RR = RR_n - RR_{n-1}$) of the time series shown in (a); (**c**) The magnitude of the increment; (**d**) The sign of the increment; (**e**) The acceleration ($\Delta^2 RR = RR_n - 2RR_{n-1} + RR_{n-2}$).

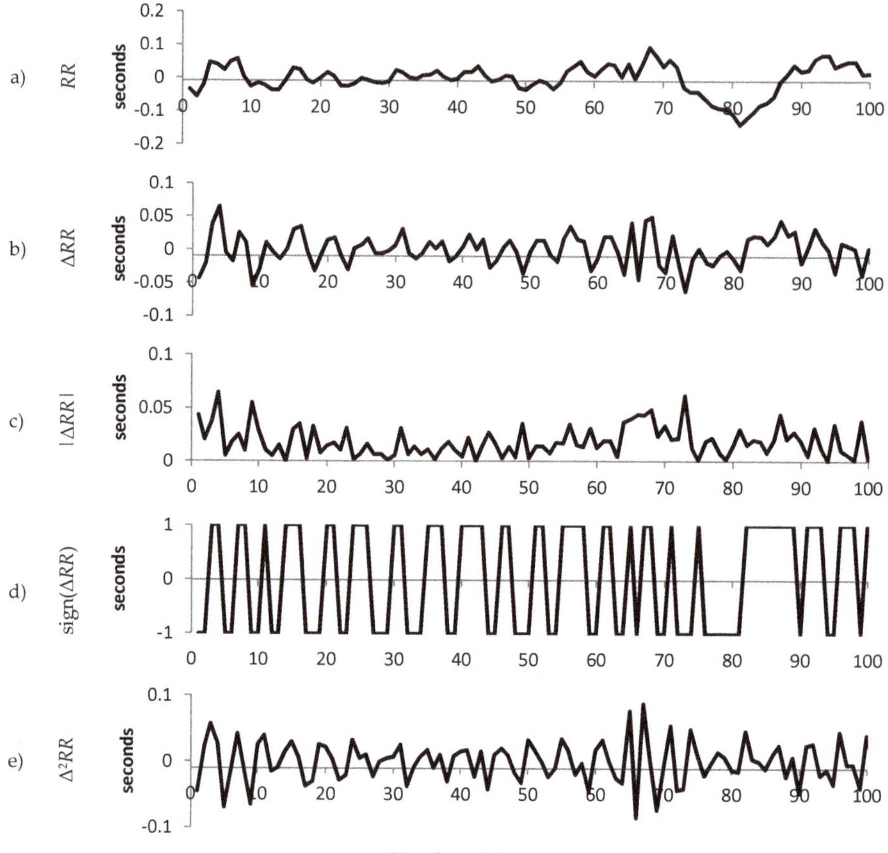

Figure 3. An illustration of the composition of the raw 100 *RR* tachogram for a participant with Early CAN. (**a**) The *RR* time series after filtering and pre-processing; (**b**) The increment ($\Delta RR = RR_n - RR_{n-1}$) of the time series shown in (a); (**c**) The magnitude of the increment; (**d**) The sign of the increment; (**e**) The acceleration ($\Delta^2 RR = RR_n - 2RR_{n-1} + RR_{n-2}$).

Figure 3 shows similar information from the representative participant with early CAN. The range of variation in *RR* interval, ΔRR and $|\Delta RR|$ can be observed to be much smaller than those in Figure 2, indicating a smaller variance in the *RR* interval. In addition, the acceleration is large compared to the representative with early CAN in Figure 3. The sign (Figure 2e) shows fewer changes in direction compared to the representative with early CAN in Figure 3.

Figure 4 shows a sample of the information from a participant with definite CAN. The difference in *RR* intervals (Figure 4b,c) can be seen to be even smaller than the example shown in Figure 2 or Figure 3, while the acceleration (Figure 4e) also has a smaller range than the example from either the Normal or Early group. The sign is different to either of the previous examples, as there appears to be more frequent reversals in sign when compared to those examples, but there is less of a mixture of fast and slow changes in sign.

In order to quantify these differences, the variation of each time series was evaluated, for each participant in the study, using the Rényi entropy. A variety of values were used for the parameters (sequence length π, exponent α and width of the kernel function σ).

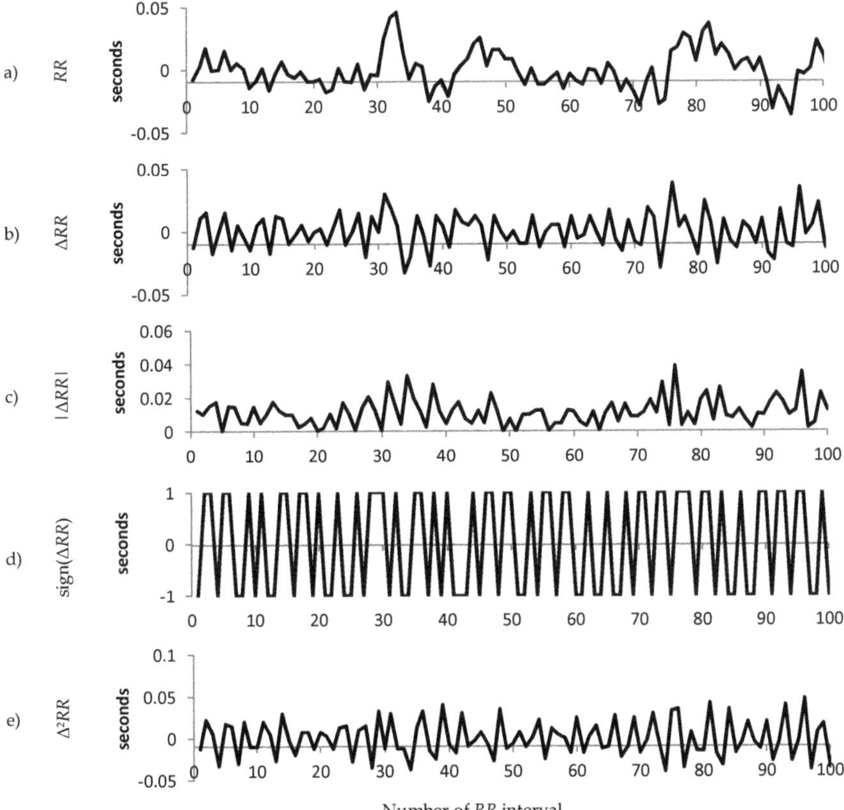

Figure 4. An illustration of the composition of the raw 100 *RR* tachogram for a person with Definite CAN. (**a**) The *RR* time series after filtering and pre-processing; (**b**) The increment ($\Delta RR = RR_n - RR_{n-1}$) of the time series shown in (a); (**c**) The magnitude of the increment; (**d**) The sign of the increment; (**e**) The acceleration ($\Delta^2 RR = RR_n - 2RR_{n-1} + RR_{n-2}$).

Our results comparing Normal (N), Early (E) and Definite (D) CAN, based on the magnitude of the difference of *RR* intervals for sequence length π set to 1, 2, 4, 8, 16, and for $\alpha = +5$ applying the Mann-Whitney tests obtained the smallest *p*-value ($p < 0.0001$) for the Definite to Normal comparison with a sequence length of $\pi = 1$ values for ΔRR. Normal versus Early was best differentiated with longer sequences of length $\pi = 4$ or $\pi = 8$, whereas for Early versus Definite the optimal sequence length was again $\pi = 1$. Extracting the sign of a ΔRR sequence provides information on the linear aspects of the traces but the separation of the classes is less pronounced as seen in the figures above and results in the tables below and hence the *p*-values are much larger, indicating a lesser role of the linear characteristics of the signals in differentiating CAN progression. Only Normal to Early CAN was significantly different for a sequence length of $\pi = 8$, suggesting that the nonlinear, fractal-like characteristics may play a larger role in CAN development.

Separating the classes based on the acceleration of ΔRR increments results in the smallest *p*-value (8.13×10^{-5}) obtained for the Mann-Whitney test comparing Definite CAN to Normal, and a sequence length $\pi = 4$. For acceleration, separation of CAN progression improves with sequence length up to $\pi = 4$, and then decreases again for all comparisons, except for Normal versus Early, where the best separation is seen using $\pi = 16$. However, the best overall comparative results were found with $\pi = 4$.

Table 1 concerns the magnitude of differences ($|\Delta RR|$) and shows p-values obtained from three Mann-Whitney tests comparing Normal to Early (NE), Early to Definite (ED) and Definite to Normal (DN), for different values of the Rényi parameters sequence length π, and exponent α. The width of the kernel function σ was always chosen as $\pi/100$. Nearly all of these p-values are significant at the $p < 0.01$ level, and some are extremely small, suggesting that these Rényi exponents show an effect for all three of these comparisons.

Table 1. Classification results based on Rényi exponents applied to the magnitude of differences ($|\Delta RR|$). Figures represent p-values for the results of Mann-Whitney tests for comparisons of Normal to Early (NE), Early to Definite (ED) and Definite to Normal (DN). Values shown in bold are the smallest p-value for each comparison (Table 1), while tests that were not significant are indicated by n.s.

	Test	$\pi = 1$	$\pi = 2$	$\pi = 4$	$\pi = 8$	$\pi = 16$
	NE	0.0001	<0.0001	**<0.0001**	0.0001	0.002
$\alpha = 1$	ED	0.004	**0.006**	0.02	n.s.	n.s.
	DN	**<0.0001**	<0.0001	0.0002	0.002	0.03
	NE	0.0002	<0.0001	**<0.0001**	<0.0001	0.0007
$\alpha = 2$	ED	0.002	**0.004**	0.01	0.05	n.s.
	DN	<0.0001	**<0.0001**	0.0001	0.001	0.02
	NE	0.0002	<0.0001	**<0.0001**	<0.0001	0.0004
$\alpha = 3$	ED	0.003	**0.005**	0.009	0.004	0.2
	DN	<0.0001	**<0.0001**	<0.0001	0.001	0.01
	NE	0.0003	<0.0001	**<0.0001**	<0.0001	0.0003
$\alpha = 4$	ED	0.002	**0.005**	0.009	n.s.	n.s.
	DN	<0.0001	**<0.0001**	<0.0001	0.0009	0.007
	NE	0.0003	<0.0001	**<0.0001**	**<0.0001**	0.0002
$\alpha = 5$	ED	**0.002**	0.005	0.009	0.02	n.s.
	DN	**<0.0001**	<0.0001	<0.0001	0.0007	0.006

In general, the smallest values are found for normal versus definite CAN as would be expected. However, the table suggests that some values of the Rényi parameters are better than others at demonstrating this effect. Generally, the sequence length of 2 provides the best separation for ED (early–definite) and DN (definite–normal), but using $\pi = 4$ provides the best separation for NE (normal–early). n.s.—not significant.

Table 2 illustrates results for acceleration and shows p-values obtained from three Mann-Whitney tests for different values of sequence length π, and exponent α. The figures show an optimum value for $\pi = 4$ and for a variety of values for α. For short sequence lengths, the p-value increases with increasing α. For longer sequences the opposite is true.

The actual values of the Rényi entropy calculated from the magnitude of the increment of RR intervals $|\Delta RR|$, using the parameters sequence length $\pi = 4$, and width of the kernel function $\sigma = 0.04$, are illustrated in Figure 5. The exponent α was varied so that $-5 \leq \alpha \leq 5$. The inset highlights details of the exponents corresponding to the positive values of α. Rényi entropy calculated for the class of Early CAN lies in between those for the Normal and Definite classes.

Rényi entropy calculated from the acceleration of the RR intervals $\Delta^2 RR$, using the parameters sequence length $\pi = 4$, and width of the kernel function $\sigma = 0.04$ indicates a better separation for positive α (Figure 6.).

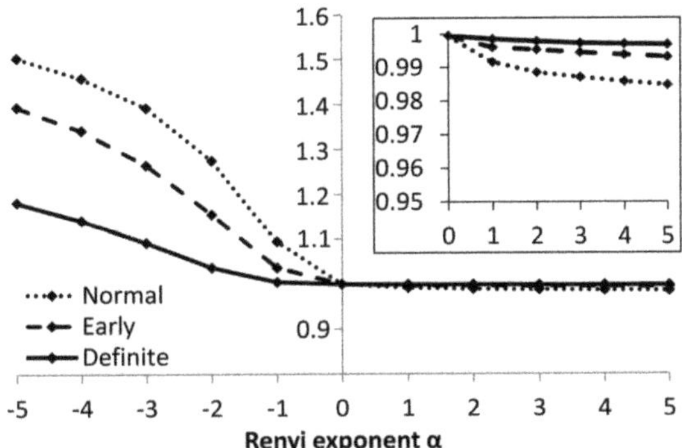

Figure 5. Values of Rényi entropy based on the magnitude of the difference between RR intervals, for sequences of 4 values.

Table 2. Classification results based on Rényi exponents applied to acceleration ($|\Delta^2 RR|$). Figures represent p-values for the results of Mann-Whitney tests for comparisons of Normal to Early (NE), Early to Definite (ED) and Definite to Normal (DN). Values shown in bold are the smallest p-value for each comparison (Table 1), while tests that were not significant are indicated by n.s.

	Test	$\pi = 1$	$\pi = 2$	$\pi = 4$	$\pi = 8$	$\pi = 16$
	NE	0.003	0.0005	**0.0002**	0.0003	0.0005
$\alpha = 1$	ED	n.s.	0.03	**0.01**	0.02	n.s.
	DN	0.00147	0.0002	**<0.0001**	0.0005	0.004
	NE	0.007	0.0008	**0.0003**	0.0004	0.0004
$\alpha = 2$	ED	n.s.	0.02	**0.009**	0.02	0.05
	DN	0.002	0.0003	**<0.0001**	0.0003	0.001
	NE	0.01	0.001	0.0003	0.0005	**0.0003**
$\alpha = 3$	ED	n.s.	0.03	**0.007**	0.02	0.04
	DN	0.002	0.0004	**<0.0001**	0.0002	0.001
	NE	0.01	0.0009	0.0004	0.0005	**0.0003**
$\alpha = 4$	ED	n.s.	0.03	**0.007**	0.02	0.03
	DN	0.003	0.0004	**<0.0001**	0.0002	0.0009
	NE	0.01	0.001	0.0004	0.0006	**0.0002**
$\alpha = 5$	ED	n.s.	0.03	**0.007**	0.02	0.03
	DN	0.0038	0.0004	**<0.0001**	0.0001	0.0006

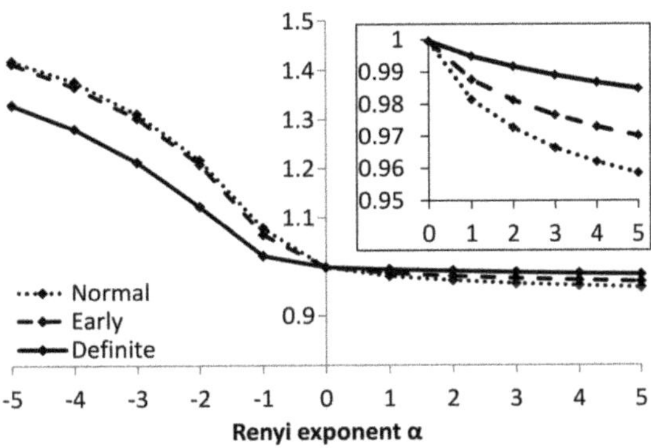

Figure 6. Values of Rényi entropy based on the acceleration of RR intervals, for sequences of 4 values.

4. Discussion and Conclusions

In physiological dynamic systems, various extended concepts of entropy, such as approximate entropy (ApEn), sample entropy (SampEn), and multi-scale entropy have been developed to quantify various physiological signals [42,43,57]. The major advantage of Rényi entropy is that it is robust for short time series, nonlinearity and nonstationarity. The Rényi entropy introduced here also has the advantage of addressing how the probabilities are calculated by applying a density method based on a Gaussian kernel rather than a histogram method, which is the standard for calculation of multiscale entropy. $H(\alpha)$ is the order of the Rényi entropy measure. As α increases, the measures become more sensitive to the values occurring at higher probability and less to those occurring at lower probability, which provides a picture of the *RR* length distribution within a signal [49]. Including acceleration in our model then adds information about the heart rate variability by providing information not only on the change in the system using the first difference but also the magnitude and direction of the change measured by the second difference (acceleration) with respect to sequence length.

Heart rhythm is characterized by a scale-invariant, nonlinear dynamics displaying long-range power-law correlations over a range of time scales [58,59] akin to 1/f [60] or fractal-like scaling [1,39,47,57,58,61–65]. Fractal-like scaling analysis has been shown to indicate risk of adverse cardiac events [66,67].

Bio signals quantifying cardiac interbeat intervals (*RR* intervals) exhibit complex dynamics that vary with age and disease and can be characterized by scaling laws [12,68]. In healthy subjects, *RR* interval time signals present large variability, which is a function of the numerous physiological processes that influence heart rhythm, including ANS and neuroendocrine factors. Nonstationarity and nonlinear dynamics characteristic of these signals are believed to be due to the complex interaction between the two branches of the ANS, endocrine factors and the intrinsic cardiac control mechanisms.

Early identification of CAN is crucial for more effective clinical outcomes. Studies have shown that the one of the earliest signs for a CAN diagnosis is the reduction of HRV. Thus, understanding of the time series characteristics and selecting an appropriate method to analyze these signals and interpret the results is paramount. A consistent finding of ours is that the most difficult two classes to separate were Definite and Early CAN. This implies that patients in the early stages of CAN have similar HRV features to those in the definite group. This may be a reflection that the existing CART criteria are somewhat conservative in identifying CAN, or the two blood pressure tests included in the CART battery indicative of sympathetic dysfunction do not clearly identify disease progression from early to definite CAN, and that sympathetic dysfunction may already be a factor in early CAN [69].

The typical *RR* tachogram consists of linear and nonlinear portions, which overlap and lead to the characteristic heart rate variability. In this work, we show that different components of the *RR* tachogram are able to differentiate between the stages of CAN progression from normal and early CAN to definite CAN. These different components rely on the fact that control of heart rate entails changes in both a positive and negative directions. In particular, the magnitude and acceleration of the changes in *RR* increments separate all three groups. Both of these series carry information on the nonlinear properties of the interbeat interval time series and indicate that fractal-like or power law dynamics within the biosignals become more prominent with disease progression. This complex behavior is further illustrated by the larger and more often occurring deviations in acceleration.

Recent NLD methods continue to shed light on HRV changes under various physiological and pathological conditions, providing valuable potential prognostic and diagnostic information and complementing traditional time- and frequency-domain analyses. With the advent of multiple tools and algorithms, it is critical to identify which of these methods should be selected and under which conditions they should be applied. Our work aligns with previous work, confirming the efficacy of complex measures for representing and quantifying heart rate variability. The current research has focused on investigating the differences in successive *RR* intervals adopting interbeat acceleration as a novel feature, which provides additional information about the nonlinearity of heartbeat regulation and hence the identification of disease. The high degree of separation obtained between classes of disease points to its diagnostic and risk stratification potential in cardiac autonomic neuropathy, and provides a much less invasive test for this disease, with the advantages of faster diagnosis, better access to treatment and more effective clinical outcomes.

Author Contributions: Conceptualization, H.F.J. and D.J.C. and K.K.; Methodology, H.F.J., D.J.C. and M.P.T.; Formal Analysis, H.P.J., D.J.C., M.T. and K.K.; Investigation, H.F.J., D.J.C., M.P.T. and K.K.; Writing-Original Draft Preparation, H.F.J., D.J.C., M.T. and K.K.; Writing-Review & Editing, H.F.J., D.J.C., M.P.T. and K.K.

Funding: MPT was (funded through an EFSD award) supported by EFSD/JDRF/Lilly.

Conflicts of Interest: The authors declare no conflict of interest.

References

1. Goldberger, A.L.; Amaral, L.A.N.; Hausdorff, J.M.; Ivanov, P.C.; Peng, C.-K.; Stanley, H.E. Fractal dynamics in physiology: Alterations with disease and aging. *Proc. Natl. Acad. Sci. USA* **2002**, *99*, 2466–2472. [CrossRef] [PubMed]
2. Costa, M.; Goldberger, A.L.; Peng, C.-K. Multiscale entropy analysis of biological signals. *Phys. Rev. E* **2005**, *71*, 021906. [CrossRef] [PubMed]
3. Peng, C.K.; Havlin, S.; Stanley, H.E.; Goldberger, A.L. Quantification of scaling exponents and cross over phenomena in nonstationary heartbeat time series analysis. *Chaos* **1995**, *5*, 82–87. [CrossRef] [PubMed]
4. Ivanov, P.C.; Rosenblum, M.G.; Peng, C.K.; Mietus, J.; Havlin, S.; Stanley, H.E.; Goldberger, A.L. Scaling behaviour of heartbeat intervals obtained by wavelet-based time-series analysis. *Nature* **1996**, *383*, 323–327. [CrossRef] [PubMed]
5. Agelink, M.W.; Malessa, R.; Baumann, B.; Majewski, T.; Akila, F.; Zeit, T.; Ziegler, D. Standardized tests of heart rate variability: Normal ranges obtained from 309 healthy humans, and effects of age, gender and heart rate. *Clin. Auton. Res.* **2001**, *11*, 99–108. [CrossRef] [PubMed]
6. Bellavere, F.; Balzani, I.; De Masi, G.; Carraro, M.; Carenza, P.; Cobelli, C.; Thomaseth, K. Power spectral analysis of heart-rate variations improves assessment of diabetic cardiac autonomic neuropathy. *Diabetes* **1992**, *41*, 633–640. [CrossRef] [PubMed]
7. Yeragani, V.K.; Srinivasan, K.; Vempati, S.; Pohl, R.; Balon, R. Fractal dimension of heart rate time series: An effective measure of autonomic function. *J. Appl. Physiol.* **1993**, *75*, 2429–2438. [CrossRef]
8. Malik, M.; Camm, J. *HRV Variability*; Futura Publishing Co.: Armonk, NY, USA, 1995.
9. Electrophysiology, Task Force of the European Society of Cardiology the North American Society of Pacing. Special report: Heart rate variability standards of measurement, physiological interpretation, and clinical use. *Circulation* **1996**, *93*, 1043–1065. [CrossRef]
10. Teich, M.C.; Lowen, S.B.; Jost, B.M.; Vibe-Rheymer, K. *Heart Rate Variability: Measures and Models*; IEEE Press: New York, NY, USA, 2001.
11. Khandoker, A.H.; Jelinek, H.F.; Moritani, T.; Palaniswami, M. Association of cardiac autonomic neuropathy with alteration of sympatho-vagal balance through heart rate variability analysis. *Med. Eng. Phys.* **2010**, *32*, 161–167. [CrossRef]
12. Pikkujämsä, S.M.; Mäkikallio, T.H.; Sourander, L.B.; Räihä, I.J.; Puukka, P.; Skyttä, J.; Peng, C.K.; Goldberger, A.L.; Huikuri, H.V. Cardiac interbeat interval dynamics from childhood to senescence: Comparison of conventional and new measures based on fractals and chaos theory. *Circulation* **1999**, *100*, 393–399. [CrossRef]
13. Schmitt, D.T.; Ivanov, P.C. Fractal scale-invariant and nonlinear properties of cardiac dynamics remain stable with advanced age: A new mechanistic picture of cardiac control in healthy elderly. *Am. J. Physiol.* **2007**, *293*, 1923–1937. [CrossRef] [PubMed]
14. Burr, R.L. Interpretation of normalised specrtal heart rate variability in sleep research: A critical review. *Sleep* **2007**, *30*, 913–919. [CrossRef] [PubMed]
15. Vanoli, E.; Adamson, P.B.; Ba, L.; Pinna, G.D.; Lazzara, R.; Orr, W.C. Heart rate variability during specific sleep stages. A comparison of healthy subjects with patients after myocardial infarction. *Circulation* **1995**, *91*, 1918–1922. [CrossRef] [PubMed]

16. Hautala, A.J.; Makikallio, T.H.; Kiviniemi, A.; Laukkanen, R.T.; Nissila, S.; Huikuri, H.V.; Tulppo, M.P. Cardiovascular autonomic function correlates with the response to aerobic training in healthy sedentary subjects. *Am. J. Physiol. Heart Circ. Physiol.* **2003**, *285*, H1747–H1752. [CrossRef] [PubMed]
17. Jelinek, H.F.; Huang, Z.Q.; Khandoker, A.H.; Chang, D.; Kiat, H. Cardiac rehabilitation outcomes following a 6-week program of PCI and CABG Patients. *Front. Physiol.* **2013**, *4*, 302. [CrossRef] [PubMed]
18. Kiviniemi, A.M.; Tulppo, M.P.; Eskelinen, J.J.; Savolainen, A.M.; Kapanen, J.; Heinonen, I.H.; Huikuri, H.V.; Hannukainen, J.C.; Kalliokoski, K.K. Cardiac autonomic fucntion and high-intensity interval training in middle-aged men. *Med. Sci. Sports Exerc.* **2014**, *46*, 1960–1967. [CrossRef] [PubMed]
19. La Rovere, M.; Mortara, A.; Sandrone, G.; Lombardi, F. Autonomic nervous system adaptations to short-term exercise training. *Chest* **1992**, *101*, 299–303. [CrossRef] [PubMed]
20. Soares-Miranda, L.; Sandercock, G.; Valente, H.; Vale, S.; Santos, R.; Mota, J. Vigorous physical activity and vagal modulation in young adults. *Eur. J. Cardiovasc. Prevent. Rehab.* **2009**, *16*, 705–711. [CrossRef]
21. Tulppo, M.P.; Mäkikallio, T.H.; Seppänen, T.; Laukkanen, R.T.; Huikuri, H.V. Vagal modulation of heart rate during exercise: Effects of age and physical fitness. *Am. J. Physiol. Heart Circ. Physiol.* **1998**, *274*, H424–H429. [CrossRef]
22. McLachlan, C.S.; Ocsan, R.; Spence, I.; Hambly, B.; Matthews, S.; Wang, L.; Jelinek, H.F. Increased total heart rate variability and enhanced cardiac vagal autonomic activity in healthy humans with sinus bradycardia. In *Baylor University Medical Center Proceedings*; Taylor & Francis: Oxford, UK, 2010; Volume 23, pp. 368–370.
23. Mäkikallio, T.H.; Huikuri, H.V.; Hintze, U.; Videbæk, J.; Mitrani, R.D.; Castellanos, A.; Myerburg, R.J.; Møller, M.; DIAMOND Study Group. Fractal analysis and time- and frequency-domain measures of heart rate variability as predictors of mortality in patients with heart failure. *Am. J. Cardiol.* **2001**, *87*, 178–182.
24. Huikuri, H.V.; Valkama, J.O.; Airaksinen, K.E.; Seppänen, T.; Kessler, K.M.; Takkunen, J.T.; Myerburg, R.J. Frequency domain measures of heart rate variability before the onset of nonsustained and sustained ventricular tachycardia in patients with coronary artery disease. *Circulation* **1993**, *87*, 1220–1228. [CrossRef] [PubMed]
25. Khandoker, A.H.; Jelinek, H.F.; Palaniswami, M. Heart rate variability and complexity in people with diabetes associated cardiac autonomic neuropathy. In Proceedings of the 2008 30th Annual International Conference of the IEEE Engineering in Medicine and Biology Society, Vancouver, BC, Canada, 20–25 August 2008; pp. 4696–4699.
26. Kemp, A.H.; Quintana, D.S.; Felmingham, K.L.; Matthews, S.; Jelinek, H.F. Heart rate variability in unmedicated depressed patients without comorbid cardiovascular disease. *PLoS ONE* **2012**, *7*, e30777. [CrossRef] [PubMed]
27. Carney, R.M.; Freedland, K.E. Depression and heart rate variability in patients with coronary artery disease. *Clev. Clin. J. Med.* **2009**, *76*, S13–S17. [CrossRef] [PubMed]
28. Barbieri, R.; Citi, L.; Valenza, G.; Guerrisi, M.; Orsolini, S.; Tessa, C.; Diciotti, S.; Toschi, N. Increased instability of heartbeat dynamics in Parkinson's disease. In *Computing in Cardiology*; IEEE: Piscataway Township, NJ, USA, 2013; Volume 40, pp. 89–92.
29. Kallio, M.; Suominen, K.; Bianchi, A.M.; Mäkikallio, T.; Haapaniemi, T.; Astafiev, S.; Sotaniemi, K.A.; Myllylä, V.V.; Tolonen, U. Comparison of heart rate variability analysis methods in patients with Parkinson's disease. *Med. Biol. Eng. Comput.* **2002**, *40*, 408–414. [CrossRef] [PubMed]
30. Vinik, A.I.; Erbas, T.; Casellini, C.M. Diabetic cardiac autonomic neuropathy, inflammtion and cariovascular disease. *J. Diabetes Investig.* **2013**, *4*, 4–8. [CrossRef] [PubMed]
31. Charles, M.; Fleischer, J.; Witte, D.R.; Ejskjaer, N.; Borch-Johnsen, K.; Lauritzen, T.; Sandbaek, A. Impact of early detection and treatment of diabetes on the 6-year prevalence of cardiac autonomic neuropathy in people with screen-detected diabetes: ADDITION-Denmark, a cluster-randomised study. *Diabetiologia* **2013**, *56*, 101–108. [CrossRef] [PubMed]
32. Hurst, H.E. Long-term storage capacity of reservoirs. *Trans. Am. Soc. Civ. Eng.* **1951**, *116*, 770–808.
33. Ashkenazy, Y.; Ivanov, P.C.; Havlin, S.; Peng, C.K.; Goldberger, A.L.; Stanley, H.E. Magnitude and sign correlations in heartbeat fluctuations. *Phys. Rev. Lett.* **2001**, *86*, 1900–1903. [CrossRef] [PubMed]
34. Ashkenazy, Y.; Ivanov, P.C.; Havlin, S.; Peng, C.K.; Yamamoto, Y.; Goldberger, A.L.; Stanley, H.E. Decomposition of heartbeat time series: Scaling analysis of the sign sequence. *Comput. Cardiol.* **2000**, *27*, 139–142.

35. Ashkenazy, Y.; Lewkowicz, M.; Levitan, J.; Moelgaard, H.; Thomsen, P.E.B.; Saermark, K. Discrimination of the healthy and sick cardiac autonomic nervous system by a new wavelet analysis of heartbeat intervals. *Fractals* **1998**, *6*, 197–203. [CrossRef]
36. Jelinek, H.F.; Tarvainen, M.P.; Cornforth, D.J. Renyi entropy in the identification of cardiac autonomic neuropathy in diabetes. *Comput. Cardiol.* **2012**, *39*, 909–911.
37. Kurths, J.; Voss, A.; Saparin, P.; Witt, A.; Kleiner, H.J.; Wessel, N. Quantitative analysis of heart rate variability. *Chaos* **1995**, *5*, 88–94. [CrossRef] [PubMed]
38. Lake, D.E. Renyi entropy measures of heart rate Gaussianity. *IEEE Trans. Biomed. Eng.* **2006**, *53*, 21–27. [CrossRef]
39. Voss, A.; Schulz, S.; Schroeder, R.; Baumert, M.; Caminal, P. Methods derived from nonlinear dynamics for analysing heart rate variability. *Phil. Trans. Math. Phys. Eng. Sci.* **2009**, *367*, 277–296. [CrossRef] [PubMed]
40. Wessel, N.; Schumann, A.; Schirdewan, A.; Voss, A.; Kurths, J. Entropy measures in heart rate variability data. In *International Symposium on Medical Data Analysis*; Springer: Berlin/Heidelberg, Germany, 2000; pp. 78–87.
41. Wessel, N.; Voss, A.; Malberg, H.; Ziehmann, C.; Voss, H.U.; Schirdewan, A.; Meyerfeldt, U.; Kurths, J. Nonlinear analysis of complex phenomena in cardiological data. *Herzschr. Elektrophys.* **2000**, *11*, 159–173. [CrossRef]
42. Pincus, S. Approximate entropy as a measure of system complexity. *Proc. Nat. Acad. Sci. USA* **1991**, *88*, 2297–2301. [CrossRef] [PubMed]
43. Richman, J.S.; Moorman, J.R. Physiological time-series analysis using approximate entropy and sample entropy. *Am. J. Physiol. Heart Circ. Physiol.* **2000**, *278*, H2039–H2049. [CrossRef]
44. Grassberger, P. Finite sample corrections to entropy and dimension estimates. *Phys. Lett. A* **1988**, *128*, 369–373. [CrossRef]
45. Grassberger, P.; Procaccia, I. Measuring the strangeness of strange attractors. *Physica* **1983**, *9*, 189–208.
46. Eckmann, J.P.; Ruelle, D. Ergodic theory of chaos and strange attractors. In *The Theory of Chaotic Attractors*; Springer: New York, NY, USA, 1985; pp. 273–312.
47. Ivanov, P.C.; Amaral, L.A.N.; Goldberger, A.L.; Havlin, S.; Rosenblum, M.G.; Struzik, Z.R.; Stanley, H.E. Multifractality in human heartbeat dynamics. *Nature* **1999**, *399*, 461–465. [CrossRef]
48. Costa, M.; Goldberger, A.L.; Peng, C.-K. Multiscale entropy analysis of complex physiological time series. *Phys. Rev. Lett.* **2002**, *89*, 068102. [CrossRef]
49. Cornforth, D.; Tarvainen, M.; Jelinek, H.F. How to Calculate Renyi Entropy from Heart Rate Variability, and Why it Matters for Detecting Cardiac Autonomic Neuropathy. *Front. Bioeng. Biotechnol.* **2014**, *2*, 34. [CrossRef] [PubMed]
50. Rényi, A. On measures of information and entropy. In Proceedings of the Fourth Berkeley Symposium on Mathematics, Statistics and Probability; The Regents of the University of California: Oakland, CA, USA, 1961; pp. 547–561.
51. Xu, Y.; Ma, Q.D.Y.; Schmitt, D.T.; Bernaola-Galván, P.; Ivanov, P.C. Effects of coarse-graining on the scaling behavior of long-range correlated and anti-correlated signals. *Phys. A Stat. Mech. Appl.* **2011**, *390*, 4057–4072. [CrossRef] [PubMed]
52. Jelinek, H.F.; Wilding, C.; Tinley, P. An innovative multi-disciplinary diabetes complications screening programme in a rural community: A description and preliminary results of the screening. *Aust. J. Prim. Health* **2006**, *12*, 14–20. [CrossRef]
53. Spallone, V.; Bellavere, F.; Scionti, L.; Maule, S.; Quadri, R.; Bax, G.; Melga, P.; Viviani, G.L.; Esposito, K.; Morganti, R.; et al. Recommendations for the use of cardiovascular tests in diagnosing diabetic autonomic neuropathy. *Nutr. Metab. Cardiovasc. Dis.* **2011**, *21*, 69–78. [CrossRef] [PubMed]
54. Pop-Busui, R.; Evans, G.W.; Gerstein, H.C.; Fonseca, V.; Fleg, J.L.; Hoogwerf, B.J.; Genuth, S.; Grimm, R.H.; Corson, M.A.; Prineas, R. The ACCORD Study Group. Effects of cardiac autonomic dysfunction on mortality risk in the Action to Control Cardiovascular Risk in Diabetes (ACCORD) Trial. *Diabetes Care* **2010**, *33*, 1578–1584. [CrossRef]
55. Flynn, A.C.; Jelinek, H.F.; Smith, M.C. Heart rate variability analysis: A useful assessment tool for diabetes associated cardiac dysfunction in rural and remote areas. *Aust. J. Rural Health* **2005**, *13*, 77–82. [CrossRef]
56. Tarvainen, M.P.; Ranta-Aho, P.O.; Karjalainen, P.A. An advanced detrending method with application to HRV analysis. *IEEE Trans. Biomed. Eng.* **2002**, *49*, 172–175. [CrossRef]
57. Costa, M.; Goldberger, A.L.; Peng, C.K. Multiscale entropy analysis: A new measure of complexity loss in heart failure. *J. Electrocardiol.* **2003**, *36*, 39–40. [CrossRef]

58. Gao, J.; Gurbaxani, B.M.; Hu, J.; Heilman, K.J.; Emauele, V.A.; Lewis, G.F.; Davila, M.; Unger, E.R.; Lin, J.M.S. Multiscale analysis of heart rate variability in nonstationary environments. *Front. Physiol.* **2013**, *4*, 119. [CrossRef]
59. Saul, J.P.; Albrecht, P.; Berger, R.D.; Cohen, R.J. Analysis of long term heart rate variability: Methods, 1/f scaling and implications. *Pharmacology* **1988**, *14*, 419–422.
60. Kobayashi, M.; Musha, T. 1/f fluctuation of heart beat period. *IEEE Trans. Biomed. Eng.* **1982**, *29*, 456–457. [CrossRef] [PubMed]
61. Struzik, Z.R.; Hayano, J.; Sakata, S.; Kwak, S.; Yamamoto, Y. 1/f Scaling in heartrate requires antagonistic autonomic control. *Phys. Rev. E* **2004**, *70*, 050901. [CrossRef] [PubMed]
62. Hu, J.; Gao, J.; Tung, W.-W.; Cao, Y. Multiscale analysis of heart rate variability: A comparison of different complexity measures. *Ann. Biomed. Eng.* **2010**, *38*, 854–864. [CrossRef] [PubMed]
63. Kiyono, A.; Struzik, Z.R.; Aoyagi, N.; Yamamoto, Y. Multiscale probability density function analysis: Non-Gaussian and scale-invariant fluctuations of healthy human HRV. *IEEE Trans. Biomed. Eng.* **2006**, *53*, 95–102. [CrossRef] [PubMed]
64. Thurner, S.; Feurstein, M.C.; Teich, M.C. Multiresolution wavelet analysis of heartbeat intervals discriminates healthy patients from those with cardiac pathology. *Phys. Rev. Lett.* **1998**, *80*, 1544–1547. [CrossRef]
65. Krstacic, G.; Krstacic, A.; Smalcelj, A.; Milicic, D.; Jembrek-Gostovic, M. The "Chaos Theory" and nonlinear dynamics in heart rate variability analysis: Does it work in short-time series in patients with coronary heart disease? *Ann. Noninvasive Electrocardiol.* **2007**, *12*, 130–136. [CrossRef]
66. Ho, K.K.; Moody, G.B.; Peng, C.K.; Mietus, J.E.; Larson, M.G.; Levy, D.; Goldberger, A.L. Predicting survival in heart failure case and control subjects by use of fully automated methods for deriving nonlinear and conventional indices of heart rate dynamics. *Circulation* **1997**, *96*, 842–848. [CrossRef]
67. Laitio, T.; Jalonen, J.; Kuusela, T.; Scheinin, H. The role of heart rate variability in risk stratification for adverse postoperative cardiac events. *Anesth. Analg.* **2007**, *105*, 1548–1560. [CrossRef]
68. Goldberger, A.L. Non-linear dynamics for clinicians: Chaos theory, fractals, and complexity at the bedside. *Lancet* **1996**, *347*, 1312–1314. [CrossRef]
69. Bellavere, F.; Bosello, G.; Fedele, D.; Cardone, C.; Ferri, M. Diagnosis and management of diabetic autonomic neuropathy. *BMJ* **1983**, *287*, 61. [CrossRef] [PubMed]

Article

On the Robustness of Multiscale Indices for Long-Term Monitoring in Cardiac Signals

Mohammed El-Yaagoubi [1,2], Rebeca Goya-Esteban [1], Younes Jabrane [2], Sergio Muñoz-Romero [1,3], Arcadi García-Alberola [4] and José Luis Rojo-Álvarez [1,3,*]

1. Department of Signal Theory and Communications, Telematics and Computing Systems, Rey Juan Carlos University, 28933 Fuenlabrada, Spain; m.elyaagoubi@alumnos.urjc.es (M.E.-Y.); rebeca.goyaesteban@urjc.es (R.G.-E.); sergio.munoz@urjc.es (S.M.-R.)
2. GECOS Lab, ENSA, Cadi Ayyad University, 40000 Marrakech, Morocco; y.jabrane@uca.ac.ma
3. Center for Computational Simulation, Universidad Politécnica de Madrid, 28040 Pozuelo, Spain
4. Hospital Clínico Universitario Virgen de la Arrixaca de Murcia, 30120 Murcia, Spain; arcadi@secardiologia.es

* Correspondence: joseluis.rojo@urjc.es; Tel.: +34-914-888-744

Received: 6 April 2019; Accepted: 14 June 2019; Published: 15 June 2019

Abstract: The identification of patients with increased risk of Sudden Cardiac Death (SCD) has been widely studied during recent decades, and several quantitative measurements have been proposed from the analysis of the electrocardiogram (ECG) stored in 1-day Holter recordings. Indices based on nonlinear dynamics of Heart Rate Variability (HRV) have shown to convey predictive information in terms of factors related with the cardiac regulation by the autonomous nervous system, and among them, multiscale methods aim to provide more complete descriptions than single-scale based measures. However, there is limited knowledge on the suitability of nonlinear measurements to characterize the cardiac dynamics in current long-term monitoring scenarios of several days. Here, we scrutinized the long-term robustness properties of three nonlinear methods for HRV characterization, namely, the Multiscale Entropy (MSE), the Multiscale Time Irreversibility (MTI), and the Multifractal Spectrum (MFS). These indices were selected because all of them have been theoretically designed to take into account the multiple time scales inherent in healthy and pathological cardiac dynamics, and they have been analyzed so far when monitoring up to 24 h of ECG signals, corresponding to about 20 time scales. We analyzed them in 7-day Holter recordings from two data sets, namely, patients with Atrial Fibrillation and with Congestive Heart Failure, by reaching up to 100 time scales. In addition, a new comparison procedure is proposed to statistically compare the poblational multiscale representations in different patient or processing conditions, in terms of the non-parametric estimation of confidence intervals for the averaged median differences. Our results show that variance reduction is actually obtained in the multiscale estimators. The MSE (MTI) exhibited the lowest (largest) bias and variance at large scales, whereas all the methods exhibited a consistent description of the large-scale processes in terms of multiscale index robustness. In all the methods, the used algorithms could turn to give some inconsistency in the multiscale profile, which was checked not to be due to the presence of artifacts, but rather with unclear origin. The reduction in standard error for several-day recordings compared to one-day recordings was more evident in MSE, whereas bias was more patently present in MFS. Our results pave the way of these techniques towards their use, with improved algorithmic implementations and nonparametric statistical tests, in long-term cardiac Holter monitoring scenarios.

Keywords: nonlinear dynamics; multiscale indices; cardiac risk stratification; Holter; long term monitoring; multiscale entropy; multifractal spectrum; multiscale time irreversibility

1. Introduction

Sudden Cardiac Death (SCD) has been defined as the unexpected natural and cardiac originated death within a short period of time, given at less than 1 h from the onset of symptoms if witnessed or within 24 h of having been observed alive if unwitnessed, in a person with no prior condition that could be considered as fatal [1–3]. There are other causes of sudden death, but cardiac is the most usual origin, specifically including: (a) Arrhythmic causes, such as ventricular tachycardia (VT), ventricular fibrillation (VF), or asystole; (b) several other structural heart disease origins, for instance, those ones corresponding to congenital heart disease; and (c) abnormal function of the autonomous nervous system (ANS), which is not itself a death cause, but it can promote others such as arrhythmic or hypertensive death [4]. The SCD mechanism in the last case is often VT or VF. Given the incidence of SCD as major cause of mortality in the world, methods have been proposed aiming to provide risk stratification tools for cardiac patients [5]. SCD episodes can happen not only in patients with coronary or cardiomyopathic disease, but also they can occur in people with no previous heart alteration, which makes the risk stratification extremely complex. The prognostic significance of noninvasive studies and the efficacy of the therapeutic actions have been pointed to be etiology dependent [6]. The most widely used SCD-risk marker in the clinical practice is the Left Ventricular Ejection Fraction (LVEF), but given its low specificity, many other techniques have been proposed. A relevant subset of them is given by the computational indices that are obtained from the signal analysis of the Electrocardiogram (ECG), including a variety of proposed biomarkers such as late potentials, heart rate variability (HRV), T-wave alternans, or deceleration capacity. The interested reader can refer to [7] for a detailed review on issues related with signal processing, technology transfer, and scientific evidence for all of them.

Many of these SCD markers are obtained in a Holter recording, which is a diagnostic tool consisting of 24 to 48 h signal registers in two or three chest leads to be subsequently processed by using a computer program, so that a variety of cardiac events can be identified by the clinician. Probably one of the most scrutinized markers of SCD risk from Holter recordings is HRV, which measures the time changes between consecutive cardiac beats [8]. Its interest partially comes from its non-invasive nature and its easy for analysis, only needing to know the time instants of the beat occurrences. The heart does not behave like a periodic oscillator, but instead its rhythm is modulated by the ANS, and the simultaneous actuation of its two branches (sympathetic and parasympathetic) causes dynamic oscillations of the cardiac frequency, producing the presence of HRV [9]. Among the many methods that have been proposed in the literature to quantify the HRV indices, the nonlinear methods extract relevant information from HRV signals in terms of their complexity. Nonlinear indices are based on the underlying idea that fluctuations in the between-beat intervals (also known as RR intervals) can exhibit characteristics that are well known from Complex Dynamic Systems Theory, and broadly speaking, healthy states are expected to correspond to more complex patterns than pathological states. However, some pathologies are associated with highly erratic fluctuations with statistical properties resembling uncorrelated noise [7], and traditional algorithms could yield higher irregularity indices for such pathological signals when compared to healthy dynamics, even though the latter represent more physiologically complex states [10]. This possible inconsistency may be due to the fact that traditional algorithms are based on single scale analysis, and they can not take into account the complex temporal fluctuations inherent to healthy physiologic control systems. It is usual that studies based on 1-day Holter monitoring [11] envision that relevant information could be obtained from longer duration recordings, however, few studies [12,13] have scrutinized nonlinear indices in several-day Holter monitoring, despite its current and increasing availability in the clinical practice. Note in the following that, whereas some authors refer to long-term Holter as those with duration about 24 h, we will use long-term to refer to the Holter recordings when measured for several days throughout this paper.

Several of the nonlinear HRV measurements (based either on Chaos Theory, Information Theory, or Fractal Theory) have been paid special attention according to the electrophysiological hypothesis that the long-term regulation is a homeostatic yet dynamical equilibrium, which can be expected to be complex and multi-cause enough to require a set of indices that should be calculated at different scales. This has motivated the extension of several of those indices to what can be called their *multiscale* versions. Remarkable examples of this effort are the Multiscale Entropy (MSE) method [14–17], the Multiscale Time Irreversibility (MTI) method [15], and the Multifractal Spectrum (MFS) method [18,19]. MSE has been applied to predict stroke-in-evolution in acute ischemic stroke patients using one-hour ECG signals during 24 h [20], and also recently to predict vagus-nerve stimulation outcome in patients with drug-resistance epilepsy, who were found to have lower preoperative HRV than controls [21]. Different preprocessing stages have been proposed for it, including the use of MSE in the first difference of RR-interval time series instead of the series itself, yielding better statistical support and discrimination capabilities between Congestive Heart Failure (CHF) and control groups [22]. Some authors consider the MSE algorithm biased for two reasons: First, the similarity criteria is fixed for all scales, whereas coarse grained time series variance has been pointed to decrease with the scale; And second, spurious oscillations are introduced due to the suboptimal procedure for eliminating the fast temporal scales of the time series. Accordingly, a modified algorithm was proposed [23], so-called the refined MSE (RMSE), in order to overcome these limitations, and it was tested in simulations and in 24-h HRV signals from aortic stenosis and control groups. The use of RMSE did not allow to make inferences that could not be made by MSE with real data, however, simulations showed that RMSE can be a more reliable method for the assessment of entropy-based irregularity. A comparative study between MSE and RMSE was performed in [24] confirming that despite the differences they both present similar tendencies with scale factor. MTI has been applied to HRV and blood preasure variability (BPV) signals concluding that TI of beat-to-beat HRV and BPV is significantly altered during orthostasis [25]. In addition, recent interest has been raised in the use of some of the multiscale indices in the analysis of atrial fibrillation (AF) dynamics, which has been scrutinized in the context of ischemic stroke prediction in patients with permanent AF [26].

We can say that cardiac long-term monitoring (LTM) has been technologically achieved, and previous studies exist which have scrutinized the value of these recordings in simple and well-known clinical indices from practice, such as the number or rate of ectopic beats or the number or rate of different cardiac events [27]. However, algorithm robustness should be paid attention if deeper physiological and pathological information is to be extracted from nonlinear multiscale indices, in order to be sure about their reliability when working with long series in populational data, and to our best knowledge, few previous works can be found noting this point with attention. Therefore, we propose here to study the robustness of nonlinear multiscale HRV measurements to characterize different cardiac health states in LTM recordings.

Specifically, we made a comparative analysis of the three mentioned multiscale methods on a database consisting of patients with 7-day Holter in two cardiac conditions, namely, CHF and AF [28]. This study aims to give basic knowledge on the usefulness and current limitations of these methods towards their future and principled use for SCD risk stratification. For this purpose, a nonparametric statistical test is proposed in order to compare and give cut-off comparison levels between two different situations in terms of the confidence intervals (CI) for the median difference across multiscale representations of poblational representations. This method can be used either for establishing comparisons among patients or subjects with different conditions, or to scrutinize the impact of preprocessing or data length on the statistical properties of the multiscale representations. The procedure can be seen as an extension of previously used statistical comparisons [20,21] in terms of nonparametric bootstrap tests for confidence bands [12,29].

The structure of the paper is as follows. In the next section, the fundamentals of the multiscale methods selected here for HRV analysis are described. Then, existing methods and the new procedure based on nonparametric bootstrap tests for median difference are provided in Section 3, together with the presentation of the available recordings in Physionet [11] used as starting benchmark for this study and the LTM-ECG databases for CHF and AF during 7 days. In Section 4, a set of experiments are conducted and results are presented on the suitability, together with some technical limitations and consistency properties, of these benchmarked multiscale algorithms. Finally, in Section 5, technical and clinical discussion is given and conclusions are drawn.

2. Fundamentals of Multiscale Methods for HRV Analysis

HRV measurements aim to give a numerical magnitude of the time fluctuations between sets of consecutive beats. The short-term recordings of HRV are usually measured about 3 to 5 min, and they have been traditionally associated with the dynamic control of the ANS on the heart rate and the cardiac properties. The long-term fluctuations of HRV have been described to have a wide physiological meaning in terms of the cardiovascular system self-regulation mechanism description. The ANS is divided into two branches, namely, the sympathetic and the vagal (parasympathetic) ones. Broadly speaking, the activation and excitation of the sympathetic branch has an accelerating effect on the cardiac cycle, whereas the vagal activation has a decelerating effect, but both subsystems are simultaneously and continuously working and compensating themselves, so that oscillations on a dynamic equilibrium are produced on the heart rate [9]. In addition, the ANS receives information through the so-called efferent pathways from a wide variety of systems and organs (heart, digestive system, kidney, respiratory system, and many others), and those influences are part of the genesis of ANS afferent pathways, in which heart rate is affected and is involved through different control mechanisms. Additional influences on the ANS such as humoral factors, night-day cycles, or environmental influences, are slower than the ANS, so that they only can influence the long-term HRV [30]. All these sources are contributors to the HRV modulation, which globally has its origin on a complex dynamic equilibrium arising from diverse mechanisms in the cardiovascular system that are taking place in the short-, the middle-, and the long-term scales.

A number of scientific, technical, and medical studies have focused on the HRV, and it well might be the most studied index in the SCD risk-stratification literature. Nonlinear methods [9] include several subfamilies, according to their calculation being based on Information Theory, on Chaos Theory, or on Fractal Theory. These methods have been paid special attention not only for their attractive theoretical foundations, but also because they seemed to have promising risk capabilities in small-sized studies with patients [7]. A usual situation in the literature of nonlinear HRV indices has been that some basic index has been first proposed, which has shown descriptive capabilities and some independence for SCD risk stratification, and then this index has been subsequently extended to a multiscale formulation, aiming to capture a richer variety of descriptions for the signal behavior. As described next, this is the case of MSE, MTI, and MFS methods for HRV analysis.

We denote the continuous-time ECG signal of a patient by $S(t)$, and we register it during an observed time interval, denoted by $t \in (t_i, t_f)$. As a result of preprocessing steps devoted to signal filtering and R-wave detection, we can detect the R wave of each beat in $S(t)$, and the time instants associated to each R-wave are denoted as t_n^R, with $n = 0, \ldots, N$, so that the detected set of R-waves can be expressed as a point process, given by

$$RR(t) = \sum_{n=0}^{N} \delta(t - t_n^R) \tag{1}$$

where $\delta(t)$ denotes the continuous-time Dirac's delta function. It is often useful to work with the so-called *normal beats*, which correspond to R-waves in beats that have been only originated in sinus rhythm conditions and where artifacts and ectopic beats are discarded. Here we assume that the R-waves correspond to cardiac beats but not to artifacts or wrong R-wave detections. The RR-tachogram [9], or just *tachogram*, can be denoted by $x[n]$ and is defined as the discrete-time series given by the indexed time difference between consecutive R-wave times (excluding artifacts and non-physiological beats, as well as conventional quality-control beat filters), this is,

$$x[n] = t_n^R - t_{n-1}^R, \quad n = 1, \ldots, N \tag{2}$$

The following algorithms and indices can be expressed and obtained in terms of the tachogram registered in patients.

2.1. MSE Analysis

The approximate entropy ($ApEn$) can be described as a nonlinear fluctuation measurement that aims to quantify the irregularity of a RR-interval time series [31]. An $ApEn$ increase is usually interpreted as an indicator of irregularity increase in the underlying cardiovascular process. The Sample Entropy ($SampEn$) was subsequently introduced [32] to solve the limitations of the $ApEn$ [33], since the latter compares each pattern in a time series with other patterns but also with itself, which leads to the overestimation of similarity existence in that time series, and hence to strong bias and to some inconsistent results. $SampEn$ is the negative of the natural logarithm of the conditional probability that two similar patterns of m point segments, $x_m(j)$ and $x_m(i)$ of tachogram $x[n]$, remain similar if we increase the number of points to $m+1$, within a tolerance r that is defined as a noise-rejection filter [32]. SampEn index reduces the statistical bias of ApEn index and it is a measurement rather independent of the data length, but its problem is that it it can be unstable when the counted events are scattered.

The new concept of multiscale analysis was proposed to overcome several limitations pointed out for $ApEn$ and $SampEn$ measurements which could be leading to clinical misinterpretation of HRV in some conditions. The MSE analysis was introduced by Costa et al. in [14], and it was also intended to provide a richer description of the cardiac dynamics in terms of a set of naturally related indices, rather than to use a single number. For a given discrete-time series, a new series is constructed in a scale τ, the terms of which are the average of the consecutive elements of the original series without overlapping. For a time series with $\tau = 1$, this corresponds to the original series, whereas for $\tau = 2$, the series is constructed with the average of the elements taken from two by two, and so on. We finally calculate the $SampEn$ for each one of these new generated series. When the obtained values are represented versus the scale factor, the dependence of the measured entropy with the time scale can be scrutinized. The maximum scale to use depends on the number of samples in the time series.

Starting with tachogram signal $x[n]$, we denote MSE as $MSE(x[n], \tau, r, m)$ to explicitly consider the dependence of its design parameters. We obtain the consecutive time series y^τ, determined by scaling factor τ as follows:

- First, the original time series is divided into non-overlapping intervals with window size of τ samples. Then the signal mean is obtained for each of the sample windows.
- Each element of the series $y^\tau[j]$ is calculated according to the equation:

$$y^\tau[j] = \frac{1}{\tau} \sum_{n=n_j}^{j\tau} x[n], \qquad 1 \leq j \leq \frac{N}{\tau} \tag{3}$$

  where $n_j = (j-1)\tau + 1$ for notation simplicity. For the first scale, the time series $y^1[j]$ is just the original time series. The length of each obtained time subseries is equal to $\frac{N}{\tau}$.
- The sample entropy index is calculated for each time series y^τ, and it is represented as a scale-factor function $SampEn(\tau)$.

The MSE analysis has been applied to a variety of cardiac and cardiopathological situations, including the analysis of CHF [14], hypertensive and sino-aortic denervated conditions in experimental studies [24], and the ANS evolution before, during, and after percutaneous transluminal coronary angioplasty [34], as well as for prediction of ischemic stroke in patients with persistent AF [26], among other things. In all these cases, the length of the analyzed signals did not exceed 24 h using 20 as a maximum scale value, which involves $x[n]$ tachograms with about or more than 20,000 samples each.

2.2. MTI Analysis

Time irreversibility has recently attracted attention in the cardiovascular-signal field. A signal is said to be time irreversible if its statistical properties change after its time reversal. The consistency loss of the statistical properties of a signal when the signal reading undergoes a change through time inversion is measured using the MTI, which represents an asymmetry index. This index is higher in healthy systems (with more complex dynamics) and it decreases in conditions like pathology or aging, as introduced in [15,16]. On the other hand, physiological time series generate complex fluctuations in multiple depending time scales, due to the existence of different hierarchical and interrelated regulatory systems.

The MTI analysis has been applied to measure nonlinear dynamics in heart-rate time series. For instance, MTI indices were computed in [35] for 20 healthy neonates to detect the presence of nonlinearity in their cardiac-rhythm control system, and temporal asymmetries were detected within their heart rate dynamics even shortly after birth.

2.3. MFS Analysis

Physiological signals have been shown to present fractal temporal structure under healthy normal conditions [12,36,37]. In particular, it was shown in [38] that time series generated by certain cardiovascular control systems in healthy conditions require a large number of exponents to adequately characterize their scaling properties, and that the nonlinear properties of this behavior are encoded in the Fourier phases. The same work used examples of CHF patients to contrast the previous finding with the loss of multifractality in this example of life-threatening condition. In this setting, RR-interval time series have been analyzed in terms of multiple scaling exponents. We next summarize the basic principles for estimating the MFS from RR signals that are usually followed in the literature. The interested reader can consult the original works on its application [19,38] and the details on the wavelet-transform modulus maxima method [39], which gives a principled calculation method for this purpose.

Whereas monofractal signals have the same scaling properties through time and they can be indexed by a single global exponent (such as Hurst exponent H) characterizing their fluctuations, other signals exhibit variations in their local Hurst exponent along time. When several subsets of a signal are characterized by the same local Hurst exponent h_o, and when each of these signals can be characterized by a fractal dimension measurement, we denote this estimated dimension as $D(h_o)$. Accordingly, $D(h)$ will have nonzero values on a set of discrete points in h for some class of signals. Local value of h is modernly estimated with Wavelet Theory [39], often using successive derivatives of the Gaussian function as the analyzing wavelet at different scales a, in order to remove polinomial trends with polinomial order up to the wavelet derivative order. In these conditions, the problem reduces to obtain the modulus of the maxima extrema of the time series wavelt transform at eath time instant.

We then estimate the partition function $Z_q(a)$, as the sumation of the qth powers of these local maxima as a function of scale, and for small scales, it is fulfilled that it has the form $Z_q(a) \sim a^{\tau(q)}$, where $\tau(t)$ are exponents that can be estimated. In monofractal signals, a linear scaling exponent spectrum is obtained, given by $\tau(q) = qH - 1$. However, for multifractal signals we obtain a nonlinear expression, and it can be shown that $\tau(q) = qh(q) - D(h)$, where $h(q) = d\tau(q)/dq$ is not constant. Accordingly, the estimated fractal dimensions $D(h)$ are obtained by the Legendre transform of $\tau(q)$, finally yielding

$$D(h) = qh - \tau(t) \tag{4}$$

Given that $h = 0.5$ can be related to uncorrelated changing time series, this representation allows to determine to what extent a process conveys anticorrelated ($h < 0.5$) or correlated ($h > 0.5$) behaviour consistently manifested through different time periods.

The multifractal structure of HRV can reflect important properties of the heart-rate autonomic regulation. Multifractal analysis is an expansion of fractal analysis since it characterizes the time series variability with a collection of scaling exponents instead of a single one, which makes possible to investigate and quantify HRV in terms of its multiexponent properties. A right shift has been revealed in the multifractal spectrum peaks for healthy subjects during meditation [40], which points to a better health condition of persons with respect to multifractal nature. Accordingly, a healthy heart-rate regulation promotes a multifractal signal.

3. Statistical Methods and ECG Databases

3.1. Previously Proposed Indices and Bootstrap Resampled Median Difference

A set of metrics were computed in order to quantify population differences in terms of the information conveyed by small, medium, and large scales of MSE and MTI indices. Namely, the area under the MSE and MTI profiles between scales 1 and 5 (so-called Area 1–5), between scales 6 and 20 (Area 6–20), and between scales 21 and 100 (Area 21–100). The area under the complete profiles (so-called Area) for MSE, MTI, and MFS was also computed. These metrics have been previously used and validated in studies comparing results of multiscale indices for different populations [20,21]. The Wilcoxon-Rank Sum Test was subsequently used to evaluate the statistical difference between populations in terms of these metrics, also according to the previous works.

As an extension of the previous existing analysis, we contribute in this paper with a statistical procedure allowing to establish simple statistical comparisons, either between two different population groups or within the same group of patients, in terms of a given multiscale representation. The procedure can be summarized and described as follows. We generically denote the scale as ν (which includes both possibilities for τ in MSE and MTI, or h in MFS), and the multiscale index as $J(\nu)$ (where J stand for either MSE or MTI or MFS representations). Let us assume that we have available a set of signals from a given patient dataset A, and this set is denoted as

$$\mathcal{S}_A \equiv \{x_i[n], i = 1, \cdots, N_A\} \tag{5}$$

and that the multiscale parameter can be obtained by using a given operator Γ for each signal in the database, i.e.,

$$J_i(\nu) = \Gamma\left(x_i[n], \theta\right) \tag{6}$$

where θ includes the set of preprocessing and processing parameters established for preprocessing and conditioning the signal under analysis.

Note that in this case $J(\nu)$ represents a random process, defined by its statistical distribution $f_{J(\nu)}\left(J(\nu)\right)$, which in general has an unspecified expression. We can define its median value and denote it as $J_M(\nu)$, which in practice can be estimated as the median of the multiscale representations obtained in a given population with a given set of parameters, and denoted as $\bar{J}_M(\nu|\mathcal{S}_A, \theta)$. Therefore, statistical differences can be calculated in two kinds of situations. First, when comparing the multiscale differences in two populations of patients, $\mathcal{S}_A$ and $\mathcal{S}_B$, we can build the statistic accounting for the difference between their corresponding poblational medians, give by

$$\Delta J_M(\nu) = \bar{J}_M(\nu|\mathcal{S}_A, \theta) - \bar{J}_M(\nu|\mathcal{S}_B, \theta) \tag{7}$$

In addition, we can have two different sets of preprocessing conditions, given by θ_1 and θ_2, and in this case the differences due to this change in those conditions can be scrutinized in terms of the difference of the median multiscale spectrum in a given population, as given by

$$\Delta J_M(\nu) = \bar{J}_M(\nu|\mathcal{S}_A, \theta_1) - \bar{J}_M(\nu|\mathcal{S}_A, \theta_2) \tag{8}$$

Hence, both representations are similar enough to provide a similar-to-handle view of the scales in which differences can be observed. The use of the median gives a robust estimator for cases where non-Gaussian distributions can be present, which was previously observed to be this case.

Since the *PDF* of the multiscale indices is often complex to estimate and to handle, we used nonparametric bootstrap resampling techniques, which provide us with an estimation of the empirical distribution of any statistical magnitude that can be built from computational media [29]. In our case, given a set of observed signals, we build a resample of this set by sampling with replacement each of the individuals in $\mathcal{S}_A$ up to B times, so that we get the so-called the bth resample of the patient population, $\mathcal{S}_A^*(b)$, where superscript * is the usual notation to point out all the bootstrap-estimated magnitudes. For this resample, we obtain an estimate of the statistical magnitude of interest, widely known as its bootstrap replication, and in our case it corresponds to weight vector $w_{(b)}^*$. By repeating the procedure B times, we get an estimate of the marginal distribution given by the empirical probability density function *(PDF)* of the bootstrap replications of each weight, this is, $\bar{J}_M(\nu|\mathcal{S}_A^*, \theta)$. The estimated distribution of the median difference statistic as a function of the scale can then be estimated from the replications of this statistic, which for the case of two different patient populations is obtained by non-paired resampling as follows:

$$\Delta J_M^*(\nu, b) = \bar{J}_M(\nu|\mathcal{S}_A^*, \theta) - \bar{J}_M(\nu|\mathcal{S}_B^*, \theta) \tag{9}$$

whereas for changes in the preprocessing conditions, paired resampling can be addressed, yielding

$$\Delta J_M^*(\nu, b) = \bar{J}_M(\nu|\mathcal{S}_A^*, \theta_1) - \bar{J}_M(\nu|\mathcal{S}_A^*, \theta_2) \tag{10}$$

The corresponding CI can be readily obtained just using sorted statistics, with significance level α yielding confidence level $1 - \alpha$ (typically $1 - \alpha = 0.95$). We expect the relevant differences to exhibit non-zero overlapping CI. In addition, the band confidence width should be consistent with the expected statistical power of the bootstrap test, hence allowing to study the consistency of the estimates when increasing the number of measured days in the Holter signals, and the estimated median average should allow us to scrutinize the presence of bias.

3.2. ECG Databases

We started by using multiscale methods to asses the variability of the RR-interval signals derived from 24-h Holter recordings from control subjects and from CHF patients. Both sets of recordings were downloaded from Physionet database [11]. The control group was obtained from 24-h Holter recordings in 72 healthy subjects (35 men and 37 women, from 20 to 76 years old). The original ECG recordings were sampled at 128 Hz. The CHF group was obtained from 24-h Holter recordings in 44 subjects (from 22 to 79 years old, including 19 men and 6 women, though gender information was not available for all the recordings). A subset of the original ECG recordings were sampled at 250 Hz (15 recordings), and the rest at 128 Hz. A number of studies have been conducted with these Databases [41–44] to determine the effect of exercise training on cardiac autonomic modulation in normal older adults using HRV, to establish normal values of RR variability for middle-aged persons, and to determine the effect of beta-blockers on parasympathetic nervous system activity.

We also used a specific LTM database, in which two sets of 7-day Holter recordings were also analyzed, one set from patients with CHF in sinus rhythm (73 recordings), and another set from patients with CHF with chronic AF (14 recordings). For short, we will denote them as CHF dataset and AF dataset, keeping in mind that both of them are CHF patients, but with different basal rhythms. The protocol to collect these recordings was carried out following the principles of Helsinki Declaration. It was approved by the Local Ethics Committee. Patients were recruited during scheduled outpatient visits to the CHF outpatient clinic in Virgen de la Arrixaca University Hospital (Murcia, Spain). From June 2007 to May 2011, patients with an established diagnosis of stable chronic CHF gave written informed consent to participate. All patients had LVEF $\leq$50% and they were clinically stable, without need for hospital admission or intravenous vasoactive agents within the past 3 months. Exclusion criteria included pacemaker-dependent patients, a serious comorbid condition with associated life expectancy $<$1 year, hospitalization for Myocardial Infarction (MI) and unstable Coronary Artery

Disease (CAD) within the past 3 months, or any cardiac revascularization procedure within 30 days before enrolment. The 7-day continuous Holter recordings were obtained using a commercially available device (Lifecard CF^{TM}, Del Mar Reynolds, Issaquah, Washington). These databases had been used [12,27,37,45] in previous studies: (a) To demonstrate that the circadian rhythms detected in 7-day recordings could not always be detected in 24-h periods; (b) to compare the diagnostic sensitivity of 1-day Holter monitoring versus 7-day Holter monitoring (7DH); (c) to detect atrial and ventricular arrhythmias in a population of stable patients with chronic HF and left ventricular dysfunction; (d) to characterize the relationship between heart rate and post-discharge outcomes in patients with hospitalization for HF with reduced ejection fraction (EF) in sinus rhythm; and (e) to characterize the infradian, circadian, and ultradian components for each patient, as well as circadian and ultradian fluctuations.

A standard Holter analysis software (ELA Medical^{TM}, Sorin Group, Paris, France) was used to process the data. When needed, a trained cardiologist performed a visual check of the QRS complex classification and every arrhythmic event, therefore, manual corrections were made. Both data sets (Physionet and LTM) were preprocessed to exclude artifacts and ectopic beats, as follows. RR intervals lower than 200 ms and greater than 2000 ms were eliminated, as well as those which differed more than 20% from the previous RR interval [9]. The nonlinear indices were computed on the resulting time series.

4. Experiments and Results

In this section, we present a series of experiments in order to analyze the consistency and robustness of the indices for multiscale characterization with 1-day and 7-days Holter registers of healthy subjects, AF patients, and CHF patients. First, a robustness analysis of all these methods is made with datasets from Physionet databases (specifically, healthy subjects and CHF patients) in 1-day Holter recordings, and patients with AF and CHF are then analyzed in 7-day Holter recordings. We also scrutinized how the results can change when using 1-day Holter registers versus 7-day Holter registers and with increased scaling factor. A study on the robustness of the indices on 7-day recordings is checked in terms of 1-day segments, and a quantitative analysis is made for all of them in terms of the confidence bands for the populational medians.

4.1. Physionet Database and 1-Day Holter Recordings

We started our analysis by scrutinizing the effect of calculating multiscale indices in 1-day Holter recordings above 20 scales, which is the usual limit value in the literature. The multiscale indices were obtained in the 72 subjects in the control set of Physionet database. Figure 1 (left panels) shows these results, where each plot represents the entropy value (vertical axis) as a function of the multiscale parameter (τ or h), for every patient in the database and with no specific ordering. Typical patterns can be observed for each multiscale index. For instance, in MSE there is often a soft curve for low scales that usually turns to near constant at larger scales. We can observe also a rift effect specially in larger scales, which indicates that the method is being sensitive to noise. We can also observe a bias effect among subjects, as some of them appear to be above increased or decreased average levels compared with others. With respect to MTI, we can see that all the cases start near zero value for low scales, and there is a general trend to increase as a soft-changing curve above zero for all the cases. No rift effect is present in this multiscale index. Finally, MFS often shows an inverted-U shape, as documented in the literature, and there is a bow about $h = 0.5$ in many cases, followed by a flat or slow decaying set of values. Note that some few cases seem to deviate from this populational behavior.

A different behavior is present when we scrutinize the three multiscale indices for the CHF patients in 1-day Holter recordings, as seen in Figure 1 (right panels). This figure and its panels allow us to check the individual profiles of a given multiscale index in a population, so that individual profiles can be checked to be consistent with their population, and also non-concordant profiles can be clearly identified. Note that the axis are similar to the control panels, and that the vertical axis have been adjusted with the same range and scale. In MSE there is a diversity of shapes in low scales, though the trend to stabilize at larger scales is again observed, as well as the rift effect. In addition, there are more cases with increased average value, and in general there is a trend to exhibit lower average cases than controls. The MTI again starts from low values for low scales, but then it smoothly and often (not always) tends to go towards negative values. With respect to the MFS, it clearly decreases its width, the bow and the flat set of indices are mostly lost, and it often deviates from 1. Some very atypical cases are present, specially in the CHF set. Several patients are extremely different from the others in MSE and MTI, and several (not few) cases in MFS seem to present a breakdown and even values above 1. We thoroughly checked in all the patients that artifacts were correctly suppressed from the signals with ad-hoc designed software to represent jointly the RR-signals and the ECG signals on a similar time basis. Figure 2 presents six example cases (three from healthy subjects and three from CHF patients) in specific cases, namely, a normal-trend patient (typical MSE, MTI, and MFS) (Panels a, b), a patient with atypical MSE and MTI profiles (Panels c, d), and a patient with atypical MFS profile (Panels e, f).

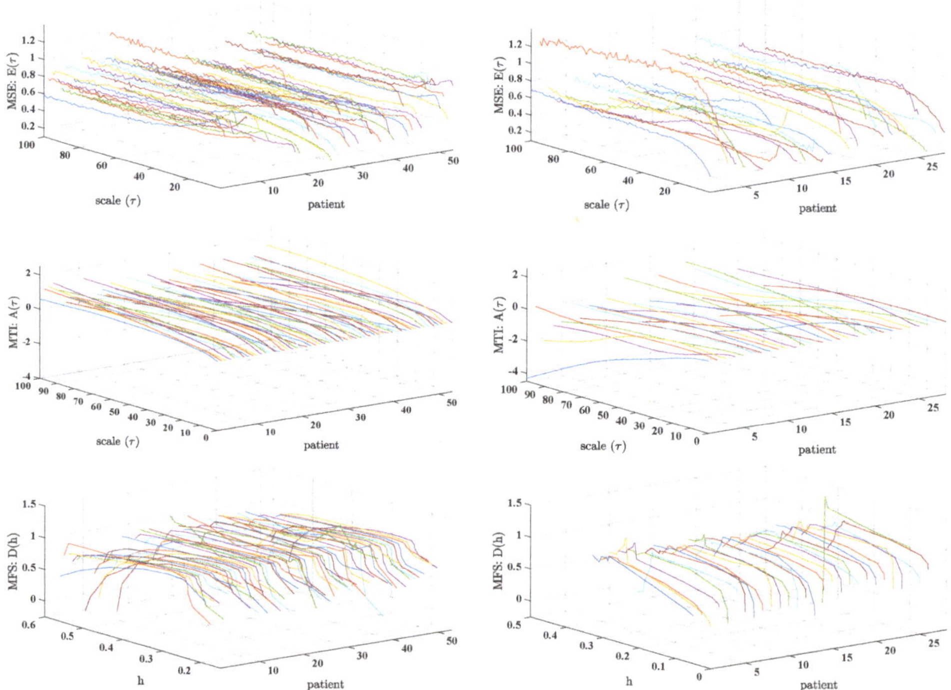

Figure 1. Results of multiscale analysis in Physionet Database: Multiscale Entropy (MSE) (**top**), Multiscale Time Irreversibility (MTI) (**middle**), and Multifractal Spectrum (MFS) (**down**) for the control database (**left**) and for the Congestive Heart Failure (CHF) patients (**right**).

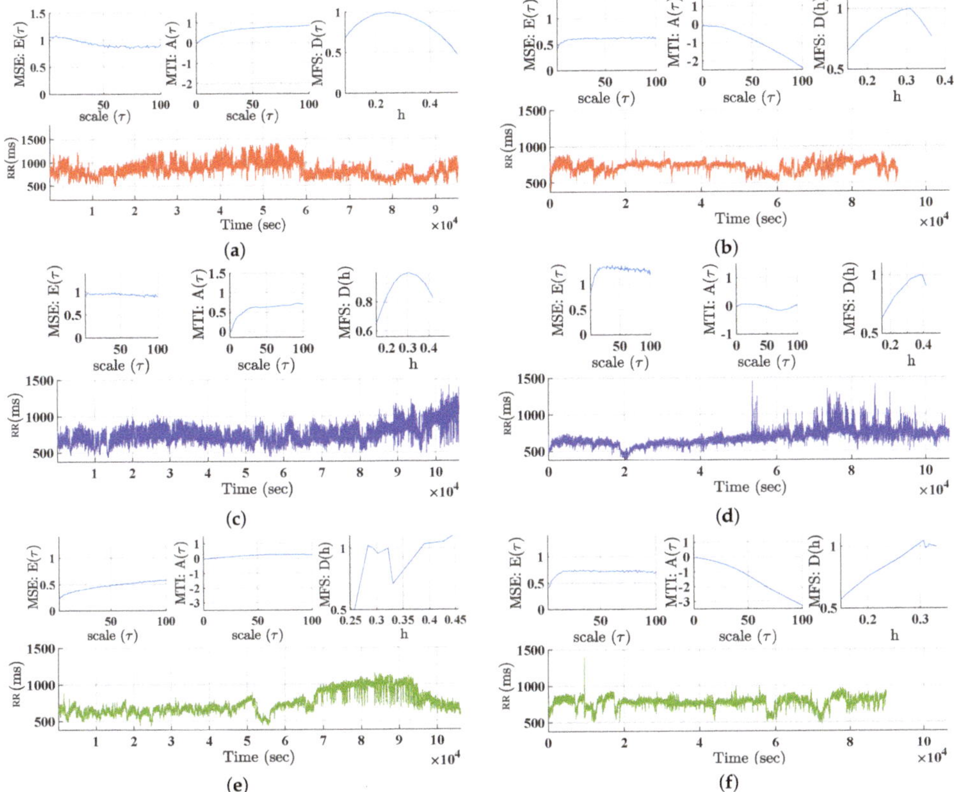

Figure 2. Results of multiscale analysis in example patients from Control Database (**left**) and from CHF Database (**right**) in Physionet: (**a**,**b**) Normal trend; (**c**,**d**) abnormal MSE and MTI profiles; (**e**,**f**) abnormal MFS profile.

4.2. Ltm Database and 7-Day Holter Recordings

Figure 3 shows the MSE, MTI, and MFS results for AF and CHF 7-day Holter recordings databases. In the right panels, we can observe typical patterns that are similar to the CHF cases in 24-h recordings. In low scales, MSE exhibits a variety of shapes that tend to stabilize at larger scales. MTI indices show decreasing values from low to large scales reaching negative values in some cases. MFS shows a shape with an increased width and a lost of the bow. In the left panels, different trends are observed for AF patients. MSE shows an inverted shape compared to CHF patients, i.e., larger values for low scales and lower values for high scales, or a decreasing soft curve for low scales that usually turns into constant at larger scales. We can also observe a deviation of some of the subjects from the general trend. Regarding MTI, for low scales, values start near zero with a generally slight-decreasing trend curve. MFS shows an inverted-U shape with a bow about $h = 0.2$ for all cases, followed by slowly decreasing values.

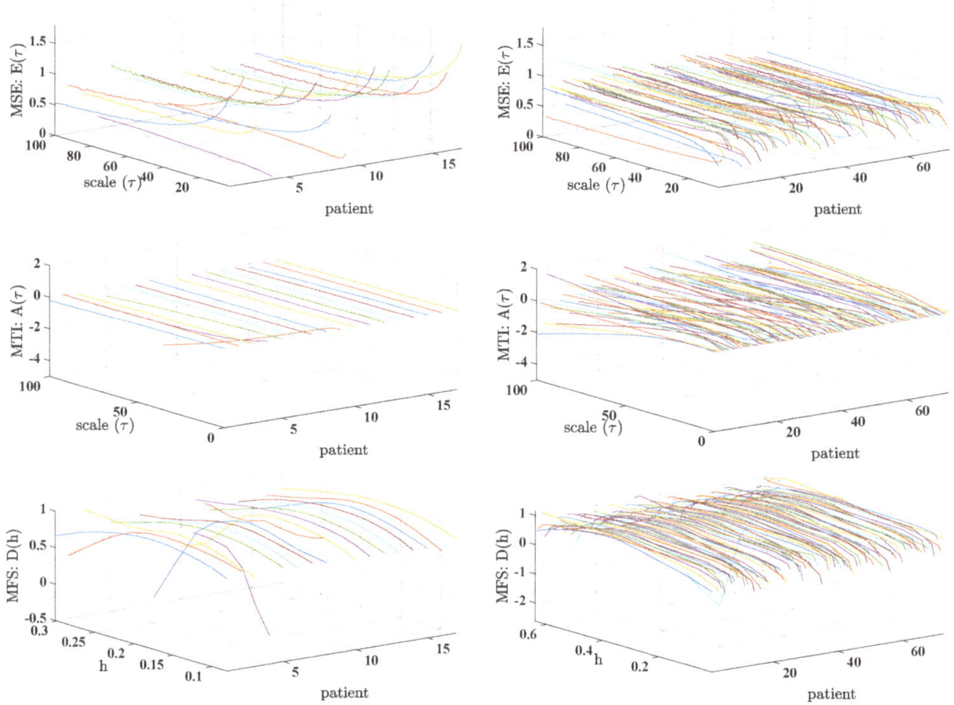

Figure 3. Results of multiscale analysis in long-term monitoring (LTM) database: MSE (**top**), MTI (**middle**), and MFS (**down**), for the atrial fibrillation (AF) database (**left**) and for the CHF database (**right**).

We designed a simple experiment accounting for windowing analysis of 7-day recordings in windows of 1-day duration. Our purpose here was to perform a robustness analysis of the 7-days Holter recordings when compared with 1-day recordings, in order to scrutinize the new knowledge provided by the scales up to 100 and to analyze the reproducibility of the multiscale indices throughout the 7 days.

Figure 4 presents a comparative analysis between 1-day and 7-day recordings by means of 3 patients with AF and 3 patients with CHF. The heart-rate signal of each patient is presented as well as the MSE, MTI, and MFS profiles. Those profiles present similar trends for every 1-day segment and for the 7-day segment with variable variance depending on the particular case. However this variance is larger in CHF examples than in AF examples for MTI in large scales and mainly for MFS.

We can also check that the consideration of larger scales in the 7-day recordings has relevant effect on the estimation of the multiscale indices. For instance, the ripple in MSE is reduced in the 7-day estimations. In addition, the variance in the 1-day estimated indices for larger scales in MSE and MTI seem to stabilize to a profile which is not just the average of the consecutive days, which implies that nonlinear irregularity effects in these larger scales are present and likely they are better accounted by the 7-day calculations. It is interesting the effect that the use of 7-day recordings has on MFS, which is different from a simple day-averaging of the windowed spectra. In some cases, a wider set of 1-day spectra is condensed into a narrower spectrum with 7-day signals. In general, larger variance seems to be present at larger scales with 1-day signals, which is reduced by 7-day based spectrum (often yielding a spectral profile lower than the individual spectra in each day). Even in one case, apparently inconsistent daily spectra turn into a well-shaped spectrum when using the 7-day recording.

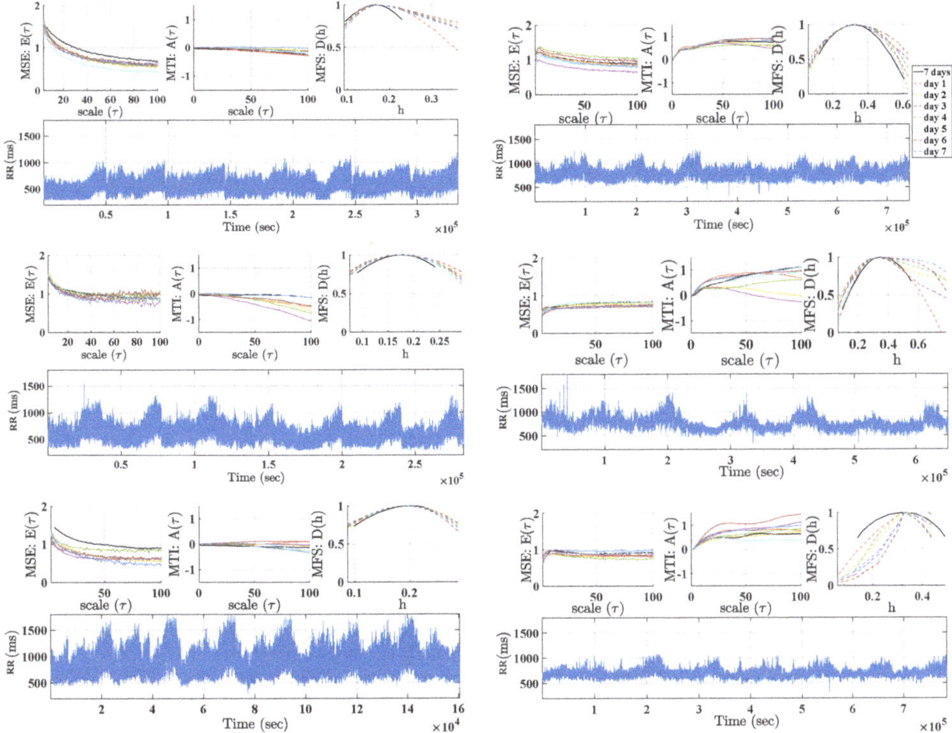

Figure 4. Examples of robustness analysis of 1-day versus 7-day Holter recordings in AF (**left**) and CHF (**right**) patients.

4.3. Statistical Analysis and Confidence Bands

Table 1 compares CHF patients and control subjects of Physionet database in terms of Area 1–5, Area 6–20, Area 21–100, and Area metrics for the multiscale indices. The Wilcoxon Rank-Sum Test reveals significant statistical differences for Area 1–5 in MSE and MTI, and for Area 6–20 in MTI, hence for small and medium scales. Table 2 shows the comparison for the two populations in the LTM database. In this case, statistical differences are present in MSE for small and medium scales, whereas in MTI these are also preset for large scales Area 21–100, and in MTI and MFS for the complete Area. The same statistical differences are observed when comparing the one day average results of LTM database (Table 3). Note that the interpretation of these differences should be taken as descriptive of the different cardiac conditions in terms of the different scales of the signal dynamics, and not as classifying characteristics for them with diagnostic purposes.

Additional details can be scrutinized from the proposed method, aiming to extend the statistical behavior for the previous indices in different scales, populations and conditions, or scale span, in terms of confidence bands width and median differences. Figure 5 shows the confidence bands for the median multiscale indices when comparing the control set with the two considered cardiac conditions, namely, CHF and AF. The former is obtained from the Physionet CHF set and the later from the LTM-AF set. Each panel shows the median and confidence bands for the first group (up, left), for the second group (up, right), and for the median difference (down). Panel (a) shows that the MFS is mainly different between both groups for low scales in MSE, whereas significant differences are present in MTI differences in all the scales. For obtaining confidence bands in MFS, an interpolation was done to a regular grid sampling using chirp interpolation, and given the different scale spam for the obtained MFS in each patient, the confidence bands in each point of the scale grid was obtained conditional to the existence of the fractal spectrum in that scale. As it can be seen in the right panel, the control set has a scale span between 0 and 1, with some exceptional case extending out of it, but the confidence band gets wider after 0.6, whereas the scale span is mainly narrower in CHF patients, though sometimes it reaches a similar set of values for different patients, as it can be observed from the wider confidence bands in the left and specially in the right of the graph. The median difference indicates a clear descending trend in the median value (blue line) from left to right, which is only non-overlapping zero at about scale 0.5. Panel (b) in Figure 5 shows the result of a similar comparison in this case between the control subjects from Physionet (1D) and the CHF set from the LTM database (7D). The relevant differences with respect to the previous comparison can be summarized as follows. For MSE, the differences in the lowest scales are less present, and there is a trend to the band to be consistently below zero, though it remains non-significantly yet borderline. For MTI, the differences between both populations show a similar trend than in the previous comparison, though the confidence bands in this case are borderline with respect to their zero overlapping. For MFS, the confidence bands in the CHF population from 7D reaches a wider interval of scale values remaining narrow, and the confidence bands for the median difference exhibit a similar trend with smoother band limits than the previous comparison. Panel (c) in Figure 5 shows the comparison between the control subjects from Physionet (1D) and the AF set from the LTM database (7D). For MSE, significant differences are present in low scales (below 60), whereas for MTI there are significant differences in all the scales. In addition, significant differences are present in MFS in almost all the scales, whereas in this case the median trend has the opposite slope than in the preceding comparisons with CHF. Whereas several of these results have been previously documented in the literature, our results are consistent with those precedents and they can be observed at a glance from the representation. It is evident that these indices are measuring different aspects of the complexity and/or nonlinear nature of the cardiac dynamics, and that they probably should be used complementarily.

Figure 6 shows the confidence bands of multiscale analyses with the aim of establishing the statistical properties that can be expected due to the use of LTM recordings. In this setting, and as it could be expected, the confidence-band widths are consistently narrower when scrutinizing paired datasets, i.e., those comparisons in which the indices estimated from 7-day recordings are compared to the indices estimated in one day (the third in each set for all of them in the 7D register) of 1D recordings for the same patients. Panel (a) shows the scale-paired confidence bands for the median differences when analyzing the CHF patient set, showing a trend to raise in the very low scales, which could mean that the significance of these scales depends on the number of days considered, when working with CHF patients. The estimated MTI remains very reproducible when obtaining it from 1D or from 7D in each patient. Note the different behavior of MSE and MTI in terms of the confidence band width in terms of increasing scales (mostly constant in MSE and increasing error standard with scale in MTI). With respect to MFS, whereas the median value difference is not significant, the most critical scale regions are again the boundaries of the population spectrum, in terms of confidence band widths. Similar conclusions can be obtained for the AF patients when compared in terms of MSE and MTI. The scale-border effects is even more critical in this case, as AF wide of this spectrum is narrower in the patients compared with control subjects and with CHF condition patients. Panels (c) and (d) show the result of comparing different conditions (CHF and AF) when using 1D or 7D in these two different sets of patients. Note that in this case the confidence bands of the differences are not very different for the cases of MSE and MTI, whereas both the confidence band widths and the scale span are clearly narrower when they are compared in terms of 7D recordings.

Table 1. Physionet database, congestive heart failure (CHF) versus Control. Area 1–5, Area 6–20, Area 21–100 and Area metrics expressed as mean ± standard deviation for the multiscale indices. Significant statistical differences given by the Wilcoxon Rank-Sum Test are indicated.

Scale Test	CHF	Control	*p*-Value
MSE (Area 1–5)	1.30 ± 0.50	1.58 ± 0.57	<0.05
MSE (Area 6–20)	9.47 ± 3.52	9.66 ± 2.75	0.53
MSE (Area 21–100)	56.45 ± 18.23	52.37 ± 11.88	0.38
MSE (Area)	68.51 ± 22.30	65.02 ± 15.27	0.69
MTI (Area 1–5)	0.09 ± 0.09	0.20 ± 0.14	<0.05
MTI (Area 6–20)	1.95 ± 1.79	4.41 ± 2.37	<0.05
MTI (Area 21–100)	50.46 ± 35.72	43.54 ± 27.76	0.35
MTI (Area)	52.77 ± 36.54	48.71 ± 29.77	0.68
MFS (Area)	0.32 ± 0.41	0.27 ± 0.09	0.62

Table 2. Long-term monitoring (LTM) database, atrial fibrillation (AF)-7-day Holter monitoring (7DH) versus heart failure (HF)-7DH. Area 1–5, Area 6–20, Area 21–100 and Area metrics expressed as mean ± standard deviation for the multiscale indices. Significant statistical differences given by the Wilcoxon Rank-Sum Test are indicated.

Scale Test	AF-7DH	HF-7DH	*p*-Value
MSE (Area 1–5)	3.01 ± 1.16	1.55 ± 0.47	<0.05
MSE (Area 6–20)	12.89 ± 4.89	10.22 ± 2.63	<0.05
MSE (Area 21–100)	52.34 ± 18.03	56.63 ± 12.44	0.79
MSE (Area)	70.16 ± 24.51	69.80 ± 15.53	0.27
MTI (Area 1–5)	0.04 ± 0.04	0.16 ± 0.13	<0.05
MTI (Area 6–20)	0.72 ± 1.61	3.13 ± 2.17	<0.05
MTI (Area 21–100)	19.74 ± 50.62	42.78 ± 25.18	<0.05
MTI (Area)	20.60 ± 52.31	46.47 ± 26.29	<0.05
MFS (Area)	0.25 ± 0.12	0.48 ± 0.91	<0.05

Table 3. LTM database, AF-1-day Holter monitoring (1DH) versus HF-1DH. Area 1–5, Area 6–20, Area 21–100 and Area metrics expressed as mean ± standard deviation for the multiscale indices. Significant statistical differences given by the Wilcoxon Rank-Sum Test are indicated.

Scale Test	AF-1DH	HF-1DH	*p*-Value
MSE (Area 1–5)	3.06 ± 1.18	1.63 ± 0.58	<0.05
MSE (Area 6–20)	12.96 ± 4.74	10.58 ± 3.24	<0.05
MSE (Area 21–100)	50.58 ± 17.37	56.68 ± 14.16	0.28
MSE (Area)	68.54 ± 23.48	70.35 ± 18.11	0.82
MTI (Area 1–5)	0.04 ± 0.06	0.18 ± 0.17	<0.05
MTI (Area 6–20)	1.12 ± 2.69	3.41 ± 2.47	<0.05
MTI (Area 21–100)	31.68 ± 87.92	45.79 ± 30.89	<0.05
MTI (Area)	33.00 ± 90.97	49.82 ± 32.48	<0.05
MFS (Area)	0.26 ± 0.11	0.45 ± 0.51	<0.05

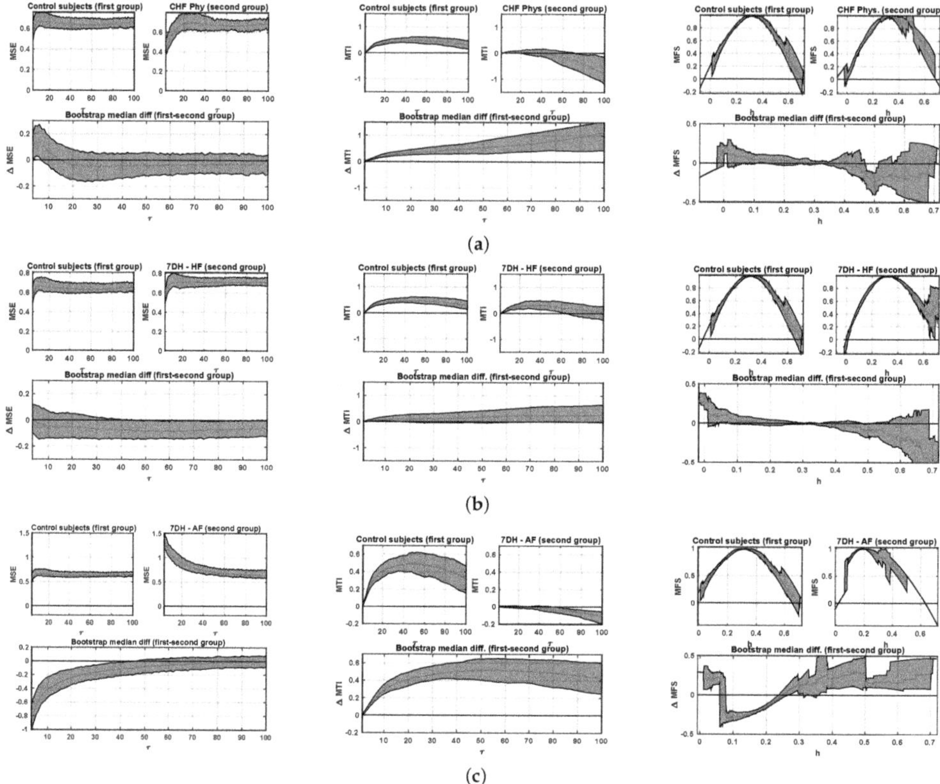

Figure 5. Confidence bands of multiscale analysis in control population vs different cardiopathy conditions: (a) Controls from Physionet (1D) vs HF patients from Physionet (1D); (b) controls from Physionet (1D) vs HF patients (7D) from LTM database; (c) controls from Physionet (1D) vs AF patients (7D) from LTM database. From left to right columns, the MSE, MTI, and MFS confidence bands and their differences are included.

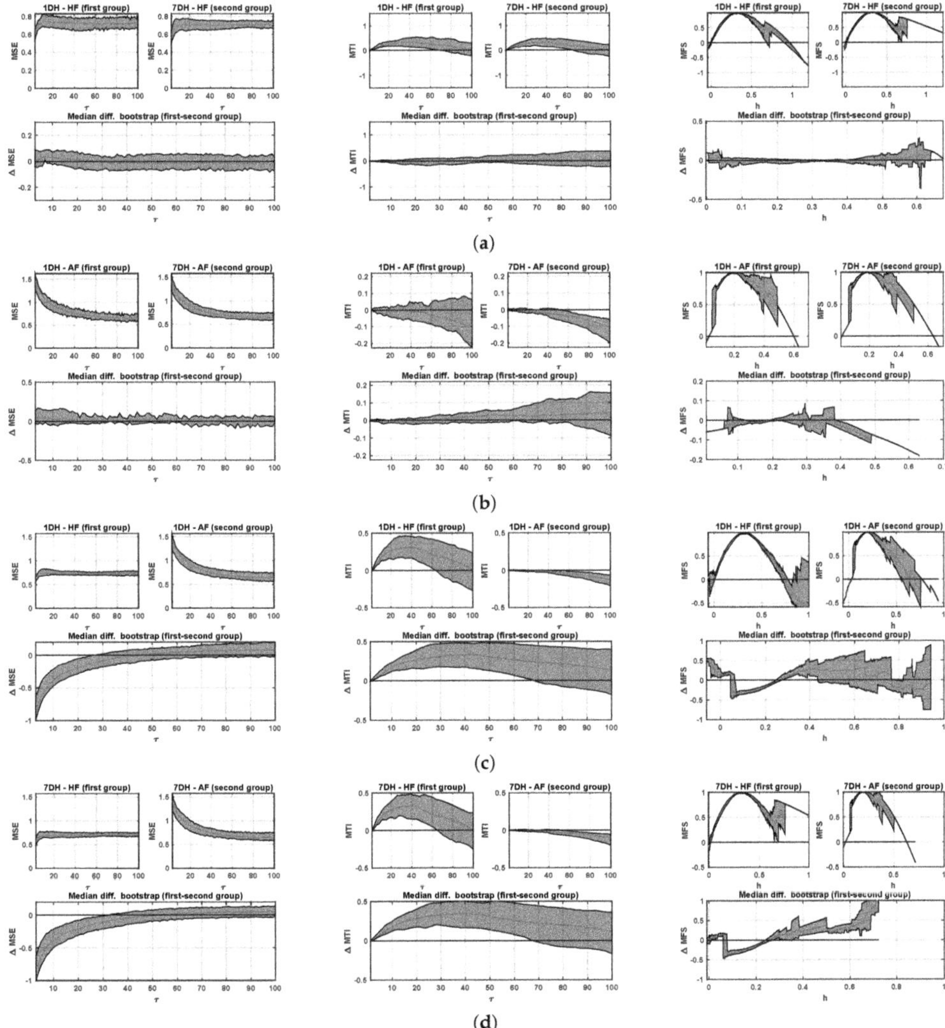

Figure 6. Confidence bands of multiscale analysis in LTM database: (**a**) Paired 1D vs 7D in HF patients; (**b**) paired 1D AF vs 7D in AF patients; (**c**) non-paired 1D HF vs 1D AF patients; (**d**) non-paired 7D HF vs 7D AF patients. From left to right columns, the MSE, MTI, and MFS confidence bands and their differences are included.

5. Discussion and Conclusions

In this work, we have addressed the calculation of three representative multiscale indices, namely, MSE, MTI, and MFS, on 1-day and 7-day Holter recordings. From our results, we can conclude that, when present, the trends are consistent between the additional scales provided by 1-day recordings and by the 7-day recordings, but the second ones can give a statistically better view of this kind of representations, specially in the MFS representations.

Contributions. Several preceding results can be found in which descriptions of 1-day estimated indices are shown, however, and to the best knowledge of the authors, few works address from this perspective the 7-days case with scales up to 100. Accordingly, and given the advances in the monitoring and well-being technology in our days, this represents a good moment to determine the possible advantages and issues of using these methods in larger observation scales. Non-linear methods used in this work (MSE, MTI, and MFS) have previously been proposed and widely used in the literature. The contributions of our work are twofold. On the one hand, we analyzed the dynamics, the consistency, and especially the robustness of these indices when applied to the new long term scenarios (7 days) and to point out that algorithms may need to be improved when using them in this area. On the other hand, to find out if further information is conveyed in the temporal scales that have not been analyzed before in shorter recordings.

Summary and Discussion of Results. Results showed significant differences between CHF and AF populations not only for short-term scales but also for long-term and very long-term scales in some indices, MSE (MTI) being higher (lower) for AF. These results are consistent with previous studies [14–16]. Here we confirmed the same trends, but we also obtained several differences in very long-term scales that had not been analyzed before. MSE was not further significantly different for very long-term scales. This may be attributed to the fact that statistical characteristics of AF signals resemble those from noise, and being MSE a entropy-based measure, it is higher for AF than for CHF for short-term and long-term scales, but not for very long-term scales, where the surrogate signals present attenuated these erratic characteristics. On the other hand, MTI shows that time irreversiblity is for all time scales lower in AF patients, i.e., for those presenting a more severe pathology.

We detected that 1-day estimations of the multiscale indices can show a distorted profile. The different estimations observed in the different time windows could be due to several reasons. In addition, this effect could have statistical roots, in terms of consistency and variance of the estimation with shorter time series. On the other hand, it also could be due to the fact that some frames had some properties in their dynamics which made the multiscale algorithm fail or disrupt. The differences in daily activities should not be a limitation, as far as the multiscale algorithm measurements are the correlations present in the signal at different time scales. In the conventionally used surrogate-signal test, which is built with lower-pass versions of the original signal, the more we increased the scales, the more the short-term relationships were eliminated to manifest the long-term. Thus, the short-term differences would be hidden for high scales. We thoroughly checked that the underlying signals did not have artifacts or aberrations with a specifically customized software to support an observer to watch at the heart-rate signals and at the original ECG signals simultaneously. Accordingly, one of our conclusions is that these multiscale algorithms, while likely informative, should trust not only on the increase of the length of the signals for their consistency, but also on improving their algorithmic robustness. While stationarity can not often be assumed for HRV signals, MSE was originally proposed and applied in 24-h Holter recordings, and later many other works have used MSE to study different dynamics in 24-h. The analysis of the underlying dynamics during the day or night can be complex with these methods. Nevertheless, we observed that the disruption effect in one day does not have any significant impact in the 7-day analysis, which promotes their use in favor of improved statistical consistency.

Usefulness of Median-difference Tests. An extension of previously proposed statistical comparisons for scales has been proposed here in terms of nonparametric bootstrap resampling, which allows us to establish poblational comparisons in terms of the median difference of the multiscale representations, either for different health conditions, or for different acquisition conditions. Our results with this method were consistent with the results identified in the literature and showed some non-observed differences, specially the ones related with AF patients, and with different statistical consistency behavior and retrieved information about the patient given by these three multiscale indices, which suggests their use jointly in these and other populations.

Related Relevant LTM Applications With the technological availability of cardiac monitoring systems for extended time periods, LTM is expected to bring interest to new applications in health, wellness, and research. For instance, it has been recently pointed out [30] that energetic environmental phenomena can affect psychophysical processes on people in different ways depending on their sensitivity, health status, and ability of the ANS to self-regulate. In that study, the HRV was recorded for 72 consecutive hours per week over a five-month period in 16 participants, in order to examine ANS responses during normal background environmental periods. Interestingly, HRV measurements were negatively correlated with the solar wind speed, and the low-frequency and high-frequency power were negatively correlated with the magnetic field. This study confirmed that the daily activity of the ANS responds to changes in geomagnetic and solar activity during periods of undisturbed normal activity, it starts at different times after changes in various environmental factors, and it persist for variable time periods. As far as the activity of the ANS reflected by HRV measurements is affected by solar and geomagnetic influences, the analysis of HRV should take into account these effects when possible. This study is focused on the ANS modulation, and hence spectral measurements were mostly used, nevertheless, this kind of data could be analyzed with multiscale indices to provide a wider view on the long-term behaviour of HRV.

On the Clinical Usefulness of Multiscale Indices. As described with detail in [7], many indices have been proposed in the last years to develop risk stratification from different kinds of analysis of electric cardiac signals. However, these techniques from the academic research world rarely are used in the clinical practice. In that reference, our team analyzes the possible reasons by decoupling the sometimes limited accuracy and the lack of consensus on the robustness with the appropriate signal processing implementations. In this line of search, the present work aims to first establish the need for robustness in the methods that are widely used nowadays, as a requirement before enrolling in risk stratification studies, which require high-cost and high effort to yield clinically useful use to these techniques. Our future research would consider first to perform the same multiscale analysis (MSE, MTI, and MFS) in LTM healthy subjects, in order to study the dynamic behavior in normal conditions when increasing the number of scales. As indicated, the development of more robust multiscale indices is a desirable target in order to continue to progress in this informative characterization of the cardiac dynamics. On the other hand, AF and CHF are different heart diseases with different dynamics, though in this study we only could analyze AF patients as a subset of CHF patients. CHF is a syndrome of the deterioration in short-term and long-term regulatory mechanisms, while AF, and especially the persistent one, is an intrinsic short-term mechanism. Whereas long-term mechanisms can be present in AF, it seems that LTM recordings should be better used to analyze their presence or deterioration.

Despite the need for more robust algorithms in long-term nonlinear indices, we still consider that it worth the effort to use these indices, as vast literature supports their informativeness in other scenarios in addition to the very long-term monitoring. In general terms, the underlying hypothesis in preceding studies is that heart rate oscillations respond to phenomena with very different characteristic times, ranging from seconds to months. The former are well evaluated in a 24-h recording, but the low frequency (such as circadian cycles, many hormonal cycles, or secondary to changes in activity during the week) would only be represented in long-term monitoring. The response to these stimuli has been moderately studied, but it could provide us with interesting information on some clinical aspects, such as the prediction of decompensation in heart failure, the evolution of cardiovascular remodelling after acute injury (e.g., in the weeks following acute myocardial infarction), the evolution of sports training, or the susceptibility to the development of malignant arrhythmias (related with sudden cardiac death). All these aspects involve hormonal processes, inflammatory processes, adaptations of different organs or systems, and they are characterized by the interaction of mechanisms with slow response times, so that they develop in days or weeks, rather than in minutes or hours. Therefore, methods of analyzing the very long-term behavior of heart rate responses could be of interest, especially in noisy rhythms such as AF, in which almost all the usual HRV parameters are difficult to interpret.

We still do not know if these methods will be widely used in future clinical practice because we are still in the phase of describing how these indices behave, their variability and their comparison with the results in 24 h, among other thing. In any case, all these efforts should be supported by robust algorithms for multiscale nonlinear indices, an idea that had not been previously paid much attention in the nonlinear HRV literature.

Author Contributions: M.E.-Y. conducted the experiments, organized and wrote the paper, and performed the analysis of experimental results. R.G.-E., Y.J., and J.-L.R.-Á. contributed to write and review some sections of the manuscript. A.G.-A. supervised the database collection and clinical orientation. S.M.-R. supported the data processing in MATLAB.

Funding: This work was partially supported by Research Grants FINALE and KERMES (TEC2016-75161-C2-1-R and TEC2016-81900-REDT) from Spanish Government.

Conflicts of Interest: The authors declare no conflict of interest.

References

1. WHO. International Classification of Diseases. *Health Care Financing Administration*; Technical Report; US Department of Health and Human Services, Centers for Disease Control and Prevention: Geneve, Switzerland, 2005.
2. WHO. Cardiovascular Disease, 2016. Available online: http://www.who.int/topics/cardiovascular_diseases/en/ (accessed on 8 July 2018).
3. Priori, S.G.; Aliot, E.; Blomstrom-Lundqvist, C.; Bossaert, L.; Breithardt, G.; Brugada, P.; Camm, A.J.; Cappato, R.; Cobbe, S.M.; Di Mario, C.; et al. Task Force on Sudden Cardiac Death of the European Society of Cardiology. *Eur. Heart J.* **2001**, *22*, 1374–1450. [CrossRef] [PubMed]
4. Zipes, D.P.; Camm, A.J.; Borggrefe, M.; Buxton, A.E.; Chaitman, B.; Fromer, M.; Gregoratos, G.; Klein, G.; Moss, A.J.; Myerburg, R.J.; et al. Guidelines for Management of Patients With Ventricular Arrhythmias and the Prevention of Sudden Cardiac Death: A Report of the American College of Cardiology/American Heart Association Task Force and the European Society of Cardiology Committee for Practice Guidelines (Writing Committee to Develop Guidelines for Management of Patients With Ventricular Arrhythmias and the Prevention of Sudden Cardiac Death). *J. Am. Coll. Cardiol.* **2006**, *48*, 247–346. [CrossRef]
5. Soguero-Ruiz, C.; Mora-Jiménez, I.; Ramos-López, J.; Fernández, T.Q.; García-García, A.; Díez-Mazuela, D.; García-Alberola, A.; Rojo-Álvarez, J.L. An Interoperable System toward Cardiac Risk Stratification from ECG Monitoring. *Int. J. Environ. Res. Public Health* **2018**, *15*, 428. [CrossRef] [PubMed]
6. Villacastín, J.; Bover, R.; Castellano, N.; Moreno, J.; Morales, R.; García-Espinosa, A. Risk Stratification and Prevention of Sudden Death in Patients With Heart Failure. *Rev. EspañOla Cardiol.* **2004**, *57*, 768–782. [CrossRef]
7. Gimeno-Blanes, F.; Blanco-Velasco, M.; Barquero-Pérez, O.; García-Alberola, A.; Rojo-Álvarez, J.L. Sudden Cardiac Risk Stratification with Electrocardiographic Indices—A Review on Computational Processing, Technology Transfer, and Scientific Evidence. *Front. Physiol.* **2016**, *7*, 1–17. [CrossRef] [PubMed]
8. Corana, A.; Bortolan, G.; Casaleggio, A. Most probable dimension value and most at interval methods for automatic estimation of dimension form time series. *Chaos Sol. Fractals* **2004**, *20*, 779–790. [CrossRef]
9. Malik, M.; Cripps, T.; Farrell, T.; Camm, A. Prognostic value of heart rate variability after myocardial infarction. A comparison of different data processing methods. *Med. Biol. Eng. Comput.* **1989**, *27*, 603–611. [CrossRef]
10. Klabunde, R.E. *Cardiovascular Physiology Concepts*; Lippincot Williams and Wilkings: Philadelphie, PA, USA, 2011; p. 243.
11. Goldberger, A.L.; Amaral, L.A.N.; Glass, L.; Hausdor, J.M.; Ivanov, P.C.; Mark, R.G.; Mietus, J.E.; Moody, G.B.; Peng, C.K.; Stanley, H.E. PhysioBank, PhysioToolkit, and PhysioNet: Components of a New Research Resource for Complex Physiologic Signals. *Circulation* **2000**, *101*, 215–220. [CrossRef]
12. Goya-Esteban, R.; Mora-Jiménez, I.; Rojo-Álvarez, J.L.; Barquero-Pérez, O.; Pastor-Pérez, F.J.; Manzano-Fernandez, S.; Pascual-Figal, D.A.; García-Alberola, A. Heart rate variability on 7-day holter monitoring using a bootstrap rhythmometric procedure. *IEEE Trans. Biomed. Eng.* **2014**, *52*, 1366–1376. [CrossRef] [PubMed]
13. El-Yaagoubi, M.; Goya-Esteban, R.; Jabrane, Y.; Muñoz-Romero, S.; Rojo-Álvarez, J.L.; García-Alberola, A. Multiscale Entropy and Multiscale Time Irreversibility for Atrial Fibrillation and Heart Failure from 7-Day Holter. *Biomed. Eng. Online* **2017**, *10*, 90.
14. Costa, M.; Goldbergez, A.L.; Peng, C.K. Multiscale entropy analysis of complex physiologic time series. *Phys. Rev. Lett.* **2002**, *89*, 68–102. [CrossRef] [PubMed]

15. Costa, M.; Goldberger, A.L.; Peng, C.K. Broken asymmetry of the human heartbeat: Lost of time irreversibility in aging and Disease. *Phys. Rev. Lett.* **2005**, *95*, 102–198. [CrossRef] [PubMed]
16. Costa, M.; Peng, C.K.; Goldberger, A.L. Multiscale analysis of heart rate dynamics: Entropy and time irreversibility measures. *Cardiovasc. Eng.* **2008**, *8*, 88–93. [CrossRef] [PubMed]
17. Aktaruzzaman, M.; Sassi, R. Parametric estimation of sample entropy in heart rate variability Analysis. *Biomed. Signal Process. Control.* **2014**, *14*, 141–147. [CrossRef]
18. Stanley, H.E.; Amaral, L.A.; Goldberger, A.L.; Havlin, S.; Ivanov, P.; Peng, C.K. Statistical physics and physiology: monofractal and multifractal approaches. *Phys. Stat. Mech. Appl.* **1999**, *270*, 309–324. [CrossRef]
19. Ivanov, P.; Luis, A.; Goldberger, A.L.; Shalomo, H.; Stanley, H.E.; Struzik, Z. From 1/f noise to multifractal cascades in heartbeat dynamics. *Chaos* **2001**, *11*, 641–652. [CrossRef] [PubMed]
20. Chen, C.H.; Huang, P.W.; Tang, S.C.; Shieh, J.S.; Lai, D.M.; Wu, A.Y.; Jeng, J.S. Complexity of Heart Rate Variability Can Predict Stroke-In-Evolution in Acute Ischemic Stroke Patients. *Nature* **2015**, *5*, 17552. [CrossRef] [PubMed]
21. Liu, H.Y.; Yang, Z.; Meng, F.G.; Guan, Y.G.; Ma, Y.S.; Yang, S.L.; Lin, J.L.; Pan, L.S.; Zhao, M.M.; QU, W.; et al. Preoperative Heart Rate Variability as Predictors of Vagus Nerve Stimulation Outcome in Patients with Drug-resistant Epilepsy. *Nature* **2018**, *8*, 3856. [CrossRef] [PubMed]
22. Liu, C.; Gao, R. Multiscale Entropy Analysis of the Differential RR Interval Time Series Signal and Its Application in Detecting Congestive Heart Failure. *Entropy* **2017**, *19*, 430. [CrossRef]
23. Valencia, J..F.; Porta, A.; Vallverdú, M.; Clarià, F.; Baranowski, R.; Orlowska-Baranowska, E.; Caminal, P. Refined multiscale entropy: Application to 24-h Holter recordings of heart period variability in healthy and aortic stenosis subjects. *IEEE Trans. Biomed. Eng.* **2009**, *56*, 2202–2213. [CrossRef]
24. Silva, L.; Lataro, R.; Castania, J.; Silva, C.D.; Valencia, J.; Murta, L.; Salgado, H.; Fazan, R.; Porta, A. Multiscale entropy analysis of heart rate variability in heart failure, hypertensive, and sinoaortic-denervated rats: Classical and refined approaches. *Am. J. Physiol. Regul. Integr. Comp. Physiol.* **2016**, *311*, 150–156. [CrossRef] [PubMed]
25. Chladekova, L.; Czippelova, B.; Turianikova, Z.; Tonhajzerova, I.; Calkovska, A.; Baumert, M.; Javorka, M. Multiscale time irreversibility of heart rate and blood pressure variability during orthostasis. *Physiol. Meas.* **2012**, *33*, 1747–1756. [CrossRef] [PubMed]
26. Watanabe, E.; Kiyono, K.; Hayano, J.; Yamamoto, Y.; Inamasu, J.; Yamamoto, M.; Ichikawa, T.; Sobue, Y.; Harada, M.; Ozaki, Y. Multiscale Entropy of the Heart Rate Variability for the Prediction of an Ischemic Stroke in Patients with Permanent Atrial Fibrillation. *PLoS ONE* **2015**, *10*, 137–144. [CrossRef] [PubMed]
27. Pastor-Pérez, F.J.; Manzano-Fernández, S.; Goya-Esteban, R.; Pascual-Figal, D.A.; Barquero-Pérez, O.; Rojo-Álvarez, J.L.; Martinez-Espejo, M.D.M.; Chavarri, M.V.; García-Alberola, A. Comparison of detection of arrhythmias in patients with chronic heart failure secondary to non-ischemic versus ischemic cardiomyopathy by 1 versus 7-day holter monitoring. *Am. J. Cardiol.* **2010**, *106*, 677–681. [CrossRef] [PubMed]
28. Mainardi, L.; Cerutti, S.; Sörnmo, L. *Understanding Atrial Fibrillation, the Signal Processing Contribution*; Morgan and Claypool: San Rafael, CA, USA, 2008.
29. Efron, B.; Tibshirani, R.J. *An Introduction to the Bootstrap*; CRC Press: Boca Raton, FL, USA, 1994.
30. Alabdulgader, A.; McCraty, R.; Atkinson, M. Long-term study of heart rate variability responses to changes in the solar and geomagnetic environment. *Sci. Rep.* **2018**, *8*, 2663. [CrossRef]
31. Singh, B.; Singh, M.; Banga, V.K. Sample Entropy based HRV: Effect of ECG Sampling Frequency. *Biomed. Sci. Eng.* **2014**, *2*, 68–72. [CrossRef]
32. Richman, J.S.; Moorman, J.R. Physiological time-series analysis using approximate entropy and sample entropy. *Heart Circ. Physiol.* **2000**, *278*, 2039–2049. [CrossRef]
33. Pincus, S.M.; Goldberger, A.L. Physiological time-series analysis: what does regularity quantify? *Am. J. Physiol.* **1994**, *266*, 1643–1656. [CrossRef]
34. Valencia, J.F.; Vallverdu, M.; Gomis, P.; Wagner, G.; Caminal, P. Multiscale Information Analysis of the Autonomous Nervous System during Myocardial Ischemia. In *Computers in Cardiology*; IEEE: Durham, NC, USA, 2007.
35. Czippelova, B.; Chladekova, L.; Uhrikova, Z.; Javorka, K.; Zibolen, M.; Javorka, M. Time irreversibility of heart rate oscillations in newborns—Does it reflect system nonlinearity? *Biomed. Signal Process. Control* **2015**, *19*, 85–88. [CrossRef]

36. Signoriniand, M.G.; Sassi, R.; Lombardi, F.; Cerruti, S. Regularity patterns in heart rate variability signal: the approximate entropy approach. In *Annual International Conference of the IEEE Engineering in Medicine and Biology Society. Vol. 20 Biomedical Engineering Towards the Year 2000 and Beyond*; IEEE: Hong Kong, China, 1998.
37. Goya-Esteban, R.; Sandberg, F.; Barquero-Pérez, O.; García-Alberola, A.; Sornmo, L.; Rojo-Álvarez, J.L. Long-term characterization of persistent atrial fibrillation: wave morphology, frequency, and irregularity analysis. *Med Biol. Eng. Comput.* **2010**, *57*, 1053–1060. [CrossRef]
38. Ivanov, P.C.; Amaral, L.A.N.; Goldberger, A.L.; Havlin, S.; Rosenblum, M.; Struzik, Z.; Stanley, H. Multifractality in Human Heartbeat Dynamics. *Nature* **1999**, *399*, 461–465. [CrossRef] [PubMed]
39. Muzy, J.F.; Bacry, E.; Arnéodo, A. Multifractal formalism for fractal signals: The structure-function approach versus the wavelet-transform modulus-maxima method. *Phys. Rev.* **1993**, *47*, 875–884. [CrossRef]
40. Lekkala, R.G.R.; Kuntamalla, S. The Effect of Chi Meditation on the Multifractal Nature of Heart Rate Variability. *Int. J. Signal Process. Syst.* **2016**, *4*, 128–132. [CrossRef]
41. Mietus, J.E.; Peng, C.K.; Henry, I.; Goldsmith, R.L.; Goldberger, A.L. The pNNx files: Re-examining a widely used heart rate variability measure. *Heart* **2002**, *88*, 378–380. [CrossRef] [PubMed]
42. Stein, P.K.; Ehsani, A.A.; Domitrovich, P.P.; Kleiger, R..E.; Rottman, J.N. Effect of exercise training on heart rate variability in healthy older adults. *Am. Heart J.* **1999**, *138*, 567–576. [CrossRef]
43. Bigger, J.T.; Fleiss, J.L.; Steinman, R.C.; Rolnitzky, L.M.; Schneider, W.J.; Stein, P.K. RR variability in healthy, middle-aged persons compared with patients with chronic coronary heart disease or recent acute myocardial infarction. *Circulation* **1995**, *91*, 1936–1943. [CrossRef] [PubMed]
44. Rochelle, G.L.; Bigger, J.T.; Bloomfield, M.D.; Krum, H.; Steinman, R.C.; Sackner-Bernstein, J.; Packer, M. Long-term carvedilol therapy increases parasympathetic nervous system activity in chronic congestive heart failure. *Am. J. Cardiol.* **1997**, *80*, 1101–1104. [CrossRef]
45. Pastor-Pérez, F.J.; Manzano-Fernández, S.; Pascual-Figal, R.G.E.D.A.; Barquero-Pérez, O.; Rojo-Álvarez, J.L.; Everss, E.; Martinez-Espejo, M.D.M.; Chavarri, M.V.; García-Alberola, A. Prognostic significance of long-period heart rate rhythms in chronic heart failure. *Circulation* **2012**, *76*, 2124–2129. [CrossRef]

Article

Complexity of Frontal Cortex fNIRS Can Support Alzheimer Disease Diagnosis in Memory and Visuo-Spatial Tests

David Perpetuini [1,*], Antonio M. Chiarelli [1], Daniela Cardone [1], Chiara Filippini [1], Roberta Bucco [2], Michele Zito [2] and Arcangelo Merla [1]

1 Institute for Advanced Biomedical Technologies, Department of Neuroscience and Imaging, University G. D'Annunzio of Chieti-Pescara, Via Luigi Polacchi 13, 66100 Chieti, Italy; antonio.chiarelli@unich.it (A.M.C.); d.cardone@unich.it (D.C.); chiara.filippini@unich.it (C.F.); arcangelo.merla@unich.it (A.M.)

2 Department of Medicine and Science of Ageing, University G. d'Annunzio of Chieti-Pescara, Via dei Vestini 31, 66100 Chieti, Italy; roberta.bucco@yahoo.it (R.B.); m.zito@dmsi.unich.it (M.Z.)

* Correspondence: david.perpetuini@unich.it; Tel.: +39-0871-3556954

Received: 6 December 2018; Accepted: 27 December 2018; Published: 1 January 2019

Abstract: Decline in visuo-spatial skills and memory failures are considered symptoms of Alzheimer's Disease (AD) and they can be assessed at early stages employing clinical tests. However, performance in a single test is generally not indicative of AD. Functional neuroimaging, such as functional Near Infrared Spectroscopy (fNIRS), may be employed during these tests in an ecological setting to support diagnosis. Indeed, neuroimaging should not alter clinical practice allowing free doctor-patient interaction. However, block-designed paradigms, necessary for standard functional neuroimaging analysis, require tests adaptation. Novel signal analysis procedures (e.g., signal complexity evaluation) may be useful to establish brain signals differences without altering experimental conditions. In this study, we estimated fNIRS complexity (through Sample Entropy metric) in frontal cortex of early AD and controls during three tests that assess visuo-spatial and short-term-memory abilities (Clock Drawing Test, Digit Span Test, Corsi Block Tapping Test). A channel-based analysis of fNIRS complexity during the tests revealed AD-induced changes. Importantly, a multivariate analysis of fNIRS complexity provided good specificity and sensitivity to AD. This outcome was compared to cognitive tests performances that were predictive of AD in only one test. Our results demonstrated the capabilities of fNIRS and complexity metric to support early AD diagnosis.

Keywords: Alzheimer disease; functional near infra-red spectroscopy; signal complexity; clock drawing test; digit span test; corsi block tapping test

---

1. Introduction

Early Alzheimer's disease (AD) is characterized by subtle impairments in executive abilities and memory [1,2]. Multiple cognitive test batteries were developed to assess these initial impairments and to allow differential diagnosis between idiopathic dementia and early AD [3,4]. The Clock Drawing Test (CDT) is a functional test that evaluates visuo-constructive and visuo-spatial abilities [5] and it is commonly employed in clinical practice for early dementia screening [6,7]. However, its sensitivity and specificity to early AD is still discussed [8,9]. In fact, novel scoring methods based on CDT outcome were proposed to improve its discrimination capabilities between Mild Cognitive Impairment (MCI) and early AD [5]. The Digit Span Test (DST) and the Corsi Block Tapping Test (CBTT) were developed to assess memory deficits. DST is widely employed in clinics to evaluate short-term memory and verbal working memory abilities [10], whereas the CBTT is employed to estimate visuo-spatial memory skills [11].

Although these tests and associated scorings were developed for sensitive and specific diagnosis of AD, performances in a single test are generally not sufficient for a definitive diagnosis of the disease, often requiring a large set of tests. Functional neuroimaging can be employed during these tests to support diagnosis [12,13]. The importance of using functional neuroimaging during the administration of these tests is widely agreed. In fact, adapted versions of CDT were administrated during a functional Magnetic Resonance Imaging (fMRI) acquisition directly in the MRI scanner [14–16]. DST and CBTT were also adapted for functional neuroimaging acquisition (either through fMRI or functional Near InfraRed Spectroscopy, fNIRS) [15–19]. Adaptation could be related to either environmental constraints (because of the fMRI acquisition) or temporal constraints (because of the need of paced tasks that allow standard functional neuroimaging signal analysis). In fact, when performed with concurrent functional brain imaging, these tests are modified from a non-paced administration of tasks and free doctor-patient interaction to a block or event related paradigm where doctor is replaced by an automated stimulation (e.g., through a screen monitor or a minidisk player) [16–19]. Although these modifications are useful from a neuroimaging standpoint, they indeed alter the ecology of the tests and the interaction between the doctor and the patient. However, these tests characteristics play a primary role in clinical practice and they should be preserved. In this perspective, one of the most suitable neuroimaging techniques is fNIRS that allows functional neuroimaging in outpatient environment.

fNIRS is a scalp-located non-invasive optical methodology able to record oscillation within the brain of oxygenated (O_2Hb) and deoxygenated (HHb) hemoglobin related to neuronal activity through the Blood Oxygen Level Dependent (BOLD) effect [20]. This technique is portable, relatively cheap, lightweight and resilient to motion artifacts with a mechanical structure resembling Electroencephalography [EEG] [20] thus being suitable for ecological measurements during the administration of clinical tests. fNIRS was indeed utilized to investigate brain functional alteration in AD during different tasks [21–23]. Studies proved the capabilities of fNIRS to investigate functional alterations between AD patients and healthy controls, as well as between AD and MCI patients. fNIRS was also employed during the administration of CDT and DST, however, it was never performed on AD patients and always through a paradigm adaptation. Shoyama and colleagues [24] employed a 52 fNIRS channels system to investigate brain functional activity of healthy subjects during CDT. Statistically significant activations in the superior temporal cortex and in the frontal cortex were detected [24]. Hoshi et al. [25] compared brain functional activation between DST and a modified DST (backward version) and found significant differences in the dorsolateral prefrontal cortex (dorsolateral PFC) [25]. Tian and colleagues [19] employed fNIRS to investigate difference in the brain activation during the administration of DST between patients with post-traumatic stress disorder and healthy controls [19]. Although both groups showed functional activation in the medium PFC, interesting lateralized differences were found. CBTT was administrated together with fNIRS monitoring to 39 healthy participants to clarify the role of the PFC during the task execution [18].

Without ad-hoc modifications of the tests structures, the ecological characteristics of these tests definitely do not allow for a standard functional neuroimaging analysis, which requires task-related design matrix to provide statistical inference about brain activity generally based on block averaging or General Linear Model (GLM) [26]. We recently implemented a novel signal analysis approach for functional data acquired through fNIRS that can assess alterations of brain signals in AD overcoming the limitations of standard analysis approaches. In particular, we suggested to estimate the complexity of cortical functional oscillations measured through fNIRS during these tests employing the Sample Entropy (SampEn) metric [27]. SampEn is the negative natural logarithm of the conditional probability that signal subseries of length m (pattern length) that match pointwise within a tolerance r (similarity factor) also match at the m+1 point and it evaluates non-linear predictability of the signal [27]. Moreover, it can be assessed at different time scales using a Multi Scale approach (MSE) [28]. The rationale behind this idea is that AD patients may present a modified structure of functional oscillations during cognitive tests that may reflect in a modified SampEn of the recorded fNIRS signals. SampEn is currently employed for biomedical signal evaluation to assess physiology and pathology. For example, it was used to investigate nonlinear properties of heart rate time series [29–31]. Moreover, it was applied to evaluate the complexity of fMRI signals in attention deficit hyperactivity disorder patients [32] and in patients affected by schizophrenia [33]. In AD, it was utilized to analyze resting-state brain activity deriving from Magnetoencephalographic [34] and EEG signals [35,36]. Concerning fNIRS, SampEn was used to investigate cortical activation in AD patients during the administration of Free and Cued Selective Reminding Test [37,38].

In the present study, we report evaluation of fNIRS SampEn in frontal cortex of early AD and healthy controls (HC) during CDT, DST and CBTT as they are performed in clinical practice. Statistical differences in signal complexities between AD and HC were assessed and the capability of fNIRS and SampEn to provide good sensitivity and specificity to the disease was investigated through a multivariate analysis.

2. Materials and Methods

2.1. Participants

Twenty-two participants were recruited in the study. The sample population was composed of eleven early AD patients (mean age $\pm$ SD: 72.2 $\pm$ 4.5 years; 7 males/4 females) and eleven HC (mean age $\pm$ SD: 67.5 $\pm$ 5.0 years; 8 males/3 females). The AD patients had a diagnosis of Mild probable Alzheimer's disease, as defined by the Diagnostic and Statistical Manual of Mental Disorders, 5th edition (DSM-5). Patients with moderate to severe cognitive impairment (Mini Mental State Examination, MMSE $< 25/30$) [39], vascular dementia, behavioral or psychiatric disorders, hydrocephalus, brain lesions or with a history of stroke or traumatic brain injury were excluded from the study. The study was approved by the Research Ethics Board of the local university, and it was conducted according to the principles described in the Declaration of Helsinki. Informed consent form was signed by all participants before the experiment and they were able to withdraw from it at any time.

2.2. Experimental Design

Experimental environment and layout are reported in Figure 1. Figure 1a shows the ecological settings of the experiment with the doctor sitting in front of the patient while freely interacting with him during the administration of the tests. Figure 1b shows the experimental paradigm. The different tasks, CDT, DST and CBTT were administered to the participants in a consecutively manner, spaced by 1-min rest periods, as they are usually performed in outpatient environment. Rest periods allowed for avoidance of overlapping effects among tasks. At the start of the experiment, the participants were instructed to relax for 1 min with their eyes closed. Afterward, CDT test started. CDT consisted in presenting to the patients a blank A4-sized paper and in asking to draw a circle with numbers representing a clock. Finally, patients were asked to draw the clock hands at a specific time. After another rest period DST was administered. The patients were asked to repeat a series of digits of increasing length (starting from two) verbally presented by the examiner at a pace of 1 s. If the participants were not able to repeat the sequence, another sequence of the same length was disclosed. The test stopped when the participants could not accurately repeat two series of the same length. After another minute of rest, CBTT was performed. In CBTT, the doctor consecutively tapped an increasing number (starting from two) of cubes located on a wooden tablet (Figure 1a) whose sequence had to be remembered by the patient. The cubes were touched with the index finger at a rate of 1 cube per second. The participants had to tap the cube sequence in the same order immediately after the doctor. The experiment temporal structure resembled that of DST. The overall experiment ended after the administration of CBTT.

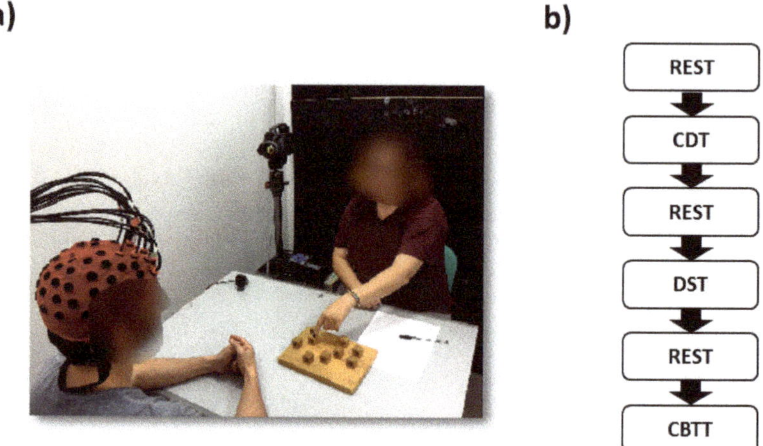

Figure 1. (**a**) Experimental environment and layout. The doctor was sitting in front of the patient while freely interacting with him during the administration of the tests. (**b**) Experimental paradigm. The different tasks, CDT, DST and CBTT were administered to the participants in a consecutively manner, spaced by 1-min rest periods, as they are usually performed in outpatient environment.

2.3. Functional Near-Infrared Spectroscopy Instrumentation and Measurement

A frequency-domain near-infrared spectroscopy (NIRS) system (Imagent, ISS Inc., Champaign, IL, USA) was used in this study. The system consisted of 32 laser diodes sources, 16 at 690 nm and 16 at 830 nm of wavelength, and 4 photomultiplier-tube (PMT) detectors. The lasers were modulated at 110 MHz and the PMTs at 110.005 MHz for heterodyne detection of modulated light. A time multiplexing scheme was utilized on the sources to prevent cross-talk. The overall instrument sampling rate (accounting for source time-multiplexing) was set at 10 Hz. The light power was lower than 4 mW/cm^2, in respect of the ANSI standard limits allowing safe measurements. NIR light was carried to the scalp using single optic fibers (0.4 mm core) and from the scalp back to the PMTs using fiber bundles (3 mm diameter). The fibers were located on the frontal and prefrontal areas and were held in place by means of an EEG helmet adapted for NIRS (Figure 1a). The fibers were placed according to the international 10–20 electrode placement system [40]. Source and detector locations were digitized with a Polhemus FastTrak 3D digitizer (Colchester, VT, USA; accuracy: 0.8 mm) using a recording stylus and three head-mounted receivers, which allowed for small movements of the head in between measurements. Optodes layout allowed to measure fNIRS signals from 21 channels (source-detector couple) with source-detector distances separation of either 3 cm or 4 cm. These distances inter-optodes bring to a depth of penetration that allow to measure cortical activity [41]. Figure 2a shows the among subjects' average channels locations after warping of the digitized sources and detectors into MNI space (Colin27) using AtlasViewerGUI of Homer2 NIRS analysis package [42]. The logarithmic channels' and subjects' average light sensitivity map (Jacobian), is displayed in Figure 2b, showing the frontal and prefrontal sensitivity of the optical probe. According to a sensitivity analysis performed in NIRS-SPM [43], the brain Brodmann Areas (BAs) investigated were numbers 8, 9, and 46. Notice that fNIRS is known to investigate regions up to ~3 cm from the scalp surface. Thus, it is indeed suited to investigate superficial cortical regions, as the BAs investigated, not being sensitive to deeper brain structures.

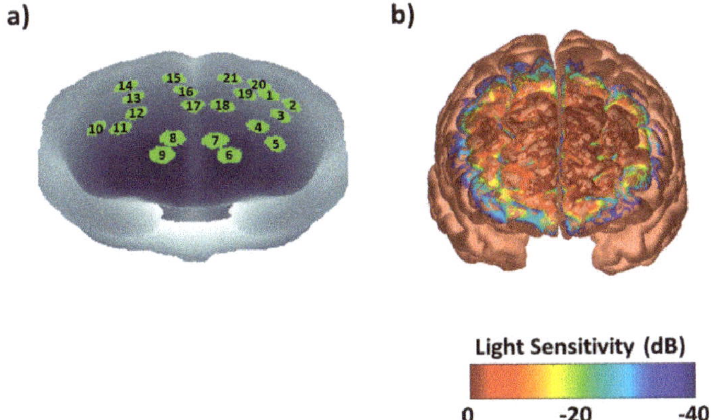

Figure 2. (**a**) Among subjects' average channels locations after warping of the digitized sources and detectors into MNI space (Colin27). (**b**) Logarithmic channels' and subjects' average light sensitivity map displayed up to an attenuation of 100 times (40 dB) showing the frontal and prefrontal sensitivity of the optical probe.

2.4. Functional Near-Infrared Spectroscopy Signal Analysis

fNIRS data were analyzed employing a standard continuous wave based fNIRS analysis by means of Homer2 NIRS Processing package [42]. Raw signal intensities were converted into optical densities (ODs). Motion artifacts were identified and removed relying on a wavelet-based algorithm [44]. Artifact free ODs were band-pass filtered using a 3rd order Butterworth filter (0.01–0.4 Hz) to remove slow drifts and physiological contaminations related to heart and breathing rates. ODs were converted into O_2Hb and HHb oscillations employing the modified Beer-Lambert law (MBLL). In order to further increase the signal to noise ratio (SNR) of functional signals, we employed an algorithm that exploit the brain activity-induced anticorrelation between the two hemoglobin forms (correlation-based signal improvement, CBSI) [45]. CBSI allows to obtain an O_2Hb signal corrected through HHb. Although this procedure increases O_2Hb SNR, it creates statistical dependencies among O_2Hb and HHb, allowing to further analyze only O_2Hb. Thus, subsequent analysis was performed on O_2Hb only.

In order to estimate O_2Hb signal complexity, the SampEn metric was evaluated for each fNIRS channel integrated over the different experimental phases.

The SampEn of a time series $\{x_1, \ldots, x_N\}$ of length N is calculated based on following set of equations [32]:

$$\mathrm{SampEn}(m, r, N) = -\ln\left[\frac{U^{m+1}(r)}{U^{m}(r)}\right] \tag{1}$$

$$U^{m}(r) = [N - mT]^{-1}\sum_{i=1}^{N-mT} C_i^{m}(r)$$

$$C_i^{m}(r) = \frac{B_i}{N-(m+1)T}$$

$$B_i = \text{number of j where } d\left|X_i, X_j\right| \leq r$$

$$X_i = \left(x_i, x_{i+T} \ldots, x_{i+(m-1)T}\right)$$

$$X_j = \left(x_j, x_{j+T} \ldots, x_{j+(m-1)T}\right)$$

$$i \leq j \leq N - mT,\ j \neq i$$

where N is total length of the time-series examined, m is the embedded dimension, r is the tolerance factor (scalar for which two subseries with distance below its value are considered identical) and T is the time delay expressed in samples. In this work, the computation of the SampEn was performed choosing a time delay T = 1, an embedded dimension of m = 2 and a similarity factor of r = 0.2·SD, where SD is the Standard Deviation of the signal evaluated. These parameters are commonly employed for complexity analysis of biological signals and they were chosen in accordance with [27].

In order to investigate signal complexity at multiple time scales we further employed the Multiscale Entropy (MSE) method [28]. MSE relies on computing coarse-grained time-series. The course-graining procedure is described by the following equation [32]:

$$y_j^{(\tau)} = \frac{1}{\tau} \sum_{i=(j-1)\tau+1}^{j\tau} x_i \; 1 \leq j \leq \frac{N}{\tau} \tag{2}$$

where $y_j^{(\tau)}$ is the coarse-grained signal for a given sample j and a scaling factor τ which is generated from an average of samples of original signal x. Afterwards, SampEn of each coarse-grained time series is estimated. In the current study only two scale factors of $\tau = 2$, $\tau = 3$ were considered, because of the limited total time of each experimental phase. Moreover, in order to eliminate a possible effect on the SampEn and MSE metric of the time series length, for each experimental phase we cut the last period of each phase for each subject to the shortest one. We obtained homogenized experimental phases lengths for CDT (370 samples), for DST (380 samples) and for CBTT (530 samples).

2.5. Statistical Inference and Multivariate Classification

Behavioral results were estimated through unpaired t-tests comparing AD with HC with a NULL hypothesis rejection set at $p < 0.05$. Receiver operating characteristics (ROC) [46] curve was evaluated to provide an estimate of the behavioral tests' sensitivity and specificity to the disease.

SampEn and MSE for O_2Hb in each channel, for each experimental phase and each subject were estimated. Average significant differences between AD and HC were tested by means of unpaired t-tests ($p < 0.05$). In order to avoid increased chance of false positive for fNIRS, multiple comparison correction, accounting for channels numerosity, was performed through the False Discovery Rate (FDR) algorithm [47]. Estimation of the BA investigated by the statistically activated channels was performed in NIRS-SPM [43].

A multivariate analysis was implemented to provide a classification of disease (AD or HC), starting from the SampEn and MSE of all the 21 channels in each experimental phase. A linear multiple regression [48] was performed on a dependent variable that labelled the presence of the disease (AD = 1, HC = 0). In order to provide an unbiased estimate of the out-of-sample performance of the classifier a leave one out cross-validation procedure was implemented. This cross-validation procedure consisted in leaving one subject out of the regression and estimating the predicted output value (between 0 and 1) on the given subject. The same procedure was iterated for all the subjects. A ROC curve analysis on the 22 out-of-sample predicted outputs provided estimates of fNIRS signal complexity sensitivity and specificity to the disease in each experimental phase.

3. Results

CDT [48] presented significant differences between AD and HC ($t = -4.20$, df = 20, $p = 4.4 \times 10^{-4}$). However, DST [49] and CBTT [50] could not discriminate between the two groups ($t = -0.31$, df = 20, p = N.S., $t = -1.45$, df = 20, p = N.S.).

Figure 3 reports statistical maps (t-scores) of AD vs. HC for signal complexity of fNIRS for the three experimental phases. Figure 3a,b reports results employing the SampEn metric, whereas Figure 3c reports results employing the MSE ($\tau = 3$).

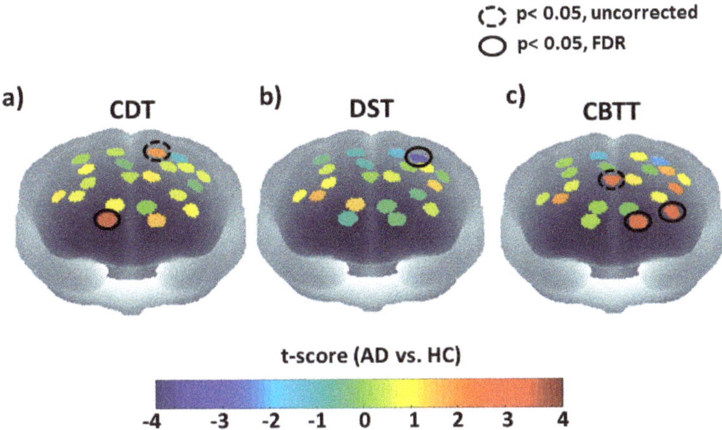

Figure 3. Statistical maps (t-scores) of AD vs. HC for signal complexity of fNIRS for the three experimental phases. Dashed black circles shows significant channels ($p < 0.05$), that did not survive FDR correction, whereas continuous black circles show channels that were still significant ($p < 0.05$) after multiple comparison correction. (**a**) t-score maps (AD vs. HC) based on the SampEn metric during CDT. (**b**) t-score maps (AD vs. HC) based on the SampEn metric during DST. (**c**) t-score maps (AD vs. HC) based on the MSE ($\tau = 3$) metric during CBTT.

In the figure we report results where we obtained statistical significance after FDR algorithm. Dashed black circles shows significant channels ($p < 0.05$), that did not survive FDR correction, whereas continuous black circles show channels that were still significant ($p < 0.05$) after multiple comparison correction. A NIRS-SPM-based evaluation of the BA investigated by the significantly activated channels (surviving multiple comparison correction) reported in Figure 3, highlighted higher SampEn for AD in BA 10 for the CDT (Channel 10, t = 3.44, df = 20, $p = 2.6 \times 10^{-3}$). Significant lower SampEn during the DST was found instead in a channel investigating BA 9 (Channel 20, t = −3.48, df = 20, $p = 2.4 \times 10^{-3}$). Higher MSE ($\tau = 3$) was finally found in two channels encompassing BA 10 and 46 during CBTT (Channel 4, t = 3.12, df = 20, $p = 5.4 \times 10^{-3}$, Channel 6, t = 2.63, df = 20, $p = 0.01$). Notice that MSE was estimated for all the different tests. However, CBTT was the only one providing statistically significant results.

ROC analysis performed on the behavioral results and the fNIRS complexity-based multiple regression analysis, described in the method section, are reported in Figure 4. We obtained significant classification outcome when behavioral tests were analyzed only for the CDT (AUC = 0.90) with a best performance sensitivity of 0.82 and specificity of 0.72. DST and CBTT were not able to provide a significant classification (AUC = 0.54 and AUC = 0.60). On the contrary, although less sensitive than CDT test, we obtained above chance classification when relying on fNIRS complexity metric for all the three tests performed with an AUC = 0.75 for CDT, AUC = 0.65 for DST and AUC = 0.71 for CBTT with a best performance sensitivity of 0.73 for all three tests and specificity of 0.82 for CDT and 0.73 for DST and CBTT. Whereas CDT and DST were evaluated on the SampEn metric, CBTT was evaluated on the MSE, which provided channels based significant average differences between AD and HC.

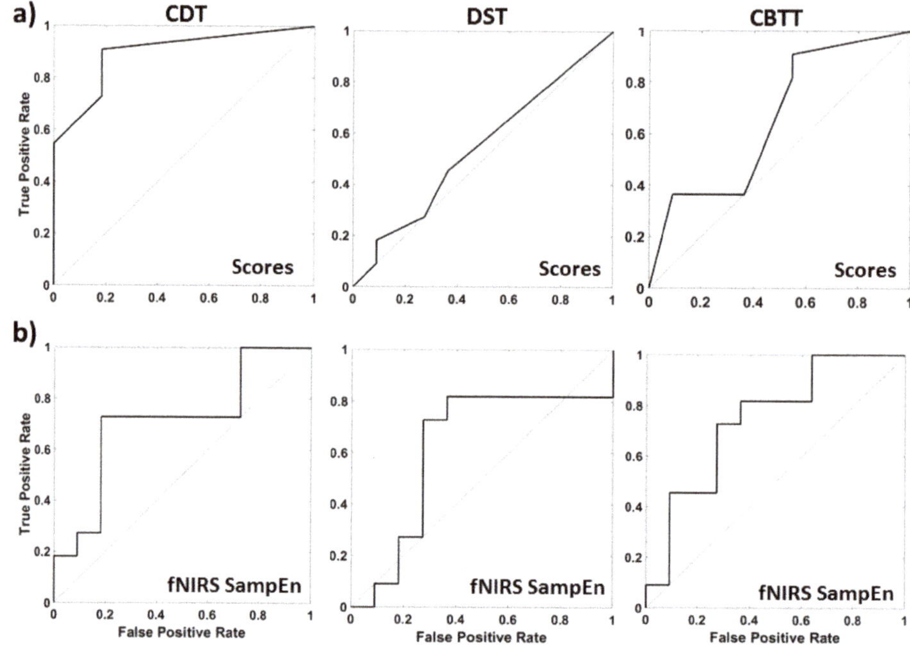

Figure 4. ROC analysis performed on the tests outcome (**a**) and on the fNIRS complexity-based multiple regression analysis (**b**).

4. Discussion

The aim of this study was to assess the capabilities of employing fNIRS during clinical tests that are commonly utilized for AD diagnosis. fNIRS could in fact support these tests by improving sensitivity and specificity to the disease.

Based on our sample population, we found that CDT scores were on average significantly different between AD and HC however DST and CBT were not, further justifying the need of a neuroimaging supporting tool.

In order to preserve the ecological conditions of the tests without altering the standard clinical protocol, we investigated whether a phase-integrated metric of complexity of hemoglobin oscillations could highlight differences between AD and HC. The complexity of the signals was evaluated by means of the SampEn and MSE algorithms.

We found fNIRS channels-based significant AD vs HC differences in SampEn and MSE of O_2Hb oscillations during CDT, DST and CBTT in brain regions (BAs) involved in visuo-spatial and mnemonic tasks [49]. As demonstrated by Vaillancourt and Newell, these results seem to support the hypothesis that disease can cause a dysregulation of brain physiology that can result in altered functional patterns identifiable through a complexity analysis of time-dependent physiological signals [50]. However, our results do not provide a definitive answer on the direction of the complexity variations due to the disease. Although an increase complexity in some channels was the dominant effect among the different tasks, we indeed had a significant complexity decrease during DST in one fNIRS channel. Previous studies evaluating complexity on electrophysiological brain recordings (EEG [51] and Magnetoencephalography, MEG [34]) showed a decrease in signal complexity with AD, however these studies were performed at rest. On the contrary, signal complexity of fNIRS was here evaluated during the execution of a visuo-spatial and working-memory tests hence strictly evaluating task-related changes induced by AD. Vaillancourt and Newell [50] showed that a pathological state can alter the complexity of an output of a biological system as a function of the input. This input dependent modulation of the alteration makes particularly unpredictable the direction of changes in fNIRS complexity as a function of channel location. In fact, the direction of change might intrinsically depend on the original task-induced activation pattern. This originally unknown activation pattern may indeed justify the variable direction of the complexity changes induced by AD during these clinical tests. In fact, because of the heterogeneous response of different brain region to a given stimulus, fNIRS signals may vary both in space and time. Due to the spatial distribution of fNIRS optodes on the frontal cortex, it is indeed plausible that different channels show a different statistic, generating a peculiar statistical map.

Inferring about the underlying causes of the found fNIRS signal complexity variations induced by AD, early findings by Hock et al. [52] identified reduced oxygenation and cerebral blood flow in AD during the execution of verbal fluency tasks. The authors justified the found alterations in oxygenation and CBF based on an impaired neurovascular coupling. In fact, neurovascular coupling is regulated by means of neurons, glia and vascular cells [51,53,54]. AD patients exhibit a deposition of amyloid β-peptide in neuropil and vessels that could impair such mechanism [55]. Moreover, an altered brain electrical activity and a loss of functional neural connectivity is observed in AD patients [55–57], that could further impact a complexity analysis of functional brain modulations.

A ROC curve analysis of clinical tests scores showed above chance classification capabilities only for CDT (AUC = 0.90) which evaluates visuo-spatial skills and apraxia [58]. On the contrary, fNIRS complexity-based multiple regression analysis showed good classification capabilities for all the three tests performed (AUC = 0.75. for CDT, AUC = 0.65 for DST and AUC = 0.71 for CBTT). These results clearly suggest that the use of fNIRS during the administration of cognitive tests may help AD diagnosis.

In fact, in clinical practice it is a common procedure to diagnose AD after long battery of tests, thus intrinsically assuming a high variability in a single test performance and across tests. The group of patients analyzed in this study had in fact a determined diagnosis based on this long battery. Although in our sample CDT seemed particularly suited for a single-test diagnosis, fNIRS entropy-based outcome seemed to provide more stable results among tests compared to cognitive performances. This means that fNIRS seems to be a good supporting tool for a standard multi-test diagnosis.

However, further studies should be performed increasing population sample size. In fact, our study was constituted by a limited number of participants and the fNIRS-based classification outcome, since it relies on a multivariate analysis of all the channels employed, might dramatically increase its performance with a large sample numerosity. In fact, although the study can be considered rather small in sample size, the investigation was conducted employing a leave-one-out cross-validation procedure (eliminating one subject at a time and testing the classifier outcome on that subject), thus intrinsically evaluating the out-of-sample performance. Thus, the results obtained are indeed generalizable. Increasing the sample size may allow further increase of the performance by decreasing a possible in-sample overfitting effect of the classifier.

Further advancement in classification procedure should employ non-linear classifiers, that were not utilizable in this work because of the small sample size and a possible over-fitting effect. Furthermore, it could be of great interest to decouple the contributions of altered brain electrical activity and impaired neurovascular coupling by combining the hemodynamic information provided by fNIRS with electrophysiological signals acquired, for example, by means of synchronous EEG measurements [59] through wearable technology [60]. These brain activity measurements could be further compared with recordings of peripheral autonomic activity (e.g., heart-rate and galvanic skin response monitoring systems) that can be themselves altered by the presence of the disease. Indeed, this study cannot provide an alternative diagnostic tool for early AD, but at least it opens the possibility of utilizing fNIRS as a supporting clinical procedure and it suggests to further explore its diagnostic potentialities.

5. Conclusions

In this study, we estimated fNIRS complexity (through SampEn and MSE) in the frontal cortex of early AD and controls during three tests that assess visuo-spatial and short-term-memory abilities (CDT, DST, CBTT). A channel-based analysis of fNIRS complexity revealed AD-induced changes in BAs 9,10 and 46 that where inhomogeneous in direction. A multivariate analysis of fNIRS task-related complexities based on multiple linear regression provided decent specificity and sensitivity to AD. This outcome was compared to test performances that were predictive of AD in one of the three tests (CDT). These findings, although preliminary, seem to confirm the hypothesis that AD may produce a dysregulation of brain electrical activity and neurovascular coupling that may present earlier than clear behavioral impairment. Our results demonstrated the capabilities of fNIRS and complexity metric to support early AD diagnosis.

Author Contributions: Conceptualization, D.P., R.B., M.Z. and A.M.; methodology, D.P.; software, D.P. and A.M.C.; validation, D.P., D.C. and C.F.; formal analysis, D.P. and A.M.C.; investigation, D.P. and R.B.; resources, M.Z. and A.M.; data curation, D.P., D.C. and C.F.; writing—original draft preparation, D.P. and A.M.C.; writing—review and editing, D.P.; visualization, D.P.; supervision, A.M.; project administration, A.M.; funding acquisition, A.M.

Funding: This research was partially funded by grant H2020, ECSEL-04-2015-Smart Health, grant n. 692470, Advancing Smart Optical Imaging and Sensing for Health (ASTONISH).

Conflicts of Interest: The authors declare no conflict of interest.

References

1. Dubois, B.; Feldman, H.H.; Jacova, C.; DeKosky, S.T.; Barberger-Gateau, P.; Cummings, J.; Delacourte, A.; Galasko, D.; Gauthier, S.; Jicha, G.; et al. Research criteria for the diagnosis of Alzheimer's disease: Revising the NINCDS–ADRDA criteria. *Lancet Neurol.* **2007**, *6*, 734–746. [CrossRef]
2. Baudic, S.; Dalla Barba, G.; Thibaudet, M.C.; Smagghe, A.; Remy, P.; Traykov, L. Executive function deficits in early Alzheimer's disease and their relations with episodic memory. *Arch. Clin. Neuropsychol.* **2006**, *21*, 15–21. [CrossRef]
3. Nitrini, R.; Caramelli, P.; Porto, C.S.; Charchat-Fichman, H.; Formigoni, A.P.; Carthery-Goulart, M.T.; Otero, C.; Prandini, J.C. Brief cognitive battery in the diagnosis of mild Alzheimer's disease in subjects with medium and high levels of education. *Dement. Neuropsychol.* **2007**, *1*, 32–36. [CrossRef] [PubMed]
4. Takada, L.T.; Caramelli, P.; Fichman, H.C.; Porto, C.S.; Bahia, V.S.; Anghinah, R.; Carthery-Goulart, M.T.; Radanovic, M.; Smid, J.; Herrera, E., Jr. Comparison between two tests of delayed recall for the diagnosis of dementia. *Arq. Neuropsiquiatr.* **2006**, *64*, 35–40. [CrossRef] [PubMed]
5. Ricci, M.; Pigliautile, M.; D'Ambrosio, V.; Ercolani, S.; Bianchini, C.; Ruggiero, C.; Vanacore, N.; Mecocci, P. The clock drawing test as a screening tool in mild cognitive impairment and very mild dementia: A new brief method of scoring and normative data in the elderly. *Neurol. Sci.* **2016**, *37*, 867–873. [CrossRef] [PubMed]
6. Shulman, K.I.; Pushkar Gold, D.; Cohen, C.A.; Zucchero, C.A. Clock-drawing and dementia in the community: A longitudinal study. *Int. J. Geriatr. Psychiatry* **1993**, *8*, 487–496. [CrossRef]
7. Ferrucci, L.; Cecchi, F.; Guralnik, J.M.; Giampaoli, S.; Noce, C.L.; Salani, B.; Bandinelli, S.; Baroni, A.; Group, F.S. Does the clock drawing test predict cognitive decline in older persons independent of the Mini-Mental State Examination? *J. Am. Geriatr. Soc.* **1996**, *44*, 1326–1331. [CrossRef] [PubMed]
8. Pinto, E.; Peters, R. Literature review of the Clock Drawing Test as a tool for cognitive screening. *Dement. Geriatr. Cogn. Disord.* **2009**, *27*, 201–213. [CrossRef] [PubMed]
9. Ehreke, L.; Luck, T.; Luppa, M.; König, H.-H.; Villringer, A.; Riedel-Heller, S.G. Clock Drawing Test–screening utility for mild cognitive impairment according to different scoring systems: Results of the Leipzig Longitudinal Study of the Aged (LEILA 75+). *Int. Psychogeriatr.* **2011**, *23*, 1592–1601. [CrossRef] [PubMed]
10. Gerton, B.K.; Brown, T.T.; Meyer-Lindenberg, A.; Kohn, P.; Holt, J.L.; Olsen, R.K.; Berman, K.F. Shared and distinct neurophysiological components of the digits forward and backward tasks as revealed by functional neuroimaging. *Neuropsychologia* **2004**, *42*, 1781–1787. [CrossRef] [PubMed]
11. Binetti, G.; Magni, E.; Padovani, A.; Cappa, S.F.; Bianchetti, A.; Trabucchi, M. Executive dysfunction in early Alzheimer's disease. *J. Neurol. Neurosurg. Psychiatry* **1996**, *60*, 91–93. [CrossRef]

12. Mueller, S.G.; Weiner, M.W.; Thal, L.J.; Petersen, R.C.; Jack, C.R.; Jagust, W.; Trojanowski, J.Q.; Toga, A.W.; Beckett, L. Ways toward an early diagnosis in Alzheimer's disease: The Alzheimer's Disease Neuroimaging Initiative (ADNI). *Alzheimers Dement.* **2005**, *1*, 55–66. [CrossRef] [PubMed]
13. Aël Chetelat, G.; Baron, J.-C. Early diagnosis of Alzheimer's disease: Contribution of structural neuroimaging. *Neuroimage* **2003**, *18*, 525–541. [CrossRef]
14. Formisano, E.; Linden, D.E.; Di Salle, F.; Trojano, L.; Esposito, F.; Sack, A.T.; Grossi, D.; Zanella, F.E.; Goebel, R. Tracking the mind's image in the brain I: Time-resolved fMRI during visuospatial mental imagery. *Neuron* **2002**, *35*, 185–194. [CrossRef]
15. Trojano, L.; Grossi, D.; Linden, D.E.; Formisano, E.; Hacker, H.; Zanella, F.E.; Goebel, R.; Di Salle, F. Matching two imagined clocks: The functional anatomy of spatial analysis in the absence of visual stimulation. *Cereb. Cortex* **2000**, *10*, 473–481. [CrossRef] [PubMed]
16. Ino, T.; Asada, T.; Ito, J.; Kimura, T.; Fukuyama, H. Parieto-frontal networks for clock drawing revealed with fMRI. *Neurosci. Res.* **2003**, *45*, 71–77. [CrossRef]
17. Kaneko, H.; Yoshikawa, T.; Nomura, K.; Ito, H.; Yamauchi, H.; Ogura, M.; Honjo, S. Hemodynamic changes in the prefrontal cortex during digit span task: A near-infrared spectroscopy study. *Neuropsychobiology* **2011**, *63*, 59–65. [CrossRef]
18. Lancia, S.; Cofini, V.; Carrieri, M.; Ferrari, M.; Quaresima, V. Are ventrolateral and dorsolateral prefrontal cortices involved in the computerized Corsi block-tapping test execution? An fNIRS study. *Neurophotonics* **2018**, *5*, 011019. [CrossRef]
19. Tian, F.; Yennu, A.; Smith-Osborne, A.; Gonzalez-Lima, F.; North, C.S.; Liu, H. Prefrontal responses to digit span memory phases in patients with post-traumatic stress disorder (PTSD): A functional near infrared spectroscopy study. *NeuroImage Clin.* **2014**, *4*, 808–819. [CrossRef] [PubMed]
20. Ferrari, M.; Quaresima, V. A brief review on the history of human functional near-infrared spectroscopy (fNIRS) development and fields of application. *Neuroimage* **2012**, *63*, 921–935. [CrossRef]
21. Hock, C.; Villringer, K.; Müller-Spahn, F.; Hofmann, M.; Schuh-Hofer, S.; Heekeren, H.; Wenzel, R.; Dirnagl, U.; Villringer, A. Near Infrared Spectroscopy in the Diagnosis of Alzheimer's Disease a. *Ann. N. Y. Acad. Sci.* **1996**, *777*, 22–29. [CrossRef] [PubMed]
22. Fallgatter, A.J.; Roesler, M.; Sitzmann, L.; Heidrich, A.; Mueller, T.J.; Strik, W.K. Loss of functional hemispheric asymmetry in Alzheimer's dementia assessed with near-infrared spectroscopy. *Cogn. Brain Res.* **1997**, *6*, 67–72. [CrossRef]
23. Herrmann, M.J.; Langer, J.B.; Jacob, C.; Ehlis, A.-C.; Fallgatter, A.J. Reduced prefrontal oxygenation in Alzheimer disease during verbal fluency tasks. *Am. J. Geriatr. Psychiatry* **2008**, *16*, 125–135. [CrossRef] [PubMed]
24. Shoyama, M.; Nishioka, T.; Okumura, M.; Kose, A.; Tsuji, T.; Ukai, S.; Shinosaki, K. Brain activity during the Clock-Drawing Test: Multichannel near-infrared spectroscopy study. *Appl. Neuropsychol.* **2011**, *18*, 243–251. [CrossRef] [PubMed]
25. Hoshi, Y.; Oda, I.; Wada, Y.; Ito, Y.; Yamashita, Y.; Oda, M.; Ohta, K.; Yamada, Y.; Tamura, M. Visuospatial imagery is a fruitful strategy for the digit span backward task: A study with near-infrared optical tomography. *Cogn. Brain Res.* **2000**, *9*, 339–342. [CrossRef]
26. Friston, K.J.; Holmes, A.P.; Worsley, K.J.; Poline, J.-P.; Frith, C.D.; Frackowiak, R.S. Statistical parametric maps in functional imaging: A general linear approach. *Hum. Brain Mapp.* **1994**, *2*, 189–210. [CrossRef]
27. Richman, J.S.; Moorman, J.R. Physiological time-series analysis using approximate entropy and sample entropy. *Am. J. Physiol.-Heart Circ. Physiol.* **2000**, *278*, H2039–H2049. [CrossRef]
28. Costa, M.; Goldberger, A.L.; Peng, C.-K. Multiscale entropy analysis of biological signals. *Phys. Rev. E* **2005**, *71*, 021906. [CrossRef]
29. Kim, W.-S.; Yoon, Y.-Z.; Bae, J.-H.; Soh, K.-S. Nonlinear characteristics of heart rate time series: Influence of three recumbent positions in patients with mild or severe coronary artery disease. *Physiol. Meas.* **2005**, *26*, 517. [CrossRef]
30. Al-Angari, H.M.; Sahakian, A.V. Use of sample entropy approach to study heart rate variability in obstructive sleep apnea syndrome. *IEEE Trans. Biomed. Eng.* **2007**, *54*, 1900–1904. [CrossRef]
31. Lake, D.E.; Richman, J.S.; Griffin, M.P.; Moorman, J.R. Sample entropy analysis of neonatal heart rate variability. *Am. J. Physiol.-Regul. Integr. Comp. Physiol.* **2002**, *283*, R789–R797. [CrossRef] [PubMed]

32. Sokunbi, M.O.; Fung, W.; Sawlani, V.; Choppin, S.; Linden, D.E.; Thome, J. Resting state fMRI entropy probes complexity of brain activity in adults with ADHD. *Psychiatry Res. Neuroimaging* **2013**, *214*, 341–348. [CrossRef]
33. Sokunbi, M.O.; Gradin, V.B.; Waiter, G.D.; Cameron, G.G.; Ahearn, T.S.; Murray, A.D.; Steele, D.J.; Staff, R.T. Nonlinear complexity analysis of brain fMRI signals in schizophrenia. *PLoS ONE* **2014**, *9*, e95146. [CrossRef] [PubMed]
34. Gómez, C.; Hornero, R.; Abásolo, D.; Fernández, A.; Escudero, J. Analysis of MEG background activity in Alzheimer's disease using nonlinear methods and ANFIS. *Ann. Biomed. Eng.* **2009**, *37*, 586–594. [CrossRef] [PubMed]
35. Abásolo, D.; Hornero, R.; Espino, P.; Alvarez, D.; Poza, J. Entropy analysis of the EEG background activity in Alzheimer's disease patients. *Physiol. Meas.* **2006**, *27*, 241–253. [CrossRef] [PubMed]
36. Escudero, J.; Abásolo, D.; Hornero, R.; Espino, P.; López, M. Analysis of electroencephalograms in Alzheimer's disease patients with multiscale entropy. *Physiol. Meas.* **2006**, *27*, 1091–1106. [CrossRef] [PubMed]
37. Perpetuini, D.; Bucco, R.; Zito, M.; Merla, A. Study of memory deficit in Alzheimer's disease by means of complexity analysis of fNIRS signal. *Neurophotonics* **2017**, *5*, 011010. [CrossRef]
38. Lemos, R.; Simões, M.R.; Santiago, B.; Santana, I. The free and cued selective reminding test: Validation for mild cognitive impairment and A lzheimer's disease. *J. Neuropsychol.* **2015**, *9*, 242–257. [CrossRef]
39. Folstein, M.F.; Folstein, S.E.; McHugh, P.R. "Mini-mental state": A practical method for grading the cognitive state of patients for the clinician. *J. Psychiatr. Res.* **1975**, *12*, 189–198. [CrossRef]
40. Homan, R.W.; Herman, J.; Purdy, P. Cerebral location of international 10–20 system electrode placement. *Electroencephalogr. Clin. Neurophysiol.* **1987**, *66*, 376–382. [CrossRef]
41. Chiarelli, A.M.; Maclin, E.L.; Low, K.A.; Mathewson, K.E.; Fabiani, M.; Gratton, G. Combining energy and Laplacian regularization to accurately retrieve the depth of brain activity of diffuse optical tomographic data. *J. Biomed. Opt.* **2016**, *21*, 036008. [CrossRef] [PubMed]
42. Huppert, T.J.; Diamond, S.G.; Franceschini, M.A.; Boas, D.A. HomER: A review of time-series analysis methods for near-infrared spectroscopy of the brain. *Appl. Opt.* **2009**, *48*, D280–D298. [CrossRef] [PubMed]
43. Ye, J.C.; Tak, S.; Jang, K.E.; Jung, J.; Jang, J. NIRS-SPM: Statistical parametric mapping for near-infrared spectroscopy. *Neuroimage* **2009**, *44*, 428–447. [CrossRef] [PubMed]
44. Molavi, B.; Dumont, G.A. Wavelet-based motion artifact removal for functional near-infrared spectroscopy. *Physiol. Meas.* **2012**, *33*, 259–270. [CrossRef] [PubMed]
45. Cui, X.; Bray, S.; Reiss, A.L. Functional near infrared spectroscopy (NIRS) signal improvement based on negative correlation between oxygenated and deoxygenated hemoglobin dynamics. *Neuroimage* **2010**, *49*, 3039–3046. [CrossRef] [PubMed]
46. Zweig, M.H.; Campbell, G. Receiver-operating characteristic (ROC) plots: A fundamental evaluation tool in clinical medicine. *Clin. Chem.* **1993**, *39*, 561–577. [PubMed]
47. Genovese, C.R.; Lazar, N.A.; Nichols, T. Thresholding of statistical maps in functional neuroimaging using the false discovery rate. *Neuroimage* **2002**, *15*, 870–878. [CrossRef]
48. Neter, J.; Kutner, M.H.; Nachtsheim, C.J.; Wasserman, W. *Applied Linear Statistical Models*; Irwin Chicago: Chicago, IL, USA, 1996; Volume 4.
49. Catani, M.; Dell'Acqua, F.; Bizzi, A.; Forkel, S.J.; Williams, S.C.; Simmons, A.; Murphy, D.G.; de Schotten, M.T. Beyond cortical localization in clinico-anatomical correlation. *Cortex* **2012**, *48*, 1262–1287. [CrossRef]
50. Vaillancourt, D.E.; Newell, K.M. Changing complexity in human behavior and physiology through aging and disease. *Neurobiol. Aging* **2002**, *23*, 1–11. [CrossRef]
51. Deng, B.; Liang, L.; Li, S.; Wang, R.; Yu, H.; Wang, J.; Wei, X. Complexity extraction of electroencephalograms in Alzheimer's disease with weighted-permutation entropy. *Chaos* **2015**, *25*, 043105. [CrossRef]
52. Hock, C.; Villringer, K.; Müller-Spahn, F.; Wenzel, R.; Heekeren, H.; Schuh-Hofer, S.; Hofmann, M.; Minoshima, S.; Schwaiger, M.; Dirnagl, U.; et al. Decrease in parietal cerebral hemoglobin oxygenation during performance of a verbal fluency task in patients with Alzheimer's disease monitored by means of near-infrared spectroscopy (NIRS)—Correlation with simultaneous rCBF-PET measurements. *Brain Res.* **1997**, *755*, 293–303. [CrossRef]
53. Girouard, H.; Iadecola, C. Neurovascular coupling in the normal brain and in hypertension, stroke, and Alzheimer disease. *J. Appl. Physiol.* **2006**, *100*, 328–335. [CrossRef]

54. Croce, P.; Zappasodi, F.; Merla, A.; Chiarelli, A.M. Exploiting neurovascular coupling: A Bayesian sequential Monte Carlo approach applied to simulated EEG fNIRS data. *J. Neural Eng.* **2017**, *14*, 046029. [CrossRef] [PubMed]
55. Pijnenburg, Y.A.L.; Vd Made, Y.; Van Walsum, A.V.C.; Knol, D.L.; Scheltens, P.; Stam, C.J. EEG synchronization likelihood in mild cognitive impairment and Alzheimer's disease during a working memory task. *Clin. Neurophysiol.* **2004**, *115*, 1332–1339. [CrossRef] [PubMed]
56. Stam, C.J.; Jones, B.F.; Nolte, G.; Breakspear, M.; Scheltens, P. Small-world networks and functional connectivity in Alzheimer's disease. *Cereb. Cortex* **2006**, *17*, 92–99. [CrossRef] [PubMed]
57. Supekar, K.; Menon, V.; Rubin, D.; Musen, M.; Greicius, M.D. Network analysis of intrinsic functional brain connectivity in Alzheimer's disease. *PLoS Comput. Biol.* **2008**, *4*, e1000100. [CrossRef]
58. Pasquier, F. Early diagnosis of dementia: Neuropsychology. *J. Neurol.* **1999**, *246*, 6–15. [CrossRef] [PubMed]
59. Chiarelli, A.M.; Zappasodi, F.; Di Pompeo, F.; Merla, A. Simultaneous functional near-infrared spectroscopy and electroencephalography for monitoring of human brain activity and oxygenation: A review. *Neurophotonics* **2017**, *4*, 041411. [CrossRef] [PubMed]
60. Chiarelli, A.M.; Libertino, S.; Zappasodi, F.; Mazzillo, M.C.; Di Pompeo, F.; Merla, A.; Lombardo, S.A.; Fallica, G.P. Characterization of a fiber-less, multichannel optical probe for continuous wave functional near-infrared spectroscopy based on silicon photomultipliers detectors: In-vivo assessment of primary sensorimotor response. *Neurophotonics* **2017**, *4*, 035002. [CrossRef] [PubMed]

Article

Multiscale Entropy Quantifies the Differential Effect of the Medium Embodiment on Older Adults Prefrontal Cortex during the Story Comprehension: A Comparative Analysis

Soheil Keshmiri [1,*,†], Hidenobu Sumioka [1,†], Ryuji Yamazaki [2] and Hiroshi Ishiguro [1,3]

1 Hiroshi Ishiguro Laboratories (HIL), Advanced Telecommunications Research Institute International (ATR), 2-2 Hikaridai Seika-cho, Kyoto 619-02, Japan; sumioka@atr.jp (H.S.); ishiguro@sys.es.osaka-u.ac.jp (H.I.)

2 School of Social Sciences, Waseda University, 1 Chome-104 Totsukamachi, Shinjuku, Tokyo 169-8050, Japan; rys@aoni.waseda.jp

3 Graduate School of Engineering Science, Osaka University, 2-1 Yamadaoka, Suita, Osaka 565-0871, Japan

* Correspondence: soheil@atr.jp

† Both authors contributed equally to this work.

Received: 23 January 2019; Accepted: 16 February 2019; Published: 19 February 2019

Abstract: Todays' communication media virtually impact and transform every aspect of our daily communication and yet the extent of their embodiment on our brain is unexplored. The study of this topic becomes more crucial, considering the rapid advances in such fields as socially assistive robotics that envision the use of intelligent and interactive media for providing assistance through social means. In this article, we utilize the multiscale entropy (MSE) to investigate the effect of the physical embodiment on the older people's prefrontal cortex (PFC) activity while listening to stories. We provide evidence that physical embodiment induces a significant increase in MSE of the older people's PFC activity and that such a shift in the dynamics of their PFC activation significantly reflects their perceived feeling of fatigue. Our results benefit researchers in age-related cognitive function and rehabilitation who seek for the adaptation of these media in robot-assistive cognitive training of the older people. In addition, they offer a complementary information to the field of human-robot interaction via providing evidence that the use of MSE can enable the interactive learning algorithms to utilize the brain's activation patterns as feedbacks for improving their level of interactivity, thereby forming a stepping stone for rich and usable human mental model.

Keywords: multiscale entropy; embodied media; tele-communication; humanoid; prefrontal cortex

---

1. Introduction

Socially assistive robotics (SAR) [1] is an emerging field of research that focuses on intelligent and interactive media to provide assistance through social than physical means [2]. SAR builds upon the behavioral and neuroscientific findings of the positive motivational effect of physically embodied media on humans' social inclinations [3,4]. For instance, it identifies that children who read with a learning companion robot consider their reading companion to support their reading comprehension and that it motivates a deepening social connection [5]. Moreover, it indicates that tele-communication through a humanoid results in the older people's brain to exhibit a similar activation pattern as in the case of in-person communication [6].

In this regard, a distinct attribute of robotic media is their physical embodiment which allows for a sense of togetherness [7]. This property can amplify human work [1] by, for example, filling the gap in human personnel shortage in elderly care facilities [8,9]. Such potentials become more intriguing by considering the positive effect of these media in robot-assistive cognitive training of the older people [10].

On the other hand, communication research suggests [11] the tendency of individuals to undermine the significant influence of the physical embodiment during verbal communication in a wide range of social [12] and behavioral domains [13]. However, almost all of these findings are based on the subjective assessments of the behavioral responses of the human subjects. This severely limits the possibility of drawing a reliable conclusion. In fact, there is a paucity of research on the neurophysiological effect of physical embodiment in communication research.

The human brain, as with any other healthy physiological system, is an inherently complex system whose dynamics strongly correlate with the productivity of its cognitive functions such as attention and language [14]. Interestingly, an increase in complexity reflects the information content of the dynamics of physiological systems, which, in turn, have a direct correspondence to the variational information in their activities [15]. Therefore, the presence or the absence of a potential shift in the dynamics of the brain activity in response to a physically embodied medium can provide evidence on the (lack of) significance of the physical embodiment during verbal communication.

In this article, we argue that if the medium reinforces the perception of the conveyed message [16] then the embodiment is inevitably an essential part of the communicated content. To verify our claim, we performed a multiscale entropy (MSE) analysis [17–20] of the older people's prefrontal cortex (PFC) responses to tele-communicated stories, in which we communicated these stories with the older people through a speaker, a video-chat system, and a humanoid. In addition, we considered face-to-face to be a control setting, through which we communicated these stories with participants in-person. We chose MSE due to its discriminative power in detecting the change in complexity of biological signals [21,22]. Furthermore, we considered older participants due to the research findings that indicate the reduction of neurophysiological signal complexity by aging [23–27]. This allowed us to utilize MSE as a reliable biomarker for detection of any potential increase in the dynamics of the older people's brain activity in response to physical embodiment. We chose storytelling since stories' scripts can be kept intact and repeated to different individuals without any change in their contents, thereby allowing for the control of such confounders as subtle differences in conveyed information. We chose PFC due to neuroscientific findings on its pivotal role in language processing [28], social cognition [29], and story comprehension [30–32]. Our objective was to verify the following hypothesis.

Hypothesis 1. *Embodied media differentially stimulate the dynamics of the older people's PFC during a tele-communicated verbal communication.*

Our contributions are threefold. First, we provide evidence that physical embodiment induces a significant increase in MSE of the older people's PFC activity. Second, we show that such a shift in dynamics of the older people's PFC activity significantly reflects their perceived feeling of fatigue that is induced by listening to the stories. Third, we show that the increase in MSE by physical embodiment significantly differentiates the individuals whose perceived feeling of fatigue are above its average perception by the older people population in our study.

Our results benefit the researchers in age-related cognitive function and rehabilitation [23] who seek for the adaptation of these media in robot-assistive cognitive training of the older people [10]. In addition, they offer a complementary information to researchers in the field of human–robot interaction for modeling the human behavior. For instance, robotic media with the ability to detect the perceived feeling of fatigue by their human companions can close the gap on attaining a sustained verbal communication with them via adapting to individuals' pace and interest in response to conversational nuances and complexity. Moreover, changes in the pattern of brain activation (as reflected by the MSE) can enable the interactive learning algorithms [33] to utilize the brain's activation patterns as feedbacks for improving their level of interactivity. This, in turn, can form a stepping stone for rich and usable model of human mental state [34]. To the best of our knowledge, this is the first study that utilized MSE to investige the effect of physical embodiment on human subjects' brain activity.

2. Results

Figure 1A shows the grand-average MSEs of the older people's left-hemispheric PFC in different media settings. In this plot, scale factors 10 and 20 correspond to the one-second and two-second data acquisition intervals, given the sampling rate of our device (i.e., 10.0 Hz). We observed an increase in MSEs between the first two scale factors that was followed by a reduced MSEs up until scale factor 8. In addition, these MSEs exhibited a subtle increase at around one second data acquisition. Moreover, they roughly followed a straight line passed the 10 scale factor. We also observed that the MSEs of the older people's PFC was, on average, highest in the case of physically embodied medium (magenta), followed by the in-person setting (red).

Friedman test identified a significant effect of media on older people's MSEs in different media settings ($p < 0.001$, $H(3, 79) = 51.12$, $r = 0.80$). Post hoc Wilcoxon test identified that face-to-face setting induced a significant increase in MSEs of older people's left-hemispheric PFC in comparison with speaker ($p < 0.05$, $W(38) = 2.12$, $r = 0.34$, $M_F = 0.67$, $SD_F = 0.10$, $M_S = 0.62$, $SD_S = 0.11$) and video-chat ($p < 0.001$, $W(38) = 3.69$, $r = 0.58$, $M_V = 0.57$, $SD_V = 0.09$). Similarly, Telenoid induced an increase in MSE values that was significantly higher than speaker ($p < 0.01$, $W(38) = 2.88$, $r = 0.46$, $M_T = 0.69$, $SD_T = 0.11$) and video-chat ($p < 0.001$, $W(38) = 3.61$, $r = 0.57$). Moreover, speaker induced a significantly higher MSE than video-chat ($p < 0.05$, $W(38) = 2.15$, $r = 0.34$). On the other hand, we observed a non-significant difference between Telenoid and face-to-face ($p = 0.063$, $W(38) = 1.85$, $r = 0.29$). Figure 1B shows these results.

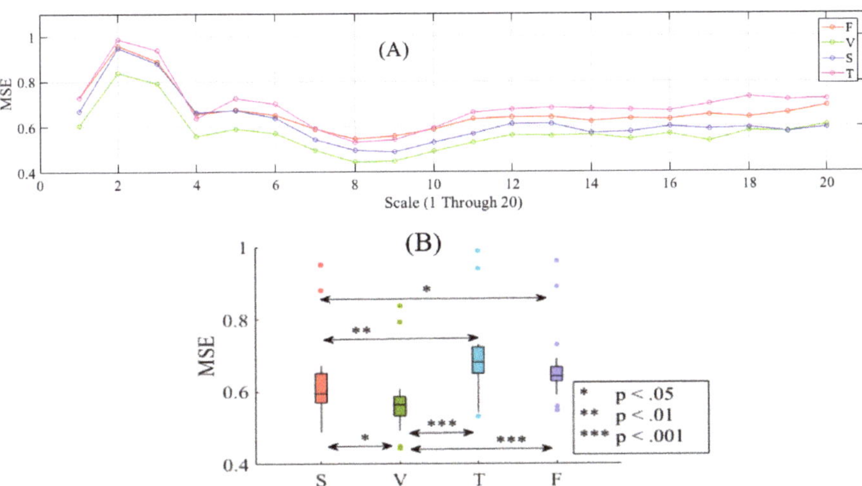

Figure 1. (**A**) Grand-average MSE of older people's Left-hemispheric PFC activation in speaker (S), video-chat (V), Telenoid (T), and face-to-face (F) settings. In these plots, scale factors 10 and 20 correspond to the one-second and two-second data acquisition intervals, given the sampling rate of our device (i.e., 10.0 Hz). (**B**) Descriptive Statistics of the older people's left-hemispheric MSE in speaker (S), video-chat (V), Telenoid (T), and face-to-face (F). Asterisks mark the significant differences between these media settings.

Whereas we observed no correlation between older people's MSEs and their self-assessed responses to the feeling of fatigue in speaker (Figure 2S) ($r = -0.05$, $p = 0.85$, $M_{Fatigue} = 3.67$, $SD_{Fatigue} = 1.80$), video-chat (Figure 2V) ($r = -0.13$, $p = 0.65$, $M_{Fatigue} = 4.20$, $SD_{Fatigue} = 2.11$), and face-to-face (Figure 2F) ($r = -0.16$, $p = 0.57$, $M_{Fatigue} = 3.80$, $SD_{Fatigue} = 2.21$), we found a significant anti-correlation in the case of Telenoid (Figure 2T) ($r = -0.57$, $p < 0.03$, $M_{Fatigue} = 4.33$, $SD_{Fatigue} = 2.12$).

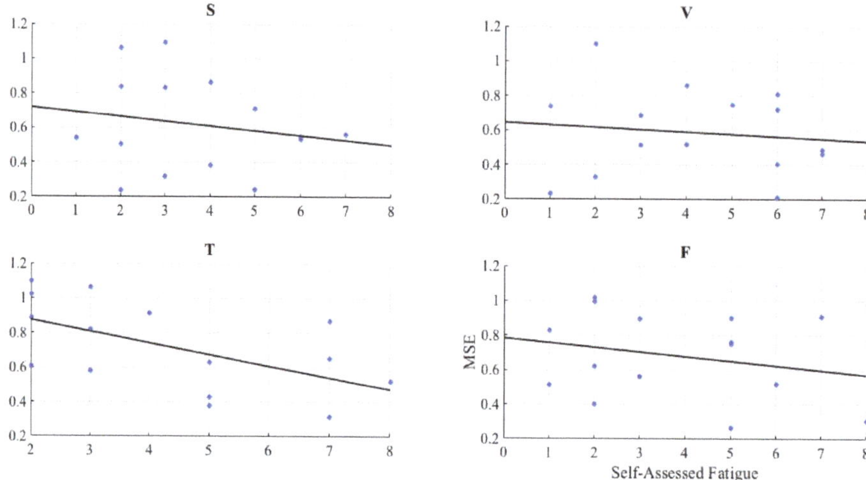

Figure 2. Spearman correlation between MSEs of the older people left-hemispheric PFC activation and their self-assessed responses to feeling of fatigue. (**S**) speaker; (**V**) video-chat; (**T**) Telenoid; (**F**) face-to-face.

Prediction of the Older People Perceived Fatigue Using MSE

We observed a significant anti-correlation between MSEs of the older people's left hemisphere and their self-assessed responses to the feeling of fatigue in the Telenoid setting. This suggested the potential utility of MSE for predicting the perceived level of fatigue by older people based on MSE of their frontal brain activity. To investigate this possibility, we interpreted the mean of the MSE clusters in different media settings as their respective decision boundaries. Figure 3 shows the MSE clusters associated with the older people's left PFC. Clusters' boundaries are depicted in black line segments in this figure.

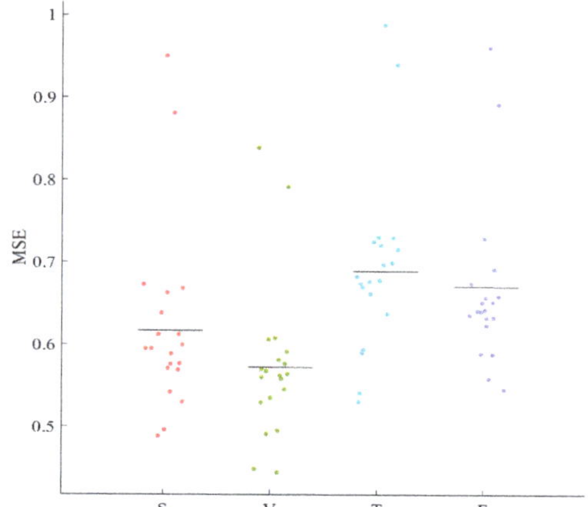

Figure 3. Older people's left-hemispheric MSE clusters. The boundaries associated with these clusters (i.e., clusters' mean) are shown (black line segments) in the figure.

Table 1 shows the metrics associated with the use of left-hemispheric MSEs of the older people for prediction of their perceived the feeling of fatigue. Significantly above chance prediction accuracy of the MSEs in the case of the Telenoid is evident in this table.

Table 1. Left PFC: Prediction accuracy, true positive (TP), true negative (TN), false positive (FP), false negative (FN), and F1-score. Significantly above chance (i.e., 50.00%, given two-class classification) prediction accuracy of the older people's feeling of fatigue in Telenoid setting is apparent in this table.

Medium	Accuracy	TP	TN	FP	FN	Precision	Recall	F1-Score
S	60.00%	4	5	2	4	0.67	0.50	0.57
V	53.33%	3	5	4	3	0.43	0.50	0.46
T	80.00%	6	6	1	2	0.86	0.65	0.80
F	46.67%	4	3	4	4	0.50	0.50	0.50

Figure 4 presents the confusion matrix pertinent to the prediction of the perceived feeling of fatigue using the left-hemispheric MSEs. We observed that the use of the MSEs achieved the highest TP and TN in case of the Telenoid. Similarly, FP and FN were smallest in the case of the Telenoid in comparison with other media settings. On other hand, speaker, video-chat, and face-to-face media settings achieved a comparable FP and FN. These results were in accord with the correlation analysis of the MSEs of the older people's left PFC.

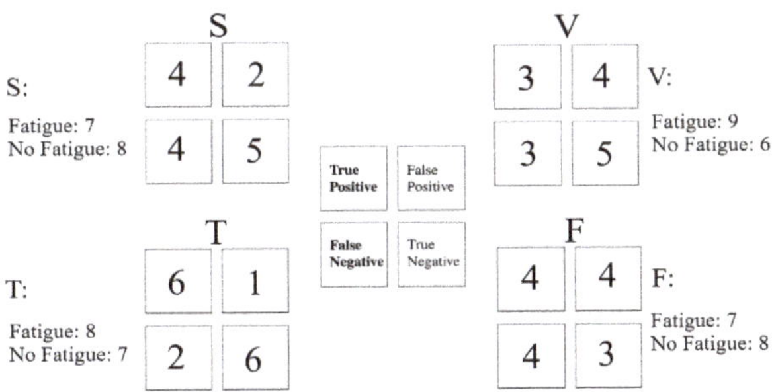

Figure 4. Left-hemispheric MSE vs. self-assessed responses to the perceived feeling of fatigue. In this figure, "Fatigue" and "No Fatigue" refer to the number of older people whose self-assessed responses to the feeling of fatigue was >4.0 and ≤4.0, respectively.

3. Discussion

We argued that the physical embodiment is inevitably an essential part of the communicated content and therefore the embodied media differentially stimulate the dynamics of the individuals' PFC during a tele-communicated verbal communication. We verified our claim through MSE analysis [17,18] of the older people's PFC responses to tele-communicated stories in which we communicated these stories with the older people through a speaker, a video-chat system, and a humanoid. In addition, we considered face-to-face to be a control setting, through which we communicated these stories with participants in-person.

We found evidence in support of our hypothesis. Precisely, the analysis of the older people's MSE identified that the physically embodied medium and the in-person setting induced an increase in the dynamics of their PFC activity. This increase that was significantly different from the speaker and the video-chat settings was non-significant between the physically embodied medium and the in-person setting. We also observed that the physically embodied medium and the in-person setting induced a bilateral PFC activation. This observation was in line with the neuroscientific findings on bilateral effect of stories on human subjects' PFC [30–32]. It also complemented these previous results by identifying that such an effect was not only present in the PFC's dynamical complexity but also the embodiment of the medium through which such contents were communicated had significant impact on the induced PFC activity.

We also observed that the older people's MSEs were higher in the finer time scales, which was accompanied with their decrease in the coarser scales. This was in accord with the findings that attribute such an effect to a shift from long-range brain regions connectivity (captured by the coarse scale MSEs) to a more local processing by aging [22,25]. Moreover, these MSEs tended to straight lines from scale factors 12 to 20. Such a tendency that is a characteristic of nonlinear systems [35,36] suggested the existence of a power law property [24]: an intrinsic complexity criterion in physiological systems [37]. In this regard, the ability of the physically embodied medium (similar to the in-person setting) to increase the complexity of the older people's brain activity implied the potential utility of embodiment in healthy cognitive aging through consideration of the correlation between increased physiological complexity and healthy condition [14,37].

Our further analyses identified that the shift in dynamics of the older people's PFC activity by physically embodied medium reflected their perceived feeling of fatigue and that such a shift was sufficient to significantly differentiate the individuals whose perceived feeling of fatigue were above the average perception of the older people population in our study. Additionally, they implied that such a differentiability was significantly stronger in the older people's left- than right-hemispheric PFC. Surprisingly, we found no correlation between the older people's MSEs and their perceived feeling of fatigue in the in-person setting, which was also in accord with the significantly low prediction accuracy in this condition. This observation appeared to undermine the effect of physical embodiment through such counterarguments as since the presence of a human is the best embodied representation for a human, the lack of correlation in this case may also suggest the observed correlation in the case of physically embodied medium to be spurious. However, several scientific studies identify the utility of tele-communication to surpass the in-person setting [38–41]. For instance, Joinson [42] noted the effect of the former in increasing the willingness of the individuals for self-disclosure in contrast with the in-person communication. Along the same direction, Zimmerman [43] concluded that, in contrast with the in-person communication, such settings as computer-mediated communication (CMC) [44] elicit emotionally rich, relationship-oriented verbal interaction among emotionally disturbed adolescents. In light of these findings, the observed phenomenon in our results may point at the utility of physically embodied medium in moderating the stress and unwillingness of individuals during their verbal communication with an unknown person. Although this interpretation finds evidence in the non-significant difference of the older people's increase in MSE in the case of physically embodied medium and in-person setting, further research is necessary to draw a more informed conclusion.

Our results contribute to such socially assistive robotics [1,2] scenarios as child education and elderly care. For instance, they suggest that the use of MSE can enable these media to determine whether their level of interaction (e.g., socialization [45], reading and comprehension [5]) is exceeding the comfort level of children, thereby allowing for modulation of their communicated contents and/or behavioral interaction. Similarly, they can advance the use of these media in elderly care facilities by enabling them to act as cognitive training mediators with exclusive access to older people's PFC dynamical changes during their cognitive training to determine their comfort in continuing their cognitive task [8,9]. The latter potential becomes more intriguing, considering the positive effect of these media in robot-assistive cognitive training of the older people [10].

Our results show a promising first step toward the use of brain information for quantification of one of the basic component of the human mental state: perceived fatigue due to the verbally communicated contents. These results benefit the researchers in age-related cognitive function and rehabilitation [23] that seek for adaptation of these media in robot-assistive cognitive training of the older people [10]. They also benefit the human–robot interaction research through such paradigms as interactive learning [33] in which such algorithms can utilize the brain's dynamical changes in the form of MSE as feedbacks for improving their level of interactivity. This, in turn, can form a stepping stone for rich and usable models of human mental state [34]. In a broader perspective, the ability to estimate the perceived feeling of fatigue during a humanoid-mediated verbal communication can contribute to the study and analysis of a robotic theory of mind (ToM) [46] through critical investigation of its implications in humans' neurological responses while interacting with their synthetic companions.

Limitations and Future Direction

Although our results indicated the significant role of the physical embodiment during verbal communication, a larger human sample is necessary for an informed conclusion on the utility of these results. Moreover, our participants were limited to older people. Therefore, it is necessary to investigate the effect of the physical embodiment in other age groups (e.g., kids, adolescents, and younger adults) to verify that our results are not affected by this factor.

Considering the utility of the physically embodied medium in differentiating the older people's perceived feeling of fatigue in conjunction with the absence of such a differentiability in the case of face-to-face setting, it is adequate to question the potential role of the novelty effect on this result. Therefore, further investigation in a longitudinal setting in which older people participate in multiple conversational sessions is necessary to verify that our findings are not affected by the long-term exposure to such media.

Moreover, our experimental setting was limited to a storytelling in which participants listened to a verbally communicated content without any requirement for their response. Therefore, it is crucial to analyze the effect of the physical embodiment in the conversational scenarios to examine the effect of such bidirectional verbal interaction on dynamical changes of the PFC.

In addition, the present study did not include other types of physically embodied media (e.g., mechanical looking robots, pet robots, etc.). Therefore, it is necessary to determine the correspondence between the media embodiment and the PFC's dynamics. It is also important to verify whether different embodiments can induce differential impact on the prediction accuracy of the feeling of fatigue.

4. Materials and Methods

4.1. Participants

Our participants consisted of twenty older people (ten females and ten males, 62–80 years old, MEAN = 70.70, SD = 4.62). All participants were free of neurological or psychiatric disorders and had no history of hearing impairment. Since we failed to collect the NIRS time series of the brain activity of four elderly adults due to recording complications, they were excluded from our analyses. We used an employment service center for older people to recruit our participants. This study was carried out in accordance with the recommendations of the ethical committee of Advanced Telecommunications Research Institute International (ATR) with written informed consent from all subjects. The protocol was approved by the ATR ethical committee (approval code:16-601-1).

4.2. Communication Media

Our experimental settings included a humanoid robotic medium (Figure 5a), an audio speaker (Figure 5c), a video-chat system (Figure 5d), and a human (i.e., in-person, Figure 5e). We chose a minimalist teleoperated android called, Telenoid R4^{TM} (Telenoid hereafter, Figure 5b) [47]. Telenoid is approximately 50.0 cm long and weighs about 3.0 kg. It comes with nine degrees-of-freedom (3 for its eyes, 1 for its mouth, 3 for its neck, and 2 for its arms) and is equipped with an audio speaker on its chest. It is primarily designed to investigate the basic and essential elements of embodiment for the efficient representation and transfer of a humanlike presence. Therefore, its design follows a minimalist anthropomorphic principle to convey a gender-and-age-neutral look-and-feel. In the present study, we chose a minimalist anthropomorphic embodiment to eliminate the projection of such physical traits as gender and age onto our robotic medium.

Telenoid conveyed the vocal information of its teleoperator through its speaker. Its motion was generated based on the operator's voice, using an online speech-driven head motion system [48]. However, its eyes and arms were motionless in this study. We placed Telenoid on a stand approximately 1.40 m (Figure 5a) from the participant's chair to prevent any confounding effect due to tactile interaction (e.g., holding, hugging, etc.). We adjusted this stand to resemble an eye-contact setting between Telenoid and the participant. We maintained the same distance in the case of the other media as well as for the in-person setting. In the in-person condition (Figure 5e), we adjusted the storyteller's seat to maintain eye-contact with the participant. For the video-chat (Figure 5d), we adjusted its placeholder in such a way that the storyteller's appearance on the screen resembled an eye-contact setting. In the speaker setting (Figure 5c), we placed the video-chat screen in front of the participant (as in the video-chat condition) and placed the speaker behind its screen.

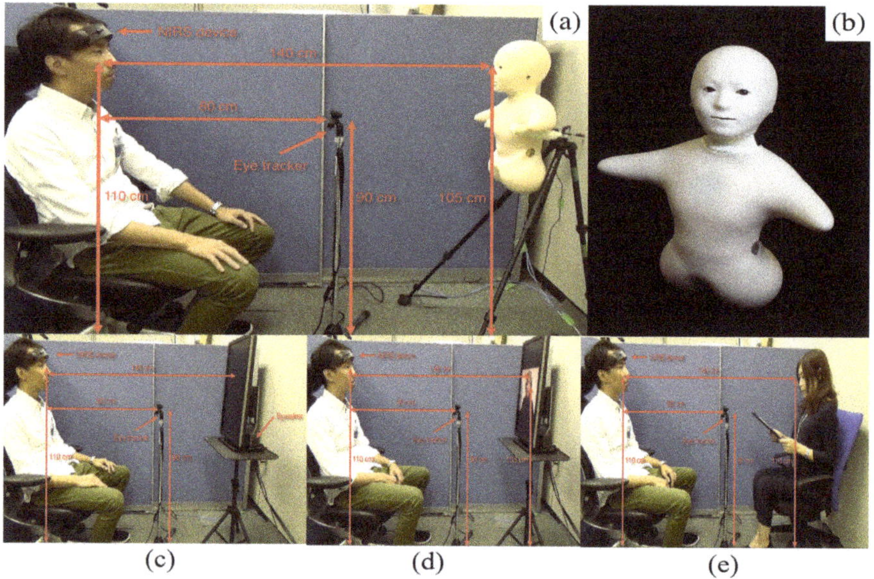

Figure 5. (**a**) Telenoid setting; (**b**) Telenoid medium; (**c**) Speaker setting (**d**) Video-chat setting; and (**e**) face-to-face setting. In these figures, an experimenter demonstrates the experimental setup.

We used the same audio device in the speaker and video-chat settings to prevent any confounding effect due to audio quality. We used the same recorded voice of a woman, who was naive to the purpose of this study, in speaker, video-chat, and Telenoid. These recordings took place in a single session in which we recorded her voice and video while telling stories. In Telenoid setting, we played back the same prerecorded voice for the speaker through the audio speaker on its chest. In the in-person setting, the same woman read the stories to the participants.

We asked our female storyteller to stay as neutral as possible while reading these stories. However, we are unable to confirm the absence of any difference in emotional impact of the stories' content on her during the in-person or the voice/video recordings.

4.3. Sensory Device

We used functional near infrared spectroscopy (fNIRS) to collect the frontal brain activity of the participants and acquired their NIRS time series data using a wearable optical topography system called "HOT-1000", developed by Hitachi High-Technologies Corp. (Figure 6). Participants wore this device on their forehead to record their frontal brain activity through detection of the total blood flow by emitting a wavelength laser light (810 nm) at a 10.0 Hz sampling rate. Data acquisition was carried out through four channels ($L1$, $L3$, $R1$, and $R3$, Figure 6). Postfix numerical values that are assigned to these channels specify their respective source-detector distances. In other words, $L1$ and $R1$ have a 1.0 cm source-detector distance and $L3$ and $R3$ have a 3.0 cm source-detector distance. Note that, whereas a short-detector distance of 1.0 cm is inadequate for the data acquisition of cortical brain activity (e.g., 0.5 cm [49], 1.0 cm [50], 1.5 cm [51], and 2.0 cm [52]), 3.0 cm is suitable [49,51].

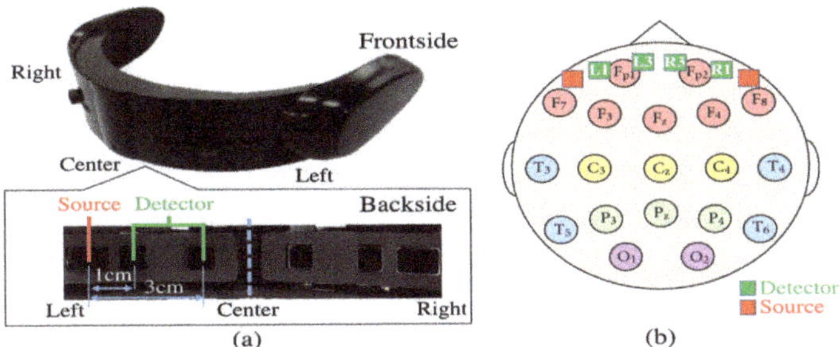

Figure 6. (**a**) fNIRS device in present study. Bottom subplot on left shows arrangement of source-detector of four channels of this device. Distances between short (i.e., 1.0 cm) and long (i.e., 3.0 cm) source and detector of left and right channels are shown. (**b**) Arrangement of 10–20 International Standard System: In this figure, relative locations of channels of fNIRS device in our study (i.e., $L1$, $L3$, $R1$, and $R3$) are depicted in red (i.e., sources) and green (i.e., detectors) squares. $L1$, $R1$, $L3$, and $R3$ are channels with short (i.e., 1.0 cm) and long (i.e., 3.0 cm) source-detector distances.

Findings on the brain activation during memory and language processing suggest a left-lateralized activation in both genders with higher specificity in females [53,54]. Therefore, we reported the results pertinent to $Left_3$ in the main body of our article (Appendix A for results on $Right_3$).

4.4. Paradigms

Our experimental paradigm consisted of storytelling sessions in which a woman narrated three-minute stories from Greek mythology in the selected media settings. This resulted in four separate sessions per participant. To prevent any confounding effect of visual distraction and to control the field of view of the participants within the same spatial limit, we placed their seat in a cubicle throughout the experimental sessions (height = 130.0 cm, width = 173.0 cm, depth = 210.0 cm). This cubicle's side wall is visible in blue in Figure 5a–e.

Every participant first gave written informed consent in the waiting room next to the experimental room. Then, a male experimenter explained the experiment's full procedure to the participants. This included the total number of session (four sessions), the duration of the narrated story in each session (i.e., three minutes), instructions about the one-minute rest period prior to the actual session (i.e., sitting still with eyes closed), and the content of the stories. The experimenter also asked the participant to focus on listening to the stories that were either narrated by a woman through a medium or in-person. Next, he led the participants to the experimental room and helped them to get seated in an armchair with proper head support in a sound-attenuated testing chamber. Then, the experimenter calibrated the eye tracker device and instructed the participants to fully relax and keep their eyes closed. Last, the experimenter verified the proper adjustment of the medium (or helped the storyteller get comfortable in the proper position for the in-person condition) and began the experimental session. In every session, we first acquired one-minute rest data. Next, the experimenter asked the participants to open their eyes and prepare to listen to the story, followed by a story session during which we recorded the NIRS time series of the frontal brain activity of the participants. During the in-person setting, we asked the storyteller to maintain as much eye-contacts with the participants as possible.

At the end of each session, all participants completed a questionnaire on their perceived (on an 8-point scale with 1 = not at all and 8 = very much) of feeling of fatigue due to listening to the story.

We provided our participants a one-minute rest period prior to the commencement of each of the storytelling sessions and asked them to keep their eyes closed. In this period, we prepared the setting for the next storytelling session. We video-recorded all the activities throughout the experiment.

Every subject participated in all four sessions. We kept the content of the stories intact in these sessions and randomized the order of the media among the participants without changing the order of the narrated stories. The entire experiment lasted about 90 minutes per participant.

4.5. Data Processing

To attenuate the effect of systemic physiological artefacts [55] (e.g., cardiac pulsations, respiration, etc.), we applied a one-degree polynomial Butterworth filter with 0.01 and 0.6 Hz for low and high bandpass which was then followed by performing the linear detrending on the data. Detrending of the signal that was adapted from signal processing and time series analysis and forecasting was a necessary step to ensure that the assumptions of stationarity and homoscedasticity (as reflected in wide spread application of linear models in analysis of fNIRS/fMRI time series) were not strongly violated (e.g., due to seasonality and/or repetitive increasing/decreasing patterns). Finally, we attenuated the effect of the skin blood flow (SBF) using an eigen decomposition technique [56]. This approach considers the first three principal components of all NIRS recorded channels of the participants' frontal brain activity during rest period to represent the SBF. Subsequently, it eliminates the SBF effect by removing these three components from participants' NIRS time series in task period. We followed the same approach and removed the first three principal components of the respective rest period of the participants from the NIRS time series of their frontal brain activity that was recorded during the task period. Similar to our NIRS recording, [56] also used 3.0 cm source-detector distance channels. Cooper et al. [57] show that this filter also attenuates the effect of motion artefact (e.g., head motion).

We used the processed time series of the older people's PFC to calculate their MSE. We used the pattern length $m = 2$, the similarity criterion $r = 0.15$, and the scale factors 1 through 20. We adapted the approach in [58] for computing the MSEs of older people's frontal brain activity.

4.6. Analysis

We computed the individuals' averaged MSE and applied Friedman test to determine any significance in different media settings. This was followed by post hoc paired Wilcoxon signed-rank test.

We also computed the Spearman correlation between the older people's averaged MSE of their frontal brain activity and their self-assessed responses to the feeling of fatigue.

The growing adaptation of physically embodied media in elderly care [8,9] along with the similarity of the older people's pattern of brain activity during in-person and physically embodied communication [6] envision their use for older people's cognitive training [10]. In fact, through such applications as brain machine interface (BMI) [59], it is foreseeable for these media to act as cognitive training mediators whose exclusive access to older people's brain activity can help determine whether these individuals have been overcome by their feeling of fatigue, thereby signaling the need for a break in the training session. Therefore, we used the MSE clusters associated with each media setting to determine whether the use of individuals' averaged MSE can predict their perceived feeling of fatigue during the storytelling. We expected that the change in PFC dynamics to be inversely proportional with such a feeling: the more tired the participants felt, the lower their MSE became. To check the validity of our expectation, we first calculated the mean of the MSE clusters in different media settings and interpreted these mean values as their respective clusters' decision boundary. Then, we calculated the true positive (TP), true negative (TN), false positive (FP), and false negative (FN) associated with each cluster's decision boundary. We considered individuals with their averaged MSEs above a given cluster's decision boundary and their self-assessed responses to the feeling of fatigue $\leq$4.0 (i.e., on an 8-point scale with 1 = not at all and 8 = very much, Section 4.4) as TP. Similarly, we considered the individuals with their averaged MSEs below the decision boundary and their self-assessed responses to the feeling of fatigue >4.0 as TN. On the other hand, we considered those individuals with their averaged MSEs above the decision boundary while their self-assessed responses to the feeling of fatigue >4.0 as FP and those with their averaged MSEs below the decision boundary while their self-assessed responses to the feeling of fatigue ≤4.0 as FN. We used these values to calculate the accuracy, precision, recall, and F1-score of the MSEs for predicting the older people's perceived feeling of fatigue. Considering our two-class paradigm (i.e., presence or absence of the feeling of fatigue by older people), the chance level was at 50.0%.

We also report the effect of the physical embodiment on the right-hemispheric PFC in Appendix A.

For Friedman test, we reported the effect size $r = \sqrt{\frac{\chi^2}{N}}$ [60] with N denoting the sample size. In the case of Wilcoxon test, we used $r = \frac{W}{\sqrt{N}}$ [61] as effect size with W denoting the Wilcoxon statistics and N the sample size. All results reported are Bonferroni corrected (i.e., multiplying the p-values with the sample size, given the use of non-parametric tests). We were unable to collect the self-assessed responses of five participants to the feeling of fatigue and therefore these participants were excluded from correlation and prediction analyses.

We used Python 2.7 for data acquisition and processing. We carried out analyses in Matlab R2016a. We used Gramm [62] for data visualization.

Author Contributions: Conceptualization, H.S. and R.Y.; formal analysis, S.K.; investigation, S.K. and H.S.; resources, H.I.; data curation, R.Y.; writing—original draft preparation, S.K.; writing—review and editing, S.K. and H.S.; supervision, H.S.; project administration, H.I.; and funding acquisition, H.S.

Funding: This research was funded by ImPACT Program of Council for Science, Technology and Innovation (Cabinet Office, Government of Japan) under Grant No.: 2014-PM11-07-01 and partially supported by JSPS KAKENHI, Grant No. 16K16480.

Conflicts of Interest: The authors declare no conflict of interest.

Abbreviations

The following abbreviations are used in this manuscript:

PFC	Prefrontal cortex
MSE	Multiscale entropy
NIRS	Near-infrared spectroscopy
S	Speaker setting
V	Video-chat setting
T	Telenoid setting
F	Face-to-face (in-person) setting
M_a	Mean of the variable "a"
SD_a	Standard deviation of the variable "a"

Appendix A. Right-Hemispheric PFC

Figure A1A shows the grand-average MSEs of the older people's right-hemispheric PFC in different media settings. In this plot, scale factors 10 and 20 correspond to the one-second and two-second data acquisition intervals, given the sampling rate of our device (i.e., 10.0 Hz). We observed an increase in MSEs between the first two scale factors that was followed by a reduced MSEs up until scale factor 8. In addition, these MSEs exhibited a subtle increase at around one second data acquisition. Moreover, they roughly followed a straight line that passed the 10 scale factor. We also observed that the MSEs of the older people's PFC was, on average, highest in the case of physically embodied medium (magenta), followed by the in-person setting (red).

Friedman test identified a significant effect of media on older people's MSEs in different media settings ($p < 0.001$, H(3, 79) = 50.10, r = 0.80). Post hoc Wilcoxon test identified that face-to-face setting induced a significant increase in MSEs of older people's right-hemispheric PFC in comparison with speaker ($p < 0.05$, W(38) = 2.07, r = 0.33, M_F = 0.66, SD_F = 0.11, M_S = 0.62, SD_S = 0.12) and video-chat ($p < 0.001$, W(38) = 3.39, r = 0.54, M_V = 0.58, SD_V = 0.14). Similarly, Telenoid induced an increase in

MSE values that was significantly higher than speaker ($p < 0.01$, W(38) = 2.83, $r = 0.45$, $M_T = 0.69$, $SD_T = 0.13$) and video-chat ($p < 0.001$, W(38) = 3.34, $r = 0.53$). On the other hand, we observed a non-significant difference between speaker and video-chat ($p = 0.06$, W(38) = 1.91, $r = 0.30$) as well as Telenoid and the face-to-face ($p = 0.11$, W(38) = 1.58, $r = 0.25$). Figure A1B shows these results.

Whereas we observed no correlation between older people's change in MSEs and their self-assessed responses to the feeling of fatigue in speaker (Figure A2S) ($r = -0.31$, $p = 0.26$), video-chat (Figure A2V) ($r = 0.13$, $p = 0.63$), and face-to-face (Figure A2F) ($r = 0.16$, $p = 0.57$), we found a significant anti-correlation in THE case of Telenoid (Figure A2T) ($r = -0.53$, $p < 0.05$).

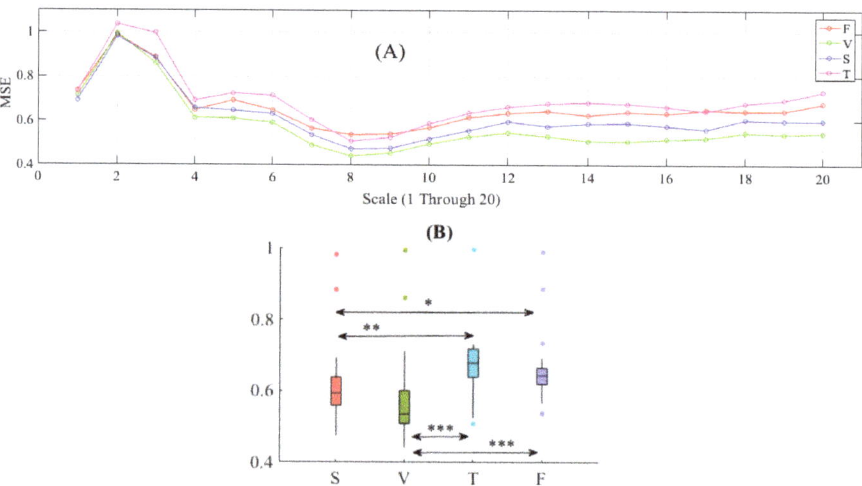

Figure A1. (**A**) Grand-average MSE of older people's right-hemispheric PFC activation in speaker (S), video-chat (V), Telenoid (T), and face-to-face (F) settings. In these plots, scale factors 10 and 20 correspond to the one-second and two-second data acquisition intervals, given the sampling rate of our device (i.e., 10.0 Hz). (**B**) Descriptive Statistics of the older people's right-hemispheric MSE in speaker (S), video-chat (V), Telenoid (T), and face-to-face (F). Asterisks mark the significant differences between these media settings.

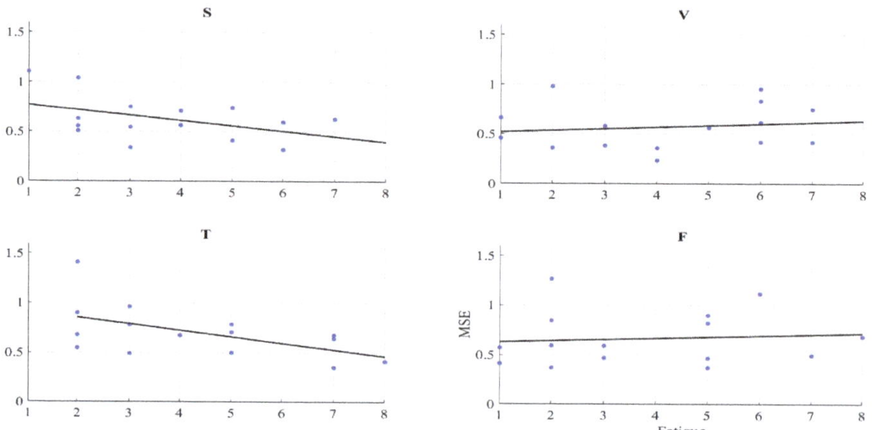

Figure A2. Spearman correlation between MSEs of the older people's right-hemispheric PFC activation and their self-assessed responses to feeling of fatigue. (**S**) speaker; (**V**) video-chat; (**T**) Telenoid; (**F**) face-to-face.

Prediction of the Older People Perceived Fatigue Using Right-Hemispheric MSE

We observed a significant anti-correlation between MSEs of the older people's right hemisphere and their self-assessed responses to the feeling of fatigue in the Telenoid setting. This suggested the potential utility of MSE for predicting the perceived level of fatigue by older people based on MSE of their frontal brain activity. To investigate this possibility, we interpreted the mean of the MSE clusters in different media settings as their respective decision boundaries. Figure A3 shows the MSE clusters associated with the older people's left PFC. Clusters' boundaries are depicted in black line segments in this figure.

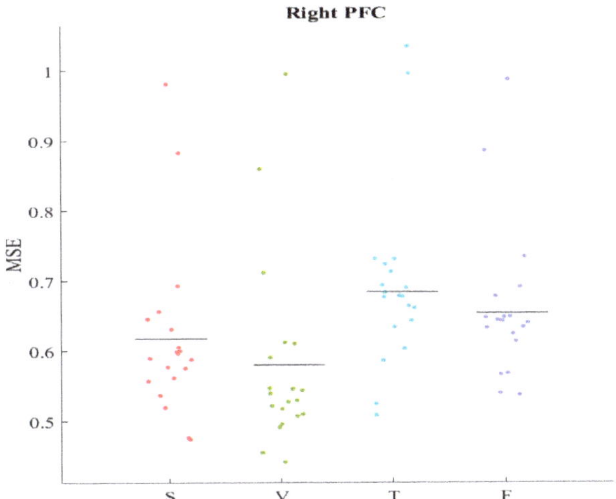

Figure A3. Older people's right-hemispheric MSE clusters. The boundaries associated with these clusters (i.e., clusters' mean) are shown (black line segments) in the figure.

Table A1 shows the metrics associated with the use of right-hemispheric MSEs of the older people for prediction of their perceived feeling of fatigue. Significantly above chance accuracy of the MSEs associated with the right PFC in THE case of Telenoid setting is apparent in this figure. However, its accuracy is significantly lower than the left-hemispheric MSEs.

Figure A4 presents the confusion matrix pertinent to the prediction of the perceived feeling of fatigue using the right-hemispheric MSEs. We observed TP and TN were highest in the Telenoid setting. In addition, Telenoid setting resulted in lowest FN. On the other hand, FP was lowest in the case of Telenoid and speaker settings alike.

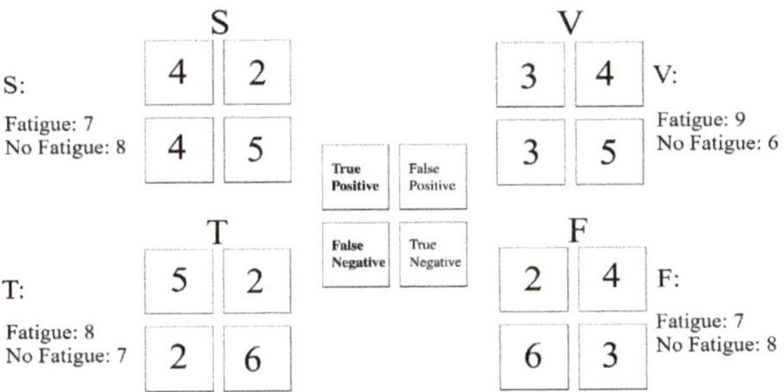

Figure A4. Right-hemispheric MSE vs. self-assessed responses to the feeling of fatigue. In this figure, "Fatigue" and "No Fatigue" refer to the number of older people whose self-assessed responses to the feeling of fatigue was >4 and ≤4.0, respectively.

Table A1. Right PFC: Prediction accuracy, true positive (TP), true negative (TN), false positive (FP), false negative (FN), and F1-score. Significantly above chance (i.e., 50.00%, given two-class classification) prediction accuracy of the older people's feeling of fatigue in Telenoid setting is apparent in this table.

Medium	Accuracy	TP	TN	FP	FN	Precision	Recall	F1-Score
S	60.00%	4	5	2	4	0.67	0.50	0.57
V	53.33%	3	5	4	3	0.43	0.50	0.46
T	73.33%	5	6	2	2	0.61	0.61	0.61
F	33.33%	2	3	4	6	0.33	0.25	0.29

References

1. Matarić, M.J. Socially assistive robotics: Human augmentation versus automation. *Sci. Robot.* **2018**, *2*, eaam5410. [CrossRef]
2. Matarić, M.J.; Scassellati, B. Socially assistive robotics. In *Springer Handbook of Robotics*; Scassellati, B., Khatib, O., Eds.; Springer International Publishing: New York, NY, USA, 2016.
3. Broadbent, E. Interactions with robots: The truths we reveal about ourselves. *Annu. Rev. Psychol.* **2017**, *68*, 627–652. [CrossRef] [PubMed]
4. Mann, J.A.; MacDonald, B.A.; Kuo, I.; Li, X.; Broadbent, E. People respond better to robots than computer tablets delivering healthcare instructions. *Comput. Hum. Behav.* **2015**, *43*, 112–117. [CrossRef]
5. Michaelis, J.E.; Mutlu, B. Reading socially: Transforming the in-home reading experience with a learning-companion robot. *Sci. Robot.* **2018**, *3*, eaat5999. [CrossRef]
6. Keshmiri, S.; Sumioka, H.; Yamazaki, R.; Okubo, M.; Ishiguro, H. Similarity of the Impact of Humanoid and In-Person Communications on Frontal Brain Activity of Older People. In Proceedings of the 2018 IEEE/RSJ International Conference on Intelligent Robots and Systems (IROS), Madrid, Spain, 1–5 October 2018.
7. Sakamoto, D.; Kanda, T.; Ono, T.; Ishiguro, H.; Hagita, N. Android as a telecommunication medium with a human-like presence. In Proceedings of the Human Robot Interaction (HRI), Arlington, VA, USA, 10–12 March 2007; Volume 3, pp. 193–200.
8. Valenti, S.M.; Agüera-Ortiz, L.; Olazarán, R.J.; Mendoza, R.C.; Pérez, M.A.; Rodriguez, P.I.; Osa, R.E.; Barrios, S.A.; Herrero, C.V.; Carrasco, C.L.; et al. Social robots in advanced dementia. *Front. Aging Neurosci.* **2015**, *7*, 133.
9. Robinson, H.; MacDonald, B.; Kerse, N.; Broadbent, E. The psychosocial effects of a companion robot: A randomized controlled trial. *J. Am. Med. Dir. Assoc. JAMA* **2013**, *14*, 661–667. [CrossRef] [PubMed]
10. Kim, G.H.; Jeon, S.; Im, K.; Kwon, H.; Lee, B.H.; Kim, G.Y.; Jeong, H.; Han, N.E.; Seo, S.W.; Cho, H.; et al. Structural brain changes after traditional and robot-assisted multi-domain cognitive training in community-dwelling healthy elderly. *PLoS ONE* **2015**, *10*, e0123251. [CrossRef] [PubMed]
11. Reeves, B.; Nass, C. *The Media Equation: How People Treat Computers, Television, and New Media Like Real People and Places*, 1st ed.; CSLI Publications: Stanford, CA, USA, 2002.
12. Nass, C.; Moon, Y.; Green, N. Are machines gender neutral? Gender-stereotypic responses to computers with voices. *J. Appl. Soc. Psychol.* **1997**, *27*, 864–876. [CrossRef]
13. Nass, C.; Moon, Y.; Carney, P. Are respondents polite to computers? Social desirability and direct responses to computers. *J. Appl. Soc. Psychol.* **1999**, *29*, 1093–1110. [CrossRef]
14. Goldberger, A.L.; Amaral, L.A.N.; Hausdorff, J.M.; Ivanov, P.C.; Peng, C.-K.; Stanley, H.E. Fractal dynamics in physiology: Alterations with disease and aging. *Proc. Natl. Acad. Sci. USA* **2002**, *99*, 2466–2472. [CrossRef] [PubMed]
15. Zhang, Y.-C. Complexity and $\frac{1}{f}$ noise. A phase space approach. *J. Phys. I* **1991**, *7*, 971–977.
16. McLuhan, M.; Quentin, F. *The Medium Is the Message: An Inventory of Effects*; Agel, J., Eds.; Pinguin Group: New York, NY, USA, 1967.
17. Costa, M.; Goldberger, A.L.; Peng, C.K. Multiscale entropy analysis of complex physiologic time series. *Phys. Rev. E* **2002**, *86*, 068102. [CrossRef] [PubMed]
18. Costa, M.; Goldberger, A.L.; Peng, C.K. Multiscale entropy analysis of biological signals. *Phys. Rev. E* **2005**, *71*, 021906. [CrossRef] [PubMed]

19. Humeau-Heurtier, A. Evaluation of systems' irregularity and complexity: Sample entropy, its derivatives, and their applications across scales and disciplines. *Entropy* **2018**, *20*, 794. [CrossRef]
20. Gao, Y.; Villecco, F.; Li, M.; Song, W. Multi-Scale permutation entropy based on improved LMD and HMM for rolling bearing diagnosis. *Entropy* **2017**, *19*, 176. [CrossRef]
21. McIntosh, A.R.; Kovacevic, N.; Itier, R.J. Increased brain signal variability accompanies lower behavioral variability in development. *PLoS Comput. Biol.* **2008**, *4*, e1000106. [CrossRef] [PubMed]
22. McDonough, I.M.; Nashiro, K. Network complexity as a measure of information processing across resting-state networks: Evidence from the Human Connectome Project. *Front. Hum. Neurosci.* **2014**, *8*, 409. [CrossRef] [PubMed]
23. Manor, B.; Lipsitz, L.A. Physiologic complexity and aging: Implications for physical function and rehabilitation. *Prog. Neuro-Psychopharmacol. Biol. Psychiatry* **2013**, *45*, 287–293. [CrossRef] [PubMed]
24. Takahashi, T.; Cho, R.Y.; Murata, T.; Mizuno, T.; Kikuchi, M.; Mizukami, K.; Kosaka, H.; Takahashi, K.; Wada, Y. Age-related variation in EEG complexity to photic stimulation: A multiscale entropy analysis. *Clin. Neurophysiol.* **2009**, *120*, 476–483. [CrossRef] [PubMed]
25. McIntosh, A.R.; Vakorin, V.; Kovacevic, N.; Wang, H.; Diaconescu, A.; Protzner, A.B. Spatiotemporal dependency of age-related changes in brain signal variability. *Cereb. Cortex* **2013**, *24*, 1806–1817. [CrossRef] [PubMed]
26. Yang, A.C.; Huang, C.C.; Yeh, H.L.; Liu, M.E.; Hong, C.J.; Tu, P.C.; Chen, J.F.; Huang, N.E.; Peng, C.K.; Lin, C.P.; et al. Complexity of spontaneous BOLD activity in default mode network is correlated with cognitive function in normal male elderly: A multiscale entropy analysis. *Neurobiol. Aging* **2013**, *34*, 428–438. [CrossRef] [PubMed]
27. Liao, F.; Cheing, G.L.; Ren, W.; Jain, S.; Jan, Y.K. Application of multiscale entropy in assessing plantar skin blood flow dynamics in diabetics with peripheral neuropathy. *Entropy* **2018**, *20*, 127. [CrossRef]
28. Gabrieli, J.D.; Poldrack, R.A.; Desmond, J.E. The role of left prefrontal cortex in language and memory. *Proc. Natl. Acad. Sci. USA* **1998**, *95*, 906–913. [CrossRef] [PubMed]
29. Forbes, C.E.; Grafman, J. The role of the human prefrontal cortex in social cognition and moral judgment, *Annu. Rev. Neurosci.* **2010**, *33*, 299–324. [CrossRef] [PubMed]
30. Mar, R.A. The neural bases of social cognition and story comprehension. *Annu. Rev. Psychol.* **2011**, *33*, 103–134. [CrossRef] [PubMed]
31. Molenberghs, P.; Johnson, H.; Henry, J.D.; Mattingley, J.B. Understanding the minds of others: A neuroimaging meta-analysis. *Neurosci. Biobehav. Rev.* **2016**, *65*, 276–291. [CrossRef] [PubMed]
32. Lerner, Y.; Honey, C.J.; Silbert, L.J.; Hasson, U. Topographic mapping of a hierarchy of temporal receptive windows using a narrated story. *J. Neurosci.* **2011**, *31*, 2906–2915. [CrossRef] [PubMed]
33. Clabaugh, C.; Matarić, M.J. Robots for the people, by the people: Personalizing human-machine interaction. *Sci. Robot.* **2018**, *3*, eaat7451. [CrossRef]
34. Yang, G.-Z.; Bellingham, J.; Dupont, P.E.; Fischer, P.; Floridi, L.; Full, R.; Jacobstein, N.; Kumar, V.; McNutt, M.; Merrifield, R.; et al. The grand challenges of Science Robotics. *Sci. Robot.* **2018**, *3*, eaar7650. [CrossRef]
35. Bak, P.; Tang, C.; Wiesenfield, K. Self-organized criticality: An explanation of the $1/f$ noise. *Phys. Rev. Lett.* **1987**, *5*, 381–384. [CrossRef] [PubMed]
36. Stanley, H.E.; Buldyrev, S.V.; Goldberger, A.L.; Goldberger, Z.D.; Havlin, S.; Mantegna, R.N. Statistical mechanics in biology: How ubiquitous are long-range correlations? *Phys. A* **1994**, *205*, 214–253. [CrossRef]
37. Goldberger, A.I.; Peng, C.K.; Lipsitz, L.A. What is physiologic complexity and how does it change with ageing and disease? *Neurobiol. Aging* **2002**, *23*, 23–26. [CrossRef]
38. Finn, K.E.; Sellen, A.J.; Wilbur, S.B. Video-mediated communication. In *Video-Mediated Communication;* Lawrence Erlbaum Associates: Hillsdale, NJ, USA, 1997.
39. de Greef, P.; Ijsselsteijn, W. Social presence in a home tele-application. *Cyberpsychol. Behav.* **2001**, *4*, 307–315. [CrossRef] [PubMed]
40. Bailenson, J.N.; Yee, N.; Merget, D.; Schroeder, R. The effect of behavioral realism and form realism of real-time avatar faces on verbal disclosure, nonverbal disclosure, emotion recognition, and copresence in dyadic interaction. *Presence* **2006**, *15*, 359–372. [CrossRef]
41. Bente, G.; Ruggenberg, S.; Kramer, N.C.; Eschenburg, F. Avatar-mediated networking: Increasing social presence and interpersonal trust in net-based collaborations. *Hum. Commun. Res.* **2008**, *34*, 287–318. [CrossRef]

42. Joinson, A.N. Self-disclosure in computer-mediated communication: The role of self-awareness and visual anonymity. *Eur. J. Soc. Psychol.* **2001**, *31*, 177–192. [CrossRef]
43. Joinson, A.N. Effects of computer conferencing on the language use of emotionally disturbed adolescents. *Behav. Res. Methods Instrum. Comput.* **1987**, *19*, 224–230.
44. Holt-Lunstad, J.; Smith, T.B.; Layton, J.B. Computer-mediated communication: Impersonal, Interpersonal, and hyperpersonal integration. *Commun. Res.* **1996**, *23*, 3–43.
45. Tanaka, F.; Cicourel, A.; Movellan, J.R. Socialization between toddlers and robots at an early childhood education center. *Proc. Natl. Acad. Sci. USA* **2007**, *104*, 17954–17958. [CrossRef] [PubMed]
46. Scassellati, B. Theory of mind for a humanoid robot. *Auton. Robots* **2002**, *12*, 13–24. [CrossRef]
47. Sumioka, H.; Nishio, S.; Minato, T.; Yamazaki, R.; Ishiguro, H. Minimal human design approach for sonzai-kan media: Investigation of a feeling of human presence. *Cogn. Comput.* **2014**, *6*, 760–774. [CrossRef]
48. Sakai, K.; Minato, T.; Ishi, C.T.; Ishiguro, H. Novel Speech Motion Generation by Modelling Dynamics of Human Speech Production. *Front. Robot. AI* **2017**, *4*, 49. [CrossRef]
49. Takahashi, T.; Takikawa, Y.; Kawagoe, R.; Shibuya, S.; Iwano, T.; Kitazawa, S. Influence of skin blood flow on near-infrared spectroscopy signals measured on the forehead during a verbal fluency task. *NeuroImage* **2011**, *57*, 991–1002. [CrossRef] [PubMed]
50. Gagnon, L.; Perdue, K.; Greve, D.N.; Goldenholz, D.; Kaskhedikar, G.; Boas, D.A. Improved recovery of the hemodynamic response in diffuse optical imaging using short optode separations and state-space modeling. *NeuroImage* **2011**, *56*, 1362–1371. [CrossRef] [PubMed]
51. Sato, T.; Nambu, I.; Takeda, K.; Aihara, T.; Yamashita, O.; Isogaya, Y.; Inoueand, Y.; Otaka, Y.; Wadaand, Y.; Kawato, M.; et al. Reduction of global interference of scalp-hemodynamic in functional near-infrared spectroscopy using short distance probes. *NeuroImage* **2016**, *141*, 120–132. [CrossRef] [PubMed]
52. Yamada, T.; Umeyama, S.; Matsuda, K. Multidistance probe arrangement to eliminate artifacts in functional near-infrared spectroscopy. *J. Biomed. Opt.* **2009**, *14*, 120–132. [CrossRef] [PubMed]
53. Li, T.; Luo, Q.; Gong, H.; Roebers, C.M.; Rammsayer, T.H. Gender-specific hemodynamics in prefrontal cortex during a verbal working memory task by near-infrared spectroscopy. *Behav. Brain Res.* **2010**, *29*, 148–153. [CrossRef] [PubMed]
54. Haut, K.; Barch, D. Sex influences on material-sensitive functional lateralization in working and episodic memory: Men and women are not all that different. *NeuroImage* **2006**, *32*, 411–422. [CrossRef] [PubMed]
55. Tak, S.; Ye, J.C. Statistical analysis of fNIRS data: A comprehensive review. *NeuroImage* **2014**, *85*, 72–91. [CrossRef] [PubMed]
56. Zhang, Y.; Brooks, D.H.; Franceschini, M.A.; Boas, D.A. Eigenvector-based spatial filtering for reduction of physiological interference in diffuse optical imaging. *J. Biomed. Opt.* **2005**, *10*, 011014. [CrossRef] [PubMed]
57. Cooper, R.; Selb, J.; Gagnon, L.; Phillip, D.; Schytz, H.W.; Iversen, H.K.; Ashina, M.; Boas, D.A. A systematic comparison of motion artifact correction techniques for functional near-infrared spectroscopy. *Front. Neurosci.* **2012**, *6*, 147. [CrossRef] [PubMed]
58. Goldberger, A.L.; Amaral, L.; Glass, L.; Hausdorff, J.M.; Ivanov, P.C.; Mark, P.G.; Moody, G.B.; Peng, C.-K.; Stanley, H.E. PhysioBank, PhysioToolkit, and PhysioNet: Components of a new research resource for complex physiologic signals. *Circulation* **2000**, *101*, e215–e220. [CrossRef] [PubMed]
59. Donoghue, J.P. Connecting cortex to machines: Recent advances in brain interfaces. *Nat. Neurosci.* **2002**, *5*, 1085. [CrossRef] [PubMed]
60. Rosenthal, R.; DiMatteo, M.R. Meta-analysis: Recent developments n quantitative methods for literature reviews. *Annu. Rev. Psychol.* **2001**, *52*, 59–82. [CrossRef] [PubMed]
61. Tomczak, M.; Tomczak, E. The need to report effect size estimates revisited. An overview of some recommended measures of effect size. *Trends Sport Sci.* **2014**, *1*, 19–25.
62. Morel, P. Gramm: Grammar of Graphics Plotting for Matlab. 2016. Available online: https://github.com/piermorel (accessed on 10 April 2018).

Article

Multiscale Entropy Analysis of Page Views: A Case Study of Wikipedia

Chao Xu [1], Chen Xu [1], Wenjing Tian [1], Anqing Hu [2] and Rui Jiang [3,*]

[1] School of Mathematics and Computer Science, Wuhan Textile University, Wuhan 430200, China; chaox@wtu.edu.cn (C.X.); christinaxuc@gmail.com (C.X.); wjtian.cs@hotmail.com (W.T.)

[2] Accounting College, Wuhan Textile University, Wuhan 430200, China; aria.aqhu@gmail.com

[3] Department of Electrical Engineering and Computer Science, University of Michigan, Ann Arbor, MI 48109-2122, USA

* Correspondence: ruij@umich.edu; Tel.: +86-134-7707-2303

Received: 18 January 2019; Accepted: 24 February 2019; Published: 27 February 2019

Abstract: In this study, the Wikipedia page views for four selected topics, namely, education, the economy/finance, medicine, and nature/environment from 2016–2018 are collected and the sample entropies of the three years' page views are estimated and investigated using a short-time series multiscale entropy (sMSE) algorithm for a comprehensible understanding of the complexity of human website searching activities. The sample entropies of the selected topics are found to exhibit different temporal variations. In the past three years, the temporal characteristics of the sample entropies are vividly revealed, and the sample entropies of the selected topics follow the same tendencies and can be quantitatively ranked. By taking the 95% confidence interval into account, the temporal variations of sample entropies are further validated by statistical analysis (non-parametric), including the Wilcoxon signed-rank test and the Mann-Whitney *U*-test. The results suggest that the sample entropies estimated by the sMSE algorithm are feasible for analyzing the temporal variations of complexity for certain topics, whereas the regular variations of estimated sample entropies of different selected topics can't simply be accepted as is. Potential explanations and paths in forthcoming studies are also described and discussed.

Keywords: human behavior; complexity; page view; multiscale entropy; sample entropy; Wikipedia

1. Introduction

We are entering an era of big data, in which the datasets we work with are characterized by the 4 Vs: volume, velocity, variety, and veracity (where veracity emphasizes the uncertainties of data) [1,2]. The statistics underlying data are consequently crucial to making the data valuable and results worthwhile, notably for large volumes of data. Traditional statistical methods, like 1st and 2nd moment statistics (the mean value and variance, respectively) or the probability density function (PDF), often ignore data's temporal and spatial characteristics, and are even invalid in some special cases. For instance, complex physiologic time series having the same mean value or variance often contain different information [3,4]. Once these methods are applied, nuance can be lost and misunderstandings may occur. For medical applications, such misunderstandings can be fatal [4].

The multiscale entropy (MSE) algorithm [5,6] was first introduced to analyze the complexity of biological time series, in which an original time series is coarsely divided into many subseries and the sample entropy [7,8] of each subseries is calculated separately. Compared to traditional statistical methods, the MSE algorithm exhibits several advances and strengths [9–11]: (1) It emphasizes the temporal correlation of series elements; (2) Multiscale processing ensures that the data is deeply mined; (3) In signal processing, the averaging process that occurs as part of coarse-grained integration of the subseries can be regarded as a low pass filter and effectively eliminates noise or interference.

Website page views, one well-known set of big data, are widely touted for their potential to reflect public interest in a subject [12,13]. However, most page view data are hidden by enterprises as commercial secrets, and are inaccessible to common users. Openly authorized page view data from organizations like Wikipedia and Google are thus frequently consulted and used for commercial purposes and data mining applications. For example, market prediction or consumption style analysis [13–16]. Daily human activities dominate page views. Thus, page view series are then endowed with a variety of temporal characteristics. Many studies have addressed the statistical properties of page views [13–16], although no research has yet highlighted the temporal characteristics of page views. Meanwhile, many widely used and well-performing techniques have been introduced to analyze the big data of web searching, for example, the clustering, SVM, etc. [17–22]. However, on this topic, the angle of entropy had never been investigated in previous studies.

Motivated by this gap in the literature, and with the aim of developing a comprehensive understanding of human behavior by taking advantages of MSE algorithms, we examine the complexity of page views in this study. Considering that website page views are commonly affected by commercial actions, as in the case of China's Internet "water armies"(people who search website for earning money or driving by commercial activities) [23], we therefore analyze Wikipedia searches, as page views of a given Wikipedia topic are highly reliable (given the importance of veracity in big data) and these searches are dominated by human intentions, rather than robotic or automatic page views. Without loss of generality, in this study, the page views (search times) of four selected topics from the years 2016–2018 were collected and then given as the input into a short-time series multiscale entropy (sMSE) algorithm to investigate their complexity.

This paper is organized as follows. Section 2 briefly introduces the basics of the sMSE algorithm. Section 3 describes the characteristics of the Wikipedia data, as well as its acquisition and processing. Section 4 calculates and discusses the sample entropies of page views of the selected topics, including their temporal characteristics, compares sample entropies across the selected topics for each year, and validates the sample entropies by using statistical analysis. Finally, Section 5 concludes the paper and suggests paths for forthcoming studies.

2. sMSE Algorithm

The original MSE algorithm contains two main procedures: coarse-grained division and sample entropy calculation [5–8]. MSE variants, in which one or both steps are modified or replaced, are widely applied to meet the needs of different series or signals [24]. In particular, the sMSE algorithm [25] selected in this study is ideal for the short length of page view series. For short time series, the modifications included in the sMSE algorithm are conducted as follows.

During coarse-grained division, factor p is induced and is effective in eliminating potential fluctuations in sample entropy [7,8]. Coarse-grained division in the sMSE algorithm is defined as

$$y_j{}^{(\tau)(p)} = \frac{1}{\tau} \sum_{i=(j-1)\tau+1+p}^{j\tau+p} x_i \quad 1 \le j \le (N-p)/\tau \tag{1}$$

in which y and x denote the elements of the coarse-grained and original series, respectively, τ is a scale factor, i and j represent the element IDs of series $\{x\}$ and $\left\{y^{(\tau)(p)}\right\}$, and N is the length of original series $\{x\}$. In addition, p fulfills $0 \le p \le \tau - 1$.

Equation (1) and the range of p show that each scale factor τ corresponds to τ coarse-grained subseries that are produced by p. The sample entropy from the original MSE algorithm is then redefined in the sMSE algorithm as

$$SE(x,\tau,m,r) = \frac{1}{\tau}\sum_{p=0}^{\tau-1} S_E(y^{(\tau)(p)},m,r) \tag{2}$$

in which SE is the sample entropy of scale factor τ and S_E denotes the sample entropy of the pth coarse-grained subseries, m is the shortest length between points, and r denotes the threshold in sample entropy algorithm. The sample entropy in Equation (2) is defined as

$$S_E(m,r) = \lim_{N\to\infty} -\ln \frac{A_m(r)}{B_m(r)} \tag{3}$$

and is generally estimated using

$$S_E(m,r) = -\ln \frac{A_m(r)}{B_m(r)} \tag{4}$$

in which $A_m(r)$ stands for the probability that two sequences match for $m+1$ points and $B_m(r)$ denotes the probability that two sequences will match for m points, with both the tolerance r and self-match sequences excluded. More details on $A_m(r)$, $B_m(r)$, and the sample entropy algorithm can be found in [7,8].

In general, r is set to 0.15 times the standard derivation of the series. However, it is worth noting that the decreasing of sample entropy in the MSE algorithm is determined by both the selection of tolerance r and the coarse-graining produced coherence of elements in data series, especially the former one. In many cases [26,27], for enhancing accuracy and eliminating potential errors introduced by selecting a certain tolerance r, the MSE algorithm is even refined by adopting an adaptive threshold r as a function of the scale factor in coarse-graining process. In this study, for simplification, and for focusing on the novel background we choose, we follow the original MSE algorithm [5–8] and set $m = 2$ and $r = 0.15 * std$.

By using the sMSE algorithm, the sample entropies of the white noise series and the $1/f$ noise series are calculated and shown in Figure 1. The independent elements in the white noise series tend correlate to each other, due to the averaging process in coarse-graining, in which the coherence between elements is consequently strengthened along with an increasing scale factor, and the sample entropies are therefore monotonically decreased with an increasing scale factor. On the other hand, the invariant sample entropies at all scale factors of the $1/f$ noise series are due to its special self-like property (fractal) [28], whose geometric shape won't be changed at all scale factors. The results in Figure 1 agree with those of the original MSE algorithm results in [5], both quantitatively and qualitatively. The correctness and accurateness of the sMSE algorithm are therefore guaranteed and validated in this study.

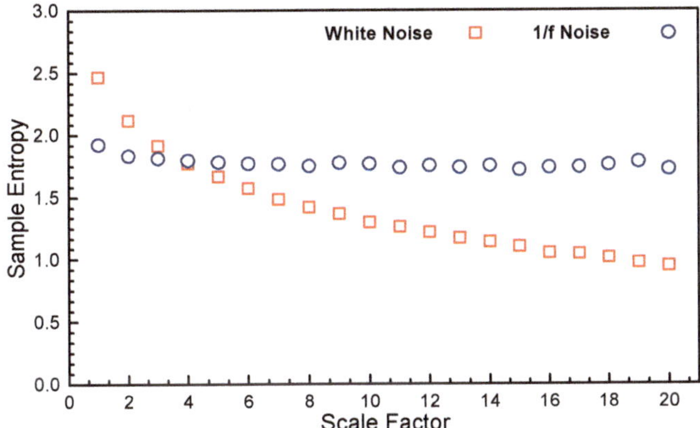

Figure 1. The calculated sample entropies of the white noise and $1/f$ noise series using the series multiscale entropy (sMSE) algorithm (series length: 1024, $m = 2$, $r = 0.15 *$ std).

3. Data Acquisition and Processing

Wikipedia content contains many topics, each divided into many subcategories, which may be further divided depending upon their intricacy. This division continues until reaching a given intricacy standard. For instance, as depicted in Figure 2, the topic of medicine is divided into many Level 1 subcategories, including clinical medicine, health insurance, and medical associations, etc. Each Level 1 category is further divided into subcategories (Level 2), until reaching given intricacy standards, and so forth. For simplification, further divisions are replaced by an ellipsis in Figure 2.

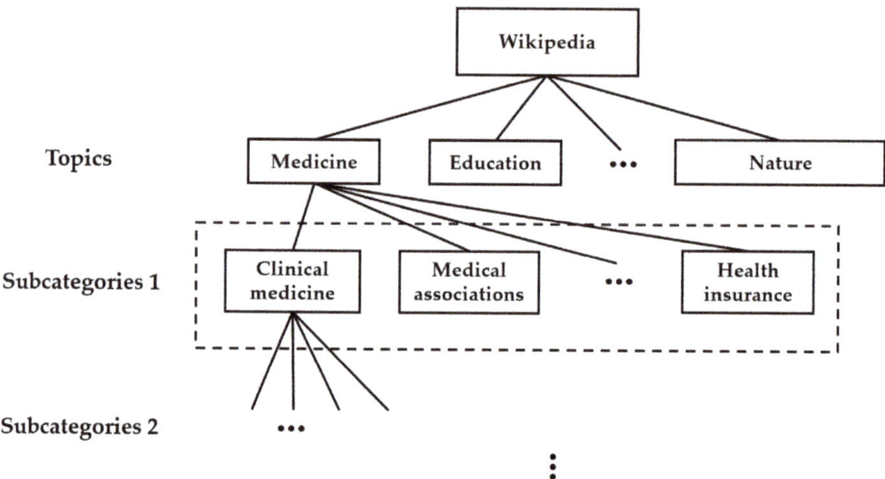

Figure 2. Data classification and the multilevel structure of Wikipedia.

In this paper, we are only concerned with the page views of the topics named in Figure 2. We find the sum of the page views of subcategories to determine the total page views of the corresponding topic. As shown in Figure 2, the page views of the medicine topic are the sum of the page views of its Level 1 subcategories, the page views of each subcategory in Level 1 are the sum of the page views of its Level 2 subcategories, and so forth. Four topics—education, economy/finance, medicine, and nature/environment—were selected in this study for their significant impact on human lives, and their wider significance to the governments of developing countries pursuing sustainable development, for example, China [29–33]. Page views of the four selected topics in three recent years (2016–2018) were downloaded from https://tools.wmflabs.org/massviews/. Table 1 presents the basic data of the four selected topics, including the number of subcategories in Level 1, the length, and the mean value of the page view series of each selected topic. Because this download was conducted on 25 December 2018, the length of page view series for 2018 is therefore 359.

To illustrate the temporal variations of the page view series, Figure 3 depicts the page views of the four selected topics as a function of days of the year. Figure 3 shows that, for each selected topic, the curves are close to each other, which make them indistinguishable at first glance, with only random outliers. This characteristic can be found quantitatively in Table 1, where for some topics, the mean values of page view series across the three years tended to be close. Despite these similar mean values, the three page view curves of each selected topic fluctuate rapidly, highlighting temporal fluctuations. The 1st or 2nd moment statistics and the temporal connections of page views may not be suitable for revealing connections. In addition, the nature/environment topic has the fewest page views, which reveals that environmental problems are rarely researched compared to the other three selected topics.

Table 1. Characteristics of the collected data.

Topics	Number of Subcategories	Year	Length	Mean Value
Medicine	28	2016	366	212,281
		2017	365	199,207
		2018	359	198,467
Education	27	2016	366	312,365
		2017	365	319,447
		2018	359	298,771
Economy/finance	32	2016	366	354,896
		2017	365	323,208
		2018	359	293,831
Nature/environment	16	2016	366	145,095
		2017	365	148,254
		2018	359	110,283

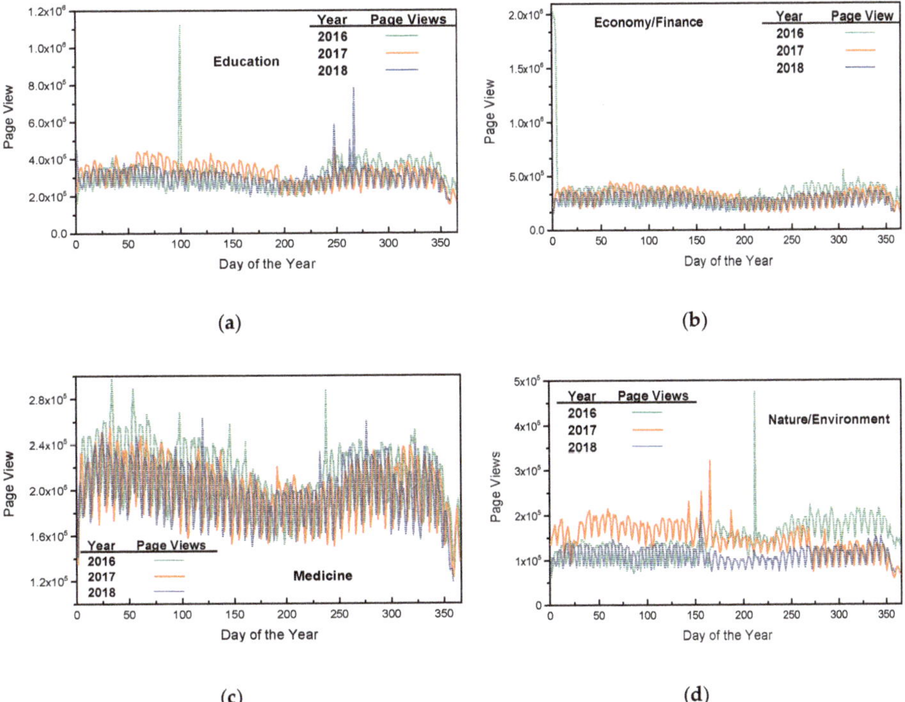

Figure 3. Page views of the four selected topics in 2016, 2017 and 2018: (**a**) Education; (**b**) Economy/ Finance; (**c**) Medicine; (**d**) Nature/Environment.

4. Results and Discussion

In biological applications of the MSE algorithm, sample entropy is physically defined as the adaptability of organisms to a certain circumstance or ecosystem [5], in which a larger sample entropy corresponds to a higher complexity and a stronger adaptability. Following this viewpoint, in this study, we define sample entropy as the complexity of human website searching activities, as well as its internal and temporal connections. That is, a larger sample entropy denotes higher complexity of human website searching activities, and elements in the page view series are less interconnected and have a weaker temporal correlation. Notice that 1st order sample entropy, or even sample entropies at small scale factors, may be invalid or insufficient to reveal hidden information conveyed by time series [3,4]. Therefore, in this section, we examine scale factors ranging from 1 to 10, as the sample entropy at scale factor 10 approaches zero. Hence, in this study, we emphasize sample entropies at large scale factors, which are also highlighted in the MSE algorithm.

For each topic, the sample entropy (with a 95% confidence interval) is first calculated using the sMSE algorithm, as depicted in Figure 4. Macroscopically speaking, in Figure 4, the sample entropies of the four selected topics exhibit the same tendencies, i.e., for each selected topic, the three years' sample entropies decrease as the scale factor increases. Compared with the sample entropies of the white noise series in Figure 1, the sample entropies of the selected topics in the three years are smaller at all scale factors and decrease more rapidly. Such a characterization fully reflects a lower complexity of the page view series of the selected topics, and the elements in the page view series are highly correlated when compared with a white noise series whose elements are independent of each other.

Specifically, for the topic of education, page views from the year 2016 appear to be the most complex, since they have the largest sample entropies at scale factors of 2 and above, whereas the page views of the year 2018 have the lowest complexity and the strongest temporal interconnections, although it has the largest 1st scale sample entropy. For the topic of economy/finance, page views from 2016 had the smallest sample entropies at small factors, but had the largest sample entropies at large scale factors (greater than 3). The page views from 2017 and 2018 had the same sample entropies at scale factors greater than 3, which somehow shows that these two years exhibited the same complexity and temporal correlations of human website searching activities. An interesting phenomenon was found for the topic of medicine: for all the years examined, the sample entropies tended to be the same at all scale factors, which reflects that in three recent years, the page view series of medicine topics have the same complexity and temporal correlations. It is not clear whether this is because the topic of medicine widely concerns all people, which would make the variations in recent years highly regularized. Lastly, for the topic of nature/environment, the sample entropies of 2016 and 2017 fluctuate as the scale factor increases, and the page views of the year 2017 had the largest sample entropies at scale factors greater than 5, whereas the page views of 2016 and 2018 had the same sample entropies at scale factors of 3 and above.

Although some selected topics, such as nature/the environment, tended to have irregular sample entropies, this could not be easily concluded when compared with other topics. However, the results in Figure 4 show that, for irregularly varied page view series, the sMSE algorithm affords an alternative analysis method and reveals the complexities and temporal correlations of page views in different years.

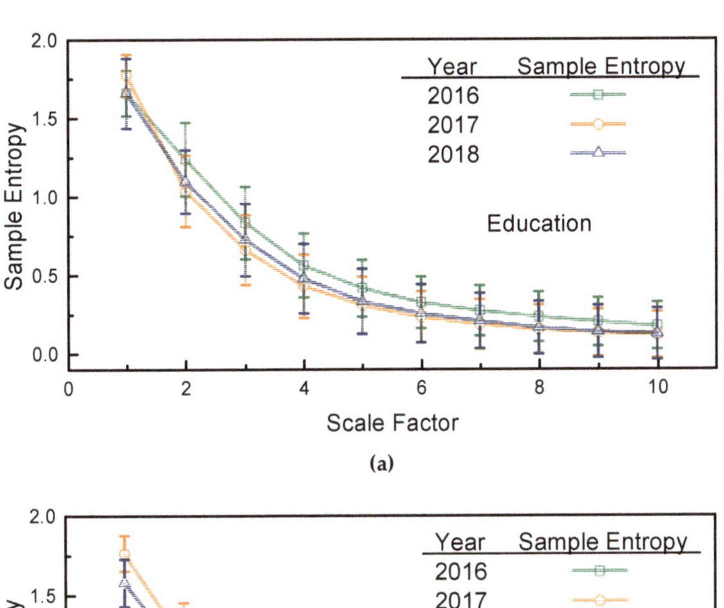

(a)

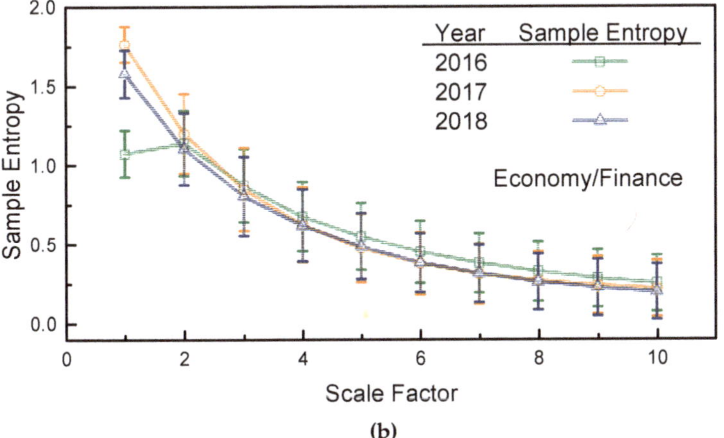

(b)

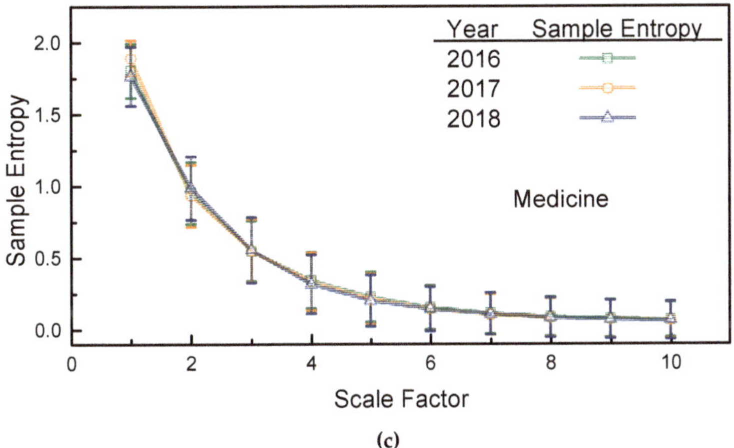

(c)

Figure 4. *Cont.*

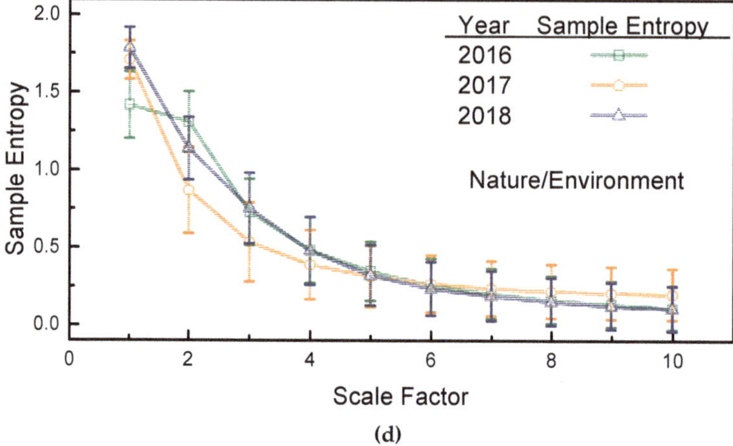

(d)

Figure 4. Sample entropies of page views of the four selected topics: (**a**) Education; (**b**) Economy/ Finance; (**c**) Medicine; (**d**) Nature/Environment.

We have focused on three years' variations in sample entropies for selected topics. A horizontal comparison of the sample entropies with a 95% confidence interval of the four selected topics was also conducted, and the results of 2016, 2017, and 2018 are presented in Figure 5, in which it can readily be seen that the sample entropies of the four selected topics exhibit the same tendencies across years. That is, at large scale factors, the sample entropies of the four selected topics are quantitatively ranked as follows: economy/finance has the largest sample entropies, followed by education, nature/environment, and medicine. Notably, in 2016, the gaps between these quantitatively ranked curves are more obvious. These results suggest that human website searching activities on economy/finance topics are the most complicated and that elements in this page view series are less temporally correlated when compared to the other three topics.

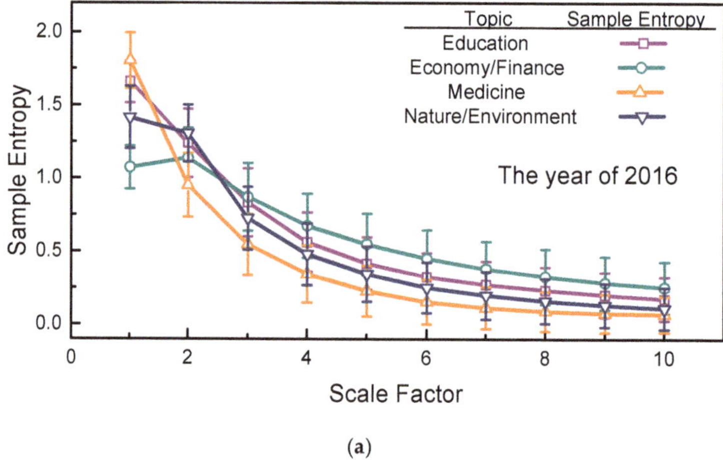

(a)

Figure 5. *Cont.*

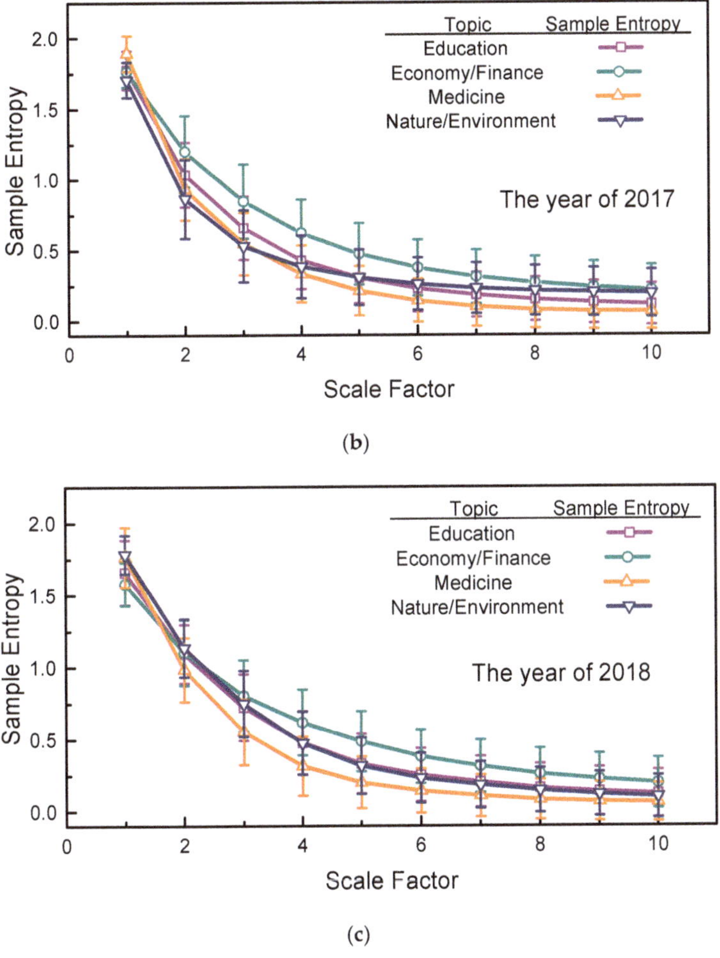

Figure 5. Comparison of sample entropies between the four selected topics: (**a**) 2016; (**b**) 2017; and (**c**) 2018.

Comparing the rate of descent of the curves in Figure 5, the sample entropies for the topic of medicine show the most rapid decrease as the scale factor increases, which shows that the elements of the page view series for medicine are the most strongly temporally correlated among the four selected topics. In Figure 3, note that no obvious outlier occurred in the page view series for medicine either, and the smoothness of these curves may accelerate their descent. Hence, this topic's smallest sample entropies at larger scale factors and its most rapid descent suggests that the human website searching activities related to medicine are more strongly temporally correlated and more regular than those of the other three selected topics. Again, as in previous discussions of Figure 4, the sMSE algorithm highlights the results shown in Figure 5 at large scale factors, in which chaotic page view series are ordered and regularized.

It should be noted that, above, the estimated sample entropies (mean value) using the sMSE algorithm are compared. Once the 95% confidence interval is involved, the comparative results should be furtherly validated using statistical analysis. In what follows, we conduct an assessment of significance of difference for the comparative results of Figures 4 and 5. For comparative results in

Figure 4, for each selected topic, the Wilcoxon signed-rank test is selected to access the significance of difference for the sample entropies with a 95% confidence interval in different years. For Figure 5, for a given year, the Mann-Whitney U-test is applied to compare different selected topics. Without loss of generality, a $p < 0.05$ is considered to be significant. Both the Wilcoxon signed-rank test and the Mann-Whitney U-test is realized by using Matlab R2018a.

For the selected topics, the p values of the Wilcoxon signed-rank test for Figure 4 with different year pairs are given in Table 2. It readily can be seen in Table 2 that, for education, the p-values of all year pairs are smaller than 0.05. The variations of sample entropy in Figure 4a are therefore considered to be significantly different, whereas for the other three topics, the p values are all larger than 0.05, except the sample entropies for economy/finance topics in the years of 2017 and 2018. Looking back to the sample entropies in Figure 4, although the variations of sample entropy of the economy/finance and the nature/environment topics are complicated, the statistical analysis results suggest the sample entropies of these two topics should be regarded as the same in the past three years. Specially, for medicine topics, the large p values in Table 2 show agreement with the invariant sample entropies in Figure 4c in the past three years. Therefore, by combining the results in Figure 4 and Table 2, the variations of sample entropy are acceptable and feasible for analyzing the complexity of the education and medicine topics, whereas for topics of economy/finance and natural/environment, the sample entropies in Figure 4 should be treated as undistinguished. Similarly to Table 2, the p value of the Mann-Whitney U-test for sample entropies in Figure 5 are presented in Tables 3–5, respectively. The p values in Tables 3–5 are all found to be larger than 0.05, which suggests that the sample entropies of different selected topics in Figure 5 are the same with regard to statistics. The complexity of different selected topics is therefore regarded to be undistinguished in Figure 5.

Table 2. p values of the Wilcoxon signed-rank test for results in Figure 4.

Year Pairs / ***p-Value*** / **Topic**	**(2016,2017)**	**(2016,2018)**	**(2017,2018)**
Education	0.0273	0.0488	0.0137
Economy/finance	0.4922	0.0840	0.0039
Medicine	0.4922	0.6250	1.0000
Nature/environment	0.8457	0.4316	0.9219

Table 3. p values of the Mann-Whitney U-test for results in Figure 5a.

Topic Paris	**(Edu, E/F)**	**(Edu, Med)**	**(Edu, N/E)**	**(E/F, Med)**	**(E/F, N/E)**	**(Med, N/E)**
p-value	0.4727	0.1859	0.5205	0.0757	0.2730	0.3847

Table 4. p values of the Mann-Whitney U-test for results in Figure 5b.

Topic Paris	**(Edu, E/F)**	**(Edu, Med)**	**(Edu, N/E)**	**(E/F, Med)**	**(E/F, N/E)**	**(Med, N/E)**
p-value	0.2123	0.3447	0.5708	0.1041	0.2703	0.2413

Table 5. p values of the Mann-Whitney U-test for results in Figure 5c.

Topic Paris	**(Edu, E/F)**	**(Edu, Med)**	**(Edu, N/E)**	**(E/F, Med)**	**(E/F, N/E)**	**(Med, N/E)**
p-value	0.4274	0.2413	0.7913	0.1212	0.2447	0.3847

By taking the results of sample entropies and statistical analysis, the sample entropies, which are estimated by the sMSE algorithm, are found to be feasible for analyzing the temporal variations of complexity of page views of certain topics, for example, education and medicine. However, it can not be simply applied when analyzing the temporal variations of the complexity of page views over different topics.

Potential explanations are, firstly, that the complexities of the selected topics in past three years are undistinguished as is. Secondly, it should be noted that we sum up all the page view data in subcategories Level 2 as the total page views of the topic in subcategories Level 1, since the page view data in subcategories Level 2 may be independent of each other. Based on the law of large numbers, once summation is conducted, the total distribution of the four selected topics in subcategories Level 1 can be regarded as normal. Lastly, as we mentioned in Section 2, the selection of tolerance r affects the sample entropy estimates clearly, and probably leads to the undistinguished sample entropy estimates over the four selected topics. We therefore suggest these paths for forthcoming studies:

(1) On data processing, the way of collecting page view data should be considered carefully, subcategories in low levels in Figure 2 should probably be investigated separately;
(2) On methodology, with regard to the MSE algorithm, the difference between two ways of selecting the threshold value r should be investigated for more accurate and robust results;
(3) On background, for topics, which can be feasibly analyzed by sample entropy, the variations and explanations of complexity may be related to certain social issues, if possible.

5. Conclusions

This paper attempts to examine the complexities and temporal correlations of page views of four selected topics on Wikipedia using an sMSE algorithm. Sample entropies of the four selected topics are compared to reveal their temporal variations, showing vivid variations between different topics in three recent years. Meanwhile, the complexities of the page views of the selected topics are investigated and regular variations in the sample entropies of different topics are also found. Statistical analysis is then conducted to validate the variations, and the results suggest the sample entropy estimated by the sMSE is feasible in analyzing the temporal variations of the complexity of page view data for some topics. However, the regular variations of sample entropy can't be simply accepted as is when different topics are compared. Potential explanations are given and discussed, and paths for forthcoming studies are also suggested.

Author Contributions: Conceptualization, C.X. (Chao Xu) and R.J.; Data curation, C.X. (Chen Xu), W.T. and A.H.; Formal analysis, C.X. (Chao Xu), C.X. (Chen Xu) and W.T.; Investigation, A.H.; Methodology, C.X. (Chao Xu) and R.J.; Software, C.X. (Chao Xu), C.X. (Chen Xu), W.T. and R.J.; Supervision, R.J.; Validation, R.J.; Writing—original draft, C.X. (Chao Xu), C.X. (Chen Xu) and W.T.; Writing—review & editing, R.J.

Funding: This research received no external funding.

Acknowledgments: The authors would sincerely like to thank Xiangpeng Li at Wuhan Textile University for affording support and suggestions on this work. We would also like to thank the anonymous reviewers for improving the quality of this paper. Additionally, we are grateful to Wikipedia for sharing data with its users.

Conflicts of Interest: The authors declare no conflict of interests.

References

1. The Four V's of Big Data. Available online: https://www.ibmbigdatahub.com/infographic/four-vs-big-data (accessed on 8 January 2019).
2. Walker, S.J. Big Data: A revolution that will transform how we live, work, and think. *Int. J. Advert.* **2014**, *33*, 181–183. [CrossRef]
3. Goldberger, A.L.; Peng, C.K.; Lipsitz, L.A. What is physiologic complexity and how does it change with aging and disease? *Neurobiol. Aging* **2002**, *23*, 23–26. [CrossRef]
4. Variability vs. Complexity. Available online: https://physionet.org/tutorials/cv/ (accessed on 8 January 2019).
5. Costa, M.; Goldberger, A.L.; Peng, C.K. Multiscale entropy analysis of complex physiologic time series. *Phys. Rev. Lett.* **2002**, *89*, 068102. [CrossRef] [PubMed]
6. Costa, M.; Goldberger, A.L.; Peng, C.K. Multiscale entropy analysis of biological signals. *Phys. Rev. E* **2005**, *71*, 021906. [CrossRef] [PubMed]
7. Richman, J.S.; Moorman, J.R. Physiological time-series analysis using approximate entropy and sample entropy. *Am. J. Physiol. Heart Circ. Physiol.* **2000**, *278*, H2039–H2049. [CrossRef] [PubMed]

8. Hansen, C.; Wei, Q.; Shieh, J.S.; Fourcade, P.; Isableu, B.; Majed, L. Sample entropy, univariate, and multivariate multi-scale entropy in comparison with classical postural sway parameters in Young healthy adults. *Front. Hum. Neurosci.* **2017**, *11*. [CrossRef] [PubMed]
9. Zhang, N.; Lin, A.; Shang, P. Symbolic phase transfer entropy method and its application. *Commun. Nonlinear Sci. Numer. Simul.* **2017**, *51*, 78–88. [CrossRef]
10. Zhang, N.; Lin, A.; Ma, H.; Shang, P.; Yang, P. Weighted multivariate composite multiscale sample entropy analysis for the complexity of nonlinear times series. *Phys. A Stat. Mech. Appl.* **2018**, *508*, 595–607. [CrossRef]
11. Zhang, N.; Lin, A.; Shang, P. Multiscale symbolic phase transfer entropy in financial time series classification. *Fluct. Noise Lett.* **2017**, *16*, 1750019. [CrossRef]
12. Mestyán, M.; Yasseri, T.; Kertész, J. Early prediction of movie box office success based on Wikipedia activity big data. *PLoS ONE* **2013**, *8*, e71226. [CrossRef] [PubMed]
13. Ciglan, M.; Norvag, K. WikiPop: Personalized event detection system based on Wikipedia page view statistics. In Proceedings of the 19th ACM International Conference on Information and Knowledge Management (CIKM'10), Toronto, ON, Canada, 26–30 October 2010; pp. 1931–1932.
14. Kämpf, M.; Tessenow, E.; Kenett, D.Y.; Kantelhardt, J.W. The detection of emerging trends using wikipedia traffic data and context networks. *PLoS ONE* **2015**, *10*, e0141892. [CrossRef] [PubMed]
15. Elshendy, M.; Colladon, A.F.; Battistoni, E.; Gloor, P.A. Using four different online media sources to forecast crude oil price. *J. Inf. Sci.* **2017**, *44*. [CrossRef]
16. Moat, H.S.; Curme, C.; Avakian, A.; Kenett, D.Y.; Stanley, H.E.; Preis, T. Quantifying Wikipedia usage patterns before stock market moves. *Sci. Rep.* **2013**, *3*. [CrossRef]
17. Wen, J.R.; Nie, J.Y.; Zhang, H.J. Clustering user queries of a search engine. In Proceedings of the 10th International Conference on World Wide Web, Hong Kong, China, 1–5 May 2001; ACM: New York, NY, USA, 2001; pp. 162–168.
18. Strehl, A.; Ghosh, J.; Mooney, R.J. Impact of similarity measures on web-page clustering. In Proceedings of the AAAI: Workshop on Artificial Intelligence for Web Search (AAAI, 2000), Austin, TX, USA, 30–31 July 2000; pp. 58–64.
19. Dong, X.; Halevy, A.; Madhavan, J.; Nemes, E.; Zhang, J. Similarity search for web services. In Proceedings of the Thirtieth international conference on Very large data bases, Toronto, ON, Canada, 31 August–3 September 2004; pp. 372–383.
20. Lukashevich, H.; Nowak, S.; Dunker, P. Using one-class SVM outliers detection for verification of collaboratively tagged image training sets. In Proceedings of the IEEE International Conference on Multimedia and Expo, New York, NY, USA, 28 June–3 July 2009; pp. 682–685.
21. Joachims, T. Optimizing search engines using clickthrough data. In Proceedings of the eighth ACM SIGKDD International Conference on Knowledge Discovery and Data Mining, Edmonton, AB, Canada, 23–26 July 2002; ACM: New York, NY, USA, 2002; pp. 133–142.
22. Steinmetz, N.; Lausen, H.; Brunner, M. Web Service Search on Large Scale. In Proceedings of the 7th International Joint Conference, ICSOC-ServiceWave 2009, Stockholm, Sweden, 24–27 November 2009; pp. 437–444.
23. Internet Water Army. Available online: https://en.wikipedia.org/wiki/Internet_Water_Army (accessed on 8 January 2019).
24. Humeau-Heurtier, A. The multiscale entropy algorithm and its variants: A review. *Entropy* **2015**, *17*, 3110–3123. [CrossRef]
25. Chang, Y.C.; Wu, H.T.; Chen, H.R.; Liu, A.B.; Yeh, J.J.; Lo, M.T.; Tsao, J.H.; Tang, C.J.; Tsai, I.T.; Sun, C.K. Application of a modified entropy computational method in assessing the complexity of pulse wave velocity signals in healthy and diabetic subjects. *Entropy* **2014**, *16*, 4032–4043. [CrossRef]
26. Valencia, J.F.; Porta, A.; Vallverdu, M.; Claria, F.; Baranowski, R.; Orlowska-Baranowska, E.; Caminal, P. Refined multiscale entropy: Application to 24-h holter recordings of heart period variability in healthy and aortic stenosis subjects. *IEEE Trans. Biomed. Eng.* **2009**, *56*, 2202–2213. [CrossRef] [PubMed]
27. Faes, L.; Porta, A.; Javorka, M.; Nollo, G. Efficient computation of multiscale entropy over short biomedical time series based on linear state-space models. *Complexity* **2017**, 1–13. [CrossRef]
28. Lowen, S.B.; Teich, M.C. Fractal renewal processes generate 1/f noise. *Phys. Rev. E* **1993**, *47*, 992. [CrossRef]
29. Georgiou, S.; Whittington, D.; Pearce, D. *Economic Values and the Environment in the Developing World*; Edward Elgar Publishing Ltd.: Cheltenham, UK, 1997; pp. 264–267, ISBN 1858985005.

30. Pearce, D.W.; Warford, J.J. *World without end: Economics, Environment and Sustainable Development*; Oxford University Press: New York, NY, USA, 1993; pp. 322–326, ISBN 0195208811.
31. Zhang, X.; Kanbur, R. Spatial inequality in education and health care in China. In *Regional Inequality in China*, 1st ed.; Routledge: London, UK, 2009; p. 19, ISBN 9781135972257.
32. Niu, D.; Jiang, D.; Li, F. Higher education for sustainable development in China. *Int. J. Sustain. High. Educ.* **2010**, *11*, 153–162. [CrossRef]
33. Liu, J.; Raven, P.H. China's environmental challenges and implications for the world. *Crit. Rev. Environ. Sci. Technol.* **2010**, *40*, 823–851. [CrossRef]

Article

Performance Evaluation of an Entropy-Based Structural Health Monitoring System Utilizing Composite Multiscale Cross-Sample Entropy

Tzu-Kang Lin * and Yi-Hsiu Chien

Department of Civil Engineering, National Chiao Tung University, Hsinchu 30010, Taiwan; s9n60208.cv05g@nctu.edu.tw

* Correspondence: tklin@nctu.edu.tw; Tel.: +886-03-571-2121-54919

Received: 14 November 2018; Accepted: 7 January 2019; Published: 9 January 2019

Abstract: The aim of this study was to develop an entropy-based structural health monitoring system for solving the problem of unstable entropy values observed when multiscale cross-sample entropy (MSCE) is employed to assess damage in real structures. Composite MSCE was utilized to enhance the reliability of entropy values on every scale. Additionally, the first mode of a structure was extracted using ensemble empirical mode decomposition to conduct entropy analysis and evaluate the accuracy of damage assessment. A seven-story model was created to validate the efficiency of the proposed method and the damage index. Subsequently, an experiment was conducted on a seven-story steel benchmark structure including 15 damaged cases to compare the numerical and experimental models. A confusion matrix was applied to classify the results and evaluate the performance over three indices: accuracy, precision, and recall. The results revealed the feasibility of the modified structural health monitoring system and demonstrated its potential in the field of long-term monitoring.

Keywords: structural health monitoring; multi-scale; composite cross-sample entropy

1. Introduction

Over the preceding few decades, structural health monitoring (SHM) techniques have been developed for early damage detection in various engineering fields. Traditional SHM techniques such as visual inspection tend to be time consuming and rely heavily on manpower; therefore, these techniques have low generalizability. Recently, novel SHM methods based on signal processing techniques have been proposed for analyzing measured responses in succession. Dynamic monitoring entails measuring the displacement, velocity or acceleration signals of structures to obtain time–frequency characteristics.

In 1999, Wahab and De Roeck [1] utilized the change in modal curvature between the undamaged and damaged conditions of simply supported and continuous beams to detect damage in prestressed concrete beams. In 2000, Maeck [2] identified the location and degree of damage in reinforced concrete (RC) beams by conducting dynamic stiffness analysis. In 2003, Chang [3] summarized the limitations and applications of vibration-based SHM methods. The impact of measuring noise, environmental and damage on the sensitivity of the damage detection was then analyzed [4]. In 2016, Amezquita-Sanchez et al. summarized the current signal processing techniques for vibration-based SHM and point out its advantages and disadvantages [5]. Opoka et al. then applied the root mean square deviation (RMSD) estimator to detect the damage by comparing the averaged frequency spectrum [6]. In 2018, Soman et al. proposed a two-step methodology based on frequency spectrum for the damage detection and localization [7]. This method used the change of the frequency response function (FRF) spectrum to distinguish the healthy and the damaged tripod structure. However, complex structural behavior is difficult to capture because of incomplete measurements and the randomness of ambient vibration [8,9].

Information theory can be applied to study the quantification, storage, and communication of information. In 1948, Shannon [10] proposed Shannon entropy as a measure of uncertainty in the outcomes of random processes. Subsequently, Kolmogorov [11] defined the notion of entropy for a new class of dynamical systems, and Sinai [12] introduced a definition of entropy that can be applied to all dynamical systems. The Kolmogorov–Sinai entropy (KS entropy) has been used to measure the complexity of time series in D-dimensional dynamic systems and is crucial in ergodic theory. Subsequent studies have demonstrated that KS entropy results are affected by various levels of noise during analysis of experimental data [13].

In 1991, Pincus [14] modified KS entropy and named the modified version approximate entropy (ApEn). ApEn was initially utilized to analyze medical data such as heart rate and served as a regularity statistic for quantifying the unpredictability of fluctuations over a time series. An and Ou [15] proposed the mean curvature difference method based on ApEn theory and successfully used the proposed method to locate damage in shear frame structures. In 2000, Richman and Moorman [16] further modified ApEn to develop sample entropy (SampEn). SampEn has two advantages over ApEn: data length independence and higher relative consistency under different parameters. Lake et al. [17] used SampEn to investigate neonatal heart rate variability and validate the characteristics of SampEn.

In 2002, Costa et al. [18] proposed multiscale entropy (MSE) to solve the problem of distinguishing between healthy individuals, patients with congestive heart failure, and patients with erratic cardiac arrhythmia on a single time scale. Consequently, the coarse-graining procedure was proposed to obtain more reliable results during entropy calculation [19]; the results demonstrated that the time-series structure exhibited complexity loss under pathologic conditions on a multiple time scale. MSE has been used extensively in not only medicine but also mechanical engineering and finance. MSE and an adaptive neuro-fuzzy inference system were utilized to distinguish between fault damage categories and identify levels of fault severity [20]. Xia and Shang [21] applied MSE to investigate degrees of self-match and measure complexity in the American, European, and Asian stock markets.

Cross-ApEn was introduced by Pincus and Singer [22] in 1996 for analyzing the degree of asynchrony between two related time series. Furthermore, Richman and Moorman [16] proposed Cross-SampEn for measuring the asynchrony and dissimilarity between two distinct time series. The results indicated that Cross-SampEn was a more consistent measure of joint synchrony between pairs of clinical cardiovascular time series. In 2013, Fabris et al. [23] extended applications of SampEn and Cross-SampEn to analyze electroglottogram and microphone signals. Healthy patients and patients with throat or voice disorders could be identified by evaluating the degree of asynchrony between two time series.

Although the MSE algorithm has been successfully applied in multiple fields, a problem has emerged in practical applications: the statistical reliability of SampEn decreases as the corresponding time scale increases. Therefore, in 2013, Wu et al. [24] introduced composite MSE (CMSE) to overcome the problem. To validate the applicability of CMSE to real data, the acceleration signals of bearing faults were analyzed to demonstrate that CMSE provided more accurate entropy values than did MSE. In 2016, Yin et al. [25] proposed composite multiscale cross-sample entropy (CMSCE) to address accuracy concerns regarding the MSCE method. Subsequently, CMSCE was applied to analyze the asynchrony between financial time series, which included six stock indices from multiple regions.

The aim of this study was to develop an entropy-based SHM system for solving the problem of unstable entropy values observed when multiscale cross-sample entropy (MSCE) is employed to assess damage in laboratory-scale structure. The flowchart of the study procedures is shown in Figure 1. The proposed system was validated by simulating the ambient vibration response of a seven-story model and conducting a steel structure experiment. The health condition was first detected by the ambient vibration response through CMSCE. Moreover, as most of the practical damages were observed on the lower part of a structure under earthquake excitation, monitoring the change of the first fundamental mode frequency is an alternative way for rapid screening. Therefore, ensemble empirical mode decomposition (EEMD) was attempted to extract the first mode of structural response to evaluate the feasibility of detecting damage according to only the first mode [26]. A previously proposed damage index (DI) was used to efficiently quantify damage, and the decision of damage location and condition was made.

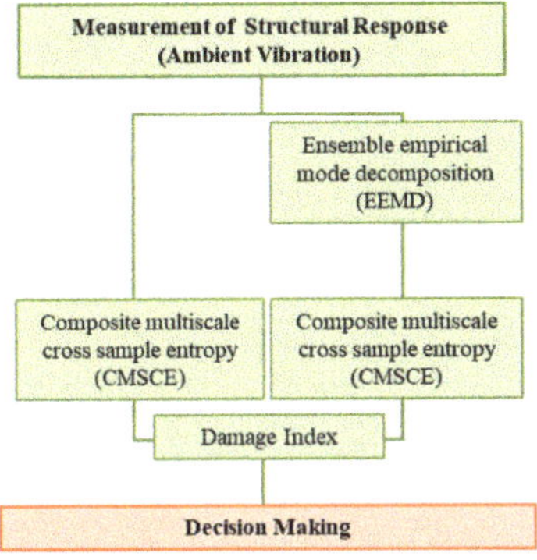

Figure 1. Flowchart of the study.

2. Methodology

2.1. Cross-SampEn Method

Cross-SampEn is used to evaluate the degree of dissimilarity between two time series from the same system. The procedure of Cross-SampEn is similar to that of SampEn and can be summarized as follows [16]. First, consider two individual time series $\{X_i\} = \{x_1, \ldots, x_i, \ldots, x_N\}$ and $\{Y_j\} = \{y_1, \ldots, y_j, \ldots, y_N\}$ with length N. Two signals are divided into templates of length m: $u_m(i) = \{x_i, x_{i+1}, \ldots, x_{i+m-1}\}$, $1 \leq i \leq N-m+1$ and $v_m(j) = \{y_j, y_{j+1}, \ldots, y_{j+m-1}\}$, $1 \leq j \leq N-m+1$. Therefore, two template spaces T_x and T_y are presented as follows:

$$T_x = \begin{bmatrix} x_1 & x_2 & \cdots & x_m \\ x_2 & x_3 & \cdots & x_{m-1} \\ \vdots & \vdots & \ddots & \vdots \\ x_{N-m+1} & x_{N-m+2} & \cdots & x_N \end{bmatrix} T_y = \begin{bmatrix} y_1 & y_2 & \cdots & y_m \\ y_2 & y_3 & \cdots & y_{m-1} \\ \vdots & \vdots & \ddots & \vdots \\ y_{N-m+1} & y_{N-m+2} & \cdots & y_N \end{bmatrix} \quad (1)$$

The number of similarities between $u_m(i)$ and $v_m(j)$ is defined as $n_i^m(r)$, which can be expressed as follows:

$$n_i^m(r) = \sum_{j=1}^{N-m} d[u_m(i), v_m(j)] \quad (2)$$

where $d[u_m(i), v_m(j)]$ is the maximum distance between two templates i and j. Under the condition that the maximum distance is within r, which is a predetermined threshold, $n_i^m(r)$ can be calculated.

$$d[u_m(i), v_m(j)] = \max\{|x(i+k) - y(j+k)|: 0 \leq k \leq m - 1\} \quad (3)$$

$$d[u_m(i), v_m(j)] \leq r, 1 \leq j \leq N - m \quad (4)$$

The similarity probability of a template is calculated using the following equation:

$$U_i^m(r)(v\|u) = \frac{n^m(r)}{(N - m)} \quad (5)$$

Subsequently, the average probability of similarity for template m is calculated as follows:

$$U^m(r)(v\|u) = \frac{1}{(N - m)} \sum_{i=1}^{N-m} U_i^m(r)(v\|u) \quad (6)$$

where $U^m(r)(v\|u)$ is the degree of asynchrony between the two template spaces, which are segmented with length m. Next, new template spaces T_x and T_y are created with different template lengths $m + 1$, and then the procedure for calculating the probability of similarity is repeated to obtain the average probability of similarity $U^{m+1}(r)(v\|u)$. Cross-SampEn is expressed as follows:

$$CS_E(m, r, N) = -\ln\left\{\frac{U^{m+1}(r)(v\|u)}{U^m(r)(v\|u)}\right\} \quad (7)$$

2.2. MSE Method

Compared with sample entropy values on a single time scale, MSE is capable of extracting more information from a time series [12,13]. New time series are constructed by performing coarse graining. The original time series is divided into nonoverlapping windows with time scale τ from 1 to N. A new time series $\left\{y_j^{(\tau)}\right\}$ is constructed by taking the arithmetic mean of each window. The equation is as follows:

$$y_j^{(\tau)} = \frac{1}{\tau}\sum_{i=(j-1)\tau+1}^{j\tau} x_i,\ 1 \le j \le N/\tau \quad (8)$$

The length of each created coarse-grained time series is N/τ. After the coarse-graining procedure, SampEn is conducted for each coarse-grained time series $\left\{y_j^{(\tau)}\right\}$. Then, the obtained entropy values are considered the multiscale sample entropy of the time series and can be plotted as a function of the time scale factor $(f(\tau) = S_E)$. The MSE error grows as the length of the time series decreases, particularly when the response of a real structure is being measured.

2.3. CMSE Method

CMSE was proposed by Wu et al. [24] for improving the accuracy of MSE. CMSE defines the kth coarse-grained time series for a scale factor of τ. Therefore, each coarse-grained time series can be obtained using the following equation:

$$y_{k,j}^{(\tau)} = \frac{1}{\tau}\sum_{i=(j-1)\tau+k}^{j\tau+k-1} x_i,\ 1 \le j \le N/\tau,\ 1 \le k \le \tau \quad (9)$$

The CMSE algorithm produces τ coarse-grained time series at scale factor τ. Subsequently, SampEn values for all coarse-grained time series are calculated, and the CMSE value is defined as the mean of τ SampEn values, as shown in Equation (10):

$$\mathrm{CMSE}(x, \tau, m, r) = \frac{1}{\tau}\sum_{k=1}^{\tau} SampEn\left(y_k^{(\tau)}, m, r\right) \quad (10)$$

Based on the concept of CMSE, CMSCE proposed by Yin et al. [25] is defined as follows:

$$\mathrm{CMSCE}(x, y, \tau, m, r) = \frac{1}{\tau}\sum_{k=1}^{\tau} SampEn\left(x_k^{(\tau)}, y_k^{(\tau)}, m, r\right) \quad (11)$$

2.4. DI Measure

Inspired by a series of biomedical studies that have used the area of the MSE curve as an index for quantifying complexity [27–29], the DI was proposed as a measure for rapidly and efficiently diagnosing the floor damage of a structure. Signals are considered more complex when their entropy values are higher than those of other signals. For a structure with F floors, the CMSCE curves under undamaged and damaged conditions can be respectively expressed as follows:

$$\mathrm{CMSCE}_{\mathrm{undamaged}} = \left\{\begin{matrix} H_1 \\ H_2 \\ \vdots \\ H_F \end{matrix}\right\} \mathrm{CMSCE}_{\mathrm{damaged}} = \left\{\begin{matrix} D_1 \\ D_2 \\ \vdots \\ D_F \end{matrix}\right\} \quad (12)$$

where H and D represent the undamaged and damaged conditions, respectively. The subscript F indicates the analyzed floor; for example, H_1 is the Cross-SampEn of signals between the foundation and the first floor under the undamaged condition. Moreover, H_1 can be expressed as matrices: $H_1 = \left\{CS_E^{H_{11}}, CS_E^{H_{12}}, CS_E^{H_{13}}, \cdots, CS_E^{H_{1\tau}}\right\}$, where $CS_E^{H_{1\tau}}$ represents the Cross-SampEn of the first floor at scale factor τ. Hence, the general expression of Cross-SampEn on floor F is as follows:

$$D_F = \left\{CS_E^{D_{F1}}, CS_E^{D_{F2}}, CS_E^{D_{F3}}, \cdots, CS_E^{D_{F\tau}}\right\} \tag{13}$$

The DI is evaluated by calculating the differences between the areas of CMSCE curves from scale 1 to τ. The DI of floor F can be expressed as follows:

$$DI_F = \sum_{q=1}^{\tau}\left(CS_E^{D_{Fq}} - CS_E^{H_{Fq}}\right) \tag{14}$$

A positive DI indicates that the floor is damaged, whereas a negative DI indicates that the floor is undamaged. Based on results from previous studies, the DI performs to a satisfactory degree in numerical models that contain 10–20% noise. Therefore, in this study, the DI is utilized to identify the damaged floor in the numerical model and experiment.

3. Feasibility Assessment

3.1. Comparison of MSCE and CMSCE

According to the experience of applying MSCE for SHM, the optimized template length m was set as 4 based on the characteristics of recorded time series [30]. However, inevitable fluctuation was occasionally observed on the calculated entropy curve, which makes the diagnosis of damage condition and location difficult. To verify the ability of CMSCE to provide accurate entropy values subject to a long template length, experimental data obtained from a shaking table test were first analyzed to compare the performance of MSCE and CMSCE in this section. The shaking table test with alternating spells of white noise and TCU052 earthquake was conducted on a two-bay three-story reinforced concrete (RC) structure, which is shown in Figure 2. The procedure is shown in Figure 3. The acceleration of each floor was measured for SHM. The entropy-based monitoring system was based on ambient vibration, which was simulated by white noise during the shaking table test. The acceleration response of the first white noise iteration was used to represent the healthy status of the RC structure; the acceleration responses of the second and third white noise iterations were considered as unknown conditions after 800 and 1000 gal earthquakes.

Figure 2. Three-story RC structure.

Figure 3. Shaking table test procedure.

The sampling rate was 200 Hz for 120 s as the shaking table was running the white noise signal. The recorded time series were analyzed using MSCE and CMSCE with identical parameters: template length $m = 4$, threshold $r = 0.1 \times$ standard deviation (SD), and $0.08 \times$ SD. Figure 4 shows the results of MSCE and CMSCE under various conditions to illustrate the necessity of using CMSCE for on-site monitoring.

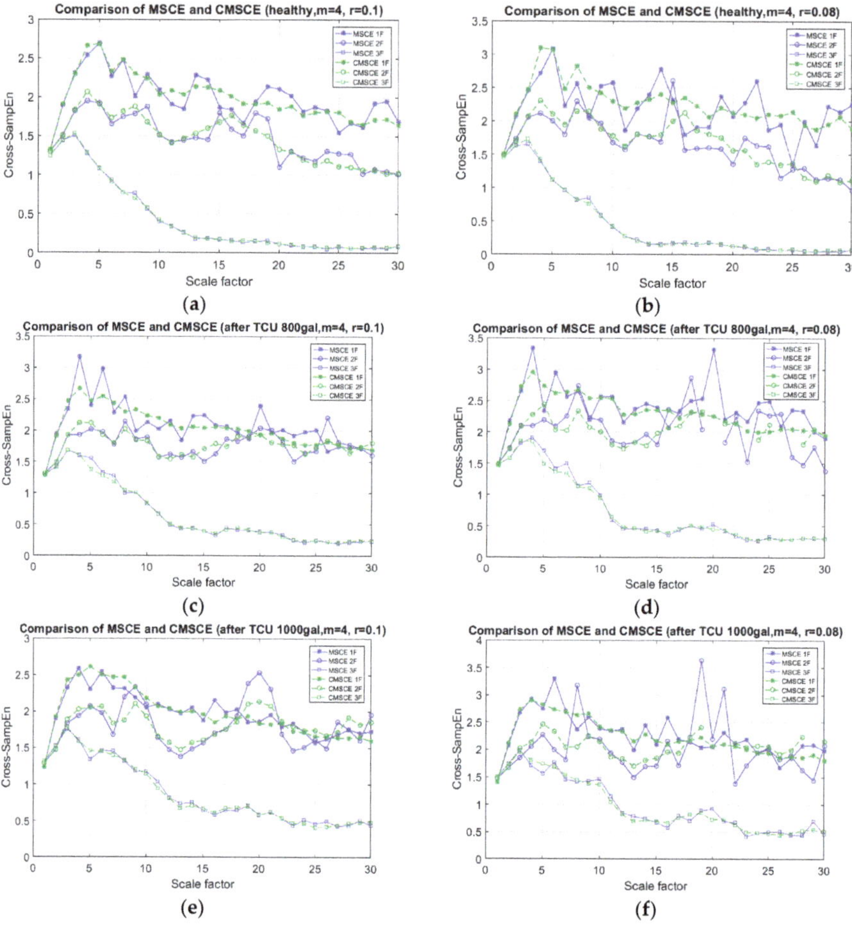

Figure 4. Comparison diagram of MSCE and CMSCE for: (**a**) healthy condition with $m = 4$, $r = 0.1$; (**b**) healthy condition with $m = 4$, $r = 0.08$; (**c**) after TCU 800 gal with $m = 4$, $r = 0.1$; (**d**) after TCU 800 gal $m = 4$, $r = 0.08$; (**e**) after TCU 1000 gal with $m = 4$, $r = 0.1$; (**f**) after TCU 1000 gal with $m = 4$, $r = 0.08$.

Diagrams obtained for the healthy condition are presented in Figure 4a,b; the diagram obtained when $m = 4$ and $r = 0.1$ is on the left, and that obtained when $m = 4$ and $r = 0.08$ is on the right. The figure shows that the results of MSCE were fluctuated markedly compared with those of CMSCE, especially on the low floor curves. Similarly, an unstable trend of MSCE was observed (Figure 4c–f), whereas CMSCE provided a more stable and reliable post-earthquake trend. The entropy values of MSCE became more undulant as the threshold r decreased. Based on the results of the RC structural experiment, CMSCE was chosen to mitigate the fluctuation concern of MSCE encountered during the analysis of the data, which contained a certain level of noise.

3.2. Numerical Simulation

After the comparison of the MSCE and CMSCE algorithms through the shaking table test, the advantage of CMSCE for SHM was validated. Subsequently, a numerical simulation was executed to assess the feasibility of using CMSCE on more complex structures. SAP2000 software was used to construct and analyze a seven-story model for numerical simulation. Regarding geometric properties, the model was a steel structure with a yield strength F_y of 2500 kg/cm^2. The height of each story was 1.06 m, and the floor widths on the x- and y-axes were 1.32m and 0.92 m, respectively. The beam was a steel plate measuring 100 mm $\times$ 70 mm. The column was a steel plate measuring 25 mm $\times$ 150 mm. Steel bracings were set up on the y-axis and selected as L-shaped steel angles of 65 mm $\times$ 65 mm $\times$ 6 mm. Moreover, an additional mass of 500 kg was added to each floor to simulate real structural behavior. Structural damage was simulated by removing the bracings symmetrically. The SAP2000 model and damage scenario are shown in Figure 5.

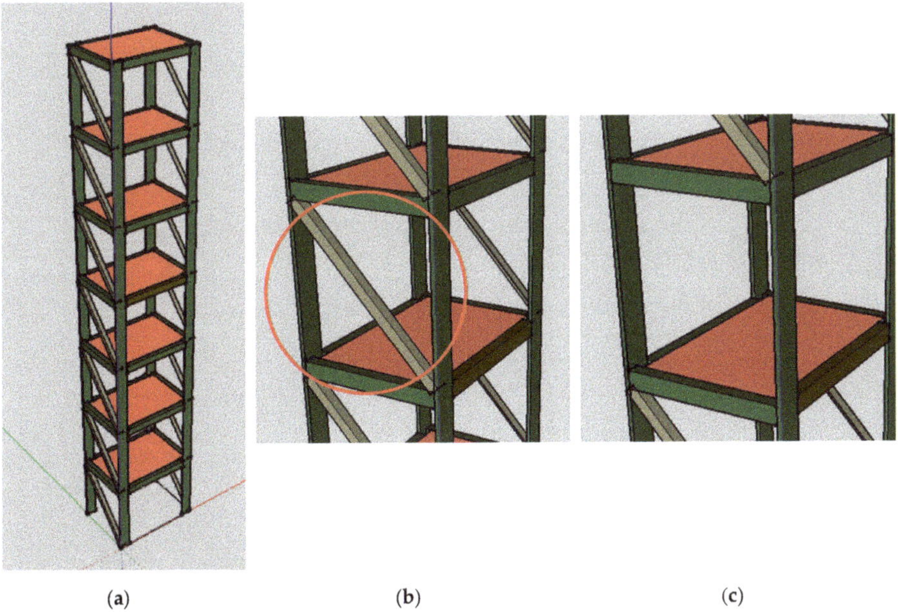

(a) (b) (c)

Figure 5. (**a**) View of the numerical model. (**b**) The undamaged scenario. (**c**) The damaged scenario.

Time history analysis was performed using a white noise signal of 1 MW power as the input acceleration to simulate the response of the structure under ambient vibration. The sampling rate was 200 Hz for 300 s. After the database had been created on the basis of the time history analysis of each damaged case, the velocity response data, which is relatively sensitive for ambient vibration, were extracted from the center of each floor on the y-axis [31]. The responses of velocity signals in the undamaged case are shown in Figure 6. All damaged cases and the modal analysis results are presented in Table 1.

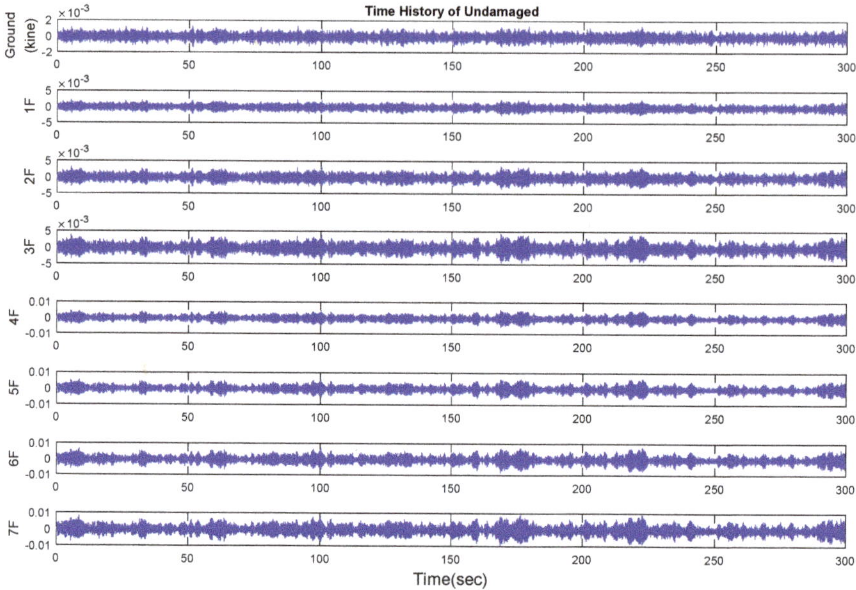

Figure 6. The time history of the velocity response of the undamaged case.

EEMD was utilized to extract the first mode of structural response for evaluating SHM performance based on the first mode. In the simulation, the results obtained with and without EEMD were displayed to evaluate the effectiveness of using the first mode to detect damage. EEMD can efficiently decompose signals into several intrinsic mode functions (IMFs) with a trend, and this is because the decomposition is based on the local characteristic time scale of the data [20,25]. The first mode of the model appeared in IMF4, and the frequencies are listed in Table 1. Subsequently, the extracted IMF4 and the original velocity response were analyzed separately to assess the damage location by using CMSCE.

Table 1. Damage cases and modal analysis.

Case Number	Damage Case	Frequency (Hz)	
		SAP2000	IMF4
1	Undamaged	3.16	3.12
2	1F	2.56	2.49
3	2F	2.55	2.49
4	3F	2.63	2.68
5	4F	2.73	2.73
6	5F	2.85	2.88
7	6F	2.99	2.98
8	7F	3.12	3.12
9	1&2F	2.15	2.15
10	3&4F	2.32	2.34
11	5&6F	2.69	2.69
12	1&2&3F	1.93	1.95
13	4&5&6F	2.38	2.34
14	1&2&3&4F	1.81	1.81
15	4&5&6&7F	2.35	2.34

3.3. Numerical Simulation Results

3.3.1. Damage Detection from the Original Velocity Response

In every damaged case, the velocity signals of two vertically adjacent floors were processed through CMSCE after the coarse-graining procedure to evaluate the degree of dissimilarity between floors. According to previous studies that have discussed detection accuracy under various parameter combinations, parameters such as the template length m, threshold r, and signal length N were optimized as 4, 0.08 × SD of the time series, and 20,000 points, respectively. Additionally, the required data length recommended by Gow et al. [32] was set between 14^m and 23^m points in the final MSE scale analysis. Therefore, the DI was conservatively calculated to time scale 10 ($\tau = 10$) for the better results. After analysis of the undamaged case and 14 damaged cases, the entropy curves for the damaged cases were compared with those for the undamaged case by calculating the DI. When a floor is damaged, the DI is positive, indicating that the signal complexity increases because of a loss of story stiffness. By contrast, a negative DI indicates a healthy floor.

Figure 7 presents the CMSCE curves for the undamaged case. In the figure, G*1F denotes the curve for the first floor and 1F*2F denotes the curve for the second floor; the curves for the remaining floors follow similar designations. Curve G*1F had the highest entropy values, indicating high complexity and low similarity; other curves ranked further down in the order, showing a decline in complexity between the two vertically adjacent floors. The case of single-floor damage is illustrated in Figure 8. The curve for the second floor was the highest among all curves. Furthermore, Figure 8a shows a substantial gap between the curve for the second floor and the remaining curves; an increase in complexity due to a loss of story stiffness could be clearly observed. According to the DI results shown in Figure 8b, the positive DI indicated that the damage occurred on the second floor.

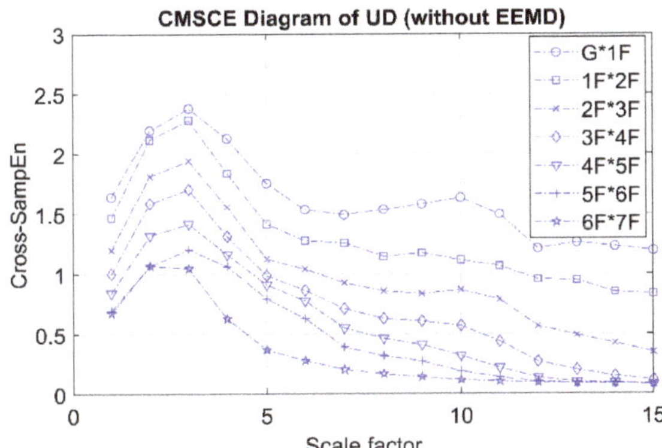

Figure 7. CMSCE diagram of undamaged case.

The results obtained for damage on the third and fourth floors are presented in Figure 9. Figure 9a shows that the curves for the third and fourth floors were higher than that for the undamaged case. Additionally, the curves for the third and fourth floors maintained higher positions than did those for the other floors at scale 5 to 15. In Figure 9b, the bars representing the third and fourth floors exhibit significantly positive values. Although extremely low positive values were observed for the first, second, and seventh floors, these could be disregarded because they were within the error tolerance range because the curves for the undamaged case maintained an almost identical level of complexity, resulting in low negative or zero-approaching values.

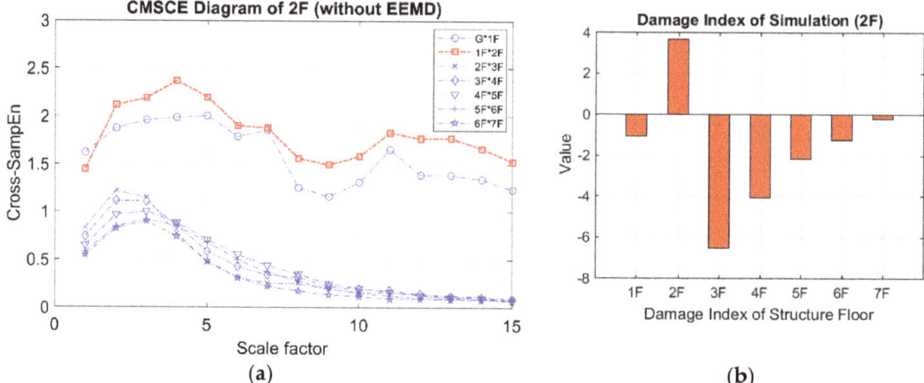

Figure 8. (**a**) CMSCE diagram of damage on the second floor (2F); (**b**) Damage index of damage on the second floor.

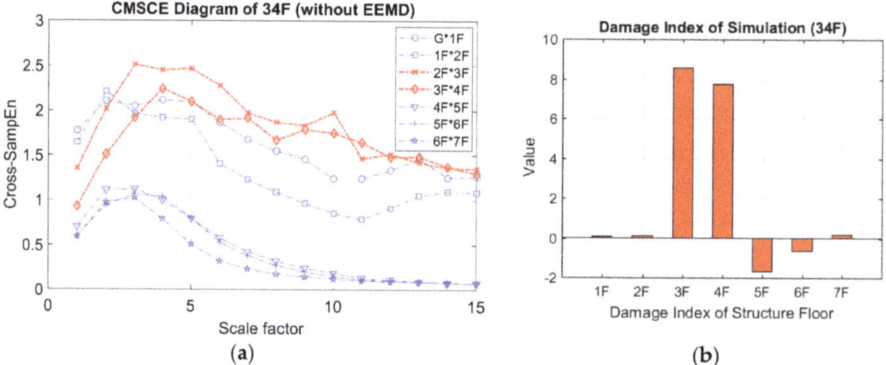

Figure 9. (**a**) CMSCE diagram of damage on the third and the fourth floor (34F); (**b**) Damage index of damage on the third and the fourth floor.

In the case of three-story damage, Figure 10a illustrates the CMSCE curves for damage on floors ranging from the fourth floor to the sixth floor, revealing that the curves for the fourth floor to the seventh floor climbed up at scale 5 to 15. Compared with the curve for the undamaged case, the curves for the fourth floor to the seventh floor increased, indicating an increase in complexity. Hence, the evident increase could be quantified through the DI analysis, as depicted in Figure 10b. The fourth floor to the sixth floor had positive values, indicating the presence of damage; however, few undefined entropies occurred on the fourth floor.

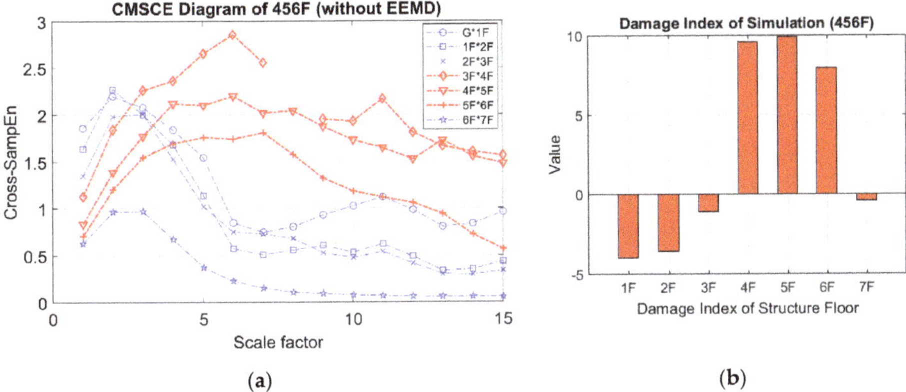

Figure 10. (**a**) CMSCE diagram of damage from the fourth to the sixth floor (456F); (**b**) Damage index of damage from the fourth to the sixth floor.

Figure 11 illustrates the curves for damage on floors ranging from the fourth floor to the seventh floor. These curves were markedly higher than those for the undamaged case at scale 5 to 15. The peak values of the damaged curves shifted slightly from scale 3 to 6, indicating that maximum complexity occurred after the coarse-graining procedure. The DI diagram for multistory damage is shown in Figure 11b; the positive indices for the fourth floor to the seventh floor can easily be identified as indicating damage.

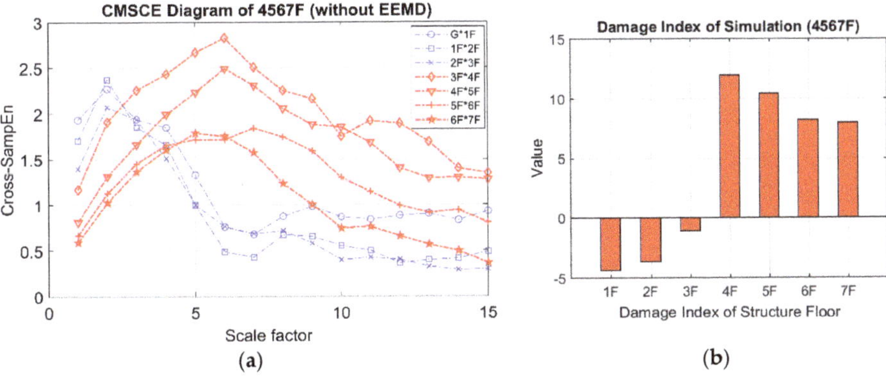

Figure 11. (**a**) CMSCE diagram of damage from the fourth to the seventh floor (4567F); (**b**) Damage index of damage from the fourth to the seventh floor.

3.3.2. Damage Detection from the Extracted First Mode Time Series (IMF4)

A study on sample entropy noted that signals with higher frequencies may have unpredictable effects on the calculations of SampEn values [32]; however, ambient noise usually contains a high percentage of high-frequency noise in real applications, and most of the practical structural damage mainly causes a frequency change on the first fundamental mode. Hence, EEMD was utilized to eliminate the influence of background noise and extract the first mode signal of the structure. IMF4 was selected to conduct CMSCE to evaluate the complexity between two signals. The template length m, threshold r, and signal length N were set to 4, $0.08 \times$ SD of the time series, and 20,000 points, respectively. Similarly, the DI was calculated to time scale 10 ($\tau = 10$). The damaged cases were compared with the undamaged case, and then the DI was applied to quantify the CMSCE results.

Figure 12 presents the CMSCE diagram obtained through EEMD for the undamaged case. G*1F denotes the curve for the first floor and 1F*2F denotes the curve for the second floor; the curves for the remaining floors follow similar designations. The entropy gradually increased with the time scale and reached a plateau at scale 10, where information richness could be accumulated if the system responded well. Moreover, the complexity rankings did not follow the order of the floors from low to high. Complexity was fairly consistent, except for the curves for the fourth and fifth floors.

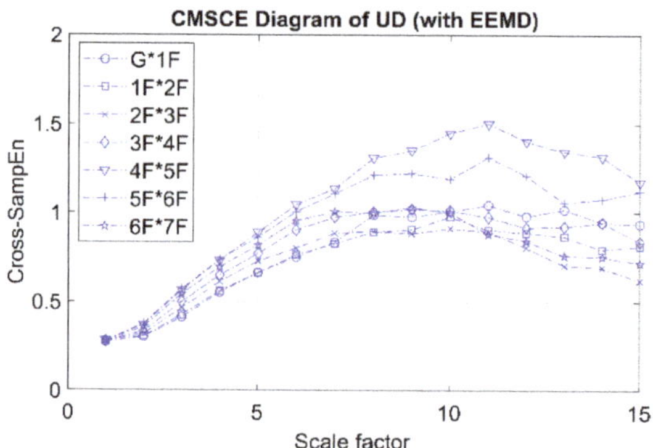

Figure 12. CMSCE diagram of undamaged case with EEMD.

Figure 13 presents the CMSCE diagram for damage on the second floor. The entropy values obtained for the second floor gradually increased to a peak at scale 10, as did the values for the first floor. All floors had almost identical complexity before time scale 5; the curves diverged between long scales. Regarding the DI results, a positive value for the second floor revealed that damage had occurred. Hence, the removal of bracings could result in significant differences among entropy curves.

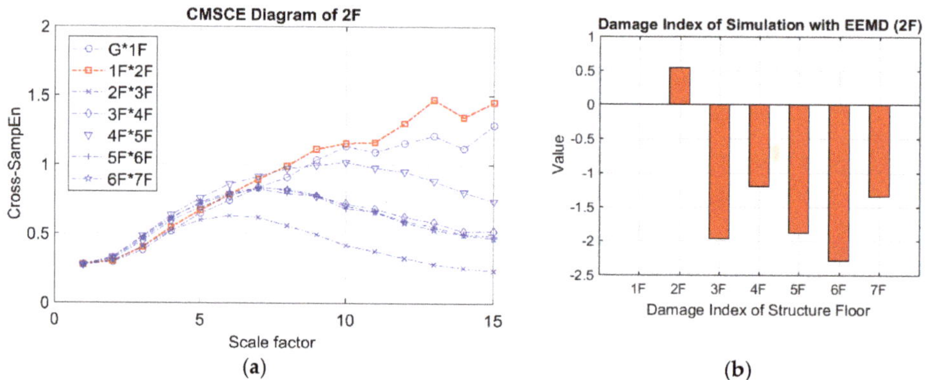

Figure 13. (**a**) CMSCE diagram of damage on the second floor (2F) with EEMD; (**b**) Damage index of damage on the second floor with EEMD.

The results of damage assessment on the third and fourth floors are presented in Figure 14. Compared with those for the damaged floors, the curves for the healthy floor above the damaged floor, the fifth to seventh floors, slowly descended after time scale 6. The curves for the third and fourth floors increased, but the increase was not evidently greater than that of the curves for the undamaged case, which caused a misclassification in the DI analysis. In addition, the same trend occurred in the case of three-story damage, as shown in Figure 15. Figure 15a shows the CMSCE diagram for damage from the fourth floor to the sixth floor, as observed through analysis with EEMD; the curve of the undamaged floor above the sixth floor was lower than those for the sixth floor. Moreover, the curves for the damaged floor sustained a stable plateau, indicating a high degree of complexity. Nevertheless, the first and second floors were misclassified as damaged in the DI analysis because this damage reduced the stability of the entire structure.

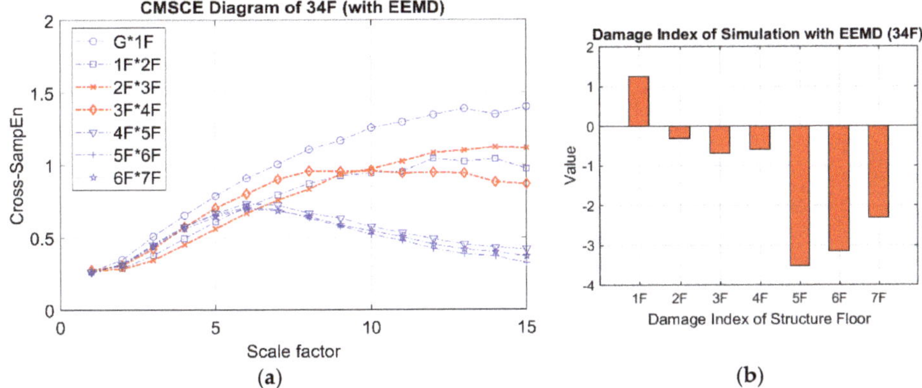

(a) (b)

Figure 14. (**a**) CMSCE diagram of damage on the third and the fourth floor (34F) with EEMD; (**b**) Damage index of damage on the third and the fourth floor with EEMD.

The CMSCE diagram for multistory damage is illustrated in Figure 16a. The curves for the fourth floor to the seventh floor gradually ascended and reached a plateau at scale 10. However, a comparison of the results in Figure 16a with those for the undamaged case (Figure 12) revealed that the overall entropy values were almost identical, signifying that the damage location could not be detected by applying the first mode signal. The obscure change in the CMSCE diagram resulted in a failure to diagnose the damaged floor in the DI results.

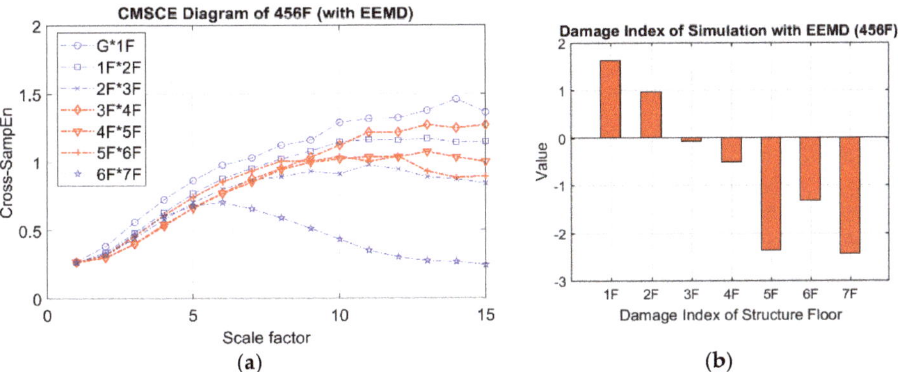

(a) (b)

Figure 15. (**a**) CMSCE diagram of damage from the fourth to the sixth floor (456F) with EEMD; (**b**) Damage index of damage from the fourth to the sixth floor with EEMD.

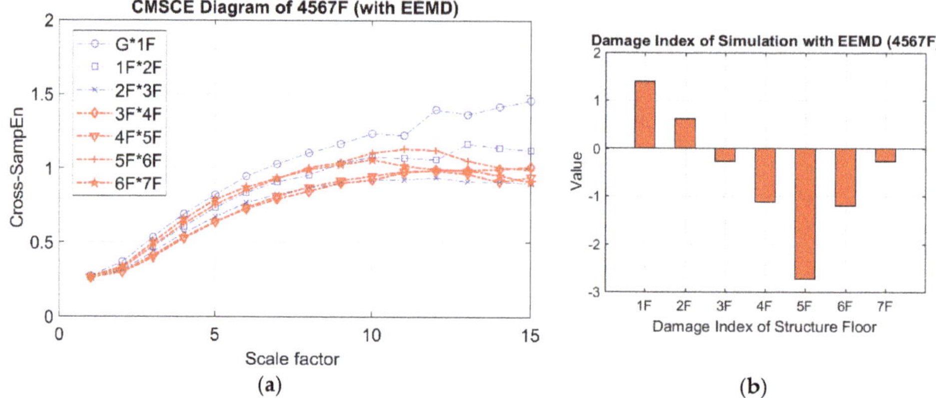

(a) (b)

Figure 16. (**a**) CMSCE diagram of damage from the fourth to the seventh floor (4567F) with EEMD; (**b**) Damage index of damage from the fourth to the seventh floor with EEMD.

3.3.3. Discussion on Numerical Simulation

A total of 15 damage cases were examined to verify the feasibility of CMSCE and evaluate the effectiveness of using EEMD to extract the structural first mode. A two-class statistical classification system, namely a confusion matrix, was applied for further performance evaluation. The entries in a confusion matrix are explained as follows: true positives (TP), referring to the number of "yes" predictions for "yes" instances; true negatives (TN), referring to the number of "no" predictions for "no" instances; false positives (FP), referring to the number of "yes" predictions for "no" instances; and false negatives (FN), referring to the number of "no" predictions for "yes" instances. On the basis of these entries, three indices can be calculated as follows:

$$\text{Accuracy} = \frac{\text{TP} + \text{TN}}{\text{TP} + \text{TN} + \text{FP} + \text{FN}} \; \text{Precision} = \frac{\text{TP}}{\text{TP} + \text{FP}} \; \text{Recall} = \frac{\text{TP}}{\text{TP} + \text{FN}} \tag{15}$$

In this study, on the basis of the definitions of the confusion matrix, the DI results were first classified into four categories: TP, indicating that the damaged floor had been correctly identified; TN, indicating that the undamaged floor had been correctly classified as healthy; FP, indicating that the healthy floor had been misclassified as damaged; and FN, indicating that the damaged floor had been misclassified as healthy. Subsequently, the three indices were calculated. The classification results are listed in Table 2. Accuracy denoted the overall accuracy of the classifier. Precision referred to the proportion of instances that had been classified as damaged and were actually damaged. "Recall" referred to the proportion of damaged instances that had been correctly classified as damaged. Hence, precision could be considered a measure of exactness, and recall could be considered a measure of completeness.

In the analysis without EEMD, a high accuracy rate of 93.9% was obtained, and a precision of 92% was obtained because two floors were misclassified as damaged. Subsequently, a recall rate of 85.2% was observed, revealing a high likelihood of detecting actual damage. By contrast, in the analysis with EEMD, the observed accuracy was 77.6% and precision was 66.7%, which was too low to precisely classify the healthy floor. Moreover, the observed recall was only 37% because the inconspicuous change in complexity resulted in errors in the detection of damaged floors. The method without EEMD was considered to be appropriate in monitoring structure.

Table 2. Results of confusion matrix for numerical model.

Two-Class Statistical Classification: Confusion Matrix									
Case Number	**Damage Floors**	**CMSCE**				**EEMD + CMSCE**			
		TP	**FP**	**TN**	**FN**	**TP**	**FP**	**TN**	**FN**
1	None								
2	1F	0	0	6	1	1	0	6	0
3	2F	1	0	6	0	1	0	6	0
4	3F	1	0	6	0	1	0	6	0
5	4F	1	0	6	0	1	0	6	0
6	5F	1	0	6	0	0	0	6	1
7	6F	1	0	6	0	0	0	6	1
8	7F	1	1	5	0	1	0	6	0
9	1&2F	1	0	5	1	1	0	5	1
10	3&4F	2	0	5	0	0	1	4	2
11	5&6F	2	0	5	0	0	0	5	2
12	1&2&3F	2	1	3	1	1	0	4	2
13	4&5&6F	3	0	4	0	0	2	2	3
14	1&2&3&4F	3	0	3	1	3	0	3	1
15	4&5&6&7F	4	0	3	0	0	2	1	4
Total		23	2	69	4	10	5	66	17
Accuracy		93.9%				77.6%			
Precision		92%				66.7%			
Recall		85.2%				37%			

3.3.4. Noise Statistical Analysis

The CMSCE without EEMD has been numerically demonstrated to be an effective SHM method. However, the basic assumption of ambient vibration condition in the methodology may be changed in practical application. In order to verify the robustness of the proposed method, different levels of noise are randomly added into the original time series for noise statistical analysis. The noise is simulated by a Gaussian white-noise. The signal-to-noise ratio (SNR) values are chosen to be 60 dB, 40 dB, and 20 dB, respectively. The damage location is diagnosed by the CMSCE and damage index methods. The symbol C indicates that the damage can be correctly identified and F represents for false identification. As shown in the Table 3, the accuracy of the method based on CMSCE remains the same for the accuracy of 78.57% under all the cases of SNR 60, 40, and 20, which shows no influence by possible external noise. The performance of the damage index has the same result of 85.71% accuracy under SNR 60 and 40. For a higher noise level (SNR 20), the damage index drops slightly on damage location assessment and has the same result of 78.57% as the CMSCE method. The result has proven that a reliable result can be provided by the proposed methods when the ambient condition is affected by possible external noise.

Table 3. The accuracy of CMSCE and damage index method for different noise levels.

Damage Location	**SNR = 60**		**SNR = 40**		**SNR = 20**	
	CMSCE	**Damage Index**	**CMSCE**	**Damage Index**	**CMSCE**	**Damage Index**
1F	C	F	C	F	C	F
2F	C	C	C	C	C	C
3F	C	C	C	C	C	C
4F	C	C	C	C	C	C
5F	C	C	C	C	C	C
6F	F	C	F	C	F	C
7F	F	C	F	C	F	F
1&2F	C	C	C	C	C	C
3&4F	C	C	C	C	C	C
5&6F	F	C	F	C	F	C
1&2&3F	C	C	C	C	C	C
4&5&6F	C	F	C	F	C	F
1&2&3&4F	C	C	C	C	C	C
4&5&6&7F	C	C	C	C	C	C
Accuracy (%)	78.57%	85.71%	78.57%	85.71%	78.57%	78.57%

Note: C = Correct, F = False.

4. Experimental Verification

4.1. Experimental Setup

To verify the practicality of the SHM system, an ambient vibration experiment for a scaled-down steel benchmark structure was conducted. The number of damaged cases was designed to be identical to that of the aforementioned numerical simulation. Moreover, the characteristics of the experimental structure were similar to those of the numerical model; the height of each floor was 1.1 m, and the widths of each floor were 1.5 and 1.1 m. An additional mass of 500 kg (Figure 17b) was added to each floor to simulate the actual structural characteristics. An L-shaped steel angle measuring 65 mm × 65 mm × 6 mm was selected as the bracing.

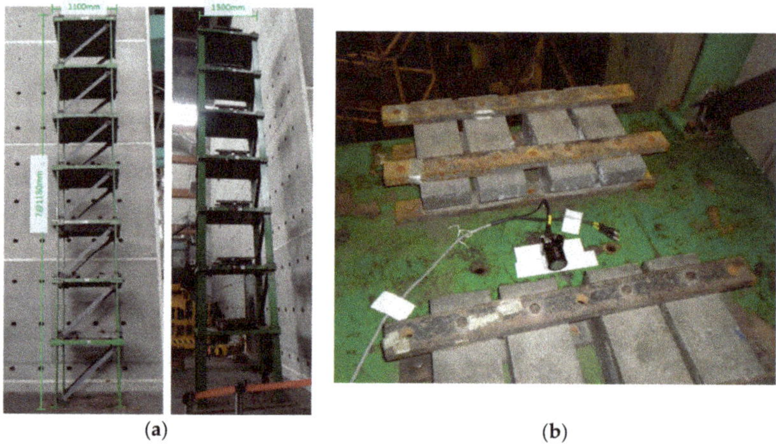

Figure 17. (**a**) A seven-story scale-down benchmark structure; (**b**) The arrangement of velocity meter and mass block.

For data acquisition, a sensitive velocity sensor VSE-15D (Tokyo Sokushin, Tokyo, Japan) was mounted on each floor to record the ambient vibration from the weak axis. Similar to the numerical analysis, damage was simulated as the removal of the bracing in the weak axis direction. The damage simulation is depicted in Figure 18. The experiment was executed at night to avoid interference from the testing field. Therefore, the velocity response under ambient vibration on each floor was recorded at a sampling rate of 200 Hz. A set of data was measured for 300 s, and then four sets were recorded to eliminate variance. The response signals of the undamaged case are shown in Figure 19.

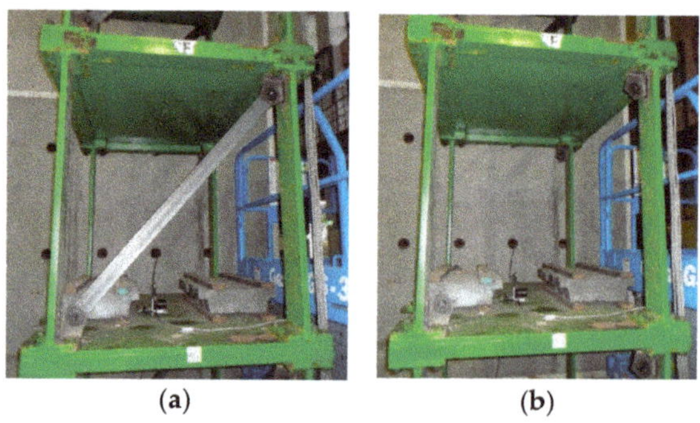

Figure 18. The damage simulation: (**a**) the healthy condition; (**b**) the damaged condition.

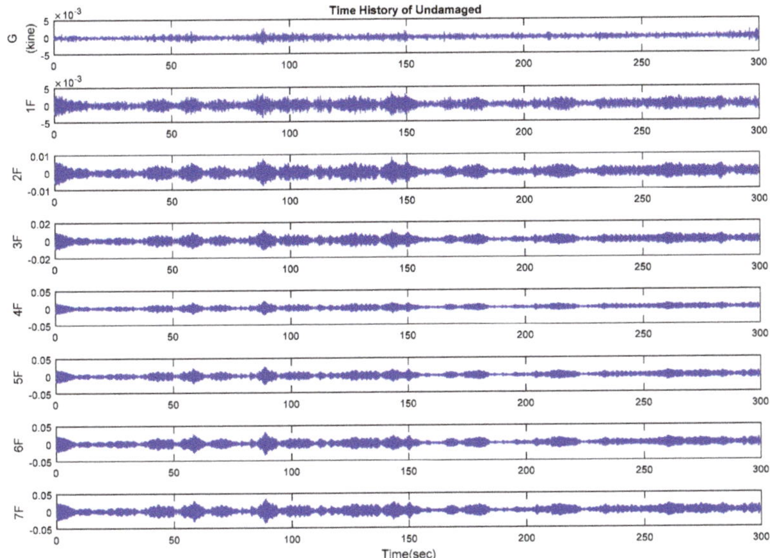

Figure 19. The velocity response of the undamaged case under ambient vibration.

The velocity responses of each case were examined using fast Fourier transform. Subsequently, the response of the seventh floor for each case was selected to examine the rationality of the signal; the results are listed in Table 4. The healthy condition had the highest frequency (3.34 Hz). Additionally, the frequency decreased markedly as damage occurred on the bottom floor, indicating that the removal of the bracing on the low floor had a relatively severe effect on the global stiffness of the structure.

Table 4. FFT results of the experimental data.

Case Number	Damage Group	Damage Floors	Frequency (Hz)
1	Undamaged	None	3.34
2	One-story damage	1F	2.08
3		2F	2.13
4		3F	2.12
5		4F	2.29
6		5F	2.61
7		6F	2.88
8		7F	3.2
9	Two-story damage	1&2F	1.64
10		3&4F	1.83
11		5&6F	2.32
12	Three-story damage	1&2&3F	1.44
13		4&5&6F	1.88
14	Multistory damage	1&2&3&4F	1.33
15		4&5&6&7F	1.86

4.2. Damage Detection Result

In all damage cases, the original velocity signals were analyzed to evaluate the complexity between floors. The parameters template length m, threshold r, and signal length N were optimized as 4, 0.08 SD of the time series, and 20,000 points, respectively. The DI range was identical to that of the simulation:

from scale 1 to 10. In addition, low positive values were excluded by a predetermined threshold value 1 during the practical application [30]. Consequently, the damaged floor could be detected.

Figure 20 presents the CMSCE diagram for the healthy condition. The curve for the first floor was the highest on all scales, indicating high complexity between the ground and first floors. The curves for the third floor maintained a gap with those for the second floor and remained constant from scale 5 to 10; this reveals that the signals tended to be similar. The healthy condition was recognized as a reference for detecting damage locations.

The CMSCE diagram for damage on the second floor is presented in Figure 21. The curve for the second floor increased evidently; however, the damage on the floor engendered an increase in the curve for the first floor. Even undefined entropies appeared from scale 12; the DI result was not influenced because the range was 1 to 10. According to the DI results, the index for the second floor was apparently higher than those for the other floors, whereas the first and fourth sets revealed an outlier on the first floor.

The CMSCE diagram for the two-story damaged case is illustrated in Figure 22a. The curves for damage on the third and fourth floors showed the highest complexity from scale 4 to 8. Compared with the curves for the undamaged case, the difference between the curves for the damaged floors was distinguishable. Moreover, the DI values for the third and fourth floors were positive, whereas the values for the other floors were close to zero. Therefore, the damage on the third and fourth floors could be detected.

Figure 23 shows the curves for three-story damaged case (i.e., damage from the fourth floor to the sixth floor). The damaged curves ascended rapidly at scale 5; however, the curve for the seventh floor slightly increased from scale 4 to 9 because of the loss of stiffness from the fourth floor to the sixth floor. The DI was used to quantify the CMSCE diagram, and the results showed that damage had occurred from the fourth floor to the seventh floor. A low positive value, which was excluded, could be observed for the seventh floor. Regarding the more severe damage condition, the results for damage from the fourth floor to the seventh floor are illustrated in Figure 24. A similar trend could be observed, revealing that the curves for damage increased rapidly at scale 4. In addition to the curves for damage, the curve for the first floor was elevated at scale 5 to 10, resulting in low positive DI values. The DI diagram is presented in Figure 24b. Apparently, the damage from the fourth floor to the seventh floor could be detected rapidly.

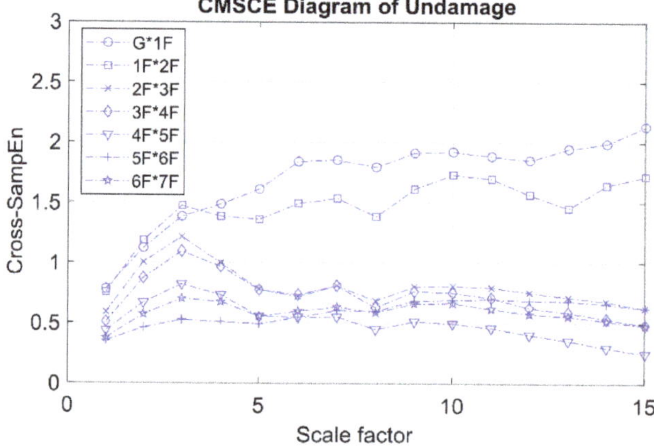

Figure 20. The experimental CMSCE diagram of undamaged case.

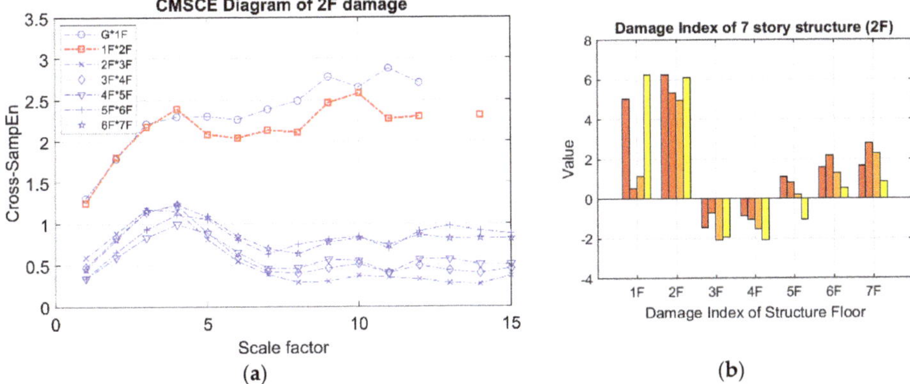

Figure 21. (**a**) The experimental CMSCE diagram of damage on the second floor (2F); (**b**) Damage index of damage on the second floor.

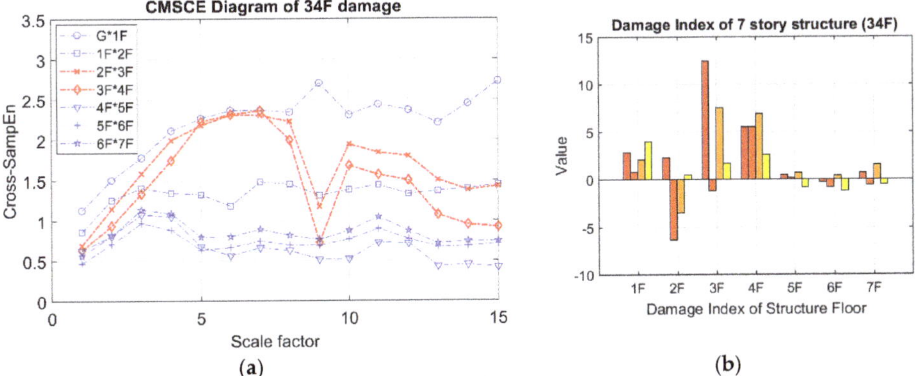

Figure 22. (**a**) The experimental CMSCE diagram of damage on the third and the fourth floor (34F); (**b**) Damage index of damage on the third and the fourth floor.

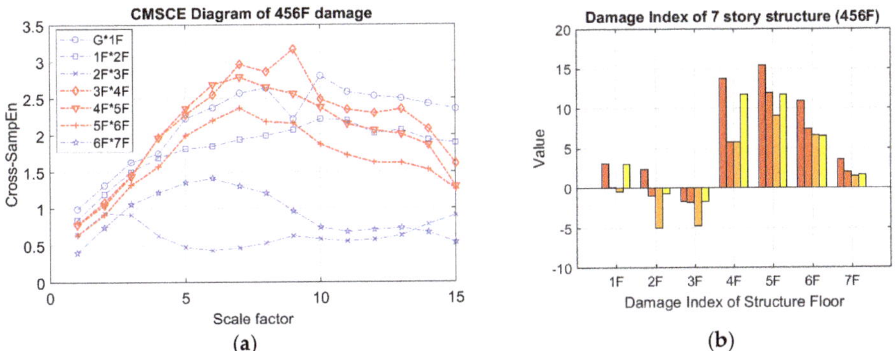

Figure 23. (**a**) The experimental CMSCE diagram of damage from the fourth to the sixth floor (456F); (**b**) Damage index of damage from the fourth to the sixth floor.

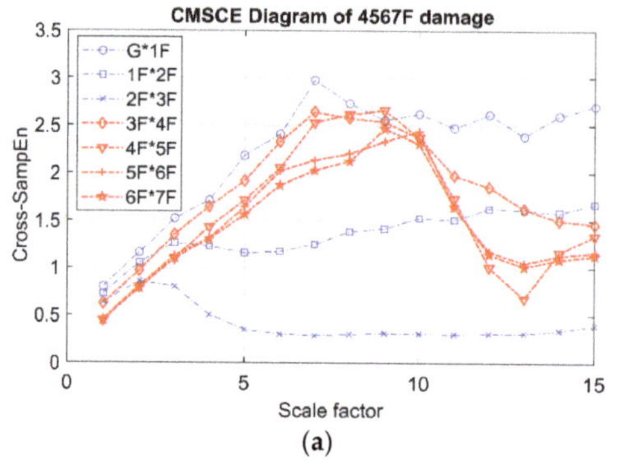

(a)

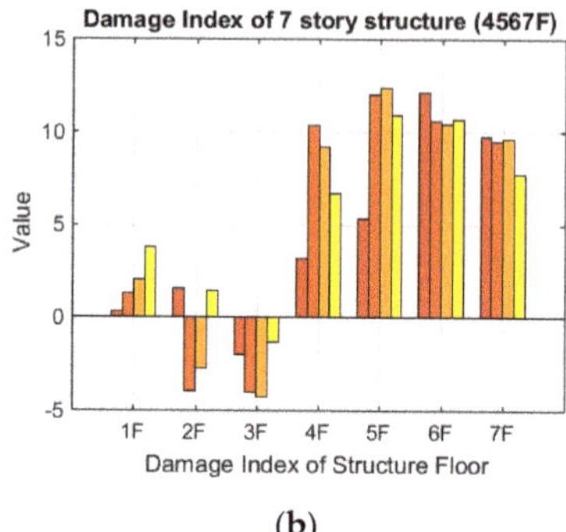

(b)

Figure 24. (**a**) The experimental CMSCE diagram of damage from the fourth to the seventh floor (4567F); (**b**) Damage index of damage from the fourth to the seventh floor.

The confusion matrix was next utilized to evaluate the performance of the DI results. First, the results were classified into four categories: TP, TN, FP, and FN. As shown in Table 5, an accuracy rate of 78.6% was obtained. Also, a precision of 56.8% was observed, indicating that the method was too conservative to classify the healthy floor. The low accuracy and precision phenomenon may be caused by the unexpected noise condition or signal interference faced in practical measurement and can be improved by slightly adjusting the threshold value. Nevertheless, a strong recall rate of 92.6% was noted, revealing that most of the floors that had been diagnosed as damaged were correctly classified as damaged floors.

Table 5. Results of the confusion matrix for experimental verification.

Two-Class Statistical Classification: Confusion Matrix					
Case Number	Damage Floors	CMSCE			
		TP	FP	TN	FN
1	None				
2	1F	1	0	6	0
3	2F	1	3	3	0
4	3F	1	4	2	0
5	4F	1	3	3	0
6	5F	1	0	6	0
7	6F	1	0	6	0
8	7F	0	1	5	1
9	1&2F	2	2	3	0
10	3&4F	2	1	4	0
11	5&6F	2	1	4	0
12	1&2&3F	2	0	4	1
13	4&5&6F	3	1	3	0
14	1&2&3&4F	4	3	0	0
15	4&5&6&7F	4	0	3	0
Total		25	19	52	2
Accuracy		78.6%			
Precision		56.8%			
Recall		92.6%			

5. Conclusions

Although entropy analysis has been applied extensively in biomedicine, finance, and mechanical engineering, few studies have utilized entropy analysis in the field of SHM. Based on the previous studies, unstable entropy values are occasionally observed when multiscale cross-sample entropy (MSCE) is employed. Therefore, the CMSCE method was utilized to enhance the reliability of entropy values on every scale. A preliminary study on a three-story RC structure has demonstrated that the CMSCE method can largely improve the performance of the original MSCE-based SHM system. In addition, as most of the practical structural damage mainly causes a frequency change on the first fundamental mode, the effectiveness of extracting first mode signals by using EEMD was also attempted. The feasibility of detecting damage locations was verified in a numerical model. It is shown that the CMSCE without EEMD method has the best overall accuracy for different levels of damage. In general, 93.9% of the floors were correctly classified and 85.2% of the actual damaged floors were correctly diagnosed. Moreover, the robustness of the SHM system was also verified through noise statistical analysis.

The proposed entropy-based SHM system for long-term monitoring was then tested by a series of experimental verification. The reliability and viability of the proposed SHM system were examined using 15 damaged cases in 5 categories representing several degrees of damage severity. The results obtained by observing the scaled-down benchmark structure reveal that 78.6% of the floors were correctly classified and 92.6% of the actual damaged floors were correctly diagnosed. Based on the outcome obtained from both numerical simulation and experimental verification, two advantages of applying CMSCE including the enhancement of system reliability on noise interference and the improvement of diagnosis accuracy have been achieved. The proposed entropy-based SHM system has been demonstrated to have high potential for practical application.

Author Contributions: T.-K.L. conceived, put forward the research ideas, and revised the paper. Y.-H.C. carried out the numerical analysis and wrote the paper.

Funding: This research received no external funding.

Conflicts of Interest: The authors declare no conflict of interest.

References

1. Wahab, M.A.; De Roeck, G. Damage detection in bridges using modal curvatures: Application to a real damage scenario. *J. Sound Vib.* **1999**, *226*, 217–235. [CrossRef]
2. Maeck, J.; Wahab, M.A.; Peeters, B.; De Roeck, G.; De Visscher, J.; De Wilde, W.; Ndambi, J.M.; Vantomme, J. Damage identification in reinforced concrete structures by dynamic stiffness determination. *Eng. Struct.* **2000**, *22*, 1339–1349. [CrossRef]
3. Chang, P.C.; Flatau, A.; Liu, S. Health monitoring of civil infrastructure. *Struct. Health Monit.* **2003**, *2*, 257–267. [CrossRef]
4. Deraemaeker, A.; Reynders, E.; De Roeck, G.; Kullaa, J. Vibration-based structural health monitoring using output-only measurements under changing environment. *Mech. Syst. Signal Process.* **2008**, *22*, 34–56. [CrossRef]
5. Amezquita-Sanchez, J.P.; Adeli, H. Signal processing techniques for vibration-based health monitoring of smart structures. *Arch. Comput. Methods Eng.* **2016**, *23*, 1–15. [CrossRef]
6. Opoka, S.; Soman, R.; Mieloszyk, M.; Ostachowicz, W. Damage detection and localization method based on a frequency spectrum change in a scaled tripod model with strain rosettes. *Mar. Struct.* **2016**, *49*, 163–179. [CrossRef]
7. Soman, R.; Mieloszyk, M.; Ostachowicz, W. A two-step damage assessment method based on frequency spectrum change in a scaled wind turbine tripod with strain rosettes. *Mar. Struct.* **2018**, *61*, 419–433. [CrossRef]
8. Kourehli, S.S.; Bagheri, A.; Amiri, G.G.; Ghafory-Ashtiany, M. Structural damage detection using incomplete modal data and incomplete static response. *KSCE J. Civ. Eng.* **2013**, *17*, 216–223. [CrossRef]
9. Siringoringo, D.M.; Fujino, Y. System identification of suspension bridge from ambient vibration response. *Eng. Struct.* **2008**, *30*, 462–477. [CrossRef]

10. Shannon, C.E. A mathematical theory of communication, Part I, Part II. *Bell Syst. Tech. J.* **1948**, *27*, 623–656. [CrossRef]
11. Kolmogorov, A.N. New metric invariant of transitive dynamical systems and endomorphisms of Lebesgue spaces. *Dokl. Russ. Acad. Sci.* **1958**, *119*, 861–864.
12. Sinai, Y.G. On the notion of entropy of a dynamical system. *Dokl. Akad. Nauk. SSSR* **1959**, *124*, 768–771.
13. Pincus, S.M.; Gladstone, I.M.; Ehrenkranz, R.A. A regularity statistic for medical data analysis. *J. Clin. Monit. Comput.* **1991**, *7*, 335–345. [CrossRef]
14. Pincus, S.M. Approximate entropy as a measure of system complexity. *Proc. Natl. Acad. Sci. USA* **1991**, *88*, 2297–2301. [CrossRef] [PubMed]
15. An, Y.-H.; Ou, J.-P. Structural damage localisation for a frame structure from changes in curvature of approximate entropy feature vectors. *Nondestruct. Test. Eval.* **2014**, *29*, 80–97. [CrossRef]
16. Richman, J.S.; Moorman, J.R. Physiological time-series analysis using approximate entropy and sample entropy. *Am. J. Physiol.-Heart Circ. Physiol.* **2000**, *278*, H2039–H2049. [CrossRef] [PubMed]
17. Lake, D.; Richman, J.S.; Griffin, M.P.; Moorman, J.R. Sample entropy analysis of neonatal heart rate variability. *Am. J. Physiol. Regul. Integr. Comp. Physiol.* **2002**, *283*, 789–797. [CrossRef] [PubMed]
18. Costa, M.; Goldberger, A.L.; Peng, C.K. Multiscale entropy analysis of complex physiologic time series. *Phys. Rev. Lett.* **2002**, *89*, 068102. [CrossRef]
19. Costa, M.; Goldberger, A.L.; Peng, C.K. Multiscale entropy analysis of biological signals. *Phys. Rev. E* **2005**, *71*, 021906. [CrossRef]
20. Zhang, L.; Xiong, G.; Liu, H.; Zou, H.; Guo, W. Bearing fault diagnosis using multi-scale entropy and adaptive neuro-fuzzy inference. *Expert Syst. Appl.* **2010**, *37*, 6077–6085. [CrossRef]
21. Xia, J.A.; Shang, P.J. Multiscale entropy analysis of financial time series. *Fluct. Noise Lett.* **2012**, *11*, 1250033. [CrossRef]
22. Pincus, S.; Singer, B.H. Randomness and degrees of irregularity. *Proc. Natl. Acad. Sci. USA* **1996**, *93*, 2083–2088. [CrossRef] [PubMed]
23. Fabris, C.; De Colle, W.; Sparacino, G. Voice disorders assessed by (cross-) sample entropy of electroglottogram and microphone signals. *Biomed. Signal Process. Control* **2013**, *8*, 920–926. [CrossRef]
24. Wu, S.D.; Wu, C.W.; Lin, S.G.; Wang, C.C.; Lee, K.Y. Time series analysis using composite multiscale entropy. *Entropy* **2013**, *15*, 1069–1084. [CrossRef]
25. Yin, Y.; Shang, P.G.; Feng, G.C. Modified multiscale cross-entropy for complex time series. *Appl. Math. Comput.* **2016**, *289*, 98–110. [CrossRef]
26. Wu, Z.H.; Huang, N.E. Ensemble empirical mode decomposition: A noise-assisted data analysis method. *Adv. Adapt. Data Anal.* **2009**, *1*, 1–41. [CrossRef]
27. Lin, Y.H.; Huang, H.C.; Chang, Y.C.; Lin, C.; Lo, M.T.; Liu, L.Y.; Tasi, P.R.; Chen, Y.S.; Ko, W.J.; Ho, Y.L.; et al. Multi-scale symbolic entropy analysis provides prognostic prediction in patients receiving extracorporeal life support. *Crit. Care* **2014**, *18*, 548. [CrossRef]
28. Lin, Y.H.; Lin, C.; Ho, Y.H.; Wu, V.C.; Lo, M.T.; Hung, K.T.; Liu, L.Y.; Lin, L.Y.; Huang, J.W.; Peng, C.K. Heart rhythm complexity impairment in patients undergoing peritoneal dialysis. *Sci. Rep.* **2016**, *6*, 280202. [CrossRef]
29. Chiu, H.C.; Ma, H.P.; Lin, C.; Lo, M.T.; Lin, L.Y.; Wu, C.K.; Chiang, J.Y.; Lee, J.K.; Hung, C.S.; Wang, T.D.; et al. Serial heart rhythm complexity changes in patients with anterior wall ST segment elevation myocardial infarction. *Sci. Rep.* **2017**, *7*, 43507. [CrossRef]
30. Lin, T.K.; Lainez, A.G. Entropy-based structural health monitoring system for damage detection in multi-bay three-dimensional structures. *Entropy* **2018**, *20*, 49.
31. Huang, C.S. Structural identification from ambient vibration measurement using the multivariate AR model. *J. Sound Vib.* **2001**, *241*, 337–359. [CrossRef]
32. Gow, B.J.; Peng, C.K.; Wayne, P.M.; Ahn, A.C. Multiscale entropy analysis of center-of-pressure dynamics in human postural control: Methodological considerations. *Entropy* **2015**, *17*, 7926–7947. [CrossRef]

Article

An Effective Bearing Fault Diagnosis Technique via Local Robust Principal Component Analysis and Multi-Scale Permutation Entropy

Mao Ge [1,2], Yong Lv [1,2,*], Yi Zhang [1,2], Cancan Yi [1,2] and Yubo Ma [1,2]

1 Key Laboratory of Metallurgical Equipment and Control Technology, Wuhan University of Science and Technology, Ministry of Education, Wuhan 430081, China; ge1656372625@gmail.com (M.G.); yizhang_de@163.com (Y.Z.); meyicancan@wust.edu.cn (C.Y.); Yubo2018.M@gmail.com (Y.M.)

2 Hubei Key Laboratory of Mechanical Transmission and Manufacturing Engineering, Wuhan University of Science and Technology, Wuhan 430081, China

* Correspondence: lvyong@wust.edu.cn; Tel.: +86-027-6886-2857; Fax: +86-027-6886-2212

Received: 11 September 2019; Accepted: 28 September 2019; Published: 30 September 2019

Abstract: The acquired bearing fault signal usually reveals nonlinear and non-stationary nature. Moreover, in the actual environment, some other interference components and strong background noise are unavoidable, which lead to the fault feature signal being weak. Considering the above issues, an effective bearing fault diagnosis technique via local robust principal component analysis (LRPCA) and multi-scale permutation entropy (MSPE) was introduced in this paper. Robust principal component analysis (RPCA) has proven to be a powerful de-noising method, which can extract a low-dimensional submanifold structure representing signal feature from the signal trajectory matrix. However, RPCA can only handle single-component signal. Therefore, in order to suppress background noise, an improved RPCA method named LRPCA is proposed to decompose the signal into several single-components. Since MSPE can efficiently evaluate the dynamic complexity and randomness of the signals under different scales, the fault-related single-components can be identified according the MPSE characteristic of the signals. Thereafter, these identified components are combined into a one-dimensional signal to represent the fault feature component for further diagnosis. The numerical simulation experimentation and the analysis of bearing outer race fault data both verified the effectiveness of the proposed technique.

Keywords: bearing fault diagnosis; weak fault; multi-component signal; local robust principal component analysis; multi-scale permutation entropy

1. Introduction

The bearing as an essential element has been widely used in rotating machinery [1,2]. Due to the severe working conditions, such as long and uninterrupted operation, alternating loads, and corrosion, the probability of bearing failure increases greatly, which may cause heavy economic losses or even serious personal injury [3]. Hence, an available diagnosis technique for bearing faults is highly valuable [4]. The bearing faults can be classified into three main types: inner race fault, outer race fault, and rolling element fault [5]. The vibration signals of the bearings are often used for fault diagnosis for their containing abundant equipment operation information [6]. When the bearing faults occur, the corresponding vibration signals will produce periodic impulses, and the feature of this signal behaves in a typical nonlinear and non-stationary nature, which increases in spectral complexity [5,7,8]. In the actual industrial production environment, the signals usually contain some interference vibrations caused by other mechanical components and strong background noise besides useful fault feature. Especially in the early stages of the bearing fault, the fault feature is weak and completely drowned by the strong noise and interferences. Therefore, in order to realize accurate diagnosis of bearing faults, suppressing the background noise and extracting weak fault features from multi-component signals are becoming an urgent work to be solved.

For the diagnosis of the bearing fault signals, some researchers have proposed many methods. Ciabattoni et al. [9] proposed a novel bearing fault classification method by adopting the empirical cumulative distribution functions (ECDFs) of the signal statistical spectral images as the fault feature vectors. The wavelet transform (WT) is the inner product operation between a translated and dialed wavelet basis function and the raw time domain signal. The different feature components and noise in the signal can be separated by the obtained wavelet coefficients [10,11]. Wang et al. [12] extracted the weak fault feature of the rolling element via wavelet packet transform method. Deng et al. [13] presented a novel fault diagnosis method for a motor bearing based on integrating empirical wavelet transform (IEWT) and fuzzy entropy. Xiao et al. [14] applied the wavelet threshold denoising method to effectively de-noise a rolling bearing signal. However, the diagnostic performance of these methods depends on the selection of the wavelet basis functions and the threshold. The Wigner–Ville distribution (WVD) [15] can extract ridges representing feature information from two-dimensional time-frequency plane of the non-stationary signals. Ming et al. [16] applied the cyclic Wiener filter to detect the rolling bearing fault, which uses the spectral coherence theory induced by the second-order cyclostationary signal to extract the weak fault feature. Nevertheless, WVD will produce cross-terms when analyzing multi-component signals. As far as the adaptive signal processing techniques, the empirical mode decomposition method (EMD) and the local mean decomposition (LMD) can be used to deal with nonlinear and nonstationary signals. LMD can decompose any signal into product functions (PFs) representing different feature components [17]. Li et al. [18] introduced a fault diagnosis scheme based on local mean decomposition and an improved multi-scale fuzzy entropy to realize the automatic identification of the bearing fault patterns. LMD is inadequate in processing the signals containing narrow bandwidth components. The multi-component signals can be decomposed into a series of intrinsic mode functions (IMFs) with physical meaning by EMD [19]. Imaouchen et al. [20] employed some demodulation analysis methods based on frequency-weighted energy operator and complementary ensemble empirical mode decompositions to identify the early weak faults of the bearing. Bustos et al. [21] successfully identified the operating state of the gears in high-speed trains through the EMD-based methodology. But the EMD and its improved version always suffer from modal aliasing and boundary effect. Moreover, they are also sensitive to noise. The above research provides rich reference information for bearing fault diagnosis.

It has become quite familiar to view the dynamic characteristics of different features of the raw system by reconstructing the observed time series from nonlinear non-stationary systems into a high dimensional phase space [22,23]. In that way, extracting fault features from high a dimensional phase space is a feasible scheme. The singular value decomposition (SVD) method [24] can decompose the signal trajectory matrix into series interpretable components. The singular values obtained can effectively display the intrinsic properties of different feature components and noise in raw signal. The fault feature can be extracted by setting the singular values representing interference components and background noise to zeros. Currently, the selection of singular values representing fault feature components still depends on experience, which may lead to considerable error. Especially for the early weak faults of the bearings, the singular values representing different feature components are almost impossible to be identified [25]. The classical manifold learning theory holds that the feature component of the signal matrix has a lower intrinsic dimension, which is distributed in a low-dimensional submanifold of a high dimensional phase space [26,27]. As a widely used dimension reduction method, the RPCA can extract this submanifold structure through a rank function constraint based on low-rank matrix approximation (LRMA) and simultaneously suppress background noise through a l_0-norm regularization strategy [28]. RPCA has proven to be a powerful de-noising tool in image processing, computer vision, and so on [29,30]. However, RPCA is inoperative for the separation of submanifold structures composed of multiple feature components; that is, it cannot process a multi-component signal. Recently, a novel, convex, locally sensitive, low rank matrix approximation (CLSLRMA) method [31] was introduced into the data completion problem, which significantly relaxes the assumption in LRMA that the feature component in matrix has a low-rank submanifold structure. CLSLRMA can decompose a matrix drawn from linear mixture of multiple low-rank manifold subspaces into their respective single subspaces. Hence, it is a feasible way to decompose the trajectory matrix composed of multi-component signals by CLSLRMA.

Permutation entropy (PE) [32] can efficiently evaluate the dynamic complexity and randomness of the signal time series through measuring similarity among the ordinal patterns extracted from the series, which has been widely used for the fault diagnosis of mechanical equipment [33,34]. The dynamic complexity of the bearings will change with the occurrence of faults, resulting in the changing PE values of the vibration signals [33]. However, because of the strong nonlinear and non-stationary characteristic of the acquired mechanical fault feature signals, their complex dynamic characteristics can usually hardly be fully displayed on the original scale, while some important information may also exit over multiple spatial-temporal levels (scales) [32]. Fortunately, based on PE, the multi-scale permutation entropy (MSPE) [32,35] has proven to be one of the most effective methods for which one can explicitly explain the characteristic information from the multiple time scales present in complex time series. Therefore, the MSPE of the signal was adopted in this paper to identify the feature component signal representing the bearing faults.

In this paper, an effective bearing fault diagnosis technique via local robust principal component analysis (LRPCA) and MSPE is introduced. Firstly, on the basis of noise suppression, we proposed an improved RPCA method to decompose the signal into several single-components, which was termed LRPCA. According to CLSLRMA, in the phase space of the weighted matrix associated with different anchor point, we assume that the signal trajectory matrix behaves as a combination of a noise component and a low-dimensional submanifold component, and those submanifold components represent different feature components in the raw signal. LRPCA shows that those submanifold components can be approximated by low-rank matrices through solving a convex program about a weighted combination of the matrix rank constraint function and the l_0-norm regularization [36,37]. After that, the MPSE was adopted to identify the low-rank matrices corresponding to the fault feature component. Finally, the identified low-rank matrices were transformed into a one-dimensional signal to represent the global approximation of the fault feature component for further diagnosis via weighted Nadaraya–Watson regression model [38]. The processing of the numerical simulation data and the experimental bearing fault data both verified that the proposed technique can provide a great diagnostic performance for bearing faults.

The rest of the paper is organized as follows: Section 2 introduces the theory description, wherein Section 2.1 defines some notations and abbreviations used in this paper; Section 2.2 illustrates the proposed LPRCA method; Section 2.3 describes the MSPE; the detailed step of the proposed bearing fault diagnosis technique is presented in Section 2.4. The analysis of the simulated signal and the experimental signal are performed in Section 3. Section 4 draws the conclusions.

2. Theory Description

2.1. Notations and Abbreviations

Throughout this paper, we use lowercase letters for scalars, e.g., x; bold and lowercase letters for vectors, e.g., $\mathbf{x} \in \mathbb{R}^{n_1}$; boldface and uppercase letters for matrices, e.g., $\mathbf{X} \in \mathbb{R}^{n_1 \times n_2}$. The anchor points in $\mathbf{X}$ are marked as $e_i = (a_i, b_i)$, $a_i = 1, \ldots, n_1$; $b_i = 1, \ldots, n_2$; $i = 1, \ldots, m$; the i-th element of $\mathbf{x}$ can be expressed as x_i, the (i,j)-th element of $\mathbf{X}$ is denoted as $\mathbf{X}(i, j)$ and the i-th row of $\mathbf{X}$ is expressed as $\mathbf{X}(i, :)$. The Hadamard product of two matrix $\mathbf{A} \in \mathbb{R}^{n_1 \times n_2}$, $\mathbf{B} \in \mathbb{R}^{n_1 \times n_2}$ is defined as $\mathbf{C} = \mathbf{A} \odot \mathbf{B} \in \mathbb{R}^{n_1 \times n_2}$ with its element $\mathbf{C}(i, j) = \mathbf{A}(i, j)\mathbf{B}(i, j)$. The l_0-norm $||\mathbf{X}||_0$ represents the sum of nonzero elements of $\mathbf{X}$ and the l_1-norm $||\mathbf{X}||_1$ the sum of absolute value of all elements in $\mathbf{X}$. The SVD of $\mathbf{X}$ is defined as $\mathbf{X} = \mathbf{U} \sum \mathbf{V}$, where **U** and **V** are the left and right singular value matrices; $\Sigma = diag(\{\sigma_i\}_{1 \le i \le r})$ represents the singular value matrix. The nuclear norm of **X** is denoted as $||\mathbf{X}||_* = \sum_i \sigma_i$.

In the rest of this paper, the following abbreviations are used: RPCA—robust principal component analysis; LRPCA—local robust principal component analysis; MSPE—multi-scale permutation entropy; CLSLRMA—convex local sensitive low rank matrix approximation; SVD—singular value decomposition; SSA—singular spectrum analysis; EMD—empirical mode decomposition; ADMM—alternating direction method of multipliers; and SNR—signal to noise ratio.

2.2. Decomposing a Signal into Single-Components via LRPCA

2.2.1. RPCA

The acquired one-dimensional bearing fault signal $\mathbf{x} \in \mathbb{R}^n$ can be converted into a high dimensional signal trajectory matrix $\mathbf{X} \in \mathbb{R}^{n_1 \times n_2}$ by phase space reconstruction, which is based on a embedding process with the parameter of the embedding dimension n_1 and the delay time τ (where $(n_1 - 1)\tau + n_2 = n$) [22]:

$$\mathbf{X} = \begin{bmatrix} x_1 & x_2 & \cdots & x_{n_2} \\ x_{1+\tau} & x_{2+\tau} & \cdots & x_{n_2+\tau} \\ \vdots & \vdots & \cdots & \vdots \\ x_{1+(n_1-1)\tau} & x_{2+(n_1-1)\tau} & \ddots & x_n \end{bmatrix} \tag{1}$$

Except for the strong background noise component, **X** is composed of multiple feature components, which include the fault feature component and the unwanted interfering components. Hence, on the basis of noise suppression, how to separate the useful feature component from these mixed multi-components and back it to the one-dimensional signal to represent the extracted fault feature component was an inevitable task for extracting weak faults in this paper.

By referring to manifold learning theory, it can be found that the feature component in **X** has a low-dimensional submanifold structure [26,27]. RPCA can extract this structure by solving the following low-rank matrix and sparse matrix decomposition model (shown as Figure 1):

$$\underset{\mathbf{L},\mathbf{S}}{\text{minarg}}\ rank(\mathbf{L}) + \alpha||\mathbf{S}||_0, s.t.\ \mathbf{X} = \mathbf{L} + \mathbf{S} \tag{2}$$

where $\alpha = 1/\sqrt{\max(n_1, n_2)}$ is the optimal weighted parameter. The first term ($\mathbf{L} \in \mathbb{R}^{n_1 \times n_2}$, $r(\mathbf{L}) \ll r(\mathbf{X})$) of this model is a rank constraint function, which uses a low-rank matrix to estimate the low-dimensional submanifold structure, and the second term ($\mathbf{S} \in \mathbb{R}^{n_1 \times n_2}$) is a regularization strategy, which is mainly used to correct the deviation of matrix data caused by noise. In fault feature extraction, **L** represents the signal feature component and **S** captures the noise [28]. Thus, RPCA can effectively separate the feature component of the signal from the noise. However, it can be seen from the theoretical basis of Equation (2) that RPCA can only deal with the matrix data composed of a single feature component and noise, which means this method cannot separate the data containing the multiple feature components.

2.2.2. LRPCA

To tackle the drawback of RPCA in processing the multi-component signals, we proposed a novel LRPCA method based on CLSLRMA to decompose **X** into several single-components. In LRPCA, the following fundamental assumption is introduced.

Fundamental Assumption: In addition to the background noise signal matrix **S**, **X** contains a feature signal matrix **L** composed of a linear mixture of several feature components. Furthermore, each feature component corresponds to a low-dimensional submanifold structure hidden in the high-dimensional phase space and has a characteristically of low-rank.

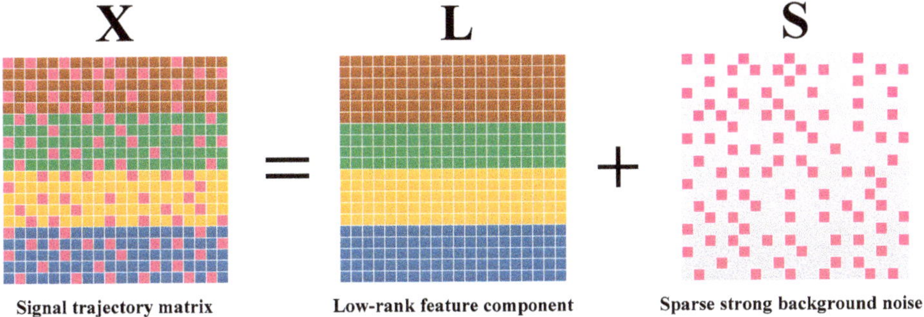

Figure 1. Illustration of the robust principal component analysis (RPCA); A signal trajectory matrix $\mathbf{X} \in \mathbb{R}^{n_1 \times n_2}$ can be decomposed into a low rank feature component $\mathbf{L} \in \mathbb{R}^{n_1 \times n_2}$ and a sparse component $\mathbf{S} \in \mathbb{R}^{n_1 \times n_2}$.

Figure 2 depicts the main ideal of the LRPCA. Specifically, each submanifold structure is generally hidden in the different high-dimensional phase space $\mathcal{T}(e_i)$ associated with the local selected anchor point ($e_i = (a_i, b_i),\ a_i = 1, \ldots, n_1;\ b_i = 1, \ldots, n_2;\ i = 1, \ldots, m$), and $\mathcal{T}(e_i)$ is derived from the weighting of $\mathbf{X}$:

$$\mathcal{T}(e_i) = \mathbf{W}_{e_i} \odot \mathbf{X} \tag{3}$$

where $\mathbf{W}_{e_i} \in \mathbb{R}^{n_1 \times n_2}$ is a local weighted coefficient matrix.

Thereafter, $\mathcal{T}(e_i)$ is decomposed into a low-rank component $\mathbf{L}_{e_i}$ and a sparse component $\mathbf{S}_{e_i}$. The resulting low-rank matrices $\mathbf{L}_{e_1}, \ldots, \mathbf{L}_{e_m}$ represent the low-dimensional submanifolds, which correspond to different feature components in raw signal, that of the fault feature component and the unwanted interfering components. Conclusively, $\mathbf{L}$ is expressed as the weighted combination of these local low-rank matrices:

$$\mathbf{L} = \mathbf{R}_1 \odot \mathbf{L}_{e_1} + \cdots + \mathbf{R}_m \odot \mathbf{L}_{e_m} \tag{4}$$

where $\mathbf{R}_i \in \mathbb{R}^{n_1 \times n_2}$ is a weighted regression matrix. Thus, the fault-related feature components can be extracted from these low-rank matrices obtained by decomposing $\mathbf{L}$. For this task, a novel local low-rank matrix and sparse matrix decomposition model was proposed to obtain these low-rank matrices from $\mathbf{X}$:

$$\mathbf{L}_{e_i} = \underset{\mathbf{L}_{e_i}, \mathbf{S}_{e_i}}{\text{minarg}}\ rank(\mathbf{L}_{e_i}) + \alpha \|\mathbf{S}_{e_i}\|_0,\ s.t.\ T(e_i) = \mathbf{L}_{e_i} + \mathbf{S}_{e_i} \tag{5}$$

The RPCA is actually a special case of this model when $\mathbf{W}_{e_i}$ is a unit matrix. Therefore, the de-noising performance of this model can be guaranteed.

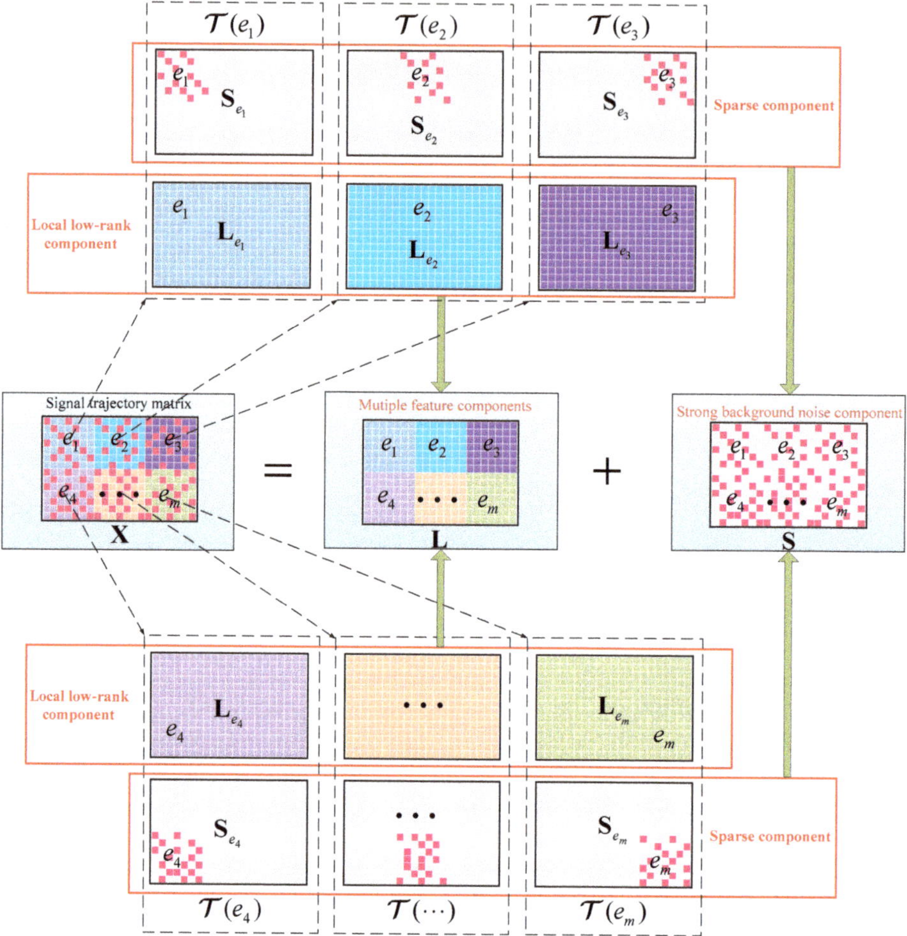

Figure 2. Illustration of the proposed LRPCA method; in the different high-dimensional phase space $\mathcal{T}(e_i)$ associated with the local selected anchor point $e_i = (a_i, b_i)$, the signal trajectory matrix $\mathbf{X} \in \mathbb{R}^{n_1 \times n_2}$ can be decomposed into a low rank component $\mathbf{L}_{e_i} \in \mathbb{R}^{n_1 \times n_2}$ and a sparse component $\mathbf{S}_{e_i} \in \mathbb{R}^{n_1 \times n_2}$.

- Model Construction and Algorithm Solving

The derivation of the phase space $\mathcal{T}(e_i)$ is the key step in LRPCA. Firstly, a distance function $d((a,b),(a',b'))$ is defined to describe the similarity between any two elements $\mathbf{X}(a,b)$ and $\mathbf{X}(a',b')$ in $\mathbf{X}$. A smaller value of d means the probability of the two elements being in the same phase space is higher. According to the theory of CLSLRMA, The standard incomplete SVD $\mathbf{X} = \mathbf{U}\mathbf{V}^T$ [39] is employed to divide d into two independent terms and the arc-cosine function is utilized to calculate these two terms:

$$d(a,a') = \arccos(\frac{\langle \mathbf{U}(a,:), \mathbf{U}(a',:)\rangle}{||\mathbf{U}(a,:)||||\mathbf{U}(a',:)||}),\ d(b,b') = \arccos(\frac{\langle \mathbf{V}(b,:), \mathbf{V}(b',:)\rangle}{||\mathbf{V}(b,:)||||\mathbf{V}(b',:)||}) \tag{6}$$

Then, the following non-parametric smoothing Epanechnikov kernel function [40] is adopted to define the local weighted coefficient matrix $\mathbf{W}_{e_i}$:

$$W((a,b),(a',b')) = W(a,a')W(b,b') = (1 - d((a,a')^2)(1 - d(b,b')^2) \tag{7}$$

The (i, j)-element of $\mathbf{W}_{e_i}$ is expressed as $W(e_i, (i, j))$. The Epanechnikov kernel function is a typical unilateral quadratic decrement function related to d. Consequently, a lager value of the weighted coefficient means that the weight for $\mathbf{X}(i, j)$ belonging to $\mathcal{T}(e_i)$ is bigger. Besides, for different local anchor point e_i, there theoretically should exist a unique corresponding phase space hidden the single submanifold. However, as the fundamental assumption states, there should be finite number of single submanifold structures in $\mathbf{X}$. Gratifyingly, this function has the mathematical property of changing slowly, which means as long as the similarity of two anchor points e_i and e_j is large enough, their corresponding phase spaces $\mathcal{T}(e_i)$ and $\mathcal{T}(e_i)$ will have quite high similarity. Moreover, both of those two phase spaces may hide the same single submanifold structure.

Through the above theoretical analysis process, the $\mathcal{T}(e_i)$ satisfying the fundamental assumption was successful created. Finally, $\mathcal{T}(e_i)$ is decomposed to extract the hidden low-rank matrix $\mathbf{L}_{e_i}$ by solving the Equation (5). Since the discrete combination nature of the l_0-norm and rank function, the solution of this model is an non-deterministic polynomial (NP)-hard problem [41]. Take this into consideration, some recent studies [36,37,42–45] pointed out that an equivalent convex optimization program of this model can be obtained from the convex hull of the two constraints; that is, the l_1-norm and the nuclear norm are employed to replace those two constraints respectively:

$$\mathbf{L}_{e_i} = \underset{\mathbf{L}_{e_i},\mathbf{S}_{e_i}}{\text{minarg}} ||\mathbf{L}_{s_i}||_* + \alpha||\mathbf{S}_{e_i}||_1, \ s.t.\ T(s_i) = \mathbf{L}_{e_i} + \mathbf{S}_{e_i} \tag{8}$$

The solution of Equation (8) is a typical convex optimization problem, whose minimizer is globally unique [43]. Algorithm 1 provides a precise and convergent solution to this equation via ADMM [44,46]. Note that steps 3 and 4 are both convex problems, they both have closed-form solutions via singular value thresholding operator [47].

Algorithm 1 solve (8) by ADMM

Input: signal trajectory matrix $\mathbf{X} \in \mathbb{R}^{n_1 \times n_2}$;
Parameter: number of anchor points: q; regularization parameter: $\alpha = 1/\sqrt{\max(n_1, n_2)}$;
for all i = 1:m, **parallel do**;
1. select $e_i(a_i, b_i)$ uniformly in $\mathbf{X}$, and calculate $\mathbf{W}_{e_i}$ by Equation (7);
2. $\mathcal{T}(e_i) = \mathbf{W}_{e_i} \odot \mathbf{X}$;
Initialize: $\mathbf{L}_{e_i}^0 = \mathbf{S}_{e_i}^0 = \mathbf{Y}^0 = 0, \gamma_0 = e^{-3}, \gamma_{\max} = e^{10}, \mu = 1.1, \varepsilon = 1e-8$;
while not converged do;
3. fix the others and update $\mathbf{L}_{e_i}^{k+1}$ by:

$$\mathbf{L}_{e_i}^{k+1} = \underset{\mathbf{L}_{e_i}}{\text{argmin}} : ||\mathbf{L}_{e_i}||_* + \frac{\gamma_k}{2}||\mathbf{L}_{e_i} + \mathbf{S}_{e_i}^k - \mathcal{T}(e_i) + \mathbf{Y}^k||_F^2$$

4. fix the others and update $\mathbf{S}_{e_i}^{k+1}$ by:

$$\mathbf{S}_{e_i}^{k+1} = \underset{\mathbf{S}_{e_i}}{\text{argmin}} : \lambda||\mathbf{S}_{e_i}||_1 + \frac{\gamma_k}{2}||\mathbf{L}_{e_i}^{k+1} + \mathbf{S}_{e_i} - \mathcal{T}(e_i) + \mathbf{Y}^k||_F^2$$

5. update Lagrange multiplier $\mathbf{Y}$: $\mathbf{Y}^{k+1} = \mathbf{Y}^k + \mathcal{T}(e_i) - \mathbf{L}_{e_i}^{k+1} - \mathbf{S}_{e_i}^{k+1}$;
6. update τ: $\tau_{k+1} = \min(\mu\tau_k, \tau_{\max})$;
7. check the convergence conditions:

$$||\mathbf{L}_{e_i}^{k+1} - \mathbf{L}_{e_i}^k||_\infty \leq \varepsilon, ||\mathbf{S}_{e_i}^{k+1} - \mathbf{S}_{e_i}^k||_\infty \leq \varepsilon, ||\mathbf{L}_{e_i}^{k+1} + \mathbf{S}_{e_i}^{k+1} - \mathcal{T}(e_i)||_\infty \leq \varepsilon$$

end;
end;
output: $\mathbf{L}_{e_1}, \cdots, \mathbf{L}_{e_m}$.

Through Algorithm 1, we can obtain m low-rank matrices corresponding to different feature components in the raw signal. Besides, it needs to be emphasized that the performance of the final fault diagnosis is highly susceptible to the location selection and the number of the anchor points. Once the chosen anchor points are inappropriate, there may be multiple low-rank matrices corresponding to the same feature component. More seriously, the fault-related low-rank matrices may be completely missed. In view of the above problems, for the selection of the anchor points, we do not have a good solution for now. But, in the process of analyzing the experimental data, the following two principles are feasible. The first one is that m should be large enough to ensure that the low-rank matrices corresponding to all feature components can be extracted specifically. This can be explained from the perspective of the probability. When the number of the anchor points is more, the probability of the extracted fault-related low-rank matrices is higher. However, this inevitably requires a lot of computing time. The other one is that these anchor points should be uniformly chosen from the elements set and the distance between any two anchor points should be made large enough. This principle is to make extracted low-rank matrices corresponding to the different feature components differ as much as possible. In the numerical simulation experiment, since the raw simulated signal contained three feature components, and we found that when m was set as six, the final decomposition performance was quite good. Additionally, when analyzing the experimental signal, it was found that $m = 6$ is also appropriate. Therefore, according to the analysis results of the experimental data, we assumed that the number of the main feature components in the acquired bearing fault signal generally would not exceed three, and set m to be twice that number; that is $m = 6$.

- Global Approximation of Fault Feature Component

These low-rank matrices can be backed to m one-dimensional component signals by inverse transform. In Section 2.2, it will show that the one-dimensional components related to fault feature can be identified from these signal through the MPSE characteristic of the signal. Thus, there are o $(o < m)$ identified one-dimensional signals and their corresponding low-rank matrices $\mathbf{L}_{e_1}, \ldots, \mathbf{L}_{e_o}$. These low-rank matrices are actually local sensitive, which can only be used to describe the fault feature information contained in the corresponding low-dimensional submanifolds. Hence, the Nadaraya-Watson regression model [38] is adopted to combine these low-rank matrices into a global approximation $\hat{\mathbf{L}}_f \in \mathbb{R}^{n_1 \times n_2}$ of the fault feature component, which can be expressed as:

$$\hat{\mathbf{L}}_f = \sum_{i=1}^{o} \frac{\mathbf{W}_{e_i}}{\sum_{j=1}^{m} \mathbf{W}_{e_j}} \odot \mathbf{L}_{e_i} \tag{9}$$

Note that if $o = m$, this equation is equivalent to Equation (4). Thus, the estimator of $\mathbf{L}$ can be obtained, which is actually a de-noising process. Thereby, we returned $\hat{\mathbf{L}}_f$ to the one-dimensional time series $\hat{\mathbf{l}}_f \in \mathbb{R}^n$, which is the expected fault feature signal. Hence, the task to suppress the strong background noise and extract the fault feature signal was completed.

2.3. Identification of Bearing Fault-Related Signal through MSPE

2.3.1. Basic Theory of MSPE

The MPSE can efficiently evaluate the dynamic complexity and randomness of the time series under different scales. The calculation of the MSPE depends on three parameters: scale factor ε, embedding dimension d and time-lag δ [32,36]. For a signal time series $\mathbf{x} = [x_1, x_2, \cdots, x_n]$, the main calculation steps can be divided as follows:

(1) Transform $\mathbf{x}$ into a successive coarse-grained time series $\mathbf{y}^\varepsilon \in \mathbb{R}^{n_\varepsilon}$ ($n_\varepsilon = [n/\varepsilon]$) by averaging the time data points in $\mathbf{x}$ with the given non-overlapping time slice of the increasing length, ε. Then, each element of $\mathbf{y}^\varepsilon$ is defined as:

$$y_i^\varepsilon = \frac{1}{\varepsilon} \sum_{j=(i-1)\varepsilon+1}^{i\varepsilon} x_j, i = 1, 2, \cdots, n_\varepsilon \tag{10}$$

(2) For each coarse-grained time series $\mathbf{y}^\varepsilon$, the PE value needs to be calculated. Firstly, $\mathbf{y}^\varepsilon$ is cut into a series of data segments through d and δ:

$$\mathbf{g}^i = [y_i^\varepsilon, y_{i+\delta}^\varepsilon, \ldots, y_{i+(d-1)\delta}^\varepsilon], i = 1, 2, \ldots, n_\varepsilon - (d-1)\delta \tag{11}$$

There will be $n_\varepsilon - (d-1)\delta$ data segments in total. Then, there are $d!$ different types of ordinal patterns ($\psi_i, i = 1, \ldots, d!$) in the data segments. Then, count the frequency of each pattern and denote them as $f(\psi_i), i = 1, \ldots, d!$. Thus, the relative frequency of each pattern can be written as:

$$p(\psi_i) = f(\psi_i)/(n_\varepsilon - (d-1)\delta) \tag{12}$$

Finally, the PE of $\mathbf{y}^\varepsilon$ is expressed as:

$$P(\varepsilon) = -\sum_{i=1}^{d!} p(\psi_i) \log_2 p(\psi_i) \tag{13}$$

For convenience, we normalize $P(\varepsilon)$ by dividing its maximum value $\log_2 d!$:

$$0 \le P(\varepsilon)/\log_2 d! \le 1 \tag{14}$$

(3) The PE values of different coarse-grained time series can be obtained and plotted as a function of the scale factors. The vector $\mathbf{P} = [P(1), P(2), \ldots]$, formed by the set of PE values, is the MSPE of the original time series. Figure 3 shows the process of coarse granulation and the data segmentation of a time series.

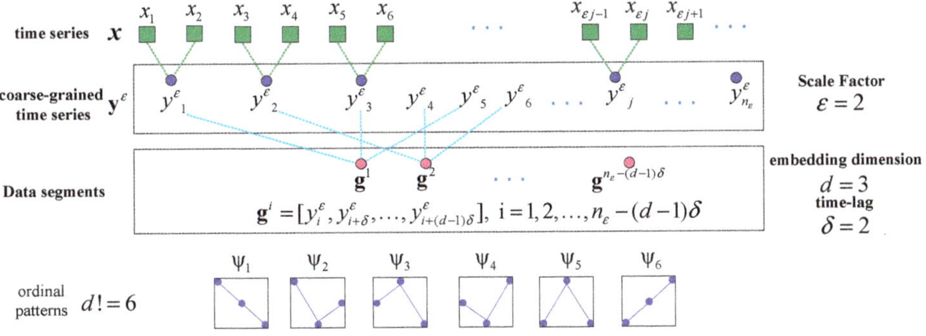

Figure 3. Illustration of the coarse-grained and data segments of the time series with $\varepsilon = 2$, $d = 3$, and $\delta = 2$, as well as the all $d! = 6$ type of ordinal patterns.

2.3.2. Mathematical Model of Bearing Fault Feature Signal

The bearing fault feature signal can be modeled as the combination of finite pulsed excitations [8,48–50]:

$$x(t) = a(t)\sum_{i=0}^{I} b_i(t)\cos(2\pi f_e t - c_i(t) + \theta_i) \tag{15}$$

where $b_i(t)$, $c(t)$, and $a_i(t)$ indicate the amplitude modulation component, the frequency modulation component, and the modulation effect caused by vibration transfer path, respectively. f_e and θ_i indicate the system resonance frequency and the initial phase, respectively.

$b_i(t)$ and $c_i(t)$ can be expressed as:

$$b_i(t) = B_i e^{-\xi(t-iT_d-v_i)} u(t - iT_d - v_i) \tag{16}$$

$$c_i(t) = \sum_{l=1}^{L} C_{il}\sin(2\pi l f_c t + \theta_{il}) \tag{17}$$

where ξ, f_c, and T_d indicate the resonance attenuation coefficient, the fault feature frequency, and the time period of fault, respectively. $u(t)$ represents a unit step function. v_i and θ_{il} represent the random slip of the i-th pulse and the initial phase, respectively. B_i and C_{il} are amplitudes.

Generally, the vibration sensor is mounted at the bearing seat, which is fixed with the outer race. In the case of the rolling element fault or the inner race fault, the vibration propagation generated by the signal transfer path from the fault location to the sensor is varies with time, resulting in an amplitude modulation effect in the signal:

$$a(t) = A[1 + \cos(2\pi f_r t)] \tag{18}$$

where f_r indicates the rotation frequency of the shaft where the fault bearing is located, and C is a constant.

In the case of the outer race fault, the vibration propagation only has a scaling effect on the signal amplitude due to the transfer path being fixed:

$$a(t) = A \tag{19}$$

One accepted approach to fault identification is to identify the fault-related frequency contents from the signal spectrum. For example, Figure 4 shows the waveform and the spectrum of a simulated bearing's inner race fault signal, and the related frequency content includes: the fault feature frequency f_c, its harmonic frequencies nf_c, the rotational frequency f_r, and the modulated side band formed by their combination $nf_c \pm f_r$; Figure 5 shows the waveform and the spectrum of a bearing outer race fault signal. The frequency content includes: the fault feature frequency f_c and its harmonic frequencies nf_c. It can be observed that nonlinear and non-stationary nature of these fault signals increases the complexity of the spectrum. Especially in the fault of inner race, where the complex modulation side band appears. As a result, the dynamic characteristic of the inner race fault should be more complex than that of outer race fault.

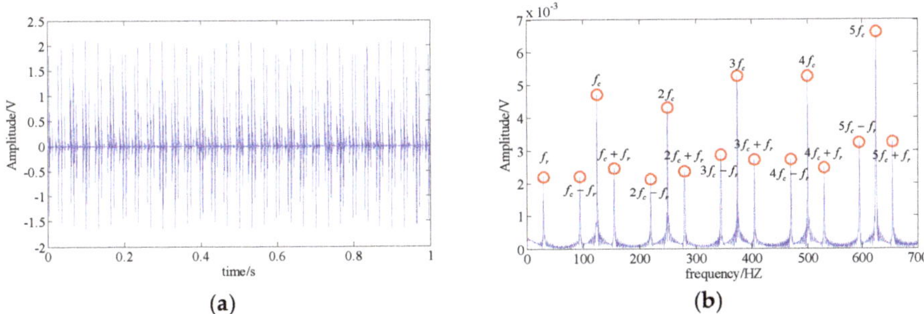

Figure 4. A simulated fault feature signal of a bearing's inner race; (a) signal waveform; (b) signal spectrum.

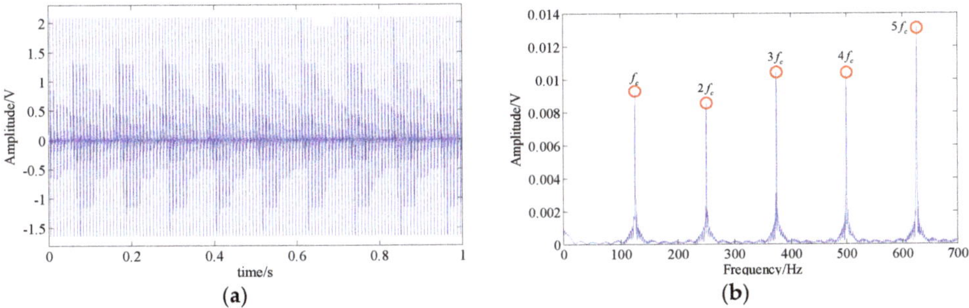

Figure 5. The simulated fault feature signal of bearing outer race; (a) signal waveform; (b) signal spectrum.

2.3.3. Identification of Fault-Related Components through MSPE

When faults occur in a mechanical system, such as gear or bearing parts, the MSPE value of the vibration signal will change, and the value varies with different type of the faults. Thus, MSPE can be adopted to identify the fault-related components.

The common interference components in bearing signals include the shock signals generated by other parts and the harmonic signals. Therefore, without loss of generality, we discuss the MSPE characteristics of different types of components bearing fault signals, including the fault feature signals of the bearing's inner race (shown as Figure 4) and the bearing outer race (shown as Figure 5), a harmonic signal, a shock signal, and a Gaussian white noise signal. The results of these five type signals are shown as Figure 6a. The results demonstrate that the MSPE value of a regular time series, such as harmonic signal or shock signal, is smaller, the examples being basically below 0.3; in contrast, since advent of the complex dynamic characteristic, the MPSE values of the bearing fault feature signals are larger, ranging from 0.4 to 0.7. In addition, it can be observed that the MSPE value of the inner race fault is larger than that of the outer race. In particular, the randomness of the noise signal is the strongest with the MSPE value more than 0.9, which proves that the MSPE is sensitive to noise.

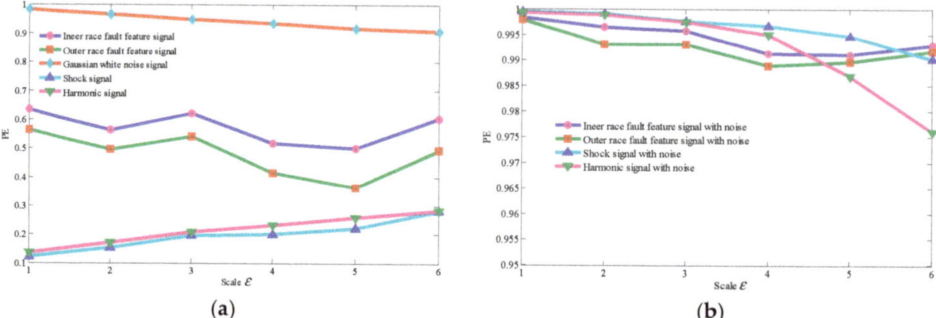

(a) (b)

Figure 6. The MSPE of different types of simulated signals; (**a**) five types of signal: the two bearing fault feature signals shown as Figures 4 and 5, a harmonic signal with the main frequency of 150 Hz, a shock signal with the main frequency of 150 Hz, and a Gaussian white noise signal; (**b**) the MSPEs of four feature signals in Figure 6a mixed with Gaussian white noise (signal to noise ratio (SNR) = −5).

According to the above analysis results, we can set a threshold range (0.5–0.85) of the MSPE value or select the larger value (but no more than 0.9) to identify the components related to bearing fault feature from the q one-dimensional component signals obtained by LRPCA.

It needs to be emphasized that although MSPE has a good anti-noise ratio, it may not be able to effectively identify the early weak fault feature signals under strong background noise. Figure 6b shows the MPSE of the four feature signals in Figure 6a, after Gaussian white noise with a SNR of −5 is added. Due to the existence of the strong background noise, the randomness of signals is greatly enhanced, resulting in the MSPE values of the four feature signals increasing to more than 0.9 and mixing together. As a result, the fault feature components are impossible to be identified. Therefore, it is an urgent problem to reduce the noise before identifying the weak fault feature components.

2.4. The Process of the Effective Fault Diagnosis Technique

Figure 7 depicts the flowchart of the proposed effective fault diagnosis technique based on the above theoretical description. And the main steps are summarized as follows:

(1) Using the proposed LRPCA method to decompose the trajectory matrix consisting of the acquired bearing fault signal into multiple low-rank matrices and to suppress the noise synchronously;
(2) Convert the low-rank matrices obtained into one-dimensional component signals by inverse transformations and identify the fault-related components from these signals through the MPSE characteristic of the signal;
(3) Using the weighted Nadaraya–Watson regression model and inverse transform to combine the low-rank matrices corresponding to identified components into a one-dimensional signal to represent the extracted fault feature component;
(4) Confirm the bearing fault by identifying the fault-related frequency contents from the signal spectrum.

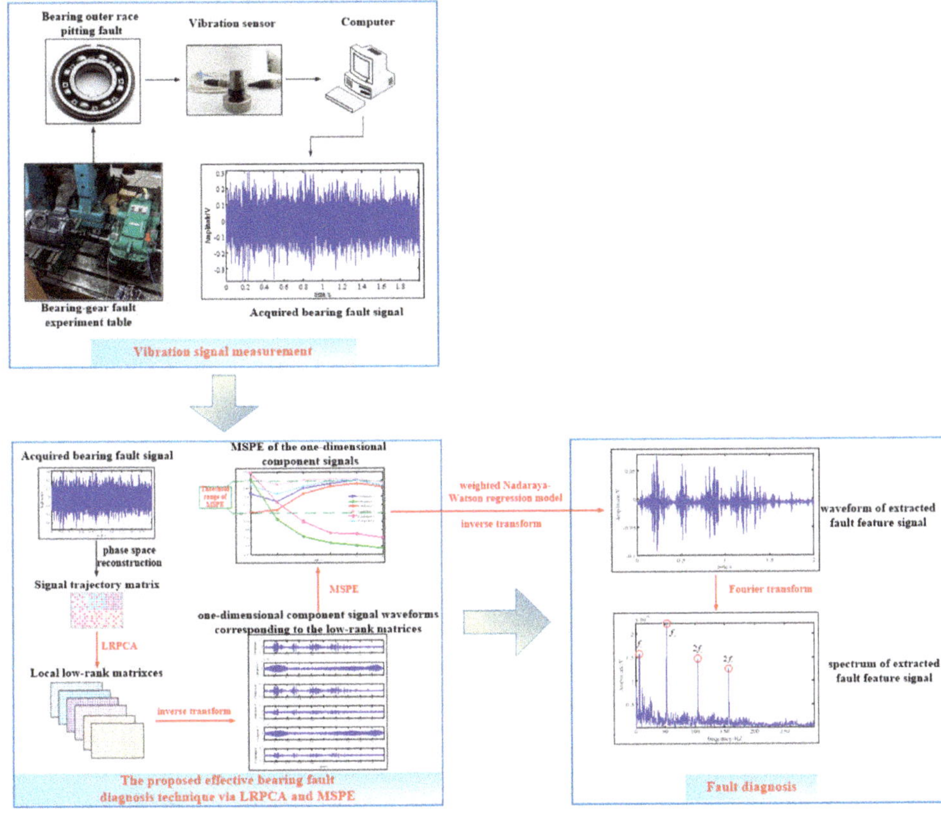

Figure 7. The flowchart of the proposed effective fault diagnosis technique via LRPCA and MSPE.

3. Experiments

3.1. Numerical Simulation Experiment

In order to verify the diagnostic performance of the proposed technique, without loss of generality, a simulated signal composed of multiple components was generated:

$$\mathbf{x} = \mathbf{x}_1(t) + \mathbf{x}_2(t) + \mathbf{x}_3(t) + \mathbf{n}(t) \tag{20}$$

where $\mathbf{x}_1(t)$ is the simulated fault feature signal of a bearing's inner race, as shown in Figure 4, and its detailed parameters are listed in Table 1. $\mathbf{x}_2(t)$ and $\mathbf{x}_3(t)$ are the interferences of a shock signal and a harmonic signal, respectively. Figure 8a,b shows the waveforms of these two signals and both of their feature frequencies are 150 Hz. $\mathbf{n}(t)$ represents the strong background white Gaussian noise. The signal sampling frequency and the sampling point are $f_s = 20{,}000$ and $n = 20{,}000$, respectively.

Table 1. Parameters in simulated fault feature signal of a bearing's inner race.

ξ	v_i	I	L	A	B_i	C_{il}	φ_i	φ_{il}	f_n	f_r	f_c
800	$0.02/f_c$	250	100	1	0.0004	$2/l^2$	0°	0°	2000 Hz	30 Hz	125 Hz

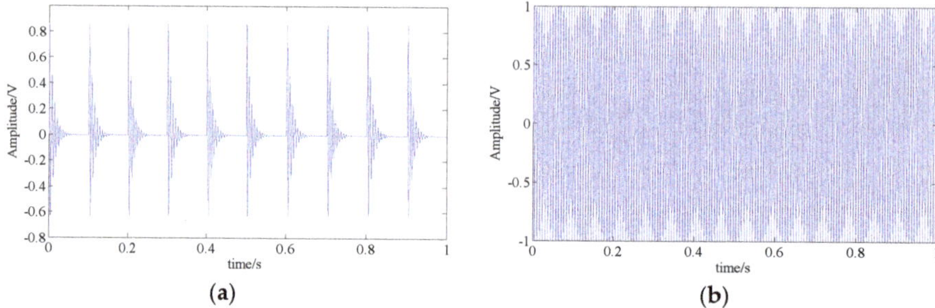

Figure 8. The simulated shock signal and harmonic signal; (**a**) shock signal waveform; (**b**) harmonic signal waveform.

Firstly, the de-noising performance of the proposed LRPCA method was tested. The strong background noise with SNR from −5 to 5 db was added to the simulated signals to imitate the early weak fault. For visually displaying the de-noising performance of LRPCA, the methods of RPCA, SSA, and EMD were employed to make a comparative analysis. The phase space reconstruction parameters used in LRPCA, RPCA, and SSA were all set as $n_1 = 200$ and $\tau = 100$. EMD employed the energy difference tracking method [51] to select desired IMF components. The hard threshold method [52] was adopted in SSA to select the best combination of singular values to reconstruct the signal. Figure 9 illustrates the de-noising result of the above four methods. In the vertical axis of the graphics, the higher SNR value indicates a better de-noising performance. It is clear that the proposed LRPCA method can provide the best noise suppression performance.

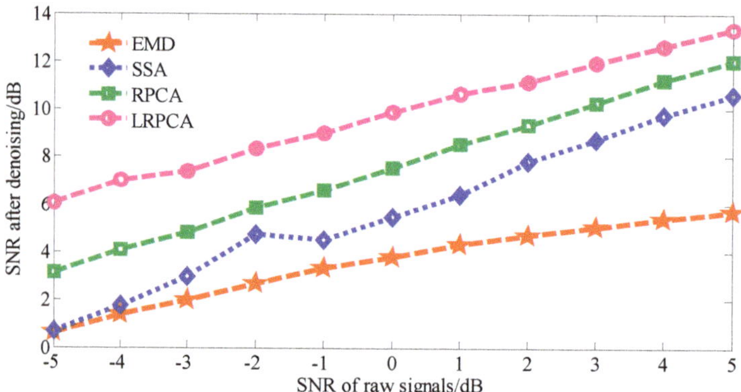

Figure 9. Comparison of the de-nosing performance of four methods when white Gaussian noise of varying SNR is added to the multi-component signal.

Then, the fault feature extraction performance of the proposed technique was investigated. In this experiment, a noise with a SNR of −5 db was added into the simulation signal. Figure 10 shows the waveform and the spectrum of the noise-signal mixture. It can be seen that the fault feature is completely submerged by noise and the other interferences, which inevitably increases the difficulty of recognizing the fault feature. The methods of wavelet shrinkage denoising [14], basis pursuit denoising [53], EMD, and SSA were selected for comparative analysis.

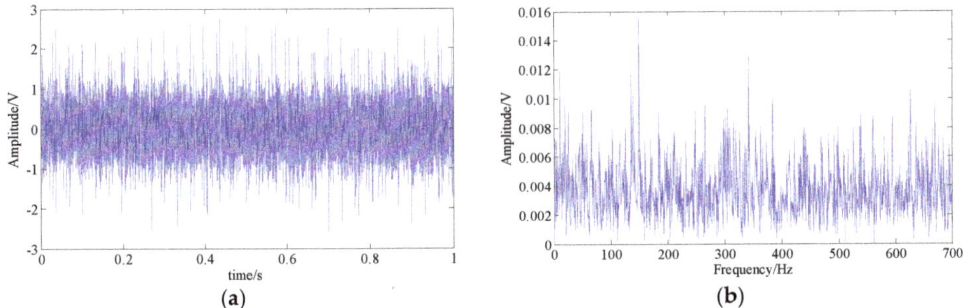

Figure 10. The simulated multi-component signal contains a strong white Gaussian noise with the SNR of −5 db; (**a**) signal waveform; (**b**) signal spectrum.

In the wavelet shrinkage denoising method, the decomposition layer was selected as 4 and the wavelet basis function was set as "db15." The waveform and the spectrum of the final extracted fault feature signal are shown in Figure 11. In the resulting spectrum, the fault feature frequency (f_c) and some of its harmonic frequencies ($2f_c$–$4f_c$) can be found. However, due to a large number of interference components and strong background noise, the identification of these fault-related frequencies was seriously affected, and the modulation sidebands were completely submerged. hence, this method was insufficient in feature extraction of a simulated signal.

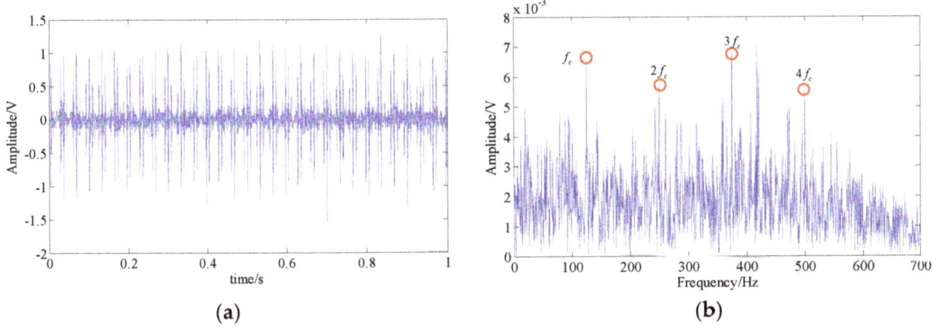

Figure 11. The analysis result of the wavelet shrinkage denoising method; (**a**) waveform of the extracted fault feature signal; (**b**) spectrum of the extracted fault feature signal.

In the basis pursuit denoising method, we adopted the compressed sensing reconstruction algorithm [54] to extract the fault feature signal and Figure 12 shows the final analysis result. In the resulting spectrum, the Fault harmonic frequencies ($3f_c$–$5f_c$) and their modulated sidebands ($3f_c \pm f_r$, $4f_c \pm f_r$, and $5f_c \pm f_r$) could be identified. But, there were still a lot of interference frequency components and strong background noise, leading to the fault of the signal not being directly determined.

In the EMD, fourteen IMFs can be obtained by decomposing the signal and the waveforms of their top twelve are shown in Figure 13a. By applying Fourier transform to these IMFs, in the resulting spectrums of IMF3 and IMF4, partial fault-related frequency contents can be found; i.e., fault harmonic frequencies ($2f_c$–$5f_c$) and the modulated sidebands ($3f_c - f_r$, $4f_c - f_r$, and $5f_c + f_r$). The waveform and the spectrum of these two IMFs are shown in Figure 13b–e. However, there were still plenty of interference frequency components and noise, which are adverse to the identification of fault feature.

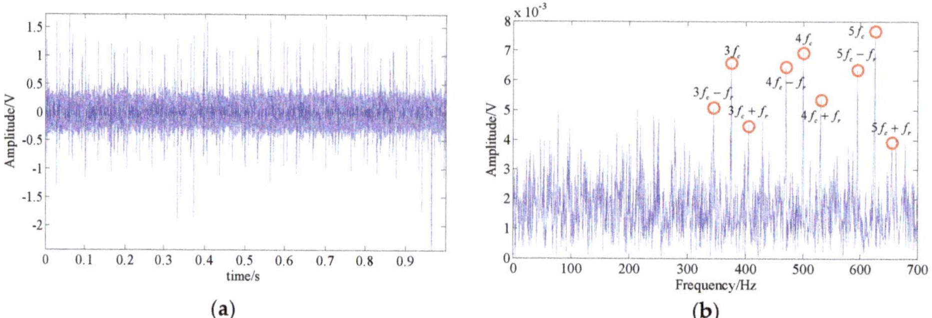

Figure 12. The analysis result of the basis pursuit denoising method; (**a**) waveform of the extracted fault feature signal; (**b**) spectrum of the extracted fault feature signal.

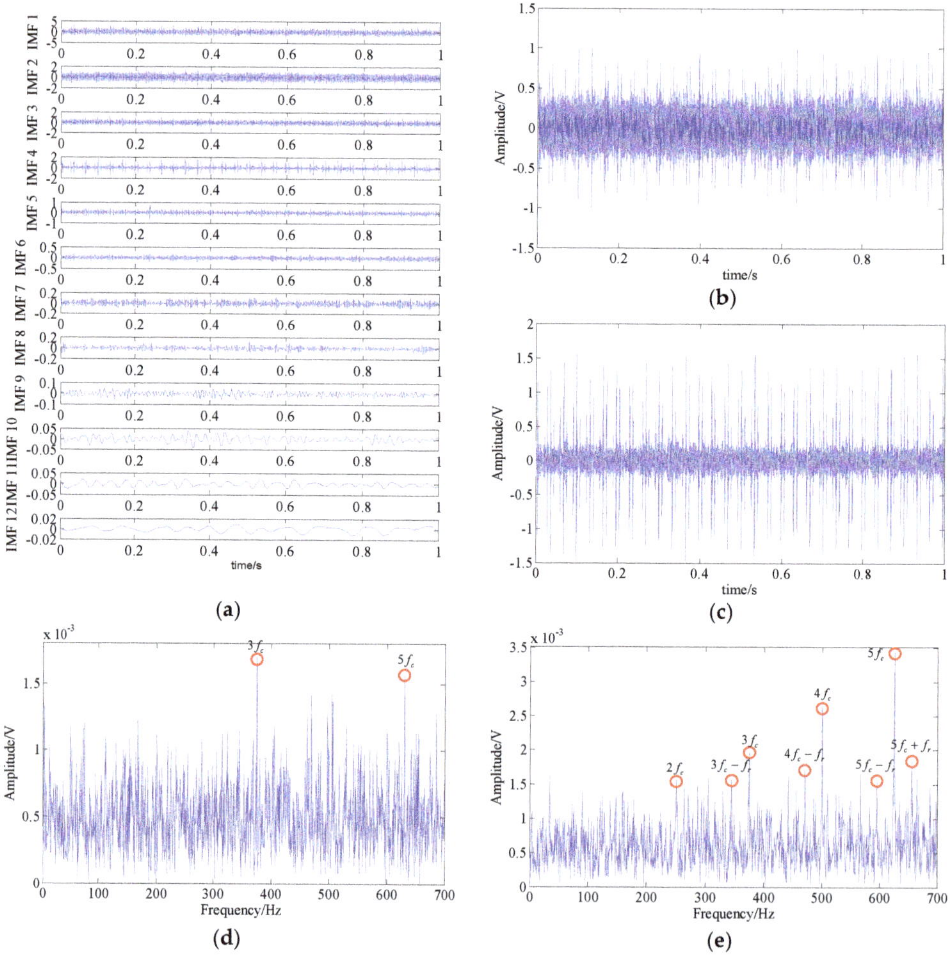

Figure 13. The analysis result of the EMD method; (**a**) waveforms of top 12 IMFs; (**b**,**d**) waveform and spectrum of IMF 3; (**c**,**e**) waveform and spectrum of IMF 4.

In the SSA, by setting the threshold value of the energy of singular value to reach 95% of the total, four singular subspaces representing different feature components can be obtained. Figure 14a illustrates the waveforms of the corresponding one-dimensional feature component signals of these four subspaces obtained through inverse transform. Then, through similarity analysis, it was concluded that components 2 and 4 had higher similarity with the raw fault feature signal. Thus, the signal obtained by adding them represents the extracted fault feature component. The waveform and the spectrum of this extracted signal are shown in Figure 14b,c. In the spectrum, some fault-related frequency contents could be found. But, similar to the analysis result of EMD, there were also some interference frequency components and noises, which may have had an adverse effect on the final diagnosed result.

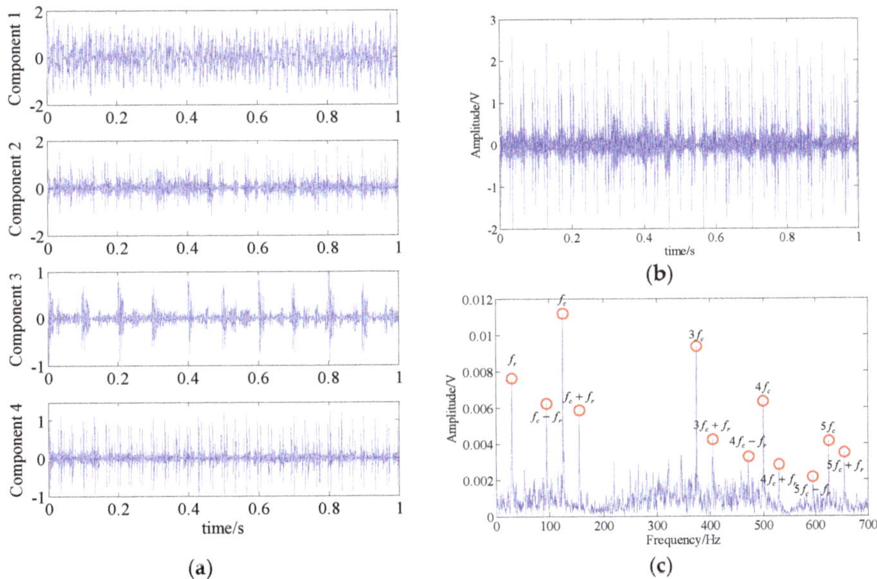

Figure 14. The analysis result of the SSA method; (**a**) waveform of the one-dimensional component signals; (**b**) waveform of the extracted fault feature signal; (**c**) spectrum of the extracted fault feature signal.

The diagnosed result of the proposed technique is shown in Figure 15. Figure 15a shows the waveforms of the one-dimensional component signals obtained by LRPCA, which correspond to the six low-rank matrices. Additionally, the MSPEs of those components are displayed in Figure 13b. It can be observed that the MSPE values of component 3 and 6 are relatively higher and evenly distributed between 0.5 and 0.8. Thus, we could combine the low-rank matrices corresponding to those two components into a one-dimensional signal to represent the extracted fault feature signal. Figure 15c,d show the waveform and the spectrum of this extracted signal. All fault related frequencies were clearly discernable in the spectrum. Furthermore, it can be observed that the interference frequency components and the background noise were basically eliminated. According to the above analysis information, we can undoubtedly confirm that the fault occurred at the inner race.

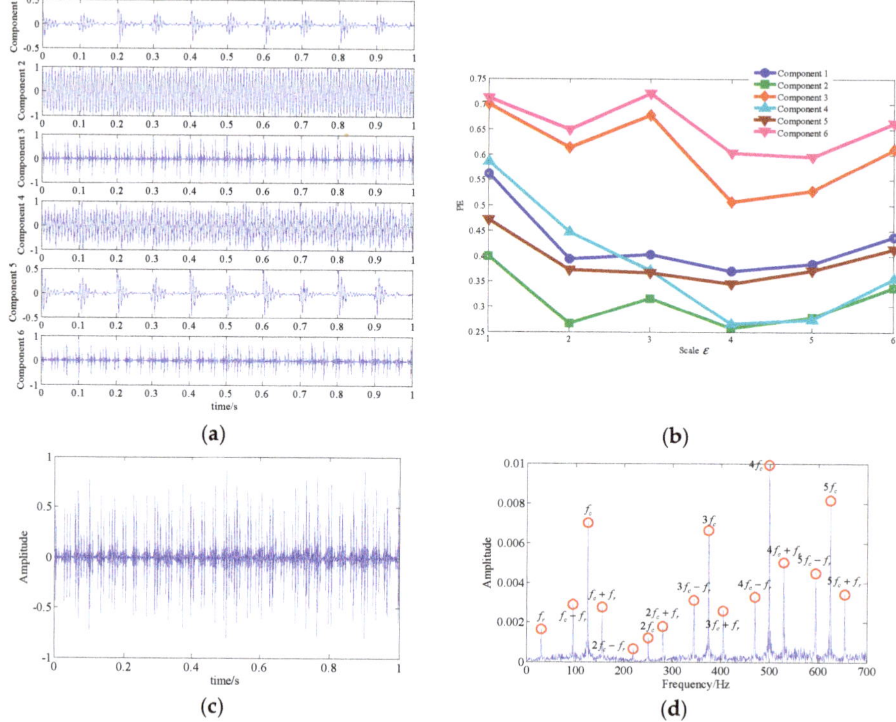

Figure 15. Analysis result of the proposed technique; (**a**) waveform of the one-dimensional component signals; (**b**) the MSPE of the components; (**c**) waveform of the extracted fault feature signal; (**d**) spectrum of the extracted fault feature signal.

The above simulation results indicate that EMD and SSA perform when decomposing a complicated multi-component signal into finite single-components. However, their anti-noise ability against strong noise is insufficient and there are still some interference components in the final extracted feature signals. On the contrary, the proposed technique is more effective in eliminating the strong background noise and extracting the weak fault feature component, and its diagnostic performance for the simulated signal is obviously better than the other two methods.

3.2. Experimental Signal Analysis

The vibration signal acquired on the spot is more complex than the simulation signal. In order to further verify the practicability of the proposed technique, a pitting fault signal of the bearing's outer race sampling from a bearing-gear fault's experimental table was analyzed. Figure 16 shows the experimental table, which consists of an AC motor, couplings, a gearbox with two pairs of meshing gears, and a magnetic powder brake. The test bearing is a single-row tapered roller bearing of the type of 32206 and its fault was handled by electrical discharge machining (EDM) method. The red arrow in Figure 16a displays the mounting position of the bearing. The experiment parameter was listed in Table 2. The vibration data of this experiment are measured by PCB acceleration sensor in the vertical direction of the bearing.

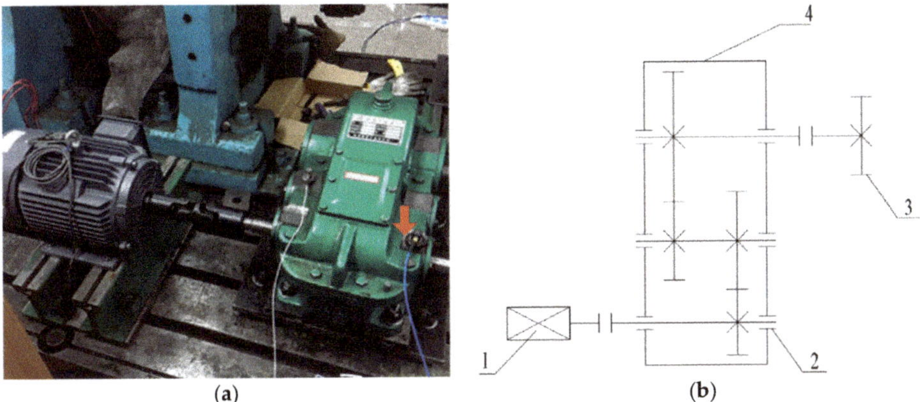

(a) (b)

Figure 16. Bearing-gear fault experiment table; (**a**) physical photograph; (**b**) structural drawing: 1—AC motor; 2—the mounting position of fault bearing; 3—magnetic powder brake; 4—gearbox.

Table 2. Experiment parameters of fault bearing.

Number of Roller Elements	Roller Diameter (mm)	Medium Diameter (mm)	Contact Angle	Rotation Frequency (Hz)	Fault Frequency (Hz)	Sampling Points	Sampling Frequency (Hz)
$z = 17$	$d = 8$	$D = 46$	$\alpha = 14.04°$	$f_r = 7.225$	$f_o = 51.05$	$N = 20{,}000$	$f_s = 10{,}000$

Figure 17 is the diagram of the waveform and the spectrum of the acquired signal. It can be seen that the weak fault feature was impossible to be identified. Then, the proposed technique was utilized to diagnose the signal, and the methods of wavelet shrinkage denoising, basis pursuit denoising, EMD, and SSA were chosen for a comparative analysis.

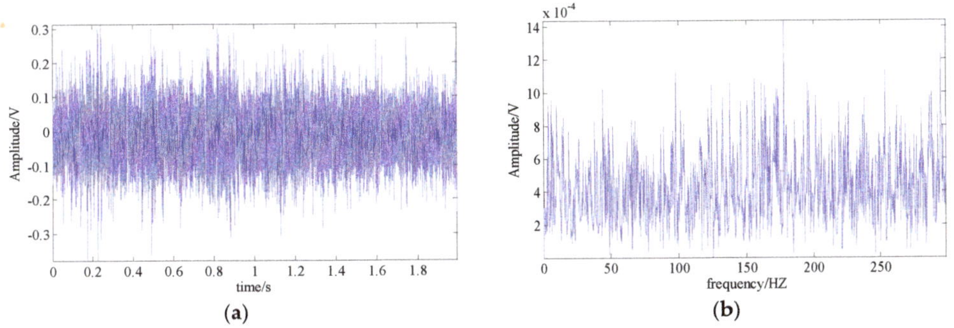

(a) (b)

Figure 17. The acquired bearing fault signal; (**a**) signal waveform; (**b**) signal spectrum.

The analysis results of the wavelet shrinkage denoising method and basis pursuit denoising method are shown as Figures 18 and 19, respectively. In their result spectrums, although the fault characteristic frequency (f_o) and its harmonic frequencies ($2f_o$–$3f_o$) can be found, the identification of these fault-related frequencies is seriously affected by a large number of interference components and strong background noise. Therefore, the fault type of bearing cannot be directly determined from these analysis results.

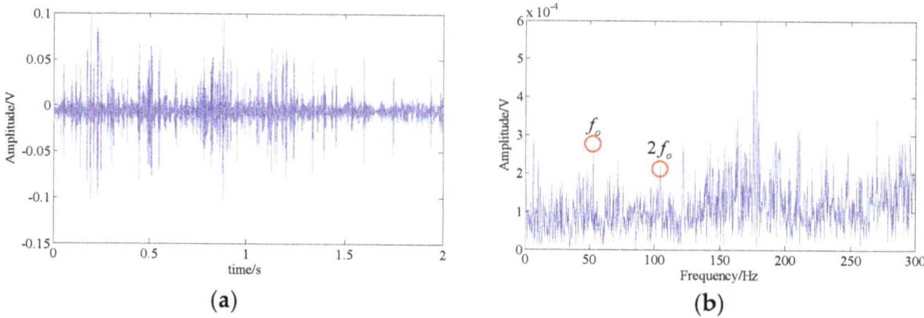

Figure 18. Analysis result of the wavelet shrinkage denoising method; (**a**) waveform of the extracted fault feature signal; (**b**) spectrum of the extracted fault feature signal.

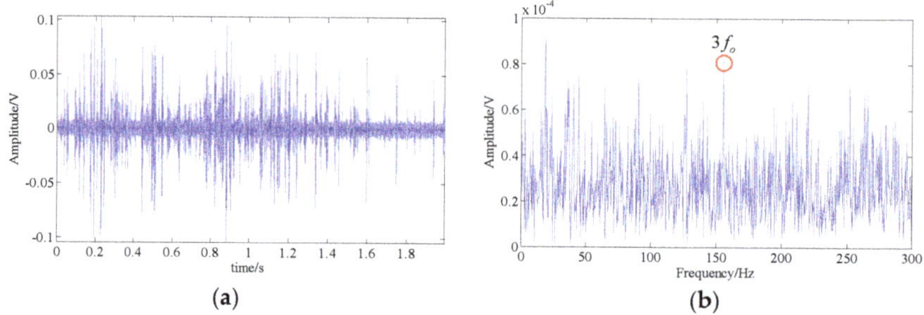

Figure 19. Analysis result of the basis pursuit denoising method; (**a**) waveform of the extracted fault feature signal; (**b**) spectrum of the extracted fault feature signal.

Figure 20a displays the waveforms of top 12 IMFs obtained through EMD. By applying the Fourier transform to them, the fault-related frequency contents (f_o and $2f_o$) can be found in the spectrums of IMF 1 and IMF 8, which are shown in Figure 20b,d,e. However, that fault feature information is hard to be recognized due to the presence of many interference frequency components and strong background noise. Therefore, it is difficult to determine whether the outer race of the bearing is faulty.

Figure 21 shows the fault feature extraction result of SSA. The peaks of fault feature frequency (f_o) and its triple frequency ($3f_o$) were obvious in the spectrum. But there are still many interference peaks and noise, which affects the identification of fault feature. These above analysis results indicate that neither EMD nor SSA can provide good fault diagnosis performances for the experimental fault signal.

The proposed technique was utilized to diagnose the signal. Figure 22a shows the waveforms of the one-dimensional component signal corresponding to the six low-rank matrices obtained by LRPCA. Furthermore, their MSPE values were displayed in Figure 22b. It can be observed that the MSPE values of component 1, 3, 4, and 6 are relatively higher. Meanwhile, the MSPE value of component 4 was basically above 0.9, which may be the feature signal of other component with more complex dynamics characteristic. The MSPE values of components 1, 3, and 6 ranged from 0.7 to 0.9, and their trends were similar, so we combined the low-rank matrices corresponding to these three components into a one-dimensional feature signal representing the extracted the fault feature signal. Figure 22c,d show the waveform and the spectrum of this extracted signal. In the spectrum, the rotation frequency (f_r), the fault feature frequency (f_o), and its double frequency ($2f_o$) and triple frequency ($3f_o$) can be easily found. Moreover, it can be seen that the energy of the noise was suppressed at a low level and there were a few interference frequency components, which had little effect on the final diagnosed result. Consequently, we could confirm that the outer race of bearing had fault. These analysis results demonstrate that the proposed technique can effectively suppress the strong background noise and extract weak fault feature component from the multi-component signal, which means the proposed technique can provide a great diagnostic performance when dealing with the experimental signal.

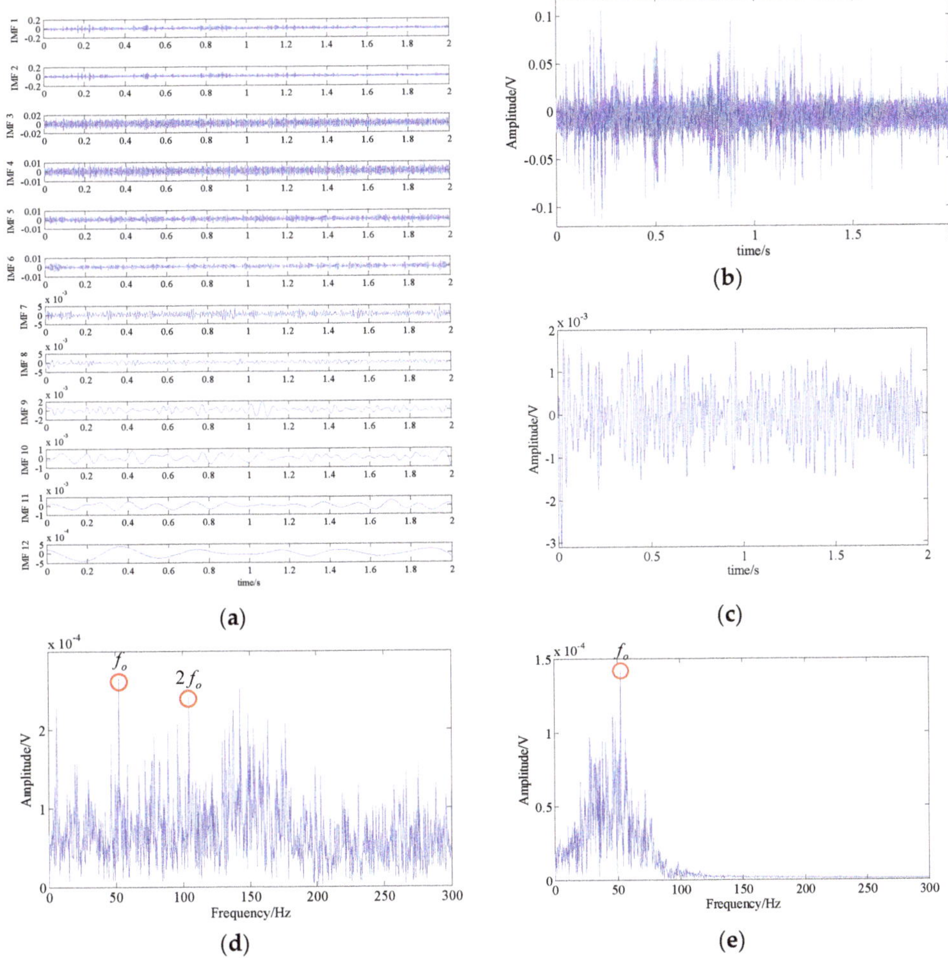

Figure 20. Analysis result of the EMD method; (**a**) waveforms of top 12 IMFs; (**b**,**d**) waveform and spectrum of IMF 1; (**c**,**e**) waveform and spectrum of IMF 8.

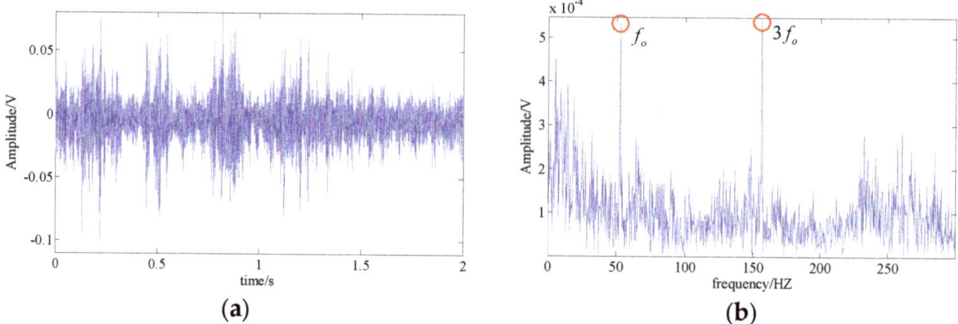

Figure 21. Analysis result of the SSA method; (**a**) waveform of the extracted fault feature signal; (**b**) spectrum of the extracted fault feature signal. The fault feature extraction results of SSA. The peaks of the fault feature frequency (f_o) and its triple frequency ($3f_o$) were obvious in the spectrum. But there are still many interference peaks and noise, which affect the identification of the fault feature. These above analysis results indicate that neither EMD nor SSA can provide a good fault diagnosis performance for the experimental fault signal.

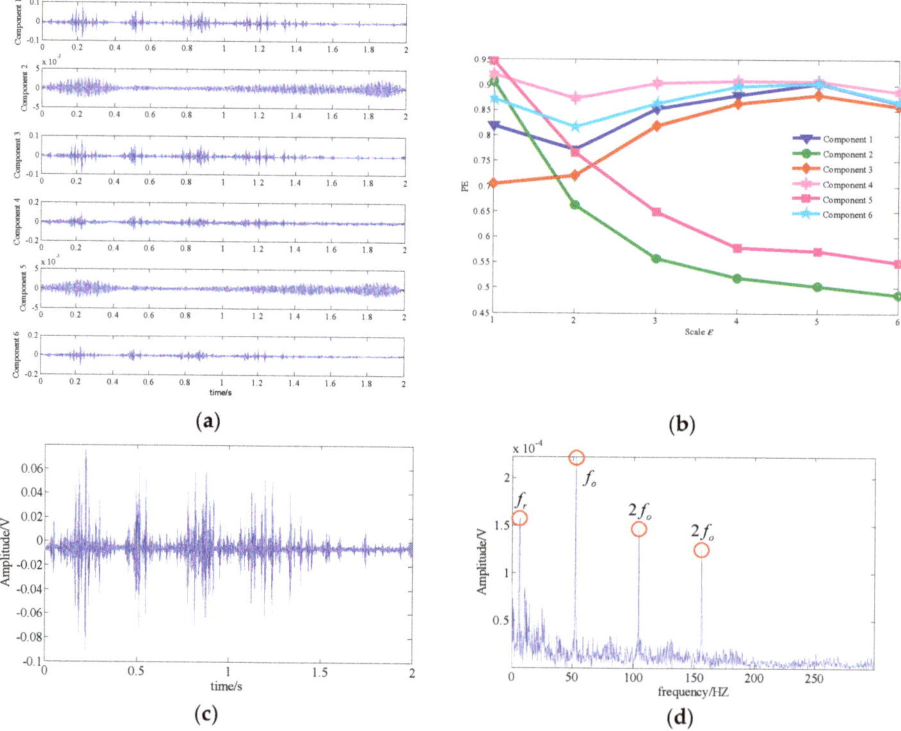

Figure 22. Analysis result of the proposed technique; (**a**) waveform of the one-dimensional component signals; (**b**) the MSPE of the components; (**c**) waveform of the extracted fault feature signal; (**d**) spectrum of the extracted fault feature signal.

4. Conclusions

In general, the bearing's weak fault feature exhibits the nature of nonlinear and non-stationary, which is hard to be extracted under the situation of existing strong background noise and interference components. In considering this problem, an effective bearing fault diagnosis technique via LRPCA and MSPE was introduced in this paper. The LRPCA can decompose the signal trajectory matrix into multiple low-rank matrices, and meanwhile, suppress the noise. The MSPE was used to identify the low-rank matrices corresponding to bearing's fault feature. Thereafter, those identified low-rank matrices were combined into a one-dimensional signal to represent the extracted fault feature component for further fault diagnosis. The principle and the effectiveness of this technique was verified by the analysis of both the simulation signal and the acquired bearing fault signal. The analysis results indicate that the proposed technique can effectively detect and locate the bearing faults accurately.

The threshold range representing the bearing fault feature component was determined based on the results of simulation experiments. However, the feature components of the acquired bearing fault signals should be much more complex than simulated signals. Therefore, we will focus on determining a more appropriate threshold range for acquired signal in future work. In addition, the de-nosing performance of proposed LRPCA method was verified by the signal simulation Equation (20), and Figure 9 shows that it performs better than other methods. In fact, the robustness of the method can be improved by statistical evidence. By extending the evaluation to greater number of rounds, the results of each round are collected, and through meaningful statistical tests, the parameters of the method can be optimized, leading to a significant improvement in noise reduction performance. Therefore, we will also carry out more deep research to that end in future work.

Author Contributions: Conceptualization: Y.L.; Methodology: Y.L. and M.G.; Investigation: C.Y. and Y.Z.; Data curation: Y.Z.; Funding acquisition: Y.L.; Project administration: C.Y.; Resources: C.Y. and Y.Z.; Software: M.G. and Y.M.; Supervision: Y.L.; Validation: M.G. and Y.M.; Writing-original draft: M.G.; Writing-review and editing: Y.L. and Y.M.

Funding: This research was funded by the National Natural Science Foundation of China under Grants No.51875416 and No.51475339, the Wuhan Science and Technology Project under Grants No.2017010201010115.

Conflicts of Interest: The authors declare that there is no conflict of interests.

References

1. Zhang, W.; Li, C.; Peng, G.; Chen, Y.; Zhang, Z. A deep convolutional neural network with new training methods for bearing fault diagnosis under noisy environment and different working load. *Mech. Syst. Signal Process.* **2018**, *100*, 439–453. [CrossRef]
2. Wang, S.; Selesnick, I.; Cai, G.; Feng, Y.; Sui, X.; Chen, X. Nonconvex sparse regularization and convex optimization for bearing fault diagnosis. *IEEE Trans. Ind. Electron.* **2018**, *65*, 7332–7342. [CrossRef]
3. Lv, Y.; Yuan, R.; Song, G. Multivariate empirical mode decomposition and its application to fault diagnosis of rolling bearing. *Mech. Syst. Signal Process.* **2016**, *81*, 219–234. [CrossRef]
4. Guo, X.; Chen, L.; Shen, C. Hierarchical adaptive deep convolution neural network and its application to bearing fault diagnosis. *Measurement* **2016**, *93*, 490–502. [CrossRef]
5. Wang, Z.; Wang, J.; Kou, Y.; Zhang, J.; Ning, S.; Zhao, Z. Weak fault diagnosis of wind turbine gearboxes based on MED-LMD. *Entropy* **2017**, *19*, 277. [CrossRef]
6. Chen, J.; Pan, J.; Li, Z.; Zi, Y.; Chen, X. Generator bearing fault diagnosis for wind turbine via empirical wavelet transform using measured vibration signals. *Renew. Energy* **2016**, *89*, 80–92. [CrossRef]
7. Bab, S.; Najafi, M.; Sola, J.F.; Abbasi, A. Annihilation of non-stationary vibration of a gas turbine rotor system under rub-impact effect using a nonlinear absorber. *Mech. Mach. Theory* **2019**, *139*, 379–406. [CrossRef]
8. Liang, M.; Bozchalooi, I.S. An energy operator approach to joint application of amplitude and frequency-demodulations for bearing fault detection. *Mech. Syst. Signal Process.* **2010**, *24*, 1473–1494. [CrossRef]
9. Ciabattoni, L.; Ferracuti, F.; Freddi, A.; Monteriu, A. Statistical Spectral Analysis for Fault Diagnosis of Rotating Machines. *IEEE Trans. Ind. Electron.* **2017**, *65*, 4301–4310. [CrossRef]

10. Tang, B.; Liu, W.; Song, T. Wind turbine fault diagnosis based on Morlet wavelet transformation and Wigner-Ville distribution. *Renew. Energy* **2010**, *35*, 2862–2866. [CrossRef]
11. Chen, J.; Li, Z.; Pan, J.; Chen, G.; Zi, Y.; Yuan, J.; Chen, B.; He, Z. Wavelet transform based on inner product in fault diagnosis of rotating machinery: A review. *Mech. Syst. Signal Process.* **2016**, *70*, 1–35. [CrossRef]
12. Wang, Y.; Xu, G.; Liang, L.; Jiang, K. Detection of weak transient signals based on wavelet packet transform and manifold learning for rolling element bearing fault diagnosis. *Mech. Syst. Signal Process.* **2015**, *54*, 259–276. [CrossRef]
13. Deng, W.; Zhang, S.; Zhao, H.; Yang, X. A novel fault diagnosis method based on integrating empirical wavelet transform and fuzzy entropy for motor bearing. *IEEE Access* **2018**, *6*, 35042–35056. [CrossRef]
14. Xiao, M.; Wen, K.; Zhang, C.; Zhao, X.; Wei, W.; Wu, D. Research on fault feature extraction method of rolling bearing based on NMD and wavelet threshold denoising. *Shock Vib.* **2018**. [CrossRef]
15. Pachori, R.B.; Nishad, A. Cross-terms reduction in the Wigner–Ville distribution using tunable-Q wavelet transform. *Signal Process.* **2016**, *120*, 288–304. [CrossRef]
16. Ming, Y.; Chen, J.; Dong, G. Weak fault feature extraction of rolling bearing based on cyclic Wiener filter and envelope spectrum. *Mech. Syst. Signal Process.* **2011**, *25*, 1773–1785. [CrossRef]
17. Liu, W.Y.; Gao, Q.W.; Ye, G.; Ma, R.; Lu, X.N.; Han, J.G. A novel wind turbine bearing fault diagnosis method based on Integral Extension LMD. *Measurement* **2015**, *74*, 70–77. [CrossRef]
18. Li, Y.; Xu, M.; Wang, R.; Huang, W. A fault diagnosis scheme for rolling bearing based on local mean decomposition and improved multiscale fuzzy entropy. *J. Sound Vib.* **2016**, *360*, 277–299. [CrossRef]
19. Lv, Y.; Yuan, R.; Wang, T.; Li, H.; Song, G. Health degradation monitoring and early fault diagnosis of a rolling bearing based on CEEMDAN and improved MMSE. *Materials* **2018**, *11*, 1009. [CrossRef] [PubMed]
20. Imaouchen, Y.; Kedadouche, M.; Alkama, R.; Thomas, M. A frequency-weighted energy operator and complementary ensemble empirical mode decomposition for bearing fault detection. *Mech. Syst. Signal Process.* **2017**, *82*, 103–116. [CrossRef]
21. Bustos, A.; Rubio, H.; Castejón, C.; García-Prada, J. EMD-based methodology for the identification of a high-speed train running in a gear operating state. *Sensors* **2018**, *18*, 793. [CrossRef] [PubMed]
22. Kennel, M.B.; Brown, R.; Abarbanel, H.D.I. Determining embedding dimension for phase-space reconstruction using a geometrical construction. *Phys. Rev. A* **1992**, *45*, 3403. [CrossRef] [PubMed]
23. Yi, C.; Lv, Y.; Dang, Z.; Xiao, H.; Yu, X. Quaternion singular spectrum analysis using convex optimization and its application to fault diagnosis of rolling bearing. *Measurement* **2017**, *103*, 321–332. [CrossRef]
24. Kouchaki, S.; Sanei, S.; Arbon, E.L.; Dijk, D.J. Tensor based singular spectrum analysis for automatic scoring of sleep EEG. *IEEE Trans. Neural Syst. Rehabil. Eng.* **2015**, *23*, 1–9. [CrossRef] [PubMed]
25. Liao, Z.; Song, L.; Chen, P.; Guan, Z.; Fang, Z.; Li, K. An effective singular value selection and bearing fault signal filtering diagnosis method based on false nearest neighbors and statistical information criteria. *Sensors* **2018**, *18*, 2235. [CrossRef] [PubMed]
26. Wang, H.; Nie, F.; Huang, H.; Makedon, F. Fast nonnegative matrix tri-factorization for large-scale data co-clustering. In Proceedings of the Twenty-Second International Joint Conference on Artificial Intelligence, Barcelona, Spain, 16–22 July 2011.
27. Lin, T.; Zha, H. Riemannian manifold learning. *IEEE Trans. Pattern Anal. Mach. Intell.* **2008**, *30*, 796–809.
28. Erichson, N.B.; Voronin, S.; Brunton, S.L.; Kutz, J.N. Randomized Matrix Decompositions Using R. *J. Stat. Softw.* **2019**, *89*, 1–48. [CrossRef]
29. Sun, W.; Du, Q. Graph-regularized fast and robust principal component analysis for hyperspectral band selection. *IEEE Trans. Geosci. Remote Sens.* **2018**, *56*, 3185–3195. [CrossRef]
30. Dang, C.; Radha, H. RPCA-KFE: Key frame extraction for video using robust principal component analysis. *IEEE Trans. Image Process.* **2015**, *24*, 3742–3753. [CrossRef]
31. Li, C.Y.; Zhu, L.; Bao, W.Z.; Jiang, Y.L.; Yuan, C.A.; Huang, D.S. Convex local sensitive low rank matrix approximation. In Proceedings of the 2017 International Joint Conference on Neural Networks (IJCNN), Anchorage, AK, USA, 14–19 May 2017.
32. Morabito, F.C.; Labate, D.; La Foresta, F.; Bramanti, A.; Morabito, G.; Palamara, I. Multivariate multi-scale permutation entropy for complexity analysis of Alzheimer's disease EEG. *Entropy* **2012**, *14*, 1186–1202. [CrossRef]
33. Yi, C.; Lv, Y.; Ge, M.; Xiao, H.; Yu, X. Tensor singular spectrum decomposition algorithm based on permutation entropy for rolling bearing fault diagnosis. *Entropy* **2017**, *19*, 139. [CrossRef]

34. Zhang, X.; Liang, Y.; Zhou, J. A novel bearing fault diagnosis model integrated permutation entropy, ensemble empirical mode decomposition and optimized SVM. *Measurement* **2015**, *69*, 164–179. [CrossRef]
35. Zheng, J.; Pan, H.; Yang, S.; Cheng, J. Generalized composite multiscale permutation entropy and Laplacian score based rolling bearing fault diagnosis. *Mech. Syst. Signal Process.* **2018**, *99*, 229–243. [CrossRef]
36. Ge, M.; Lv, Y.; Yi, C.; Zhang, Y.; Chen, X. A Joint Fault Diagnosis Scheme Based on Tensor Nuclear Norm Canonical Polyadic Decomposition and Multi-Scale Permutation Entropy for Gears. *Entropy* **2018**, *20*, 161. [CrossRef]
37. Candès, E.J.; Recht, B. Exact matrix completion via convex optimization. *Found. Comput. Math.* **2009**, *9*, 717. [CrossRef]
38. Cai, Z. Weighted nadaraya–watson regression estimation. *Stat. Probab. Lett.* **2001**, *51*, 307–318. [CrossRef]
39. Lee, J.; Kim, S.; Lebanon, G.; Singer, Y.; Bengio, S. LLORMA: Local low-rank matrix approximation. *J. Mach. Learn. Res.* **2016**, *17*, 442–465.
40. Jones, M.C. Simple boundary correction for kernel density estimation. *Stat. Comput.* **1993**, *3*, 135–146. [CrossRef]
41. Liu, G.; Lin, Z.; Yan, S.; Sun, J.; Yu, Y.; Ma, Y. Robust recovery of subspace structures by low-rank representation. *IEEE Trans. Pattern Anal. Mach. Intell.* **2012**, *35*, 171–184. [CrossRef] [PubMed]
42. Zhang, H.; Lin, Z.; Zhang, C.; Gao, J. Robust latent low rank representation for subspace clustering. *Neurocomputing* **2014**, *145*, 369–373. [CrossRef]
43. Wright, J.; Ganesh, A.; Rao, S.; Peng, Y.; Ma, Y. Robust principal component analysis: Exact recovery of corrupted low-rank matrices via convex optimization. In Proceedings of the Advances in Neural Information Processing Systems 2009, Vancouver, BC, Canada, 7–10 December 2009; pp. 2080–2088.
44. Candès, E.J.; Li, X.; Ma, Y.; Wright, J. Robust principal component analysis? *J. ACM (JACM)* **2011**, *58*, 11. [CrossRef]
45. Lv, Y.; Ge, M.; Zhang, Y.; Yi, C.; Ma, Y. A Novel Demodulation Analysis Technique for Bearing Fault Diagnosis via Energy Separation and Local Low-Rank Matrix Approximation. *Sensors* **2019**, *19*, 3755. [CrossRef] [PubMed]
46. Hong, M.; Luo, Z.Q. On the linear convergence of the alternating direction method of multipliers. *Math. Program.* **2017**, *162*, 165–199. [CrossRef]
47. Lin, Z.; Chen, M.; Wu, L.; Ma, Y. *The Augmented Lagrange Multiplier Method for Exact Recovery of Corrupted Low-Rank Matrices*; Coordinated Science Laboratory Report no. UILU-ENG-09-2215, DC-247; Coordinated Science Laboratory: Urbana, IL, USA, 2009.
48. Feng, Z.; Ma, H.; Zuo, M.J. Amplitude and frequency demodulation analysis for fault diagnosis of planet bearings. *J. Sound Vib.* **2016**, *382*, 395–412. [CrossRef]
49. Randall, R.B.; Antoni, J.; Chobsaard, S. The relationship between spectral correlation and envelope analysis in the diagnostics of bearing faults and other cyclostationary machine signals. *Mech. Syst. Signal Process.* **2001**, *15*, 945–962. [CrossRef]
50. Feng, Z.; Zuo, M.J.; Qu, J.; Tian, T.; Liu, Z. Joint amplitude and frequency demodulation analysis based on local mean decomposition for fault diagnosis of planetary gearboxes. *Mech. Syst. Signal Process.* **2013**, *40*, 56–75. [CrossRef]
51. Cheng, J.S.; Yu, D.J.; Yu, Y. Research on the intrinsic mode function (IMF) criterion in EMD method. *Mech. Syst. Signal Process.* **2006**, *20*, 817–824.
52. Tanner, J.; Wei, K. Normalized iterative hard thresholding for matrix completion. *SIAM J. Sci. Comput.* **2013**, *35*, S104–S125. [CrossRef]
53. Chen, S.S.; Donoho, D.L.; Saunders, M.A. Atomic decomposition by basis pursuit. *SIAM Rev.* **2001**, *43*, 129–159. [CrossRef]
54. Donoho, D.L. Compressed sensing. *IEEE Trans. Inf. Theory* **2006**, *52*, 1289–1306. [CrossRef]

Article

Partial Discharge Fault Diagnosis Based on Multi-Scale Dispersion Entropy and a Hypersphere Multiclass Support Vector Machine

Haikun Shang [1,*], Feng Li [2] and Yingjie Wu [3]

1 College of Electrical Engineering, Northeast Electric Power University, Jilin 132012, China
2 State Grid Electric Power Research Institute, Xinjiang 830011, China; kasuo5215@126.com
3 College of Automation Engineering, Northeast Electric Power University, Jilin 132012, China; wuyingjie6668@163.com
* Correspondence: shanghk@neepu.edu.cn; Tel.: +86-432-6480-6691

Received: 24 December 2018; Accepted: 15 January 2019; Published: 17 January 2019

Abstract: Partial discharge (PD) fault analysis is an important tool for insulation condition diagnosis of electrical equipment. In order to conquer the limitations of traditional PD fault diagnosis, a novel feature extraction approach based on variational mode decomposition (VMD) and multi-scale dispersion entropy (MDE) is proposed. Besides, a hypersphere multiclass support vector machine (HMSVM) is used for PD pattern recognition with extracted PD features. Firstly, the original PD signal is decomposed with VMD to obtain intrinsic mode functions (IMFs). Secondly proper IMFs are selected according to central frequency observation and MDE values in each IMF are calculated. And then principal component analysis (PCA) is introduced to extract effective principle components in MDE. Finally, the extracted principle factors are used as PD features and sent to HMSVM classifier. Experiment results demonstrate that, PD feature extraction method based on VMD-MDE can extract effective characteristic parameters that representing dominant PD features. Recognition results verify the effectiveness and superiority of the proposed PD fault diagnosis method.

Keywords: PD; fault diagnosis; variational mode decomposition; multi-scale dispersion entropy; HMSVM

1. Introduction

Partial discharge (PD) is an important symptom of insulation degradation for electrical equipment. PD fault diagnosis plays an irreplaceable role in the evaluation of insulation condition [1]. PD feature extraction is an important step in insulation fault diagnosis. The common methods include statistical atlas (SA) [2], wave analysis (WA) [3] and wavelet transform (WT) [4]. However, SA has the limitations of high request of sampling rate, large data size and slow speed of data processing which are not suitable for on-line monitoring. Besides, it is difficult to extract PD phase information during statistical atlas construction. WA is easily influenced by electromagnetic interference. WT has some inherent limitations such as the difficulty of selection of the wavelet basis, wavelet thresholds, decomposition levels, and so on [5].

Empirical mode decomposition (EMD), as an adaptive signal processing method that decomposes a time series into some limited intrinsic mode functions (IMFs). It is widely used in the areas of fault detection, signal processing and data compression [6–8]. However, due to the problems of ending effects and mode mixing in non-stationary signal decomposition, EMD is limited in practical applications. Variational mode decomposition (VMD) is a new signal decomposition method, which is widely applied in electrical fault feature extraction [9]. It is a non-recursive variational decomposition model. In VMD, the central frequency and bandwidth of each mode are determined by searching the optimal solution of the variation model. VMD can solve the problems of mode mixing and ending effects in traditional EMD methods [10]. In this paper, VMD is employed for PD signal decomposition to extract effective IMFs from PD signals.

In order to quantify the PD feature information extracted by VMD, entropy theory is introduced. Entropy, as a measure of uncertainty or irregularity, was widely applied in fault diagnosis recently [11]. It was first introduced by Shannon in 1948 [12]. Afterwards, approximate entropy (AE) was put forward by Pincus [13], which provided one dimensionless index representing signal features. It was suitable for both deterministic and random signals. However, AE is heavily relied on the data length. Moreover, its estimated value is uniformly lower than expected ones when processing the short dataset [14]. To overcome the weakness of AE, Richman and Moorman proposed sample entropy (SE) [15]. Due to the insensitivity to the data length and immunity to the noise in data, SE has attracted a great deal of attention. However, SE is not fast enough for some real-time applications, especially for long signals [16]. Another widely used regularity indicator is permutation entropy (PE), which is based on the order relations among values of a signal [17]. Although PE is conceptually simple and computationally fast, the method does not consider the mean value of amplitudes and differences between amplitude values [18]. In this paper, a new irregularity indicator is introduced, named dispersion entropy (DE) [19]. The method tackles the abovementioned PE and SE limitations [20]. Because of the relevance and the possible usefulness of DE in several signal analyses, it is important to understand the behavior of the technique for various kinds of classical signal concepts such as amplitude, frequency, noise power, and signal band-width. However, DE estimates the complexity at a single scale [21], which gives rise to unacceptable result when applied to analyze the multiple time scales data [22]. Regarding this disadvantage, a multi-scale dispersion entropy (MDE) procedure was put forward to estimate the complexity of the original time series over a range of scales [23]. In this work, MDE is employed to quantify the PD feature information.

In recent years, a great number of intelligent algorithms have been used in PD fault diagnosis. Support vector machine (SVM) [24], as a learning machine based on kernel functions, that has the property of global optimization and strong generalization ability. However, using hyperplane recognition model, SVM can't accurately classify the samples with nonuniform state distribution. In addition, SVM is restricted in practical application for its inherent binary classification properties [25].

Hypersphere Support Vector Machine (HSSVM), based on SVM, was first proposed by Scholkopf [26]. Instead of the hyperplane, HSSVM uses a hypersphere for pattern recognition. HSSVM can not only separate two different classes, but also divide the sample space into two different parts [27]. Moreover, in order to overcome the limitations of inherent binary classification properties, hypersphere multiclass SVM (HMSVM) was introduced [28]. In HMSVM classification, the samples in same class are assigned to a hypersphere, therefore, the data space is composed of several hyperspheres [29]. Using HMSVM, the multi-class classification is realized directly. The quadratic programming calculation of HMSVM is less than that of one-class SVM, which causes shorter training and testing time. In this paper, particle swarm optimization (PSO) [30] is employed for parameter selection in HMSVM. Then the optimized classification model is applied to PD fault pattern recognition, using extracted PD features.

In this work, the proposed PD fault diagnosis method is combined with the excellent properties of both VMD and MDE. The characteristic parameters representing dominant PD features are effectively extracted. Besides, it can solve the problems in traditional PD feature extraction methods, such as the limitations of high request of sampling rate, slow speed of data processing, difficulties to extract PD phase information, influences by electromagnetic interference, difficulty of selection of wavelet basis, and so on. Finally, HMSVM is employed for PD pattern recognition with extracted parameters. To verify the effectiveness and superiority of the proposed method, different PD feature extraction methods and diverse classifiers are introduced. Results verify the exactness of the conclusion.

The rest of this paper is organized as follows: Section 2 describes the theories of VMD, MDE and HMSVM, and presents the PD fault diagnosis procedure. Section 3 presents a brief introduction to the experimental setup used to generate PD signals. In Section 4 we show the results with their validation. The paper ends with conclusions in Section 5.

2. PD Fault Diagnosis Based on VMD-MDE and HMSVM

2.1. VMD Algorithm

VMD decomposes one real signal into K independent sub-signal u_k, which has specific sparsity. This procedure gets the minimum bandwidth estimation of each modal [31]. The procedure of signal decomposition is to solve the variational problem. The variational model with constraint condition is as follows:

$$\begin{cases} \min_{\{u_k\},\{w_k\}} \{\sum_k \left\| \partial_t[(\delta(t) + \frac{j}{\pi t})u_k(t)]e^{-jw_k t} \right\|_2^2\} \\ \text{s.t.} \sum_k u_k = f \end{cases} \tag{1}$$

where $\{u_k\} = \{u_1, u_2, \cdots, u_K\}$ demonstrates the modal components, $\{w_k\} = \{w_1, w_2, \cdots, w_K\}$ is the center frequency of each modal component, $\delta(t)$ represents impulse function, ∂_t means the partial derivatives of t, and f is the original signal.

In order to obtain the optimal solution of such constrained variational problem, Lagrangian multiplier $\lambda(t)$ is introduced. The constrained variational problem is transformed into non-constrained problem:

$$L(\{u_k\}, \{\omega_k\}, \lambda) = \alpha \sum_k \left\| \partial_t[(\delta(t) + \frac{j}{\pi t})u_k(t)]e^{-jw_k t} \right\|_2^2 + \left\| f(t) - \sum_k u_k(t) \right\|_2^2 + \left\langle \lambda(t), f(t) - \sum_k u_k(t) \right\rangle \tag{2}$$

where α is the quadratic penalty factor. Alternate direction method of multipliers (ADMM) is introduced to obtain the saddle point of such Lagrangian function, which is the optimal solution.

The procedure of VMD can be summarized in the following steps:

(1) Initialize each modal component $\{u_k^1\}$, center frequency $\{\omega_k^1\}$ and operators $\{\lambda^1\}$. Set $n = 0$.

(2) Update u_k in non-negative frequency intervals:

$$\hat{u}_k^{n+1}(\omega) \leftarrow \frac{\hat{f}(\omega) - \sum_{i<k} \hat{u}_i^{n+1}(\omega) - \sum_{i>k} \hat{u}_i^n(\omega) + \frac{\hat{\lambda}^n(\omega)}{2}}{1 + 2\alpha(\omega - \omega_k^n)^2} \tag{3}$$

(3) Update ω_k.

$$\omega_k^{n+1} \leftarrow \frac{\int_0^\infty \omega \left| \hat{u}_k^{n+1}(\omega) \right|^2 d\omega}{\int_0^\infty \left| \hat{u}_k^{n+1}(\omega) \right|^2 d\omega} \tag{4}$$

(4) Update λ in non-negative frequency intervals:

$$\hat{\lambda}^{n+1} \leftarrow \hat{\lambda}^n + \tau(\hat{f}(\omega) - \sum_k \hat{u}_i^{n+1}(\omega)) \tag{5}$$

(5) For a given precision $\varepsilon > 0$, if $\frac{\sum_k \|\hat{u}_k^{n+1} - \hat{u}_k^n\|_2^2}{\|\hat{u}_k^n\|_2^2} < \varepsilon$, then stop iteration. Otherwise, return to (2).

2.2. Theory of Multiscale Dispersion Entropy

2.2.1. Dispersion Entropy

For a univariate signal $x = x_1, x_2, \cdots, x_N$, dispersion entropy method can be described in following steps [32]:

(1) Map $x_j(j = 1,2,\cdots,N)$ into $y = \{y_1, y_2, \cdots, y_N\}$ from 0 to 1 with the normal cumulative distribution function:

$$y_j = \frac{1}{\sigma\sqrt{2\pi}} \int_{-\infty}^{x_j} e^{\frac{-(t-\mu)^2}{2\sigma^2}} dt \tag{6}$$

where σ and μ represent the standard deviation and mean of x, respectively.

(2) Assign each y_j to an integer from Label 1 to c using a linear algorithm. The mapped signal can be defined as follows:

$$z_j^c = round(c.y_j + 0.5) \tag{7}$$

(3) Define embedding vector $z_i^{m,c}$ with embedding dimension m and time delay d as:

$$z_i^{m,c} = \left\{z_i^c, z_{i+d}^c, \cdots, z_{i+(m-1)d}^c\right\}, i = 1,2,\cdots,N-(m-1)d \tag{8}$$

Each time series $z_i^{m,c}$ is mapped to a dispersion pattern $\pi_{v_0 v_1 \cdots v_{m-1}}$, where:

$$z_i^c = v_0, z_{i+d}^c = v_1. \cdots, z_{i+(m-1)d}^c = v_{m-1}$$

(4) For each dispersion pattern, the relative frequency can be obtained as:

$$p(\pi_{v_0 v_1 \cdots v_{m-1}}) = \frac{Number\{i | i \le N-(m-1)d, z_i^{m,c}\ has\ \text{type}\ \pi_{v_0 v_1 \cdots v_{m-1}}\}}{N-(m-1)d} \tag{9}$$

where $p(\pi_{v_0 v_1 \cdots v_{m-1}})$ represent the number of dispersion pattern $\pi_{v_0 v_1 \cdots v_{m-1}}$, which is assigned to $z_i^{m,c}$ divided by the total number of embedding signals with embedding dimension m.

(5) Based on Shannon's definition of entropy, dispersion entropy with embedding dimension m, time delay d, and the number of classes c can be defined as

$$DE(x,m,c,d) = -\sum_{\pi=1}^{c^m} p(\pi_{v_0 v_1 \cdots v_{m-1}}) \cdot \ln(p(\pi_{v_0 v_1 \cdots v_{m-1}})) \tag{10}$$

2.2.2. Multiscale Dispersion Entropy

Multiscale Dispersion Entropy (MDE) is the combination of the coarse-graining with dispersion entropy. In MDE, the original signal $x = x_1, x_2, \cdots, x_N$ of length N is first divided into non-overlapping scale factor τ. Then the new coarse-grained signals can be shown as follows:

$$x_j^{(\tau)} = \frac{1}{\tau} \sum_{i=(j-1)\tau+1}^{j\tau} x_i, 1 \le j \le N/\tau \tag{11}$$

Calculate the entropy value of each coarse-grained signal of length N/τ with dispersion entropy method:

$$MDE(x,\tau,m,c,d) = DE(x^{(\tau)},m,c,d) \tag{12}$$

2.3. Theory of HMSVM

2.3.1. HMSVM

HMSVM can classify the samples directly. Each type of samples needs only one-hypersphere training. All training samples are mapped into high-dimension space. Each type of training samples searches for one hypersphere that has small radius and more target samples. HMSVM classification model is shown in Figure 1.

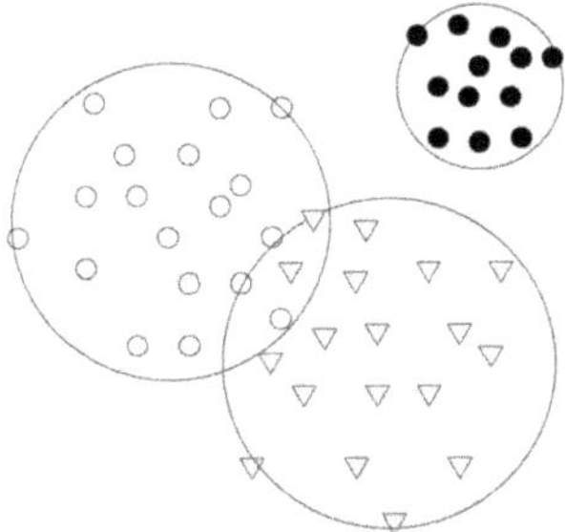

Figure 1. Classification model of HMSVM.

For an M-class problem, a collection of elements X_m (m = 1, 2, ... , M) is given. Assume that each X_m contains m-dimension sample x_{mi}, i = 1, 2 ... l_m, which represents i-th element in m-class.

Assign one hypersphere (a_m,R_m) for each sample X_m, where a_m is the center of sphere, R_m is the radius of suprasphere. The objective function of m-th suprasphere can be defined as follows:

$$\begin{aligned} &\min_{R_m}(R_m^2 + C_m \sum_{i=1}^{l_m} \xi_{m,i}) \\ &s.t. \|\Phi(x_{m,i}) - a_m\| \leq R_m^2 + \xi_{m,i}, \xi_{m,i} \geq 0 \end{aligned} \tag{13}$$

where C_m is the penalty factor, representing the trade-off between R_m and target samples. $\xi_{m,i}$ is the slack variable of HMSVM allowing remote samples staying outside the sphere.

Lagrange function can be obtained after Lagrange multiplier is introduced:

$$L(R, a, \xi, \alpha, \gamma) = {R_m}^2 + C_m \sum_{i=1}^{l_m} \xi_{mi} - \sum_{i=1}^{l_m} \alpha_i \{R^2 + \xi_{mi} - (\left\|x^2\right\| - 2a \cdot x_i + \left\|a^2\right\|)\} - \sum_{i=1}^{l_m} \gamma_i \xi_{mi} \tag{14}$$

The derivative operation of Equation (14) is processed to obtain the dual optimization problem as follows:

$$\min_{a_m} \sum_i \sum_j \alpha_{m,i} \alpha_{m,j} K(x_{m,i}, x_{m,j}) - \sum_{i=1}^{l_m} \alpha_{m,j} K(x_{m,i}, x_{m,j}) \tag{15}$$

The restricting condition that the target function should satisfy is shown as follows:

$$\sum_{i=1}^{l_m} \alpha_{m,i} = 1,\ 0 \leq \alpha_{m,i} \leq C_m \tag{16}$$

For an unknown fault sample d, we first calculate the square of the distance between d and a_m using the formula below:

$$D^2(d) = \|d - a_m\|^2 = (d \cdot d) - 2\sum_{i=1}^{l_m} \alpha_i (d \cdot x_i) + \sum_{i=1}^{l_m} \sum_{j=1}^{l_m} \alpha_i \alpha_j (x_i \cdot x_j) \tag{17}$$

The radius of the suprasphere is defined as $R_m = D(x_i)$, where x_i represents the support vector. Therefore, the category assigned to the unknown sample d can be determined according to the comparison between R_m and $D(d)$.

2.3.2. Kernel Function Selection

Due to the complexity among different PD fault samples, the spherical distribution will not appear in low-dimensional space. PD fault samples need to be mapped into high-dimension space using kernel functions to obtain the optimal hypersphere. In recent time, the common kernel functions include radial basic function (RBF) [33], polynomial kernel function and sigmoid function. After repeating tests, RBF shows outstanding performance. Therefore, RBF is selected as the kernel function for HMSVM. It can be defined in Equation (18):

$$K(x, x_i) = \exp\{-\frac{|x - x_i|^2}{\sigma^2}\} \tag{18}$$

2.4. PD Fault Diagnosis Based on VMD-MDE and HMSVM

In this paper, the proposed PD fault diagnosis method combines feature extraction and pattern recognition. Firstly, the original PD signal is decomposed using VMD to obtain the intrinsic mode functions. Secondly MDE value of each intrinsic mode function is calculated. And then principal component analysis (PCA) [34] is introduced to select principal components of MDE as PD feature vectors. Finally, the extracted vectors are sent to HMSVM pattern classifier to recognize different PD faults. The fault diagnosis procedure is as follows:

Step 1: Extract different types of PD signals in experimental environment, including floating discharge (FD), needle-surface discharge (ND), ball-surface discharge (BD) and corona discharge (CD).

Step 2: Select proper initial number of IMF according to the center frequency observation and decompose PD signals using VMD into intrinsic mode functions with different characteristic scales.

Step 3: Calculate the correlation coefficients between each IMF and original PD signal to select effective IMFs [35,36]. If the coefficient is greater than the threshold value, then keep the IMF as effective one. Otherwise, abandon the IMF. In this paper, the threshold value of the correlation coefficient is set to 0.3.

Step 4: Fix the decomposition scale for IMF and calculate the MDE value of extracted IMFs as original PD feature vectors.

Step 5: Analyze the PD vectors by PCA and extract fewer representative principal components as PD characteristic parameters.

Step 6: Send extracted PD characteristic parameters into HMSVM classifier to diagnose different PD fault modes and obtain the final diagnosis result.

The flow chart of PD fault diagnosis with proposed method is shown in Figure 2.

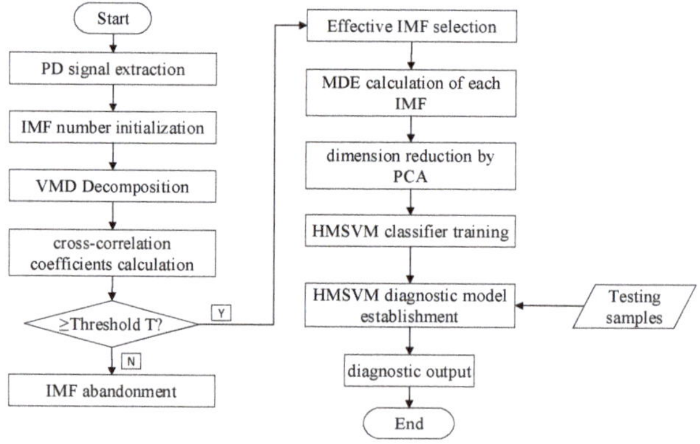

Figure 2. PD fault diagnosis procedure based on VMD-MDE and HMSVM.

3. Experiments and Analysis

3.1. Experimental Setup

Different PD types can produce different effects in insulation materials, but the range may be diverse. To analyze the characteristics of different PD types, PD signals of different models are extracted in the laboratory [37]. According to the inner insulation structure of power transformers, there are four possible different PD types, including FD, ND, BD and CD. PD models are shown in Figure 3. The experimental setup is shown in Figure 4.

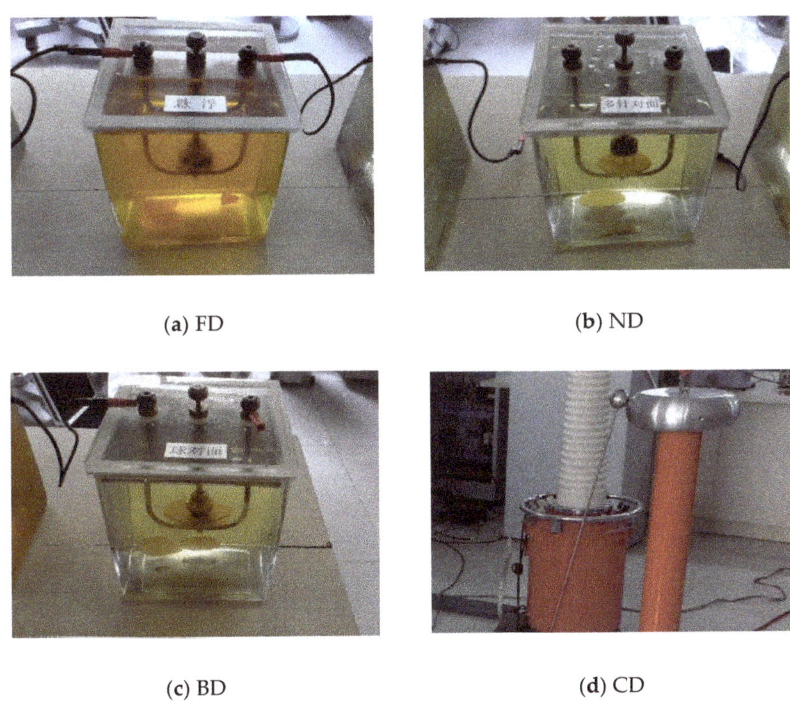

(**a**) FD (**b**) ND (**c**) BD (**d**) CD

Figure 3. PD models.

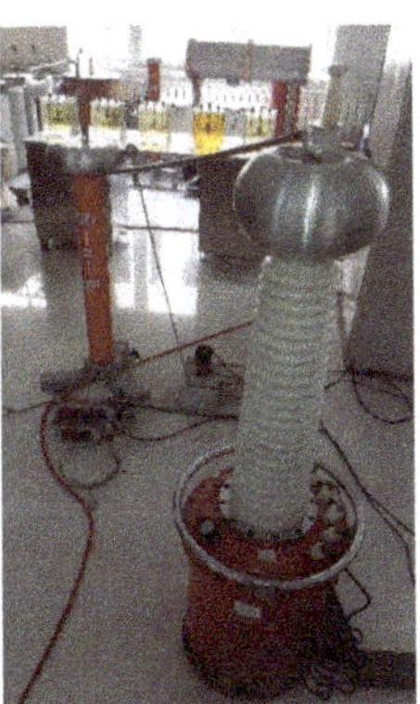

Figure 4. Photograph of experimental setup.

PD signals are detected in the simulated transformer tank in the laboratory. The pulse current is collected by a current sensor with a 500 kHz–16 MHz bandwidth. The UHF signal is extracted by a UHF sensor with a 10–1000 MHz bandwidth. The signal received is imported into the PD analyzer. The test condition is shown in Table 1 and the experimental connection diagram is shown in Figure 5.

Table 1. Test condition of PD models.

PD Types	Initial Voltage/kV	Breakdown Voltage/kV	Testing Voltage/kV	Sample Number
FD	2	7	3/4	50/50
ND	8.8	12	9/10	50/50
BD	3.5	10	5/6	50/50
CD	4.5	10	6/7	50/50

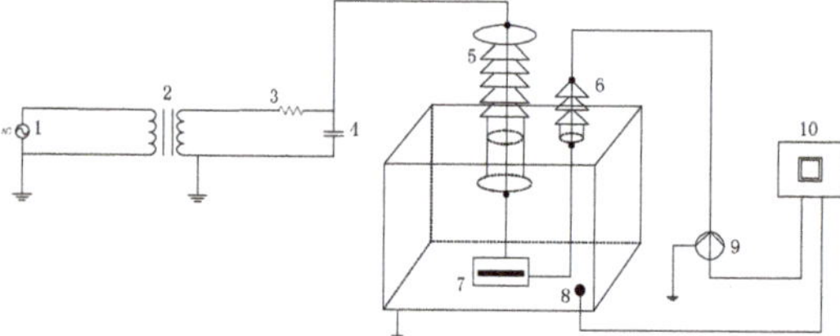

Figure 5. The connection diagram of PD experiment. 1—AC power source; 2—step up transformer; 3—resistance; 4—capacitor; 5—high voltage bushing; 6—small bushing; 7—PD model; 8—UHF sensor; 9—current sensor; 10—console.

3.2. Signal Extraction

In this paper, four different types of PD signals are extracted with above experimental setup. The extracted PD waveforms are shown in Figure 6.

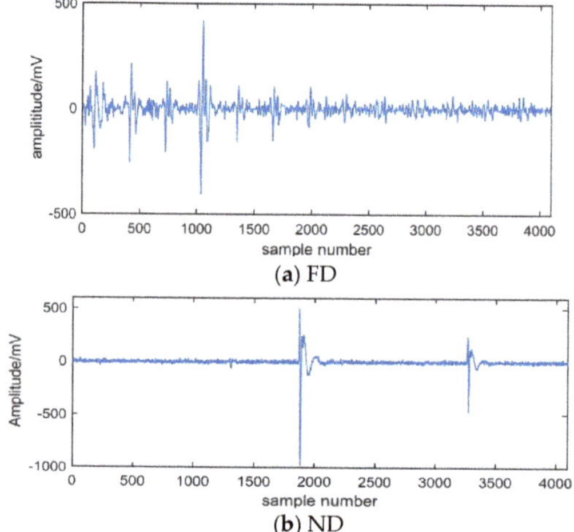

(**a**) FD

(**b**) ND

Figure 6. *Cont.*

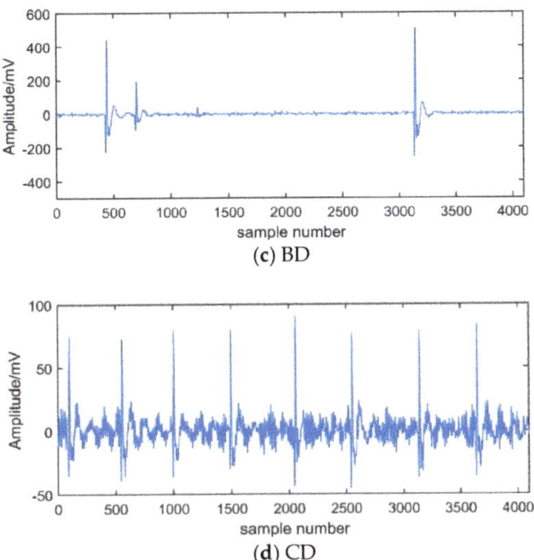

(c) BD

(d) CD

Figure 6. PD signals.

4. Results and Analysis

4.1. VMD Decomposition

In this paper, float discharge is taken as an example for VMD decomposition. The number of IMFs, represented as K, is determined according to the central frequency observation. The central frequency of IMF with the variation of K is shown in Table 2.

Table 2. Central frequency.

Number of IMFs	Central Frequency/MHz						
2	0.0079	7.3682					
3	0.0073	6.9573	12.3268				
4	0.0059	6. 8232	11.9803	13.2581			
5	0.0055	6. 8041	12.0256	13.1263	13.3572		
6	0.0059	6. 7855	11.7785	13.5579	13.2602	13.9348	
7	0.0053	6. 8034	12.1379	13.7877	13.9021	13.9975	14.2814

Table 2 shows that the IMFs with similar central frequency arise from $K = 5$, which means excessive decomposition. Therefore $K = 4$ is selected as the number of IMF. In this paper, the balancing parameter $\alpha = 2000$ and bandwidth parameter $\tau = 0.1$. The decomposition results with EMD and VMD are shown in Figures 7 and 8.

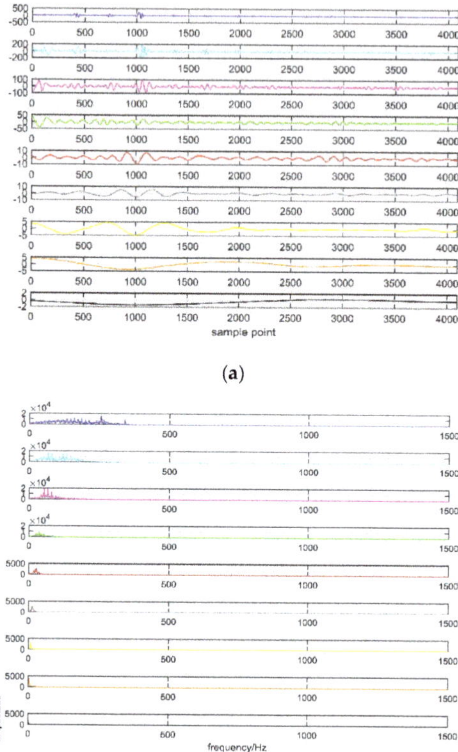

(a)

(b)

Figure 7. Results of EMD decomposition. (**a**) IMF of decomposition; (**b**) Frequency spectrum of decomposition.

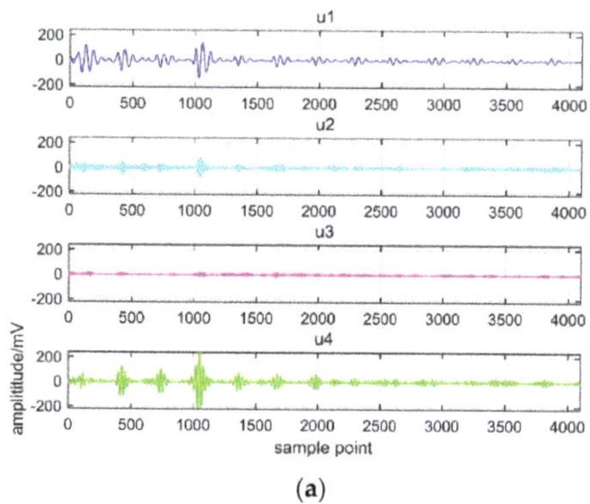

(a)

Figure 8. *Cont.*

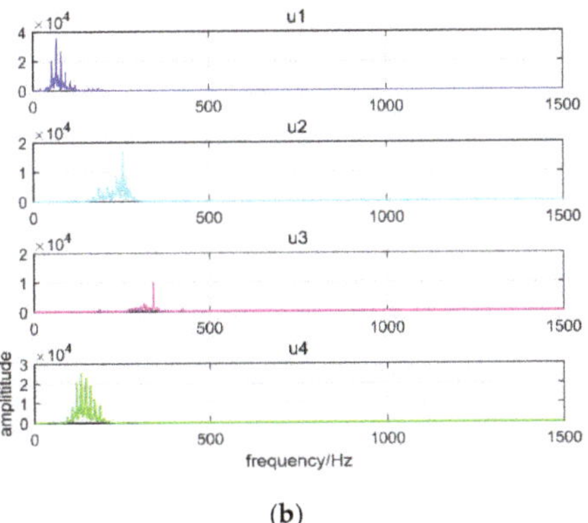

(b)

Figure 8. Results of VMD decomposition. (**a**) IMF of decomposition; (**b**) Frequency spectrum of decomposition.

Figure 7 shows the EMD decomposition results containing IMF components and frequency spectrum. From the figure we can see that eight IMF components and one remaining component are obtained. However, the problem of mode mixing occurs in EMD decomposition. Besides, IMF component in each decomposition level is different from that of original signal. Figure 8 describes the results of VMD decomposition. It can be seen from this figure that the modal components in VMD approach to the real signal. Figures 7 and 8 verify the effectiveness of VMD and the superiority over EMD. It can be concluded that VMD is more suitable for PD signal decomposition.

4.2. IMF Selection

In order to obtain the effective IMF, the correlation coefficient (CC) between each IMF and original PD signal is calculated. Given a threshold T, if the CC is greater than T, the IMF will be selected as effective component; otherwise it will be regarded as false component and abandoned. In this work, T is set to 0.3. The CC values of IMF for VMD and EMD are shown as Table 3.

Table 3. CC values.

	$u1$	$u2$	$u3$	$u4$	$u5$	$u6$	$u7$	$u8$	$u9$
VMD	0.6809	0.5129	0.3583	0.0083	-	-	-	-	-
EMD	0.7362	0.6035	0.4231	0.3026	0.2092	0.1123	0.0365	0.0086	0.0025

Table 3 shows that the CC value of first three IMFs is larger than the given threshold, which means these IMFs could represent the real components of PD signals. Therefore, the first three IMFs are selected and analyzed for VMD decomposition. Similarly, we can see that the CC value is smaller than the threshold from the fourth IMF, which means these IMFs contain less information of PD signals. Consequently, the first four IMFs are kept for EMD decomposition.

4.3. Feature Extraction

In this paper four different types of PD signals are decomposed using VMD method. The VMD decomposition parameters are shown in Table 4. K_s is the number of effective IMFs calculated as described in Section 4.2.

Table 4. VMD decomposition parameters.

PD Type	K	α	τ	K_s
FD	4	2000	0.1	3
ND	5	2000	0.1	3
BD	4	2000	0.1	4
CD	4	2000	0.1	4

Using the above parameters, the corresponding IMFs of different types of PD are obtained by VMD decomposition. Then the MDE value of each IMF is calculated. During MDE calculation, some preset parameters need to be given, including scale factor s, number of classification c, time delay d and embedded dimension m. But considering that aliasing may occur when $d > 1$, d is set to 1 as recommended. In order to avoid the trivial case of only one dispersion pattern, c is set to 2. For better detection on dynamic change of signals, m is set to 6. To analyze the variation of MDE values with different scales, s is set to 20. With above parameters, MDE values of four different types of PD signals extracted in the laboratory are calculated. For each type of PD, MDE values are averaged with different IMFs, shown in Figure 9.

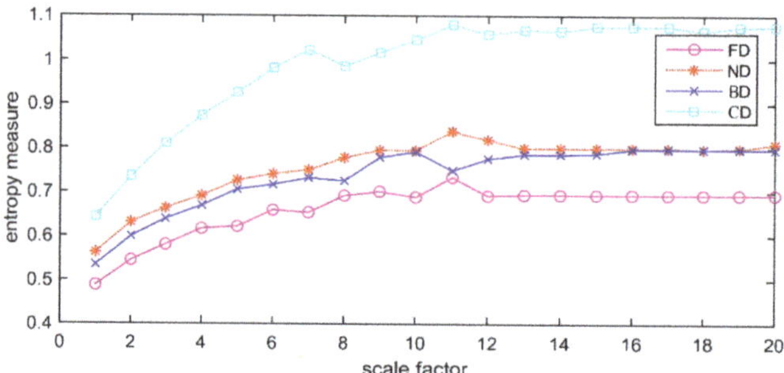

Figure 9. MDE variation with scale factors.

Figure 9 shows that different types of PD signals have diverse MDE values with variations of scale factors. The reason is that the randomness of PD signals is changing when PD fault occurs, which could change the MDE values. It also indicates that a single scale cannot completely reflect all the signal information and much more important information distributes in other scales. MDE can effectively detect the dynamic variation of PD signals which represent the fault characteristics with different scales. It can be found from the figure that MDE values start to level off after Scale 12. Therefore, the scale factor is set to 12 in this paper. In the case of FD, MDE values of IMFs using VMD and EMD are shown in Figure 10.

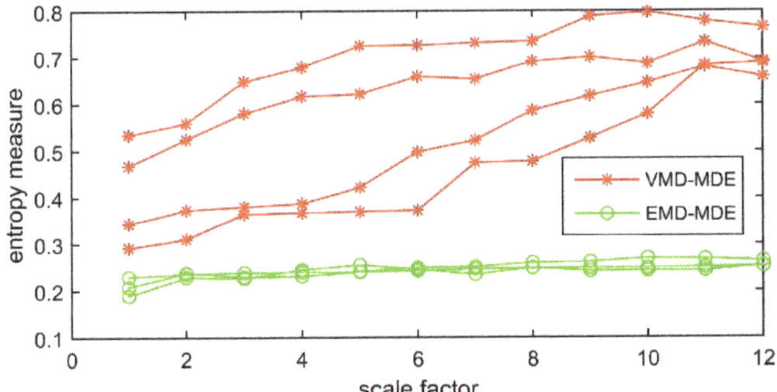

Figure 10. MDE values of IMFs using VMD and EMD.

Figure 10 shows that with the variation of scales, MDE values extracted by VMD are different. However, MDE values extracted by EMD seems to be same with the increase of decomposition scales which makes it difficult to distinguish different IMFs. The initial FD feature vectors combined with the MDE of all IMFs using VMD decomposition are shown in Table 5.

Table 5. Initial feature vectors.

IMF	Vectors
K_1	$O_1, O_2, O_3, O_4, O_5, O_6, O_7, O_8, O_9, O_{10}, O_{11}, O_{12}$
K_2	$P_1, P_2, P_3, P_4, P_5, P_6, P_7, P_8, P_9, P_{10}, P_{11}, P_{12}$
K_3	$Q_1, Q_2, Q_3, Q_4, Q_5, Q_6, Q_7, Q_8, Q_9, Q_{10}, Q_{11}, Q_{12}$
K_4	$R_1, R_2, R_3, R_4, R_5, R_6, R_7, R_8, R_9, R_{10}, R_{11}, R_{12}$

4.4. PCA-Based Dimension Reduction

Due to the high dimension of extracted feature vectors, it will cause big burden for pattern classifiers which can directly affect the recognition accuracy. In this paper, the PCA method is employed for dimension reduction of initial feature vectors. In the case of K_1, the covariance matrix is constructed to obtain the principal components. The eigenvalue and eigenvector of the covariance matrix are solved for linear transformation of original vectors. To achieve the goal of dimension reduction, those factors whose eigenvalues are greater than 1 are selected as principal components. The eigenvalue and corresponding contribution rates of the covariance matrix are shown in Table 6.

Table 6. Eigenvalues and corresponding contribution rates.

Vectors	Eigenvalue	Contribution Rate/%	Accumulated Contribution Rate/%
O_1	3.732	66.738	66.738
O_2	2.169	25.843	92.581
O_3	0.852	3.560	96.141
O_4	0.603	1.435	97.576
O_5	0.304	1.064	98.64
O_6	0.124	0.626	99.266
O_7	0.102	0.441	99.707
O_8	0.075	0.152	99.859
O_9	0.052	0.086	99.945
O_{10}	0.036	0.027	99.972
O_{11}	0.029	0.024	99.996
O_{12}	0.003	0.004	100.00

Table 6 shows that first two eigenvalues are greater than 1, and the accumulated contribution rate is larger than 90%. The contribution rate changes with the variation of principle components, shown in Figure 11.

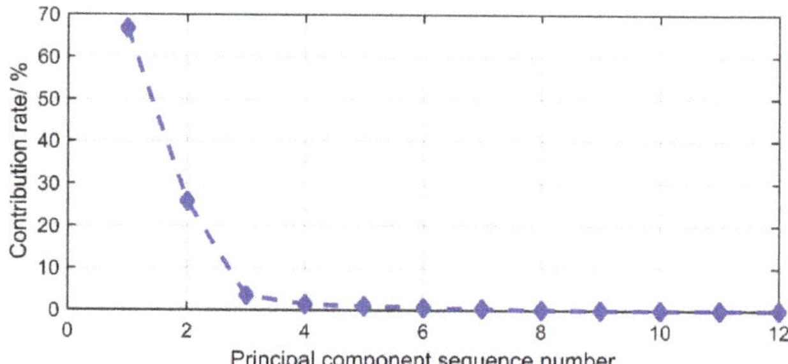

Figure 11. The variation of contribution rate with principle components.

It can be concluded from above figure that, the contribution rate from the third principle component starts to level off. In addition, the contribution rates are decreasing gradually which can be ignored. Therefore, first two principle components are suitable for further analysis which represent most of the vector information. To do so, the original 12 indicators are reduced to 2 new ones. With a similar method, the principle components of K_2, K_3 and K_4 can be obtained, shown in Table 7.

Table 7. Principle components with different IMFs.

IMF	KMO	Contribution Rate/%	Principle Component
K_1	0.852	92.581	O_1, O_2
K_2	0.767	88.379	P_1, P_2
K_3	0.734	80.232	Q_1, Q_2, Q_3
K_4	0.752	83.368	R_1, R_2

It can be seen from Table 7 that nine principle components factors are extracted from 48 feature vectors. And the contribution rate in each IMF is greater than 80%. Given the above, the dimension of feature vectors is reduced to nine after dimension reduction using PCA. Similarly, with above procedure, the calculated PD parameters of different PD types are shown in Table 8.

Table 8. Principle components with different IMFs.

PD Type	Parameters				
	K_1	K_2	K_3	K_4	K_5
FD	O_1, O_2	P_1, P_2	Q_1, Q_2, Q_3	R_1, R_2	-
ND	O_1, O_2	P_1, P_2	Q_1, Q_2	R_1, R_2	S_1, S_2
BD	O_1, O_2, O_3	P_1, P_2	Q_1, Q_2	R_1, R_2	-
CD	O_1, O_2	P_1, P_2, P_3	Q_1, Q_2	R_1, R_2	-

4.5. PD Pattern Recognition

In this paper, 400 PD samples, including FD, ND, BD and CD, are extracted in the laboratory containing 100 samples in each PD type. MDE values of four different PD types are calculated and 50 samples in each type constitute the initial feature vectors. To verify the effectiveness and superiority of the proposed method, the feature extraction methods based on multi-scale sample entropy (MSE) and multi-scale permutation entropy (MPE) are introduced. The calculation method of MSE and MPE is similar with that of MDE. Firstly, PD signals are decomposed using EMD or VMD. After that

MSE or MPE values of extracted IMFs are calculated. Finally, PCA is applied to dimension reduction. The parameters during signal decomposition are shown in Table 9.

Table 9. Parameters selection.

	EMD Decomposition			VMD Decomposition		
	Level	**Scale**	**Principle Components Number**	**Level**	**Scale**	**Principle Components Number**
MSE	4	14	10	3	12	8
MPE	3	10	8	3	10	8
MDE	3	12	9	4	12	9

PD feature vectors extracted with the above three methods are sent to the HMSVM classifier. Due to the big impact on the fault diagnosis result, HMSVM parameters need optimal configuration with PSO. In the case of VMD-MDE method, first of all, PD samples are divided into training and testing samples. After multiple experimental trials, the number of particle population is set to 20, $w = 1$, $c_1 = 2$, $c_2 = 2$, the maximum number of iterations $N = 200$. The penalty parameter C is between 1/n and 1, while the searching range of the kernel parameter σ is between 1 and 100. The optimum fitness reaches the maximum value of 96.98% after 19 iterations, when $\sigma = 12.26$ and $C = 0.35$. Similarly, HMSVM parameters with different feature extraction methods are obtained as follows.

Using the parameters in Table 10, HMSVM classifier is constructed for fault diagnosis of three different PD features. The recognition results with EMD and VMD decomposition are shown in Figures 12 and 13.

Table 10. HMSVM parameters.

	EMD-MSE	EMD-MPE	EMD-MDE	VMD-MSE	VMD-MPE	VMD-MDE
C	0.43	0.31	0.27	0.46	0.33	0.35
σ	10.38	11.86	10.19	12.05	9.37	12.26

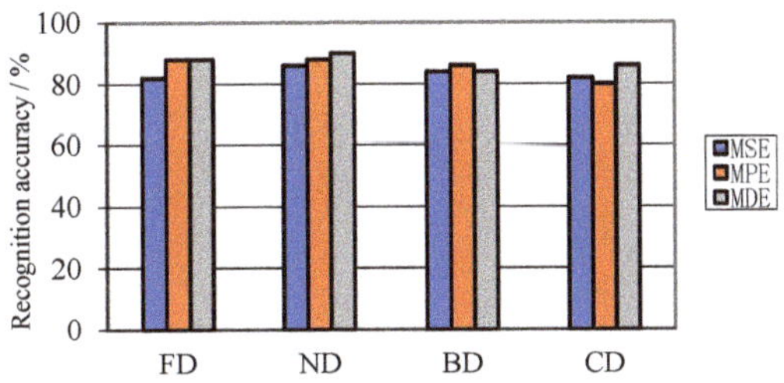

Figure 12. Recognition results using EMD decomposition.

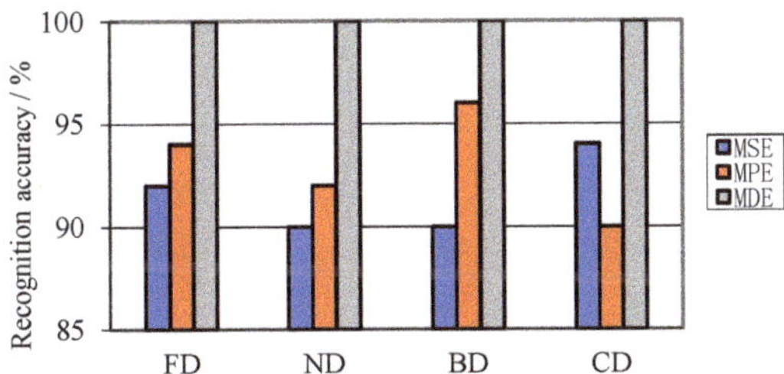

Figure 13. Recognition results using VMD decomposition.

Figures 12 and 13 demonstrate that the recognition result using EMD decomposition is significantly different with that using VMD decomposition. Figure 12 illustrates that the recognition accuracy in each PD type is not less than 80% but no more than 90%, which means, using EMD decomposition, extracted PD features cannot represent most of signal characteristics. In contrary, Figure 13 shows that the recognition accuracy in each PD type is no less than 90%. Moreover, in each PD type, there's no misjudged sample with MDE. This means that, with VMD decomposition, PD features can effectively represent most of signal information. Besides, from above two figures, it gets a satisfactory result with MDE parameters.

To compare the diagnosis results of PD features with different classifiers, artificial neural network (ANN) [38] and support vector machine (SVM) classifiers are introduced for PD pattern recognition. In ANN, back-propagation network is employed as the recognition model, which trains the weight with differentiable nonlinear functions. The classifier parameters are shown in Table 11. σ is the kernel parameter of RBF and C is the penalty factor in SVM.

Table 11. Parameters of ANN and SVM.

Classifier	Type	EMD-MSE	EMD-MPE	EMD-MDE	VMD-MSE	VMD-MPE	VMD-MDE
SVM	C	0.25	0.28	0.45	0.44	0.38	0.46
	σ	8.39	10.57	8.32	9.18	8.25	10.22
ANN	Input	10	8	9	8	8	9
	Output	4	4	4	4	4	4
	Hidden layer	16	12	14	12	10	12

With the parameters shown in Tables 10 and 11, ANN, SVM and HMSVM classifiers are constructed for PD pattern recognition. Using diverse classifiers, the recognition result with VMD-MDE can be seen in Figure 14. Table 12 shows the integrative result using different PD features, in which running time means the time used for PD fault diagnosis.

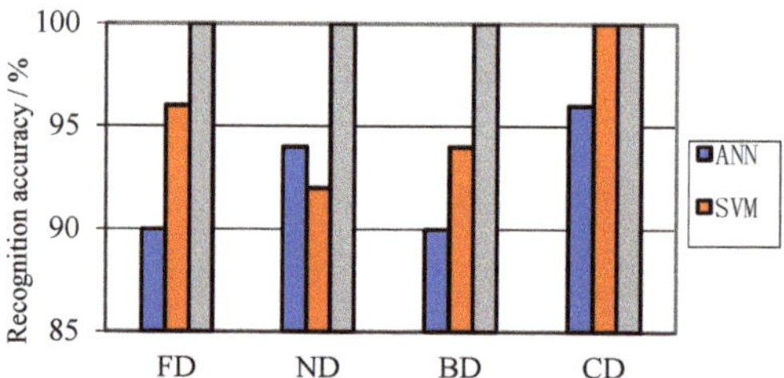

Figure 14. Recognition results using VMD-MDE method.

Table 12. Recognition result with different PD features.

Feature Types	ANN		SVM		HMSVM	
	Recognition Accuracy/%	Running Time/s	Recognition Accuracy/%	Running Time/s	Recognition Accuracy/%	Running Time/s
EMD- MSE	86.00	6.88×10^{-4}	88.50	6.92×10^{-4}	86.50	6.75×10^{-4}
EMD- MPE	86.50	3.45×10^{-3}	84.00	3.21×10^{-3}	86.00	3.51×10^{-3}
EMD- MDE	88.00	5.39×10^{-4}	90.50	5.36×10^{-4}	91.50	1.68×10^{-3}
VMD- MSE	95.00	8.16×10^{-4}	96.50	7.29×10^{-4}	97.50	7.80×10^{-4}
VMD- MPE	98.00	7.45×10^{-4}	97.50	7.12×10^{-4}	99.00	7.42×10^{-4}
VMD- MDE	98.00	5.36×10^{-4}	99.00	5.32×10^{-4}	100.00	5.27×10^{-4}

As can be illustrated in Figure 14, using the same PD feature extraction method, the recognition results with different classifiers are significantly different. The average classification accuracy achieved using HMSVM is 100.00%. HMSVM shows great advantages over ANN and SVM. Table 12 shows diverse diagnostic results with different PD features. Compared with different PD feature types,

VMD-MDE gives less running time and higher recognition accuracy. It means parameters using VMD-MDE can represent most of PD signal components. The quadratic programming calculation of HMSVM is less than that of SVM, which causes shorter training and testing time. In addition, HMSVM shows better classification ability than other two classifiers, ANN and SVM.

5. Conclusions

In this paper, a novel PD fault diagnosis method is proposed. This method combines PD feature extraction based on VMD-MDE and PD pattern recognition based on HMSVM. First of all, four types of PD signals are extracted in the experimental environment, including FD, ND, BD and CD. Then VMD is employed for PD signal decomposition. Secondly, proper IMFs are selected according to central frequency observation and MDE values in each IMF are calculated. Afterwards PCA is introduced to select effective principle components in MDE as final PD characteristic parameters. Finally, the extracted principle factors are used as PD features and sent to the HMSVM classifier. Experiment results show the following advantages: the proposed method can extract effective IMFs according to VMD decomposition. PD feature information in IMFs can be quantified successfully with MDE. Using PCA, the principle components which represent prominent characteristics are effectively selected. With small data size and low computational complexity, this approach overcomes the limitations in traditional PD feature extraction methods. Compared with PD feature extraction methods based on EMD-MSE, EMD-MPE, EMD-MDE, VMD-MSE and VMD-MPE, this proposed approach based on VMD-MDE achieves higher recognition accuracy and needs less running time, which can improve the diagnosis efficiency to satisfy real time requirements.

HMSVM uses one hypersphere for pattern recognition. HMSVM can not only separate two different classes, but also divide the sample space into two different parts. Using HMSVM, the classification of multi-classes was realized directly. Compared with ANN and SVM classifiers, HMSVM obtains higher recognition rate and improves the accuracy and efficiency in PD fault diagnosis. On the whole, this proposed method provided a new scheme for PD fault diagnosis. For further consideration, the proposed fault diagnosis method can be employed in PD on-line monitoring and diagnosis.

Author Contributions: Y.W. and H.S. conceived and designed the experiments; H.S. performed the experiments; F.L. analyzed the data and contributed to analysis tools; H.S. wrote the paper. All authors have read and approved the final manuscript.

Funding: This research was funded by the Science and Technology Project of the State Grid Corporation of China (SGLNDK00KJJS1500008), and the Doctoral Scientific Research Foundation of Northeast Electric Power University (No. BSJXM-201406), China.

Conflicts of Interest: The authors declare no conflict of interest.

References

1. Firuzi, K.; Vakilian, M.; Darabad, V.P.; Phung, B.T.; Blackburn, T.R. A novel method for differentiating and clustering multiple partial discharge sources using S transform and bag of words feature. *IEEE Trans. Dielectr. Electr. Insul.* **2018**, *24*, 3694–3702. [CrossRef]
2. Zhou, Z.L.; Zhou, Y.X.; Huang, X.; Zhang, Y.X. Feature Extraction and Comprehension of Partial Discharge Characteristics in Transformer Oil from Rated AC Frequency to Very Low Frequency. *Energies* **2018**, *11*, 1702. [CrossRef]
3. Hammarstrom, T.J.A. Partial discharge characteristics within motor insulation exposed to multi-level PWM waveforms. *IEEE Trans. Dielectr. Electr. Insul.* **2018**, *25*, 559–567. [CrossRef]
4. Mota, H.D.O.; Vasconcelos, F.H.; Castro, C.L.D. A comparison of cycle spinning versus stationary wavelet transform for the extraction of features of partial discharge signals. *IEEE Trans. Dielectr. Electr. Insul.* **2016**, *23*, 1106–1118. [CrossRef]
5. Castillo, J.; Mocquet, A.; Saracco, G. Wavelet transform: A tool for the interpretation of upper mantle converted phases at high frequency. *Geophys. Res. Lett.* **2018**, *28*, 4327–4330. [CrossRef]

6. Li, Y.B.; Xu, M.Q.; Liang, X.H.; Huang, W.H. Application of Bandwidth EMD and Adaptive Multiscale Morphology Analysis for Incipient Fault Diagnosis of Rolling Bearings. *IEEE Trans. Ind. Electron.* **2017**, *64*, 6506–6517. [CrossRef]
7. Bustos, A.; Rubio, H.; Castejon, C.; Garcia-prada, J.C. EMD-Based Methodology for the Identification of a High-Speed Train Running in a Gear Operating State. *Sensors* **2018**, *18*, 793. [CrossRef]
8. Xiao, B.; Fang, L.J.; Li, J.F.; Qi, X.S.; Bai, Y.R. An EMD Method for Ascertaining Maximal Value of Cellular Load in Spatial Load Forecasting. *J. Northeast Electr. Power Univ.* **2018**, *38*, 8–14.
9. Dragomiretskiy, K.; Zosso, D. Variational Mode Decomposition. *IEEE Trans. Signal Process.* **2014**, *62*, 531–544. [CrossRef]
10. Yao, J.C.; Xiang, Y.; Qian, S.; Wang, S. Noise source identification of diesel engine based on variational mode decomposition and robust independent component analysis. *Appl. Acoust.* **2017**, *116*, 184–194. [CrossRef]
11. Zhang, L.; Veitch, D. Learning Entropy. *Lect. Notes Comput. Sci.* **2017**, *6640*, 15–27.
12. Shannon, C.E. A mathematical theory of communications. *Bell Syst. Tech. J.* **1948**, *27*, 379–423. [CrossRef]
13. Pincus, S. Approximate entropy (ApEn) as a complexity measure. *Chaos* **1995**, *5*, 110–117. [CrossRef] [PubMed]
14. Wu, H.T.; Yang, C.C.; Lin, G.M.; Haryadi, B. Multiscale Cross-Approximate Entropy Analysis of Bilateral Fingertips Photoplethysmographic Pulse Amplitudes among Middle-to-Old Aged Individuals with or without Type 2 Diabetes. *Entropy* **2017**, *19*, 145. [CrossRef]
15. Richman, J.S.; Moorman, J.R. Physiological time-series analysis using approximate entropy and sample entropy. *Am. J. Physiol. Heart Circ. Physiol.* **2000**, *278*, H2039. [CrossRef] [PubMed]
16. George, M.; Md, A.; Roberto, S. Low Computational Cost for Sample Entropy. *Entropy* **2018**, *20*, 61. [CrossRef]
17. Bandt, C.; Pompe, B. Permutation entropy: A natural complexity measure for time series. *Phys. Rev. Lett.* **2002**, *88*, 174102. [CrossRef]
18. Zhou, S.H.; Qian, S.L.; Chang, W.B.; Xiao, Y.Y. A Novel Bearing Multi-Fault Diagnosis Approach Based on Weighted Permutation Entropy and an Improved SVM Ensemble Classifier. *Sensors* **2018**, *18*, 1934. [CrossRef]
19. Azami, H.; Rostaghi, M.; Fernandez, A.; Escudero, J. Dispersion entropy for the analysis of resting-state MEG regularity in Alzheimer's disease. In Proceedings of the International Conference of the IEEE Engineering in Medicine and Biology Society, Orlando, FL, USA, 16–20 August 2016; p. 6417.
20. Baldini, G.; Giuliani, R.; Steri, G.; Neisse, R. Physical layer authentication of Internet of Things wireless devices through permutation and dispersion entropy. In Proceedings of the Global Internet of Things Summit, Geneva, Switzerland, 6–9 June 2017; pp. 1–6.
21. Azami, H.; Rostaghi, M.; Abasolo, D.; Escudero, J. Refined Composite Multiscale Dispersion Entropy and its Application to Biomedical Signals. *IEEE Trans. Bio-Med. Eng.* **2017**, *99*, 1. [CrossRef]
22. Goldberger, A.L.; Bruce, A.; Peng, C.K.; Costa, M. Multiscale entropy analysis of biological signals. *Phys. Rev. E Stat. Nonlinear Soft Matter Phys.* **2005**, *71*, 1–9.
23. Azami, H.; Escudero, J. Coarse-Graining Approaches in Univariate Multiscale Sample and Dispersion Entropy. *Entropy* **2018**, *20*, 138. [CrossRef]
24. Vapnik, V. *Statistical Learning Theory*; Wiley: New York, NY, USA, 1998.
25. Velazquez-Pupo, R.; Sierra-Romero, A.; Torres-Roman, D.; Romero-Delgado, M. Vehicle detection with occlusion handling, tracking, and OC-SVM classification: A high performance vision-based system. *Sensors* **2018**, *18*, 374. [CrossRef] [PubMed]
26. Scholkopf, B.; Smola, A. *Kernel Methods and Support Vector Machines;* Academic Press Library in Signal Processing: Amsterdam, The Netherlands, 2003; pp. 857–881.
27. Ai, Q.; Wang, A.; Wang, Y.; Sun, H.J. Improvements on twin-hypersphere support vector machine using local density information. In *Progress in Artificial Intelligence*; Springer: Berlin, Germany, 2018; pp. 1–9.
28. Xu, T.; He, D.K. Theory of hypersphere multiclass SVM. *Control Theory Appl.* **2009**, *26*, 1293–1297.
29. Guo, Y.; Xiao, H. Multiclass multiple kernel learning using hypersphere for pattern recognition. *Appl. Intell.* **2017**, *48*, 1–9. [CrossRef]
30. Kennedy, J.; Eberhart, R. Particle Swarm Optimization. In Proceedings of the IEEE International Conference on Neural Networks, Perth, Western Australia, 27 November–1 December 1995; pp. 1942–1948.
31. Wang, Z.; Jia, L.; Qin, Y. Adaptive Diagnosis for Rotating Machineries Using Information Geometrical Kernel-ELM Based on VMD-SVD. *Entropy* **2018**, *20*, 73. [CrossRef]

32. Rostaghi, M.; Azami, H. Dispersion Entropy: A Measure for Time-Series Analysis. *IEEE Signal Process. Lett.* **2016**, *23*, 610–614. [CrossRef]
33. Chen, S.; Mclaughlin, S.; Mulgrew, B. Complex-valued radial basic function network, Part I: Network architecture and learning algorithms. *Signal Process.* **1994**, *35*, 19–31. [CrossRef]
34. Pearson, K. On Lines and Planes of Closest Fit to Systems of Points in Space. *Philos. Mag.* **1901**, *2*, 559–572. [CrossRef]
35. Peng, Z.K.; Tse, P.W.; Chu, F.L. A comparison study of improved Hilbert-Huang transform and wavelet transform: Application to fault diagnosis for rolling bearing. *Mech. Syst. Signal Process.* **2005**, *19*, 974–988. [CrossRef]
36. Mostafizur, R.M.; Anowarul, F.S. Mental Task Classification Scheme Utilizing Correlation Coefficient Extracted from Interchannel Intrinsic Mode Function. *BioMed Res. Int.* **2017**, 1–11. [CrossRef]
37. Shang, H.K.; Kwok, L.; Li, F. Partial Discharge Feature Extraction Based on Ensemble Empirical Mode Decomposition and Sample Entropy. *Entropy* **2017**, *19*, 439. [CrossRef]
38. Folkes, S.R.; Lahav, O.; Maddox, S.J. An artificial neural network approach to the classification of galaxy spectra. *Mon. Notices R. Astron. Soc.* **2018**, *283*, 651–665. [CrossRef]

MDPI
St. Alban-Anlage 66
4052 Basel
Switzerland
Tel. +41 61 683 77 34
Fax +41 61 302 89 18
www.mdpi.com

Entropy Editorial Office
E-mail: entropy@mdpi.com
www.mdpi.com/journal/entropy

www.ingramcontent.com/pod-product-compliance
Lightning Source LLC
LaVergne TN
LVHW071620170726
843515LV00009B/2438

* 9 7 8 3 0 3 9 4 3 3 4 0 7 *